TECHNOLOGY OF MACHINE TOOLS

Third Edition

S. F. Krar

J. W. Oswald

J. E. St. Amand

Gregg Division/McGraw-Hill Book Company

New York   Atlanta   Dallas   St. Louis   San Francisco
Auckland   Bogotá   Guatemala   Hamburg   Johannesburg   Lisbon
London   Madrid   Mexico   Montreal   New Delhi   Panama   Paris
San Juan   São Paulo   Singapore   Sydney   Tokyo   Toronto

Sponsoring Editor: D. Eugene Gilmore
Editing Supervisor: Iris Wheat
Design and Art Supervisor: Patricia Lowy
Production Supervisor: Priscilla Taguer

Text Designer: Sharkey Design
Cover Photograph: The Image Bank/Gabe Palmer

Library of Congress Cataloging in Publication Data

Krar, Stephen F.
  Technology of machine tools.

  Includes index.
  1. Machine-tools.   2. Machine-shop practice.
I. Oswald, James William.   II. St. Amand, J. E.
III. Title.
TJ1185.K668   1984   621.9'02   83-19902
ISBN 0-07-035425-1

TECHNOLOGY OF MACHINE TOOLS, THIRD EDITION

   3 4 5 6 7 8 9 0   VNHVNH   8 9 1 0 9 8 7 6

ISBN 0-07-035425-1

CONTENTS

PREFACE

No single invention since the Industrial Revolution has left such an impact on society as the computer. During the last two or three decades, basic computers have been applied to machine tools to program and control various machine operations. Computers have steadily improved until there are now highly sophisticated units capable of controlling the operation of a single machine, a group of machines, or soon even a complete manufacturing plant. Since the use of computers in the machine tool industry continues to progress at an ever accelerating rate, this third edition contains a section titled "Computer-Age Machining," Section 15, which introduces and deals with computers and computer controls. The wide scope of this text makes it suitable for use in vocational and secondary schools, community colleges, and apprenticeship training and adult retraining programs.

This book is based on the authors' many years of practical experience as skilled workers in the trade and as specialists in teaching. To keep abreast of rapid technological change, the authors have researched the latest technical information available, and have visited industries which are leaders in their field. Many sections of this book were reviewed by key personnel in various manufacturing firms, so that the most accurate and up-to-date information is presented. The authors appreciated the opportunity to incorporate into the text the suggestions and recommendations made by these people.

The third edition of *Technology of Machine Tools* is presented in unit form with each unit introduced with a set of objectives followed by related theory and operational sequence. Each operation is explained in a step-by-step procedure which students can readily follow. Advanced operations are introduced by problems, followed by step-by-step solutions and machining procedures. Review questions at the end of each unit can be used for review or for homework assignments to prepare students for subsequent operations. To make this text easily understood, each unit contains many new illustrations and photographs. Color has been used throughout to emphasize important points and to make the illustrations more meaningful.

Throughout the text, dual dimensioning (U.S. Customary: inch; SI: metric) is used. The need for a working knowledge of the metric system cannot be overstressed, because most of the countries in the world are already on or in the process of converting to the metric system. Since those involved in the machine shop trade will have to be familiar with both the metric and inch systems during the changeover period, wherever possible, metric tools and information are included.

The purpose of this text is to assist instructors in giving students basic training in the operation of machine tools, and in helping students understand the latest machining processes and developments. The material is organized so that instructors may easily select those topics that are most suitable for class projects, or that suit the individual differences of students.

A technician in the machine shop trade should be neat, develop sound work habits, and have a good knowledge of mathematics, print reading, and computers. To keep up to date on technological changes, technicians must continue to expand their knowledge by reading specialized texts, trade literature, and magazine articles in this field.

Steve F. Krar
J. William Oswald
Joseph E. St. Amand

ABOUT THE AUTHORS

STEVE F. KRAR majored in Machine Shop Practice and spent fifteen years in the trade, first as a machinist and finally as a tool and diemaker. After this period he entered Teachers' College and graduated from the University of Toronto with a Specialist's Certificate in Machine Shop Practice. During his nineteen years of teaching, Mr. Krar was active in Vocational and Technical education and served on the executive of many educational organizations. For ten years he was on the summer staff of the University of Toronto, involved in teacher training programs. Active in machine tool associations, Steve Krar has been a senior member of the Society of Manufacturing Engineers for over twenty-five years. He is a co-author of the McGraw-Hill text *Machine Tool Operations*. He is also co-author of the following McGraw-Hill Ryerson Ltd. publications: *Machine Shop Training, Machine Shop Operations,* and the overhead transparency kits: *Machine Tools, Measurement and Layout, Threads and Testing Equipment,* and *Cutting Tools.*

J. WILLIAM OSWALD served an apprenticeship in machine shop, and after sixteen years in the trade attended Teachers' College at the University of Toronto. After graduation, he received a Specialist's Certificate in Machine Shop Practice and taught machine shop work for twenty-five years. During this time he attended several up-grading courses in the operation of the latest machine shop and testing equipment. For several years Mr. Oswald served on the teacher-training staffs at the University of Toronto and the University of Western Ontario. He had also worked with various technical educational committees and organizations. He is a co-author of the McGraw-Hill text *Machine Tool Operations.* Mr. Oswald is also a co-author of *Machine Shop Operations,* and the overhead transparency kits: *Machine Tools, Measurement and Layout, Threads and Testing Equipment,* and *Cutting Tools.*

JOSEPH E. ST. AMAND served his apprenticeship in general machine shop work, jigs and fixtures, and tool and die work. After twelve years in industry he entered Teachers' College and graduated with a Specialist's Certificate in Machine Shop Practice from the University of Toronto. During his thirty-three years of teaching, Joe St. Amand served on several educational committees which were involved with technical research, courses, and standardized examinations. For more than twenty-five years he has been active in chapter work for the Society of Manufacturing Engineers. He is a co-author of the McGraw-Hill text *Machine Tool Operations.* He is also co-author of the texts: *Machine Shop Training, Machine Shop Operations,* and Machine Shop Transparency Kits: *Machine Tools, Measurement and Layout, Threads and Testing Equipment,* and *Cutting Tools.*

ACKNOWLEDGMENTS

The authors wish to express their sincere thanks and appreciation to Alice H. Krar for her untiring devotion in reading, typing, and checking the manuscript for this text. Without her supreme efforts, this text could not have been produced.

Special thanks are due to Mr. Oliver Coley, formerly Technical Director, Garson-Falconbridge Secondary School; Mr. Nelson Durst, formerly of the Arthur Voaden Vocational School, St. Thomas; Mr. J. Demaline of the Hagersville Secondary School; Mr. H. Bacsu, Westlane Secondary School, Niagara Falls; Don Matthews, San Joaquin Delta College, Stocton, California; and to all the teachers who offered suggestions which we were happy to include.

Our deep thanks go to the following firms which reviewed sections of the manuscript and offered suggestions which were incorporated to make this text as accurate and up to date as possible: American Superior Electric Co., Bendix Corporation, Carborundum Co., FAG Bearing Co. Ltd., Federal Products Corporation, Norton Co., and Moore Special Tool Co. Cincinnati Milacron Inc. was most helpful in offering advice and in reviewing several sections on milling and special processes.

We are grateful to the following firms who have assisted in the preparation of this text by supplying illustrations and technical information.

Allen, Chas. G. & Co.
American Chain and Cable Co. Inc., Wilson
   Instrument Division
"American Machinist"
American Superior Electric Co. Ltd.
Ametek Testing Equipment
Armstrong Bros. Tool Co.
Ash Precision Equipment Inc.
Atlas Press Co., Clausing Division
Avco-Bay State Abrasive Company
Bausch & Lomb Inc.
Bendix Corporation, Automation Group
Bethlehem Steel Corporation
Bickle Co. H. W.
Boston Gear Works
Bridgeport Machines Division of Textron Inc.
Brown & Sharpe Manufacturing Co.
Buffalo Forge Co.
Butterfield Division, Union Twist Drill Co.
Canadian Blower and Forge Co. Ltd.
Canadian Tap and Die Co. Ltd.
Carborundum Company
Cincinnati Lathe and Tool Co.
Cincinnati Milacron Inc.
Cincinnati Shaper Co.
Clausing Corporation
Cleveland Tapping Machine Co.
Cleveland Twist Drill (Canada) Ltd.
Colchester Lathe & Tool Co.
Coleman Engineering Co. Inc.
Covel Manufacturing Co.
Delmar Publishers Inc.
Delta File Works
DeVlieg Machine Co.
Dillon, W. C. and Co. Inc.
DoALL Company

Elliott Machine Tools
Emco Maier Corp.
Enco Manufacturing Co.
Everett Industries Inc.
Ex-Cell-O Corp.
Explosive Fabricators Division, Tyco Corp.
FAG Bearing Co. Ltd.
Federal Products Corp.
Firth-Brown Tools (Canada) Ltd.
Frontier Equipment Ltd.
General Electric Co. Ltd.
General Motors of Canada
Greenfield Tap and Die Co.
Hones, Charles H. Inc.
Inland Steel Co.
Jacobs Manufacturing Co.
Jones and Lamson Division of Waterbury Farrel
Kaiser Steel Corp.
Kostel Enterprises Ltd.
Kalamazoo Machine Tool Co.
KTS Industries
LeBlond, R. K. Machine Tool Co.
Lionite Abrasives Ltd.
Lodge & Shipley
Mahr Gage Co. Inc.
Mobil Oil Corporation
Moore Special Tool Co.
Morse Twist Drill and Machine Co.
MTI Corporation
National Broach & Machine Division, Lear Siegler Inc.
National Twist Drill & Tool Co.
Neill, James & Co. (Sheffield) Ltd.
Nicholson File Co. of Canada Ltd.
Norton Company of Canada Ltd.
Powder Metallurgy Parts Manufacturers' Association
Pratt & Whitney Co. Inc., Machine Tool Division

Precision Diamond Tool Co.
Retor Developments Ltd.
Rockford Machine Tool Co.
Rockwell International
Shore Instrument & Mfg. Co. Inc.
Slocomb, J. T. Co.
South Bend Lathe Inc.
Standard-Modern Tool Co. Ltd.
Standard Tool Co.
Stanley Tools Division, Stanley Works
Starrett, L. S. Co.
Sun Oil Co.
Taft-Peirce Manufacturing Co.
Taper Micrometer Corp.

Thompson Grinder Co.
Union Carbide Corp., Linde Division
United States Steel Corporation
Volstro Manufacturing Co. Inc.
Watts Bros. Tool Works
Weldon Tool Co.
Wells Manufacturing Corp.
Whitman & Barnes
Wickman, A. C. Ltd.
Wilkie Brothers Foundation
Williams, A. R., Machinery Co. Ltd.
Williams, J. H. & Co.
Wilton Machinery Co.
Woodworth, W. J., and J. D. Woodworth

INTRODUCTION TO MACHINE TOOLS

SECTION 1

The progress of humanity throughout the ages has been governed by the types of tools available. Ever since primitive people used rocks as hammers or as weapons to kill animals for food, tools have governed our standard of living. The use of fire to extract metals from ore led to the development of newer and better tools. The harnessing of water led to the development of hydropower, which greatly improved humanity's well-being.

With the industrial revolution in the mid-18th century, early machine tools were developed and were continually improved. The development of machine tools and related technologies showed rapid advances during and immediately after World Wars I and II. Since World War II, processes such as numerical control, electromachining, and powder metallurgy greatly altered manufacturing methods.

Today we are living in a society greatly affected by the development of the computer. Computers affect us from the growing and sale of food to manufacturing processes, and entertainment. Although the computer is influencing our everyday lives, it is still important that the student or apprentice be able to perform basic operations on standard machine tools. This knowledge will provide the necessary background for a person seeking a career in the machine tool trade.

HISTORY OF MACHINES
UNIT 1

The high standard of living we enjoy today did not just happen. It has been the result of the development of highly efficient machine tools over the past several decades. Processed foods, automobiles, telephones, televisions, refrigerators, clothing, books and practically all the things we use are produced by machinery.

OBJECTIVES

After completing this unit you will be familiar with:
1. The development of tools throughout history
2. The standard types of machine tools used in shops
3. The newly developed space-age machines and processes

The history of machine tools began during the stone age (over 5000 years ago), when the only tools were hand tools made of wood, animal bones, or stone (Fig. 1-1).

Between 2000 and 2500 B.C., stone spears and axes were replaced with copper and bronze implements and power supplied by humans was in a few cases replaced with animal power. It was during this bronze age that human beings first enjoyed ''power-operated'' tools.

Around 1000 B.C., the iron age dawned, and most bronze tools were replaced with more durable iron imple-

| Beginning Over 50,000 Yrs. Ago | Beginning Over 6500 Yrs. Ago | Beginning About 3400 Yrs. Ago | Beginning About 150 Yrs. Ago |

STONE AGE — BRONZE AGE — IRON AGE — MACHINE AGE

Sawing · Drilling · Chopping · Slicing · Burnishing · Scraping · Measuring

Fig. 1-1 The development of hand tools over the years.

ments. After the smiths learned to harden and temper iron, its use became widespread. Tools and weapons were greatly improved and animals were domesticated to provide power for some of these tools, such as the plow. During the iron age, all commodities required by humans, such as housing and shipbuilding materials, wagons, and furniture, were all handmade by the skilled craftsmen of the era.

About 300 years ago, the iron age became the machine age. In the 17th century, people began exploring new sources of energy. Water power began to replace human and animal power. With the new power came improved machines and, as production increased, more products became available. Machines continued to be improved, and the boring machine made it possible for James Watt to produce the first steam engine in 1776, beginning the industrial revolution. The steam engine made it possible to provide power to any area where it was needed. With quickening speed, machines were improved and new ones invented. Newly designed pumps reclaimed thousands of acres of the Netherlands from the sea. Mills and plants which had depended on water power were converted to steam power to produce flour, cloth, and lumber more efficiently. Steam engines replaced sails and steel replaced wood in the shipbuilding industry. Railways sprung up unifying the countries and steamboats connected the continents. Steam-driven tractors and improved farm machinery lightened the task of the farmer. As machines improved, further sources of power were developed. Generators were made to produce electricity and diesel and gasoline engines were developed.

With further sources of energy available, industry grew and new and better machines were built. Progress continued slowly during the first part of the 20th century except for spurts during the two world wars. The Second World War in the 1940s sparked an urgent need for new and better machines, which resulted in more efficient production (Fig. 1-2).

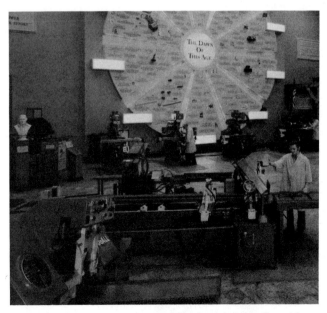

Fig. 1-2 New machine tools were developed during the mid-20th century.

Since the 1950s, progress has been rapid and we are now in the space age. Calculators, computers, robots, and automated machines and plants are commonplace. The atom has been harnessed and nuclear power is used to produce electricity and to drive ships. We have traveled to the moon and outer space, all because of the fantastic technological developments. Machines can mass produce parts to millionths of an inch accuracy. The fields of measurement and metallurgy have become very sophisticated. All of these factors have produced a high standard of living for us. Now all people, regardless of their occupation or status, are dependent on machines and/or their products (Fig. 1-3).

Through constant improvement, modern machine tools have become more accurate and efficient. Improved production and accuracy have been made possible through the application of hydraulics and electronic devices such as numerical control to basic machine tools.

COMMON MACHINE TOOLS

Machine tools are generally power-driven metal-cutting or -forming machines used to shape metals by:

☐ The removal of chips
☐ Pressing, drawing, or shearing
☐ Controlled electrical machining processes

Any machine tool generally has the capability of:

☐ Holding and supporting the workpiece
☐ Holding and supporting a cutting tool
☐ Imparting a suitable movement (rotating or reciprocating) to the cutting tool or the work
☐ Feeding the cutting tool or the work so that the desired cutting action and accuracy will be achieved

The machine tool industry is divided into several different categories such as the general machine shop, the toolroom, and the production shop. The machine tools found in the metal trade fall into three broad categories:

1. *Chip-producing machines* which form metal to size and shape by cutting away the unwanted sections. These machine tools generally alter the shape of steel products by casting, forging, or rolling in a steel mill.
2. *Non-chip-producing machines* which form metal to size and shape by pressing, drawing, or shearing. These machine tools generally alter the shape of sheet steel products and also produce parts which need little or no machining by compressing granular or powdered metallic materials.
3. *New-generation machines* which were developed to perform operations which would be very difficult, if not impossible, to perform on chip- or non-chip-producing machines. Electro-discharge and electro-chemical machines, for example, use either electrical and/or chemical energy to form metal to size and shape.

A general machine shop contains a number of standard machine tools that are basic to the production of a variety of metal components. Operations such as turning, boring, threading, drilling, reaming, sawing, milling, filing, and grinding are most commonly performed in a machine shop.

HISTORY OF MACHINES

Fig. 1-3　Machine tools produce tools and machines for all types of products.　(Courtesy DoAll Company.)

Fig. 1-4　Common machine tools found in a machine shop. (Courtesy DoAll Company.)

Fig. 1-5　A standard upright drill press.

Machines such as the drill press, engine lathe, power saw, shaper, milling machine, and grinder are usually considered the *basic machine tools* in a machine shop (Fig. 1-4).

STANDARD MACHINE TOOLS

DRILL PRESS

The drill press or drilling machine (Fig. 1-5), probably the first mechanical device developed prehistorically, is used primarily to produce round holes. Drill presses range from the simple hobby type to the more complex automatic and numerically controlled machines used for production purposes. The function of a drill press is to grip and revolve the cutting tool (generally a twist drill) so that a hole can be produced in a piece of metal or other material. Operations such as drilling, reaming, spot facing, countersinking, counterboring, and tapping are commonly performed on a drill press.

ENGINE LATHE

The engine lathe (Fig. 1-6) is used to produce round work. The workpiece, held by a work-holding device mounted on

4

Fig. 1-6 An engine lathe is used to produce round work. (Courtesy Monarch Machine Tool Company.)

the lathe spindle, is revolved against a cutting tool which produces a cylindrical form. Straight turning, tapering, facing, drilling, boring, reaming, and thread cutting are some of the common operations performed on a lathe.

METAL SAW

The metal-cutting saws are used to cut metal to the proper length and shape. There are two main types of metal-cutting saws: the bandsaw (horizontal and vertical) and the reciprocating cut-off saw. On the vertical bandsaw (Fig. 1-7) the workpiece is held on the table and brought into contact with the continuous-cutting saw blade. It can be used to cut work to length and shape. The horizontal bandsaw and the reciprocating saw are used to cut work to length only. The material is held in a vise and the saw blade is brought into contact with the work.

SHAPER

The shaper (not illustrated) is generally used for producing flat, curved, or angular surfaces on metal workpieces. The cutting tool moves back and forth (reciprocates) in a horizontal plane across the face of the work, which may be held in a vise or fastened to the table. The work is moved across for successive cuts either by hand or automatic feed. Shapers are manufactured with either the crank-type or the hydraulic-driving mechanism.

MILLING MACHINE

The horizontal milling machine (Fig. 1-8) and the vertical milling machine are two of the most useful and versatile machine tools. Both machines use one or more rotating milling cutters having single or multiple cutting edges. The workpiece, which may be held in a vise, fixture, accessory, or fastened to the table, is fed into the revolving cutter. Equipped with proper accessories, milling machines are capable of performing a wide variety of operations such as drilling, reaming, boring, counterboring, and spot facing,

Fig. 1-7 A contour-cutting bandsaw. (Courtesy DoAll Company.)

Fig. 1-8 A horizontal milling machine. (Courtesy Cincinnati Milacron Inc.)

and of producing flat and contour surfaces, grooves, gear teeth, and helical forms.

GRINDER

Grinders use an abrasive cutting tool to bring a workpiece to an accurate size and produce a high surface finish. In the grinding process, the surface of the work is brought into contact with the revolving grinding wheel. The most common types of grinders are the surface, cylindrical, cutter and tool, and the bench or pedestal grinder.

Surface grinders (Fig. 1-9) are used to produce flat, angular, or contoured surfaces on a workpiece.
Cylindrical grinders are used to produce internal and external diameters which may be straight, tapered, or contoured.
Cutter and tool grinders are generally used to sharpen milling machine cutters.
Bench or pedestal grinders are used for offhand grinding and the sharpening of cutting tools such as chisels, punches, drills, and lathe and shaper tools.

Fig. 1-9 A surface grinder is used to grind flat surfaces. (Courtesy DoAll Company.)

Fig. 1-10 Robots are finding ever-increasing applications in industry. (Courtesy Cincinnati Milacron Inc.)

SPECIAL MACHINE TOOLS

Special machine tools are designed to perform all the operations necessary to produce a single component. Some special-purpose machine tools include gear-generating machines; centerless, cam, and thread grinders; turret lathes; and automatic screw machines.

SPACE-AGE MACHINES AND PROCESSES

The introduction of *electro-discharge machining, electro-chemical machining,* and *electrolytic grinding* has made it possible to machine new space-age materials and to produce shapes which were difficult or often impossible to produce by other methods.

Numerical-controlled and *computer-controlled* machines, more recent additions to the array of machine tools, have greatly increased the production and improved the quality of the finished product. Consistent accuracy over many hundreds of parts is one of the features of these machines.

The numerical control principle has also been applied to *robots,* which are now capable of handling materials and changing machine tool accessories as easily and probably more efficiently than a person can (Fig. 1-10). *Robotics* has become the fastest-growing phase of the manufacturing industry.

Since its development, the *laser* has been applied to several areas in the manufacturing industry. Lasers are now used to an increasing extent for cutting and welding all types of metals—even those which have been impossible by other methods. Laser beams can pierce diamonds or any known material and are also used in extremely accurate measuring and surveying devices.

With the introduction of the numerous special machines and special cutting tools, production has increased tremendously over standard machine methods. Many products are produced automatically by a continuous flow of finished parts from these special machines. With product control and high production rates, everyone can enjoy the pleasures and conveniences of the automobile, power lawn mowers, automatic washers, stoves, and scores of other modern products. Without the basic machine tools to produce the first pieces required to develop these ideas of production and automation, the costs of many luxuries that we now enjoy would be prohibitive.

REVIEW QUESTIONS

1. Briefly trace the development of tools from the stone age to the industrial revolution.
2. Why are machine tools so important to our society?
3. How have improved production and accuracy been achieved with basic machine tools?
4. Name three categories of machine tools used in the metalworking trade.
5. List five operations which can be performed on each of the following:
   (a) drill press
   (b) lathe
   (c) milling machine
6. Name four types of grinders found in a machine shop.
7. What is the importance of the electro-machining processes?
8. What effect has numerical and computer control had on manufacturing?
9. State two applications of robots.
10. What is the importance of lasers in modern industry?

MACHINE TRADE
OPPORTUNITIES

SECTION 2

Almost all products used by people, whether in farming, mining, manufacturing, construction, transportation, communication, or the professions, are dependent on machine tools for their manufacture. Constant improvements to and efficient use of machine tools affect the standard of living of any nation. Only through machine tools have we been able to enjoy such pleasures as the automobile, airplane, television, home furnishings, appliances, and many other items on which we rely in our daily lives. Through constant improvement, modern machine tools have become more accurate and efficient. Improved production and accuracy have been made possible through the application of hydraulics and electronic devices such as numerical control to basic machine tools.

CAREERS IN THE METALWORKING INDUSTRY
UNIT 2

The advancing technology, new ideas, new products, and special processes and manufacturing techniques are creating new and more specialized jobs. To advance in the machine trade, it is necessary to keep up-to-date with modern technology. A young person leaving school may be employed in an average of five jobs during his or her lifetime, three of which do not even exist today. Industry is always on the lookout for bright young people who are precise and do not hesitate to assume responsibility. To be successful, do your job to the best of your ability and never be satisfied with inferior workmanship. Always try to produce quality products, at a reasonable price, in order to compete with the foreign products which are such a serious concern for North American industry.

OBJECTIVES

After completing this unit you will:
1. Know the various types of jobs available in the machine tool industry
2. Know the type of work each job entails

There are many different careers available in the metalworking industry depending on the skill, initiative, and qualifications of an individual. The metalworking trade will offer exciting opportunities to any ambitious young person who is willing to accept the challenge of working to close tolerances and producing intricate parts. To be successful in this trade, other characteristics such as care of self, orderliness, accuracy, confidence, and safe work habits are essential.

APPRENTICESHIP TRAINING

One of the best ways to learn a skilled trade is through an apprenticeship program. An apprentice (Fig. 2-1) is one who is employed to learn a trade under the guidance of skilled tradespeople. The apprenticeship program is set up in conjunction with and under the supervision of the company, the Department of Labor, and the trade union. It is usually about two- to four-years' duration and includes on-the-job training and related theory or classroom work. This period of time may be reduced by the completion of approved courses or because of previous experience in the trade.

To qualify for an apprenticeship, it is advantageous to have completed a high school program or its equivalent. Mechanical ability with a good standing in mathematics, science, English, and mechanical drawing is very desirable.

Fig. 2-1 An apprentice learns the trade under the guidance of a skilled tradesperson. (Courtesy DoAll Company.)

Apprentices earn as they learn; the wage scale increases periodically during the training program.

Upon completion of an apprenticeship program, a certificate is granted which qualifies a person to apply as a journeyman in the trade. Further opportunities in the trade are limited only by the initiative and interest of a person. It is quite possible for an apprentice eventually to become an engineer.

MACHINE OPERATOR

Machine tool operators are classified as semiskilled tradespeople. They (Fig. 2-2) are usually rated and paid according to their job classification, skill, and knowledge. The class A operator possesses more skill and knowledge than class B and C operators. For example, a class A operator should be able to operate the machine and:

- [ ] Make necessary machine setups
- [ ] Adjust cutting tools
- [ ] Calculate cutting speeds and feeds
- [ ] Read and understand drawings
- [ ] Read and use precision measuring tools

With the continued advancement in tape-controlled machines and programmable robots, gradually operators' jobs will be minimized. However, some machine tool operators may qualify with advanced technology courses and remain employed as operators on computer numerical control (CNC) turning centers, machining centers, and CNC robots.

MAINTENANCE MACHINIST

The maintenance machinist trade is a combination of mechanical, rigging, and carpentry skills. The time for apprenticeship varies but usually ranges from two to four years and includes the necessary related theory for the job. During the apprenticeship the trainee works with qualified tradespeople. A maintenance machinist may be required to:

- [ ] Move and/or install machinery including production lines
- [ ] Read drawings and calculate sizes, fits, and tolerances of machine parts
- [ ] Repair machines by replacing and fitting new parts
- [ ] Dismantle and install equipment

To become a maintenance machinist apprentice, a student should have good technical training and be a high school graduate. A general knowledge of electricity, carpentry, sheet metal, and the machine trade is very helpful.

The job outlook and forecasts are good. It is expected that most large industries will employ a work force of maintenance machinists to maintain and install machinery and production lines.

MACHINIST

Machinists (Fig. 2-3) are skilled workers who can efficiently operate all standard machine tools. Machinists must be able to read drawings and use precision measuring instruments and hand tools. They must have acquired enough knowledge and developed sound judgment to perform any bench, layout, or machine tool operation. In addition, they should be capable of making mathematical calculations required for setting up and machining any part. Machinists should have a thorough knowledge of metallurgy and heat-treating. They should also have a basic understanding of welding, hydraulics, electricity, pneumatics, and a familiarization with computer technology.

TYPES OF MACHINE SHOPS

A machinist may qualify to work in a variety of shops. The three most common are maintenance, production, and jobbing shops.

A *maintenance shop* is usually connected with a manufacturing plant, lumber mill, or a foundry. A maintenance machinist generally makes and replaces parts for all types of setup and cutting tools and production machinery. The machinist must be able to operate all machine tools and be

Fig. 2-2　A machine operator generally operates only one type of machine.　(Courtesy Cincinnati Milacron Inc.)

Fig. 2-3　A machinist is skilled in the operation of all machines. (Courtesy Cincinnati Lathe & Tool Company.)

familiar with bench operations such as layout, fitting, and assembly.

A *production shop* may be connected with a large factory or plant which makes many types of identical machined parts such as pulleys, shafts, bushings, motors, and sheet-metal pieces. A person working in a production shop generally operates one type of machine tool and often produces identical parts (Fig. 2-4).

A *jobbing shop* is generally equipped with a variety of standard machine tools and perhaps a few production machines such as a turret lathe and punch and shear presses. A jobbing shop may be required to do a variety of tasks usually under contract to another company. This work may involve the production of jigs, fixtures, dies, molds, tools, or short runs of special parts. A person working in a jobbing shop generally is a qualified machinist or toolmaker and would be required to operate all types of machine tools and measuring equipment.

TOOL AND DIEMAKER

A *tool and diemaker* is a highly skilled craftsman who must be able to make different types of dies, molds, cutting tools, jigs, and fixtures. These tools may be used in the mass production of metal, plastic, or other parts. For example, to make a die to produce a 90° bracket in a punch press, the tool and diemaker must be able to select, machine, and heat treat the steel for the die components. He or she should also know what production method will be used to produce the part since this information will help to produce a better die. For a mold used to produce a plastic handle in an injection molding machine, the tool and diemaker must know the type of plastic used, finish required, and the process used in production.

To qualify as a tool and diemaker, a person should serve an apprenticeship, have above-average mechanical ability, and be able to operate all standard machine tools. This person also requires a broad knowledge of shop mathematics, print reading, machining operations, metallurgy, heat-treating, computers, and space-age machining processes.

TECHNICIAN

A *technician* is a person who works at a level between the professional engineer and the machinist. The technician may assist the engineer in making cost estimates of products, preparing technical reports on plant operation, or program a numerically controlled machine.

A technician should have completed high school education and have at least two years of postsecondary education in a community college, technical institute, or a university. A technician must also possess a good knowledge of drafting, mathematics, and technical writing.

Opportunities for technicians are becoming more plentiful because of the development of machine tools such as numerical control, turning centers, and electro-machining processes. Technicians are usually trained in only one area

Fig. 2-4 This production driller produces many identical parts.

of technology such as electrical, manufacturing, machine tool, or metallurgy. Some technicians of one technology may require a knowledge of more than their specialty field. For example, a machine tool technician (Fig. 2-5) should have a knowledge of industrial machines and manufacturing processes in order to know the best method of manufacturing a product. However, it is not necessary for a technician to perform as a skilled machinist. A technician may qualify as a technologist after at least one year of on-the-job training under a technologist or engineer.

TECHNOLOGIST

A *technologist* works at a level between a graduate engineer and a technician. Most technologists are three- or four-year graduates from a community or technical college. Their studies generally include physics, advanced mathematics, chemistry, engineering graphics, computer programming, business organization, and management.

An *engineering technologist* may do many jobs performed by an engineer such as design studies, production planning, laboratory experiments, and supervision of technicians. This permits the engineer to work in other important areas. Technologists are often employed to serve in a middle management position within a large company. A technologist may be employed in many areas such as quality and cost control, production control, labor relations, training, and product analysis. A technologist may qualify as an engineer by taking further education at university level and by passing a qualifying examination.

Fig. 2-5 A technician is often required to check the setup and operation of a machine program. (Courtesy Cincinnati Milacron Inc.)

INSPECTORS (QUALITY CONTROL)

An *inspector* (Fig. 2-6) checks and examines machined parts to determine if they meet the specifications on the drawing. If the parts are not within the limits on the drawing, they will not fit together when assembled. This is very important since parts made in one country may be assembled in another or be interchanged with worn or broken parts.

An inspector should have a technical or vocational education in order to be familiar with measuring tools and inspection processes. The on-the-job training may take from several weeks to several years, depending on the job, the items to be inspected, and the technical knowledge required. An inspector may require varying degrees of skill depending on the size, cost, type, and tolerances required on the finished workpiece. A good inspector should:

☐ Understand and read mechanical drawings
☐ Make basic mathematical calculations
☐ Use micrometers, gages, comparators, and precision measuring instruments

INSTRUMENT MAKERS

Instrument makers are highly skilled tool and diemakers who work directly with scientists and engineers. Instrument makers should be able to operate all precision machine tools in order to make measuring instruments, gages, and special machines for testing purposes. Generally an instrument maker requires more training than a machinist or a tool and diemaker. They must work to closer tolerances and limits than a machinist. Also, they usually must perform all the work on the instrument or gage being produced.

Instrument makers generally serve four or five years of apprenticeship. They may also learn the craft by transferring from a toolmaker's or machinist trade and further training on the job.

With new processes in manufacturing, the demand for instrument makers will remain fairly close to the demand

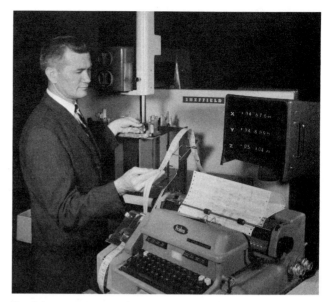

Fig. 2-6 An inspector checks the dimensions of finished part and products. (Courtesy The Bendix Corporation.)

for skilled tradespeople. Instrument makers are employed by research centers, scientific laboratories, manufacturers of gages and measuring tools, and government and standards departments.

PROFESSIONS

There are many areas open to the engineering graduate. Teaching is one of the most satisfying and challenging professions. Graduation from a college in a teacher training course is required along with on-the-job industrial experience. Although industrial experience is not a prerequisite, it will prove helpful in teaching. Some states and/or some schools require on-the-job experience to qualify for technical and vocational certification. With industrial experience a person may teach industrial arts or in designated subject areas in a technical or vocational school.

Engineers in industry are responsible for the design and development of new products and production methods and for redesigning and improving existing products. Most engineers specialize in a certain phase of engineering such as:

☐ Civil
☐ Metallurgical
☐ Aerospace
☐ Mechanical
☐ Electrical
☐ Electronics

A bachelor's degree in engineering is usually required to enter the profession; however, some classes of engineering, such as tool and manufacturing, progress through a practical experience program and obtain certificates after passing qualifying examinations. Because of the variety of engineering jobs available, women as well as men are entering the many phases of this profession.

NUMERICAL CONTROL PROGRAMMERS

Tool programmers for numerically controlled machines must be thoroughly familiar with all machining techniques. A programmer should be familiar with a skilled machine operator's job and know all the sequences and procedures for producing a machined part. In order to program a machine to produce a part, an operator must be able to:

☐ Read working drawings
☐ Select the best tools for the machining operation
☐ Calculate speeds and feeds for different materials and types of cutting tools
☐ Have a knowledge of production costs and machine tool processes

To obtain the necessary programming background and knowledge, many vocational schools, technical schools, and universities offer CNC and NC (numerical control) programming courses. Computer-assisted machining (CAM) is perhaps one of the fastest-growing specialized areas in the machining industry. To learn programming on a specialized turning center is only the beginning of what industry will be offering in the future. A good education is essential along with related information in electronics, mathematics, and manufacturing processes.

REVIEW QUESTIONS

1. Define an apprentice.
2. Name three desirable qualities a person should have for an apprenticeship training.
3. Explain the difference between a machinist and a machine operator.
4. Briefly explain the difference between a jobbing and a production shop.
5. Define a tool and diemaker.
6. How can a person become a tool and diemaker?
7. Explain the difference between a technician and a technologist.
8. Briefly define the duties of an inspector and an instrument maker.
9. What qualifications are required to become a technical teacher?
10. List four areas in industry that require an engineer's qualifications.

GETTING THE JOB
UNIT 3

After graduation or leaving school, probably the most important task is that of finding a full-time job. *Choosing the right job is very important.*

After consulting with the school guidance counselor, the state employment services, and any other agency which may be helpful, start looking into the job which appeals to you. In most areas aptitude tests are available to help you decide on the career you may wish to explore.

ASSESS YOUR ABILITIES

In order to help determine where your interests lie, ask yourself:

☐ What type of work do I like?
☐ What type of work do I not like?
☐ What jobs have I done with some success?
☐ What skills have I acquired at school?
☐ What have I done in my part-time work that has been outstanding?
☐ Do I enjoy taking apart and repairing appliances and items which are not working?

EXPLORE YOUR INTERESTS

After having narrowed the field, then—

☐ Read any books on the chosen subject(s) (Fig. 3-1).
☐ Talk to people who do the type of work that you are considering. Ask them about the job and job opportunities.
☐ Have further talks with the school guidance counselor.
☐ Consult the state or federal employment services.

When you have gathered enough information to discuss reasonably the type of work you are interested in, check the help-wanted ads in the newspapers.

Fig. 3-1 Read as much as possible about a job which might appeal to you.

MAKING A RÉSUMÉ

When you have decided on the job you wish to apply for, prepare a résumé listing the following facts in a logical order.

☐ Your full name.
☐ Address, area or zip code, telephone number.
☐ Social security number (if any).
☐ Father's and mother's names and address.
☐ Education—schools attended and grade or level completed.
☐ Other special training which may be helpful; for example, first-aid courses.
☐ Special interests and hobbies.
☐ Sports in which you are active or interested.
☐ Any organizations to which you belong or are active in.
☐ Names of firms and places where you have worked. (List in reverse chronological order.) Include the following information:
    Period and dates of employment.
    Type of work and equipment operated (if any).
    Supervisor's name.
    Salary.
    Reason for leaving. (Be honest because this information is usually available if the prospective employer wishes to phone your former employer.)
☐ A list of at least three persons who may be contacted for character references. Include addresses and telephone numbers. Be sure to ask permission to use their names *before* listing them on an application.

Many state employment services and government agencies have pamphlets which can be used as a guide to prepare résumés. Also available is information on how to get and keep a job which can be valuable to a person looking for a job.

FACTS ABOUT INTERVIEWS

ARRANGING THE INTERVIEW

After completing the résumé, submit it with a covering letter to the personnel manager of the company involved. Be sure to include a request for an interview. In many cases you may phone the company and make an appointment for an interview. In this case, leave the résumé with the person who interviews you.

QUESTIONS COMMONLY ASKED DURING AN INTERVIEW

Prior to the interview, consider the following questions, often asked by employers and the probable reason for asking them. You should be prepared with satisfactory answers.

Question: Why would you like to work here?
  Reason: To see if you have gathered any information about the company before the interview.
Question: What were your best subjects at school?
  Reason: This will reveal some of your interests and abilities.
Question: What sports or activities did you participate in when attending school?
  Reason: To find out the interests and abilities outside of school. Also to see if you can work as a team.
Question: What type of job do you hope to have five years from now?
  Reason: To assess your ambition and initiative.
Question: At what salary would you expect to start?
  Reason: To see if you are familiar with the "going rates" and to see how you assess your abilities.
Question: What do you have to offer for the job?
  Reason: To give you a chance to outline your abilities.
Question: What type of books or plays are you interested in?
  Reason: To assess your interests and often your environment.
Question: How did you get along with your previous employer?
  Reason: Your answer may reveal if you are a complainer or one that is hard to get along with.
Question: Why are you applying for this job?
  Reason: To see if you have checked into this particular job and not just looking for any job.

THE INTERVIEW

In preparation for the interview you should consider the following:

☐ Be sure of the address, the office, and the time.
☐ Know the name and position of the person that you are to see. This can be obtained by a phone call to the company prior to the interview.
☐ Be neatly dressed and groomed. Remember, a neat applicant usually commands more attention (Fig. 3-2).

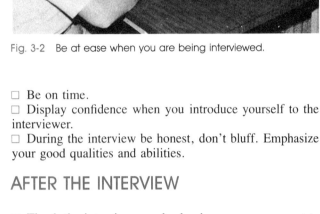

Fig. 3-2   Be at ease when you are being interviewed.

Fig. 3-3   Always thank the interviewer—it may pay off!

□ Be on time.
□ Display confidence when you introduce yourself to the interviewer.
□ During the interview be honest, don't bluff. Emphasize your good qualities and abilities.

AFTER THE INTERVIEW

□ Thank the interviewer and ask when you may expect to hear from him or her (Fig. 3-3).
□ If you are offered a job, accept it (or reject it) as soon as possible. Never leave the prospective employer "hanging on," awaiting your decision. If you are not interested, explain why.
□ The next day, send the interviewer a short letter expressing your appreciation for the valuable time he or she took for the interview.
□ If you do not hear from the company in a reasonable time (7 to 10 days) call and ask for the person that interviewed you. Ask if there has been any decision as yet and when you might expect one.
□ If you don't get the first job, apply to other plants. Don't stop looking.
□ Try to use each interview as a learning experience which will eventually help you to get the right job.
□ After several unsuccessful interviews, you might seek professional counseling from the Department of Labor, a vocational school, or a community college.

POINTS TO REMEMBER

□ The job will not find you. You must find the job.
□ Know the type of work that you want and don't offer to take any job.
□ Look for the type of work which you feel will be interesting. You will be much more successful if you like your work.

REVIEW QUESTIONS

1. List four things that should be considered when you are attempting to assess your abilities for a job.
2. Name four ways to obtain further information about a trade or job.
3. Assume you are applying for a job; prepare a personal résumé that you would submit to an employer.
4. Name three methods of arranging an interview.
5. List four important points that must be considered in preparation for an interview.
6. Name four important points that should be followed after an interview.

SAFETY
SECTION 3

There has never been a more meaningful saying than "Safety Is Everyone's Business." To become a skilled craftsman, it is very important to learn to work safely taking into consideration not only your own safety but the safety of your fellow workers.

In general, everyone has a tendency to be careless about safety many times. We take chances every day by not wearing seat belts, walking under ladders, cluttering the work area, and doing many other careless and unsafe things. It seems that people tend to feel that accidents always happen to others. However, be sure to remember that a moment of carelessness can result in an accident which can affect you for the rest of your life. A loss of eyesight due to not wearing glasses, or the loss of a limb due to loose clothing caught in a machine can seriously affect or end your career in the machine tool trade. *THINK SAFE, WORK SAFE,* and *BE SAFE.*

SAFETY IN THE MACHINE SHOP

UNIT 4

All hand and machine tools can be dangerous if used improperly or carelessly. Working safely is one of the first things a student or apprentice should learn because the safe way is usually the most correct and efficient way. A person learning to operate machine tools *must first* learn the safety regulations and precautions for each tool or machine. Far too many accidents are caused by carelessness in work habits. It is easier and much more sensible to develop safe work habits than to suffer the consequences of an accident. *SAFETY IS EVERYONE'S BUSINESS AND RESPONSIBILITY.*

OBJECTIVES

After completing this unit you will be able to:
1. Recognize safe and unsafe work practices in a shop
2. Identify and correct hazards in the shop area
3. Perform the job in a manner which is safe to you and other workers

SAFETY ON THE JOB

The safety programs initiated by accident prevention associations, safety councils, governmental agencies, and industrial firms are constantly attempting to reduce the number of industrial accidents. *Nevertheless,* each year accidents which could have been avoided result not only in millions of dollars' worth of lost time and production but also in a great deal of pain, many lasting physical handicaps, or even the death of a worker. Modern machine tools are equipped with safety features, but it is still the operator's responsibility to use these machines wisely and safely.

Accidents don't just happen; they are caused. The cause of an accident can usually be traced to carelessness on someone's part. Accidents can be avoided, and a person learning the machine shop trade must first develop safe work habits.

A safe worker should:

☐ Be neat, tidy, and safely dressed for the job he or she is performing
☐ Develop a responsibility for personal safety and the safety of fellow workers
☐ Think safety and work safely at all times

SAFETY IN THE SHOP

Safety in a machine shop may be divided into two broad categories:

☐ Those practices which will prevent injury to the workers.
☐ Those practices which will prevent damage to machines

and equipment. Too often damaged equipment results in personal injuries.

In considering these categories, one must consider personal grooming, proper housekeeping (including machine maintenance), safe work practices, and fire prevention.

PERSONAL GROOMING

The following rules should be observed when working in a machine shop.

1. Always wear approved safety glasses in any area of the machine shop. Most plants now insist that all employees and visitors wear safety glasses or some eye protection device when entering a shop area. There are several types of eye protection devices available for use in the machine shop.

a. The most common are the *plain safety glasses* with side shields (Fig. 4-1A). These offer sufficient eye protection when operating any machine or when performing any bench or assembly operation. The lenses are made of shatterproof glass and the side shields protect the sides of the eyes from flying particles.

b. *Plastic safety goggles* (Fig. 4-1B) are generally worn by anyone wearing prescription eyeglasses. These are soft, flexible plastic and fit closely around the upper cheeks and forehead. Unfortunately, these goggles have a tendency to fog up when it is warm.

c. *Face shields* (Fig. 4-1C) may also be used by those wearing prescription glasses. The plastic shield gives full face protection and permits air circulation between the face and the shield, thus preventing fogging up in most situations. These shields, as well as approved pro-

Fig. 4-1 Types of safety glasses. (A) Plain; (B) plastic goggles; (C) Face shields.

Fig. 4-2 Loose clothing can easily be caught in moving parts of machinery.

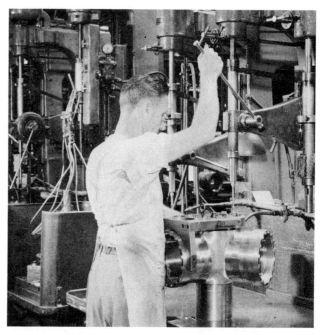

Fig. 4-3 Tie aprons behind your back to keep the ties from being caught in machinery.   (Courtesy Whitman & Barnes.)

Fig. 4-4 Wearing rings and watches can be the cause of serious injuries.

tective clothing and gloves, *must* be worn when heating and quenching materials during heat-treating operations or where there is any danger of hot flying particles. In industry, some companies provide their employees with prescription safety glasses, which eliminate the need for protective goggles or shields.

NOTE: Never think that because you are wearing glasses your eyes are safe. If the lenses are not made of approved safety shatterproof glass, serious eye injury can result.

2. Never wear loose clothing when operating any machine (Fig. 4-2).
  a. Always roll up your sleeves or wear short sleeves.
  b. Clothing should be made of hard, smooth material that will not catch easily in a machine. Loose-fitting sweaters should not be worn for this reason.
  c. Remove or tuck in a necktie before starting a machine. If you want to wear a tie, make it a bow tie.
  d. When wearing a shop apron, *always tie it at the back* and never in front of you so that the apron strings will not get caught in rotating parts (Fig. 4-3).
3. Remove wrist watches, rings, and bracelets; these can get caught in the machine causing painful and often serious injury (Fig. 4-4).
4. Never wear gloves when operating a machine.
5. Long hair must be protected by a hair net or approved protective shop cap (Fig. 4-5). One of the most common accidents on a drill press is caused by long, unprotected hair getting caught in a revolving drill.

Fig. 4-5 Long hair must be protected by a hair net or approved shop cap.

6. Canvas shoes or open-toe sandals must *never* be worn in a machine shop since they offer no protection to the feet against sharp chips or falling objects. In industry, most companies make it mandatory for the employees to wear safety shoes.

HOUSEKEEPING

The operator should remember that *good housekeeping will never interfere with safety or efficiency;* therefore, the following points should be observed.

Fig. 4-6 Grease and oil on the floor can cause dangerous falls.

1. *ALWAYS STOP THE MACHINE BEFORE YOU ATTEMPT TO CLEAN IT.*
2. Always keep the machine and hand tools clean. Oily surfaces can be dangerous. Metal chips left on the table surface may interfere with the safe clamping of a workpiece.
3. Always use a brush and not a cloth to remove any chips.
4. Oily surfaces should be cleaned with a cloth.
5. Do not place tools and materials on the machine table—use a bench near the machine.
6. Keep the floor free from oil and grease (Fig. 4-6).
7. Sweep up the metal chips on the floor frequently. They become embedded in the soles of shoes and can cause dangerous slippage if a person walks on a terrazzo or concrete floor. A scraper, mounted on the floor near the door, should be used to remove these chips before leaving the shop (Fig. 4-7).
8. Never place tools or materials on the floor close to a machine where they will interfere with the operator's ability to move safely around the machine (Fig. 4-8).
9. Return bar stock to the storage rack after cutting off the required length (Fig. 4-9).
10. Never use compressed air to remove chips from a machine. Not only is it a dangerous practice because of flying metal chips, but small chips and dirt can become wedged between machine parts and cause undue wear.

SAFE WORK PRACTICES

1. *Do not operate any machine before understanding its mechanism and knowing how to stop it quickly.* Knowing how to stop a machine quickly can prevent a serious injury.
2. Before operating any machine, be sure that the safety devices are in place and in working order. Remember, safety devices are for the operator's protection and should not be removed.
3. Always disconnect the power and lock it off at the switch box when making repairs to any machine (Fig. 4-10). A sign should be placed on the machine noting that it is out of order.
4. Always be sure that the cutting tool and the workpiece are properly mounted before starting the machine.
5. *Keep hands away from moving parts.* It is dangerous practice to "feel" the surface of the revolving work or to stop a machine by hand.

Fig. 4-7 Remove the chips from the soles of your shoes before leaving the shop.

Fig. 4-8 Poor housekeeping can lead to accidents.

Fig. 4-9 Store stock safely in a material stock rack.

6. *Always stop a machine before measuring, cleaning, or making any adjustments.* It is dangerous to do any type of work around moving parts of a machine (Fig. 4-11).

7. *Never use a rag near the moving parts of a machine.* The rag may be drawn into the machine, along with the hand that is holding it.

8. *Never have more than one person operate a machine at the same time.* Not knowing what the other person would or would not do has caused many accidents.

9. *Get first aid immediately for any injury, no matter how small.* Report the injury and be sure that the smallest cut is treated to prevent the chance of a serious infection.

10. Before you handle any workpiece, remove all burrs and sharp edges with a file.

11. Do not attempt to lift heavy or odd-shaped objects which are difficult to handle on your own.

12. For heavy objects, follow safe lifting practices.

    a. Assume a squatting position with your knees bent and back straight.

    b. Grasp the workpiece firmly.

    c. Lift the object by straightening the legs and keeping the back straight (Fig. 4-12). This procedure uses the leg muscles and prevents injury to the back.

13. Be sure the work is clamped securely in the vise or to a machine table.

14. Whenever work is clamped, be sure the bolts are placed closer to the workpiece than to the clamping blocks.

15. Never start a machine until you are sure that the cutting tool and machine parts will clear the workpiece (Fig. 4-13).

16. Use the proper wrench for the job, and replace nuts with worn corners.

17. It is safer to pull on a wrench than to push on it.

FIRE PREVENTION

1. Always dispose of oily rags in proper metal containers.

2. Be sure of the proper procedure before lighting a gas furnace.

3. Know the location and the operation of every fire extinguisher in the shop.

Fig. 4-10 Power switches must be locked off before making repairs or adjustments to a machine.

Fig. 4-12 Follow recommended lifting procedures to avoid back injury.

Fig. 4-11 The machine must be stopped before the work is measured.

Fig. 4-13 Make sure the cutting tool and parts of the machine will clear the work.

4. Know the location of the nearest fire exit from the building.
5. Know the location of the nearest fire-alarm box and its operating procedure.
6. When using a welding or cutting torch, be sure to direct the sparks away from any combustible material.

REVIEW QUESTIONS

1. What must be learned before operating a machine tool for the first time?
2. List three qualities of a safe worker.

PERSONAL GROOMING

3. Name three types of eye protection which may be found in a shop.
4. State four precautions which must be observed in regard to clothing worn in a shop.
5. Why should gloves not be worn when operating a machine?

6. How must long hair be protected?

HOUSEKEEPING

7. Why must a cloth not be used to remove chips?
8. Why should shoe soles be scraped before leaving the shop?
9. State two reasons why compressed air should not be used for cleaning machines.

SAFE WORK PRACTICES

10. State three precautions to observe before operating any machine.
11. Describe the procedure to follow when lifting a heavy object.
12. What should you do immediately after receiving any injury?

FIRE PREVENTION

13. What three fire prevention factors should everyone become familiar with before starting to work in a machine shop?

JOB PLANNING
SECTION 4

Machine shop work consists of machining a variety of parts (round, flat, contour) and either assembling them into a unit or using them separately to perform some operation. It is important that the sequence of operations be carefully planned in order to produce a part quickly and accurately. Improper planning or following a wrong sequence of operations can often result in spoiled work.

PLUNGE CUT

THREAD CUTTING

TAPER TURNING

Programming time: approx. 30 min
Input time (manually): approx. 17 min
Input time (via tape): approx. 30 s
Machining time: 15 min (incl. 6 tool changes)

TURNING OF RADIUS

TURNING OF RADIUS

PARTING OFF

ENGINEERING DRAWINGS
UNIT 5

Engineering drawing is the common language by which draftspersons, tool designers, and engineers indicate to the machinist and toolmaker the physical requirements of a part. Drawings are made up of a variety of lines which represent surfaces, edges, and contours of a workpiece. By adding symbols, dimension lines and sizes, and word notes, the draftsperson can indicate the exact specifications of each individual part. The student must be familiar with this language of industry so that he or she can interpret lines and symbols in order to produce the parts quickly and accurately.

A complete product is usually shown on an *assembly drawing* by the draftsperson. Each part or component of the product is then shown on a *detailed drawing,* which is reproduced in copies called *prints.* The prints are used by the machinist or toolmaker to produce the individual parts which eventually will make up the complete product. Some of the more common lines and symbols will be reviewed briefly. However, it must be assumed that the student already has the basic knowledge of print reading or that a good book on this subject is available for review purposes.

OBJECTIVES

After completing this unit you will be able to:
1. Understand the meaning of the various lines used in engineering drawings
2. Recognize the various symbols used to convey information
3. Read and understand engineering drawings or prints

TYPES OF DRAWINGS AND LINES

Orthographic Projection In order to describe the shape of noncylindrical parts accurately on a drawing or print, the draftsperson uses the orthographic view or projection method. The orthographic view shows the part from three sides: the front, the top, and the right-hand side (Fig. 5-1). These three views enable the drafter to describe a part or object so completely that the machinist knows exactly what is required.

Cylindrical parts are generally shown on prints in two views: the front and the right side (Fig. 5-2). However, if a part contains many details, it may be necessary to use the top, bottom, or left-side views to describe the part accurately to the machinist.

Section Views In many cases, complicated interior forms are difficult to describe in the usual manner by a draftsper-

Fig. 5-1 Three views of orthographic projection make it easier to describe the details of the part.

son. Whenever this occurs, a sectional view, which is obtained by making an imaginary cut through an object, is made. This imaginary cut can be made in a straight line in any direction to best expose the interior contour or form of a part (Fig. 5-3).

Types of Lines A wide variety of standard lines are used in engineering drawings in order for the designer to indicate to the machinist exactly what is required. Thick, thin, broken, wavy, and section lines are used on shop or engineering drawings. See Table 5-1 for an example, description, and purpose of some of the more common lines used on shop drawings.

DRAFTING TERMS AND SYMBOLS

Common drafting terms and symbols are used in shop and engineering drawings in order for the designer to describe each part accurately. If it were not for the universal use of terms, symbols, and abbreviations, the designer would have to make extensive notes describing exactly what is required. These would not only be cumbersome but could be misunderstood and therefore result in costly errors. Some of the common drafting terms and symbols are explained below.

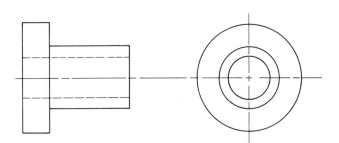

Fig. 5-2 Cylindrical parts are generally shown in two views.

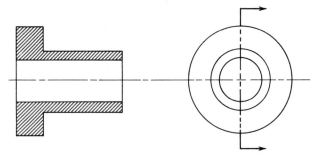

Fig. 5-3 Section views are used to show complicated interior forms.

Table 5-1 Common Lines Used on Shop Drawings			
EXAMPLE	**NAME**	**DESCRIPTION**	**USE**
a	Object lines	Thick black lines approximately ¹⁄₃₂ in. wide (the width may vary to suit drawing size).	Indicate the visible form or edges of an object
b	Hidden lines	Medium-weight black lines of ⅛-in. long dashes and ¹⁄₁₆-in. spaces.	Indicate the hidden contours of an object.
c	Center lines	Thin lines with alternating long lines and short dashes. Long lines from ½ to 3 in. long. Short dashes ¹⁄₁₆ to ⅛ in. long, spaces ¹⁄₁₆ in. long.	Indicate the centers of holes, cylindrical objects, and other sections.
d	Dimension lines	Thin black lines with an arrowhead at each end and a space in the center for a dimension.	Indicate the dimensions of an object.
e	Cutting-plane lines	Thick black lines make up a series of one long line and two short dashes. Arrowheads show the line of sight from where the section is taken.	Show where a section is imagined to be cut.
f	Cross-section lines	Fine evenly spaced parallel lines at 45°. Line spacing is in proportion to the part size.	Show the surfaces exposed when a section is cut.

Fig. 5-4 Limits show the largest and smallest size of a part.

Fig. 5-5 Tolerance is the permissible variation of a size.

Fig. 5-6 Allowance is the intentional difference in the sizes of mating parts.

Limits (Fig. 5-4) are the largest and the smallest permissible dimensions of a part (the maximum and minimum dimensions) on a shop drawing both sizes would be given.

EXAMPLE

0.751 largest dimension
0.749 smallest dimension

Tolerance (Fig. 5-5) is the permissible variation of the size of a part. On a drawing the basic dimension is given plus or minus to variation allowed.

EXAMPLE

$$0.750 \begin{array}{c} +0.001 \\ -0.003 \end{array}$$

The tolerance in this case would be 0.004 (the difference between $+0.001$ oversize or -0.003 undersize).

Allowance (Fig. 5-6) is the intentional difference in the sizes of mating parts such as the diameter of a shaft and the size of the hole. On a shop drawing both the shaft and the hole would be indicated with maximum and minimum sizes in order to produce the best fit.

Fit is the range of tightness between two mating parts. There are two general classes of fits:

1. *Clearance fits,* whereby a part may revolve or move in relation to a mating part
2. *Interference fits,* whereby two parts are forced together to act as a single piece

Scale size is used in most shop or engineering drawings since it would be impossible to draw parts to the exact size because some drawings would be too large while others would be too small. The scale size of a drawing is generally found in the title block section and it indicates the scale to which the drawing has been made which is a representative measurement.

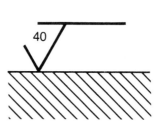

Fig. 5-7 Surface finish symbols indicate the type and finish of the surface.

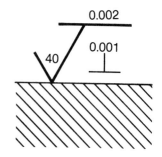

Fig. 5-8 Surface finish specifications.

SCALE	DEFINITION
1:1	Drawing is made to the actual size of part or full scale
1:2	Drawing is made one-half the actual size of the part
2:1	Drawing is made twice the actual size of the part

SYMBOLS

Some of the symbols and abbreviations used on shop drawings indicate the surface finish, type of material, roughness symbols, and common machine shop terms and operations.

Surface Symbols

Surface finish is the deviation from the nominal surface caused by the machining operation. Surface finish includes roughness, waviness, lay, and flaws, and is measured by a surface finish indicator in microinches.

The surface finish mark, used in many cases, indicates which surface of the part must be finished. The number in the $\sqrt{}$ indicates the quality of finish required on the surface (Fig. 5-7). In this case the *roughness height* or the measurement of the finely spaced irregularities caused by the cutting tool cannot exceed 40 microinches (μ in.).

If the surface of a part must be finished to exact specifications, each part of the specification is indicated on the symbol (Fig. 5-8) as follows:

40	Surface finish in microinches
0.002	Waviness height in thousandths of an inch
0.001	Roughness width in thousandths of an inch
⊥	Machining marks run perpendicular to the boundary of the surface indicated

The following symbols indicate the direction of lay (marks produced by machining operations on work surfaces)

=	Parallel to the boundary line of the surface indicated by the symbol
X	Angular in both directions on the surface indicated by the symbol
M	Multidirectional
C	Approximately circular to the center of the surface indicated by the symbol
R	Approximately radial to the center of the surface indicated by the symbol

Material Symbols

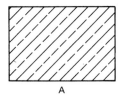

Represents copper, brass, bronze, etc.

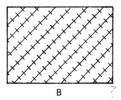

Represents aluminum, magnesium, and their alloys.

Represents steel and wrought iron.

Represents cast iron and malleable iron.

Fig. 5-9   Symbols used to indicate types of material.

Common Machine Shop Abbreviations

CBORE	Counterbore
CSK	Countersink
DIA	Diameter
Ø	Diameter
HDN	Harden
L	Lead
LH	Left hand
mm	Millimeter
NC	National coarse
NF	National fine
P	Pitch
R	Radius
Rc	Rockwell hardness test
RH	Right hand
THD	Thread or threads
TIR	Total indicated run-out
TPI	Threads per inch
UNC	Unified national coarse
UNF	Unified national form

REVIEW QUESTIONS

1. How can a drafter indicate the exact specifications required on a part?
2. What is the purpose of
   (a) An assembly drawing?
   (b) A detailed drawing?
3. What is the purpose of an orthographic view?
4. Why are section views shown?
5. What lines are used to show:
   (a) The form of a part?
   (b) The centers of holes, objects, or sections?
   (c) The exposed surfaces of where a section is cut?
6. Define:
   (a) Limits
   (b) Tolerance
   (c) Allowance
7. How is half scale indicated on an engineering drawing?
8. Define each part of the surface-finish symbol

9. What do the following abbreviations mean?
   (a) CBORE
   (b) HDN
   (c) mm
   (d) THD
   (e) TIR

MACHINING PROCEDURES FOR VARIOUS WORKPIECES

UNIT 6

Planning the procedures for machining any part so that it can be machined accurately and as quickly as possible is very important. Many parts have been spoiled because the incorrect sequence has been followed in the machining process. Although it would be impossible to list the exact sequence of operations that would apply to every type and shape of workpiece, some general rules should be followed to machine a part accurately and in the shortest time possible.

OBJECTIVES

After completing this unit you should be able to:
1. Plan the sequence of operations and machine round work mounted between lathe centers
2. Plan the sequence of operations and machine round work mounted in lathe chuck
3. Plan the sequence of operations and machine flat workpieces

MACHINING PROCEDURES FOR ROUND WORK

Most of the work produced in a machine shop is round and is turned to size on a lathe. In industry much of the round work is held in a chuck, while a larger percentage of work in school shops is machined between centers because of the need to reset work more often. In either case it is important to follow the correct machining sequence of operations in order to prevent spoiling work, which often happens when incorrect procedures are followed.

GENERAL RULES FOR ROUND WORK

1. Rough-turn all diameters to within $\frac{1}{32}$ inch (in.) [0.79 millimeters (mm)] of the size required.
   — Machine the largest diameter first and progress to the smallest.
   — If the small diameters are rough-turned first, it is quite possible that the work would bend when the large diameters are machined.
2. Rough-turn all steps and shoulders to within $\frac{1}{32}$ in. (0.79 mm) of the length required (Fig. 6-1).

— Be sure to measure all lengths from one end of the workpiece.
— If all measurements are not taken from the end of the workpiece, the length of each step would be $\frac{1}{32}$ in. (0.79 mm) shorter than required. If four steps were required, the length of the fourth step would be $\frac{1}{8}$ in. (3.17 mm) shorter than required (4 × $\frac{1}{32}$ in. or 4 ×

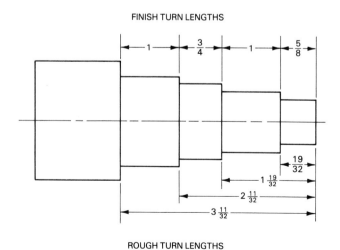

FINISH TURN LENGTHS

ROUGH TURN LENGTHS

Fig. 6-1 A sample part showing rough and finish turn lengths.

0.79 mm) and would leave too much material for the finishing operation.

3. If any special operations such as knurling or grooving are required, they should be done next.

4. Cool the workpiece before starting the finishing operations.

— Metal expands from the friction caused by the machining process and any measurements taken while work is hot are incorrect.

— When the workpiece is too cold, the diameters of round work will be smaller than required.

5. Finish-turn all diameters and lengths.

— Finish the largest diameters first and work down to the smallest diameter.

— Finish the shoulder of one step to the correct length and then cut the diameter to size.

WORKPIECES REQUIRING CENTER HOLES

There are times when it is necessary to machine the entire length of a round workpiece. When this is required, usually on shorter workpieces, center holes are drilled in each end. The workpiece shown in Fig. 6-2 is a typical part which can be machined between the centers on a lathe.

Machining Sequence

1. Cut off a piece of steel ⅛ in. (3.17 mm) longer and ⅛ in. (3.17 mm) larger in diameter than required.

— In this case the diameter of the steel cut off would be 1⅝ in. (41.27 mm) and its length would be 16⅛ in (409.57 mm).

2. Hold the work in a three-jaw chuck, face one end square, and then drill the center hole.

3. Face the other end to length and then drill the center hole.

4. Mount the workpiece between the centers on a lathe.

5. Rough-turn the largest diameter to within 1/32 in. (0.79 mm) of finish size or 1¹⁷/₃₂ in. (38.89 mm).

NOTE: The purpose of the rough cut is to remove excess metal as quickly as possible.

6. Finish-turn the diameter to be knurled.

NOTE: The purpose of the finish cut is to cut work to the required size and produce a good surface finish.

7. Knurl the 1½-in. (38.1-mm) diameter.

8. Machine the 45° chamfer on the end.

9. Reverse the work in the lathe, being sure to protect the knurl from the lathe dog with a piece of soft metal.

10. Rough-turn the 1¼-in. (31.75-mm) diameter to 1⁹/₃₂ in. (32.53 mm) (Fig. 6-3).

— Be sure to leave the length of this section ⅛ in. (3.17 mm) short [12⅞ in. (327.02 mm) from the end] to allow for finishing the ⅛-in. (3.17 mm) radius.

11. Rough-turn the 1⅛-in. (28.57-mm) diameter to 1⁵/₃₂ in. (29.36 mm).

Fig. 6-2 A sample of a round shaft which can be machined.

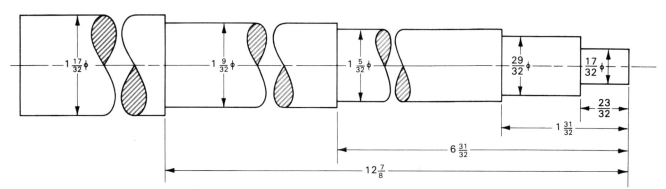

Fig. 6-3 The rough-turned diameters and lengths on a shaft.

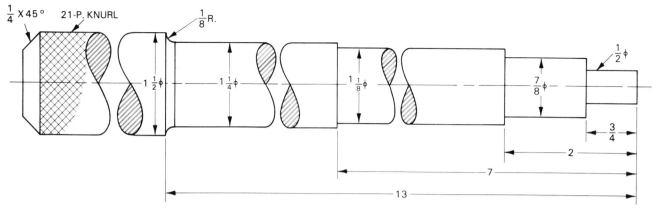

Fig. 6-4 The shaft-turned to diameter and length.

— Leave the length of this section ¹/₃₂ in. (0.79 mm) short [6³¹/₃₂ in. (176.98 mm) from the end] to allow for finishing the shoulder.

12. Rough-turn the ⁷/₈-in. (22.22-mm) diameter to ²⁹/₃₂ in. (23.01 mm).

— Leave the length of this section ¹/₃₂ in. (0.79 mm) short [1³¹/₃₂ in. (49.98 mm) from the end] to allow for finishing the shoulder.

13. Rough-turn the ½-in. (12.7-mm) diameter to ¹⁷/₃₂ in. (13.48 mm).

— Machine the length of this section to ²³/₃₂ in. (18.23 mm).

14. Cool the work to room temperature before starting the finishing operations.

15. Finish-turn the 1¼-in. (31.75-mm) diameter to 12⅞ in. (327.02 mm) from the end.

16. Mount a ⅛-in. (3.17-mm)-radius tool and finish the corner to the correct length (Fig. 6-4).

17. Finish-turn the 1⅛-in. (28.57-mm) diameter to 7 in. (177.8 mm) from the end.

18. Finish-turn the ⁷/₈-in. (22.22-mm) diameter to 2 in. (50.8 mm) from the end.

19. Set the compound rest to 7° and machine the taper to size.

20. Finish-turn the ½-in. (12.7-mm) diameter to ¾ in. (19.05 mm) from the end.

21. With a cut-off tool cut the groove at the end of the ½-in. (12.7-mm) diameter (Fig. 6-5).

22. Chamfer the end of the section to be threaded.

23. Set the lathe for threading and cut the thread to size.

WORKPIECES HELD IN A CHUCK

The procedure for machining the external surfaces of round workpieces held in a chuck (three-jaw, four-jaw, collet, etc.) is basically the same as for machining work held between lathe centers. However, if both external and internal surfaces must be machined on work held in a chuck, the sequence of some operations is changed.

Fig. 6-5 Special operations completed on a shaft.

Fig. 6-6 A round part requiring internal and external machining.

Whenever work is held in a chuck for machining, it is very important that the workpiece be held short for rigidity and to prevent accidents. Never have work extending more than *three times its diameter* beyond the chuck jaws unless it is supported by some means such as a steady rest or center.

MACHINING EXTERNAL AND INTERNAL DIAMETERS IN A CHUCK

To machine the part shown in Fig. 6-6, the following sequence of operations is suggested.

1. Cut off a piece of steel ⅛ in. (3.17 mm) larger in diameter and ½ in. (12.7 mm) longer than required.

— In this case the rough diameter would be 2⅛ in. (53.98 mm).

— The length would be 3⅞ in. (98.4 mm); the extra length is to allow the piece to be gripped in the chuck.

2. Mount and center the workpiece in a four-jaw chuck, gripping only 5/16 to 3/8 in. (7.9 to 9.5 mm) of the material in the chuck jaws.
— A three-jaw chuck would not hold this size workpiece securely enough for the internal and external machining operations.

3. Face the end of the work square.
— Remove only the minimum amount of material required to square the end.

4. Rough-turn the three external diameters, starting with the largest and progressing to the smallest to within 1/32 in. (0.79 mm) of size and length.

5. Mount a drill chuck in the tailstock spindle and center-drill the work.

6. Drill a ½-in. (12.7-mm)-diameter hole through the work.

7. Mount a 15/16-in. (23.81-mm) drill in the tailstock and drill through the work.

8. Mount a boring bar in the toolpost and bore the 1-in. (25.4-mm) ream hole to 0.968 in. (24.58 mm) in diameter.

9. Bore the 1¼ in. 7 UNC threaded section to the tap-drill size, which is 1.107 in (28.12 mm).

$$TDS = D - P$$
$$D = diameter$$
$$P = pitch$$

10. Cut the groove at the end of the section to be threaded to length and a little deeper than the major diameter of the thread.

11. Mount a threading tool in the boring bar and cut the 1¼ in. 7 UNC thread to size.

12. Mount a 1-in. (25.4-mm) reamer in the tailstock and cut the hole to size.

13. Finish-turn the external diameters to size and length starting with the largest and working down to the smallest.

14. Reverse the workpiece in the chuck and protect the finish diameter with a piece of soft metal between it and the chuck jaws.

15. Face the end surface to the proper length.

MACHINING FLAT WORKPIECES

Since there are so many variations in size and shape of flat workpieces, it is difficult to give specific machining rules to follow that would apply to each. Some general rules are listed; however, these may have to be modified to suit particular workpieces.

1. Select and cut off the material a little larger than the size required.

2. Machine all surfaces to size in a shaper or milling machine.

3. Lay out the physical contours of the part such as angles, steps, radii, and so on.

4. Lightly prick-punch the layout lines which indicate the surfaces to be cut.

5. Remove large sections of the workpiece on a contour bandsaw.

6. Machine all forms such as steps, angles, radii, and grooves.

Fig. 6-7 A typical flat part which must be laid out and machined.

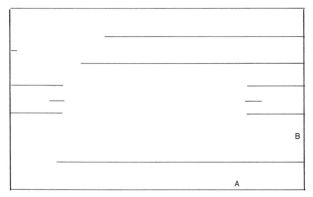

Fig. 6-8 Layout all horizontal lines using edge A as the reference surface.

7. Lay out all hole locations and with dividers scribe the reference circle.
8. Drill all holes and tap those which require threads.
9. Ream holes.
10. Surface-grind any surfaces which require it.

OPERATIONS SEQUENCE FOR A SAMPLE FLAT PART

The part shown in Fig. 6-7 is used only as an example to set forth a sequence of operations which should be followed when machining similar parts. These are not meant to be hard-and-fast rules but act only as a guide for the student.

The sequence of operations suggested for the sample part shown in Fig. 6-7 is different from those suggested for machining a block square and parallel as outlined in Unit 65 because:

1. The part is relatively thin and has a large surface area.
2. Since at least ⅛ in. (3.17 mm) of work should be above the vise jaws, it would be difficult to use a round bar between the work and movable jaw for machining the large flat surfaces.
3. A small inaccuracy (out-of-squareness) on the narrow edge would create a greater error when the large surface was machined.

Procedure

1. Cut off a piece of steel ⅝ in. (15.87 mm) × 3⅜ in. (85.72 mm) × 5⅝ in. (142.87 mm) long.
2. In a milling machine, finish one of the larger surfaces (face) first.
3. Turn the workpiece over and machine the other face to ½ in. (12.7 mm) thick.
4. Machine one edge square with the face.
5. Machine an adjacent edge square (at 90°) with the first edge.
6. Place the longest finished edge (A) down in the machine vise and cut the opposite edge to 3¼ in. (82.55 mm) wide.

NOTE: Leave 0.010 in. (0.25 mm) on each surface to be ground.

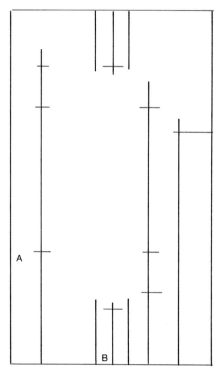

Fig. 6-9 Layout all vertical lines using edge B as the reference surface.

7. Place the narrower finished edge (B) down in the machine vise and cut the opposite edge to 5½ in. (139.7 mm) long.
8. With edge A as a reference surface, lay out all the horizontal dimensions with an adjustable square, surface gage, or height gage (Fig. 6-8).
9. With edge B as a reference surface, lay out all the vertical dimensions with an adjustable square, surface gage, or height gage (Fig. 6-9).
10. Use a bevel protractor to lay out the 30° angle on the upper left-hand edge.
11. With a divider set to ¼ in. (6.35 mm), draw the arcs for the two center slots.
12. With a sharp prick punch, lightly mark all the surfaces to be cut and the centers of all hole locations.
13. Center-punch and drill ½-in. (12.7-mm) diameter holes for the two center slots.
14. On a vertical bandsaw, cut the 30° angle to within 1/32 in. (0.79 mm) of the layout line.
15. Place the workpiece in a vertical mill and machine the two ½-in. (12.7-mm) slots.
16. Machine the step on the top edge of the workpiece.
17. Set the work to 30° in the machine vise and finish the 30° angle.
18. Prick-punch the hole locations, scribe reference circles, and then center-punch all hole centers.
19. Center drill all hole locations.
20. Drill and counterbore the holes for the ¼ in.-20 NC screws.
21. Tap drill the 5/16 in.-18-thread holes. (F drill or 6.52 mm)

22. Drill the ¼-in. (6.35-mm) ream holes to ¹⁵⁄₆₄ in. (5.96 mm).

23. Countersink all holes to be tapped slightly larger than their finished size.

24. Ream the ¼-in. (6.35-mm) holes to size.

25. Tap the ⁵⁄₁₆-in.—18 UNC holes.

REVIEW QUESTIONS

1. Why is it not advisable to rough-turn small diameters first?

2. To what size should work be rough-turned?

3. Why must all measurements be taken from one end of a workpiece?

4. Why is it important that the workpiece be cooled before finish-turning?

5. Why can a workpiece be bent during the knurling operation?

6. Why is the work cut off longer than required when machining in a chuck?

7. How far should a 6-in.-long workpiece, 1 in. in diameter, extend beyond the chuck jaws?

8. How deep should the groove be cut for the internal section to be threaded?

9. How is a finish diameter protected from the chuck jaws?

10. How much material should be left on a flat surface for grinding?

11. When using a bandsaw to remove excess material, how close to the layout line should the cut be made?

12. When machining flat surfaces, what surface should be machined first?

MEASUREMENT
SECTION 5

The world has depended on some form of measurement system since the beginning of civilization. The Egyptians, for example, used a unit of length called the cubit, a unit equal to the length of the forearm from the middle finger to the elbow. James Watt, on the other hand, improved his steam engine by maintaining its tolerances to the thickness of a thin shilling, an English coin. The days of such crude measurements, however, have gone. Today we live in a demanding world where products must be built to very precise tolerances. These same products may start out as components built by several sub-industries and find their way to other industries which use them in the manufacture of final consumer products. From start to finish the product may cover several industries located in widely separate places, often nations away, and then be sold at still another location. *Interchangeable manufacture,* world trade, and the need for high precision have all contributed to the need for a *highly accurate international* measurement system. Consequently, in 1960 the International System of Units (SI) was developed to satisfy this need.

Presently there are two major systems of measurement used in the world. The inch system, often called the English system of measurement, is still widely used in the United States and Canada. However, since over 90 percent of the world's population uses some form of metric measurement, the need for universal adoption of metric measurement is evident.

METRIC (DECIMAL) SYSTEMS

On December 8, 1975, the United States Senate passed Metric Bill S100 "to facilitate and encourage the substitution of metric measurement units for customary measurement units in education, trade, commerce and all other sections of the economy of the United States. . . ." On January 16, 1970, the Canadian government adopted SI for implementation throughout Canada by 1980.

Although both the United States and Canada are now reasonably committed to conversion to the metric system as rapidly as possible, it is likely to be some years before all machine tools and measuring devices are redesigned or converted. The change to the metric system in the machine shop trade will be gradual because of the long life expectancy of the costly machine tools and measuring equipment involved. It is probable, therefore, that people involved in the machine shop trade will have to be familiar with both the metric and the inch systems during the long changeover period.

THE CHANGEOVER PERIOD

Although precision tool manufacturers are producing most of the measuring tools in metric sizes, the use of these tools is not widespread because of the reluctance of industry to make the costly changeover. Consequently, many present-day machinists will probably have to be familiar with both the inch and the metric systems of measurement.

Since the changeover will be gradual, it is safe to assume that students graduating from schools for the next several years must be taught a dual system of measurement (inch and metric) until the majority of industries have changed to the metric system. With this in mind, we have used dual measurement throughout this book so that the students will be able to work effectively in both systems.

To accommodate these problems, the following policy has been adopted throughout this book to enable you to work effectively in both systems now, while permitting an easy transition to full metric as the new materials and tools become available.

1. Where general measurements or references to quantity are not related specifically to metric standards, tools, or products, inch units are given with the metric equivalent in parentheses.
2. Where the student may be exposed to equipment designed to both metric and inch standards, separate information is given on both types of equipment in *exact* dimensions.
3. Where only inch standards, tools, or products exist at the present time, inch measurements are given with a metric conversion, to two decimal places, provided in parentheses.

SYMBOLS FOR USE WITH SI

Here is a list of some common SI quantities, names, and symbols which you are likely to encounter in your work in the machine shop.

QUANTITY	NAME	SYMBOL
length*	meter	m
volume*	liter	ℓ and l
mass*	gram	g
time	minute	min
	second	s
force	newton	N
pressure, stress*	pascal	Pa
temperature	degree Celsius	°C
area*	square meter	m²
velocity (speed)	meters per minute and	m/min and m/s
	meters per second	
angles	degrees	°
	minutes	'
	seconds	"
electric potential	volt	V
electric current	ampere	A
frequency	hertz	Hz
electric capacitance	farad	F

Here is a list of prefixes often used with the quantities indicated by * above.

PREFIX	MEANING	MULTIPLIER	SYMBOL
micro	one-millionth	0.000 001	μ
milli	one-thousandth	0.001	m
centi	one-hundredth	0.01	c
deci	one-tenth	0.1	d
deka	ten	10	da
hecto	one hundred	100	h
kilo	one thousand	1 000	k
mega	one million	1 000 000	M

BASIC MEASUREMENT
UNIT 7

Basic measurement can be termed as those measurements taken by use of a rule or any other nonprecision measuring tool, whether it be in the inch or metric standard.

OBJECTIVES

After completing this unit you should be able to:
1. Identify several types of steel rules
2. Measure round and flat work to 1/64-in. accuracy with a rule
3. Measure with spring calipers and a rule

INCH SYSTEM

The unit of length in this system is the inch, which may be divided into fractional or decimal fraction divisions. The fractional system is based on the binary system having base 2. The binary fractions commonly used in this system are 1/2, 1/4, 1/8, 1/16, 1/32, and 1/64. The decimal-fraction system has a base 10 whereby any number may be written as a product of ten and/or a fraction of ten.

VALUE	FRACTION	DECIMAL
one-tenth	1/10	0.1
one-hundredth	1/100	0.01
one-thousandth	1/1000	0.001
one ten-thousandth	1/10,000	0.0001
one hundred-thousandth	1/100,000	0.00001
one millionth	1/1000,000	0.000001

METRIC SYSTEM

Linear metric dimensions are expressed in multiples and submultiples of the meter. In the machine shop trade, the millimeter is used for the expression of most metric dimensions. Fractions of the millimeter are expressed in decimals.

A brief comparison of common inch and metric equivalents are:

$$
\begin{array}{lll}
1 \text{ yd} & = & 36 \text{ in.} \\
1 \text{m} & = & 39.37 \text{ in.} \\
1000 \text{ m} & = & 1 \text{ km} \\
1 \text{ km} & = & 0.621 \text{ mi} \\
1 \text{ mi} & = & 1.609 \text{ km}
\end{array}
$$

Table 7-1 shows inch-metric comparisons for common inch/metric system measurements.

NOTE: In machine shop metric measurements, most dimensions will be given in millimeters (mm). Very large dimensions would be given in meters (m) and millimeters (mm). For metric-inch conversion tables and decimal equivalents, see the Appendix of Tables in the back of the book.

FRACTIONAL MEASUREMENT

Fractional dimensions, often called *scale dimensions,* can be measured with such instruments as rulers or calipers. The steel rules used in machine shop work are graduated either in binary-fractional divisions of 1, 1/2, 1/4, 1/8, 1/16, 1/32, and 1/64 of an inch (Fig. 7-1) or in decimal fractional divisions, decimeters, centimeters, millimeters, and half-millimeters (Fig. 7-2). Divisions of 1/64 of an inch or 0.50 mm are about as fine as can be seen on a rule without the use of a magnifying glass. Precision measuring instruments such as micrometers, verniers, etc., are required when metric-drawn prints show any dimensions of less than 0.50 mm or when inch-drawn prints show any dimensions in decimals.

STEEL RULES

METRIC STEEL RULES

Metric steel rules (Fig. 7-3), usually graduated in millimeter and half-millimeter, are used for making linear metric measurements which do not require great accuracy. A wide

Table 7-1

INCH SIZES	METRIC SIZES			
	MILLIMETER (mm)	CENTIMETER (cm)	DECIMETER (dm)	METER (m)
1 in.	25.4	2.54	0.254	0.0254
1 ft.	304.8	30.48	3.048	0.3048
1 yd.	914.4	91.44	9.144	0.9144

Fig. 7-1 Fractional divisions of an inch.

Fig. 7-2 Metric rules are usually graduated in millimeters and half-millimeters. (Courtesy Kostel Enterprises Limited.)

Fig. 7-3 A 15-cm metric rule. (Courtesy Kostel Enterprises Limited.)

Fig. 7-4 A spring-tempered (quick-reading) 6-in. rule.

Fig. 7-5 A hook rule is used to make accurate measurements from an edge or shoulder. (Courtesy The L. S. Starrett Company.)

Fig. 7-6 Short-length rules are used for measuring small openings. (Courtesy The L. S. Starrett Company.)

variety of metric rules are available in lengths from 15 cm to 1 m. The 15-cm rule shown in Fig. 7-3 is 2.4 mm or about 3/32 in. shorter than a standard 6-in. rule.

FRACTIONAL STEEL RULES

The common binary fractions found on inch steel rules are 1/64, 1/32, 1/16, and 1/8 of an inch. Several varieties of inch steel rules may be used in machine shop work, such as *spring-tempered, flexible, narrow,* and *hook.* Lengths range from 1 to 72 in. Again, these rules are used for measurements which do not require great accuracy. Spring-tempered quick-reading 6-in. rules (Fig. 7-4) with No. 4 graduations are the most frequently used inch rules in machine shop work. These rules have four separate scales, two on each side. The front is graduated in eighths and sixteenths and the back is graduated in thirty-seconds and sixty-fourths of an inch. Every fourth line is numbered to make reading in thirty-seconds and sixty-fourths easier and quicker.

Hook rules (Fig. 7-5) are used to make accurate measurements from a shoulder, step, or edge of a workpiece.

They may also be used to measure flanges, circular pieces, and for setting inside calipers to a dimension.

Short-length rules (Fig. 7-6) are useful in measuring small openings and hard-to-reach locations where an ordinary rule cannot be used. Five small rules come to a set; they range between 1/4 and 1 in. in length and can be interchanged in the holder.

Decimal rules (Fig. 7-7) are most often used when it is necessary to make linear measurements smaller than 1/64 in. Since linear dimensions are sometimes specified on drawings in decimals, these rules are very useful to the machinist. The most common graduations found on decimal rules are 0.100 (1/10 of an inch), 0.050 (1/20 of an inch), 0.020 (1/50 of an inch), and 0.010 (1/100 of an inch). A 6-in. decimal rule is shown in Fig. 7-8.

MEASURING LENGTHS

With a reasonable amount of care, fairly accurate measurements can be made using steel rules. Whenever possible, it is advisable to butt the end of a rule against a shoulder or step (Fig. 7-9) to assure an accurate measurement.

Through constant use, the end of a steel rule becomes worn. Measurements taken from the end are, therefore, often inaccurate. Fairly accurate measurements of flat work can be made by placing the 1-in. or 1-cm graduation line on the edge of the work, taking the measurement, and subtracting 1 in. or 1 cm from the reading (Fig. 7-10). When measuring flat work, be sure that the edge of the rule is parallel to the edge of the work. If the rule is placed at an angle to the edge (Fig. 7-11), the measurement will not be accurate. When you measure the diameter of round stock, start from the 1-in. or 1-cm graduation line.

THE RULE AS A STRAIGHTEDGE

The edges of a steel rule are ground flat. The rule may therefore be used as a straightedge to test the flatness of workpieces. The edge of a rule should be placed on the work surface, which is then held up to the light. Inaccuracies as small as a few thousandths of an inch or 0.05 mm can easily be seen by this method.

OUTSIDE CALIPERS

Outside calipers are tools used to measure the outside surface of either round or flat work. They are made in several styles such as *spring joint* and *firm joint calipers*. The spring joint caliper consists of two curved legs, a spring, and an adjusting nut. The outside spring joint caliper is most commonly used because it can easily be adjusted to size. The caliper itself cannot be read directly and therefore must be set to a steel rule or a standard-size gage. Therefore, calipers should not be used when an accuracy of less than 0.007 to 0.008 in. (0.18 to 0.20 mm) is required.

SETTING AN OUTSIDE CALIPER TO A RULE

When calipers are set to a rule, it is important that the end of the rule be in good condition and not worn or damaged.

Procedure

1. Place the rule in the left hand with the forefinger extending slightly beyond the end of the rule (Fig. 7-12).
2. Hold the caliper in the right hand and place one leg of the caliper over the end of the rule supporting it with the left forefinger.

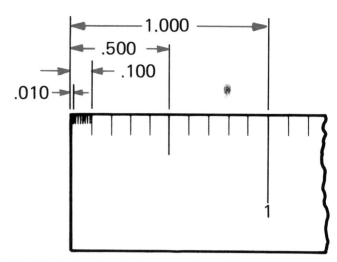

Fig. 7-7 Decimal graduations on a rule provide an accurate and simple form of measurement.

Fig. 7-8 Common graduations found on a 6-in. decimal rule. (Courtesy L. S. Starrett Company.)

Fig. 7-9 Butting a rule against a shoulder.

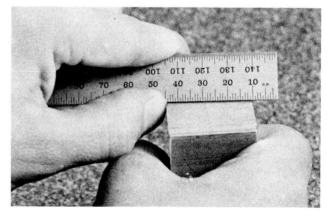

Fig. 7-10 Measuring with a rule starting at the 1-in. or 1-cm line. (Courtesy Kostel Enterprises Limited.)

Fig. 7-11 The rule must be held parallel to the edge of the work otherwise the measurement will not be correct.

Fig. 7-12 Setting an outside caliper to size with a rule.

Fig. 7-13 Checking a diameter with an outside caliper.

Fig. 7-14 Adjusting an inside caliper to the size of a hole.

3. Hold both legs of the caliper parallel to the edge of the rule and turn the adjusting nut until the end of the lower caliper leg splits the desired graduation on the rule.

MEASURING WITH OUTSIDE CALIPERS

Never attempt to measure work while it is moving or revolving. Not only is it a dangerous practice which could result in an accident, but any measurements taken will not be accurate.

Procedure

1. Place the caliper on the work with both legs at right angles to the centerline of the work (Fig. 7-13).
2. Hold the top of the caliper (spring) lightly between the thumb and forefinger.
3. The diameter is correct when the caliper will just slide over the work by its own weight.

NOTE: *Never* force the caliper over the diameter.

INSIDE CALIPERS

Inside calipers are used to measure the diameter of holes or the width of keyways and slots. They are made in several styles, such as the *spring joint* and the *firm joint calipers*.

MEASURING AN INSIDE DIAMETER

Fairly accurate measurements of holes and slots may be made using an inside caliper and a rule.

Procedure

1. Place one leg of the caliper near the bottom edge of the hole (Fig. 7-14).
2. Hold the caliper leg in this position with a finger.
3. Keep the caliper legs vertical or parallel to the hole.
4. Move the top leg in the direction of the arrows and turn

the adjusting nut until a slight drag is felt on the caliper leg.

5. Find the size of the setting by placing the end of a rule and one leg of the caliper against a flat surface.

6. Hold the legs of the caliper parallel to the edge of the rule and note the reading on the rule.

TRANSFERRING MEASUREMENTS

When an accurate measurement is required, the caliper setting should be checked with an outside micrometer.

Procedure

1. Hold the micrometer in the right hand so that it can be easily adjusted with the thumb and forefinger (Fig. 7-15).

2. Place one leg of the caliper on the micrometer anvil and hold it in position with a finger.

3. Rock the top leg of the caliper in the direction of the arrows.

4. Adjust the micrometer thimble until *only a slight drag* is felt as the caliper leg passes over the measuring face.

Fig. 7-15   An inside caliper setting being checked with a micrometer.

REVIEW QUESTIONS

1. Name two systems of measurement presently used in North America.

2. What is the common unit of length in the SI system?

3. How are metric rules usually graduated?

4. Name four types of steel rules used in machine shop work.

5. Describe a rule with No. 4 graduations.

6. State the purpose of:
   (a) hook rules
   (b) decimal rules

7. Name and describe two types of outside calipers.

8. What is the procedure for setting an outside caliper to a size?

9. Explain how you would know when the work is the same size as the caliper setting.

10. Explain the procedure for setting an inside caliper to the size of a hole.

11. In point form, list the procedure for checking an inside caliper setting with a micrometer.

SQUARES AND SURFACE PLATES

UNIT 8

The square is a very important tool used by the machinist for layout, inspection, and setup purposes. Squares are manufactured in various degrees of accuracy ranging from semiprecision to precision squares. Precision squares are hardened and accurately ground.

OBJECTIVES

After completing this unit you will be able to identify and know the uses of:
1. The machinist's combination square
2. Three types of solid and adjustable squares
3. Two types of surface plates

STEEL RULE

PROTRACTOR HEAD

CENTER HEAD

SQUARE HEAD

SPIRIT LEVEL

SCRIBER

Fig. 8-1 The combination set may be used for laying out and checking work.

Fig. 8-2 The combination (adjustable) square being used as a depth gage.

MACHINIST'S COMBINATION SQUARE

The combination square is a basic tool used by the machinist for quickly checking 90 and 45° angles. It is part of a combination set (Fig. 8-1) which includes the square head, the center head, the bevel protractor, and a graduated grooved rule to which the various heads may be attached. In addition to laying out and checking angles, the combination square may also be used as a depth gage (Fig. 8-2) or to measure the length of work to reasonable accuracy. Further uses of the combination set will be discussed in layout work (Unit 19).

PRECISION SQUARES

Precision squares are used chiefly for inspection and setup purposes. They are hardened and accurately ground and must be handled carefully to preserve their accuracy. A great variety of squares are manufactured for specific purposes. The squares are all variations of the *solid square* or the *adjustable square*.

BEVELED-EDGE SQUARES

The better-quality standard squares used in inspection have a beveled-edge blade which is hardened and ground. The beveled edge allows the blade to make a line contact with the work, thereby permitting a more accurate check. Two methods of using a beveled-edge square for checking purposes are illustrated in Figs. 8-3 and 8-4. In Fig. 8-3, if

PAPER FEELERS

Fig. 8-3 Using paper between the blade of the square and the work to check for squareness.

the work is square (90°), both pieces of paper will be tight between the square and the work. In Fig. 8-4, the light is shut out only where the blade makes line contact with the surface of the work. Light shows through where the blade of the square does not make contact with the surface being checked.

43

Fig. 8-4 The light shines through where the blade of the square does not make line contact with the surface.

Fig. 8-5 A toolmaker's surface plate square.

Fig. 8-6 A direct-reading cylindrical square being used to check a part for squareness.

Fig. 8-7 A diemaker's square is useful for checking the clearance on dies.

Fig. 8-8 The direct-reading diemaker's square indicates the angle at which the blade is set.

TOOLMAKER'S SURFACE PLATE SQUARE

The *toolmaker's surface plate square* (Fig. 8-5) provides a convenient method of checking work for squareness on a surface plate. Since it is of one-piece construction, there is little chance of any inaccuracy developing, as is the case with a blade and beam square.

CYLINDRICAL SQUARES

Cylindrical squares are commonly used as master squares against which other squares are checked. The square consists of a thick-walled alloy steel cylinder which has been hardened, ground, and lapped. The outside diameter is a nearly true cylinder, and the ends are ground and lapped square with the axis. The ends are recessed and notched to

decrease the inaccuracy from dust and to reduce friction. When a cylindrical square is used, it must be set carefully on a clean surface plate and rotated slightly to force particles of dust and dirt into the end notches; the square can then make the proper contact with surface plate. Cylindrical squares provide a perfect line contact with the part being checked.

Another type of cylindrical square is the *direct-reading cylindrical square* (Fig. 8-6), which indicates directly the amount that the part is out of square. One end of the cylinder is lapped square with the axis, while the other end is ground and lapped slightly out of square. The circumference is etched with several series of dots which form elliptical curved lines. Each curve is numbered at the top to indicate the amount, in ten-thousandths of an inch (0.0001), that the workpiece is out of square over the length of the square. Absolute squareness, or *zero deviation,* is indicated by an etched, vertical, dotted line on the square.

When used, the square is carefully placed in contact with the work and turned until no light is seen between it and the part being inspected. The uppermost curved line in contact with the work is noted and followed to the top where the number shows the amount the work is out of square. This square may also be used as a conventional cylindrical square if the opposite end, which is ground and lapped square with the axis, is used.

ADJUSTABLE SQUARES

The *adjustable square,* while not providing the accuracy of a good solid square, is used by the toolmaker where it would be impossible to use a fixed square.

A *diemaker's square* (Fig. 8-7) is used to check the clearance angle on dies. The blade is adjusted to the angle of the workpiece by means of a blade-adjusting screw. This angular setting must then be checked with a protractor. Another form of diemaker's square is the *direct-reading* type which indicates the angle at which the blade is set (Fig. 8-8).

ADJUSTABLE MICROMETER SQUARE

The *adjustable micrometer square* (Fig. 8-9) may be used to check a part for squareness accurately. When a piece of work is being checked and light shows between the blade and the work, turn the micrometer head until the full length of the blade, which may be tilted, touches the work. The amount the part is out of square may be read from the micrometer head. When the micrometer head is set at zero, the blade is perfectly square with the beam.

STRAIGHTEDGES

A *straightedge* is used to check surfaces for flatness and to act as a guide when scribing long, straight lines in layout work. Straightedges are generally rectangular bars of hardened and accurately ground steel, having both edges flat and parallel. They are supplied with either plain or beveled edges. Long straightedges are generally made of cast iron with ribbed construction.

SURFACE PLATES

A *surface plate* is a rigid block of granite or cast iron, the flat surface of which is used as a reference plane for layout, setup, and inspection work. Surface plates generally have a three-point suspension to prevent rocking when mounted on an uneven surface.

Cast-iron plates are well ribbed and supported to resist deflection under heavy loads. They are made of close-grained cast iron, which has high strength and good wear-

Fig. 8-9 The amount a part is out of square can be read on an adjustable micrometer square. (Courtesy Ash Precision Equipment Inc.)

Fig. 8-10 Granite surface plates are not affected by humidity and temperature change.

resistance qualities. After a cast-iron surface plate has been machined, its surface must be scraped by hand to a flat plane. This operation is long and tedious; therefore, the cost of these plates is high.

Granite surface plates (Fig. 8-10) have many advantages over the cast-iron types and are replacing them in many shops. They may be manufactured from gray, pink, or black granite and are obtainable in several degrees of accuracy. Extremely flat finishes are produced by lapping.

The advantages of granite plates are:

1. They are not appreciably affected by temperature change. Granite will not burr as does cast iron; therefore the accuracy is not impaired.
2. These plates are nonmagnetic.
3. They are rustproof.
4. Abrasives will not embed themselves as easily in the surface; thus they may be used near grinding machines.

CARE OF SURFACE PLATES

1. Keep surface plates clean at all times, and wipe them with a dry cloth before using.
2. Clean them occasionally with solvent or surface plate cleaner to remove any film.
3. Protect them with a wooden cover when not in use.
4. Use parallels whenever possible to prevent damage to plates by rough parts or castings.
5. Remove burrs from the workpiece before placing it on the plate.

6. Slide heavy parts onto the plate rather than place them directly on the plate, since a part might fall and damage the plate.

7. Remove all burrs from cast-iron plates by honing.

8. When they are not in regular use, cover cast-iron plates with a thin film of oil to prevent rusting.

REVIEW QUESTIONS

PRECISION SQUARES

1. Name two types of solid squares and state the advantage of each.

2. Why are beveled-edge squares used in inspection work?

3. What procedure should be followed when using a cylindrical square?

4. State the purpose of a diemaker's square.

5. How can the angle of the workpiece be determined using each type of diemaker's square?

SURFACE PLATES

6. What is the purpose of a surface plate?

7. Name three types of granite used in making surface plates.

8. State five advantages of granite over cast-iron surface plates.

9. List eight points necessary for the care of surface plates.

PRECISION MEASURING TOOLS— UNITS 9–18

Although fixed gauges are convenient for checking the upper and lower limits of external and internal dimensions, they do not measure the actual size of the part. The machinist must use some form of precision measuring instrument to obtain this desired size. Precision measuring tools may be divided into five categories, namely, tools used for outside measurement, inside measurement, depth measurement, thread measurement, and height measurement.

MICROMETERS
UNIT 9

The *micrometer caliper*, usually called the *micrometer*, is the most commonly used measuring instrument when accuracy is required. The *standard inch micrometer*, shown in a cutaway view in Fig. 9-1, measures accurately to 0.001 in. Since many phases of modern manufacturing require greater accuracy, the *vernier micrometer*, capable of even finer measurements, is being used to an increasing extent.

The standard metric micrometer (Fig. 9-5) measures in hundredths of a millimeter, whereas the vernier metric micrometer measures up to 0.002 mm.

The only difference in construction and reading between the standard inch and the vernier micrometer is the addition of the vernier scale on the sleeve above the index or centerline.

After completing this unit you should be able to:
1. Identify the most common types of outside micrometers and their uses
2. Measure the size of a variety of objects to within .001 in. accuracy
3. Read vernier micrometers to 0.0001-in. accuracy
4. Measure the size of a variety of objects to within 0.01-mm accuracy

PRINCIPLE OF THE INCH MICROMETER

In order to understand the principle of the inch micrometer, the student should be familiar with two important thread terms:

☐ Pitch: The distance from a point on one thread to a corresponding point on the next thread. For inch threads this is expressed as 1/N (number of threads). For metric threads it is expressed in millimeters.
☐ Lead: The distance a screw thread advances axially in one complete revolution or turn.

Since there are 40 threads per inch on the micrometer, the pitch is ¹⁄₄₀ (0.025) in. Therefore, one complete revolution of the spindle will either increase or decrease the distance between the measuring faces by ¹⁄₄₀ (0.025) in. The 1-in. distance marked on the micrometer sleeve is divided into 40 equal divisions, each of which equals ¹⁄₄₀ (0.025) in.

If the micrometer is closed until the measuring faces just touch, the zero line on the thimble should line up with the index line on the sleeve (barrel). If the thimble is revolved counterclockwise one complete revolution, it will be noted that one line has appeared on the sleeve. Each line on the sleeve indicates 0.025 in. Thus, if three lines were showing on the sleeve (or barrel), the micrometer would have opened 3 × 0.025, or 0.075 in.

Every *fourth* line on the sleeve is longer than the others and is numbered to permit easy reading. Each numbered line indicates a distance of 0.100 in. For example, #4 showing on the sleeve indicates a distance between the measuring faces of 4 × 0.100, or 0.400 in.

The thimble has 25 equal divisions about its circumference. Since one turn moves the thimble 0.025 in., one

Fig. 9-2   An inch micrometer reading of 0.288 in.   (Courtesy Kostel Enterprises Limited.)

division would represent ¹⁄₂₅ of 0.025 or 0.001. Therefore, each line on the thimble represents 0.001.

TO READ A STANDARD INCH MICROMETER

1. Note the last number showing on the sleeve. Multiply this by 0.100.
2. Note the number of small lines visible to the right of the last number shown. Multiply this number by 0.025.
3. Add the number of divisions on the thimble from zero to the line that coincides with the index line on the sleeve. In Fig. 9-2

☐ #2 is shown on the sleeve   2 × 0.100 = 0.200
☐ 3 lines are visible
past the number   3 × 0.025 = 0.075
☐ #13 line on thimble
coincides with the index line   13 × 0.001 = 0.013
Total reading   0.288 in.

VERNIER MICROMETER

The *vernier inch micrometer* (Fig. 9-3) has, in addition to the graduations found on a standard micrometer, a *vernier scale* on the sleeve. This vernier scale consists of 10 divisions which run *parallel to and above* the index line. It will be noted that these 10 divisions on the sleeve occupy the same distance as 9 divisions (0.009) on the thimble. One division on the vernier scale, therefore, represents ¹⁄₁₀ × 0.009, or 0.0009 in. Since one graduation on the thimble

Fig. 9-1   Cutaway view of a standard micrometer with a ratchet stop.

Fig. 9-3 An inch vernier micrometer caliper with a friction thimble.

Fig. 9-4 An inch vernier micrometer reading of 0.2363 in. (Courtesy Kostel Enterprises Limited.)

represents 0.001 or 0.0010 in., the difference between one thimble division and one vernier scale division represents 0.0010 − 0.0009, or 0.0001. Therefore, each division on the vernier scale has a value of 0.0001 in.

TO READ A VERNIER MICROMETER

1. Read the micrometer as you would a standard micrometer.
2. Note the line on the vernier scale that coincides with a line on the thimble. This line will indicate the number of ten-thousandths that must be added to the above reading.

Referring to Fig. 9-4A, the reading of the vernier micrometer is as follows:

☐ #2 is shown on the sleeve 2 × 0.100 = 0.200
☐ 1 line is visible past the
number 1 × 0.025 = 0.025
☐ #11 line on the thimble
is just past the index line 11 × 0.001 = 0.011
☐ In Fig. 9-4B #3 line on the vernier scale
coincides with a line on the
thimble 3 × 0.0001 = 0.0003
☐ Total reading 0.2363 in.

METRIC MICROMETER

The *metric micrometer* (Fig. 9-5) is similar to the inch micrometer with two exceptions. The two basic differences are the pitch of the spindle screw and the graduations on the sleeve and thimble.

The pitch of the screw is 0.5 mm; therefore, a complete revolution of the thimble increases or decreases the distance between the measuring faces 0.5 mm. Above the index line on the sleeve, the graduations are in millimeters (from 0 to 25) with every fifth line being numbered. Below the index line, each millimeter is subdivided into two parts of 0.5 mm, which corresponds to the pitch of the thread. It is

Fig. 9-5 A metric micrometer measures in hundreds of a millimeter.

apparent, therefore, that two turns of the thimble will be required to move the spindle 1 mm.

The circumference of the thimble is divided into 50 equal divisions, with every fifth line being numbered. Since one revolution of the thimble advances the spindle 0.5 mm, each graduation on the thimble equals 1/50 × 0.5 mm = 0.01 mm.

TO READ A METRIC MICROMETER

1. Note the number of the last main division showing *above the line* to the left of the thimble. Multiply this by 1 mm.
2. If there is a half-millimeter line showing *below the index line,* between the whole millimeter and the thimble, then add 0.5 mm.
3. Add the number of the line on the thimble that coincides with the index line.

In Fig. 9-6 there are:

☐ 17 lines above the index line 17 × 1   = 17.00
☐ 1 line below the index line   1 × 0.5 = 0.50
☐ 11 lines on the thimble   11 × 0.01 = 0.11
      Total reading   17.61 mm

METRIC VERNIER MICROMETER

The *metric vernier micrometer,* in addition to the graduations found on the standard micrometer, has five vernier divisions on the barrel, each representing 0.002 mm. In the vernier micrometer reading illustrated in Fig. 9-7, each major division (below the index line) has a value of 1 mm. Each minor division (above the index line) has a value of 0.50 mm. There are 50 divisions around the thimble, each having a value of 0.01 mm.

TO READ A METRIC VERNIER MICROMETER

1. Read the micrometer as you would for a standard metric micrometer.
2. Note the line on the vernier scale that coincides with a line on the thimble. This line will indicate the number of two-thousandths of a millimeter that must be added to the above reading.

Referring to Fig. 9-7, the reading of the metric vernier micrometer would be:

Major divisions
  (below the index line)
                (10 × 1.00) = 10.00
Minor divisions
  (above the index line)
                (1 × 0.50) = 0.50
  Thimble divisions
                (16 × 0.01) = 0.16
  The 2nd vernier division
coincides with a thimble
line   (2 × 0.002) = 0.004
      Reading   10.664 mm

COMBINATION INCH-METRIC MICROMETER

With the gradual shift to metric measurement, there will be a need for a dual dimension system for some time. The combination inch-metric micrometer (Fig. 9-8A) will give readings in both inch and metric sizes. It has a digital reading for one system and a standard barrel and thimble reading for the other.

Another type of inch-metric micrometer (Fig. 9-8B) has dual scales on the sleeve and the barrel. A horizontal scale on sleeve (usually black) and the left-hand scale on the

Fig. 9-6 A metric micrometer reading of 17.61 mm. (Courtesy Kostel Enterprises Limited.)

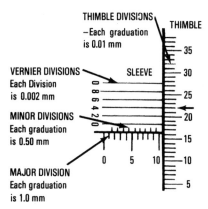

Fig. 9-7 A metric vernier micrometer reading of 10.164 mm.

Fig. 9-8 (A) A combination inch-metric micrometer with a digital reading for one system. (Courtesy Mitutoyo/MTI Canada Ltd.) (B) A combination inch-metric micrometer which has dual scales on the sleeve and the barrel.

barrel represent inch readings. The angular scale on the sleeve (usually red) and the right-hand scale on the barrel represent metric readings.

MICROMETER ADJUSTMENTS

Proper care and use of a micrometer is necessary if its accuracy is to be preserved and adjustments kept to a minimum. Minor adjustments to micrometers can easily be made; however, it is extremely important that all parts of the micrometer be kept free from dust and foreign matter during any adjustment.

TO REMOVE PLAY IN THE MICROMETER THREADS

To remove play (looseness) in the spindle threads due to wear:

1. Back off the thimble as shown in Fig. 9-9.
2. Insert the C-spanner into the slot or hole of the adjusting nut.
3. Turn the adjusting nut clockwise until play between the threads has been eliminated.

NOTE: After the micrometer has been adjusted, it should advance the spindle freely while turning the ratchet stop or friction thimble.

TESTING THE ACCURACY OF MICROMETERS

The accuracy of a micrometer should be tested periodically to ensure that the work produced is the size required. Always make sure that both measuring faces are clean before checking a micrometer for accuracy.

To test a micrometer, first clean the measuring faces and then turn the thimble using the friction thimble or ratchet stop until the measuring faces contact each other. If the zero line on the thimble coincides with the center (index) line on the sleeve, the micrometer is accurate. Micrometers can also be checked for accuracy by measuring a gage block or known standard.

The reading of the micrometer must be the same as the gage block or standard. Any micrometer which is not accurate should be adjusted by a qualified person.

TO ADJUST THE ACCURACY OF A MICROMETER

Should the accuracy of a micrometer require adjustment:

1. Clean the measuring faces and inspect them for damage.
2. Close the measuring faces carefully by turning the ratchet stop or friction thimble.
3. Insert the C-spanner into the hole or slot provided in the sleeve (Fig. 9-10).

Fig. 9-9   Removing the play in the micrometer spindle screw threads.   (Courtesy L. S. Starrett Company.)

Fig. 9-10   Resetting the accuracy of the micrometer.   (Courtesy L. S. Starrett Company.)

4. Carefully turn the *sleeve* until the index line on the sleeve coincides with the zero line on the thimble.
5. Recheck the accuracy of the micrometer by opening the micrometer and then closing the measuring faces by turning the ratchet stop or friction thimble.

MICROMETER ACCESSORIES

Although the design of most micrometers is fairly standard, certain desired refinements may be added to the basic design. Items such as the lock ring, ratchet, friction thimble, carbide measuring faces, and anvil extensions increase the accuracy and range of these instruments. Figure 9-11A–C illustrates several models of outside micrometers.

Fig. 9-11 (A) Direct-reading micrometer; (B) micrometer with carbide measuring faces; (C) tubular frames are used in large-size micrometers to reduce weight. (Courtesy The L. S. Starrett Company.)

INDICATING MICROMETER

Figure 9-12 shows an *indicating micrometer,* employing an indicating dial and a moving anvil, which permits accurate measurements to 0.0001 in. (0.002 mm). This type of micrometer may also be used as a comparator by setting it to the desired size with gage blocks or a standard and locking the spindle. The tolerance arms are then set to the required limits. As each piece of work is inserted, the relieving button is depressed. After the button is released, a comparative or a limit reading will be indicated on the dial.

SCREW THREAD MICROMETERS

Sharp-V, American National, Unified, and International Organization for Standardization (ISO) threads may be measured with reasonable accuracy with a screw thread mi-

Fig. 9-12 An indicating micrometer used for inspection work.

Fig. 9-13 Screw thread micrometer and insert showing the measuring faces.

crometer (Fig. 9-13). These micrometers have a pointed spindle and a double-V swivel anvil which are shaped to contact the pitch diameter of the thread being measured (Fig. 9-13). The micrometer reading indicates the pitch diameter of the thread, which is equal to the outside diameter less the depth of one thread.

Each thread micrometer is limited to measuring a certain range of threads; this range is stamped on the micrometer frame. One-inch thread micrometers are manufactured in four ranges to cover the following range of threads per inch (TPI):

  8 to 13 TPI
14 to 20 TPI
22 to 30 TPI
32 to 40 TPI

Metric-thread micrometers are available in sizes from 0 to 25 mm, 25 to 50 mm, 50 to 75 mm, and 75 to 100 mm. A set of 12 anvil and spindle inserts is available for thread pitches from 0.4 to 6 mm.

To check the accuracy of a thread micrometer, carefully bring the measuring faces into light contact; the micrometer reading for this setting should be zero.

Fig. 9-14 Threads may be compared to a gage or standard with a screw-thread-comparator micrometer.

When measuring threads, the micrometer gives a slightly distorted reading because of the helix angle of the thread. In order to overcome this inaccuracy, set the thread micrometer to a thread plug gage or to a thread which must be duplicated.

TO MEASURE WITH A THREAD MICROMETER

1. Select the correct thread micrometer to suit the number of threads per inch of the workpiece.
2. Thoroughly clean the measuring surfaces.
3. Check the micrometer for accuracy by bringing the measuring faces together; the reading should be zero.
4. Clean the thread to be measured.
5. Set the micrometer to the required thread plug gage and note the reading.
6. Fit the swivel anvil onto the threaded workpiece.
7. Adjust the spindle until the point just bears against the opposite side of the thread.
8. Carefully roll the micrometer over the thread to get the proper feel.
9. Note the readings and compare them to the micrometer reading of the thread plug gage.

Threads may also be checked by the *screw-thread-comparator micrometer* (Fig. 9-14), which has two conical measuring surfaces. Since it does not measure the pitch diameter, it is important to set this instrument to a thread plug gage before measuring a threaded workpiece. The screw micrometer is used for quick comparison of threads, as well as for checking small grooves and recesses where regular micrometers cannot be used.

When thread micrometers or comparators are not available, threads may be accurately checked by the *three-wire method,* which is fully discussed in Unit 49.

REVIEW QUESTIONS

MICROMETERS

1. How many threads per inch are there on a standard inch micrometer?

2. What is the value of:
   (a) each line on the sleeve?
   (b) each numbered line on the sleeve?
   (c) each line on the thimble?
3. Read the following standard micrometer settings.

4. Describe briefly the principle of the vernier micrometer.
5. Describe the procedure for reading a vernier micrometer.
6. Read the following vernier micrometer settings.

7. Explain how to adjust a micrometer
   (a) to remove play in the spindle threads
   (b) for accuracy

METRIC MICROMETERS

8. What are the basic differences between a metric and an inch micrometer?
9. What is the value of one division on:
   (a) the sleeve above the index line?
   (b) the sleeve below the index line?
   (c) the thimble?

10. Read the following metric micrometer settings.

11. State two uses for an indicating micrometer.
12. What is the purpose of the relieving button on the indicating micrometer?

SCREW THREAD MICROMETERS

13. Describe the construction of the contact points of a screw thread micrometer.
14. What dimension of the thread is indicated on a screw thread micrometer reading?
15. List the four ranges covered by screw thread micrometers.
16. How may threads be measured *accurately* with a screw thread micrometer?

VERNIER CALIPERS

UNIT 10

Vernier calipers are precision measuring tools used to make accurate measurements to within 0.001 in. for inch verniers or to 0.02 mm for metric verniers. The bar and the movable jaw may be graduated on both sides or both edges. One side is used to take outside measurements; the other side is for taking inside measurements. Vernier calipers are available in inch and metric graduations; however, some types have both inch and metric graduations on the same caliper.

OBJECTIVES

After completing this unit you should be able to:
1. Measure workpieces to within an accuracy of 0.001 in. with a 25-division inch vernier caliper
2. Measure workpieces to within an accuracy of 0.001 in. using a 50-division inch vernier caliper
3. Measure workpieces to within an accuracy of 0.02 mm using a metric vernier caliper

VERNIER CALIPERS

Vernier calipers (Fig. 10-1) are precision tools used to make accurate measurements to within 0.001 in. or 0.02 mm, depending on whether they are inch or metric.

PARTS OF THE VERNIER CALIPER

The vernier caliper, regardless of the standard of measurement used, consists of an L-shaped frame and a movable jaw.

The L-shaped frame consists of a *bar,* which shows the *main scale graduations,* and the *fixed jaw.* The *movable jaw,* which slides along the bar, contains the *vernier scale.* Adjustments for size are made by means of an *adjusting nut.* Readings may be locked in place by means of the *clamp screws.*

Most bars are graduated on both sides or on both edges, one for outside measurements and the other set for inside measurements. The outer tips of the jaws are cut away to form nibs which permit inside measurements to be taken. Inch vernier calipers are manufactured with both 25- and

CLAMP SCREWS

BAR

VERNIER
SCALE

FINE ADJUSTING NUT

FIXED
JAW

MOVABLE
JAW

Fig. 10-1 Parts of a vernier caliper.

50-division vernier scales. The 50-division scale is much easier to read than the 25-division scale. Metric vernier calipers graduated in millimeters are also available.

Some vernier calipers are provided with two small indentations, or points, on the bar and a movable jaw which may be used to set dividers accurately to a specific dimension or radius.

The bar of the vernier caliper with the 25-division vernier scale on the movable jaw is graduated exactly the same as a micrometer. Each inch is divided into 40 equal divisions, each having a value of 0.025 in. Every fourth line, representing $\frac{1}{10}$ or 0.100, is numbered. The vernier scale on the movable jaw has 25 equal divisions, each representing 0.001. The 25 divisions on the vernier scale, which are 0.600 in. in length, are equal to 24 divisions on the bar. The difference between *one* division on the bar and one vernier division equals 0.025 − 0.024, or 0.001 in. Therefore, only one line of the vernier scale will line up exactly with a line on the bar at any one setting.

MEASURING A WORKPIECE WITH A 25-DIVISION INCH VERNIER CALIPER

1. Remove all burrs from the workpiece and clean the surface to be measured.
2. Open the jaws enough to pass over the work.
3. Close the jaws against the work and lock the right-hand clamp screw.
4. Turn the adjusting screw until the jaws *just touch* the work surface. Be sure that the jaws are in place by attempting to move the bar slightly sideways and vertically while turning the adjusting nut.
5. Lock the clamp screw on the movable jaw.
6. Read the measurement shown in Fig. 10-2 as follows:

Fig. 10-2 A 25-division inch vernier caliper reading of 1.436 in. (Courtesy L. S. Starrett Company.)

☐ The large #1 on the bar = 1.000
☐ The small #4 past the #1 4 × 0.100 = 0.400
☐ One line is visible past
the #4 1 × 0.025 = 0.025
☐ The 11th line of the vernier
scale coincides with a line
on the bar 11 × 0.001 = 0.011
    Total reading 1.436 in.

THE 50-DIVISION INCH VERNIER CALIPER

Because 25-division vernier calipers are often difficult to read, many vernier calipers are now manufactured with 50 divisions (equal to 49 on the main scale) on the vernier

Fig. 10-3 A 50-division inch vernier caliper reading of 1.464 in.

Fig. 10-4 A vernier caliper with inch and metric readings.

Fig. 10-5 A metric vernier caliper reading of 43.18 mm.

sliding or *vernier scale* with every fifth one being numbered. These 50 graduations occupy the same space as 49 graduations on the main scale (49 mm). Therefore,

$$1 \text{ vernier division} = \frac{49}{50}$$
$$= 0.98 \text{ mm}$$

The difference between 1 main scale division and 1 vernier scale division is

$$1 - 0.98 = 0.02 \text{ mm}$$

TO READ A METRIC VERNIER CALIPER

1. The last numbered division on the bar to the left of the vernier scale represents the number of millimeters multiplied by 10.
2. Note how many full graduations are showing between this numbered division and the zero on the vernier scale. Multiply this number by 1 mm.
3. Find the line on the vernier scale which coincides with a line on the bar. Multiply this number by 0.02 mm.
In Fig. 10-5:

☐ The large #4 graduation on the bar
(4 × 10 mm) = 40
☐ Three full lines past the #4 graduation
(3 × 1 mm) = 3
☐ The 9th line on the vernier scale
coincides with a line on the bar (9 × 0.02) = <u>0.18</u>
Total reading   43.18 mm

DIRECT-READING DIAL CALIPER

An instrument which resembles the vernier caliper is the *direct-reading dial caliper* (Fig. 10-6), capable of reading to 0.001 in. (0.02 mm). A dial indicator, the hand of which is attached to a pinion, is mounted on the sliding jaw. For the inch dial caliper, one revolution of the hand represents 0.200 in. of travel. The dial is divided into 200 divisions, each representing 0.001 in. The pinion meshes with a rack attached to the bar which is graduated in inches (1.000) and tenths of an inch (0.100). Dial calipers are quickly set and more easily read than vernier calipers. Most direct-reading calipers have a narrow sliding blade attached to the sliding jaw (and dial). This narrow blade permits the dial caliper to be used as an efficient and accurate depth gage. See Unit 11.

scale of the movable jaw. Each of these scales on the bar and movable jaw is equal to 2.450 in. in length. Each division on the bar then equals 2.450 divided by 49 divisions, or 0.050 in. in length. Each division on the vernier scale would equal 2.450 divided by 50 divisions, or 0.049 in. in length. The *difference in length* between one main scale division and one vernier division equals 0.050 − 0.049, or 0.001 in.

Each line on the main scale of a 50-division vernier caliper has a value of 0.050 in. Each line on the vernier scale has a value of 0.001 in. In Fig. 10-3,

☐ The large #1 on the bar   = 1.000
☐ The small #4 past the #1   4 × 0.100 = 0.400
☐ 1 line is visible past #4   1 × 0.050 = 0.050
☐ The 14th line on
the vernier scale
coincides with a line
on the bar   14 × 0.001 = <u>0.014</u>
☐ Total reading   1.464 in.

THE METRIC VERNIER CALIPER

Vernier calipers are also made with metric readings and many have both metric and inch graduations on the same instrument (Fig. 10-4). The parts of metric vernier calipers are the same as those of the inch vernier.

The *main scale* is graduated in millimeters and every main division is numbered. Each numbered division has a value of 10 mm, for example, #1 represents 10 mm, #2 represents 20 mm, etc. There are 50 graduations on the

Fig. 10-6   A direct-reading dial caliper.   (Courtesy L. S. Starrett Company.)

GEAR-TOOTH VERNIER CALIPER

The *gear-tooth vernier caliper* (Fig. 10-7) consists basically of a vernier caliper and a vernier depth gage mounted at right angles to each other. This caliper is used to measure the thickness of gear teeth at the pitch-circle line. It may be used also to measure the width of Acme threads at the pitch line. Inch instruments are manufactured in two ranges: 20-diametral pitch to 2-diametral pitch and 10-diametral pitch to 1-diametral pitch, measuring in thousandths of an inch.

Metric gear-tooth vernier calipers are graduated in steps of 0.02 mm and come in the 1.25- to 12.00-mm range and the 2.5- to 25.00-mm range.

Corrected Addendum   When measuring gear teeth, it is necessary to set the gear-tooth vernier caliper to the *corrected addendum,* which is a point slightly lower than the *true addendum* of the gear. If the vernier depth slide is set to the true addendum, the caliper will measure the sides of the tooth at a point above the pitch circle and not give a true measurement (Fig. 10-8A and B). Most handbooks contain tables giving the corrected addendum of gear teeth. If tables are not available, the following formula may be used to calculate the corrected addendum.

$$CA = A + \left[\tfrac{1}{2}\, PD \left(1 - \cos \frac{90°}{N}\right) \right]$$

CA = corrected addendum
A = addendum
PD = pitch diameter
N = number of teeth in the gear

TO CHECK GEAR-TOOTH DIMENSIONS

1. Set the vertical slide to the *corrected addendum* of the gear.
2. Place the slide on top of the gear tooth to be measured.
3. Using the adjusting nut, set the horizontal slide to the thickness of the tooth.
4. Note the reading on the horizontal scale and check it with the required thickness.

Fig. 10-7   A gear-tooth vernier caliper.

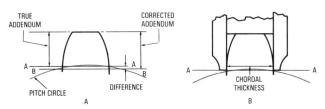

Fig. 10-8   Gear-tooth measurement showing the (A) true and (B) corrected addendum.

REVIEW QUESTIONS

VERNIER CALIPER

1. Describe the principle of
   (a) the 25-division vernier
   (b) the 50-division vernier
2. Describe the procedure for reading a vernier caliper.
3. What are the following vernier caliper readings?

4. Describe the principle of the metric vernier caliper.
5. Read the following metric caliper settings.

A

B

GEAR-TOOTH VERNIER CALIPER

6. A 16-diametral pitch gear has 54 teeth.
   (a) Calculate the corrected addendum.
   (b) Explain how to measure the gear tooth.

INSIDE-, DEPTH-, AND HEIGHT-MEASURING INSTRUMENTS

UNIT 11

Because of the great variety of measurements required in machine shop work, many types of measuring tools are available. These permit the machinist to measure not only outside sizes but also inside diameters, depths, and heights. Direct-reading instruments are most commonly used and generally the most accurate; however, because of the shape or size of the part, transfer-type instruments may be required.

OBJECTIVES

After completing this unit you should be able to:
1. Measure hole diameters to within 0.001-in. (0.02-mm) accuracy using inside micrometers and inside micrometer calipers
2. Measure depths of slots and grooves to an accuracy of 0.001 in. (0.02 mm)
3. Measure heights to an accuracy of 0.001 in. (0.02 mm) using a vernier height gage.

INSIDE-MEASURING INSTRUMENTS

All inside-measuring instruments fall into two categories: direct-reading and transfer-type.

With *direct-reading instruments* the size of the hole can be read on the instrument being used to measure the hole. The most common direct-reading instruments are the inside micrometer, the Intrimik, and the vernier caliper.

Transfer-type instruments are set to the diameter of the hole and then this size is transferred to an outside micrometer to determine the actual size. The most common transfer-type instruments are inside calipers, small hole gages, and telescopic gages.

DIRECT-READING INSTRUMENTS

Inside Micrometer Calipers

The inside micrometer caliper (Fig. 11-1) is designed for measuring holes, slots, grooves, etc. from 0.200 to 2.000 in. for the inch-designed instruments, or 5 to 50 mm for metric instruments. The nibs or ends of the jaws are hardened and ground to a small radius to permit the accurate measurement of holes. A locking nut provided with this micrometer can be used to lock the micrometer at any desired size.

The inside micrometer caliper employs the same principle as a standard micrometer, except that the barrel readings on some calipers are reversed (as shown in Fig. 11-1). Extreme care must be taken in reading this type of instrument. Other inside micrometer calipers have the readings on the spindle and are read in the same manner as a standard outside micrometer. Inside micrometer calipers are special-purpose tools and are not used in mass production measurement.

To Use an Inside Micrometer Caliper

1. Adjust the jaws to slightly less than the diameter to be measured.
2. Hold the fixed jaw against one side of the hole and adjust the movable jaw until the proper "feel" is obtained.

NOTE: Move the movable jaw back and forth to ensure that the measurement taken is across the true diameter.

3. Set the lock nut, remove the instrument, and check the reading.

Inside Micrometers

For internal measurements larger than 1½ in. or 40 mm, inside micrometers (Fig. 11-2) are used. The inside micrometer set consists of a micrometer head, having a range ½ or 1 in.; several extension rods of different lengths which may be inserted into the head; and a ½-in. spacing collar. These sets cover a range from 1½ to over 100 in., or from 40 to 1000 mm for metric tools. Sets that are used for the

Fig. 11-1 An inside micrometer caliper.

Fig. 11-2 An inside micrometer set can measure a large range of sizes.

larger ranges generally have hollow tubes rather than rods for greater rigidity.

The inside micrometer is read in the same manner as the standard micrometer. Since there is no locking nut on the inside micrometer, the thimble nut is adjusted to a tighter fit on the spindle thread to prevent a change in the setting while it is being removed from the hole.

To Measure with an Inside Micrometer

1. Measure the size of the hole with a rule.
2. Insert the correct micrometer extension rod after having carefully cleaned the shoulders of the rod and the micrometer head.
3. Align the zero marks on the rod and micrometer head.
4. Hold the rod firmly against the micrometer head and tighten the knurled set screw.
5. Adjust the micrometer to slightly less than the diameter to be measured.
6. Hold the head in a fixed position and adjust the micrometer to the hole size while moving the rod end in the direction of the arrows (Fig. 11-3).
7. When a micrometer is properly adjusted to size, there should be a slight drag or feel when the rod end is moved past the center line of the hole.
8. Carefully remove the micrometer and note the reading.
9. To this reading, add the length of the extension rod and collar.

Fig. 11-3 Using an inside micrometer to measure the size of a hole.

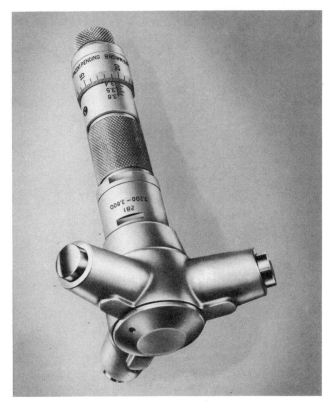

Fig. 11-4 The Intrimik, which has three contact points, measures holes accurately.

Intrimik

A difficulty encountered in measuring hole sizes with instruments employing only two measuring faces is that of properly measuring the diameter and not a chord of the circle. An instrument which eliminates this problem is the *Intrimik* (Fig. 11-4).

The Intrimik consists of a head with three contact points spaced 120° apart; this head is attached to a micrometer-type body. The contact points are forced out to contact the inside of the hole by means of a tapered or conical plug attached to the micrometer spindle (Fig. 11-5). The construction of a head with three contact points permits the

Fig. 11-5 The construction of the Intrimik head.

Intrimik to be self-centering and self-aligning. It is more accurate because it provides a direct reading, eliminating the necessity of transferring measurements to determine hole size as with telescope or small hole gages.

The range of these instruments is from 0.275 to 12.000 in., and the accuracy varies between 0.0001 and 0.0005 in., depending on the head used. Metric Intrimiks have a range from 6 to 300 mm, with graduations in 0.001 mm. The accuracy of the Intrimik should be checked periodically with a setting ring or master ring gage.

TRANSFER-TYPE INSTRUMENTS

Transfer measurement is where the size of an object is taken with an instrument which is not capable of giving a direct reading. The size is then determined by measuring the setting of the instrument with a direct-reading instrument or gage of a known size.

Small Hole Gages

Small hole gages are available in sets of *four*, covering a range from ⅛ to ½ in. (3.1 to 12.7 mm). They are manufactured in *two* types (Fig. 11-6A and B).

The small hole gages shown in Fig. 11-6A have a small, round end, or ball, and are used for measuring holes, slots, grooves, and recesses which are too small for inside calipers or telescope gages. Those shown in Fig. 11-6B have a flat bottom and are used for similar purposes. The flat bottom permits the measurement of shallow slots, recesses, and holes impossible to gage with the rounded type.

Both types are of similar construction and are adjusted to size by turning the knurled knob on the top. This draws up a tapered plunger, causing the two halves of the ball to open up and contact the hole.

To Use a Small Hole Gage Small hole gages require extreme care in setting since it is easy to get an incorrect setting when checking the diameter of a hole.

1. Measure the hole to be checked with a rule.

INSIDE-, DEPTH-, AND HEIGHT-MEASURING DEVICES

Fig. 11-6 (A) Small hole gages with a hardened-ball end; (B) small hole gages with a flat-bottom end. (Courtesy L. S. Starrett Company.)

2. Select the proper small hole gage.
3. Clean the hole and the gage.
4. Adjust the gage until it is slightly smaller than the hole and insert into the hole.
5. Adjust the gage until it can be felt just touching the sides of the hole or slot.
6. Swing the handle back and forth, and adjust the knurled end until the proper "feel" is obtained across the widest dimension of the ball.
7. Remove the gage and check the size with an outside micrometer.

NOTE: It is important that when transferring the measurement, the same "feel" is obtained as when adjusting the gage to the hole.

Telescope Gages

Telescope gages (Fig. 11-7) are used for obtaining the size of holes, slots, and recesses from $\frac{5}{16}$ to 6 in. (8 to 152 mm). They are T-shaped instruments, each consisting of a pair of telescoping tubes or plungers connected to a handle. The plungers are spring-loaded to force them apart. The knurled knob on the end of the handle locks the plungers into position when turned in a clockwise direction.

NOTE: In some sets, only one plunger moves.

To Measure Using a Telescope Gage

1. Measure the hole size and select the proper gage.
2. Clean the gage and the hole.
3. Depress the plungers until they are slightly smaller than the hole diameter and lightly tighten the knurled knob.
4. Insert it into the hole and, with the handle tilted upward slightly, loosen the knurled knob to release the plungers.
5. *Lightly* snug up the knurled knob.
6. Hold the bottom leg of the telescope gage in position with one hand.
7. Move the handle downward through the center while slightly moving the top leg from side to side (Fig. 11-8).

Fig. 11-7 A set of telescope gages. (Courtesy L. S. Starrett Company.)

8. Tighten the knurled knob to lock the plungers in position.
9. Recheck the "feel" on the gage by testing it in the hole again.
10. Check the gage size with outside micrometers, maintaining the same "feel" as in the hole (Fig. 11-9).

Dial Bore Gages

A quick and accurate method of checking hole diameters and bores for size, out-of-round, taper, bell-mouth, hourglass, or barrel shapes, is by means of the *dial bore gage* (Fig. 11-10).

Fig. 11-8   Setting a telescope gage to the hole diameter.

Fig. 11-9   Measuring the telescope gage setting with a micrometer.

Gaging is accomplished by three spring-loaded centralizing plungers in the head, one of which actuates the dial indicator, graduated in ten-thousandths of an inch, or in 0.01-mm graduations for metric tools.

These instruments are available in six sizes to cover a range from 3 to 12 in. or from 75 to 300 mm. Each instrument is supplied with extensions to increase its range. The dial bore gage must be set to a size with a master gage, and then the hole size is compared to the gage setting. Should the hole size vary, it is not necessary to adjust the gage as long as the size remains within the range of the gage.

DEPTH MEASUREMENT

Although rules and various attachments can be used for measuring depth, the depth micrometer and the depth vernier are most commonly used where accuracy is required.

MICROMETER DEPTH GAGE

Micrometer depth gages are used for measuring depth of blind holes, slots, recesses, and projections. Each gage consists of a flat base attached to a micrometer sleeve. An extension rod of the required length fits through the sleeve and protrudes through the base (Fig. 11-11). This rod is held in position by a threaded cap on the top of the thimble.

Micrometer extension rods are available in various lengths, providing a range up to 9 in. or 225 mm for metric tools. The micrometer screw has a range of ½ or 1 in., or up to 25 mm for metric tools. Depth micrometers are available with both round or flat rods which are *not interchangeable* with other depth micrometers. The accuracy of these micrometers is controlled by a nut on the end of each extension rod which can be adjusted if necessary.

To Measure with a Micrometer Depth Gage

1. Remove burrs from the edge of the hole and the face of the workpiece.

Fig. 11-10   Dial bore gages provide a quick and accurate method of measuring hole diameters.

Fig. 11-11   A micrometer depth gage and extension rods. (Courtesy L. S. Starrett Company.)

Fig. 11-12 Measuring the depth of a shoulder.

Fig. 11-13 Graduations on a depth micrometer are reversed from those on an outside micrometer.

2. Clean the work surface and the base of the micrometer.
3. Hold the micrometer base firmly against the surface of the work (Fig. 11-12).
4. Rotate the thimble lightly with the tip of one finger in a clockwise direction until the bottom of the extension rod touches the bottom of the hole or recess.
5. Recheck the micrometer setting a few times to make sure that not too much pressure was applied in the setting.
6. Carefully note the reading.

NOTE: *The numbers on the thimble and the sleeve are the reverse of those on a standard micrometer* (Fig. 11-13).

VERNIER DEPTH GAGES

The depths of holes, slots, and recesses may also be measured by a vernier depth gage. This instrument is read in the same manner as a standard vernier caliper. Figure 11-14 illustrates how the toolmaker's button may be set up with this instrument.

Depth measurements may also be made with certain types of vernier or dial calipers which are provided with a thin sliding blade or depth gage attached to the movable jaw (Fig. 11-15). The blade protrudes from the end of the bar opposite the sliding jaw. The caliper is placed vertically over the depth to be measured, and the end of the bar is held against the shoulder while the blade is inserted into the hole to be measured. Depth readings are identical to standard vernier readings.

HEIGHT MEASUREMENT

Accurate height measurement is very important in layout and inspection work. With the proper attachments, the

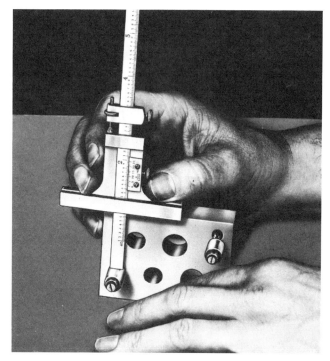

Fig. 11-14 Checking the position of toolmaker's buttons using a vernier depth gage.

vernier height gage is a very useful and versatile tool for these purposes. Where extreme accuracy is required, gage blocks or a precision height gage may be used.

VERNIER HEIGHT GAGE

The vernier height gage is a precision instrument used in toolrooms and inspection departments on layout and jig-

Fig. 11-15 A vernier dial caliper with a depth gage blade. (Courtesy L. S. Starrett Company.)

Fig. 11-16 A vernier height gage is used for layout and inspection work.

BEAM

CLAMP SCREW

VERNIER SLIDE

VERNIER SLIDE

SCRIBER

BASE

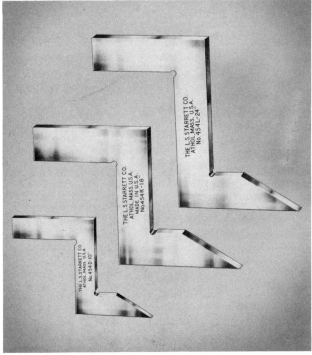

Fig. 11-17 Offset scribers are used with a vernier height gage for accurate layout work.

Fine adjustments are made by means of an adjusting nut. The vernier height gage is read in the same manner as a vernier caliper.

The vernier height gage is very well suited to accurate layout work and may be used for this purpose if a scriber is mounted on the movable jaw (Fig. 11-16). The scriber height may be set either by means of the vernier scale or by setting the scriber to the top of a gage block buildup of the desired height.

The *offset scriber* (Fig. 11-17) is a vernier height gage attachment which permits the setting of heights from the face of the surface plate. When using this attachment, it is not necessary to consider the height of the base or the width of the scriber and clamp.

A *depth gage attachment* may be fastened to the movable jaw, permitting the measurement of height differences which may be difficult to measure by other methods.

Another important use for the vernier height gage is in inspection work. A dial indicator may be fastened to the movable jaw of the height gage (Fig. 11-18), and distances between holes or surfaces can be checked to within an

and fixture work to measure and mark off distances accurately. These instruments are available in a variety of sizes from 12 to 72 in. or from 300 to 1000 mm and can be accurately set at any height to within 0.001 in. or 0.02 mm, respectively. Basically, a vernier height gage is a vernier caliper with a hardened, ground, and lapped base instead of a fixed jaw and is always used with a surface plate or an accurate flat surface. The sliding jaw assembly can be raised or lowered to any position along the beam.

Fig. 11-18 Using a height gage and dial indicator to check a height.

accuracy of 0.001 in. (0.02 mm) on the vernier scale. If greater accuracy (0.0001 in. or less) is required, the indicator may be used in conjunction with gage blocks.

In Fig. 11-18, the height gage is being used to check the location of reamed holes in relation to the edges of the plate and to each other.

To Measure with a Vernier Height Gage and Dial Indicator

1. Thoroughly clean the surface plate, height gage base, and work surface.
2. Place a finished edge of the work on the surface plate and clamp it against an angle plate if required.
3. Insert a snug-fitting plug in the hole to be checked with about ½ in. (12.7 mm) projecting beyond the work.
4. Mount the dial indicator on the movable jaw of the height gage.
5. Adjust the movable jaw until the indicator almost touches the surface plate.
6. Lock the upper slide of the height gage and use the adjusting nut to move the indicator until the dial registers about one-quarter turn.
7. Set the indicator dial to zero.
8. Record the reading of the vernier height gage.
9. Adjust the vernier height gage until the indicator registers zero on the top of the plug. Record this vernier height gage reading.
10. From this reading, subtract the initial reading plus half the diameter of the plug. This will indicate the distance from the surface plate to the center of the hole.
11. Check other hole heights using the same procedure.

To Measure Heights Using Gage Blocks

When hole locations must be accurate to 0.0005 in. (0.010 mm) or less, a gage block buildup is made for the proper dimension from the surface plate to the top of a plug fitted in the hole.

Fig. 11-19 Checking hole locations with a precision height gage and a height transfer gage.   (Courtesy ExCello Corporation.)

1. Prepare the required gage block buildup (the center of the hole height plus one-half the hole diameter). (See Unit 12.)
2. Mount a suitable dial indicator on a surface or vernier height gage.
3. Set the dial indicator to register zero on the top of the gage blocks.
4. Move the indicator over the top of the plug. The difference between the gage block buildup and the top of the plug will be registered on the dial indicator (Fig. 11-18).

PRECISION HEIGHT GAGE

The precision height gage (Fig. 11-19) provides a quick and accurate method of setting any height within the range of the instrument, eliminating the need for calculating and assembling specified gage blocks for comparative measurements. A surface plate is used as the reference surface.

The precision height gage is made from a hardened and ground round steel bar, with ground and lapped measuring steps or disks spaced exactly at 1-in. (25.4-mm) intervals. The measuring bar or column is raised or lowered by turning the large micrometer thimble which is graduated by steps of 0.0001 in., or 0.002 mm if the instrument is metric. The column may be raised or lowered a full inch or 25 mm, permitting any reading from zero to the range of the instrument in increments of 0.0001 in. or 0.002 mm, respectively. Some models have a vernier scale below the micrometer thimble; readings in increments of 0.000 010 in. or 0.000 25 mm are then possible. It is important that the accuracy of precision height gages be checked periodically with a master set of gage blocks.

Height gages are available in models of 6, 12, 24, and 36 in., and their range may be increased by the use of riser blocks under the base. Metric gages range from 300 to 600 mm.

To Use a Precision Height Gage

1. Clean the surface plate and the feet of the height gage.
2. Clean the bottom of the work to be checked and place it on the surface plate, using parallels and an angle plate if required.
3. Insert plugs into the holes to be checked.
4. Mount a dial indicator on the movable jaw of a vernier height gage.
5. Adjust the height gage until the dial indicator registers approximately 0.015 in. (0.40 mm) across the top of the plug.
6. Turn the dial of the indicator to zero.
7. Move the dial indicator over the nearest disk of the precision height gage and raise the column by turning the micrometer until the dial indicator reads zero.
8. Check the micrometer reading. This reading will indicate the distance from the surface plate to the top of the plug.
9. Subtract half the diameter of the plug from this reading.

NOTE: If the work is set on parallels, this height must be subtracted from the precision height gage reading.

REVIEW QUESTIONS

INSIDE MICROMETER CALIPERS

1.
   (a) On what type of inside micrometer calipers are the readings reversed to an outside micrometer?
   (b) What type are read the same as an outside micrometer?
2. What precautions must be observed in taking a measurement with an inside micrometer caliper?

INSIDE MICROMETERS

3. What construction feature compensates for a lock nut on inside micrometers?
4. What precautions must be taken when:

   (a) assembling the inside micrometer and extension rod?
   (b) using the inside micrometer?
5. What is the correct feel with an inside micrometer?

SMALL HOLE GAGES

6. Name two types of small hole gages and state the purpose of each.
7. What precaution must be observed when using a small hole gage to obtain a dimension?

TELESCOPE GAGES

8. List the steps required to measure a hole with a telescope gage.

DIAL BORE GAGES

9. What hole defects may be conveniently measured with a dial bore gage?

INTRIMIK

10. Why is an Intrimik particularly suited to measure hole sizes?
11. Why is an Intrimik more accurate than a telescope gage for measuring a hole size?

MICROMETER DEPTH GAGE

12. How is the accuracy of a micrometer depth gage adjusted?
13. How must the workpiece be prepared prior to measuring the depth of a hole or slot with a micrometer depth gage?
14. Explain the procedure for measuring a depth with a depth micrometer.
15. How does the reading of a depth micrometer differ from that of a standard outside micrometer?

VERNIER HEIGHT GAGE

16. State the two main applications for the vernier height gage.
17. What accessories are required for a vernier height gage to check *accurately* the height of a workpiece?

PRECISION HEIGHT GAGE

18. What are the advantages of using a precision height gage in lieu of a gage block buildup?
19. What dimension(s) must be subtracted from the reading so that the correct reading for the height of a hole being checked will be obtained?

GAUGE BLOCKS.

...g ...e u.iii you will be able to:

1. Understand the use and application of gage blocks
2. Calculate inch and metric gage block buildups
3. Prepare and wring gage blocks together

GAGE BLOCK MANUFACTURE

Gage blocks are rectangular blocks of hardened and ground alloy steel which have been stabilized through alternate subjections of extreme heat and cold until the crystalline structure of the metal is without strain. The two measuring surfaces are lapped and polished to an optically flat surface and to a specific size accurate within a range of 2 to 8 millionths of an inch (50 to 200 millionths of a millimeter). The size of each block is stamped on one of its surfaces. Chrome-plated gage blocks are also available and when long wear is desirable, carbide blocks are used. Great care is exercised in their manufacture; the final calibration is made under ideal conditions where the temperature is maintained at 68°F (20°C). Therefore, an inch or a centimeter is accurate to size *only* when measured at the standard temperature of 68°F (20°C).

USES

Industry has found gage blocks to be invaluable tools. Because of their extreme accuracy, they are used for the following purposes.

1. To check the dimensional accuracy of fixed gages to determine the extent of wear, growth, or shrinkage
2. To calibrate adjustable gages, such as micrometers and vernier calipers, imparting accuracy to these instruments
3. To set comparators, dial indicators, and height gages to exact dimensions
4. To set sine bars and sine plates when extreme accuracy is required in angular setups
5. For precision layout with the use of attachments
6. To make machine tool setups
7. To measure and inspect the accuracy of finished parts in inspection rooms

GAGE BLOCK SETS

INCH-STANDARD GAGE BLOCKS

Gage blocks are manufactured in a variety of sets that vary from a few blocks to as many as 115 in a set. The most commonly used is the 83-piece set (Fig. 12-1) from which it is possible to make over 120,000 different measurements ranging from one hundred-thousandths of an inch to over 25 inches. The blocks which comprise an 83-piece set are listed in Table 12-1.

Fig. 12-1 An 83-piece set of gage blocks.

Two *wear* blocks are supplied with an 83-piece set. Some manufacturers supply 0.050-in. wear blocks, while others have 0.100-in. wear blocks. They should be used at each end of a combination, especially if the blocks will be in contact with hard surfaces or abrasives. In this way, the wear which occurs during use will be on the two wear blocks only, rather than on many blocks, and the useful life and accuracy of the set will be prolonged. During use, it is considered good practice always to expose the same face of the wear block to the work surface. A good habit to acquire is always to have the words ''Wear Block'' appear on the outside of the combination. This way, all the wear will be on one surface and the wringing quality of the other surface will be preserved.

METRIC GAGE BLOCKS

Metric gage blocks are supplied by most manufacturers in sets of 47-, 88-, and 113-piece blocks. The most common set is the 88-piece set (Table 12-2). Each of the above sets

TABLE 12-1 Sizes in an 83-Piece Set of Inch Standard Gage Blocks

FIRST: 0.0001-IN. SERIES—9 BLOCKS

0.1001	0.1002	0.1003	0.1004	0.1005	0.1006	0.1007	0.1008	0.1009

SECOND: 0.001-IN. SERIES—49 BLOCKS

0.101	0.102	0.103	0.104	0.105	0.106	0.107	0.108	0.109
0.110	0.111	0.112	0.113	0.114	0.115	0.116	0.117	0.118
0.119	0.120	0.121	0.122	0.123	0.124	0.125	0.126	0.127
0.128	0.129	0.130	0.131	0.132	0.133	0.134	0.135	0.136
0.137	0.138	0.139	0.140	0.141	0.142	0.143	0.144	0.145
0.146	0.147	0.148	0.149					

THIRD: 0.050-IN. SERIES—19 BLOCKS

0.050	0.100	0.150	0.200	0.250	0.300	0.350	0.400	0.450	0.500
0.550	0.600	0.650	0.700	0.750	0.800	0.850	0.900	0.950	

FOURTH: 1.000-IN. SERIES—4 BLOCKS

1.000	2.000	3.000	4.000

TWO 0.050-in. WEAR BLOCKS

TABLE 12-2 Sizes in an 88-Piece Set of Metric Gage Blocks

0.001-MM SERIES—9 BLOCKS

1.001	1.002	1.003	1.004	1.005	1.006	1.007	1.008	1.009

0.01-MM SERIES—49 BLOCKS

1.01	1.02	1.03	1.04	1.05	1.06	1.07	1.08	1.09
1.10	1.11	1.12	1.13	1.14	1.15	1.16	1.17	1.18
1.19	1.20	1.21	1.22	1.23	1.24	1.25	1.26	1.27
1.28	1.29	1.30	1.31	1.32	1.33	1.34	1.35	1.36
1.37	1.38	1.39	1.40	1.41	1.42	1.43	1.44	1.45
1.46	1.47	1.48	1.49					

0.5-MM SERIES—1 BLOCK

0.5

0.5-MM SERIES—18 BLOCKS

1	1.5	2	2.5	3	3.5	4	4.5	5
5.5	6	6.5	7	7.5	8	8.5	9	9.5

10-MM SERIES—9 BLOCKS

10	20	30	40	50	60	70	80	90

TWO 2-mm WEAR BLOCKS

contains a pair of 2-mm wear blocks. These blocks are used at each end of the buildup to prolong the accuracy of the other blocks in the set.

ACCURACY

Gage blocks in inch and metric standards are manufactured in three common degrees of accuracy, depending on the purpose for which they are used.

1. The *Class AA* set, commonly called a *laboratory* or *master* set, is accurate to ±0.000 002 in. in the inch standard. The metric gage block set is accurate to ±0.000 05 mm. These gage blocks are used in temperature-controlled laboratories as references to compare or check the accuracy of working gages.
2. The *Class A* set is used for inspection purposes and is accurate to ±0.000 004 in. in the inch standard. The metric gage block set is accurate to +0.000 15 mm and −0.000 05 mm.
3. The *Class B* set, commonly called the *working* set, is accurate to ±0.000 008 in. in the inch standard. The metric gage block set is accurate to +0.000 25 mm and −0.000 15 mm. These blocks are used in the shop for machine tool setups, layout work, and measurement.

THE EFFECT OF TEMPERATURE

While the effect of temperature on ordinary measuring instruments is negligible, changes in temperature are important when precision gage blocks are handled. Gage blocks have been calibrated at 68°F (20°C), while the human body temperature is around 98°F (37°C). A 1°F (0.5°C) rise in temperature will cause a 4-in. (100-mm) stack of gage blocks to expand approximately 0.000 025 in. (0.0006 mm); therefore it is important that these blocks be handled as little as possible. The following suggestions are offered to eliminate as much error as possible due to temperature change.

1. Handle gage blocks only when they must be moved.
2. Hold them by hand for as little time as possible.
3. Hold them between the tips of the fingers so that the area of contact is small, or use insulated tweezers.
4. Have the work and gage blocks at the same temperature. If a temperature-controlled room is not available, both the work and gage blocks may be placed in kerosene until both are at the same temperature.
5. Where extreme accuracy is necessary, use insulating gloves and tweezers to prevent temperature change during handling.

GAGE BLOCK BUILDUPS

Gage blocks are manufactured to great accuracy; they adhere to each other so well when wrung together properly that they can withstand a 200-pound (lb) [890-newton (N)] pull. Many theories have been advanced to explain this adhesion. Scientists have felt that it may be atmospheric pressure, molecular attraction, the extremely flat surfaces of the blocks, or a minute film of oil which gives the blocks this quality. Possibly a combination of any of these could be responsible.

When the blocks required to make up a dimension are being calculated, the following procedure should be followed to save time, reduce the chance of error, and use as few blocks as possible. For example, if a measurement of 1.6428 in. is required:

Step	Procedure	
1	Write the dimension required on the paper	1.6428
2	Deduct the size of two wear blocks: 2 × .050 in.	0.100
	Remainder	1.5428
3	Use a block which will eliminate the right-hand digit	0.1008
	Remainder	1.4420
4	Use a block which will eliminate the right-hand digit and at the same time bring the digit to the left of it to a zero (0) or a five (5)	0.142
	Remainder	1.300
5	Continue to eliminate the digits from the right to the left until the dimension required is attained	0.300
	Remainder	1.000
6	Use a 1.000-in. block	1.000
	Remainder	0.000

To eliminate the possibility of error in subtraction while making a buildup, it is good practice to use two columns for this calculation. As the following example illustrates, the left-hand column is where the gage blocks are subtracted from the original dimension, and the right-hand column is used as a check column. For example, to build up a dimension of 3.8716 in.:

		Procedure Column	Check Column
1.	Two wear blocks (2 × .050 in.)	3.8716 in. 0.100 3.7716	0.100
2.	Use 0.1006 in.	0.1006 3.6710	0.1006
3.	Use 0.121 in.	0.121 3.550	0.121
4.	Use 0.550 in.	0.550 3.000	0.550
5.	Use 3.000 in.	3.000 0.000 in.	3.000 3.8716 in.

When using metric blocks for buildup of 57.150 mm, proceed as follows:

Step	Procedure	
1	Write the dimension required on paper	57.150
2	Deduct the size of 2 wear blocks (2 × 2 mm)	4.000
	Remainder	53.150
3	Use a block which will eliminate the right-hand digit	1.050
	Remainder	52.100
4	Use a block to eliminate the right-hand digit	1.100
	Remainder	51.000
5	Use a 1-mm block	1.000
	Remainder	50.000
6	Use a 50-mm block	50.000
	Remainder	0.000 mm

For a metric buildup of 27.781 mm, proceed as follows:

	Procedure Column	Check Column
1. Two wear blocks (2 × 2 mm)	27.781 mm 4.000 23.781	4.000
2. Use 1.001 mm	1.001 22.780	1.001
3. Use 1.080 mm	1.080 21.700	1.080
4. Use 1.700 mm	1.700 20.000	1.700
5. Use 20.000 mm	20.000 0.000 mm	20.000 27.781 mm

To Wring Blocks Together

It is important when wringing blocks together to take care not to damage them. The sequence of movement to wring blocks together correctly is illustrated in Fig. 12-2.

1. Clean the blocks with a clean, soft cloth.
2. Wipe each of the contacting surfaces on the clean palm of the hand or on the wrist. This procedure removes any dust particles left by the cloth and also applies a light film of oil.
3. Place the end of one block over the end of another block as shown in Figure 12-2.
4. While applying pressure on the two blocks, slide one block over the other.
5. If the blocks do not adhere to each other, it is generally because the blocks have not been thoroughly cleaned.

Fig. 12-2 Procedure for wringing gage blocks.

Care of Gage Blocks

1. Gage blocks should always be protected from dust and dirt by being kept in a closed case when not in use.
2. Gages should not be handled unnecessarily since they absorb the heat of the hand. Should this occur, the gage blocks must be permitted to return to room temperature before use.
3. Fingering of lapped surfaces should be avoided to prevent tarnishing and rusting.
4. Care should be taken that gage blocks are not dropped or their lapped surfaces scratched.
5. Immediately after use, each block should be cleaned, oiled, and replaced in the storage case.
6. Before gage blocks are wrung together, their faces must be free of oil and dust.
7. Gage blocks should never be left wrung together for any length of time. The slight moisture between the blocks can cause rusting which will permanently damage the block.

REVIEW QUESTIONS

1. How are gage blocks stabilized, and why is this necessary?
2. State five general uses for gage blocks.
3. For what purpose are wear blocks used?
4. How should wear blocks always be assembled into a buildup?
5. State the difference between a master set and a working set of gage blocks.
6. What precautions are necessary when handling gage blocks in order that the effect of heat on the blocks is minimal?
7. List five precautions necessary for the proper care of gage blocks.
8. Calculate the gage blocks required for the following buildups. (Use the check column for accuracy.)

(a) 2.1743 in.	(c) 7.8923 in.	(e) 74.213 mm
(b) 6.2937 in.	(d) 32.079 mm	(f) 89.694 mm

Precise angular measurement and setups constitute an important phase of machine shop work. The most commonly used tools for accurately laying out and measuring angles are the universal bevel protractor, sine bar, and sine plate.

OBJECTIVES

After completing this unit you will be able to:
1. Make angular measurements to an accuracy of 5' (minutes) of a degree using a universal bevel protractor
2. Make angular measurements to less than 5' of a degree using a sine bar, gage blocks, and a dial indicator

THE UNIVERSAL BEVEL PROTRACTOR

The *universal bevel protractor* (Fig. 13-1) is a precision instrument capable of measuring angles to within 5'

(0.083°). It consists of a *base* to which a *vernier scale* is attached. A *protractor dial*, graduated in degrees, with every tenth degree being numbered, is mounted onto the circular section of the base. A *sliding blade* is fitted into this dial; it may be extended in either direction and set at any angle to the base. The blade and the dial are rotated as a unit. Fine adjustments are obtained with a small knurled-headed pinion which when turned engages with a gear attached to the blade mount. The protractor dial may be locked in any desired position by means of the *dial clamp nut*.

The vernier protractor (Fig. 13-2) is being used to measure an obtuse angle, or an angle greater than 90° but less than 180°. An *acute angle attachment* is fastened to the

Fig. 13-1 The universal bevel protractor can measure accurate angles.

Fig. 13-2 Measuring an obtuse angle using a universal bevel protractor.

vernier protractor to measure angles less than 90° (Fig. 13-3).

The vernier protractor dial, or main scale, is divided into two arcs of 180°. Each arc is divided into two quadrants of 90° and has graduations from 0 to 90° to the left and right of the zero ("0") line, with every tenth degree being numbered.

The vernier scale is divided into 12 spaces on each side of the "0" line which occupy the same space as 23° on the protractor dial. By simple calculation, it is easy to prove that one vernier space is 5', less than two graduations on the main scale. If the zero on the vernier scale coincides with a line on the main scale, the reading will be in degrees only. However, if any other line on the vernier scale coincides with a line on the main scale, the number of vernier graduations beyond the zero should be multiplied by 5 and added to the number of full degrees indicated on the protractor dial.

TO READ A VERNIER PROTRACTOR

1. Note the number of whole degrees between the zero on the main scale and the zero on the vernier scale.
2. Proceeding in the *same direction* beyond the zero on the vernier scale, note which vernier line coincides with a main scale line.
3. Multiply this number by 5 and add it to the number of degrees on the protractor dial.

In Fig. 13-4, the angular reading is calculated as follows. The number of degrees indicated on the main scale is 50 plus. The fourth line on the vernier scale *to the left* of the zero coincides with a line on the main scale. Therefore, the reading is:

Number of full degrees	= 50°
Value of vernier scale (4 × 5') =	20'
Reading	= 50°20'

NOTE: A double check of the reading would locate the vernier scale line on the other side of zero which coincides with a protractor scale line. This line should always equal the complement of 60'. In Fig. 13-4, the 40' line to the right of the zero coincides with a line on the protractor scale. This reading, when added to the 20' on the left of the scale, is equal to 60', or 1°.

THE SINE BAR

A *sine bar* (Fig. 13-5) is used when the accuracy of an angle must be checked to less than 5' or work must be located to a given angle within close limits. The sine bar consists of a steel bar with two cylinders of equal diameter fastened near the ends. The centers of these cylinders are on a line exactly parallel with the edge of the bar. The distance between the centers of these lapped cylinders is usually 5 or 10 in. Sine bars are generally made of stabilized tool steel, hardened, ground, and lapped to extreme accuracy. They are used on surface plates, and any desired

Fig. 13-3 Measuring an acute angle.

Fig. 13-4 A vernier protractor reading of 50°20'.

Fig. 13-5 A 5-in. sine bar with gage block buildup is used to set up work to an angle.

angle can be set by raising one end of the bar to a predetermined height with gage blocks.

Sine bars are generally made 5 in. or multiples of 5 in. in length. That is, the lapped cylinders are 5 in. ±0.0002 or 10 in. ±0.00025 between centers. The face of the sine bar is accurate to within 0.00005 in. in 5 in. In theory, the sine bar is the hypotenuse of a right-angle triangle. The gage block buildup forms the side opposite, while the face of the surface plate forms the side adjacent in the triangle.

Using trigonometry, it is possible to calculate the side opposite or gage block buildup for any angle between 0 and 90° as follows.

$$\text{Sine of a given angle} = \frac{\text{side opposite}}{\text{hypotenuse}}$$
$$= \frac{\text{gage-block buildup}}{\text{length of sine bar}}$$

When using a 5-in. sine bar, this would become:

$$\text{Sine of the angle} = \frac{\text{buildup}}{5}$$

Therefore, by transposition, the gage block buildup for any required angle with a 5-in. bar is as follows.

$$\text{Buildup} = 5 \times \text{sine of the required angle}$$

EXAMPLE

Calculate the gage block buildup required to set a 5-in. sine bar to an angle of 30°.

$$\begin{aligned} \text{Buildup} &= 5 \sin 30° \\ &= 5 \times .5000 \\ &= 2.5000 \text{ in.} \end{aligned}$$

NOTE: This formula is applied only to angles up to 60°.

When an angle greater than 60° is to be checked, it is better to set up the work using the complement of the angle (Fig. 13-6A). The angle plate is then turned 90° to produce the correct angle (Fig. 13-6B). The reason is that when the sine bar is in a near-horizontal position, a small change in the height of the buildup will produce a smaller change in the angle than when the sine bar is in the near-vertical position. This change in gage block height may be shown by calculating the buildups required for 75° and also for the complementary angle of 15°.

Buildup required for:

$$\begin{aligned} 75°1' &= 5 \sin 75°1' \ (0.9660) \\ &= 4.8300 \text{ in.} \\ 75° &= 5 \sin 75° \ (0.96592) \\ &= 4.82960 \text{ in.} \\ \text{Difference in the buildup for } 1' \\ &= 0.00040 \text{ in.} \end{aligned}$$

Buildup required for:

$$\begin{aligned} 15°1' &= 5 \sin 15°1' \ (0.25910 \text{ in.}) \\ &= 1.29550 \text{ in.} \\ 15° &= 5 \sin 15° \ (0.25882) \\ &= 1.29410 \text{ in.} \\ \text{Difference in buildup for } 1' \\ &= 0.00140 \text{ in.} \end{aligned}$$

This example shows that exactly 3.5 times the buildup is required to produce a change of 1' at 15°, than is required for 1' at 75°. Therefore, a small inaccuracy in setup would result in a smaller error at a smaller angle than it would at a larger one. If the complementary angles of 80 and 10° are used, this ratio increases to over 5:1.

Fig. 13-6 Setting up for an angle greater than 60°. (A) Set the sine bar to the complement of the angle; (B) turn the angle plate 90° on its side. (Courtesy Kostel Enterprises.)

When small angles are to be checked, it is sometimes impossible to get a buildup small enough to place under one end of the sine bar. In such situations, it will be necessary to place gage blocks under both rolls of the sine bar, having a net difference in measurement equal to the required buildup. For example, the buildup required for 2° is 0.1745 in. Since it is impossible to make this buildup, it is necessary to place the buildup for 1.1745 in. under one roll and a 1.000-in. block under the other roll, giving a net difference of 0.1745 in.

Before the sine bar is used to check a taper, it is necessary to calculate the angle of the taper so that the proper gage block buildup may be made. Fig. 13-7A and B illustrates how this is done.

In the right-angled triangle ABD:

$$\begin{aligned} \text{Tan } \frac{a}{2} &= \frac{\frac{1}{2}}{12} \\ &= 0.04166 \\ &= 2°23'10'' \\ \therefore a &= 4°46'20'' \end{aligned}$$

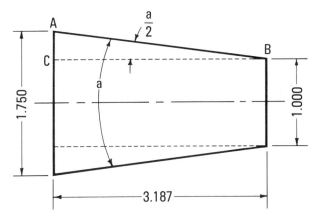

Fig. 13-7 The angle formed by a taper of 1 in./ft.

Fig. 13-8 Work may be clamped to a sine plate. (Courtesy The Taft-Pierce Manufacturing Company.)

Fig. 13-9 A hinged sine plate may be clamped to the machine table.

From this solution, the following formula for solving the included angle, when the taper per foot (TPF) is known, is derived.

$$\text{Tan } \frac{a}{2} = \frac{\text{TPF}}{24}$$

NOTE: When calculating the angle of a taper *do not* use the formula tan a = TPF/12 since the triangle ABC is not a right-angled triangle.

By transposition, if the included angle is given, the tpf may be calculated as follows:

$$\text{TPF} = \tan \frac{1}{2} a \times 24$$

If the tpf is not known, the angle may be calculated as shown in Fig. 13-7B.

SOLUTION

$$AC = \frac{1.750 - 1.000}{2}$$

$$= \frac{0.750}{2}$$

$$= 0.375$$

$$\text{Tan } \frac{a}{2} = \frac{0.375}{3.187}$$

$$= 0.11766$$
$$= 6°42'22''$$
$$\therefore a = 13°24'44''$$

To check the accuracy of this taper using a 5-in. sine bar, it is necessary to calculate the buildup as follows.

Buildup = 5 sin 13°24'44''
    = 5 × 0.23196
    = 1.1598 in.

Metric tapers are expressed as a ratio of 1 mm per unit of length; for example, a taper having a ratio of 1:20 would taper 1 mm in diameter in 20 mm of length. (See Unit 48.)

Tapers can be checked conveniently and accurately with a *taper micrometer*. This measuring instrument measures work to sine bar accuracy while the work is still in the machine. See ''Checking a Taper'' in Unit 48.

The *sine plate* (Fig. 13-8) uses the same principle as the sine bar and is similar in construction except that it is wider. Sine bars are up to 1 in. in width, while sine plates are generally more than 2 in. wide. They have several tapped holes in the surface which permit the work to be clamped to the surface of the sine plate. An end stop on a sine plate prevents the workpiece from moving during machining.

Sine plates may be hinged to a base (Fig. 13-9) and are often called *sine tables*. Both types are supplied in 5- and 10-in. lengths and have a step or groove of 0.100 or 0.200

in. deep ground in the base to permit the buildup for small angles to be placed under the free roll.

THE COMPOUND SINE PLATE OR TABLE

The *compound sine plate* (Fig. 13-10) consists of one sine plate superimposed on another sine plate. The lower plate is hinged to a base and may be tilted to any angle from 0 to 60° by placing gage blocks under the free roll or cylinder. The upper base is hinged to the lower base so that its cylinder and hinge are at right angles to those of the lower plate. The upper plate may also be tilted to any angle up to 60°. This feature permits the setting of compound angles (angles in two directions).

Compound sine plates have a side and end plate on the top table to facilitate setting up work square with the table edge and to prevent movement of the work during machining. A step or groove of 0.100 or 0.200 in. deep ground in the base and in the lower table permits the setup for small angles, as the gage block buildup may be placed in the groove.

Since most angles are machined in fixtures, sine plates are not used until the finishing operation, which is generally grinding. To facilitate the holding of parts, both simple and compound sine plates are available with built-in magnetic chucks.

Fig. 13-10   A compound sine plate allows the setting of angles in two directions.

REVIEW QUESTIONS

UNIVERSAL BEVEL PROTRACTOR

1. Name the parts of a universal bevel protractor and state the purpose of each.
2. Describe the principle of the vernier protractor.
3. Sketch a vernier protractor reading of:
   (a) 34°20′
   (b) 17°45′

SINE BAR

4. Describe the construction and principle of a sine bar.

5. What are the accuracies of the 5-in. and the 10-in. sine bars?
6. Calculate the gage block buildup for the following angles using a 5-in. sine bar:
   (a)   7°40′
   (b) 25°50′
   (c) 40°10′
7. What procedure should be followed to check an angle of 72° using a sine bar and gage blocks? Why is this procedure recommended?
8. In calculating the angle of a taper, why is the formula $\text{Tan } \frac{1}{2}a = \dfrac{\text{TPF}}{24}$ used, rather than

$$\text{Tan } a = \frac{\text{TPF}}{12} ?$$

Illustrate by means of a suitable sketch.

SINE PLATE

9. Describe a sine plate and state its purpose.
10. What is the advantage of a hinged sine plate?
11. What is the purpose of a compound sine plate?

UNIT 14

Although modern production processes have reached a high degree of precision, producing a part to an *exact* size would be far too costly. Consequently, industrial production methods or interchangeable manufacture permit certain variations from the exact dimensions to ensure that the component part will fit into the unit on assembly.

To determine the size of the various parts, the inspector generally uses some form of gaging. Gages used in industry vary from the simpler type of fixed gage to the sophisticated electronic and laser devices used to measure extremely fine variations.

To ensure that the part will meet the specifications, certain basic terms are used and these are usually added onto the drawing of the part. These terms apply to all forms of measurement and inspection and should be understood by both the machinist and the inspector. (See Table 14-1.)

OBJECTIVES

After completing this unit you will be able to:
1. Recognize and know the uses of three types of plug and ring gages
2. Check the accuracy of a part with a plug or ring gage
3. Check the accuracy of a part using a snap gage

TABLE 14-1 An Example of Limits and Tolerances		
Nominal size	3 in.	
Basic dimension	3.000	Decimal equivalent of nominal size
Basic dimension and amount of bilateral tolerance permitted	3.000 ±0.002	
Limits	3.002	Largest size permitted
	2.998	Smallest size permitted
Tolerance	0.004	Difference between minimum and maximum limits

Basic dimension is the exact size of a part from which all limiting variations are made.

Limits are the maximum and minimum dimensions of a part (the high and low dimensions).

Tolerance is the permissible variation of a part. This is often shown on the drawing by the basic dimension plus or minus the amount of variation allowed. If a part on a drawing was dimensioned to 3.00 in. ±0.02 (76.2 mm ±0.05), the tolerance would be 0.04 in. (0.10 mm).

If the tolerance is in one direction only, that is plus *or* minus, it is said to be *unilateral tolerance*. However, if the tolerance is both plus *and* minus, it is referred to as *bilateral tolerance*.

Allowance is the intentional difference in the dimensions of mating parts, for example, the difference between the maximum diameter of the shaft and the minimum diameter of the mating bore.

FIXED GAGES

Fixed gages are used for inspection purposes because they provide a quick means of checking a specific dimension. These gages must be easy to use and accurately finished to the required tolerance. They are generally finished to one-tenth the tolerance they are designed to control. For example, if the tolerance of a piece being checked is to be maintained at 0.001 in. (0.02 mm), then the gage must be finished to within 0.0001 in. (0.002 mm) of the required size.

CYLINDRICAL PLUG GAGES

Plain *cylindrical plug gages* (Fig. 14-1) are used for checking the inside diameter of a straight hole and are generally of the "go" and "no-go" variety. This type consists of a handle and a plug on each end ground and/or lapped to a specific size. The smaller-diameter plug, or the "go" gage, checks the lower limit of the hole. The larger-diameter plug, or the "no-go" gage, checks the upper limit of the hole (Fig. 14-2). For instance, if a hole size is to be maintained at 1.000 in. ±0.0005 (25.4 mm ±0.012) the "go" end of the gage would be designed to fit into a hole 0.9995 in. (25.38 mm) in diameter. The larger end ("no-go") would not fit into any hole smaller than 1.0005 in. (25.41 mm) in diameter.

The dimensions of these gages are usually stamped on the handle at each end adjacent to the plug gage. "Go" gages are made longer than the "no-go" end for easy identification. Sometimes a groove is cut on the handle near the "no-go" gage to distinguish it from the "go" end.

Due to the wear caused by the constant use of plug gages, many of them are made with carbide tips which greatly increase their life.

TO USE A CYLINDRICAL PLUG GAGE

1. Select a plug gage of the correct size and tolerance for the hole being checked.
2. Clean both ends of the gage and the hole in the workpiece with a clean, dry cloth.
3. Check the gage (both ends) and the workpiece for nicks and burrs.
4. Wipe both ends of the gage with an oily cloth to deposit a thin film of oil on the surfaces.
5. Start the "go" gage *squarely* into the hole (Fig. 14-3). If the hole is within the limits, the gage will enter easily. DO NOT FORCE OR TURN IT. The plug should enter the hole for the full length, and there should be no excessive play between the plug and the part.

NOTE: If the gage enters only part way, there is a taper in the hole. Excessive play or looseness in one direction indicates that the hole is elliptical (out of round).

6. After the hole has been checked with the "go" gage, it should be checked with the "no-go" end. This gage should not begin to enter the hole. An entry of more than ¹⁄₁₆ in. (1.58 mm) indicates an oversize, bell-mouth, or tapered hole.

PLAIN RING GAGES

Plain ring gages, used to check the outside diameter of pieces, are ground and lapped internally to the desired size.

Fig. 14-1   A cylindrical plug gage is used to check hole sizes.

Fig. 14-2   The go end of the gage checks the minimum tolerance limit, while the no-go end checks the maximum tolerance. (Courtesy Bendix Corporation.)

Fig. 14-3   Checking a hole size with a plug gage.

Fig. 14-4   Plain ring gages are used to check the diameter of round work.

The size is stamped on the side of the gage. The outside diameter is knurled, and the "no-go" gage is identified by an annular groove on the knurled surface (Fig. 14-4). The precautions and procedures for using a ring gage are similar to those outlined for a plug gage and should be followed carefully.

TAPER PLUG GAGES

Taper plug gages (Fig. 14-5), made with standard or special tapers, are used to check the size of the hole and the accuracy of the taper. Some of these gages used for inspection purposes have "go" and "no-go" rings scribed on them. If the gage fits into the hole between these two rings, the hole is then within the required tolerances. Other taper plug gages have steps ground on the large end to indicate the limits. The rings or steps measure hole size limits only. An incorrect taper is evidenced by a wobble between the plug gage and the hole.

TO CHECK AN INTERNAL TAPER USING A TAPER PLUG GAGE

1. Select the proper taper gage for the hole being checked.
2. Wipe the gage and the hole with a clean, dry cloth.
3. Check both the gage and hole for nicks and burrs.
4. Apply a *thin* coating of Prussian blue to the surface of the plug gage.
5. Insert the plug gage into the hole as far as it will go (Fig. 14-6).
6. Maintaining light end pressure on the plug gage, rotate it in a *counterclockwise* direction for approximately one-quarter turn.
7. Check the diameter of the hole. A proper size is indicated when the edge of the workpiece lies between the limit steps or lines on the gage.
8. Check the taper of the hole by attempting to move the gage radially in the hole. Any error in the taper will be indicated by play at either end between the hole and the gage. Movement or play at the large end indicates excessive taper; movement at the small end indicates insufficient taper.
9. Remove the gage from the hole to see if the bluing has rubbed off evenly along the length of the gage, a result which would indicate a proper fit. A poor fit is evident if the bluing has been rubbed off more at one end than the other.

TAPER RING GAGES

Taper ring gages (Fig. 14-5) are used to check both the accuracy and the outside diameter of the taper. Ring gages often have scribed lines or a step ground on the small end to indicate the "go" and "no-go" dimensions.

For a taper ring gage, the precautions and procedures are similar to those outlined for a taper plug gage. However,

Fig. 14-5 Taper ring and plug gages.

Fig. 14-6 Checking a tapered hole with a taper plug gage.

when work which has not been ground or polished is being checked, three equally spaced chalk lines around the circumference and extending for the full length of the tapered section may be used to indicate the accuracy of the taper (Fig. 14-7). If the work has been ground or polished, it is advisable to use three thin lines of Prussian blue.

Fig. 14-7 Checking the accuracy of a taper using chalk lines.

CARE OF PLUG AND RING GAGES

Gage life is dependent on the following factors.

1. Materials from which the gage is made
2. Material of the part being checked
3. Class of fit required
4. Proper care of the gage

In order to preserve the accuracy and life of gages, observe the following points.

1. Store gages in divided wooden trays to protect them from being nicked or burred.
2. Check them frequently for size and accuracy.
3. Correctly align gages with the workpiece to prevent binding.
4. Do not force or twist a plain plug or ring gage. Forcing or twisting will cause excessive wear.
5. Clean the gage and workpiece thoroughly before checking the part.
6. Use a light film of oil on the gage to help prevent binding.
7. Make provision for air to escape when gaging blind holes with a plug gage.
8. Have gages and work at room temperature to ensure accuracy and prevent damage to the gage.
9. Never use an inspection gage as a working gage.

THREAD PLUG GAGES

Internal threads are checked with *thread plug gages* (Fig. 14-8) of the "go" and "no-go" variety and employ the same principle as cylindrical plug gages.

When a thread plug gage is used, the go end, which is the longer end, should be turned in flush to the bottom of the hole. The no-go end should just start into the hole and become quite snug before the third thread enters.

Since thread plug gages are quite expensive, certain precautions should be observed in their use.

1. Thread plug gages have a chip groove cut along the thread to clear loose chips. Do not depend on this for removing burrs or loose chips. To prolong the life of the

Fig. 14-8 Thread plug gages are used to check the size and accuracy of an internal thread. (Courtesy Taft-Pierce Manufacturing Company.)

Fig. 14-9 Go and no-go thread ring gages in a holder. (Courtesy Bendix Corporation.)

gage, it is advisable to remove burrs and loose chips (wherever possible) by means of an old tap.
2. Before using the thread plug gage, apply a little oil to its surface.
3. Never force the gage.

THREAD RING GAGES

The most popular gage of this type is the *adjustable thread ring gage*. These gages, used to check the accuracy of an external thread, have a threaded hole in the center with three radial slots and a setscrew to permit small adjustment. The outside diameter is knurled, and the "no-go" gage is identified by an annular groove cut on the knurled surface. Both the "go" and "no-go" gages are generally assembled in one holder for checking the part easily (Fig. 14-9).

When these gages are used, the thread being checked should fully enter the go gage, but should not enter the no-go gage by more than 1½ turns. Before checking a thread, remove any dirt, grit, or burrs. A little oil will help to prolong the life of the gage.

SNAP GAGES

Snap gages, one of the most common types of comparative measuring instruments, are faster to use than micrometers but are limited in their application. They are used to check diameters within certain limits by comparing the part size to the preset dimension of the snap gage. Snap gages generally have a C-shaped frame with adjustable gaging anvils or rolls which are set to the "go" and "no-go" limits of the part. These gages are supplied in several styles, some of which are shown in Fig. 14-10.

TO USE A SNAP GAGE TO CHECK A DIMENSION

Proper use of a snap gage is required to prevent springing the gage and marring the work surface.

1. Thoroughly clean the anvils of the gage.
2. Set the "go" and "no-go" anvils to the required limits using gage blocks or a standard.
3. Lock the anvils in position and recheck the accuracy of the settings.

NOTE: If a dial-indicator snap gage is used, set the bezel (outer ring) of the indicator to read zero and lock it into position.

4. Clean the surface of the work.
5. Hold the gage in the right hand, keeping it square with the work.

6. With the left hand, hold the lower anvil in position on the workpiece.
7. Push the gage over the work surface with a rolling motion. Only light hand pressure should be used to pass the go pins.

NOTE: Do not force the gage; if the work is the correct size, the gage should pass easily over the work.

8. Advance the gage until the "no-go" anvils or rolls contact the work. If the gage stops at this point, the work is within the limits.

REVIEW QUESTIONS

1. For each of the following dimensions, indicate the basic diameter, upper limit, lower limit, and tolerance:

(a) 1.750 + 0.002 in.
       − 0.000 in.
(b) 0.625 + 0.0015 in.
       − 0.0000 in.

Fig. 14-10 Various types of snap gages. (A) Adjustable snap gage   (Courtesy Taft-Pierce Manufacturing Company); (B) adjustable roll snap gage   (Courtesy Bendix Corporation); (C) dial indicator snap gage   (Courtesy Taft-Pierce Manufacturing Company.)

(c) 12.50 ± 0.02 mm
(d) 20.500 + 0.000 mm
           − 0.015 mm
(e) 0.500 ± 0.005 mm

2. State whether the tolerances for each size in question 1 are unilateral or bilateral.

FIXED GAGES

3. What purpose do fixed gages serve in industry?
4. To what tolerance are fixed gages finished?
5. If a hole size is to be maintained at 1.750 in. ±0.002 (44.45 mm ±0.05) what would be the sizes of the "go" and "no-go" gages?

CYLINDRICAL PLUG GAGES

6. How are the "go" and "no-go" ends of a cylindrical plug gage identified?
7. What precautions must be observed when a cylindrical plug gage is used?

PLAIN RING GAGES

8. How is a "no-go" ring gage distinguished from a "go" ring gage?

TAPER PLUG AND RING GAGES

9. How may the limits of a taper plug gage be indicated on the gage?
10. List the precautions to observe when checking with a taper plug or ring gage.
11. When may chalk be used to check an external taper and when should bluing be used?
12. Why should a taper plug or ring gage be rotated no more than one-quarter of a turn when checking a taper?

THREAD PLUG AND RING GAGES

13. What parts of a thread are checked with a "go" gage? A "no-go" gage?
14. List three precautions to observe when using a thread plug gage.
15. How should an external thread of the proper size fit into the thread ring gage?

SNAP GAGES

16. Describe a snap gage.
17. What advantage has the dial indicator snap gage over an adjustable snap gage?
18. List the precautions necessary when a workpiece is checked with a snap gage.

COMPARISON MEASUREMENT

UNIT 15

Manufacturing processes have now become so precise that component parts are often made in several areas, then shipped to a central location for final assembly. In order for this process of *interchangeable manufacture* to be economical, there must be some assurance that these parts will fit on assembly. The components are therefore made to within certain limits, and further inspection or *quality control* ensures that only properly sized parts will be used.

Much of this inspection is done rapidly, accurately, and economically by a process called *comparison measurement.* This consists of comparing the measurement of the part to a known standard or master of the exact dimension required. Basically, *comparators* are gages that incorporate some means of amplification to compare the part size to a set standard, usually gage blocks.

Mechanical, optical, mechanical-optical comparators, and air, electrical, and electronic gages all employ the same principle of comparision measurement.

After completing this unit you will be able to:
1. Understand the principle of comparison measurement
2. Identify four types of comparators and know their use
3. Measure to within 0.0005-in. (0.01-mm) accuracy with a dial indicator, mechanical and optical comparator, or air and electronic gages

COMPARATORS

A comparator may be classified as any instrument which is used to compare the size of a workpiece to a known standard. The simplest form of comparator is a dial indicator mounted on a surface gage. All comparators are provided with some means of amplification by which variations from the basic dimensions are noted easily.

DIAL INDICATORS

Dial indicators are used to compare sizes and measurements to a known standard and to check the alignment of machine tools, fixtures, and workpieces prior to machining.

Many types of dial indicators operate on a gear and rack principle (Fig. 15-1). A *rack* cut on the *plunger or spindle* is in mesh with a *pinion* which in turn is connected with a *gear train*. Any movement of the spindle is then magnified and transmitted to a *hand or pointer* over a *graduated dial*. Inch-designed dials may be graduated in thousandths of an inch or less. The dial, attached to a *bezel*, may be adjusted and locked in any position.

During use, the contact point on the end of the spindle bears against the work and is held in constant engagement with the work surface by the rack spring. A hair spring is attached to the gear that meshes with the center pinion. This flat spiral spring takes up the backlash from the gear train and prevents any lost motion from affecting the accuracy of the gage.

Dial indicators are generally available in two types: the continuous-reading dial indicator and the dial test indicator.

The *continuous-reading dial indicator* (Fig. 15-2), numbered clockwise for 360°, is available in two types: the

regular-range and the long-range indicator. The regular-range dial indicator has only about 2½ revolutions of travel. It is generally used for comparison measurement and setup purposes. The long-range dial indicator (Fig. 15-2) is often used to indicate table travel or cutting tool movement on machine tools. It has a second smaller hand that indicates the number of revolutions that the large hand has traveled.

Dial test indicators (Fig. 15-1) may have a balanced-type dial, that is, one which reads both to the right and left from zero and indicates a plus or minus value. Indicators of this type have a total spindle travel of only 2½ revolutions. These instruments may be equipped with tolerance pointers to indicate the permissible variation of the part being measured.

Perpendicular dial test indicators or back plunger indicators have the spindle at right angles (90°) to the dial. They are used extensively in setting up lathe work and for machine table alignment.

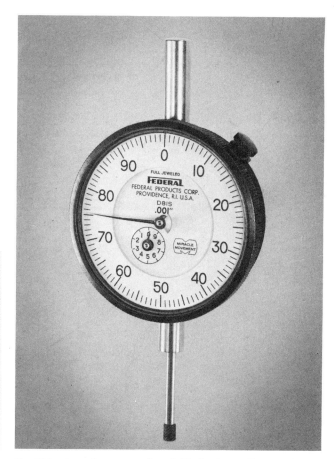

Fig. 15-1 A balanced-type dial test indicator showing the internal mechanism. (Courtesy Federal Products Corporation.)

Fig. 15-2 A continuous-reading dial indicator. (Courtesy Federal Products.)

The *universal* dial test indicator (Fig. 15-3) has a contact point that may be set at several positions through a 180° arc. This type of indicator may be conveniently used to check internal and external surfaces. Figure 15-4A and B illustrates typical applications of this type of indicator.

Metric dial indicators (Fig. 15-5) are available in both the balanced and continuous-reading types. The type used for inspection purposes is usually graduated in 0.002 mm and has a range of 0.5 mm. The regular indicators are usually graduated in 0.01 mm and have a range up to 25 mm.

TO MEASURE WITH A DIAL TEST INDICATOR AND HEIGHT GAGE

1. Clean the face of the surface plate and the vernier height gage.
2. Mount the dial test indicator on the movable jaw of the height gage (Fig. 15-4B).

Fig. 15-3 The universal dial test indicator.

3. Lower the movable jaw until the indicator point just touches the top of a gage block resting on the surface plate.
4. Tighten the upper locking screw on the vernier and loosen the lower locking screw.
5. Carefully turn the adjusting nut until the indicator needle registers approximately one-quarter turn.
6. Turn the bezel to set the indicator to zero.
7. Note the reading on the vernier and record it on a piece of paper.
8. Raise the indicator to the height of the first hole to be measured.
9. Adjust the vernier until the indicator reads zero.
10. Note the vernier reading again and record it.
11. Subtract the first reading from the second and add the height of the gage block.
12. Proceed in this manner to record the location of all other holes.

MECHANICAL COMPARATORS

The *mechanical comparator* (Fig. 15-6) consists of a base, a column, and a gaging head. Mechanical comparators operate on several principles. Some employ the gear and rack principle used in some dial indicators while others use a system of levers similar to the universal dial indicator (Fig. 15-4). The mechanical comparator (Fig. 15-6) uses two twisted steel ribbons which are tensioned when the comparator is set to a standard. Any variance from the standard size changes the tension of the ribbons and the position of the pointer. All types are used in inspection to check the size of a part against a master gage. The variation between the part and the master is shown in a scale as a plus or minus quantity.

Fig. 15-4 (A) A universal dial test indicator being used to center a workpiece with the machine spindle; (B) checking measurements with a dial test indicator and height gage. (Courtesy Federal Products Corporation.)

Fig. 15-5   A metric dial indicator with a balanced dial.

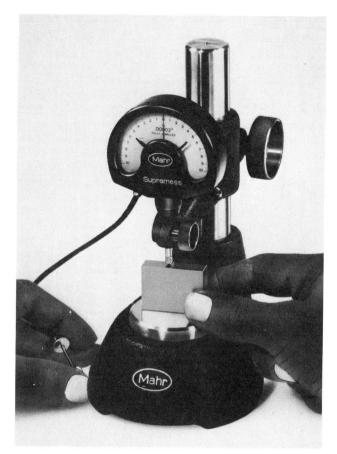

Fig. 15-6   A mechanical comparator showing tolerance pointers to indicate upper and lower limits.

TO MEASURE WITH A MECHANICAL COMPARATOR

1. Clean the anvil and master gage of the required size.
2. Place the master on the anvil.
3. Carefully lower the gaging head until the stylus touches the master and indicates a movement of the needle.
4. Lock the gaging head to the column.
5. Adjust the needle to zero using the fine adjustment knob, and set the limit pointers on the face.
6. Recheck the setting by removing the master and replacing it.
7. Set the tolerance pointers to the upper and lower tolerances of the part being checked.
8. Substitute the work being gaged for the master and note the reading. If the reading is to the right of zero, the work is too large; if to the left, it is too small. If the needle stops between the tolerance pointers, the work is within the permitted limits.

OPTICAL COMPARATORS

An *optical comparator* or *shadowgraph* (Fig. 15-7) projects an enlarged shadow onto a screen where it may be compared to lines or to a master form which indicates the limits of the dimensions or the contour of the part being checked.

Fig. 15-7   Intricate forms can be easily checked with an optical comparator.

The optical comparator is a fast, accurate means of measuring or comparing the workpiece with a master. It is often used when the workpiece is difficult to check by other methods. Optical comparators are particularly suited to the checking of extremely small or odd-shaped parts which would be difficult to inspect without the use of expensive gages.

Optical comparators are available in bench and floor models, identical in principle and operation (Fig. 15-8). Light from a *lamp* passes through a *condenser lens* and is projected against the workpiece. The shadow caused by the workpiece is transmitted through a *projecting lens* system which magnifies the image and casts it onto a *mirror*. The image is then reflected to the *viewing screen* and is further magnified in this process.

The extent of the image magnification depends on the lens used. Interchangeable lenses for optical comparators are available in the following magnifications: 5X, 10X, 31.25X, 50X, 62.5X, 90X, 100X, and 125X.

A comparator chart or master form mounted on the viewing screen is used to compare the accuracy of the enlarged image of the workpiece being inspected. Charts are usually made of translucent material, such as cellulose acetate or frosted glass. There are many different charts available for special jobs, but the most commonly used are linear-measuring, radius, and angular charts. A vernier protractor screen is also available for checking angles. *Since charts are available in several magnifications, care must be taken to use a chart of the same magnification as the lens mounted on the comparator.*

Many accessories are available for the comparator, increasing the versatility of the machine. Some of the most common are *tilting work centers* which permit the workpiece to be tilted to the required helix angle when checking threads; a *micrometer work stage* which permits quick and accurate measuring of dimensions in both directions; and *gage blocks, measuring rods,* and *dial indicators* used on comparators for checking measurement. The surface of the workpiece may be checked by a *surface illuminator* which lights up the face of the workpiece adjacent to the projecting lens system and permits this image to be projected onto the screen.

TO CHECK THE ANGLE OF A 60° THREAD USING AN OPTICAL COMPARATOR (FIG. 15-9)

1. Mount the correct lens in the comparator.
2. Mount the tilting centers on the micrometer cross-slide stage.
3. Set the tilting centers to helix angle of the thread.
4. Set the workpiece between centers.
5. Mount the vernier protractor chart and align it horizontally on the screen.
6. Turn on the light switch.
7. Focus the lens so that a clear image appears on the screen.
8. Move the micrometer cross-slide stage until the thread image is centralized on the screen.

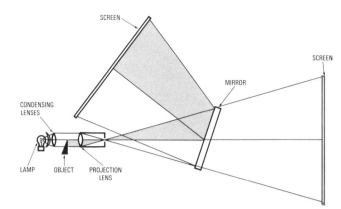

Fig. 15-8   The principle of the optical comparator.

Fig. 15-9   Checking a thread form on an optical comparator.

9. Revolve the vernier protractor chart to show a reading of 30°.
10. Adjust the cross-slides until the image coincides with the protractor line.
11. Check the other side of the thread in the same manner.

NOTE: If the thread angle is not correct or square with the centerline, adjust the vernier protractor chart to measure the angle of the thread image.

Other dimensions of the thread, such as depth, diameters, and width of flats, may be measured with the micrometer-measuring stages or devices such as rods, gage blocks, and indicators.

MECHANICAL-OPTICAL COMPARATORS

The *mechanical-optical comparator* (Fig. 15-10), or the *reed*-type comparator, combines a reed mechanism with a light beam to cast a shadow on a magnified scale to indicate

Fig. 15-10 A reed-type comparator uses both the mechanical and optical principles of measurement. (Courtesy The Bendix Corporation.)

the dimensional variation of the part. It consists of a base, a column, and a gaging head which contains the reed mechanisms and light source.

THE REED MECHANISM

Figure 15-11 illustrates the principle of the reed mechanism. A fixed steel block A and a movable block B have two pieces of spring steel, or reeds, attached to them (Fig. 15-12A). The upper ends of the reeds are joined and connected to a pointer. Since block A is fixed, any movement of the spindle attached to block B will move this block up or down, causing the pointer to move a much greater distance to the right or left (Fig. 15-12B).

A light beam passing through an aperture illuminates the scale (Fig. 15-13). The pointer, with a *target* attached, is located below the aperture so that a movement of block B will cause the target to interrupt the light beam, casting a highly amplified shadow on the scale. The movement of

Fig. 15-11 Construction and principles of the reed mechanism (mechanical lever).

Fig. 15-12 (A) Cutaway section showing the blocks and the reed mechanism; (B) the upper part of the reed deflects a greater distance than the lower part.

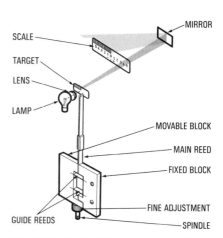

Fig. 15-13 The light beam or optical lever amplifies the target movement.

the pointer and target is obviously greater than the movement of the spindle. Also, the shadow cast on the scale will be larger than the movement of the target. Therefore, the measurement on the scale will be much greater than the movement of the spindle.

To illustrate the total magnification of this instrument, suppose that the ratio of target movement to the stylus is 25:1; the ratio to the light beam lever is 20:1. By combining these two movements, the shadow cast would then be 25 × 20 or 500 times as large as the movement of the stylus. This large amplification would permit extremely accurate gaging to be performed on this type of instrument. Reed-type comparators are manufactured with magnifications from 500:1 to 20 000:1.

The scales for these instruments are graduated in plus and minus with zero being in the center of the scale. The value for each graduation is marked on the scale of every machine.

TO MEASURE WITH A REED COMPARATOR

1. Raise the gaging head above the required height, and clean the anvil and master thoroughly.
2. Place the master gage or gage block buildup on the anvil.
3. Carefully lower the gaging head until the end of the spindle *just touches the master.*

NOTE: A shadow will begin to appear on the left side of the scale.

Fig. 15-14 A flow- or column-type air gage.

4. Clamp the gaging head to the column.
5. Turn the adjusting sleeve until the shadow coincides with the zero on the scale.
6. Remove the gage blocks and carefully slide the workpiece between the anvil and the spindle.
7. Note the reading. If the shadow is to the right of the zero, the part is oversize; if to the left, it is undersize.

AIR GAGES OR PNEUMATIC COMPARATORS

Air gaging, a form of comparison measurement, is used to compare workpiece dimensions with those of a master gage by means of air pressure or flow.

Air gages are of two types: the *flow or column type* (Fig. 15-14), which indicates air velocity, and the *pressure type* (Fig. 15-16), which indicates air pressure in the system.

COLUMN-TYPE AIR GAGE

After air has been passed through a filter and a regulator, it is supplied to the gage at about 10 psi (68.9 kPa) (Fig. 15-15). The air flows through a transparent tapered tube in which a float is suspended as a result of this air flow. The top of this tube is connected to the gaging head by a plastic tube.

The air flowing through the gage exhausts through the passages in the gaging head into the clearance between the head and the workpiece. The rate of flow is proportional to the clearance indicated by the position of the float in the column. The gage is set to a master and the float is then positioned by means of an adjusting knob. The upper and lower limits for the workpiece are then set. If the hole in the workpiece is larger than the hole size of the master, more air will flow through the gaging head, and the float will rise higher in the tube. Conversely, if the hole is smaller than the master, the float will fall in the tube. Amplification from 1000:1 to 40,000:1 may be obtained

Fig. 15-15 Principle of the column-type air gage.

Fig. 15-16 Gaging a hole using a pressure-type air gage.

Fig. 15-17 Operation of the pressure-type air gaging system.

PRESSURE-TYPE AIR GAGE

In the pressure-type air gage (Fig. 15-16), air passes through a filter and regulator and is then divided into two channels (Fig. 15-17). The air in the *reference channel* escapes to atmosphere through a zero-setting valve. The air in the *measuring channel* escapes to atmosphere through the gage head jets. The two channels are connected by an extremely accurate differential pressure meter.

The master is placed over the gaging spindle and the zero-setting valve is adjusted until the gage needle indicates zero. Any deviation in the workpiece size from the master size changes the reading. If the workpiece is too large, more air will escape through the gaging plug; therefore, pressure in the measuring channel will be less and the dial gage hand will move counterclockwise, indicating how much the piece is oversize. A diameter smaller than the master gage indicates a reading on the right side of the dial. Amplification from 2500:1 to 20,000:1 may be obtained with this type of gage. Pressure-type air gages may also use plug, ring, or snap gaging heads for a wide variety of measuring jobs.

Air gages are widely used since they have several advantages over other types of comparators.
1. Holes may be checked for taper, out-of-roundness, concentricity, and irregularity more easily than with mechanical gages (Fig. 15-18).
2. The gage does not touch the workpiece; therefore there is little chance to mar the finish.

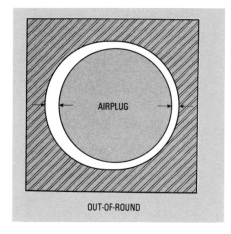

Fig. 15-18 Irregular-shaped holes may be easily checked with an air gage. (Courtesy Federal Products Corporation.)

3. Gaging heads last longer than fixed gages since wear is reduced between the head and the workpiece.

4. Less skill is required to use this type of gaging equipment than other types.

5. Gages may be used at a machine or bench.

6. More than one diameter may be checked at the same time.

ELECTRONIC COMPARATORS

The *electronic comparator* (Fig. 15-19), a highly accurate form of comparator, uses the Wheatstone bridge circuit to transform minute changes in spindle movement into a relatively large needle movement on the gage. This degree of magnification is controlled by a selector on the front of the *amplifier*. The widely spaced graduations represent values from 0.0001 to .000 01 in. (0.002 mm to 0.0002 mm), depending on the scale selected.

When it is not necessary to know the exact dimension of a part, but rather if it falls within the required limits, a signal light attachment may be installed on the gage. When a workpiece is tested, an amber light indicates that it is within the prescribed limits. A red light indicates that the workpiece is too small, and a blue light indicates that it is too large.

Electronic units may also be used as height gages by mounting a rectangular gage head on a height gage stand (Fig. 15-20). This method is particularly suited to the checking of soft, highly polished surfaces because of the light gaging pressure required.

Fig. 15-19   The value of the scale divisions on an electronic comparator can be readily changed to suit the accuracy required.

Fig. 15-20   The electronic gage with a rectangular head is used to accurately check heights.   (Courtesy Bendix Corporation.)

REVIEW QUESTIONS

COMPARISON MEASUREMENT

1. Describe
  (**a**) quality control
  (**b**) comparison measurement

DIAL INDICATORS

2. What is the difference between a regular-range dial indicator and a long-range indicator?

3. Compare a perpendicular dial indicator with a dial test indicator.

4. How are metric dial indicators usually graduated?

COMPARATORS

5. Define a comparator.

6. List three principles used in mechanical comparators.

7. Why is high amplification necessary in any comparison measurement process?

8. Describe the procedure for measuring a workpiece with a reed-type comparator.

OPTICAL COMPARATORS

9. List the advantages of an optical comparator.

10. Describe the principle of an optical comparator. Illustrate by means of a suitable sketch.

11. What precautions are necessary when charts are used on an optical comparator?

AIR GAGES

12. Describe the principle of the column type air gage. Illustrate by means of a suitable sketch.

13. Describe and neatly illustrate the principle of the pressure-type air gage.

14. List six advantages of air gages.

ELECTRONIC COMPARATORS

15. What circuit is employed in electronic comparators?

16. Describe the operation of the signal light attachment for electrical or electronic gages.

THE COORDINATE MEASURING SYSTEM

UNIT 16

Coordinate measurement is a method used to speed up layout, machining, and inspection by having all measurements dimensioned from three rectangular coordinates (surfaces of the workpiece) *x, y,* and *z.* Coordinate measuring systems are capable of measuring to 0.000 050 in. or 0.001 mm. This high degree of accuracy is achieved by the use of a series of dark and light bands called the *moiré fringe pattern.* This system has been applied extensively to machine tools such as lathes, milling machines, and jig borers (Fig. 16-1A and B) to speed up the measurement process, thus reducing the overall machining time. Coordinate measuring machines (Fig. 16-1C), widely used in the field of measurement and inspection, provide a quick and accurate means of checking machined parts prior to assembly.

Various units may be mounted on a machine tool to give readings for the *x* (length), *y* (width), and *z* (depth) axes. A unit providing direct readout in degrees, minutes, and seconds, is available for angular positioning and measurement.

Fig. 16-1 (A) A lathe equipped with a coordinate measuring system; (B) the digital readout shows the exact location of the cutting tool or workpiece; (C) the coordinate measuring system being used for inspection purposes.   (Courtesy The Bendix Corporation.)

After completing this unit you will know:
1. The purpose and application of the coordinate measuring system
2. The main components and operation of the unit
3. The advantages of the coordinate measuring system

PARTS OF THE MEASURING UNIT

The measuring unit (Fig. 16-2) has three basic components: *a machined spar with a calibrated grating* (A), *a reading head* (B), and a counter with *a digital readout display* (C).

The main element in this system is an accurately ruled grating (Fig. 16-2A) of the desired length of travel. The face of this grating for systems having 0.0001-in. (0.002-mm) resolution, is etched for its entire length with lines spaced 0.001 in. (0.02 mm) apart. For machines capable of greater accuracy, the grating spar is etched with 2500 lines per inch, so that the digital readout resolution is 0.000 050 in. (0.001 mm). Metric readings are obtained by pushing a button on the side of the box.

A *transparent index grating,* having the same graduations as the grating spar, is mounted in the reading head (Fig. 16-2B). The index grating is positioned so that its lines are at a slight angle to the lines on the spar. The reading head is mounted so that the index grating is positioned just 0.002 in. (0.05 mm) above the main grating. Mounted in the reading head are a small light, a collimating lens, and four photoelectric cells (Fig. 16-3).

NOTE: This electro-optical system ''counts'' the moiré fringes, and produces the high resolution measurement accuracy of the Sheffield Cordax Measuring System.

PRINCIPLE OF THE MOIRÉ FRINGE

To illustrate the moiré fringe pattern, draw a series of equally spaced lines on two pieces of plastic sheeting (Fig. 16-4).

One sheet should then be placed over the other at an angle as in Fig. 16-4. Where these lines cross, dark bands appear. If the top sheet is moved to the right, the position of these bands will shift down and the pattern appear as a series of dark and light bands moving vertically on the sheet. Thus, any longitidinal movement will produce a vertical movement of the light bands. This principle can also be illustrated by placing one comb upon another at a small angle and moving one across the face of the other.

OPERATION OF THE MEASURING UNIT

Since the etched lines on the reading-head grating are at an angle to the lines on the main spar, a series of bands would

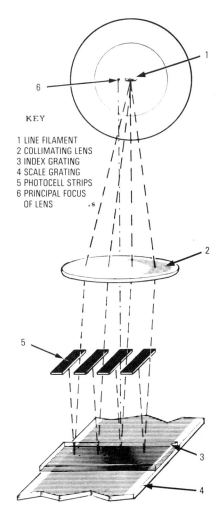

KEY

1 LINE FILAMENT
2 COLLIMATING LENS
3 INDEX GRATING
4 SCALE GRATING
5 PHOTOCELL STRIPS
6 PRINCIPAL FOCUS
   OF LENS

Fig. 16-3 Principle of the coordinate measuring system.

PRESET SWITCHES

x-AXIS READOUT

+ 12.4792

X

− 09.3485

Y

y-AXIS READOUT
C

ZERO RESET

READING HEAD
B

SPAR
A

READOUT CONSOLE

Fig. 16-2 Components of a coordinate measuring system.

Fig. 16-4 Principle of the moiré fringe pattern. (Courtesy Bendix Corporation.)

PHOTOCELLS

1/4 LINE TRAVEL

1/2 LINE TRAVEL

3/4 LINE TRAVEL

LIGHT INTENSITY ON PHOTOCELLS 3 AND 4

MAX.

MIN.

1/4   1/2   3/4   1

LINE DISPLACEMENT

Fig. 16-5 Lateral movement of the fringe pattern.

appear when viewed from directly above. Let us consider one band only, to explain the operation of the system.

Light from the lamp (Fig. 16-3) passes through the collimating lens and is converted into a parallel beam of light. This beam strikes the fringe pattern and is reflected back up to one of four photoelectric cells which convert the fringe pattern into an electrical signal. As the reading head is moved longitudinally, the band moves vertically on the face of the spar and will be picked up by the next photoelectric cell creating another signal (Fig. 16-5). These signals from the photoelectric cells (output signals) are transmitted to the digital readout display box where they indicate accurately the head travel at any point. A signal from the next photoelectric cell will increase or decrease the reading on the digital readout box by 0.0001 in. (0.002 mm), depending on the direction of the movement.

NOTE: The fringe pattern (Fig. 16-4) shifts laterally and continuously across the grating path.

ADVANTAGES OF THE COORDINATE MEASURING SYSTEMS

1. Readout boxes provide clear, visible numbers which eliminate the possibility of misreading a dial gage.
2. It provides a constant readout of the tool (or table) position.
3. The reading indicates the exact position of the tool (or table) and is not affected by machine or lead screw wear.
4. This system eliminates the need for gage blocks and measuring rods on jig borers and vertical milling machines.
5. The need for operator calculations and the inherent possibility of errors are eliminated.
6. Machine setup time is greatly reduced.
7. Production is increased since the workpiece need only be checked for one size. For example, when a workpiece having several diameters is to be machined in a lathe, it is necessary to measure the first diameter only. Once the machine has been set to this size, all other diameters will be accurate.

8. The need for operator skill is reduced.
9. Scrap and rework are virtually eliminated.
10. Most machine tools can be fitted with this system.

REVIEW QUESTIONS

1. What is the principle of coordinate measurement?
2. Why have coordinate measuring systems been so widely accepted by industry?
3. State two different areas where coordinate measuring systems may be used.
4. Name the three main parts of the measuring unit.
5. Describe the operation of this measuring system.
6. State seven important advantages of a coordinate measuring system.

Two of the most precise measuring methods are those using optical flats and the laser. Although each uses a different principle, both use a source of monochromatic light to produce highly accurate measurements.

OBJECTIVES

After completing this unit you will be able to:
1. Check pieces for size, flatness, and parallelism, using optical flats
2. Understand the operation of a laser interferometer
3. Understand the applications of the laser in the measurement field

MEASURING WITH OPTICAL FLATS

One of the most accurate and reliable means of measuring is by the use of light waves.

Optical flats (Fig. 17-1), used with a monochromatic light, employ this principle to check work for flatness (Fig. 17-4), parallelism (Fig. 17-5), and size (Fig. 17-6).

Optical flats are disks of clear fused quartz, lapped to within a few millionths of an inch of flatness. They are generally used with a helium light source which produces a greenish-yellow light, having a wavelength of 23.1323 μin. (0.587 56 μm).

The optical flat, a perfectly flat, transparent disk, is placed on the surface of the work to be checked. The functioning surface of the optical flat is the surface adjacent to the workpiece. It is transparent and capable of reflecting light; therefore all light waves that strike this surface are split into two parts (Fig. 17-2). One part is reflected back by the lower surface of the flat. The other part passes through this surface and is reflected by the upper surface of the work. Whenever the reflected split portions of two light waves cross each other, or *interfere,* they become visible and produce dark interference bands or fringe lines. This happens whenever the distance between the lower surface of the flat and the upper surface of the workpiece is *only one-half of a wavelength* or multiples thereof (Fig. 17-2).

Since the wavelength of helium light is 23.1323 μin. (0.587 56 μm or millionths of a meter), each half-wavelength will represent 11.6 μin. (0.293 μm). Each dark band then represents a progression of 11.6 μin. (0.293 μm) above the point of contact between the workpiece and the optical flat. Therefore, when a height is checked, the number of bands between two points on a surface multiplied by 0.0000116 (0.293 μm) will indicate the height difference between two surfaces.

For comparison measurements, the difference in height between a master block and the workpiece can be determined as shown in Fig. 17-3. This illustrates the method used for checking accurately the height of an unknown surface by comparing it with a gage block of a known

Fig. 17-1 Checking gage block using an optical flat and a helium light source.

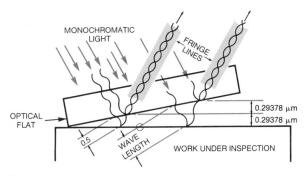

Fig. 17-2   The principle of the optical flat.

2 BANDS .0000232 in.

1 BAND .0000116 in.

Fig. 17-3   Checking the height of a block with a master block.

height. It is necessary first to know which block is larger before the unknown block can be measured. To determine the larger block, apply finger pressure to points X and Y. If this pressure at X makes no change in the band pattern and the pressure at Y causes the bands to separate, the master block (M) is larger. If the opposite is true, the unknown block (U) is larger. In Fig. 17-3, two bands appear on the low block; therefore the unknown block is two bands, or 2 × 11.6 = 23.2 μin. or 2 × 0.293 = 0.586 μm below the master.

HOW TO INTERPRET THE BANDS

Referring to Fig. 17-4, since the lines show only a slight curve, the block would be convex or very slightly higher in the center. The block is 2 × 11.6 or 0.000 0232 in. out of flat.

In Fig. 17-5, since two bands appear on the master block, the workpiece is 2 × 11.6 or 23.2 μin. smaller or larger than the master. If pressure on the master causes the spacing of the bands to widen, the part is smaller. Note the curve in the band shows that the workpiece is not exactly parallel. It is about one-half band or 5.8 μin. out of parallel.

In Fig. 17-6, the master block is on the left-hand side. Note that there are three bands on the master block and six bands on the unknown size on the right. The unknown block then has three more bands than the master. (The block with more bands is always smaller.) Also note that the lines on the unknown block are straight and evenly spaced. This indicates that it is parallel over its length. Since the lines slope down toward the line of contact, the left side of the unknown block is lower by one-half band or 5.8 μin. (11.6 ÷ 2).

MEASUREMENT WITH LASERS

Besides its many other uses in medicine and industry, the laser provides one of the most accurate means of measure-

Fig. 17-4   Checking fitness with an optical flat.

Fig. 17-5   Checking parallelism with an optical flat.

Fig. 17-6   Checking size with an optical flat.

Fig. 17-7 The principle of the laser interferometer.

Fig. 17-8 The laser beam's wave components are in-phase.

Fig. 17-9 The laser beam's wave components are out-of-phase.

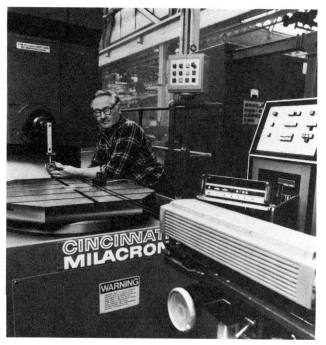

Fig. 17-10 A laser interferometer being used to check the alignment of machine parts.

ment. The laser-light measuring device, called an *interferometer* (Fig. 17-7), measures changes in position (alignment) by means of light-wave interference.

Referring to Fig. 17-7, the laser beam is split into two parts by the *beam splitter*. One of these beams (red) is transmitted through the beam splitter to a *motion-sensitive mirror* back to the beam splitter at the point X. From this point where the two beams join, the recombined beams are transmitted to the *detector*.

If there is no movement, both portions of the beam stay in the same phase and the light reaching the detector will remain constant (Fig. 17-8). If there is any movement at the sensitive mirror, the beam (black) reflected from that mirror will be altered and it will fluctuate in and out of phase with the other (red) beam (Fig. 17-9). When this beam (black) is out-of-phase, it then cancels the other beam (red) causing the light to fluctuate. This fluctuating light pattern registers on the beam detector where the number of fluctuations are computed relative to the laser wavelength and the precise movement is displayed on a readout box.

Because of the very thin, straight beam produced by a laser interferometer, these devices are used for very precise linear measurement and alignment in the production of large machines. They are also used to calibrate precision machines and measuring devices. Laser devices may also be used to check machine setups (Fig. 17-10). A laser beam is projected against the work and measurements are made by the beam and displayed on a digital readout panel.

Because of the very thin, straight beam characteristics, lasers are used extensively in the construction and surveying industries. They may be used to indicate the exact location when positioning girders on tall buildings or for establishing directional lines for a tunnel being constructed under a river. They are also widely used for establishing grades for sewer and drainage systems.

Lasers have become an established part of the space program. Laser beams are used to indicate how far the spaceship is from a given planet. They are also used for military purposes for range findings, guidance, and tracking of missiles.

The operation of the laser and its other industrial applications will be covered in Unit 93.

REVIEW QUESTIONS

OPTICAL FLATS

1. Describe an optical flat.
2. What light source is used with optical flats and what is its wavelength?
3. Describe, and illustrate in detail, the principle of an optical flat.
4. When measuring the height of a block using a gage block and an optical flat, how is it possible to determine the higher block?

LASERS

5. Name five applications of the interferometer in measurement.
6. Name the four main parts of an interferometer.
7. Briefly describe the operation of the interferometer.

SURFACE FINISH MEASUREMENT

UNIT 18

Modern technology has demanded improved surface finishes to assure proper functioning and long life of machine parts. Pistons, bearings, and gears depend to a great extent on a good surface finish for proper functioning and therefore require little or no break-in period. Finer finishes often require additional operations, such as lapping or honing. These higher finishes are not always required on an article and only result in higher production costs. To prevent overfinishing a part, the desired finish is indicated on the print. This information specifying the degree of finish is conveyed to the machinist by a system of symbols devised by the American Standard Association (ASA). This provides a standard system of determining and indicating surface finish. The inch unit of surface finish measurement is the *microinch* (μin.). The metric unit for surface finish is the *micrometer* (μm).

OBJECTIVES

After completing this unit you will be able to:
1. Understand the surface finish symbols which may be found on a drawing
2. Use a surface finish indicator to measure the surface finish of a part

The most common instrument used to measure finish is the *surface indicator* (Fig. 18-1).

This device consists of a *tracer head* and an *amplifier*. The tracer head houses a diamond stylus, having a point radius of 0.0005 in. (0.013 mm), which bears against the surface of the work. It may be moved along the work surface by hand or it may be motor-driven. Any movement of the stylus caused by surface irregularities is converted into electrical fluctuations by the tracer head. These signals are magnified by the amplifier and registered on the meter by a hand or needle. The reading shown on the meter indicates the *average* height of surface roughness or the departure of this surface from the reference (center) line.

Readings may be in either *arithmetical average (AA)* or *root mean square (rms)*. A highly magnified cross section of a workpiece would appear as in Fig. 18-2 with "hills and valleys" above and below the centerline. To calculate the surface finish without a surface indicator, the height of these deviations must be measured and recorded as shown. The AA or the rms could then be calculated as in Fig. 18-2. The rms is considered the better method of determining surface roughness since it emphasizes extreme deviations.

Fig. 18-1 Using a surface indicator to check the surface finish of a part.

a = 3	a² = 9
b = 19	b² = 361
c = 22	c² = 484
d = 15	d² = 225
e = 30	e² = 900
f = 19	f² = 361
g = 27	g² = 729
h = 19	h² = 361
i = 30	i² = 900
j = 12	j² = 144
k = 22	k² = 484
l = 14	l² = 196
m = 5	m² = 25
Totals 237	5179

$$\text{ARITHMETICAL AVERAGE} = \frac{237}{13} = 18.2\ \mu\ \text{in.}$$

$$\text{ROOT MEAN SQUARE AVEARGE} = \sqrt{\frac{5179}{13}} = 19.9\ \mu\ \text{in.}$$

Fig. 18-2 Calculation of arithmetical average and root mean square.

Fig. 18-3 Surface characteristics and symbols.

For accurate determination of the surface finish, the indicator must first be calibrated by setting it to a precision reference surface on a test block calibrated to ASA standards.

The symbols used to identify surface finishes and characteristics are shown in Fig. 18-3.

SURFACE FINISH DEFINITIONS

Surface deviations—any departures from the nominal surface in the form of waviness, roughness, flaws, lay, and profile.

Waviness—surface irregularities which deviate from the mean surface in the form of waves; they may be caused by vibrations in the machine or work. These are generally widely spaced.

Waviness height—The peak-to-valley distance in inches or millimeters.

Waviness width—The distance between successive waviness peaks or valleys in inches or millimeters.

Roughness—relatively finely spaced irregularities superimposed on the waviness pattern and caused by the cutting tool or the abrasive grain action and the machine feed. These irregularities are much narrower than the waviness pattern.

Roughness height—The AA deviation measured normal to the centerline in microinches or micrometers.

Roughness width—The distance between successive roughness peaks parallel to the nominal surface in inches or millimeters.

Roughness width cutoff—The greatest spacing of repetitive surface irregularities to be included in the measurement of roughness height. It must always be greater than the roughness width. Standard values are 0.003, 0.010, 0.030, 0.100, 0.300 and 1 in. (0.075, 0.25, 0.76, 2.54, 7.62, and 25.4 mm).

Flaws—irregularities such as scratches, holes, cracks,

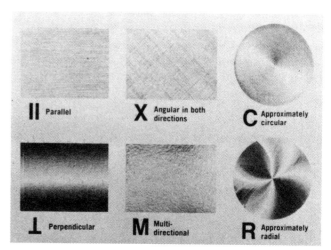

Fig. 18-4 Surface symbols used to designate the direction of the lay. (Courtesy GAR Electroforming Limited.)

ridges, or hollows that do not follow a regular pattern as in the case of waviness and roughness.

Lay—the direction of the predominant surface pattern caused by the machining process.

Profile—the contour of a specified section through a surface.

Microinch or micrometer—the unit of measurement used to measure the surface finish. The microinch is equal to 0.000 001 in.; the micrometer to 0.000 001 m.

The following symbols indicate the direction of the lay (Fig. 18-4).

‖ parallel to the boundary line of the surface indicated by the symbol

⊥ perpendicular to the boundary line of the surface indicated by the symbol

X angular in both directions on the surface indicated by the symbol

M multidirectional
C approximately circular to the center of the surface indicated by the symbol
R approximately radial in relation to the center of the surface indicated by the symbol

Average Surface Roughness Produced by Standard Machining Processes		
	MICROINCHES	MICROMETERS
Turning	100—250	2.5—6.3
Drilling	100—200	2.5—5.1
Reaming	50—150	1.3—3.8
Grinding	20—100	0.5—2.5
Honing	5—20	0.13—0.5
Lapping	1—10	0.025—0.254

TO MEASURE SURFACE FINISH WITH A SURFACE INDICATOR

1. Turn the switch on and allow the instrument to warm up for approximately 3 min.
2. Check the machine calibration by moving the stylus over the 125-μin. (3.2-μm) test block at approximately ⅛ in./s (3 mm/s).
3. If necessary, adjust the calibration control so that the instrument registers the same as the test block.
4. Unless otherwise specified, use the 0.030-in. (0.76-mm) cutoff range for surface roughness of 30 μin. (0.76 μm) or more. For surfaces less than 30 μin. (0.76 μm) use the 0.010-in. (0.25-mm) cutoff range.

NOTE: When measuring a surface where the roughness is unknown, it is good practice to set the range switch at a high setting to avoid damage to the instrument. After an initial test, the range switch may be turned to a finer setting for an accurate surface reading.

5. Thoroughly clean the surface to be measured to ensure accurate readings and reduce wear on the rider cap protecting the stylus.
6. With a smooth, steady movement of the stylus, trace the work surface at approximately ⅛ in./s (3 mm/s).
7. Note the reading from the meter scale.

A more elaborate device for measuring surface finish is the *surface analyzer*. This incorporates a recording device which produces a graph ink line record of the surface irregularities on a graduated chart.

Although the surface indicator is the most common, other methods may be used to measure surface finish with reasonable accuracy during machining processes.

1. *Comparison blocks* are used for comparing the finish on the workpiece with the calibrated finish on a test block using the fingernail test.
2. *Commercial sets of standard finished specimens.* Standard sets having up to 25 different surface finish samples are available for checking surface finish. They consist of

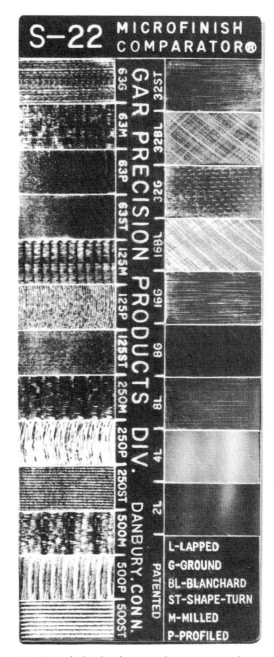

Fig. 18-5 A visual surface roughness comparator gage. (Courtesy GAR Electroforming Limited.)

blocks or plates having surfaces varying from the smoothest to the roughest likely to be required (Fig. 18-5).

When using these specimens, the finish of the machined part is checked against the sample finish to determine approximately the finish produced on the part. It is often difficult to determine the finish visually. In such cases, the surfaces may be compared by moving the tip of a fingernail over the two surfaces and comparing them.

Table 18-1 shows the results obtained on pieces of round and flat metal by various machining operations. A model B-1 110 Brush surface analyzer was used to obtain the readings. The speeds, feeds, and tool radii used are those recommended for a high-speed steel toolbit.

TOOL	OPERATION	MATERIAL	SPEED	FEED	TOOL	ANALYZER SETTING		SURFACE FINISH RMS
						CUTOFF	RANGE	
Cutoff saw	Sawing	2½-in.-diameter aluminum	320 ft/min (97.5 m/min)	—	10-pitch saw	.030 in. (0.76 mm)	1000	300–400
Shaper	Shaping a flat surface	Machine steel	100 ft/min (30.5 m/min)	0.005 in. (0.13 mm)	3/64 in. radius HSS	.030 in. (0.76 mm)	300	225–250
Vertical milling machine	Fly cutting (flat surface)	Machine steel	820 r/min	0.015 in. (0.38 mm)	1/16 in. radius Stellite	.030 in. (0.76 mm)	300	125–150
Horizontal milling machine	Slab milling	Cast aluminum	225 r/min	2½ in./min (63.5 mm/min)	Slab cutter, 4 in. diameter HSS	.030 in. (0.76 mm)	100	40–50
Lathe	Turning	2½-in.-diameter aluminum	500 r/min	0.010 in. (0.25 mm)	3/64 in. radius HSS	.030 in. (0.76 mm)	300	100–200
	Turning	2½-in.-diameter aluminum	500 r/min	0.007 in. (0.18 mm)	5/64 in. radius HSS	.030 in. (0.76 mm)	100	50–60
	Facing	2-in.-diameter aluminum	600 r/min	0.010 in. (0.25 mm)	1/32 in. radius HSS	0.30 in. (0.76 mm)	300	200–225
	Facing	2-in.-diameter aluminum	800 r/min	0.005 in. (0.13 mm)	1/32 in. radius HSS	.030 in. (0.76 mm)	100	30–40
	Filing	3/4-in.-diameter machine steel	1200 r/min	—	10-in. lathe file	.010 in. (0.25 mm)	100	50–60
	Polishing	3/4-in.-diameter machine steel	1200 r/min	—	#120 abrasive cloth	.010 in. (0.25 mm)	30	13–15
	Machine reaming	Aluminum	500 r/min	—	Machine reamer HSS, 3/4 in. diameter	.030 in. (0.76 mm)	100	25–32
Surface grinder	Grinding a flat surface	Machine steel	—	0.030 in. (0.76 mm)	60-grit grinding wheel	.003 in. (0.076 mm)	10	7–9
Cutter and tool grinder	Cylindrical grinding	1-in. machine steel	—	Hand (slow)	46-grit grinding wheel	.010 in. (0.25 mm)	30	12–15
Lapping	Flat lapping	7/8 in. × 5½ in. tool steel (hardened)	—	Hand	600-grit abrasive	.010 in. (0.25 mm)	10	1–2
	Cylindrical lapping	½-in.-diameter tool steel (hardened)	—	Hand	600-grit abrasive	.010 in. (0.25 mm)	10	1–2

TABLE 18-1   Surface Finishes Obtained by Various Machining Operations*

* Metric figures given represent soft conversions.

REVIEW QUESTIONS

SURFACE FINISH MEASUREMENT

1. Explain why present-day standards for surface finish are very important to industry.

2. Define the following surface finish terms: microinch, lay, flaw, roughness, waviness, root mean square (rms), arithmetical average (AA).

3. Explain each symbol and number (inches) as it applies to surface finish:

$$50\sqrt{\frac{0.002 - 2}{}} = 0.020$$

4. Explain what the following lay symbols represent: $=$, $\perp$, X, M, C, R.

5. Describe briefly the principle and operation of a surface indicator and a surface analyzer in measuring surface finish.

LAYOUT TOOLS AND PROCEDURES
SECTION 6

Laying out is the process of scribing or marking center points, circles, arcs, or straight lines on metal to indicate the shape of the object, the amount of metal to be removed during the machining process, and the position of the holes to be drilled. The layout will help the machinist determine the amount of material to be removed, although the size for rough and finish cuts must be checked by actual measurement.

All layouts should be made from a *baseline* or finished surface to ensure an accurate layout, correct dimensions, and proper location of holes. The importance of proper layout cannot be over-emphasized. The accuracy of the finished product depends greatly on the accuracy of the layout. The drawing does not show how much material must be removed from each surface of the casting or workpiece; it merely shows which surfaces are to be finished.

The layout for holes, whether in cored or solid material, is as important as the layout for other dimensions of the workpiece.

Layouts may be of two types—*basic or semiprecision* and *precision.* A semiprecision layout may involve the use of basic measuring and layout tools such as a rule and surface gage. It is generally not as accurate as a precision layout which requires the use of more accurate equipment such as the vernier height gage. If the part does not have to be precise, time should not be spent in making a precision layout. It is advisable, therefore, to keep the layout as simple as the workpiece requirements permit.

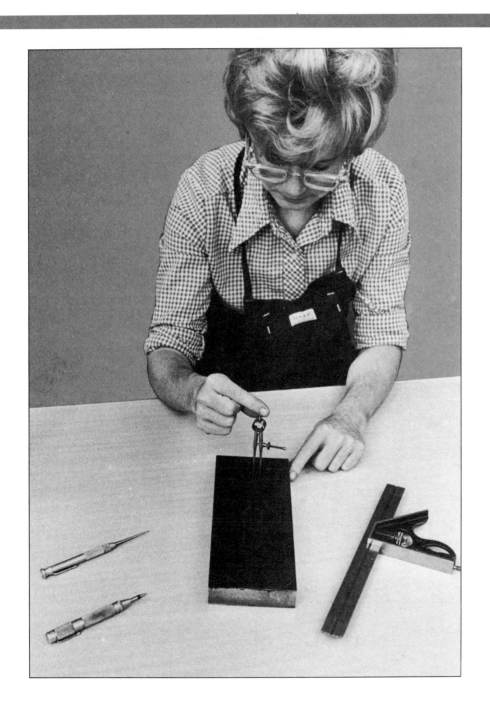

BASIC LAYOUT MATERIALS, TOOLS, AND ACCESSORIES

UNIT 19

The accuracy of a layout is very important to the accuracy of the finished product. If the layout is not correct, it is only right to assume that the workpiece will not be usable. The student should therefore realize that a good layout will entail the proper and careful use of all layout tools.

OBJECTIVES

After completing this unit you will be able to:
1. Prepare a work surface for layout
2. Use and care for various types of surface plates
3. Identify and use the main basic layout tools and accessories

LAYOUT SOLUTIONS

The surface of the metal is usually coated with a layout solution to make layout lines visible. There are several types of layout solutions available. Regardless of the type used, the surface should be clean and free of grease.

The most commonly used layout solution is *layout dye*, or *bluing* (Fig. 19-1). This quick-drying solution, when coated lightly on the surface of any metal, will produce a background for sharp, clear-cut lines. Layout dye may be applied with a cloth, a brush, a dauber, or sprayed on the work surface.

A copper-colored surface can be produced if the clean surface of a steel workpiece is coated with a copper sulfate ($CuSO_4$) solution to which a few drops of sulfuric acid have been added. When this solution is used, the surface of the workpiece must be absolutely clean and free from grease and finger marks.

NOTE: Copper sulfate should be used on ferrous metal only.

Steel may be blued by heating the surface of the workpiece with a torch until it turns blue and then quenching the workpiece to prevent further color change. Care must be taken to avoid distortion when heating the workpiece.

A mixture of vermilion powder and shellac is often used for aluminum since some layout compounds corrode the aluminum. Alcohol should be used to thin this solution or to remove it from the workpiece.

The surfaces of castings and hot-rolled steel are often prepared for layout by merely chalking the surface. A mixture of lime and alcohol, which readily clings to the rough surface of the castings, is often used for this purpose.

LAYOUT TABLES AND SURFACE PLATES

Layout work may be performed on a layout table (Fig. 19-2), or on a surface plate (Fig. 19-3) made of granite or cast iron. Granite layout tables and plates are considered better than cast-iron ones because they:

☐ Do not become burred.
☐ Do not rust.
☐ Are not affected by temperature change.
☐ Do not have internal stresses and therefore will not warp or distort.
☐ Are nonmagnetic.
☐ Can be used for checking near grinding machines since abrasive particles will not embed in the surface.
☐ Are cheaper than a similar sized cast-iron plate.

Granite plates are available in three colors: black, pink, and gray. Black granite plates are considered superior because they are harder, denser, less porous, and therefore less liable to absorb moisture.

CARE OF SURFACE PLATES AND LAYOUT TABLES

Although surface plates are rugged, their accuracy can be easily destroyed. The following points should be observed in regard to surface plates.

☐ Keep the working surface clean.
☐ Cover the plate or table when not in use.
☐ Carefully place the work on the surface plate—do not drop it onto the plate.
☐ Use parallels under the workpiece whenever possible.

Fig. 19-1 The work surface should be coated with layout dye before starting a layout.

Fig. 19-2 A granite layout table provides an accurate reference surface for layout work.

Fig. 19-3 A cast-iron surface plate.

☐ Never hammer or punch any layout on a surface plate.
☐ Remove burrs from cast-iron plates and always protect the surface with a thin film of oil and a cover when not in use.

SCRIBERS

The *scriber* (Fig. 19-4) has a hardened steel point, or points, and may be used in conjunction with a combination square, a rule, or a straightedge to draw straight lines. On

Fig. 19-4 A pocket and double-end scriber.

Fig. 19-5 Sharpening a scriber on an oilstone.

Fig. 19-6 A spring divider can be used to scribe arcs and circles.

some scribers, one end is bent at an angle to allow marking lines in hard-to-reach places. To be accurate, any layout requires fine lines; therefore the scriber point must always be sharp. The points of scribers, hermaphrodite calipers, dividers, and trammels should be honed frequently on a fine oilstone (Fig. 19-5) for sharp points to be maintained. When extremely fine lines are required, knife-edge scribers should be used.

DIVIDERS AND TRAMMELS

Dividers are used for scribing arcs and circles on a layout and for transferring measurements. The spring divider (Fig. 19-6), the most common type, is available in sizes from 3

103

Fig. 19-7 Trammels are used to scribe large arcs and circles.

Fig. 19-9 Locating the center of round work with a hermaphrodite caliper.

Fig. 19-8 Scribing a line parallel to an edge with a hermaphrodite caliper.

to 12 in. Larger circles and arcs may be scribed with a trammel (Fig. 19-7).

A trammel consists of a *beam* on which *two sliding or adjustable heads* with *scriber points* are mounted. Some trammels may have an *adjusting screw* for fine adjustment. Rods or beams of different lengths can be used to increase the capacity of the trammel. When a circle is laid out from the hole, a ball attachment may be substituted for one of the scriber points.

HERMAPHRODITE CALIPERS

The hermaphrodite caliper (Fig. 19-8) has one curved leg and one straight leg which contains a scriber point. It is used for laying out lines parallel to an edge (Fig. 19-8) or for locating the center of round or irregular-shaped stock (Fig. 19-9).

When setting the hermaphrodite caliper to a size, place the bent leg on the end of the rule and adjust the other leg until the scriber point is at the desired graduation. When scribing parallel lines with this tool, be sure that the scriber is always held at 90° to the edge of the workpiece.

SQUARES

Squares are used to lay out lines at right angles (90°) to a machined edge, to test the accuracy of surfaces which must be square (90° to each other), and to set up work for machining.

Adjustable squares are used for general-purpose work. The *solid square* (Fig. 19-10) is used where greater accuracy is required. It is made up of two parts, the beam (A) and the blade (B) (Fig. 19-10). Extremely accurate solid squares called *master squares* are used to check the accuracy of other squares.

THE COMBINATION SET

The *combination set* (Fig. 19-11), used extensively in layout work, consists of a steel rule, square head, bevel protractor, and center head.

The *steel rule* may be fitted to the other three parts of the combination set for various layout, setup, or checking operations.

The *square head* and the rule or *combination square* may be used to lay out lines parallel to an edge (Fig. 19-12). It is also used to lay out angles at 45 and 90° to an edge (Fig. 19-13). The square head may be moved along the rule to any position. Other applications of the square head are for checking 45 and 90° angles and measuring depths.

Fig. 19-10 The solid square is used for inspection purposes.

Fig. 19-11 The combination set is used for laying out and checking work.

Fig. 19-12 Scribing a line parallel to an edge.

Fig. 19-13 Scribing a line at 90° to an edge.

The *bevel protractor* when mounted on the rule is used to lay out and check various angles. The protractor can be adjusted to any angle from 0 to 180°. The accuracy of this protractor is ±0.5° (30'). A universal bevel protractor may be used if an accuracy of 5' is required.

The *center head* forms a center square when mounted on the rule. It may be used for locating the centers on the ends of round, square, and octagonal stock.

SURFACE GAGE

The surface gage (Fig. 19-14) is used with a surface plate or any flat surface to scribe layout lines on a workpiece. It consists of a *base*, *spindle*, and *scriber*.

The surface gage may be set to the required dimension by using a combination square (Fig. 19-14). This method is usually accurate enough for most layout work. A surface gage may be used on a surface plate to scribe parallel lines on a workpiece (Fig. 19-15). When the workpiece is fas-

Fig. 19-14 Setting a surface gage to a dimension with a combination square.

BASIC LAYOUT MATERIALS, TOOLS, AND ACCESSORIES

Fig. 19-15 Using a surface gage to lay out lines parallel to the top of a surface plate.

Fig. 19-16 A prick punch and a center punch are used in layout work.

Fig. 19-17 An automatic center punch produces uniform indentations on a layout.

Fig. 19-18 Angle plates have sides at 90° to each other.

tened to an angle plate, horizontal and vertical lines can be laid out in one setup by merely rotating the angle plate 90° on the surface plate and scribing the line(s).

Most surface gages have two pins in the base which, when pushed down, are used to guide the surface gage along the edge of the workpiece or surface plate. Some surface gages have a V-groove machined in the base which allows it to be used on cylindrical work.

LAYOUT OR PRICK PUNCHES AND CENTER PUNCHES

The *layout or prick punch* and the *center punch* (Fig. 19-16) differ only in the angle of the point. The *prick punch* is ground to an angle of 30 to 60° and is used to mark permanently the location of layout lines. The narrower angle of this punch makes a smaller and neater indentation in the metal surface.

The *center punch* is ground to an angle of 90° and is used to mark the location of the centers of holes. The wider indentation permits easier and more accurate starting of a drill point.

After the layout lines have been scribed on the workpiece, they should be permanently marked by means of layout or prick-punch marks. This step will ensure that the layout line location will still be visible should the line be rubbed off through handling. The intersection of the centerlines of a circle should be carefully prick-punched and then enlarged with a center punch.

NOTE: Extreme care should be taken when the intersections of layout lines are being prick-punched. Regardless of how accurate the layout is, it is impossible to mark locations with a prick punch to closer than 0.003 to 0.004 in. (0.07 to 0.10 mm).

Uniform layout punch marks may be obtained with an automatic center punch (Fig. 19-17). This punch contains a striking block in the body which is released by downward pressure; the block then strikes the punch proper, causing it to indent the metal. The impression size can be changed by adjusting the tension on the screw cap on the upper end of the tool. This type of punch produces marks uniform in size which improve the appearance of the workpiece. Automatic center punches may also have a spacing attachment which provides uniformity in the spacing of the layout punch marks.

LAYOUT ACCESSORIES

In addition to regular layout tools, certain accessories are helpful in layout work. When lines are required on the face of a plate, it is customary to clamp the work to an *angle*

Fig. 19-19 (A) Scribing horizontal lines with a surface gage; (B) the angle plate turned on its side to scribe vertical lines. (Courtesy Kostel Enterprises Limited.)

Fig. 19-20 Parallels keep the bottom surface of the work parallel with the surface plate.

plate (Fig. 19-18) with a *toolmaker's clamp*. This will hold the work in a vertical plane so that the layout lines will be accurately positioned. Since an accurate angle plate will have all adjacent surfaces at 90° to each other, it is possible to scribe lines accurately which intersect at 90°. This is achieved by scribing all the horizontal lines on the workpiece then turning the angle plate on its side and scribing the intersecting lines (Fig. 19-19A and B).

Parallels (Fig. 19-20) may be used when it is necessary to raise the workpiece to a desired height and to maintain the work surface parallel to the top of the surface plate.

V-blocks are used to hold round work for layout and for inspection. They may be used singly or in pairs. Some blocks are so constructed that they may be rotated 90° on their sides without the work having to be removed. This feature permits the laying out of lines at 90° on a shaft without a change in the position of the work in the V-block (Fig. 19-21).

Fig. 19-21 Lines may be conveniently scribed at 90° with a special V-block.

Fig. 19-22 Keyseat rules are used to scribe lines parallel to the axis of a cylinder. (A) Solid keyseat rule; (B) keyseat rule and clamp.

Keyseat rules are used to lay out keyseats on shafts or to draw lines parallel to the axis of the shaft. A solid keyseat rule resembles two straightedges machined at 90° to each other (Fig. 19-22A). *Keyseat clamps* when attached to a rule or straightedge will convert it to a keyseat rule (Fig. 19-22B).

REVIEW QUESTIONS

1. State two reasons why a layout is necessary.
2. Why should the layout be as simple as possible?

LAYOUT MATERIALS

3. What is the purpose of a layout solution?
4. Name four layout solutions and state one application of each.
5. State two methods of preparing the surface of a casting prior to laying out the workpiece.

LAYOUT TABLES AND SURFACE PLATES

6. State five reasons why granite surface plates are considered better than cast-iron ones.
7. List five points to be observed in the care of surface plates.

SCRIBERS

8. Why must the point of a scriber always be sharp?

DIVIDERS AND TRAMMELS

9. For what purpose are dividers used when laying out a workpiece?
10. What is the purpose of a trammel?
11. How can a circle be laid out concentric with a hole?

HERMAPHRODITE CALIPERS

12. List two uses for hermaphrodite calipers.

SQUARES

13. Name two types of squares used in layout work.
14. List four main parts of a combination set.
15. State three uses for the combination square.
16. What is the accuracy of the bevel protractor?
17. For what purpose is the center head used?

SURFACE GAGE

18. Name the three main parts of the surface gage.
19. What is the purpose of the two pins in the base of the surface gage?

LAYOUT OR PRICK PUNCHES AND CENTER PUNCHES

20. How may the layout be made ''permanent''?
21. What is the purpose of:
    (a) a prick punch?
    (b) a center punch?
22. To what angle is each ground?

LAYOUT ACCESSORIES

23. List four layout accessories and state the purpose of each.

BASIC OR SEMIPRECISION LAYOUT

UNIT 20

Basic or semiprecision layout requires the use of the basic tools described in Unit 19. Remember that the layout should be kept as simple as possible in order to save time and reduce the chance of error. Since the accuracy of the layout will affect the accuracy of the finished workpiece, care should be exercised in the layout procedure.

Although the layout required will naturally not be the same for each workpiece, there are certain procedures which should be followed in any layout. The following jobs are intended to acquaint the reader with basic layout procedures.

OBJECTIVES

After completing this unit you will be able to:
1. Lay out a workpiece to an accuracy of ±0.007 in.
2. Lay out straight lines using the combination square and surface gage
3. Lay out hole centers, arcs, and circles

TO LAY OUT HOLE LOCATIONS, SLOTS, AND RADII

1. Study Fig. 20-1 and select the proper stock.
2. Cut off the stock, allowing enough material to square the ends if required.
3. Remove all burrs.
4. Clean the surface thoroughly and apply layout dye.
5. Place a suitable angle plate on a surface plate.

NOTE: Clean both plates (Fig. 20-2).

6. Clamp the work to the angle plate with a finished edge (A) of the part against the surface plate or on a parallel. Leave one end of the angle plate protruding beyond the workpiece.
7. With the surface gage set to the proper height, scribe a centerline for the full length of the workpiece (Fig. 20-3A).
8. Using the centerline as a reference, set the surface gage for each horizontal line and scribe the center lines for all hole and radii locations (Fig. 20-3B).
9. With the work still clamped to the angle plate, turn the angle plate 90° with edge B down and scribe the baseline at the bottom of the workpiece (Fig. 20-4A).
10. Using the baseline as a reference line, locate and scribe the other centerlines for each hole or arc (Fig. 20-4B).

NOTE: All measurements for any location must be taken from the baseline or finished edge.

11. Locate the starting points for the angular layout (Fig. 20-4A).
12. Remove the workpiece from the angle plate.
13. Carefully prick-punch the center of all hole or radii locations.
14. Using a divider set as required, scribe all circles and arcs (Fig. 20-5A).

Fig. 20-1 A layout exercise.

A

Fig. 20-2 Place a suitable angle plate on the clean surface.

B

Fig. 20-3 (A) Centerlines are scribed parallel to the base; (B) using a surface gage to scribe parallel lines.

BASE LINE

A

B

Fig. 20-4 (A) Centerlines scribed. (B) The angle plate turned to 90° to scribe vertical lines.

15. Scribe any lines required to connect the arcs or circles (Fig. 20-6).

16. Draw in the angular lines.

TO LAY OUT A CASTING HAVING A CORED HOLE

When a casting which requires a hole in it is molded in a foundry, a core is used to produce the rough hole which may have to be machined later. Often the core shifts out of place and the hole is cast off center as shown in Fig. 20-7. If the hole must be machined concentric with the outside of the casting, it may be necessary to lay out the location of the hole.

PROCEDURE

1. Grind the scale off the surface to be laid out.

2. Tap a tightly fitting wooden piece into the cast hole (Fig. 20-7).

3. Coat the surface to be laid out and the wooden piece with a solution of slaked lime and alcohol or layout dye.

4. With the hermaphrodite calipers scribe four arcs as

A

B

Fig. 20-5 (A) Arcs and circles are scribed with a divider; (B) arcs and circles are scribed.

Fig. 20-6 Arcs and circles are connected.

shown, using the outside diameter of the shoulder as the reference surface.

5. Using the intersection of these arcs as a center, scribe a circle of the required diameter on the casting. The hole should be concentric with the outside of the casting (Fig. 20-7).

6. Prick-punch the layout line at about eight equidistant points around the layout circle.

Fig. 20-7 Method of centering a hole location on a casting.

Fig. 20-8 Keyseat layouts.

TO LAY OUT A KEYSEAT IN A SHAFT

A keyseat is a recessed groove cut in a shaft and into which a key is fitted to prevent a mating part, such as a pulley or gear, from turning on the shaft. Laying out a keyseat requires much care, particularly if the mating parts must maintain a certain relative position.

Procedure

1. Apply layout dye to the end of the shaft and to the area where the keyseat is to be laid out.
2. Mount the workpiece in a V-block.

Fig. 20-9 Aligning the keyseat layout in a vise.

3. Set the surface gage scriber to the center of the shaft.
4. Scribe a line across the end and continue it along the shaft to the keyseat location (Fig. 20-8).
5. Rotate the work in the V-block and mark the length and the position of the keyseat on the shaft.
6. Set the dividers to half the width of the keyseat, and scribe a circle at each end of the layout (Fig. 20-8).
7. Using a keyseat rule and scriber, connect the circles with a line on either side of the centerline and tangent to the circles.

NOTE: If a keyseat rule is not available, the circles may be connected using the surface gage.

8. Prick-punch the layout of the keyseat, and center-punch the centers of the circles.
9. If it is necessary to drill holes at the end of the keyseat, it is advisable to set up the shaft by aligning the end layout line in a vertical position with a square (Fig. 20-9).

REVIEW QUESTIONS

1. Outline the main steps that should be followed in making a layout consisting of straight lines, slots, and radii.
2. Name three common tools used to make basic or semi-precision layouts.
3. Describe how to layout a cored hole concentric with the outside shoulder of a casting.
4. List the main steps for laying out a keyseat on a shaft.

UNIT 21

The accuracy of the finished workpiece is generally determined by the accuracy of the layout; therefore it is very important that great care be used when laying out. In order to make a precision layout, a person must be able to read and understand drawings, select and use the proper layout tools for the job, and accurately transfer measurements from the drawing to the workpiece. After a layout is completed, it is wise to check all layout work against the sizes on the working drawings to make sure that the layout is accurate. When the layout lines must be accurate to within 0.001 in. (0.02 mm), a *vernier height gage* may be used (Fig. 21-1).

When hole locations and dimension lines are made on a layout, they are generally made from two machined edges called *reference surfaces* using *X* and *Y* coordinates. Any layout composed of holes, angles, and lines may be calculated using trigonometry to determine the coordinate measurements. Once the coordinates have been determined, they can be used for setting up the workpiece and accurately positioning the holes for machining. Another method of calculating hole locations is by means of the Woodworth Tables of Coordinate Factors and Angle Calculations found in the Appendix of Tables in this book.

OBJECTIVES

After completing this unit you will be able to:
1. Make a precision layout using the vernier height gage
2. Use the Woodworth Coordinate Factors and Angles tables to calculate equidistant hole locations
3. Make precision layouts using the sine bar and gage blocks

THE VERNIER HEIGHT GAGE

The vernier height gage may be used to measure or mark off vertical distances to ±0.001 in. (0.02 mm) accuracy.

The main parts of the vernier height gage (Fig. 21-2) are the *base, beam, vernier slide,* and the *scriber,* which is attached to the vernier slide when making layouts. Other accessories such as a dial indicator or a depth gage attachment may be added to the slide for measurement and inspection work. The graduations on the beam and the vernier slide are the same as those on a vernier caliper and the readings are made in the same manner as with a vernier caliper.

TO MAKE A PRECISION LAYOUT USING A VERNIER HEIGHT GAGE

It is required to lay out the position of five equally spaced holes on a 5-in.-diameter circle located in the center of a 7-in.-*square* steel plate (See Fig. 21-3).

Fig. 21-1 A vernier height gage is used when an accurate layout is required.

BEAM

CLAMP SCREW

VERNIER SLIDE

VERNIER SLIDE

SCRIBER

BASE

Fig. 21-2 The main parts of a vernier height gage.

Fig. 21-4 Offset scribers.

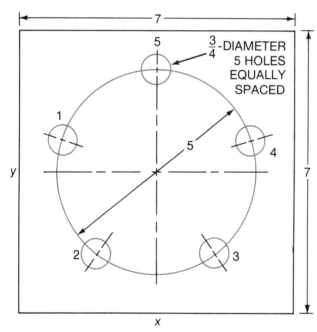

Fig. 21-3 Five equally spaced holes are required on circle.

Procedure

1. Refer to the drawing of the required workpiece (Fig. 21-3).

2. Remove all burrs from the workpiece.

3. Apply layout dye to the surface and mount it on an angle plate.

4. Clean the surface of the layout table, the angle plate, and the base of the height gage.

5. Mount an offset scriber (Fig. 21-4) on the vernier slide and clamp it in position.

6. Move the vernier slide and scriber down until the scriber touches the top of the surface plate.

7. Check the reading on the vernier scale. The zero mark on the vernier should align exactly with the zero graduation on the beam. If the zero on the vernier does not coincide with the zero on the beam, recheck the assembly of the scriber and the vernier slide.

8. Refer to the Coordinate Factors and Angles tables in the Appendix for the coordinates for five equidistant holes.

9. Calculate the location of all five holes as follows:

Hole 1
Horizontal distance from left-hand edge Y
= (diameter of circle × factor for A) + 1.000
= 5 × 0.024472 + 1.000
= 0.12236 + 1.000
= 1.122

Vertical distance from upper edge X
= (diameter × factor for B) + 1.000
= (5 × 0.345492) + 1.000
= 1.72746 + 1.000
= 2.727

Hole 2
Distance from edge Y
= (5 × 0.206107) + 1.000
= 1.030535 + 1.000
= 2.031

Distance from edge X
= (5 × 0.904508) + 1.000
= 4.52254 + 1.000
= 5.523

Hole 3
Distance from edge Y
= (5 × 0.793893) + 1.000
= 3.969465 + 1.000
= 4.969

Distance from edge X
= (5 × 0.904508) + 1.000
= 4.52254 + 1.000
= 5.523 (same as for hole 2)

Hole 4

Distance from edge Y
= (5 × 0.975528) + 1.000
= 4.87764 + 1.000
= 5.878

Distance from edge X
= (5 × 0.345492) + 1.000
= 1.72746 + 1.000
= 2.727   (same as for hole 1)

Hole 5

Distance from edge Y
= (5 × .5000) + 1.000
= 2.5000 + 1.000
= 3.500

Distance from edge X
= (5 × 0.000) + 1.000
= 0.000 + 1.000
= 1.000

10. Place edge Y on the layout table surface.

11. Set the height gage to 1.122 and scribe a line by *drawing* the scriber across the face of the workpiece at the location for hole 1.

12. Set the height gage to each of the following settings. After each setting is made, scribe the line for the appropriate hole location (Fig. 21-5).

Hole 2—2.031
Hole 3—4.969
Hole 4—5.878
Hole 5—3.500

13. Rotate the angle plate and the work 90° and place the edge (X) on the layout table.

14. Set the vernier height gage to 2.727.

15. Scribe the intersecting lines at the centers of holes 1 and 4.

16. Set the height gage to 5.523.

17. Scribe the intersecting lines for the centers of holes 2 and 3.

18. Set the height gage to 1.000 and scribe the intersecting line for hole 5 (Fig. 21-6).

19. Remove the workpiece from the angle plate and place it on the bench with the layout surface up.

20. Using a sharp prick punch and a magnifying glass, carefully mark the centers of the holes at the intersecting lines.

21. Set the dividers to ⅜ in. and scribe the five ¾-in. circles.

22. Carefully prick-punch the circumference of each circle at four equidistant points to ensure the permanency of the layout (Fig. 21-7).

TO MAKE A PRECISION LAYOUT USING A SINE BAR, GAGE BLOCKS, AND A VERNIER HEIGHT GAGE

If a more accurate layout is required for hole positions and angles (Fig. 21-8), it may be done by using a sine bar, gage blocks, and height gage to accurately establish the hole locations and their X and Y axes. The use of coordinates is preferred to locate hole centers since the work can be set up on a jig borer or vertical mill using these same coordinate measurements to position the holes for machining. This type of layout example is shown in Fig. 21-9.

1. Check the drawing for the required dimensions (Fig. 21-8).

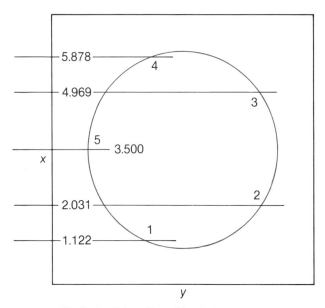

Fig. 21-5   The horizontal position of five holes.

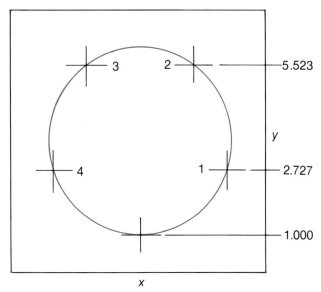

Fig. 21-6   The vertical distance of each hole.

2. Machine and grind a plate square to size 4.750 in. × 4.750 in.

3. Clean the surface of the workpiece, and coat it with layout dye.

4. Set the work on edge on a surface plate, and clamp it to an angle plate.

5. Using a vernier height gage, scribe the centerlines for hole A (Fig. 21-9).

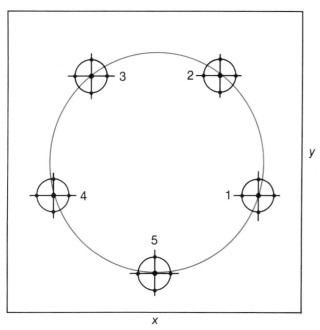

Fig. 21-7 Prick-punch marks on the circles ensure the permanency of the layout.

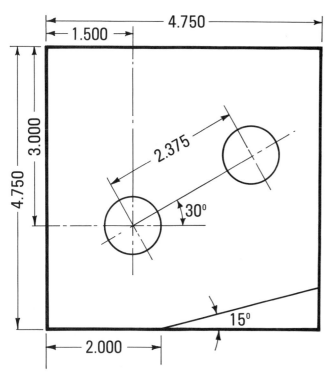

Fig. 21-8 An accurate layout for precision work.

6. Calculate the position of hole B as follows:

Length of side X

$$\frac{X}{2.375} = \cos 30°$$
$$X = 2.375 \cos 30°$$
$$= 2.375 \times 0.86603$$
$$= 2.0568 \text{ in.}$$

Length of side Y

$$\frac{Y}{2.375} = \sin 30°$$
$$Y = 2.375 \sin 30°$$
$$= 2.375 \times 0.5000$$
$$= 1.1875 \text{ in.}$$

The position of the center of hole B is then 2.0568 in. to the right of the centerline for hole A and 1.1875 in. above the centerline for hole A. It is now possible to position the hole using the coordinate method.

The vertical center line for hole B would then be located 1.500 + 2.0568 = 3.5568 in. along the X axis.

The horizontal centerline for hole B would be located 3.000 − 1.1875 = 1.8125 in. along the Y axis.

7. Using a vernier height gage, mark off the centerline for hole B.

8. Remove the workpiece from the angle plate.

9. Carefully prick-punch the intersection of the centerlines of these holes using a sharp punch and a magnifying glass.

10. To lay out the 15° angle at the corner of the plate, calculate the buildup for 15° using a 5-in. sine bar.

$$\text{Buildup} = 5 \sin 15°$$
$$= 5 \times 0.25882$$
$$= 1.2941 \text{ in.}$$

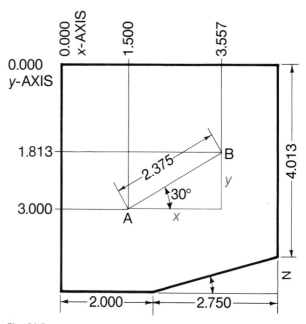

Fig. 21-9 The position of hole B is calculated by trigonometry.

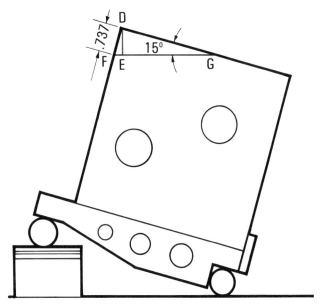

Fig. 21-10 To locate the line *EG* accurately the distance *DE* must be known.

11. Place the buildup under one end of a sine bar (on a surface plate).

12. Place the workpiece, angled edge up, on a sine bar and clamp to an angle plate.

13. Calculate the length of *DF* (Fig. 21-10) as follows:

$$\frac{DF}{2.750} = \tan 15°$$

$$\frac{DF}{2.750} = 0.26795$$

$$DF = 0.26795 \times 2.750$$

$$DF = 0.7368 \text{ in.}$$

14. Calculate the length of the line *DE* (Fig. 21-10). In order to scribe the line accurately at 15° as required in Fig. 21-8, it is necessary first to calculate the length of the line

DE since this is the vertical distance below *D* that the line must be scribed. By previous calculations it was determined that the line *DF* was 0.737 in. In the triangle *DEF* the angle *FDE* is 15°.

$$\therefore \frac{DE}{DF} = \cos 15°$$

$$DE = \cos 15° \times DF$$
$$= 0.96592 \times 0.737$$
$$= 0.7118$$
$$= 0.712 \text{ in.}$$

15. Set the scriber on the vernier height gage to the uppermost corner of the plate (point *D*).

16. Lower the scriber 0.712 in. and scribe the line *GF*. This will locate the point *G* at a position 2.000 in. from the side of the plate as required in Fig. 21-8.

17. Remove the workpiece from the angle plate.

18. Lightly prick-punch the layout line using a magnifying glass and a sharp prick punch.

REVIEW QUESTIONS

1. Name three requirements a machinist must have in order to make a precision layout.

2. List four main parts of a vernier height gage.

3. How can the vertical and horizontal distances of equally spaced holes on a circle be calculated from the edges of a workpiece?

4. Calculate the vertical and horizontal distances for three equally spaced holes on a 4-in.-diameter circle located in the center of a 6-in.-square plate.

5. How can the hole centers at intersecting lines be accurately marked?

6. How can accurate angular lines be laid out?

7. Calculate the gage block buildup for setting a sine bar to an 18° angle.

HAND TOOLS AND BENCH WORK

SECTION 7

The machine trade may be divided into two categories: hand tool and machine tool operations.

Although this era is looked upon as the machine age, the importance of hand tool operations or bench work should not be overlooked. Bench work may refer to the operations of laying out, fitting, and assembling. These operations may involve sawing, chipping, filing, polishing, scraping, reaming, and threading. A good machinist should be capable of using all hand tools skillfully. Effective selection and use of these tools is possible only with continued practice.

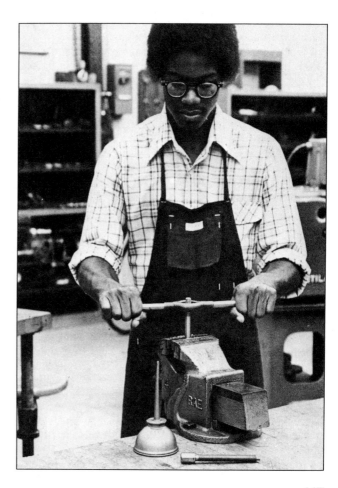

HOLDING, STRIKING, AND ASSEMBLING TOOLS

UNIT 22

Hand tools may be divided into two classes: *noncutting* and *cutting types.* Noncutting tools include such items as vises, hammers, screwdrivers, wrenches, and pliers which are used basically for holding, assembling, or dismantling parts.

OBJECTIVES

After completing this unit you will be able to:
1. Select various tools used for holding, assembling, or dismantling workpieces
2. Properly use these tools for holding, assembling, and dismantling workpieces

THE BENCH VISE

The machinist's, or bench, vise (Fig. 22-1) is used to hold small work securely for sawing, chipping, filing, polishing, drilling, reaming, and tapping operations.

Vises are mounted close to the edge of the bench; they permit long work to be held in a vertical position. Vises may be made of cast iron or cast steel. The vise size is determined by the width of the jaws.

A machinist's vise may be of the solid-base or swivel-base type. The swivel-base vise (Fig. 22-1) differs from the solid-base type by having a swivel plate added to the bottom of the vise. This allows the vise to be swung in any circular position. When gripping finished work of soft materials, use jaw caps made of brass, aluminum, or copper to protect the work surface from being marred or damaged.

Fig. 22-1 A swivel bench vise can be rotated to any position.

HAMMERS

There are many types of hammers used by the machinist, the most common being the *ball-peen hammer* (Fig. 22-2). The larger striking surface is called the *face,* and the smaller, rounded end is the *peen.* Ball-peen hammers are made in a variety of sizes, with head masses ranging from approximately 2 ounces (oz) to 3 pounds (lb) (55 to 1400 g). The smaller sizes are used for layout work, and the larger ones are for general work. The peen is generally used in riveting or peening operations.

Soft-faced hammers (Fig. 22-3A) have heads made of plastic, rawhide, copper, or lead. These heads are fastened to a steel body and can be replaced when worn. Soft-faced hammers are used in assembling or dismantling parts so the finished surface of the work will not be marred. Lead

PEEN

FACE

Fig. 22-2 A ball-peen hammer. (Courtesy Stanley Tools Division, Stanley Works)

118

Fig. 22-3 (A) A soft-faced hammer; (B) cracked hammer handles are dangerous to use.

Fig. 22-4 A standard screwdriver.

Fig. 22-5 A Phillips screwdriver.

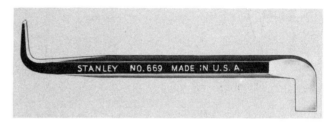

Fig. 22-6 An offset screwdriver is often used in confined spaces.

Fig. 22-7 The socket in the screw suits the tip of the Phillips screwdriver.

hammers are often used to seat the workpiece properly on parallels when setting up work in a vise for milling or shaping operations.

When using a hammer, always grasp it at the end of the handle to provide better balance and greater striking force. This also tends to keep the hammer face flat on the work and reduces the chance of damage to the workface.

The following safety precautions should always be observed when using a hammer.

1. Be sure that the handle is solid and not cracked (Fig. 22-3B).
2. See that the head is tight on the handle and secured with a proper wedge to keep the handle expanded in the head.
3. Never use a hammer with a greasy handle or when your hands are greasy.
4. Never strike two hammer faces together. The faces have been hardened and a metal chip may fly off, causing an injury.

SCREWDRIVERS

Screwdrivers are manufactured in a variety of shapes, types, and sizes. The two most common types used in a machine shop are the standard or flat blade (Fig. 22-4) and the Phillips screwdriver (Fig. 22-5). Both of these types are manufactured in various sizes and styles such as standard shank, stubby shank, and offset (Fig. 22-6).

Phillips-type screwdrivers have a +-shaped tip for use with Phillips-type recessed screw heads (Fig. 22-7). These screwdrivers are manufactured in four sizes, #1, #2, #3, and #4 to suit the various-sized recesses in the heads of the fasteners. When using this type of screwdriver, care must be taken to use the proper size screwdriver. Too small a driver will damage both the tip and the recess in the screw head. It is important that the screwdriver be held firmly in the recess and square with the screw.

Blades for smaller standard screwdrivers are generally made of round stock, and blades for larger ones are often square so that a wrench may be applied for leverage.

CARE OF A SCREWDRIVER

1. Choose the correct size of screwdriver for the job. If too small a screwdriver is used, both the screw slot and the tip of the driver may become damaged.
2. Do not use the screwdriver as a pry, chisel, or wedge.
3. When the tip of a standard screwdriver becomes worn or broken, it should be redressed to shape (Fig. 22-8).

REGRINDING A STANDARD SCREWDRIVER BLADE

When regrinding the tip, make the sides of the blade slightly concave by holding the side of the blade tangential to the

Fig. 22-8   Regrinding a standard screwdriver blade.

SINGLE-END WRENCH

Fig. 22-9   A single-end open-end wrench.

DOUBLE-END WRENCH

Fig. 22-10   A double-end open-end wrench.

Fig. 22-11   A box or 12-point wrench.

periphery of the grinding wheel (Fig. 22-8). Grind an equal amount off each side of the blade. This shape will enable the blade to maintain a better grip in the slot. Care should be taken that the original taper, width, and thickness of the tip are retained and that the end is ground square with the centerline of the blade.

NOTE: When grinding, remove a minimum amount of metal so as not to grind past the hardened zone in the tip. Quench the tip frequently in cold water so as not to draw the temper from the blade.

WRENCHES

Many types of wrenches are used in machine shop work, each being suited for a specific purpose. The name of the wrench is derived from either its use, its shape, or its construction. The more common types of wrenches used in a machine shop are as follows:

Open-end wrenches may be single-ended (Fig. 22-9) or double-ended (Fig. 22-10). The openings on these wrenches are usually offset at a 15° angle to permit turning the nut or bolt head in limited spaces by ''flopping'' the wrench.

Double-ended wrenches usually have a different-size opening at each end to accommodate two different size bolt heads or nuts. These wrenches are available in both inch and metric sizes.

Box-end 12-point wrenches (Fig. 22-11) completely surround the nut and are useful in close quarters where only a small rotation of the nut can be obtained at one time. The box end has 12 precisely cut notches around the inside face which fit closely over the points on the outside of the nut. Because this wrench cannot slip when the proper size is used, it is preferred over most other styles of wrenches. The wrenches usually have a different size at each end and are available in inch and metric sizes.

Socket wrenches (Fig. 22-12) are similar to box wrenches in that they are usually made with 12 points and surround the nut. These sockets are also available in inch and metric sizes. Several types of drives, including ratchet and torque-wrench handles, are available for the various sockets. When nuts or bolts must be tightened to within certain limits to

Fig. 22-12   A set of socket wrenches.

prevent warping, socket wrenches are used in conjunction with a torque-wrench handle.

Adjustable wrenches (Fig. 22-13) may be adjusted to within a certain range to fit several sizes of nuts or bolt heads. This wrench is particularly useful for odd-size nuts or when another wrench of the proper size is not available. Unfortunately, this type of wrench can slip when not properly adjusted to the flats of the nut. This may result in injury to the operator and damage to the corners of the nut.

When using an adjustable wrench, it should be tightened securely to the faces of the nut and the turning force applied in the direction indicated in Fig. 22-13.

Allen setscrew wrenches (Fig. 22-14), commonly called *Allen* wrenches, are hexagonal and fit into the recesses of socket head setscrews. They are made of tool steel and are available in sets to suit the wide variety of screw sizes. The indicated size of the wrench is same as the distance across the flats of the wrench. Usually this distance is one-half the outside diameter of the Allen setscrew in which it is used. These wrenches are available in both inch and metric sizes.

Pin spanner wrenches are specialized wrenches generally supplied by the machine tool manufacturer for use on specific machines. They are supplied in various types. Fixed-face spanners and adjustable-face spanners (Fig. 22-15A)

Fig. 22-13   The correct method of using an adjustable wrench.

ALLEN WRENCH

Fig. 22-14   An Allen wrench.

PIN SPANNER WRENCH

Fig. 22-15   (A) A fixed-face spanner; (B) a hook-pin spanner.

COMBINATION

Fig. 22-16   A slip-joint or combination plier.

SIDE CUTTING

Fig. 22-17   Side-cutting pliers.

are positioned in two holes on the face of a special nut or threaded fitting on a machine.

A *hook-pin spanner* (Fig. 22-15B) is used on the circumference of a round nut, the pin of the spanner fitting into a hole in the periphery of the nut.

HINTS ON USING WRENCHES

1. Always select a wrench which fits the nut or bolt properly. A wrench that is too large may slip off the nut and possibly cause an accident.
2. Whenever possible, it is advisable to *pull* rather than push on a wrench in order to avoid injury if the wrench should slip.
3. Always be sure that the nut is fully seated in the wrench jaw.
4. Use a wrench in the same plane as the nut or bolt head.
5. When tightening or loosening a nut, a sharp quick jerk is more effective than a steady pull.
6. A drop of oil on the threads, when assembling a bolt and nut, will ensure easier removal later.

PLIERS

Pliers are useful for gripping and holding small parts for certain machining operations (such as drilling small holes)

or when assembling parts. Pliers are made in many types and sizes and are named by their shape, their function, or their construction. The more common types of pliers used in a machine shop are:

Combination or slip-joint pliers (Fig. 22-16) are adjustable to grip both large and small workpieces. They may be used to grip certain work when small holes must be drilled or for bending or twisting light, thin materials.

Side-cutting pliers (Fig. 22-17) are used mainly for cutting, gripping, and bending of small-diameter ($\frac{1}{8}$ in. or less) rods or wire.

Needle-nose pliers (Fig. 22-18) are available in both straight and bent-nose types. They are useful for holding very small parts, for positioning them in hard-to-get-at places, and for bending or forming wire.

HOLDING, STRIKING, AND ASSEMBLING TOOLS

NEEDLE-NOSE

Fig. 22-18   Needle-nose pliers.

DIAGONAL

Fig. 22-19   Diagonal cutters.

Fig. 22-20   A vise-grip plier.

Diagonal cutters (Fig. 22-19) are used solely for cutting wire and small pieces of soft metal.

Vise-grip pliers (Fig. 22-20) provide an extremely high gripping power because of the adjustable lever action. The screw in the handle provides adjustment for various sizes of work. This type of plier is available in several different styles such as standard jaws, needle jaws, and C-clamp jaws.

HINTS ON USING PLIERS

The following points should be observed if pliers are to give the proper service.

1. Never use a plier instead of a wrench.
2. Never attempt to cut large-diameter or heat-treated ma-

terial with pliers. This may cause the jaws to distort or the handle to break.
3. Always keep the pliers clean and lubricated.

REVIEW QUESTIONS

THE BENCH VISE

1. What is the advantage of the swivel-base vise over the solid-base vise?
2. How may finished work be held in a vise without the surface being marred or damaged?

HAMMERS

3. Describe the most common hammer used by a machinist.
4. For what purpose are soft-faced hammers used?
5. State three safety rules which should be observed when using a hammer.

SCREWDRIVERS

6. List three important points which should be observed in the care of a screwdriver.
7. Explain the procedure for regrinding the tip of a screwdriver blade.
8. List two precautions that should be observed when using a Phillips screwdriver.

WRENCHES

9. Why are open-end wrenches offset about 15° to the handle?
10. Why is the proper size box wrench preferred to other types of wrenches?
11. What advantage does a socket wrench have over a box wrench?
12. What precaution should be observed when using an adjustable wrench?
13. What will happen if excess pressure is applied to an adjustable wrench or pressure applied on the wrong jaw?
14. What is the cross-sectional shape of an Allen wrench and for what purpose is it used?
15. Where is a hook-pin spanner used?
16. State four hints which you feel are useful when using any wrench.

PLIERS

17. Name four types of pliers and state one use for each.
18. What advantage do vise grips have over other types of pliers?

HAND-TYPE CUTTING TOOLS
UNIT 23

Although most metal cutting can be done more easily, quickly, and accurately on a machine, it is often necessary to do certain metal-cutting operations at the bench or on a job. Such operations include sawing, filing, scraping, reaming, and tapping. It is therefore important that the prospective machinist know how to use hand-type cutting tools properly.

OBJECTIVES

After completing this unit you should be able to:
1. Select and use the proper hacksaw blade for sawing a variety of materials
2. Select and use a variety of files to perform various filing operations
3. Identify and know the purpose of rotary files, ground burrs, and scrapers

SAWING, FILING, AND SCRAPING

Hacksaws, files, and scrapers are very common tools in the machine shop and often the most incorrectly used and abused. The proper use of these tools will not come immediately. It is only through practice that the student or apprentice will become proficient in the use of these tools.

THE HAND HACKSAW

The *pistol-grip hand hacksaw* (Fig. 23-1) is composed of three main parts: the *frame*, the *handle*, and the *blade*. The frame can be of either the solid or adjustable type. The solid frame is more rigid and will accommodate blades of only one specific length. The adjustable frame is more commonly used and will take blades which range from 10 to 12 in. (250 to 300 mm) long. A wing nut at the back of the frame provides adjustment for blade tensioning.

Hacksaw blades are made of high-speed molybdenum or tungsten-alloy steel that has been hardened and tempered. There are two types: the solid, or all-hard, blade and the flexible blade. Solid blades are hardened throughout and are very brittle. They break easily if not used properly. Only the teeth of the flexible blade are hardened, while the back of the blade is soft and flexible. Although this type of blade will stand more abuse than the all-hard blade, it will not stand up as long for general use.

Solid blades are usually used on brass, tool steel, cast iron, and larger sections of mild steel since they do not run out of line when pressure is applied. Flexible blades may be used on channel iron, tubing, copper, and aluminum since they do not break as easily on material with thin cross sections.

Blades are manufactured in various pitches (number of teeth per inch), such as 14, 18, 24, and 32. The pitch is the most important factor to consider when selecting the proper blade for a job. An 18-tooth blade (18 teeth per inch) is recommended for general use. When selecting a blade, choose as coarse a blade as possible in order to provide plenty of chip clearance and to cut through the work as quickly as possible. The blade selected should have at least *two teeth in contact with the work* at all times. This will prevent the work from jamming between the teeth and stripping the teeth from the blade. Fig. 23-2 will provide a guide for proper blade selection.

TO USE THE HAND HACKSAW

1. Check that the blade is of the proper pitch for the job and that the teeth point away from the handle.

Fig. 23-1 Parts of a hand hacksaw.

123

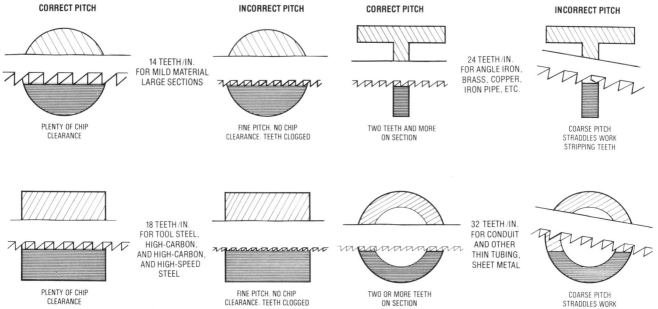

| CORRECT PITCH | INCORRECT PITCH | | CORRECT PITCH | INCORRECT PITCH |

14 TEETH/IN. FOR MILD MATERIAL LARGE SECTIONS

PLENTY OF CHIP CLEARANCE

FINE PITCH. NO CHIP CLEARANCE. TEETH CLOGGED

24 TEETH/IN. FOR ANGLE IRON, BRASS, COPPER, IRON PIPE, ETC.

TWO TEETH AND MORE ON SECTION

COARSE PITCH STRADDLES WORK STRIPPING TEETH

18 TEETH/IN. FOR TOOL STEEL, HIGH-CARBON, AND HIGH-CARBON, AND HIGH-SPEED STEEL

PLENTY OF CHIP CLEARANCE

FINE PITCH. NO CHIP CLEARANCE. TEETH CLOGGED

32 TEETH/IN. FOR CONDUIT AND OTHER THIN TUBING, SHEET METAL

TWO OR MORE TEETH ON SECTION

COARSE PITCH STRADDLES WORK

Fig. 23-2 The selection of the proper blade pitch is very important.

2. Adjust the blade tension so that the blade cannot flex or bend.

3. Mount the stock in the vise so that the cut will be about ¼ in. (6.35 mm) from the vise jaws.

4. Grasp the hacksaw as shown in Fig. 23-3. Assume a comfortable stance, standing erect with the left foot slightly ahead of the right foot.

5. Start the saw cut just outside and parallel to a previously scribed line.

NOTE: File a V-shaped nick at the starting point to help start the file at the right spot.

6. After the cut has started, apply pressure only on the forward stroke. Use about 50 strokes per minute.

7. When cutting thin material, hold the saw at an angle to have at least two teeth in contact with the work at all times. Thin work is often clamped between two pieces of wood, and the cut is made through all pieces (Fig. 23-3).

8. When nearing the end of the cut, slow down to control the saw as it breaks through the material.

NOTE: If a saw blade breaks or becomes dull in a partly finished cut, replace the blade and rotate the work one-half turn so that the old cut is at the bottom. A new blade will bind in an old cut and the "set" of the new teeth will be ruined quickly.

FILES

A *file* is a hand cutting tool made of high-carbon steel, having a series of teeth cut on the body by parallel chisel cuts. The parts of a file are shown in Fig. 23-4. Files are used to remove surplus metal and to produce finished surfaces. Files are manufactured in a variety of types and shapes, each for a specific purpose. They may be divided into two classes: single cut and double cut (Fig. 23-5).

WOOD PIECES

THIN METAL

Fig. 23-3 The correct method of holding a hacksaw.

TANG

LENGTH

HEEL   FACE   EDGE   POINT

Fig. 23-4 The main parts of a file.

Single-cut files have a single row of parallel teeth running diagonally across the face. They include *mill, long-angle lathe,* and *saw files.* Single-cut files are used when a smooth finish is desired or when hard materials are to be finished.

Double-cut files have two intersecting rows of teeth. The first row is usually coarser and is called the *overcut.* The

SINGLE-CUT

DOUBLE-CUT

BASTARD-CUT

SECOND-CUT

SMOOTH-CUT

Fig. 23-5 File classification.

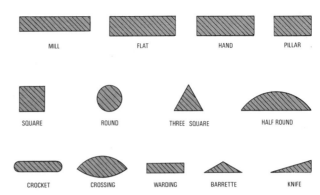

MILL   FLAT   HAND   PILLAR

SQUARE   ROUND   THREE SQUARE   HALF ROUND

CROCKET   CROSSING   WARDING   BARRETTE   KNIFE

Fig. 23-6 Cross-sectional views of machinist files.

Fig. 23-7 Cleaning a file with a file brush.

other row is called the *upcut*. These intersecting rows produce hundreds of cutting teeth which provide for fast removal of metal and easy clearing of chips.

DEGREES OF COARSENESS

Both single- and double-cut files are manufactured in various degrees of coarseness, such as *rough, coarse, bastard, second cut, smooth,* and *dead smooth*. Those most commonly used by the machinist are the bastard, second cut, and smooth (Fig. 23-5).

MACHINISTS' FILES

The types of files most commonly used are the *flat, hand, round, half-round, square, pillar, three-quarter* (triangular), *warding,* and *knife* (Fig. 23-6).

Care of Files

Because files are relatively inexpensive hand tools, they are often abused. Proper care, selection, and use are most important if good results are to be obtained with files. The following points should be observed in the care of files:

1. Do not store files where they will rub together. Hang or store them separately.
2. Never use a file as a pry or a hammer. Since the file is

hard, it snaps easily, often causing small pieces to fly, which may result in a serious eye injury.
3. Do not knock a file on a vise or other metallic object to clean it. Always use a file card or brush for this purpose (Fig. 23-7).
4. Apply pressure only on the forward stroke when filing. Pressure on the return stroke will dull the file.
5. Do not press too hard on a new file. Too much pressure tends to break off the cutting edges and shorten the life of the file.
6. Too much pressure also results in "pinning" (*small particles being wedged between the teeth*) which causes scratches on the surface of the work. Keep the file clean. A piece of brass, copper, or wood pushed through the teeth will remove the "pins." Applying chalk to the face of the file will lessen the tendency for the file to become clogged (pinned).

FILING PRACTICE

Filing is an important hand operation and one that can be mastered only through patience and practice. The following points should be observed when cross filing:

1. Never use a file without a handle. Ignoring this rule is a dangerous practice. Serious hand injury may result should the file slip.

Fig. 23-8 The file must be level when filing.

Fig. 23-9 Cross-filing will show any high spots on the work surface.

2. Fasten the work to be filed, at about elbow height, in a vise.

3. To produce a flat surface, the right hand, right forearm, and left hand should be held in a horizontal plane (Fig. 23-8). Push the file across the work face in a straight line and do not rock the file.

4. Apply pressure only on the forward stroke.

5. Never rub the fingers or hand across a surface being filed. Grease or oil from the hand causes the file to slide over instead of cutting the work. Oil will also clog the file.

6. Keep the file clean by using a file card frequently.

For rough filing, use a double-cut file and cross the stroke at regular intervals to help keep the surface flat and straight (Fig. 23-9). When finishing, use a single-cut file and take shorter strokes to keep the file flat.

Fig. 23-10 Draw filing is used to produce a flat, smooth surface.

Test the work for flatness occasionally by laying the edge of a steel rule across its surface. A steel square should be used to test the squareness of one surface to another.

DRAW FILING

Draw filing is used to produce a smooth, flat surface on the workpiece. This method of filing removes file marks and scratches left by cross filing.

NOTE: When draw filing, hold the file as in Fig. 23-10 and move the file back and forth along the length of the work.

POLISHING

After a surface has been filed, it may be finished with abrasive cloth to remove small scratches left by the file. This may be done with a piece of abrasive cloth held under the file which is moved back and forth along the work.

SPECIAL FILES

Long-angle lathe files are used for filing on a lathe as they provide a better shearing action than mill files. The long angle of the teeth tends to clean the file, helps eliminate chatter, and reduces the possibility of tearing the metal.

Aluminum files are designed for soft, ductile metals, such as aluminum and white metal, because regular files tend to clog quickly when used on this type of material.

The modified tooth construction on aluminum files tends to reduce clogging. The upcut tooth is deep, and the overcut is fine. This produces small scallops on the upcut which breaks up the chips and permits them to clear more easily.

Brass files have a small upcut angle and a fine, long-angle overcut which produces small, easily cleared chips. The almost straight upcut prevents grooving the surface of the work.

Shear tooth files combine a long angle and a single-cut

ROUND

HALF ROUND

FLAT

CROSSING

KNIFE

SQUARE

THREE SQUARE

EQUALING

BARRETTE

JOINT (2 Round Edges)

SLITTING

MARKING

Fig. 23-11 Needle files are used for intricate work. (Courtesy Nicholson File Co. of Canada Ltd.)

Fig. 23-12 The broken-line teeth of rotary files tend to dissipate heat quickly. (Courtesy Nicholson File Co. of Canada Ltd.)

Fig. 23-13 Ground burrs are used on nonferrous metals. (Courtesy Nicholson File Co. of Canada Ltd.)

coarse tooth for filing materaials such as brass, aluminum, copper, plastics, and hard rubber.

PRECISION FILES

Precision files comprise swiss pattern, needle, and riffler files. *Swiss pattern* and *needle files* (Fig. 23-11) are small files having very fine tooth cuts and round integral handles. They are made in several shapes and are generally used in tool and die shops for finishing delicate and intricate pieces. *Die sinker rifflers* are curved up at the ends to permit filing the bottom surface of a die cavity.

ROTARY FILES AND BURRS

The increased use of portable electric and pneumatic power tools has developed the widespread application of rotary files and burrs. The wide range of shapes and sizes available makes these tools particularly suitable for metal pattern making and die sinking.

Rotary file teeth (Fig. 23-12) are cut and form broken lines in contrast to the unbroken flutes of the ground burr (Fig. 23-13). The teeth of the rotary file tend to dissipate the heat of friction, a feature which makes this tool particularly useful for work on tough die steels, forgings, and scaly surfaces.

Ground burrs (Fig. 23-13) may be made of high-speed steel or carbide. The flutes of a burr are generally machine ground to a master burr to ensure uniformity of tooth shape and size. Ground high-speed burrs are used more efficiently on nonferrous metals, such as aluminum, brass, bronze, and magnesium, since they have better chip clearance than rotary files.

Carbide burrs may be used on hard or soft materials with good results and will last up to 100 times longer than a high-speed steel burr.

Using High-Speed Steel Rotary Files and Burrs

For best results, the following points should be observed when using rotary files or burrs.

1. Move the file or burr at an even rate to produce a smooth surface. An uneven rate of pressure produces work with ridges and hollows.
2. Use the proper speed for the burr diameter or file as recommended by the manufacturer.
3. Use only sharp burrs or files.
4. For more accurate control of the burr or file, grip the grinder as close as possible to the end.
5. Medium-cut burrs and files generally provide satisfactory metal removal and finish for most jobs. If greater stock removal is required, use a coarse burr or file. For an extra-smooth finish, use a fine burr or file.

SCRAPERS

When a truer surface is required than can be produced by machining, the surface may be finished by scraping. However, this is a long and tedious process. Most bearing surfaces (flat and curved) are now finished by grinding, honing, or broaching.

Scraping is a process of removing small amounts of metal from specific areas to produce an accurate bearing surface. It is used to produce flat surfaces on cast-iron surface plates or in fitting brass and babbitt bearings to shafts.

Fig. 23-14   A set of hand scrapers.

Scrapers are made in various shapes, depending on the surface to be scraped (Fig. 23-14). They are generally made of high-grade tool steel, hardened and tempered. Carbide-tipped scrapers are very popular because they maintain the cutting edge longer than other types.

REVIEW QUESTIONS

THE HAND HACKSAW

1. Compare the flexible blade and the solid, or all-hard, hacksaw blade.

2. What pitch hacksaw blade should be selected to cut:
   (a) tool steel?
   (b) thin-wall tubing?
   (c) angle iron and copper?
3. What procedure is recommended if a saw blade breaks or becomes dull in a partially finished cut?

FILES

4. Describe and state the purpose of:
   (a) single-cut files
   (b) double-cut files
5. Name the most commonly used degrees of coarseness in which files are manufactured.
6. List four important points which should be observed in the care of files.
7. How can pinning of a file be kept to a minimum?
8. Describe and state the purpose of:
   (a) long-angle lathe files
   (b) aluminum files
   (c) shear tooth files
9. Describe and state the purpose of:
   (a) swiss pattern files
   (b) die sinker rifflers
10. Compare rotary files and ground burrs.
11. List three important points which should be observed when rotary files or ground burrs are used.
12. Describe the correct procedure for filing a flat surface.

THREAD-CUTTING TOOLS AND PROCEDURES

UNIT 24

Threads may be cut internally using a tap and externally using a die. The proper selection and use of these threading tools is an important phase of machine shop work.

OBJECTIVES

After completing this unit you will be able to:
1. Calculate the tap drill size for inch and metric taps
2. Cut internal threads using a variety of taps
3. Know the methods used to remove broken taps from a hole
4. Cut external threads using a variety of dies

HAND TAPS

Taps are cutting tools used to cut internal threads. They are made from high-quality tool steel, hardened and ground. Two, three, or four flutes are cut lengthwise across the threads to form cutting edges, provide room for the chips, and admit cutting fluid to lubricate the tap. The end of the shank is square so that a tap wrench (Fig. 24-1A, B) can be used to turn the tap into a hole. For inch taps, the major diameter, number of threads per inch, and type of thread are usually found stamped on the shank of a tap. For example, ½ in. − 13 UNC represents:
1. ½ in = major diameter of the tap
2. 13 = number of threads per inch
3. UNC = Unified National Coarse (a type of thread)

Hand taps are usually made in sets of three, called *taper, plug,* and *bottoming* (Fig. 24-2A).

TAPER   PLUG   BOTTOMING

A

A

B

Fig. 24-1 (A) A T-handled tap wrench; (B) a double-end adjustable tap wrench.

B

Fig. 24-2 (A) A set of hand taps; (B) taps may be identified by an annular ring on the shank.

A *taper tap* is tapered from the end approximately six threads and is used to start a thread easily. It can be used for tapping a hole which goes *through* the work, as well as for starting a *blind* hole (one that does not go through the work).

A *plug tap* is tapered for approximately three threads. Sometimes the plug tap is the only tap used to thread a hole going through a piece of work.

A *bottoming tap* is not tapered but chamfered at the end for one thread. It is used for threading to the bottom of a blind hole. When tapping a blind hole, first use the taper tap, then the plug tap, and complete the hole with a bottoming tap.

Taps may also be identified by an annular ring or rings cut around the shank of the tap. One ring around the shank indicates that it is a taper tap, two rings indicate a plug tap, and three rings indicate a bottoming tap (Fig. 24-2B).

TAP DRILL SIZE

Before a tap is used, the hole must be drilled to the correct tap drill size (Fig. 24-3). This is the drill size that would leave the proper amount of material in the hole for a tap to cut a thread. The tap drill is always smaller than the tap and leaves enough material in the hole for the tap to produce 75 percent of a full thread.

When a chart is not available, the tap-drill size for any American, National, or Unified thread can be found easily by applying this simple formula:

$$TDS = D - \frac{1}{N}$$

TDS = tap drill size
D = major diameter of tap
N = number of threads per inch

EXAMPLE

Find the tap drill size for a ⅞ in. − 9 NC tap.

$$
\begin{aligned}
TDS &= 7/8 - 1/9 \\
&= 0.875 - 0.111 \\
&= 0.764 \text{ in.}
\end{aligned}
$$

The nearest drill size to 0.764 in. is 0.765 in. (⁴⁹⁄₆₄). Therefore, ⁴⁹⁄₆₄ in. is the tap drill size for a ⅞ in. − 9 NC tap.

METRIC TAPS

Although there are several thread forms and standards in the metric thread system, the International Standards Organization (ISO) has adopted a standard metric thread which will be used throughout the United States, Canada, and many other countries throughout the world. This new series will have only 25 thread sizes ranging from 1.6 to 100 mm diameter. See Table 6 in the Appendix for the size and pitch of the threads in this series. Also see Unit 49 for the thread form and dimensions of the ISO metric thread.

Like inch taps, metric taps are available in sets of three: the *taper, plug,* and *bottoming* tap. They are identified with

Fig. 24-3 A cross section of a tapped hole. (A) Body size; (B) tap drill size; (C) minor diameter.

the letter M followed by the nominal diameter of the thread in millimeters times the pitch in millimeters.

A tap with the markings M 4 × 0.7 would indicate

M—a metric thread
4—the nominal diameter of the thread in millimeters
0.7—the pitch of the thread in millimeters

TAP DRILL SIZES FOR METRIC TAPS

The tap drill size for metric taps is calculated in the same manner as for U.S. Standard threads.

$$TDS = \text{major diameter (mm)–pitch (mm)}$$

EXAMPLE

Find the tap drill size for a 22–2.5 mm thread.

$$
\begin{aligned}
TDS &= 22–2.5 \\
&= 19.5 \text{ mm}
\end{aligned}
$$

TAPPING A HOLE

Tapping is the operation of cutting an internal thread using a tap and tap wrench. Because taps are hard and brittle, they are easily broken. *Extreme* care must be used when tapping a hole to prevent breakage. A broken tap in a hole is difficult to remove and often results in scrapping the work.

To Tap a Hole by Hand

1. Select the correct taps and tap wrench for the job.
2. Apply a suitable cutting fluid to the tap.

NOTE: No cutting fluid is required when tapping brass or cast iron.

3. Place the tap in the hole as vertically as possible, press downward on the wrench applying equal pressure on both handles, and turn clockwise (for right-hand thread) for about two turns.
4. Remove the tap wrench and check the tap for squareness.

NOTE: Check at two positions at 90° to each other (Fig. 24-4).

5. If the tap has not entered squarely, remove it from the hole and restart it by applying pressure in the direction from which the tap leans. *Be careful* not to exert too much pressure in the straightening process.
6. When a tap has been properly started, feed it into the hole by turning the tap wrench.
7. Turn the tap clockwise one-quarter turn, and then turn it backward about one-half turn to break the chip. This must be done with a steady motion to prevent the tap from breaking.

NOTE: When tapping blind holes, use all three taps in order: taper, plug, and then the bottoming tap. Before using the bottoming tap, remove all the chips from the hole and be careful not to hit the bottom of the hole with the tap.

REMOVING BROKEN TAPS

If extreme care is not used when cutting a thread, particularly in a blind hole, the tap may break in the hole and considerable work will be required to remove it. In some cases, it may not be possible and another piece of work must be started.

Several methods may be used to remove a broken tap; some may be successful, others will not.

Tap Extractor

This tool (Fig. 24-5) has four fingers that slip into the flutes of a broken tap. It is adjustable in order to support the fingers close to the broken tap, even when the broken end is below the surface of the work. A wrench is fitted to the extractor and turned counterclockwise to remove a right-hand tap. Tap extractors are made to fit all sizes of taps.

To Remove a Broken Tap
Using a Tap Extractor:

1. Select the proper extractor for the tap to be removed.
2. Slide *collar A,* to which the fingers are attached, down *body B* so that the fingers project well below the end of the body.
3. Slide the fingers into the flutes of the broken tap, making sure they go down into the hole as far as possible.
4. Slide the body down until it rests on top of the broken tap. This will give the maximum support to the fingers.
5. Slide the *collar C* down until it rests on top of the work. This also provides support for the fingers.

Fig. 24-4 Checking a tap for squareness while the work is held in the vise.

Fig. 24-5 A tap extractor being used to remove a broken tap.

6. Apply a wrench to the square section on the top of the body.
7. Turn the wrench *gently* in a counterclockwise direction.

NOTE: Do not force the extractor as this will damage the fingers. It may be necessary to turn the wrench back and forth carefully to free the tap sufficiently to back it out.

Drilling

If the broken tap is made of carbon steel, it may be possible to drill it out.

131

Procedure:

1. Heat the broken tap to a bright red color, and allow it to cool *slowly*.
2. Center-punch the tap as close to the center as possible.
3. Using a drill considerably smaller than the distance between opposite flutes, proceed *carefully* to drill a hole through the broken tap.
4. Enlarge this hole to remove as much of the metal between the flutes as possible.
5. Collapse the remaining part with a punch and remove the pieces.

Acid Method

If the broken tap is made of high-speed steel and cannot be removed with a tap extractor, it is sometimes possible to remove it by the acid method.

Procedure:

1. Dilute one part nitric acid with five parts water.
2. Inject this mixture into the hole. The acid will act upon the steel and cause the tap to become loose.
3. Remove the tap with an extractor or a pair of pliers.
4. Wash the remaining acid from the thread with water so that it will not continue to act on the threads.

Tap Disintegrators

Taps may sometimes be successfully removed by a tap disintegrator which may be held in the spindle of a drill press. The disintegrator employs the *electrical discharge principle* to cut its way through the tap, using a hollow brass tube as an electrode. Taps may also be removed using the same method on any electrical discharge machine (see Unit 91).

THREADING DIES

Threading dies are used to cut external threads on round work. The most common threading dies are the sloid, adjustable split, and the adjustable and removable screw plate die.

The *solid die* (Fig. 24-6) is used for chasing or recutting damaged threads and may be driven by a suitable wrench. It is not adjustable.

The *adjustable split die* (Fig. 24-7) has an adjusting screw which permits an adjustment over or under the stan-

dard depth of thread. This type of die fits into a die stock (Fig. 24-8).

The *adjustable screw plate die* (Fig. 24-9) is probably a more efficient die since it provides for greater adjustment than the split die. Two die halves are held securely in a collet by means of a threaded plate which also acts as a guide when threading. The plate, when tightened into the collet, forces the die halves with tapered sides into the tapered slot of the collet. Adjustment is provided by means of two adjusting screws which bear against each die half. The threaded section at the bottom of each die half is tapered to provide for easy starting of the die. Note that the upper side of each die half is stamped with the manufacturer's name, while the lower side of each is stamped with the same serial number. Care should be taken in assembling the die that both serial numbers are facing down. *Never* use two die halves with different serial numbers.

TO THREAD WITH A HAND DIE

1. Chamfer the end of the workpiece with a file or on the grinder.
2. Fasten the work securely in a vise.
3. Select the proper die and die stock.
4. Lubricate the tapered end of the die with a suitable cutting lubricant.
5. Place the tapered end of the die squarely on the work (Fig. 24-10).
6. Press down on the die stock handles and turn clockwise several turns.
7. Check the die to see that it has started squarely with the work.
8. If it is not square, remove the die from the work and restart it squarely.
9. Turn the die forward one turn and then reverse it approximately one-half of a turn to break the chip.
10. During the threading process apply cutting fluid frequently.

CAUTION: When cutting a long thread, keep the arms and hands clear of the sharp threads coming through the die.

If the thread must be cut to a shoulder, remove the die and restart it with the tapered side of the die facing up. Complete the thread, being careful not to hit the shoulder, otherwise the work may be bent and the die broken.

Fig. 24-6 A solid die nut.

Fig. 24-7 An adjustable, round split die.

Fig. 24-8 A die stock is used to turn the die onto the workpiece.

Fig. 24-9   An adjustable screw plate die.

Fig. 24-10   Start the tapered end of the die on the work.

REVIEW QUESTIONS

HAND TAPS

1. Name, describe, and state the purposes of the three taps in a set.
2. Define tap drill size.
3. Use the formula and calculate the tap drill size for:
   (a) ½ in.—13 NC tap
   (b) M 42—4.5 mm tap
4. Why should care be used when a hole is being tapped?

5. Explain the procedure for correcting a tap which has not started squarely.
6. Briefly outline the procedure for tapping a hole using the tap and reamer aligner.
7. Briefly explain the method of removing a broken tap using a tap extractor.

THREADING DIES

8. For what purpose are threading dies used?
9. State the purpose of the adjustable split die and the solid die.
10. Explain the procedure for starting a die on the work.
11. What procedure should be followed when it is necessary to cut a thread to a shoulder?

FINISHING PROCESSES— REAMING, BROACHING, AND LAPPING

UNIT 25

Hand cutting tools are generally used to remove only small amounts of metal and are designed to do specific operations.

Reamers, available in a wide variety of types and sizes, are used to bring a hole to size and produce a good finish.

Broaches, when used in a machine shop, are generally used with an arbor press to produce special shapes in the workpiece. The broach, which is a multitooth cutting tool of the exact shape and size, is forced through a hole in the workpiece to reproduce its shape in the metal.

Lapping is a process whereby very fine abrasive powder, embedded in a proper tool, is used to remove minute amounts of material from a surface.

After completing this unit you should be able to:
1. Identify and know the purpose of several types of hand reamers
2. Ream a hole accurately with a hand reamer
3. Cut a keyway in a workpiece using a broach and arbor press
4. Lap a hole or an external diameter of a workpiece to size and finish

HAND REAMERS

A hand reamer is a tool used to finish drilled holes accurately and provide a good finish. Reaming is generally performed by machine, but there are times when a hand reamer must be used to finish a hole. Hand reamers, when used properly, will produce holes accurate to size, shape, and finish.

TYPES OF HAND REAMERS

The *solid hand reamer* (Fig. 25-1) may be made of carbon steel or high-speed steel. These straight reamers are available in inch sizes from ⅛ to 1½ in. in diameter and in metric sizes from 1 to 26 mm in diameter. For easy starting, the cutting end of the reamer is ground to a slight taper for a distance equal to the diameter of the reamer. Solid reamers are not adjustable and may have straight or helical flutes. Straight-fluted reamers should not be used on work with a keyway or any other interruption since chatter and poor finish will result. Since hand reamers are designed to remove only small amounts of metal, no more than 0.005 to 0.010 in. (0.12 to 0.25 mm) should be left for reaming, depending on the diameter of the hole. A square on the end of the shank provides the means of driving the reamer with a tap wrench.

The *expansion hand reamer* (Fig. 25-2) is designed to permit an adjustment of approximately 0.006 in. (0.15 mm) above the nominal diameter. The reamer is made hollow and has slots cut along the length of the cutting section. A tapered threaded plug, fitted into the end of the reamer, provides for limited expansion. If the reamer is expanded too much, it will be easily broken. On *inch expansion hand reamers,* the limit of adjustment is up to 0.006 in. over the nominal size on reamers up to ½ in. and about 0.015 in. on reamers over ½ in. *Metric expansion hand reamers* are available in sizes from 4 to 25 mm. The maximum amount of expansion on these reamers is 1 percent over the nominal size. For example, a 10-mm-diameter reamer can be expanded to 10.01 mm (10 + 1 percent). The cutting end of the reamer is ground to a slight taper for easy starting.

The *adjustable hand reamer* (Fig. 25-3) has tapered slots cut in the body for the entire length. The inner edge of the cutting blades have a corresponding taper so that the blades remain parallel for any setting. The blades are adjusted to size by upper and lower adjusting nuts.

The blades on inch adjustable hand reamers have an adjustment range of 1/32 in. on the smaller reamers to almost 5/16 in. on the larger ones. They are manufactured in sizes ¼ to 3 in. in diameter. *Metric adjustable hand reamers* are available in sizes from #000 (adjustable from 6.4 to 7.2 mm) to #16 (adjustable from 80 to 95 mm).

Taper reamers are made to standard tapers and are used to finish tapered holes accurately and smoothly. They may be made with either spiral or straight teeth. Because of its shearing action and its tendency to reduce chatter, the spiral-fluted reamer is superior to the straight one. A *roughing reamer* (Fig. 25-4), with nicks ground at intervals along the teeth, is used for more rapid removal of surplus metal. These nicks or grooves break up the chips into smaller sections; they prevent the tooth from cutting and overloading along its entire length. When a roughing reamer is not available, an old taper reamer is often used prior to finishing the hole with a finishing reamer.

The *finishing taper reamer* (Fig. 25-5) is used to finish the hole smoothly and to size after the roughing reamer. This reamer, which has either straight or left-hand spiral flutes, is designed to remove only a small amount of metal [about 0.010 in. (0.25 mm)] from the hole. Since taper reamers do not clear themselves readily, they should be removed frequently from the hole and the chips cleared from the flutes.

Fig. 25-1 A solid hand reamer. (Courtesy Cleveland Twist Drill Company.)

Fig. 25-2 An expansion hand reamer.

Fig. 25-3 An adjustable hand reamer.

Fig. 25-4 A roughing taper reamer. (Courtesy Whitman and Barnes.)

Fig. 25-5 A finishing taper reamer. (Courtesy Whitman and Barnes.)

REAMING PRECAUTIONS

1. Never turn a reamer backward (counterclockwise) as it will dull the cutting teeth.
2. Use a cutting lubricant where required.
3. Always use a helical-fluted reamer in a hole that has a keyway or oil groove cut in it.
4. Never attempt to remove too much material with a hand reamer; about 0.010 in. (0.25 mm) maximum.
5. Frequently clear a taper reamer (and the hole) of chips.

TO REAM A HOLE WITH A STRAIGHT HAND REAMER

1. Check the size of the drilled hole. It should be between 0.004 and 0.010 in. (0.10 and 0.25 mm) smaller than the finished hole size.
2. Place the end of the reamer into the hole and place the tap wrench on the square end of the reamer.
3. Rotate the reamer in a clockwise direction to allow it to align itself with the hole (Fig. 25-6).
4. Check the reamer for squareness with the work by testing it with a square at several points on the circumference.
5. Brush cutting fluid over the end of the reamer if required.
6. Rotate the reamer slowly in a clockwise direction and apply downward pressure. Feed should be fairly rapid and steady to prevent the reamer from chattering.

NOTE: The rate of feed should be about one-quarter the diameter of the reamer for each turn.

BROACHING

Broaching is a process in which a special tapered multi-toothed cutter is forced through an opening or along the outside of a piece of work to enlarge or change the shape of the hole or to form the outside to a desired shape.

Fig. 25-6 Turn the reamer *clockwise* when starting it in a hole.

Broaching was first used for producing internal shapes, such as keyways, splines, and other odd internal shapes (Fig. 25-7). Its application has been extended to exterior surfaces, such as the flat face on automotive engine blocks and cylinder heads. Most broaching is now performed on special machines which either pull or push the broach through or along the material. Hand broaches are used in the machine shop for such operations as keyway cutting.

The cutting action of a broach is performed by a series of successive teeth, each protruding about 0.003 in. (0.07 mm) further than the preceding tooth (Fig. 25-8). The last three teeth are generally of the same depth and provide the finish cut.

Broaching has many advantages and an extremely wide range of applications.

1. It is possible to machine almost any irregular shape, providing it is parallel to the broach axis.
2. It is rapid; the entire machining process is usually completed in one pass.
3. Roughing and finishing cuts are generally combined in the same operation.
4. A variety of forms, either internal or external, may be cut simultaneously and the entire width of a surface may be machined in one pass, thus eliminating the need for a machining operation.

CUTTING A KEYWAY WITH A BROACH

Keyways may be cut by hand in the machine shop quickly and accurately by means of a broach set and an arbor press (Fig. 25-9).

A broach set (Fig. 25-10) covers a wide range of keyways and is a particularly useful piece of equipment when many keyways must be cut. The equipment necessary to cut a keyway is a bushing (A) to suit the hole size in the workpiece, a broach (B) the size of the keyway to be cut, and shims (C) to increase the depth of the cut of the broach.

PROCEDURE

1. Determine the keyway size required for the size of the workpiece.

Fig. 25-7 Examples of internal broaching.

Fig. 25-8 The cutting action of a broach.

135

Fig. 25-9 Using an arbor press to cut a keyway with a broach.

Fig. 25-10 A broach set for cutting internal keyways.

Fig. 25-11 Two shims are used when making the final pass with a broach.

2. Select the proper broach, bushing, and shims.

3. Place the workpiece on the arbor press. Use an opening on the base smaller than the opening in the workpiece so that the bushing will be properly supported.

4. Insert the bushing and the broach into the opening. Apply cutting fluid if the workpiece is made of steel.

5. Check the broach to be sure that it has started squarely in the hole.

6. Press the broach through the workpiece, maintaining constant pressure on the arbor-press handle.

7. Remove the broach, insert one shim, and press the broach through the hole.

8. Insert the second shim, if required, and press the broach through again. This will cut the keyway to the proper depth (Fig. 25-11).

9. Remove the bushing, broach, and shims.

LAPPING

Lapping is an abrading process used to remove minute amounts of metal from a surface which must be flat, accurate to size, and smooth. Lapping may be performed for any of the following reasons:

1. To increase the wear life of a part
2. To improve accuracy and surface finish
3. To improve surface flatness
4. To provide better seals and eliminate the need for gaskets.

Lapping may be performed by hand or machine, depending on the nature of the job. Lapping is intended only to remove small amounts of material, usually about 0.0005 in. (0.01 mm). Lapping by hand is a long, tedious process and should be avoided unless absolutely necessary because of the time required to remove a small amount of material.

LAPPING ABRASIVES

Both natural and artificial abrasives are used for lapping. Flour of emery and fine powders made of silicon carbide or aluminum oxide are used extensively. Abrasives used for rough lapping should be no coarser than 150 grit; fine powders used for finishing run up to about 600 grit. For fine work, diamond dust, generally in paste form, is used.

TYPES OF LAPS

Laps may be used to finish flat surfaces, holes, or the outside of cylinders. In each case, the lap material must be *softer* than the workpiece.

Fig. 25-12 (A) A roughing lapping plate; (B) a finishing lapping plate.

Flat Laps

Laps for producing flat surfaces are made from close-grained cast iron. For the roughing operation or "blocking down," the lapping plate should be scored with narrow grooves about ½ in. (12.7 mm) apart, both lengthwise and crosswise or diagonally to form a square or diamond pattern (Fig. 25-12A). Finish lapping is done on a smooth cast-iron plate (Fig. 25-12B).

Charging the Flat Lapping Plate Spread a thin coating of abrasive powder over the surface of the plate and press the particles into the surface of the lap with a hardened steel block or roll. Rub as little as possible. When the entire surface appears to be charged, clean the surface with varsol and examine it for bright spots. If any bright spots appear, recharge the lap and continue until the entire surface assumes a gray appearance after it has been cleaned.

Lapping a Flat Surface If work is to be roughed down, oil should be used on the roughing plate as a lubricant. As the work is rubbed over the lap, the abrasive powder will be washed from the grooves and act between the surface of the work and the lap. If the work has been surface-ground, rough lapping or "blocking down" is not required.

PROCEDURE

1. Place a little varsol on a finish-lapping plate which has been properly charged.
2. Place the work on top of the plate, and gently push it back and forth over the full surface of the lap using an irregular movement. *Do not dwell in one spot.*
3. Continue this movement with a light pressure until the desired surface finish is obtained.

PRECAUTIONS TO BE OBSERVED FOR FLAT LAPPING

1. Do not dwell in one area; cover the full surface of the lap.
2. Never add a fresh supply of loose abrasive. If this is required, recharge the lap.
3. Never press too hard on the work because the lap will become "stripped" in places.
4. Always keep the lap moist.

Internal Laps

Holes may be accurately finished to size and smoothness by lapping. Internal laps may be made of brass, copper, or lead and may be of three types.

The *lead lap* (Fig. 25-13A) is made by pouring lead around a tapered mandrel which has a groove cut along its length. The lap is turned to a running fit into the hole, and it is then sometimes slit on the outside to trap the loose abrasive during the lapping operation. Adjust this type by lightly tapping the large end of the mandrel on a soft block. This will cause the lead sleeve to move along the mandrel and expand.

The *internal lap* (Fig. 25-13B) may be made of copper, brass, or cast iron. A threaded-taper plug fits into the end of the lap which is slit for almost its entire length. The lap diameter may be adjusted by the threaded-taper plug.

The *adjustable lap* (Fig. 25-13C) may be made from copper or brass. The lap is split for almost its full length, but both ends remain solid. Slight adjustment is provided by means of two setscrews in the center section of the lap.

Charging and Using an Internal Lap Before charging, the lap should be a running fit in the hole.

Fig. 25-13 Various types of internal laps. (A) Lead lap; (B) copper lap; (C) adjustable lap.

1. Sprinkle some lapping powder evenly on a flat plate.
2. Roll the lap over the powder, applying sufficient pressure to embed the abrasive into the surface of the lap.
3. Remove any excess powder.
4. Mount a lathe dog on the end of the lap.
5. Fit the workpiece over the end of the lap.

NOTE: The lap should now be a wringing fit in the hole of the work and about 2.5 times the length of the workpiece.

6. Place some oil or varsol on the lap.
7. Mount the lap and the work between lathe centers.
8. Set the machine to run at a slow speed, 150 to 200 r/min for a 1 in. (25.4 mm) diameter.
9. Hold the work securely and start the machine.
10. Run the work back and forth along the entire length of the lap.
11. Remove the work and rinse in varsol to remove the abrasive and to bring it to room temperature.
12. Gage the hole for size.

NOTE: Always keep the lap moist and never add loose abrasive to the lap. This will cause the work to become bell-mouthed at the ends. If more abrasive is necessary, recharge the lap and adjust as required.

External Laps

External laps are used to finish the outside of cylindrical workpieces. They may be of several forms (Fig. 25-14); however, the basic design is the same. External laps may be made of cast iron or may have a split brass bushing mounted inside by means of a setscrew. There must be some provision for adjusting the lap.

Charging and Using an External Lap

1. Mount the workpiece in a three-jaw chuck on the lathe or drill press.
2. Adjust the lap until it is a running fit on the workpiece.
3. Grip the end of the lap in a vise.
4. Sprinkle abrasive powder in the hole.
5. With a hardened steel pin, roll the abrasive evenly around the inside surface of the lap.

Fig. 25-14   External laps.

6. Remove any excess lapping powder.

7. Place the lap on the workpiece. It should now be a wringing fit.

8. Set the machine to run at a slow speed [150 to 200 r/min for a 1 in.-diameter (25.4-mm) piece of work].

9. Add some varsol to the workpiece and the lap.

10. Hold the lap securely and start the machine.

11. Move the lap back and forth along the work.

NOTE: Always keep the lap moist.

12. In order to gage the work, remove the lap and clean the workpiece with varsol.

REVIEW QUESTIONS

HAND REAMERS

1. What is the purpose of a hand reamer?
2. Describe and state the purpose of:
   (a) the solid hand reamer
   (b) the expansion hand reamer
   (c) the taper finishing reamer
3. How much metal should be removed with a hand reamer?
4. List four important precautions which should be observed while reaming.

BROACHING

5. Define the term *broaching*.
6. Describe the cutting action of a broach.
7. State three advantages of broaching.
8. Briefly describe the procedure for broaching a keyway on an arbor press.

LAPPING

9. State three reasons for lapping.
10. What abrasives are generally used for lapping?
11. Why must the lap be softer than the workpiece?
12. Explain the procedure for charging a flat lapping plate.
13. Briefly describe the process of lapping a flat surface.
14. How are internal laps charged?

BEARINGS
UNIT 26

Bearings contribute to the smooth operation of rotating parts of machinery and motors. They are used to support and position a shaft and to reduce the friction created by the rotating part, particularly when under load. They must also be capable of absorbing and transmitting loads for the required speeds and temperatures. Bearings may be divided into two general classifications: *plain* and *rolling*.

OBJECTIVES

After completing this unit you should be able to:
1. Identify and know the purpose of the different types of bearings
2. Install a ball or roller bearing so that it operates properly

PLAIN-TYPE BEARINGS

Plain-type bearings operate on the oil film principle; that is, when rotating, the shaft is actually supported on a thin film of oil. Plain bearings are generally used in machines or components running at slower speeds. The bearing material is usually of a different type from the shaft, although great success has been claimed with hardened steel shafts rotating in hardened steel bearings at relatively high speeds. Plain bearings may be of the *solid, split,* or *thrust* type.

Solid bearings (Fig. 26-1) are of the sleeve type and often found in electric motors. They may be made of bronze, sintered bronze, or even cast iron in slowly rotating equipment. Straight sleeve bearings have no provision for adjustment and often present problems when not kept properly lubricated. Sleeve bearings on some machines have a slight taper on the outside to provide for wear adjustment. Bearings of sintered bronze are made by the powdered metallurgy process and often contain graphite between the bronze particles, which aids in the lubrication of the bearing. The porosity created by the powdered metallurgy process also provides a reservoir for oil. This tends to create a much longer lasting bearing than the plain bronze bushing. Sleeve bearings are usually made to standard sizes and are easily replaceable.

Split bearings are often used on larger machines which operate at slower speeds. They may be made of bronze, bronze with babbitt, or have a babbitt metal lining. Adjustment is usually provided for by means of laminated shims between the upper and lower bearing halves. Oil grooves chipped or machined in the bearing provide a channel for lubrication.

Thrust bearings are used to take the longitudinal thrust of a shaft. They may be flat, kidney-shaped, babbitt-faced pieces called *shoes.* They bear against a collar or collars on a shaft which rotates in oil. The bearings opererate on the oil wedge principle whereby the oil is drawn up by the rotating shaft and forms a wedge between the bearing and the collar of the shaft. These bearings may be used on large equipment where there is considerable end thrust such as on a boat propeller shaft. They prevent damage to the equipment and maintain the position of the rotating part.

ROLLING-TYPE (ANTIFRICTION) BEARINGS

Rolling bearings (ball and roller) are used in preference to plain bearings for several reasons.

1. They have a lower coefficient of friction, especially at start-up.
2. They are compact in design.
3. They have high running and dimensional accuracy.
4. The wear is not as great as on plain-type bearings.
5. They are easily replaced because of standardized sizes.

The following rolling bearing types are commonly available.

1. Radial bearings
   a. *Ball bearings*
   (1) single-row deep-groove ball bearing
   (2) double-row deep-groove ball bearing
   (3) single-row angular contact ball bearing
   (4) double-row angular contact ball bearing
   (5) self-aligning ball bearing
   b. *Roller bearings*
   (1) cylindrical roller bearing
   (2) double-row spherical roller bearing
   (3) single-row spherical roller bearing
   (4) tapered roller bearing
   (5) needle bearing

2. Thrust bearings
   (1) ball thrust bearing
   (2) cylindrical roller thrust bearing
   (3) spherical roller thrust bearing
   (4) taper roller thrust bearing

TYPES OF RADIAL BEARINGS

Ball bearings (Fig. 26-2) are widely used because of their low coefficient of friction and their suitability for high speeds. They are capable of absorbing medium to high radial and thrust loads. Ball bearings may be of single-row design for light to medium loads and of double-row design for heavy loads. There is no means of adjustment in this type of bearing.

Single-row angular contact ball bearings (Fig. 26-3A) are designed for high radial and high thrust loads in one direction. *Double-row angular contact ball bearings* (Fig. 26-3B) are designed for high radial and high thrust loads

Fig. 26-1 A plain-type sleeve bearing has no adjustment.

Fig. 26-2 Single- and double-row ball bearings.

Fig. 26-3 Angular contact bearings provide for a high radial and axial thrust. (A) Single-row bearings; (B) double-row bearings.

Fig. 26-4 Types of roller bearings. (A) Single- and double-row cylindrical; (B) double-row self-aligning; (C) taper. (Courtesy Fisher Bearing Manufacturing Limited.)

Fig. 26-5 Ball (A) and roller (B) thrust bearings.

in both directions. Both types are suitable for high speeds. When mounting single-row angular contact ball bearings, take care to have the thrust applied to the proper side of the bearing.

TYPES OF ROLLER BEARINGS

When it is necessary to support heavy loads under medium to high speeds, roller bearings are used since their basic dynamic load rating is higher than that of ball bearings of the same dimensions. The higher basic dynamic load rating is achieved by the greater area of contact between the rolling elements and the bearing raceways.

Cylindrical roller bearings (Fig. 26-4A) may be single or double row and are designed for heavy radial loads at medium to high speeds.

Double-row self-aligning roller bearings (Fig. 26-4B) consist of two rows of rollers, either spherical or barrel-shaped, revolving in a spherical outer race. They are capable of absorbing very high radial loads and moderate thrust loads in both directions. The self-aligning feature of these bearings permits shaft deflection up to 0.5° without impairing the basic dynamic load rating.

Tapered roller bearings (Fig. 26-4C) are manufactured as single row, double row, or four row and are capable of absorbing high radial and high thrust loads at moderate speeds. Provision for adjustment makes these bearings versatile. They are used extensively in machine tool manufacture.

Needle bearings employ long cylindrical rollers of small diameter. They are used where a longer bearing surface is required and where limited installation space is available. Needle bearings are designed for radial loads and moderate speeds.

TYPES OF THRUST BEARINGS

Thrust bearings consist of a set of balls or rollers held in a retainer ring between two races or washers. They are designed for heavy thrust loads and, in some instances, for combined high thrust and medium radial load.

Ball thrust bearings (Fig. 26-5A) are designed for medium to high thrust loads at moderate speeds. No radial load may be applied.

Roller thrust bearings may be of two types: cylindrical roller and spherical roller.

1. *Cylindrical roller thrust bearings* (Fig. 26-5B) are designed for high thrust load and low speed. No radial load can be applied.
2. *Spherical roller thrust bearings* are capable of absorbing very high thrust loads and moderate radial loads at low speeds. The track in the housing washer is spherical, which permits a shaft misalignment of 3°.

BEARING INSTALLATION

When a bearing must be replaced, it is advisable to select a duplicate bearing for replacement. If another make of bearing is chosen, the maker's catalogue should be consulted so that a bearing of the same specifications is used. Once the replacement bearing has been obtained, the following precautions should be observed in the installation of rolling bearings.

1. Check the shaft and housing tolerances.
2. Make sure tolerances are within the range recommended by the bearing supplier.
3. Clean the installation area and mating parts.
4. Do not unwrap a bearing until it is required for installation.
5. Do not expose a bearing to dust or dirt.
6. Do not wash a new bearing as this will remove the protective film.
7. Under no condition mount the bearing by exerting force over or through the rolling elements.

NOTE: The outer ring is generally a hand push fit into the housing while the inner ring has a light to heavy interference fit on the shaft (depending on the application).

LUBRICATION OF BALL AND ROLLER BEARINGS

Bearings operating under moderate speed and temperature conditions are generally lubricated with grease, since grease can easily be retained in the bearings and housings. It also tends to create a seal to keep out dirt and foreign matter. Oil lubrication is generally used on bearings operating at high speeds and temperatures and also in such mechanisms as closed gear trains.

Double-sealed or shielded bearings are lubricated by the manufacturer and, unless special provisions have been made, may not be relubricated.

PRECAUTIONS IN HANDLING BEARING LUBRICANTS

Because of the danger of any small particles of dirt or grit entering the bearings, grease or oil should be stored in covered containers. Grease fittings and oil caps should be wiped clean before lubricant is added. Overlubrication of ball and roller bearings must be avoided since this may

result in high operating temperatures, rapid deterioration of the lubricant, and premature bearing failure.

REVIEW QUESTIONS

1. Explain the difference between a plain-type and an antifriction-type bearing.
2. Why are antifriction bearings preferred over plain-type bearings?
3. Name three types of ball bearings and state the purpose of each.
4. Describe and state the purpose of any three types of roller bearings.
5. List three important precautions which should be observed when bearings are being installed.

PHYSICS OF
METAL CUTTING

SECTION 8

Throughout history people have continually used new developments without fully understanding the theory behind them. The cave dwellers used fire to cook food but had no knowledge of pyrology; in early battles, the boughs of trees were used to hurl large stones although we did not have a knowledge of physics or trajectory. Many hundreds of years ago, iron was mined, smelted, and forged into swords by the artisans of Damascus although metallurgy was then unknown. Without knowledge of fundamentals, we progress at a slow rate, but progress accelerates tremendously as these fundamentals are understood.

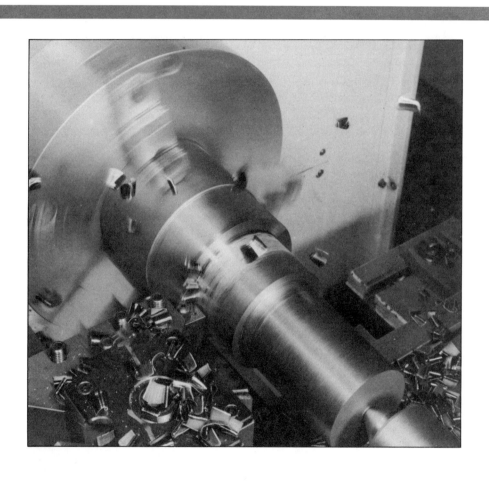

PRINCIPLES OF
METAL CUTTING
UNIT 27

We have been using tools to cut metal for hundreds of years without really understanding how the metal was cut or what was occurring where the cutting tool met the metal. For many years, it was felt that the metal ahead of the cutting tool split in a manner similar to the way that wood splits in front of an ax (Fig. 27-1). This, according to the original theory, accounted for the wear which occurred on the cutting tool face some distance away from the cutting edge. An early cutting fluid advertisement illustrated this same theory by showing the metal splitting in front of a cutting tool during a lathe-turning operation (Fig. 27-2).

OBJECTIVES

After completing this unit you should be able to:
1. Understand the various terms which apply to metal cutting
2. Understand the flow patterns of metal as it is being cut
3. Recognize the three types of chips produced from various metals

NEED FOR METAL-CUTTING RESEARCH

The manufacture of dimensionally accurate, closely fitting parts is essential to interchangeable manufacture. The accuracy and wearability of mating surfaces is directly proportional to the surface finish produced on the part. Every year in the United States alone, over 15 million tons (13.6 million tonnes) of metal are cut into chips at a cost of over $10 billion. To reduce the cost of machining, prolong the life of cutting tools, and maintain high surface finishes, it was essential for research to be done in the area of metal cutting. Since the Second World War, a great deal of research has been conducted in areas such as the theory of metal cutting, measurement of cutting forces and temperatures, machinability of metals, machining economics, and the theory of cutting fluid action. This research has found that metal of the workpiece, instead of rupturing or breaking a little before the cutting tool, is compressed and then flows up the face of the cutting tool. New cutting tools, speeds and feeds, cutting tool angles and clearances, and cutting fluids have been developed as a result of this research. All of these new developments have greatly assisted in the economical machining of metals; however, much work remains to be done before all the various factors affecting surface finish, tool life, and machine output are controlled.

METAL-CUTTING TERMINOLOGY

A number of terms resulted from the research conducted on metal cutting, and it may be wise to clearly define these terms.

A built-up edge is a layer of compressed metal from the material being cut which adheres to and piles up on the cutting tool edge during a machining operation (Fig. 27-3).

Fig. 27-1 The simile of an ax splitting wood was often used incorrectly to illustrate the action of a cutting tool.

Fig. 27-2 A false conception of metal cutting.

144

The chip-tool interface is that portion of the face of the cutting tool upon which the chip slides as it is cut from the metal (Fig. 27-3).

Crystal elongation is the distortion of the crystal structure of the work material which occurs during a machining operation (Fig. 27-4).

The deformed zone is the area in which the work material is deformed during a cutting action.

Plastic deformation is the deformation of the work material occurring in the shear zone during a cutting action (Fig. 27-4).

Plastic flow is the flow of metal occurring on the shear plane which extends from the cutting tool edge A to the corner between the chip and the work surface B in Fig. 27-4.

A rupture is the tear that occurs when brittle materials, such as cast iron, are cut and the chip breaks away from the work surface. This generally occurs when discontinuous or segmented chips are produced.

The shear angle or plane is the angle of the area of material where plastic deformation is occurring (Fig. 27-3).

The shear zone is the area where plastic deformation of the metal occurs. It is along a plane from the cutting edge of the tool to the original work surface.

PLASTIC FLOW OF METAL

In order to understand more fully what occurred to the metal while it was being deformed in the shear zone, researchers made many tests on various types of materials. Flat punches were used on ductile material to study the stress pattern, direction of material flow, and the distortion created in the metal. To observe what occurs when pressure is applied to one spot, blocks of photoelastic materials, such as celluloid and Bakelite, were used in tests. Stress lines created when pressure was exerted on the punch were observed by polarized light, through a suitable analyzer, as a series of colored bands known as *isochromatics*.

Three different types of punches (flat, narrow-faced, and knife-edge) were used on the photoelastic material to observe the various stresses created.

FLAT PUNCH

When a flat punch is forced into a block of photoelastic material, the lines of constant maximum shearing stress appear, indicating the distribution of stress. In Fig. 27-5, the shape of these stress lines, or isochromatics, appears as a family of curves almost passing through the corners of the flat punch. The greatest concentration of stress lines occurs at each corner of the punch, and larger circular stress lines appear further away from the punch. The spacing of the isochromatics is relatively wide.

NARROW-FACED PUNCH

When a narrow-faced punch is forced into a block of photoelastic material, the stress lines are still concentrated at the punch corners and where the punch meets the top surface

of the work. As can be noted in Fig. 27-6, the isochromatics are spaced closer than with the flat punch.

KNIFE-EDGE PUNCH

When a knife-edge punch is forced into the block of photoelastic material (Fig. 27-7), the isochromatics become a series of circles tangent to the two faces of the punch. In this case, the flow of material occurs upward from the point toward the free area along the faces of the punch.

When a cutting tool engages a workpiece, this flow of material takes place, and the compressed material escapes up the tool face. As the tool advances, opposition to the upward flow of material creates stresses in the material ahead of the tool due to the friction of the chip flow up the tool face. These stresses are somewhat relieved by the plastic flow or rupture of the material along a plane leading from the cutting edge of the tool to the surface of the unmachined metal. From Figs. 27-5 to 27-7 it can be concluded that internal stresses are created during metal-cutting operations.

Fig. 27-3   Chip-tool interface.   (Courtesy Cincinnati Milacron Incorporated.)

Fig. 27-4   A photomicrograph of a chip showing the crystal elongation and plastic deformation.

Fig. 27-5   The stress distribution created by a flat punch in photoelastic material. (Courtesy Cincinnati Milacron Incorporated.)

Fig. 27-6   The stress distribution created by a narrow-faced punch.   (Courtesy Cincinnati Milacron Incorporated.)

Fig. 27-7   The stress lines created by a knife-edge punch. (Courtesy Cincinnati Milacron Incorporated.)

1. Because of the forces exerted by the cutting tool, a compression of metal occurs in the material.

2. As the cutting tool or work moves forward during a cut, the stress lines concentrate at the cutting tool edge and radiate from here to the material (Fig. 27-7).

3. This concentration of stresses causes the chip to shear from the material and flow along the chip-tool interface.

4. By either plastic flow or rupture, the metal tries to flow along the chip-tool interface. Since most metals are ductile to some degree, a plastic flow generally occurs.

Whether plastic flow or rupture occurs as the metal flows along the chip-tool interface will determine the type of chip which will be produced. When brittle materials such as cast iron are being cut, the metal has a tendency to rupture and produce discontinuous or segmented chips. When relatively ductile metals are being cut, a plastic flow occurs, and continuous or flow-type chips are produced.

CHIP FORMATION

Machining operations performed on lathes, shapers, milling machines, or similar machine tools produce chips which fall into three basic types: discontinuous chip (Fig. 27-8), continuous chip (Fig. 27-9), and continuous chip with a built-up edge (Fig. 27-10).

TYPE 1—DISCONTINUOUS (SEGMENTED) CHIP

Discontinuous or segmented chips (Fig. 27-8) are produced when brittle metals such as cast iron and hard bronze or some ductile metals are cut under poor cutting conditions. As the point of the cutting tool contacts the metal, some compression occurs, as can be noted in Fig. 27-11B and C, and the chip begins flowing along the chip-tool interface. As more stress is applied to brittle metal by the cutting action, the metal compresses until it reaches a point where rupture occurs (Fig. 27-11D) and the chip separates from the unmachined portion. This cycle is repeated indefinitely during the cutting operation with the rupture of each segment occurring on the shear angle or plane. Generally, as a result of these successive ruptures, a poor surface is produced on the workpiece.

Excessive machine chatter sometimes causes discontinuous chips to be produced when ductile metal is cut.

The following conditions would be favorable to the production of a Type 1 discontinuous chip:

1. brittle work material
2. small rake angle on the cutting tool
3. large chip thickness (coarse feed)
4. low cutting speed

TYPE 2—CONTINUOUS CHIP

The Type 2 chip is a continuous ribbon produced when the flow of metal next to the tool face is not greatly retarded by a built-up edge or friction at the chip-tool interface. The continuous ribbon chip is considered ideal for efficient cutting action because it results in better surface finishes.

When ductile materials are cut, plastic flow in the metal takes place by the deformed metal sliding on a great number of crystallographic slip planes. As is the case with the Type 1 chip, fractures or ruptures do not occur because of the ductile nature of the metal.

In Fig. 27-9 it can be seen that the crystal structure of the ductile metal is elongated when it is compressed by the

Fig. 27-11 Formation of a discontinuous chip.

Fig. 27-8 A discontinuous chip. (Courtesy Cincinnati Milacron Incorporated.)

Fig. 27-9 A continuous chip. (Courtesy Cincinnati Milacron Incorporated.)

Fig. 27-10 A continuous chip with a built-up edge. (Courtesy Cincinnati Milacron Incorporated.)

PHYSICS OF METAL CUTTING

action of the cutting tool and as the chip separates from the metal. The process of chip formation occurs in a single plane extending from the cutting tool to the unmachined work surface. The area where plastic deformation of the crystal structure and shear occurs is called the *shear zone*. The angle on which the chip separates from the metal is called the *shear plane* or *shear angle*.

The mechanics of chip formation can best be understood with the aid of the schematic diagram in Fig. 27-12. As the cutting tool or work progresses, the metal immediately ahead of the cutting tool is compressed with resultant deformation or elongation of the crystal structure. This elongation does not take place in the direction of shear. As this process of compression and elongation continues, the material above the cutting edge is forced along the chip-tool interface and away from the work.

Machine steel generally forms a continuous (unbroken) chip with little or no built-up edge when machined with a cemented-carbide cutting tool. In order to reduce the amount of resistance occurring as the compressed chip slides along the chip-tool interface, a suitable rake angle is ground on the tool, and cutting fluid is used during the cutting operation. This allows the compressed chip to flow relatively freely along the chip-tool interface. A shiny layer on the back of a continuous-type chip indicates ideal cutting conditions with little resistance to chip flow.

The conditions favorable to producing a Type 2 chip are:

1. Ductile work material
2. Small chip thickness (relatively fine feeds)
3. Sharp cutting tool edge
4. A large rake angle on the cutting tool
5. High cutting speeds
6. The cutting tool and work kept cool by use of cutting fluids
7. A minimum of resistance to chip flow by:
   a. A high polish on the cutting tool face
   b. Use of cutting fluids to prevent the formation of a built-up edge
   c. Use of cutting tool materials, such as cemented carbides, which have a low coefficient of friction.

TYPE 3—CONTINUOUS CHIP WITH A BUILT-UP EDGE

Low-carbon machine steel, when cut with a high-speed steel cutting tool without the use of cutting fluids, generally produces a continuous-type chip with a built-up edge.

The metal ahead of the cutting tool is compressed and forms a chip which begins to flow along the chip-tool interface (Fig. 27-13). As a result of the high temperature, the high pressure, and the high frictional resistance against the flow of the chip along the chip-tool interface, small particles of metal begin adhering to the edge of the cutting tool while the chip shears away. As the cutting process continues, more particles adhere to the cutting tool; a larger buildup results, which affects the cutting action. The built-up edge increases in size and becomes more unstable; eventually a point is reached where fragments are torn off. Portions of these fragments which break off stick to both the chip and the workpiece. The buildup and breakdown of the built-up edge occur rapidly during a cutting action and cover the machined surface with a multitude of built-up fragments. These fragments adhere to and score the machined surface; the result is a poor surface finish.

The continuous chip with the built-up edge, as well as creating a poor surface finish on the workpiece, also shortens the cutting tool life. Two factors have an effect on the cutting tool life with this type of chip.

1. The fragments of the built-up edge cause an abrasion of the tool flank as they escape with the workpiece and chip.
2. A cratering effect is caused a short distance back from the cutting edge where the chip contacts the tool face. As this cratering continues, it eventually extends closer to the cutting edge until fracture or breakdown occurs.

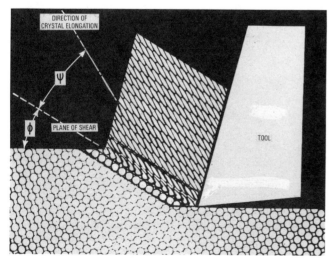

Fig. 27-12   A schematic diagram to illustrate the formation of a continuous chip and show the deformation of the crystal structure.   (Courtesy Cincinnati Milacron Incorporated.)

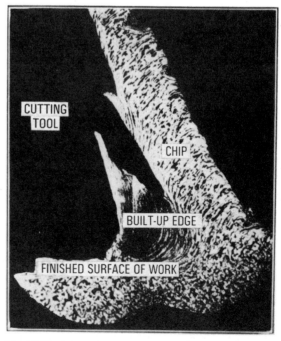

Fig. 27-13   A Type 3 continuous chip with a built-up edge being formed.   (Courtesy Cincinnati Milacron Incorporated.)

1. Explain the original theory of what occurred during a metal-cutting operation.

2. Why was extensive research carried out in the area of metal cutting?

3. Explain the new theory of what occurs when a cutting tool removes a chip from a piece of metal.

METAL-CUTTING TERMINOLOGY

4. Define the following metal-cutting terms:
   (a) built-up edge
   (b) chip-tool interface
   (c) plastic deformation
   (d) shear angle or plane

PLASTIC FLOW OF METAL

5. Why was research conducted to determine the plastic flow in metal?

6. Briefly describe what occurs when the following are forced into a block of photoelastic material:
   (a) a flat punch
   (b) a narrow-faced punch
   (c) a knife-edge punch

7. Describe the two fundamental processes which are involved in metal cutting.

CHIP FORMATION

8. Describe briefly how each of the chip types are produced:
   (a) discontinuous
   (b) continuous
   (c) continuous with a built-up edge

9. Which is the most desirable chip? Give reasons for your answer.

10. What conditions must be present in order to produce the Type 2 chip?

11. Explain how a built-up edge is formed, and state its effect on the cutting tool life.

MACHINABILITY OF METALS
UNIT 28

Machinability describes the ease or difficulty with which a metal can be machined. Such factors as cutting tool life, surface finish produced, and power required must be considered. Machinability has been measured by the length of the cutting tool life in minutes or by the rate of stock removal in relation to the cutting speed employed, that is, depth of cut. For finish cuts, machinability refers to the life of the cutting tool and the ease with which a good surface finish is produced.

OBJECTIVES

After completing this unit you should be able to:
1. Know the factors which affect the machinability of metals
2. Understand the effect of positive and negative rake angles on a cutting tool
3. Know the effects of temperature and cutting fluids on the surface finish produced

GRAIN STRUCTURE

The machinability of a metal is affected by its microstructure and will vary if the metal has been annealed. The ductility and shear strength of a metal can be modified greatly by operations such as annealing and normalizing.

Certain chemical and physical modifications of steel will improve their machinability. Free-machining steels have generally been modified in the following manner by:

1. The addition of sulfur
2. The addition of lead
3. The addition of sodium sulfite
4. Cold-working, which modifies the ductility

By making these (free-machining) modifications to the steel, three main machining characteristics become evident:

1. Tool life is increased.
2. A better surface finish is produced.
3. Lower power consumption is required for machining.

LOW-CARBON (MACHINE) STEEL

The microstructure of low-carbon steel may have large areas of ferrite (iron) interspersed with small areas of pearlite (iron) interspersed with small areas of pearlite (Fig. 28-1A and B). Ferrite is soft with high ductility and low strength, while pearlite, a combination of ferrite and iron carbide, has low ductility and high strength. When the amount of ferrite in steel is greater than pearlite or the ferrite is arranged in alternate layers with pearlite (Fig. 28-1C and D). the amount of power required to remove material increases and the surface finish produced is poor. Fig. 28-2 illustrates a more desirable microstructure in steel because the pearlite is well distributed and is therefore better for machining purposes.

HIGH-CARBON (TOOL) STEEL

A greater amount of pearlite is present in high-carbon (tool) steel because of the higher carbon content. The greater the amount of pearlite (low ductility and high strength) present in the steel, the more difficult it becomes to machine the steel efficiently. It is, therefore, desirable to anneal these steels to alter their microstructure and, as a result, improve their machining qualities.

ALLOY STEEL

Alloy steels, which are a combination of two or more metals, are generally a little more difficult to machine than low- or high-carbon steels. In order to improve their machining qualities, combinations of sulfur and lead or sulfur and manganese in proper proportions are sometimes added to alloy steels. A combination of normalizing and annealing is also used with some types of alloy steels to create desirable machining characteristics. The machining of *stainless steel,* generally difficult because of its work-hardening qualities, can be greatly eased by the addition of selenium.

CAST IRON

Cast iron, consisting generally of ferrite, iron carbide, and free carbon, form an important group of materials used by industry. The microstructure of cast iron can be controlled by the addition of alloys, the method of casting, the rate of cooling, and by heat treating. *White cast iron* (Fig. 28-3A), cooled rapidly after casting, is usually hard and brittle because of the formation of hard iron carbide. *Gray cast iron* (Fig. 28-3B) is cooled gradually; its structure is composed of compound pearlite, a mixture of fine ferrite and iron carbide, and flakes of graphite. Because of the gradual cooling, it is softer and therefore easier to machine.

Iron carbide and the presence of sand on the outer surface of the casting generally make cast iron a little difficult to machine. Through annealing, the microstructure is altered.

FERRITE   PEARLITE

PEARLITE   FERRITE

QUENCHED AND TEMPERED STRUCTURE

FERRITE   PEARLITE

Fig. 28-2A, B, Photomicrographs showing desirable microstructures in steel. (Courtesy Cincinnati Milacron Incorporated.)

FERRITE   PEARLITE

PEARLITE   FERRITE

GRAPHITE   LAMELLAR PEARLITE   PHOSPHIDE EUTECTIC

MOTTLED GRAY AND WHITE IRON

Fig. 28-1A, B, C, D Photomicrographs indicating undesirable steel microstructures. (Courtesy Cincinnati Milacron Incorporated.)

Fig. 28-3 (A) The microstructure of white cast iron; (B) the microstructure of grey cast iron. (Courtesy Cincinnati Milacron Incorporated.)

The iron carbide is broken down into graphitic carbon and ferrite; the cast iron is thus made easier to machine. The addition of silicon, sulfur, and manganese gives cast iron different qualities and improves its machinability.

CUTTING TOOL SHAPE

The shape of the cutting tool is very important to the efficient removal of metal. Every time a machine must be stopped to recondition or replace a worn cutting tool, production rates decrease. The life of a cutting tool is generally reported as:

1. The number of minutes that the tool has been cutting
2. The length cut
3. The number of cubic inches or cubic centimeters (cm^3) that has been removed
4. In the case of drills, the number of inches or millimeters of hole depth that has been drilled

To prolong cutting tool life, reduce the friction between the chip and the tool as much as possible. This can be accomplished by providing the cutting tool with a suitable rake angle and by highly polishing the cutting tool face with a honing stone. The polished cutting face reduces the friction on the chip-tool interface, reduces the size of the built-up edge, and generally results in better surface finish. The rake angle on cutting tools allows chips to flow away freely and reduces friction and the amount of power required for the machining operation.

The rake angle of the cutting tool also affects the shear angle or plane of the metal, which in turn determines the area of plastic deformation. If a *large rake angle* is ground on the cutting tool, a large shear angle is created in the metal during the cutting action (Fig. 28-4A). The results of a large shear angle are:

1. A thin chip is produced.
2. The shear zone is relatively short.
3. Less heat is created in the shear zone.
4. Good surface finish is produced.
5. Less power is required for the machining operation.

A *small or negative rake angle* on the cutting tool (Fig. 28-4B) creates a small shear angle in the metal during the cutting process with the following results:

1. A thick chip is produced.
2. The shear zone is long.
3. More heat is produced.
4. The surface finish is not quite as good as with large rake angle cutting tools.
5. More power is required for the machining operation.

Two types of rake angles are found on cutting tools:

□ *Positive rake* (Fig. 28-4A), where the point of the cutting tool and the cutting edge contact metal first and the chip moves *down the face* of the toolbit.
□ *Negative rake* (Fig. 28-4B), where the face of the cutting tool contacts the metal first and the chip is forced *up the face* of the toolbit.

Fig. 28-4 (A) Large rake angle ground in a cutting tool; (B) Negative rake angle ground on a cutting tool. (Courtesy Cincinnati Milacron Incorporated.)

Each type of rake angle serves a specific purpose. The type used depends upon the machining operation being performed and the characteristics of the work material. Rake angles can be ground on cutting tools or, in the case of cutting tool inserts, they can be held in suitable holders which provide the rake angle desired.

POSITIVE RAKE ANGLE

A positive rake angle (Fig. 28-4A) is considered best for the efficient removal of metal. It creates a large shear angle at the shear zone, reduces friction and heat, and allows the chip to flow freely along the chip-tool interface. Positive-rake-angle cutting tools are generally used for continuous cuts on ductile materials which are not too hard or abrasive in character. Even though positive-rake-angle tools remove metal efficiently, they are not recommended for all work materials or cutting applications. The following factors must be considered when the type and the amount of rake angle for a cutting tool are being determined.

1. The hardness of the metal to be cut
2. The type of cutting operation (continuous or interrupted)
3. The material and shape of the cutting tool
4. The strength of the cutting edge

NEGATIVE RAKE ANGLE

A negative rake angle (Fig. 28-4B) is used for interrupted cuts and when the metal is tough or abrasive in quality. A negative rake angle on the tool creates a small shear angle at the rear zone; therefore more friction and heat are created. Although the increase in heat may seem to be a disadvantage, it is desirable when tough metals are machined with carbide cutting tools. Face-milling cutters with carbide tool inserts are a good example of the use of negative rake for interrupted and high-speed cutting.

The advantages of negative rake on cutting tools are:

1. The shock of the work meeting the cutting tool is on the face and not the point or edge, which prolongs the life of the tool.
2. The hard outer scale on the metal does not contact the cutting edge.

3. Surfaces with interrupted cuts can be readily machined.
4. Higher cutting speeds can be employed.

The shape of a chip can be altered in a number of ways to improve the cutting action and reduce the power required. A continuous *straight* ribbon chip on a lathe can be changed to a continuous curled ribbon by:

1. Changing the *angle of the keenness* (the included angle produced by grinding the side rake and side clearance) on a toolbit.
2. A chip breaker ground behind the cutting edge of the toolbit.

A helix angle on a milling cutter affects the cutting performance by providing a shearing action when the chip is removed.

TOOL LIFE

The wear or abrasion of the cutting tool will determine its life. Three types of wear are generally associated with cutting tools: *flank wear, nose wear,* and *crater wear* (Fig. 28-5).

Flank wear is the wear occurring on the side of the cutting edge as a result of friction between the side of the cutting tool edge and the metal being machined. Too much flank wear increases friction and makes more power necessary for machining. When the flank wear is 0.015 to 0.030 in. (0.38 to 0.76 mm) long, the tool requires regrinding.

Nose wear is the wear occurring on the nose or point of the cutting tool as a result of friction between the nose and the metal being machined. Wear on the nose of the cutting tool affects the quality of the surface finish on the workpiece.

Crater wear is the wear occurring a slight distance away from the cutting edge as a result of the chips sliding along the chip-tool interface due to a built-up edge on the cutting tool. Too much crater wear eventually breaks down the cutting edge.

The following factors affect the life of a cutting tool:

1. The type of material being cut
2. The microstructure of the material

Fig. 28-5 The tool wear of a cutting tool. (1) Flank wear; (2) nose wear; (3) crater wear.

3. The hardness of the material
4. The type of surface on the metal (smooth or scaly)
5. The material of the cutting tool
6. The profile of the cutting tool
7. The type of machining operation being performed
8. Speed, feed, and depth of cut

THE EFFECTS OF TEMPERATURE AND FRICTION

In the process of cutting metals, heat is created by:

1. The plastic deformation occurring in the metal during the process of forming a chip
2. The friction created by the chips sliding along the cutting tool face

The cutting temperature varies with each type of metal and increases with the cutting speed employed and the rate of metal removal. The *greatest heat* is generated when cutting ductile material of high tensile strength, such as steel. The *lowest heat* is generated when cutting soft material of low tensile strength, such as aluminum. The maximum temperature attained during the cutting action will affect the cutting tool life, quality of the surface finish, rate of production, and accuracy of the workpiece.

At times, the temperature of metal immediately ahead of the cutting tool comes close to the melting temperature of the metal being cut. This great heat affects the life of the cutting tool. High-speed steel cutting tools cannot withstand the same high temperatures that cemented-carbide tools can without the cutting edges breaking down.

High-speed steel cutting tools are capable of cutting metal even when the cutting tools turn red due to the cutting action. This property is known as *red hardness* and occurs at temperatures above 900°F (482°C). However, when the temperature exceeds 1000°F (538°C), the edge of the cutting tool will begin to break down.

Cemented-carbide cutting tools can be used efficiently at temperatures up to 1600°F (871°C). The carbide tools are harder than high-speed steel tools but also have a greater wear-resistance. Therefore, much higher cutting speeds can be used with carbide tools than with high-speed steel tools.

FRICTION

For efficient cutting action, it is important that the friction between the chip and tool face be kept as low as possible. As the coefficient of friction increases, the more possibility there is of a built-up edge forming on the cutting edge. The larger the built-up edge, the more friction is created, which results in the breakdown of the cutting edge and poor surface finish. Every time the machine must be stopped to regrind or replace a cutting tool, production rates decrease.

The temperature created by the friction also affects the accuracy of the machined part. Even though the workpiece does not reach the same temperature as the cutting tool point, it is still high enough to cause the metal to expand.

If a part that has been heated by the cutting action is machined to size, the part will be smaller than required when it cools to room temperature. A good supply of cutting fluid will help reduce friction at the chip-tool interface and also help maintain efficient cutting temperatures.

SURFACE FINISH

Many factors affect the surface finish produced by a machining operation, the most common being the feed rate, the nose radius of the tool, the cutting speed, and the temperature generated during the machining process.

If a high temperature is created during the cutting action, there is a marked tendency for a rough surface finish to result. The reason for this is that at high temperatures metal particles tend to adhere to the cutting tool and form a built-up edge. A direct relationship between the temperature of the workpiece and the quality of the surface finish is illustrated in Fig. 28-6.

Figure 28-6A shows the results of machining a piece of aluminum without cutting fluid at 200°F (93°C). The rough surface finish indicates the presence of a built-up edge on the cutting tool. The same piece of aluminum was machined under the same conditions, but at a room temperature of 75°F (24°C) (Fig. 28-6B). A considerable improvement can be noted between the surface finishes of the samples in Figs. 28-6A and B. When the piece of aluminum was cooled to −60°F (−50°C) and machined, a further improvement in the surface finish was evident. By cooling the work material to −60°F (−50°C), the temperature of the cutting tool edge was considerably reduced and resulted in a much better surface finish than that which was produced at 200°F (93°C).

EFFECTS OF CUTTING FLUIDS

Cutting fluids are important to most machining operations in that they make it possible to cut metals at higher rates of speed. They perform two important functions.

1. They reduce the temperature of the cutting action.
2. They reduce the friction of the chips sliding along the tool face.

Some cutting fluids form a nonmetallic film on the metal surface which prevents the chip from sticking to the cutting edge. This prevents a built-up edge from forming and as a result a better surface finish is produced. The surface finish of most metals can be improved considerably by the use of the proper cutting fluids.

Cutting fluids are generally used when machining steel, alloy steel, brass, and bronze with high-speed steel cutting tools. As a rule, cutting fluids are not generally used with cemented-carbide tools unless a great quantity of cutting fluid can be applied to ensure uniform temperatures, to prevent the carbide inserts from cracking. Cast iron, aluminum, and magnesium alloys are generally machined dry;

A

B

C

Fig. 28-6 (A) Aluminum machined at 200°F (93°C); (B) aluminum machined at 75°F (24°C); (C) aluminum machined after cooling to −60°F (−50°C).

however, cutting fluids have been used with good results in some cases.

REVIEW QUESTIONS

MACHINABILITY OF METALS

1. Define *machinability*.
2. What factors affect the machinability of a metal?
3. Compare the microstructure of low-carbon and high-carbon steels with respect to their machinability.
4. How can the machining qualities of alloy steels be improved?

CUTTING TOOL SHAPE

5. Name two methods which can be used to improve the life of a cutting tool.
6. What results can be expected from grinding a large rake angle on the cutting tool?
7. How does negative rake on a tool affect the cutting process?

8. Define *flank wear, nose wear,* and *crater wear.*
9. List six important factors that affect the life of a cutting tool.

EFFECTS OF TEMPERATURE AND FRICTION

10. Name two methods by which heat is created during machining.
11. How does high temperature affect a machining operation?
12. Why is it important that friction between the chip and tool be kept to a minimum?

SURFACE FINISH

13. What common factors determine surface finish?
14. Why does high temperature affect the surface finish produced?

EFFECTS OF CUTTING FLUIDS

15. List four ways in which cutting fluids assist the machining of metals.
16. What precaution should be taken when cutting fluids are used with carbide tools?

METAL-CUTTING SAWS

SECTION 9

Archaeological discoveries show that the development of the first crude saw closely followed the origin of the stone ax and knife. The sharp edges of stones were serrated or toothed. This instrument cut by scraping away particles of the object being cut. A great improvement in the quality of saws followed the appearance of copper, bronze, and ferrous metals. With the modern steels and hardening methods of today, a wide variety of saw blades are available to suit hand hacksaws and machine power saws.

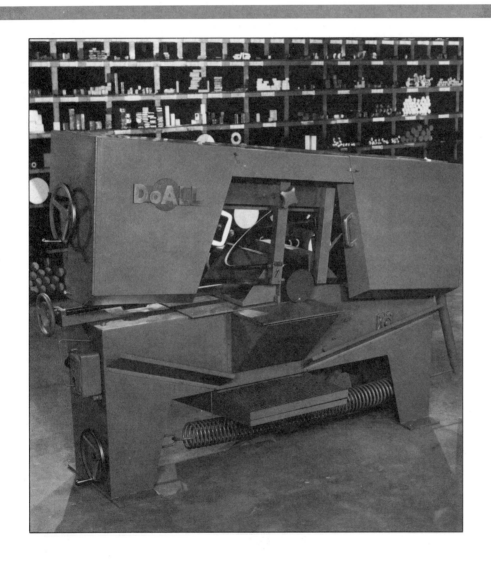

UNIT 29

Metal-cutting saws are available in a wide variety of models to suit various cutting-off operations and materials.

OBJECTIVES

After completing this unit you will be able to:
1. Name five types of cutting-off machines and state the advantage of each
2. Select the proper blade to use for cutting various cross sections
3. Install a saw band on a horizontal bandsaw
4. Use a bandsaw to cut off work to an accurate length

Five of the most common methods of cutting off material are *hacksawing, bandsawing, abrasive cutting, cold sawing,* and *friction sawing*. A brief description of each method and its advantages follows.

The *power hacksaw,* which is a reciprocating type of saw, is usually permanently mounted to the floor. The saw frame and blade travel back and forth, with pressure being applied automatically only on the forward stroke. The power hacksaw finds limited use in machine shop work since this saw cuts only on the forward stroke with resulting considerable wasted motion.

The *horizontal bandsaw* (Fig. 29-1) has a flexible, belt-like "endless" or "one-way" blade which cuts continuously in one direction. The thin continuous blade travels over the rims of two pulley wheels and passes through roller guide brackets which support the blade and keep it running true. Horizontal bandsaws are available in a wide variety of types and sizes and are becoming increasingly popular because of their high production and versatility.

The *abrasive cutoff saw* (Fig. 29-2) cuts material by a thin abrasive wheel which revolves at high speed. This type of saw is especially well suited for cutting most metals, and materials such as glass and ceramics. It can cut material to close tolerances and hardened metal does not have to be annealed in order to be cut. Abrasive cutoff can be performed under dry conditions or with a suitable cutting fluid. The use of cutting fluid keeps the work and saw cooler and produces a better surface finish.

The *cold circular cutoff saw* (Fig. 29-3) uses a circular blade which is similar to the one used on a wood-cutting table saw. The saw blade is generally made of chrome-vanadium steel, but carbide-tipped blades are used for some applications. Cold circular saws produce very accurate cuts and are especially suited for cutting aluminum, brass, copper, machine steel, and stainless steel.

Friction sawing is a burning process where a saw band, with or without saw teeth, is run at high speeds [10,000 to 25,000 surface feet per minute (sf/min)] to burn or melt its way through the metal. Friction sawing cannot be used on solid metals because of the amount of heat generated; however, it is excellent for cutting structural and honeycombed parts of machine or stainless steel.

HORIZONTAL BANDSAW PARTS

The horizontal bandsaw is the most popular machine used to cut off work. The main operative parts of this saw (Fig. 29-4) are as follows:

Fig. 29-1 A horizontal bandsaw cuts continuously in one direction.

Fig. 29-2 The abrasive cutoff saw will cut hardened metals, glass, and ceramics. (Courtesy Everett Industries, Incorporated.)

Fig. 29-3 A cold circular saw is used for cutting soft or unhardened metals. (Courtesy Everett Industries, Incorporated.)

Fig. 29-4 The main parts of a horizontal bandsaw.

The *saw frame*, hinged at the motor end, has two pulley wheels mounted on it over which the continuous blade passes.

The *step pulleys* at the motor end are used to vary the speed of the continuous blade to suit the type of material being cut.

The *roller guide brackets* provide rigidity for a section of the blade and can be adjusted to accommodate various widths of material. These brackets should be adjusted to just clear the width of the work being cut.

The *blade-tension handle* is used to adjust the tension on the saw blade. The blade should be adjusted to prevent it from wandering or twisting.

The *vise,* mounted on the table, can be adjusted to hold various sizes of workpieces. It can also be swiveled for making angular cuts on the end of a piece of material.

SAW BLADES

High-speed tungsten and high-speed molybdenum steel are commonly used in the manufacture of saw blades, and for the power hacksaw they are usually hardened completely. Flexible blades used on bandsaws have only the saw teeth hardened.

Saw blades are manufactured in various degrees of coarseness ranging from 4 to 14 pitch. When cutting large sections, use a coarse or 4-pitch blade, which provides the greatest chip clearance and helps to increase tooth penetra-tion. For cutting tool steel and thin material, a 14-pitch blade is recommended. A 10-pitch blade is recommended for general-purpose sawing. *Metric saw blades* are now available in similar sizes but in teeth per 25 mm of length rather than teeth per inch. Therefore, the pitch of a blade having 10 teeth per 25 mm would be 10 ÷ 25 mm or 0.4 mm. Always select a saw blade as coarse as possible, but make sure that *two teeth* of the blade will be in contact with the work at all times. If fewer than two teeth are in contact with the work, the work can be caught in the tooth space (gullet) which would cause the teeth of the blade to strip or break.

INSTALLING A BLADE

When replacing a blade, always make sure that the teeth are pointing in the direction of saw travel or toward the

157

motor end of the machine. The blade tension should be adjusted to prevent the blade from twisting or wandering during a cut. If it is necessary to replace a blade before a cut is finished, rotate the work one-half of a turn in the vise. This will prevent the new blade from jamming or breaking in the cut made by the worn saw.

To Install a Saw Blade

1. Loosen the blade-tension handle (Fig. 29-5).
2. Move the adjustable pulley wheel forward slightly.
3. Mount the new saw band over the two pulleys.

NOTE: *Be sure that the saw teeth are pointing toward the motor end of the machine.*

4. Place the saw blade between the rollers of the guide brackets (Fig. 29-5).
5. Tighten the blade-tension handle only enough to hold the blade on the pulleys.
6. Start and quickly stop the machine in order to make the saw blade revolve a turn or two. This will seat (track) the blade on the pulley.
7. Tighten the blade-tension handle as tightly as possible with *one hand*.

SAWING

For the most efficient sawing, it is important that the correct type and pitch of saw blade be selected and that it be run at the proper speed for the material being cut. Use finer-tooth blades when cutting thin cross sections and extra-hard materials. Coarser-tooth blades should be used for thick cross sections and material which is soft and stringy. The blade speed should suit the type and thickness of the material being cut. Too fast a blade speed or excessive feeding pressure will dull the saw teeth quickly and cause an inaccurate cut.

TO SAW WORK TO LENGTH

1. Check the solid vise jaw with a square to make sure it is at right angles to the saw blade.
2. Place the material in the vise, supporting long pieces with a floor stand (Fig. 29-6).
3. Lower the saw blade until it just clears the work. Keep it in this position by engaging the ratchet lever or by closing the hydraulic valve.
4. Adjust the roller guide brackets until they *just clear* both sides of the material to be cut (Fig. 29-5).
5. Hold a steel rule against the edge of the saw blade and move the material until the correct length is obtained.
6. Always allow 1/16 in. (1.58 mm) for each 1 in. (25.4 mm) of thickness longer than required to compensate for any *saw run-out* (slightly angular cut caused by hard spots in steel or a dull saw blade).
7. Tighten the vise and recheck the length from the blade to the end of the material to make sure that the work has not moved.

Fig. 29-5   Installing a new blade on a horizontal bandsaw.

Fig. 29-6   A floor stand is used to support long pieces while being cut.

8. Raise the saw frame slightly, release the ratchet lever or open the hydraulic valve, and then start the machine.
9. Lower the blade slowly until it just touches the work.
10. When the cut has been completed, the machine will shut off automatically.

SAWING HINTS

1. Never attempt to mount, measure, or remove work unless the saw is stopped.
2. Guard long material at both ends to prevent anyone from coming in contact with it.
3. Use cutting fluid whenever possible to help prolong the life of the saw blade.
4. When sawing thin pieces, hold the material flat in the vise to prevent the saw teeth from breaking.
5. Caution should be used when applying extra force to the saw frame, as this generally causes work to be cut out of square.

Fig. 29-7 A stop gage is used when many pieces of the same length must be cut.

6. When several pieces of the same length are required, set the stop gage which is supplied with most cutoff saws (Fig. 29-7).
7. When holding short work in a vise, be sure to place a short piece of the same thickness in the opposite end of the vise. This will prevent the vise from twisting when it is tightened (Fig. 29-8).

REVIEW QUESTIONS

1. Name and describe the cutting action of five methods of cutting off material.

HORIZONTAL BANDSAW PARTS

2. What is the purpose of the roller guide brackets?
3. How tight should the blade-tension handle be adjusted?

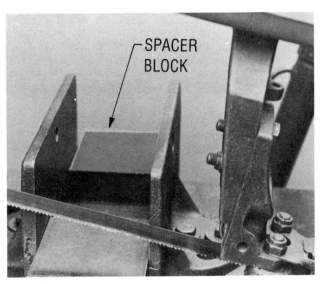

Fig. 29-8 A spacer block should be used when clamping short work in a vise.

SAW BLADES

4. Of what material are saw blades made?
5. What pitch blade is recommended for:
   (a) large sections?
   (b) thin sections?
   (c) general-purpose sawing?
6. Why should two teeth of a saw blade be in contact with the work at all times?

INSTALLING A BLADE

7. In what direction should the teeth of a saw blade point?
8. How should the workpiece be set when sawing partially cut work with a new saw blade?

SAWING

9. What may happen if too fast a blade speed is used?
10. How can a vise be checked for squareness?
11. How should the roller guide brackets be set when cutting work?
12. Explain how thin work should be held in the vise.
13. What precaution should be used when holding short work in a vise?

CONTOUR BANDSAW PARTS AND ACCESSORIES

UNIT 30

The vertical bandsaw is the latest basic machine tool to be developed. Since its development in the early 1930s, it has been widely accepted by industry as a fast and economical method of cutting metal and other materials.

Band cutoff saws differ from power hacksaws in that they have a continuous cutting action on the workpiece, while the latter cuts only on the forward stroke and on a limited section of the blade.

The contour-cutting bandsaw (vertical bandsaw) offers several features not found on other metal-cutting machines. These advantages are illustrated in Fig. 30-1.

OBJECTIVES

After completing this unit you should be able to:
1. Name the main operative parts of a contour-cutting bandsaw and state the purpose of each
2. Select the proper tooth form and set for any cutting application
3. Calculate the length of a saw band for a two-pulley machine

UNRESTRICTED MACHINING GEOMETRY

NO LIMITATION ON ANGLE, DIRECTION, OR LENGTH OF CUT

BUILT-IN TOOL HOLDER

SIMPLE FIXTURING

DOWNWARD CUTTING ACTION

CUTTING FORCE HOLDS WORK TO TABLE

HOLDS SHARPNESS

WEAR DISTRIBUTED OVER MANY TEETH

UNIFORM CHIP LOAD PER TOOTH

CONTINUOUS CUTTING

CHIP REMOVAL IS FAST AND ACCURATE

EACH TOOTH A PRECISION CUTTING TOOL

LESS POWER

NARROW TOOTH KERF

MINIMUM OF MATERIAL REDUCED TO CHIPS

LEAST MATERIAL WASTE

REMOVES WHOLE SECTIONS

CUT DIRECTLY TO FINISH LINE

Fig. 30-1   Advantages of a contour bandsaw.

CONTOUR BANDSAW PARTS

The construction of contour-band machines differs from most other machines in that the band machine is generally fabricated from steel rather than being of cast construction. The contour bandsaw consists of three basic parts: *base*, *column*, and *head* (Fig. 30-2).

BASE

The base of the contour bandsaw supports the column and houses the drive assembly which provides a drive for the saw blade.

☐ The *lower pulley* is driven by a variable-speed pulley which can be adjusted to various speeds by the *variable-speed handwheel*.
☐ The *table* is attached to the base by means of a *trunnion* (Fig. 30-3). It can be tilted 10° to the left and 45° to the right for making angular cuts by turning the *table tilt handwheel*.

☐ The *lower saw guide* is attached to the trunnion and it supports the blade to keep it from twisting.
☐ A removable *filler plate slide* and *center plate* are mounted in the table.

COLUMN

The column supports the *head*, the *left-hand blade guard*, *welding unit*, and the *variable-speed handwheel*.

☐ The *variable-speed handwheel* is used to regulate the speed of the bandsaw blade.
☐ The *blade-tension indicator* and the *speed indicator* are also located in the column.
☐ The *welding unit* is used to weld, anneal, and grind the saw blade.

HEAD

The parts found in the head of a saw are generally used to guide or support the saw band.

Fig. 30-2   A contour bandsaw provides an economical means of cutting metals to shape.

Fig. 30-3   A three-carrier-wheel contour bandsaw with table removed.   (Courtesy DoAll Company.)

□ *The upper saw pulley* supports the saw band which is adjusted by the tension and tracking controls.

□ The *upper saw guide*, attached to the saw guidepost, supports and guides the saw blade to keep it from twisting. It can be adjusted vertically to accommodate various sizes of work.

□ The *saw guard* and the *air nozzle* for keeping the area being cut free of chips are also found in the head.

Band machines may be of two types. The machine used in many toolrooms has two band carrier wheels, while the larger-capacity saws have three carrier wheels (Fig. 30-3). On the larger-capacity saws, both upper wheels may be tilted so that the band will track properly. When the blade in this type of machine becomes too short, it need not be discarded. It may be shortened to fit over the upper and lower band carrier wheels (Fig. 30-3); the capacity of the machine is reduced, but this adjustment permits the economical use of the saw blade.

BANDSAW APPLICATIONS

There are many operations which may be performed faster and easier on the contour bandsaw than on any other machine. In addition to saving time, there is also a saving in material since large sections of a workpiece can be removed as a solid instead of being reduced into chips as on conventional machines. Some of the more common operations are shown in Fig. 30-4A to F.

Notching (Fig. 30-4A) Sections of metal can be removed in one piece rather than in chips.

Slotting (Fig. 30-4B) This operation can be done quickly and accurately without the need of expensive fixtures.

Three-Dimensional Shaping (Fig. 30-4C) Complicated shapes may be cut; simply follow the layout lines.

Fig. 30-4 (A) Notching; (B) slotting; (C) three-dimensional shaping; (D) radius cutting; (E) splitting; (F) angular cutting.   (Courtesy DoAll Company.)

Radius Cutting (Fig. 30-4D)   Internal or external contours may be cut easily. Internal sections are generally removed in one piece as shown.

Splitting (Fig. 30-4E)   This can be accomplished quickly with a minimum waste of material.

Angular Cutting (Fig. 30-4F)   The work may be clamped at any angle and fed through the saw. The table may be tilted for compound angles.

COOLANTS

Some machines, particularly the power-feed models, have a cooling system which circulates and discharges coolant against the faces of the blade and work.

In fixed-table machines where coolant is required, a mist coolant system is generally employed. The mist system uses air to atomize the coolant and direct it onto the faces of the blade and the work. This method is very efficient and is recommended for the high-speed machining of non-ferrous metals such as aluminum and magnesium alloys. Tough, hard-to-machine alloys can also be cut successfully using the mist coolant system.

Grease-type lubricants and coolants may be applied directly to the blade to assist in cutting on machines having no coolant system.

POWER FEED

Some of the heavier band cutting machines are equipped with power-feed tables. The work and the table are fed toward the blade by means of a hydraulic system.

On fixed-table machines, power feeding is accomplished by means of a device which uses gravity to provide a steady mechanical feeding pressure. This allows the operator to use both hands to guide the work into the saw. The workpiece is held against a work jaw and is forced into the blade by means of cables and pulleys.

The force applied to the blade can be varied up to about 80 lb (356 N). However, for regular sawing, the feeding force of about 30 to 40 lb (133 to 178 N) should be used.

Greater feed force may be used for sawing straight lines than for cutting contours. In order to determine the feed to use for any particular job or operation, consult the job selector dial on the bandsaw.

BANDSAW BLADE TYPES AND APPLICATIONS

There are three kinds of blades commonly used in bandsawing: carbon-alloy, high-speed steel, and tungsten-carbide–tipped blades. In order to obtain the best results from any bandsaw, it is necessary to select the proper blade for the job. Consideration must be given to the kind of saw-

blade material, tooth form, pitch set, width, and gage for the material being cut.

TOOTH FORMS

Carbon and high-speed steel blades are available in three types of tooth forms (Fig. 30-5).

Precision or Regular Tooth   This is the most generally used type of tooth. It has a zero rake angle and about 30° back clearance angle. It is used when a fine finish and an accurate cut are required.

Claw or Hook Tooth   This tooth form has a positive rake on the cutting face and slightly less back clearance than the precision or buttress blade. It has the same general application as the buttress-tooth form. It is faster-cutting and longer lasting than the buttress tooth but will not produce as smooth a finish.

Buttress or Skip Tooth   This tooth form is similar to the precision tooth; however, the teeth are spaced further apart to provide more chip clearance. The tooth angles are the same as on the precision type. Buttress, or skip-tooth, blades are used to advantage on thick work sections and on deep cuts in soft material.

PITCH

Each of the tooth forms is available in various pitches, or numbers of teeth per standard reference length. Inch saw-blade pitch is determined by the number of teeth per inch; metric blade pitch by the number of teeth per 25 mm.

The thickness of the material to be cut determines the pitch of the blade to be used. When cutting thick materials, use a coarse-pitch blade; thin materials require a fine-pitch blade. It is well to remember when the proper pitch is being selected that there must be at least two teeth of the saw blade in contact with the material being cut at all times.

SET

The set of a blade is the amount that the teeth are offset on either side of the center to produce clearance for the

Fig. 30-5   Bandsaw blade tooth forms.

back of the band or blade. There are three common set patterns (Fig. 30-6).

Raker set has one tooth offset to the right, one to the left, while the third tooth is straight. This is the most common pattern and is used for most sawing applications.

Wave set has a group of teeth offset to the right and the next group to the left, a pattern which produces a wave-like appearance. Wave-set blades are generally used when the cross-section of the workpiece changes, such as on structural-steel sections or on pipe.

Straight set has one tooth offset to the right and the next to the left. It is used for cutting light nonferrous castings, thin sheet metal, tubing, and Bakelite.

WIDTH

When making straight, accurate cuts, it is advisable to select a wide blade. Narrow blades are used to cut small radii. Radius charts, advising the proper width blade to use for contour sawing, are generally found on all bandsaws. When selecting a blade for contour cutting, it is advisable to select the widest blade which can cut the smallest radius on the workpiece.

GAGE

The gage is the thickness of the saw blade. This has been standardized according to the width of the blade. Blades up to ½ in. (12.7 mm) wide are 0.025 in. (0.64 mm) thick, ⅝-in. (15.88-mm) and ¾-in. (19.05-mm) blades are 0.032 in. (0.81 mm) thick, while 1-in.-wide (25.4-mm) blades are 0.035 in. (.89 mm) thick. Since thick blades are stronger than thin blades, the thickest blade possible should be used for sawing tough material.

JOB REQUIREMENTS

The bandsaw operator should be familiar with the various types of blades and be able to select the one which will do the job to the specified requirements of finish and accuracy

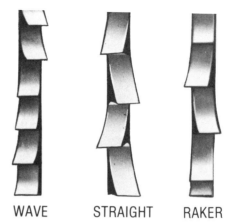

WAVE STRAIGHT RAKER

Fig. 30-6 Common set patterns. (A) Wave; (B) straight; (C) raker. (Courtesy DoAll Company.)

at the lowest cost. Table 30-1 will serve as a guide to more efficient cutting.

BLADE LENGTH

Metal-cutting saw bands are usually packaged in coils about 100 to 150 ft (30.5 to 45.7 m) in length. The length required to fit each machine is cut from the coil and its two ends are then welded together to form a continuous band.

To calculate the length required for a two-pulley bandsaw, take twice the center distance (CD) between each pulley and to it add the circumference of one pulley (PC). This is the total length of the saw band (Fig. 30-7).

EXAMPLE

Calculate the length of a saw blade for a bandsaw which has:

(a) Two 24-in.-diameter pulleys and a center to center distance of 48 in.

(b) Two 600-mm-diameter pulleys and a center-to-center distance of 1200 mm.

TABLE 30-1 Job Requirement Chart									
	TRY ONE OR MORE OF THE FOLLOWING								
TO INCREASE	FASTER TOOL VELOCITY (more teeth per min)	SLOWER TOOL VELOCITY (less teeth per min)	FINER-PITCH BAND TOOL (more teeth and smaller gullets)	COARSER-PITCH BAND TOOL (less teeth and larger gullets)	SLOWER FEEDING RATE (decrease chip load)	FASTER FEEDING RATE (increases chip load)	MEDIUM FEEDING RATE	CLAW TOOTH (positive rake angle)	PRECISION & BUTTRESS (0° rake angle)
Cutting rate	✔			✔		✔		✔	
Tool life		✔	✔				✔	✔	
Finish	✔		✔		✔				✔
Accuracy	✔				✔				

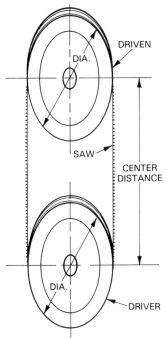

DRIVEN

DIA.

SAW

CENTER
DISTANCE

DIA.

DRIVER

Fig. 30-7 Dimensions required to calculate the length of a saw blade.

SOLUTION

(a) Blade length $= 2\ (CD) + PC$
$= 2\ (48) + (24 \times 3.1416)$
$= 96 + 75.4$
$= 171.4$ in.
(b) Blade length $= 2\ (CD) + PC$
$= 2\ (1200) + (600 \times 3.1416)$
$= 2400 + 1885$
$= 4285$ mm

REVIEW QUESTIONS

1. List six advantages of the contour bandsaw.

BANDSAW PARTS

2. How is the blade speed adjusted on a contour bandsaw?
3. What supports and guides the blade to keep it from twisting?

SAW TYPES

4. For what purpose are the following types of saw blades used?
   (a) precision
   (b) claw
   (c) buttress
5. What general rule applies when selecting the pitch of a saw blade?
6. Describe the following blade sets.
   (a) raker
   (b) wave
   (c) straight

JOB REQUIREMENTS

7. How can the following sawing factors be increased?
   (a) tool life
   (b) accuracy

BLADE LENGTH

8. Calculate the length of saw band required for the following contour bandsaws:
   (a) Two 30-in.-diameter pulleys with a center-to-center distance of 50 in.
   (b) Two 750-mm-diameter pulleys with a center-to-center distance of 1250 mm.

CONTOUR BANDSAW OPERATIONS

UNIT 31

The contour bandsaw provides a machinist with the ability to cut material close to the form required quickly while at the same time removing large sections which can be used for other jobs. The versatility of a bandsaw can be increased by using various attachments and cutting tools. Operations such as sawing, filing, polishing, grinding, and friction and high-speed sawing are all possible on a bandsaw with the proper attachments and cutting tools.

After completing this unit you should be able to:
1. Set up the machine and saw external sections to within 1/32 in. (0.8 mm) of layout lines
2. Saw internal sections to within 1/32 in. (0.8 mm) of layout lines
3. Set up a contour bandsaw to file to a layout
4. Know the purpose of special cutting tools used on a contour bandsaw

SAWING EXTERNAL SECTIONS

With the proper machine setups and attachments, a wide variety of operations can be performed on a contour bandsaw. The most common operation is that of sawing external sections. In order for any operation to be performed quickly and accurately, it is important that a person be able to select, weld, and mount the correct saw to suit the size and type of work material.

TO MOUNT A SAW BAND

1. Select and mount the proper saw inserts in the upper and lower saw guides using the proper gage for the thickness of the blade being used (Fig. 31-1). Allow 0.001- to 0.002-in. (0.02 to 0.50-mm) clearance to ensure that the blade will not bind.
2. Lower the upper saw band carrier wheel to ensure that the blade will slide over the wheels when installed.
3. Mount the blade over the upper and lower wheels with the teeth pointing down toward the table.
4. Adjust the upper wheel until some tension is registered on the blade-tension gage.
5. Set the gearshift lever in the neutral position and turn the upper saw band wheel by hand to see that the saw band rides on the center of the crown. If the saw is not tracking properly, the upper wheel must be tilted until the saw band rides on the center of the crown. When the blade is tracking properly, it should be very close to, but not touching, the back-up bearings when the saw is not cutting.
6. Re-engage the gearshift lever and close the doors on the upper and lower carrier housing.
7. Replace the filler plate in the table.
8. Lower the upper saw guide as close as possible to the work in order to ensure a straight and accurate cut.
9. Start the machine and adjust the saw band to the proper tension. This is indicated on the band-tension gage and chart.

TO CUT AN EXTERNAL SECTION

1. Study the print and check the layout on the workpiece for accuracy.
2. Check the job selector (Fig. 31-2) and determine the proper blade to use for the job. Consideration must be given to the material, thickness, type of cut (straight or curved), and the finish desired.

NOTE: When making curved cuts, it is advisable to use as wide a blade as possible.

3. Mount the proper saw blade and guides on the machine.
4. Place the work on the table and lower the upper saw guide until it just clears the work by 1/4 in. (6.35 mm). Clamp the guide in place.

NOTE: It is good practice to drill a hole at every point where sharp turns must be made to allow the workpiece to be turned easily.

5. Start the machine and see that the band is tracking properly.
6. Consult the job selector dial and set the proper speed.

SAW GUIDES

TRUNNION

Fig. 31-1   Upper and lower saw guides of bandsaw with table removed. Note table support or trunnion.

7. Place the material against the work-holding jaw or against a block of wood (Fig. 31-3).

8. Carefully bring the workpiece up to the blade and start the cut.

NOTE: When the workpiece is to be finished by some other machining operation, the cut should be made about ⅟₃₂ in. (0.794 mm) outside the layout line.

9. Carefully feed the workpiece into the saw. *Do not use too much force. Keep fingers clear of the moving blade.*

10. Saw to the layout lines.

Fig. 31-2 The job selector dial indicates the correct saw speed for the material being cut. (Courtesy DoAll Company.)

ACCURACY AND FINISH

As with any other machine, the accuracy and finish produced on a bandsaw depend upon the operator's ability to set up and operate the machine properly. Table 31-1 indicates some of the most common sawing problems which may be encountered and their possible causes.

SAWING INTERNAL SECTIONS

The bandsaw is well suited for removing internal sections of a workpiece. A starting hole must be drilled through the section to be removed to allow the saw blade to be inserted and welded. It is also good practice to drill a hole at every point where sharp turns must be made to allow the workpiece to be turned easily. The saw cut should be made close to the layout line, leaving enough material for the finishing operation.

Fig. 31-3 The work-holding jaw is used to feed the work into the saw.

TABLE 31-1 Sawing Problems and Possible Causes

PROBLEM	POSSIBLE CAUSE											
	TOO HEAVY FEED	TOO LIGHT FEED	IMPROPER BLADE TRACKING	IMPROPER BLADE TENSION	SAW GUIDES TOO FAR APART	INCORRECT SAW SPEED	PITCH TOO COARSE	PITCH TOO FINE	WRONG TYPE OF BLADE	BLADE DULL ON ONE SIDE	BLADE DULL ON BOTH SIDES	MACHINE TOO LIGHT
Blade wanders	X		X	X	X					X		
Blade not cutting		X		X	X	X		X	X		X	
Blade dulls quickly	X					X			X			
Poor finish	X			X	X	X	X		X	X	X	X
Severe diagonal waviness	X		X	X	X	X						X
Saw teeth chipping	X						X		X			
Saw teeth clogging						X		X	X			

The butt welder (Fig. 31-4) adds greatly to the versatility of the bandsaw. It permits the blade to be conveniently welded for the removal of internal sections. Blades can be cut from coil stock and welded into a continuous band, while broken blades may be welded and used again. Welders on vertical bandsaws are resistance-type welders which fuse the ends of the blade together when the correct current is applied and the blade is held properly.

In addition to knowing how to select the proper blade for the job, the operator must also be able to weld the blade. Improper welds will cause the blade to break, and it will require rewelding which is costly and time-consuming. Unless the weld is as strong as the band itself, it is not a good weld. In making a weld, the butted ends of the blade must not overlap in width, set, or pitch of the teeth.

TO WELD A BANDSAW BLADE

1. Select the proper blade for the job by checking the chart on the machine.

2. Determine the length of the blade required. The blade length for a two-wheel machine is determined by adding twice the center-to-center distance of the wheels plus the circumference of one wheel. If the upper wheel is extended the full distance, deduct 1 in. (25.4 mm) to allow for the stretch of the blade.

3. Place the blade in the cutoff shear, and cut it to the required length. Make sure that the blade is held straight and up against the blade-squaring bar on the shear.

4. If the ends of the blade are straight but not square, hold them firmly with the teeth reversed, as in Fig. 31-5, and grind both ends in one operation. If the ends are not square after grinding, it will not matter since when the blade ends are placed together for welding, the ends will match perfectly.

5. In order to get the proper tooth spacing after the blade has been welded, it is necessary to grind off some teeth at the ends of the blade to the depth of the gullet prior to welding. This is necessary since the welding operations on a DoALL welder consume about ¼ in. (6.35 mm) of the blade length. Other types of welders may vary in the blade length consumed by welding. The number of teeth that are ground off will depend on the pitch of the blade. Figure 31-6 illustrates the amount to grind off each type of blade from 4 to 10 pitch.

6. Clean the welder jaws and position the inserts for the blade size and pitch.

7. Adjust the jaw pressure by turning the selector handle for the width of the blade being welded.

8. Clamp the blade as shown in Fig. 31-7. Make sure that the blade is against the aligning surface on the back of the jaws and centered between the two jaws. If the ends of the blades do not butt against each other for the full width, they should be removed and reground.

9. Depress the welding switch or lever and hold it until the weld has cooled.

NOTE: Stand to one side and wear safety glasses to avoid injury from the welding flash.

10. Release the movable jaw clamp and then release the welding lever.

Fig. 31-4 Parts of a butt welder. (Courtesy DoAll Company.)

Fig. 31-5 Method of grinding blade ends prior to welding. (Courtesy DoAll Company.)

11. Remove the blade and check the weld for the following points.

    a. The flash material should be uniform on both sides.

    b. The spacing of the teeth should be uniform.

    c. The weld should be in the center of the gullet.

    d. The back of the blade should be straight.

If the blade does not meet all of these requirements, it should be broken, prepared again, and rewelded.

12. Move the reset lever to the anneal position and clamp the blade, with the weld in the center of the jaws and the teeth to the rear and clearing the jaws.

13. Set the selector switch at the proper setting for the

Fig. 31-6 Amounts to grind off various blades to achieve proper tooth spacing.

BLADE CLAMPED
PROPERLY FOR WELDING

Fig. 31-7 Blade clamped properly for welding.

Fig. 31-8 A thickness gage is used to check the thickness of the blade at the weld.

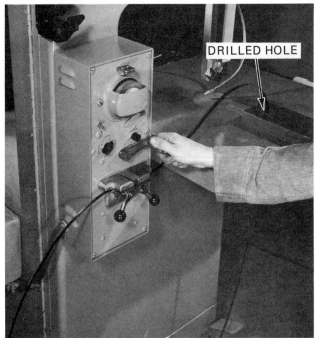

Fig. 31-9 Welding a blade for sawing internal sections.

blade to be annealed. Push in the anneal switch button and jog it intermittently until the band reaches a dull red color.

CAUTION: Do not permit the blade to become too hot at this time as it will air-harden on cooling. As the weld starts to cool, jog the anneal switch occasionally to permit the blade to cool slowly.

14. Remove the blade and grind off the welding flash. Grind the weld to the same thickness as the blade, being careful not to grind the teeth. Continually check the thickness of the blade in the thickness gage located on the welder. When properly ground, it should just slide through the thickness gage (Fig. 31-8).

15. After the blade has been ground to the proper thickness, it is advisable to anneal it to a blue color.

TO SAW INTERNAL SECTIONS

1. Drill a hole slightly larger than the width of the saw blade near the edge of the section to be cut.

NOTE: It is good practice to drill a hole at every point where sharp turns must be made to allow the workpiece to be turned easily.

2. Cut the saw blade and thread it through one of the drilled holes in the workpiece (Fig. 31-9).

3. Weld the blade and then grind off the weld bead to fit the saw thickness gage.

4. Anneal the weld section to remove the brittleness and prevent the blade from breaking.

5. Mount the saw band on the upper and lower pulleys and apply the proper tension for the size of the blade.

6. Insert the table filler plate.

7. Set the machine to the proper speed for the type and thickness of material being cut.

8. Cut out the internal section staying within $\frac{1}{32}$ in. (0.8 mm) of the layout line.

9. Remove the saw band from the pulleys.

10. Cut the blade at the weld point on the cutoff shear.

11. Remove the workpiece and the blade.

Fig. 31-10 The internal section can be used to make a die punch. (Courstey DoAll Company.)

TRIMMING AND BLANKING DIES

It is possible to use this internal cutting technique to make short-run trimming and blanking dies. By this process, the internal section, or slug, becomes the punch, while the external material forms the die (Fig. 31-10). When doing work of this type, it is necessary to tilt the table to provide the proper clearance for the die.

FRICTION SAWING

Friction sawing (Fig. 31-11) is the fastest means of sawing ferrous metals up to 1 in. (25.4 mm) in thickness. In this process, the metal is fed into a bandsaw which is traveling at a high velocity [up to 15,000 sf/min (4572 m/min)]. The tremendous heat generated by friction brings the metal immediately ahead of the saw teeth to a plastic state, and the teeth easily remove the softened metal. Since the thermal conductivity of steel is very low, the depth to which the metal is softened is only about 0.002 in. (0.50 mm).

The temperature of the saw blade remains quite low, since each tooth is only momentarily in contact with the metal and has time to cool as it travels around the carrier wheels before contacting the metal again. This method of sawing leaves a small burr which is easily removed from the workpiece.

Friction sawing is used on hardened ferrous alloys, armor plate, and parts having a thin wall or section which would be damaged by other sawing methods. Friction sawing is particularly suited to the cutting of stainless-steel alloys since this metal is difficult to cut because it work-hardens so quickly. It is not suitable for sawing most cast irons because grains break off before the metal softens. Aluminum and brass cannot be cut by friction sawing because

the high thermal conductivity of the metal does not let the heat concentrate just ahead of the blade. These metals also melt almost immediately after softening and will weld to the blade and clog the teeth. Most thermoplastics react in the same manner as the nonferrous metals.

Friction-sawing machines resemble the standard vertical band machines but are of heavier construction in the frame, bearings, spindles, and guides, to withstand the vibrations created by the high speeds required. The saw band is almost completely covered to protect the operator from the shower of sparks produced by friction sawing.

Friction-sawing bands are made of standard carbon-alloy steel but are thicker than standard blades to provide greater strength. They may be obtained in widths of ½, ¾, and 1 in. (12.7, 19.05, and 25.4 mm) and in 10 and 14 pitch with raker set only.

The teeth on a standard saw blade traveling at the high velocity required for friction sawing would dull very quickly; therefore, the teeth on friction saws are not sharpened. Since dull teeth create more friction than sharp teeth, they are more efficient for friction sawing.

The procedure for setting up a machine for friction sawing is basically the same as for conventional sawing.

HIGH-SPEED SAWING

High-speed bandsawing is performed at speeds ranging from 2000 to 6000 sf/min (609 to 1827 m/min). It is merely a standard sawing procedure performed at higher than standard speeds on nonferrous metals such as aluminum, brass, bronze, magnesium, and zinc, and other materials such as wood, plastic, and rubber.

Fig. 31-11 Friction sawing uses a high saw band speed. (Courtesy DoAll Company.)

The same machine setups and procedures apply as for conventional sawing. In high-speed sawing, the chips must be removed rapidly; consequently, buttress, or claw-tooth blades are the most efficient for this operation.

BAND FILING

When a better finish than that produced by conventional sawing is required on the edge of the workpiece, it may be produced by means of a *band file*.

The band file consists of a steel band onto which are riveted a number of short, interlocking file segments (Fig. 31-12). The ends of the band are locked together to form a continuous loop. Band files may be obtained in flat, oval, and half-round cross sections, in bastard and medium cuts, and in widths of ¼, ⅜, and ½ in. (6.35, 9.53, and 12.7 mm). Special file guides and a file-adaptor table-filler plate are used for band filing.

TO SET UP FOR BAND FILING

1. Select the proper band file for the job by consulting the job selector dial. Consideration must be given to material being filed, the shape, size, and cut of the file.
2. Set the gearshift lever into neutral position.
3. Remove the saw guides and filler plate.
4. Mount the proper file guide and backup support.
5. Lock the ends of the file blade together.
6. Mount the file band, with the teeth pointing in the proper direction, on the bandsaw carrier wheels (Fig. 31-13).

NOTE: On some makes of machines, or when filing internal sections, it may be necessary to mount the file band, and then join the ends.

Fig. 31-12 Segments of a band file.

7. Lightly add tension to the band.
8. Check the alignment and tracking of the file band.
9. Lower the upper guidepost to the proper work thickness. The distance should not exceed 2 in. (51.4 mm) for a ¼-in. (6.35-mm) file band, and 4 in. (101.6 mm) for a ⅜- or ½-in. (9.53- or 12.7-mm) band.
10. Mount the proper table-filler plate.
11. Set the gearshift lever into low gear and start the machine.
12. Adjust the band file to the proper tension.
13. Adjust the machine to the proper speed for the material being filed.

TO FILE ON A CONTOUR BANDSAW

1. Consult the job selector (Fig. 31-2) and set the machine to the proper speed. The best filing speeds are between 50 and 100 ft/min (15 and 30 m/min).
2. Apply light work pressure to the file band. It not only gives a better finish but prevents the teeth from becoming clogged.
3. Keep moving the work back and forth against the file to prevent filing grooves in the work.
4. Use a file card to keep the file clean. Loaded files cause bumpy filing and scratches in the work.

NOTE: Be sure that the machine is stopped before attempting to clean the file.

Fig. 31-13 Mounting a file band on a contour bandsaw.

171

Although the saw blade and band file are the most commonly used, several other band tools make this machine particularly versatile.

Knife-edge blades (Fig. 31-14) are available with knife, wavy, and scalloped edges, and are used for cutting soft, fibrous materials such as cloth, cardboard, cork, and rubber. Scalloped-edge blades are particularly suited for cutting thin corrugated aluminum. Special guides must be used with knife-edge blades.

Spiral-edge blades (Fig. 31-15) are round and have a continuous helical cutting edge around the circumference. This provides a cutting edge of 360° and permits the machining of intricate contours and patterns (Fig. 31-16) without the workpiece having to be turned.

Spiral-edge blades are made in two types: the spring-tempered blade which is used for plastics and wood, and the all-hard blade which is used for light metals. These blades are manufactured in diameters of 0.020, 0.040, 0.050, and 0.074 in. (0.51, 1.02, 1.27, and 1.88 mm). Special guides (Fig. 31-16) are used on the machine with this blade. When spiral blades are welded, sheet copper is used to protect the cutting edges from the welder jaws.

Line-grinding bands (Fig. 31-17) have an abrasive (either aluminum oxide or silicon carbide) bonded to the thin edge of the steel band. These bands are used to cut hardened steel alloys and other materials such as brick, marble, and glass, which could not be cut by bandsawing.

This type of machining requires a high speed—3000 to 5000 sf/min (914 to 1524 m/min)—and the use of coolant because of the heat generated. A diamond dressing stick is used to dress line-grinding bands.

Diamond-edge blades (Fig. 31-18) are used to cut superhard space-age materials, as well as ceramics, glass, silicon, and granite. This type of blade has diamond particles fused to the edges of the saw teeth. These blades operate at about 3000 sf/min (914 m/min) and generally require coolant for most efficient operation.

Although diamond-edge bands are very expensive, they will outlast 200 steel blades when cutting asbestos-cement pipe.

Polishing bands are used to remove burrs and provide a good finish to surfaces which have been sawed or filed. They may also be used for sharpening carbide toolbits. A polishing band is a continuous loop of 1-in.-wide (25.4-mm) abrasive cloth manufactured to a specific length to fit the machine. They are available in several grain sizes in both aluminum-oxide and silicon-carbide abrasive.

The polishing band is mounted in the same manner as a saw band. The special polishing guide uses the same backup support as the file bands. A special polishing-band center plate is used during band polishing. Most polishing bands are marked with an arrow on the back to indicate the direction of travel.

Electro-band machining (Fig. 31-19) is the latest development in band machining and is used to machine such

Fig. 31-14 Knife-edge blades.

Fig. 31-15 Spiral edge blades. (Courtesy DoAll Company.)

Fig. 31-16 Intricate forms may be cut with spiral edge blades. Note the special guides.

Fig. 31-17 Line grinding bands are used to cut hardened metals. (Courtesy DoAll Company.)

Fig. 31-18 Diamond edge blades are used to cut super hard materials. (Courtesy DoAll Company.)

Fig. 31-19 Cutting a heat exchanger by electro-band machining. (Courtesy DoAll Company.)

Fig. 31-20 A work-holding jaw is used to guide the work into a saw blade. (Courtesy DoAll Company.)

Fig. 31-21 The disc cutting attachment is used to cut circular shapes.

materials as thin-wall tubing, stainless steel, aluminum, and titanium honeycombing.

By this process, a low-voltage, high-amperage current is fed into the saw blade. The workpiece is connected to the opposite pole of the circuit. When the work comes close to the fast-moving band 6000 sf/min (1827 m/min), a continuous electric spark passes from the knife edge of the saw to the work. This arc acts on the material and disintegrates it. The blade does not touch the work. Coolant is flooded onto the cutting area to prevent damage to the material by the heat created. Power feed must be used in this type of machining operations.

BANDSAW ATTACHMENTS

Several standard attachments can be obtained which will increase the scope of the band machine. Some of the most common follow.

The work-holding jaw (Fig. 31-20) is a device used by the operator to hold and guide the work into the saw. Since it is usually connected to the weight-type power feed, the operator merely steers the work with the work-holding jaw and does not have to apply any feed force. *The disk-cutting attachment* (Fig. 31-21) permits the cutting of accurate circles from approximately 2.5 to 30 in. (63.5 to 760 mm) in diameter.
The cutoff and mitering attachment is used to support the work when square or angular cuts are being made.
The ripping fence provides a means for cutting long sections of flat bar stock or plate into narrow parallel sections.

When special work-holding devices are required, they are generally made in the shop to suit the specific job. These devices are usually attached to the machine table and are called *fixtures*.

REVIEW QUESTIONS

1. Name five operations which can be performed on a contour bandsaw.

SAWING EXTERNAL SECTIONS

2. Describe how to make the blade track properly.
3. Why must the upper saw guide be close to the top of the work?
4. What factors should one consider when selecting a blade for a contour cut?
5. How should the work be fed into a revolving saw?

ACCURACY AND FINISH

6. How can the following bandsawing problems be corrected?
   (a) blade wander
   (b) poor finish

SAWING INTERNAL SECTIONS

7. How is the length of a blade calculated when the upper carrier wheel is extended the full distance?
8. Make a sketch to show how the ends of the bandsaw are positioned for grinding prior to welding.
9. In order to obtain the correct tooth spacing, how many teeth should be ground off each end of a 10-pitch blade? A 14-pitch blade?
10. What are the characteristics of a good weld on a bandsaw blade?

11. List the main steps required to remove an internal section from a piece of work.

FRICTION AND HIGH-SPEED SAWING

12. Describe the principle of friction sawing.
13. Why is friction sawing particularly suited to cutting stainless steel?
14. How does high-speed sawing differ from friction sawing?

BAND FILING

15. List the main steps required to set up the machine for band filing.
16. What is the recommended filing speed?
17. How can the band file be kept clean?

ADDITIONAL BAND TOOLS

18. Name five materials that can be cut satisfactorily with knife-edge bands.
19. Name two types of spiral-edge bands and state two uses for each.
20. Describe a line-grinding band and state its purpose.
21. Name five materials which may be cut with a diamond-edge band.
22. Describe the principle of electro-band machining.

BANDSAW ATTACHMENTS

23. Describe and state the purpose of:
   (a) a work-holding jaw
   (b) a disk-cutting attachment
   (c) a mitering attachment

DRILLING MACHINES
SECTION 10

Probably one of the first mechanical devices developed prehistorically was a drill to bore holes in various materials. A bow string was wrapped around an arrow and then rapidly sawed back and forth. This process not only produced fire but it also wore a hole in the wood. The principle of a rotating tool making a hole in various materials is the one on which all drill presses operate. From this basic principle evolved the drill press—one of the most common and useful machines in industry for producing, forming, and finishing holes. Modern drilling machines are made in several types and sizes from sensitive hand-fed drill presses to the sophisticated tape-controlled production machine of the computer age.

The *drilling machine* or *drill press* is essential in any metal-working shop. Fundamentally, a drilling machine consists of a spindle (which turns the drill and which can be advanced into the work, either automatically or by hand) and a work table which holds the workpiece rigidly in position as the hole is drilled. A drilling machine is used primarily to produce holes in metal; however, operations such as tapping, reaming, counterboring, countersinking, boring, and spot-facing can also be performed.

OBJECTIVES

After completing this unit you will be able to:
1. Identify six standard operations which may be performed on a drill press
2. Identify six types of drill presses and their purposes
3. Name and state the purpose of the main parts of an upright or radial drill

STANDARD OPERATIONS

Drilling machines may be used for performing a variety of operations besides drilling a round hole. A few of the more standard operations, cutting tools, and work setups will be briefly discussed.

Drilling (Fig. 32-1A) may be defined as the operation of producing a hole by removing metal from a solid mass using a cutting tool called a *twist drill*.

Countersinking (Fig. 32-1B) is the operation of producing a tapered or cone-shaped enlargement to the end of a hole.

Reaming (Fig. 32-1C) is the operation of sizing and producing a smooth round hole from a previously drilled or bored hole with the use of a cutting tool having several cutting edges.

Boring (Fig. 32-1D) is the operation of truing and enlarging a hole by means of a single-point cutting tool which is usually held in a boring bar.

Spot-facing (Fig. 32-1E) is the operation of smoothing and squaring the surface around a hole to provide a seat for the head of a cap screw or a nut. A boring bar, with a pilot section on the end to fit into the existing hole, is generally fitted with a double-edged cutting tool. The pilot on the bar provides rigidity for the cutting tool and keeps it concentric with the hole. For the spot-facing operation, the work being machined should be securely clamped and the machine set to approximately one-quarter of the drilling speed.

Tapping (Fig. 32-1F) is the operation of cutting internal threads in a hole with a cutting tool called a tap. Special machine or gun taps are used with a tapping attachment when this operation is performed by power in a machine.

Counterboring (Fig. 32-1G) is the operation of enlarging the top of a previously drilled hole to a given depth to provide a square shoulder for the head of a bolt or capscrew.

PRINCIPAL TYPES OF DRILLING MACHINES

A wide variety of drill presses are available, ranging from the simple sensitive drill to the highly complex automatic and numerically controlled machines.

The size of a drill press may be designated in different ways by different companies. Some companies state the size as the distance from the center of the spindle to the column of the machine. Others specify the size by the diameter of the largest circular piece that can be drilled in the center.

SENSITIVE DRILL PRESSES

The simplest type of drilling machine is the sensitive drill press (Fig. 32-2). This type of machine has only a hand feed mechanism which enables the operator to "feel" how the drill is cutting and to control the downfeed pressure accordingly. Sensitive drill presses are generally light, high-speed machines and are manufactured in bench and floor models.

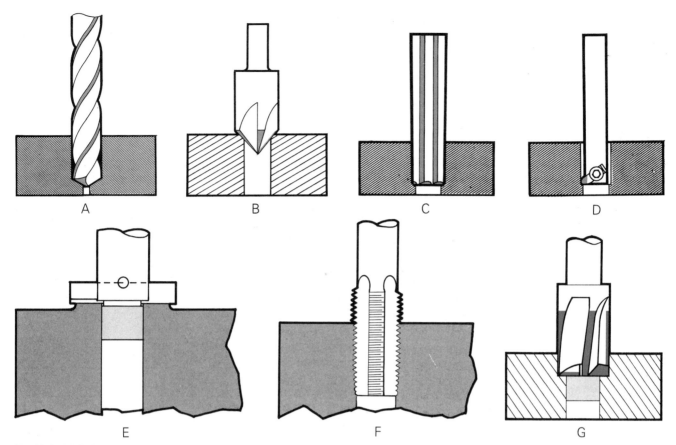

Fig. 32-1 (A) Drilling produces a straight hole; (B) countersinking produces a cone-shaped hole; (C) reaming is used to finish a hole; (D) boring is used to true and enlarge a hole; (E) Spot-facing produces a square surface; (F) tapping produces internal threads; (G) counter-boring produces square shoulders in a hole.

Fig. 32-2 A bench-type sensitive drill press.

Sensitive Drill Press Parts

Although drill presses are manufactured in a wide variety of types and sizes, all drilling machines contain certain basic parts. The main parts on the bench and floor models are *base, column, table,* and the *drilling head* (Fig. 32-2). The floor model is larger and has a longer column than the bench type.

Base The base, usually made of cast iron, provides stability for the machine and also rigid mounting for the column. The base is usually provided with holes so that it may be bolted to a table or bench. The slots or ribs in the base allow the work-holding device or the workpiece to be fastened to the base.

Column The column is an accurate cylindrical post which fits into the base. The table, which is fitted to the column, may be adjusted to any point between the base and head. The drill press head is mounted near the top of the column.

Table The table, either round or rectangular in shape, is used to support the workpiece to be machined. The table, whose surface is at 90° to the column, may be raised, lowered, and swiveled around the column. On some models it is possible to tilt the table in either direction for drilling holes on an angle. Slots are provided in most tables to allow jigs, fixtures, or large workpieces to be clamped directly to the table.

Drilling Head The head, mounted close to the top of the column, contains the mechanism which is used to revolve the cutting tool and advance it into the workpiece. The *spindle,* which is a round shaft that holds and drives the cutting tool, is housed in the *spindle sleeve* or *quill.* The spindle sleeve does not revolve but slides up and down inside the head to provide a downfeed for the cutting tool. The end of the spindle may have a tapered hole to hold taper shank tools, or may be threaded or tapered for attaching a *drill chuck* (Fig. 32-2).

The *hand feed lever* is used to control the vertical movement of the spindle sleeve and the cutting tool. A *depth stop,* attached to the spindle sleeve, can be set to control the depth that a cutting tool enters the workpiece.

UPRIGHT DRILLING MACHINE

The standard *upright drilling machine* (Fig. 32-3) is similar to the sensitive-type drill except that it is larger and heavier. The basic differences are:

1. It is equipped with a gearbox to provide a greater variety of speeds.
2. The spindle may be advanced by three methods.
   a. manually with a hand lever
   b. manually with a handwheel
   c. automatically by the feed mechanism

3. The table may be raised or lowered by means of a table-raising mechanism.
4. Some models are equipped with a reservoir in the base for the storage of coolant.

Gang drills (Fig. 32-4) are drilling machines equipped with more than one work or drilling head mounted on a single table. They are used when several operations must be performed on a single job; for example, a drill, reamer, and tap may be mounted on successive spindles so that the work may be advanced quickly from one operation to the next.

For high-speed production work, a number of spindles may be mounted on a single head. This *multispindle head* (Fig. 32-5) may incorporate 20 or more spindles on a single head driven by a drilling-machine spindle. Several heads equipped with multispindle attachments may be combined and controlled automatically to drill as many as 100 holes in a single operation. This type of automated drilling is used, for example, by the automotive industry for the drilling of engine blocks.

RADIAL DRILLING MACHINE

The *radial drilling machine* (Fig. 32-6), sometimes called a radial-arm drill, has been developed primarily for the

Fig. 32-3 A standard upright drilling machine with a square or production-type table.

Fig. 32-4 A gang drilling machine.   (Courtesy Buffalo Forge Co.)

Fig. 32-5 A multispindle drilling head.

handling of larger workpieces than is possible on upright machines. The advantages of this machine over the upright drill are:

1. Larger and heavier work may be machined.
2. The drilling head may be easily raised or lowered to accommodate various heights of work.
3. The drilling head may be moved rapidly to any desired location, while the workpiece remains clamped in one position; this feature permits greater production.
4. The machine has more power; thus, larger cutting tools can be used.
5. On universal models, the head may be swiveled so that holes can be drilled on an angle.

Radial Drilling Machine Parts

Base The base is made of heavy, box-type ribbed cast iron or of welded steel. The base, or pedestal, is used to bolt the machine to a floor or bench and also to provide a coolant reservoir. Large work may be clamped directly to the base for drilling purposes. For convenience in drilling smaller work, a *table* may be bolted to the base.

Fig. 32-6 A radial drilling machine permits large parts to be drilled.   (Courtesy Cincinnati Gilbert.)

Column The column is an upright cylindrical member fitted to the base which supports the radial arm at right angles.

Radial Arm The arm is attached to the column and may be raised and lowered by means of a *power-driven elevating screw*. The arm may also be swung about the column and may be clamped in any desired position. It also supports the drive motor and drilling head.

Drilling Head The drilling head is mounted on the arm and may be moved along the length of the arm by means of a *traverse handwheel*. The head may be clamped at any position along the arm. The head houses the change gears and controls for the spindle speeds and feeds. The drill spindle may be raised or lowered manually by means of the spindle-feed handles. When the spindle-feed handles are brought together, automatic feed is provided to the drill spindle.

NUMERICALLY CONTROLLED DRILLING MACHINE

This machine (Fig. 32-7) is a relatively new advance in production drilling. The spindle and table movements are controlled electronically by means of a punched tape. A tape reader decodes the information and passes it on to the machine tool. Here the table (and work) is positioned as required and the cutting tool needed to perform the operation is indexed.

After the speed and feed of the cutting tool have been set automatically, the machine starts and the drill or cutting tool enters the workpiece. When it has cut to the proper depth, the cutting tool will retract and another operation such as tapping or reaming may be performed. When the work cycle is complete, another workpiece is positioned under the spindle and the cycle repeated automatically with such precision that the hole positions on every workpiece will be accurate to within ±0.001 in. (0.025 mm).

REVIEW QUESTIONS

1. Define drilling, boring, reaming.
2. What is the difference between spot-facing and counterboring?

Fig. 32-7 A numerically controlled drilling machine performs automatic drilling cycles by means of punched tape. (Courtesy Cincinnati Milacron Inc.)

TYPES OF DRILLING MACHINES

3. State two methods by which the size of a drill press may be determined.
4. Compare a sensitive drill press with an upright drilling machine.
5. Describe a gang drill and state the purpose for which it is used.
6. Describe and state the purpose of the following parts of a sensitive drill press.
    (a) base
    (b) column
    (c) drilling head
    (d) table
    (e) depth stop
7. What are the advantages of a radial drilling machine?
8. Compare the construction of the radial drill press to that of a standard upright drill press.
9. Briefly describe the principle of a numerically controlled drilling machine.

DRILLING MACHINE ACCESSORIES

UNIT 33

The versatility of the drill press is greatly increased by the various accessories which are available. Drill press accessories fall into two categories:

1. *Tool-holding devices,* which are used to hold or drive the cutting tool
2. *Work-holding devices,* which are used to clamp or hold the workpiece

OBJECTIVES

After completing this unit you will be able to:
1. Identify and use three types of drill holding devices
2. Identify and use work-holding devices for drilling
3. Set up and clamp work properly for drilling

TOOL-HOLDING DEVICES

The drill press spindle provides a means of holding and driving the cutting tool. It may have a tapered hole to accommodate taper shank tools or its end may be tapered or threaded for mounting a drill chuck. Although there is a variety of tool-holding devices and accessories, the most common found in a machine shop are drill chucks, drill sleeves, and drill sockets.

DRILL CHUCKS

Drill chucks are the most common devices used on a drill press for holding straight-shank cutting tools. Most drill chucks contain three jaws that move simultaneously (all at the same time) when the outer sleeve is turned, or on some types of chucks when the outer collar is raised. The three jaws hold the straight shank of a cutting tool securely and cause it to run accurately. There are generally two common types of drill chucks: the key type and the keyless type.

CHUCKS

Straight-shank drills are held in a drill chuck. This chuck may be mounted on the drill press spindle by means of a taper (Fig. 33-1A) or a thread (Fig. 33-1B). Chucks used in larger drill presses are usually held in the spindle by means of a self-holding taper.

There are several types of drill chucks, all of which when properly used will retain their accuracy for years.

Key-type drill chucks (Fig. 33-2) are the most common. They have three jaws that move in or out simultaneously when the outer sleeve is turned. The drill is placed in the chuck and the outer sleeve turned by hand until the jaws are snug on the drill shank. The sleeve is then tightened with the key causing the drill to be held securely and accurately.

Keyless drill chucks are used more in production work

A   B

Fig. 33-1   Methods of mounting drill chucks. (A) Taper-mounted. (Courtesy The Cleveland Twist Drill Company.) (B) Thread-mounted. (Courtesy Jacobs Manufacturing Company.)

Fig. 33-2 A key-type drill chuck.

Fig. 33-3 A keyless drill chuck.

Fig. 33-5 (A) A drill sleeve; (B) a drill socket. (Courtesy The Cleveland Twist Drill Company.)

Fig. 33-4 A Jacobs impact keyless chuck. (Courtesy Jacobs Manufacturing Company.)

Fig. 33-6 Removing a taper shank drill with a drill drift. Notice the board to prevent damage to the table if the drill drops. (Courtesy Kostel Enterprises Limited.)

since the chuck may be loosened or tightened without a key.

The precision keyless chuck (Fig. 33-3) is designed to hold smaller drills accurately. The drill is changed by turning the outer knurled sleeve.

The *Jacobs impact keyless chuck* (Fig. 33-4) will hold small or large drills (within the range of the chuck) securely and accurately by means of Rubber-Flex collets. The drill is gripped or released quickly and easily by means of a built-in impact device in the chuck.

DRILL SLEEVES AND SOCKETS

The size of the tapered hole in the drill press spindle is generally in proportion to the size of the machine; the larger the machine, the larger the spindle hole. The size of the tapered shank on cutting tools is also manufactured in proportion to the size of the tool. *Drill sleeves* (Fig. 33-5A) are used to adapt the cutting tool shank to the machine spindle if the taper on the cutting tool is smaller than the tapered hole in the spindle.

A *drill socket* (Fig. 33-5B) is used when the hole in the spindle of the drill press is too small for the taper shank of

the drill. The drill is first mounted in the socket and then the socket is inserted into the drill press spindle. Drill sockets may also be used as extension sockets to provide extra length.

A flat wedge-shaped tool called a *drill drift* is used to remove tapered-shank drills or accessories from the drill press spindle. When using a drill drift (Fig. 33-6) always place the rounded edge up so that this edge will bear against the round slot in the spindle. A hammer is used to tap the drill drift and loosen the tapered drill shank in the spindle. A board or a piece of masonite should be used to protect the table in case the drill drops when it is being removed.

WORK-HOLDING DEVICES

All workpieces must be fastened securely before cutting operations are performed on a drilling machine. If the work moves or springs during drilling, the drill usually breaks. Serious accidents can be caused by work becoming loose

Fig. 33-7 A drill vise should be clamped to the table when larger holes are being drilled. (Courtesy Rockwell Tool Division Company.)

Fig. 33-8 An angle vise permits holes to be drilled at an angle in workpieces.

Fig. 33-9 A contour vise automatically adjusts to the shape of the workpiece. (Courtesy Volstro Manufacturing Company.)

A

B

Fig. 33-10 (A) V-blocks are used to hold round work for drilling. (B) A workpiece clamped in V-blocks. (Courtesy L. S. Starrett Company.)

and spinning around during a drilling operation. Some of the commonly used work-holding devices used on drill presses are listed.

A drill vise (Fig. 33-7) may be used to hold round, rectangular, square, and odd-shaped pieces for any operation that can be performed on a drill press. It is a good practice to clamp or bolt the vise to the drill table when drilling holes over ⅜ in. (9.5 mm) in diameter or to provide a table stop to prevent the vise from moving during the drilling operation.

An angle vise (Fig. 33-8) has an angular adjustment on its base to allow the operator to drill holes at an angle without tilting the drill press table.

A *contour vise* (Fig. 33-9) has special movable jaws consisting of several free-moving interlocking segments which automatically adjust to the shape of odd-shaped workpieces when the vise is tightened. These vises are valuable when one operation must be performed on many similar odd-shaped workpieces.

V-blocks (Fig. 33-10A), made out of cast iron or hardened steel, are generally used in pairs to support round work for drilling. A U-shaped strap may be used to fasten the work in a V-block, or work may be held with a T-bolt and a strap clamp (Fig. 33-10B).

Step blocks (Fig. 33-11) are used to provide support for strap clamps when work is being fastened for drilling machine operations. They are made in various sizes and steps to accommodate different work heights.

Fig. 33-11 Step blocks support the end of the clamp used to hold the work. (Courtesy Northwestern Tools, Incorporated.)

A

ANGLE PLATE

SCREW JACK

B

Fig. 33-12 (A) Work may be clamped to an angle plate for drilling; (B) a workpiece clamped to an angle plate and supported by a screw jack. (Courtesy Kostel Enterprises Limited.)

Fig. 33-13 Identical parts are drilled quickly and accurately in a drill jig.

A

B

C

Fig. 33-14 Types of clamps used to fasten work on a drill press. (A) Finger clamp; (B) U-clamp; (C) straight clamp. (Courtesy J. W. Williams & Company.)

The angle plate (Fig. 33-12A) is an L-shaped piece of cast iron or hardened steel machined to an accurate 90° angle. It is made in a variety of sizes and has slots or holes (clearance and tapped) which provide a means for fastening work for drilling. The angle plate may be bolted or clamped to the table (Fig. 33-12B).

Drill jigs (Fig. 33-13) are used in production for drilling holes in a large number of identical parts. They eliminate the need of laying out a hole location, avoid incorrectly located holes, and allow holes to be drilled quickly and accurately.

Clamps or straps (Fig. 33-14), used to fasten work to the drill table or an angle plate for drilling, are made in various sizes. They are usually supported at the end by a step block and bolted to the table by a T-bolt that fits into the table T-slot. It is good practice to place the T-bolt in the clamp or strap as close to the work as possible so that pressure will be exerted on the workpiece. Modifications of the above clamps are the *double-finger* and *gooseneck* clamps.

CLAMPING STRESSES

Whenever work is clamped for any machining operation, stresses are created. It is important that these clamping stresses should not be great enough to cause springing or distortion of the workpiece.

When work is to be held for drilling, reaming, or any machining operation, it is important that the workpiece be held securely. The clamps, bolts, and step blocks should be properly located and the work clamped firmly enough to prevent movement, yet not enough to cause the workpiece to spring or distort. It is important that the clamping pressures be applied to the work, not to the packing or step block.

Figure 33-15A illustrates the correct clamping procedure, the main pressure being applied to the workpiece.

NOTE: The step block is slightly higher than the workpiece and the bolt is close to the work.

Fig. 33-15B shows a piece of work incorrectly clamped. The bolt is located close to the step block, which is slightly lower than the workpiece. The main clamping pressure with this type of setup is applied to the step block, not the workpiece.

CLAMPING HINTS

The following suggestions should be observed when work is being clamped so that good clamping pressure will be obtained and work distortion avoided.

1. Always place the bolt as close as possible to the workpiece (Fig. 33-16).
2. Have the packing or step block slightly higher than the work surface being clamped.
3. Insert a piece of paper between the machine table and the workpiece to prevent the work from shifting during the machining process.
4. Place a metal shim between the clamp and workpiece to spread the clamping force over a wider area.
5. It is wise to use a sub-base or liner under a rough casting to prevent damage to the machine table.
6. Parts which do not lie flat on a machine table should be shimmed to prevent the work from rocking. This will prevent distortion when the work is clamped.

A

B

Fig. 33-15   (A) Work is correctly when the bolt is close to the work; (B) Work incorrectly clamped; the clamping pressure is on the step block.

Fig. 33-16   The bolts and the packing blocks correctly placed for maximum clamping pressure.

REVIEW QUESTIONS

TOOL-HOLDING DEVICES

1. What is the purpose of a drill chuck?
2. How may chucks be secured to the spindle in
   (a) smaller drill presses?
   (b) large drill presses?
3. Name three types of drill chucks
4. What is the purpose of a
   (a) drill sleeve?
   (b) drill socket?
5. How is a tapered shank drill removed from a drill press spindle?

6. Name three types of drill vises and state the purpose of each.
7. What is the purpose of
 (a) V-blocks?   (b) step blocks?
8. Describe an angle plate and state its purpose.
9. What is the advantage of a drill jig?

10. Why is it important that work be clamped properly for any machining operation?
11. Explain the procedure for clamping a workpiece properly.
12. List four important clamping hints.

TWIST DRILLS
UNIT 34

Twist drills are end-cutting tools used to produce holes in most types of material. On standard drills, two helical grooves, or flutes, are cut lengthwise around the body of the drill. They provide cutting edges and space for the cuttings to escape in the drilling process. Since drills are one of the most efficient cutting tools, it is necessary to know the main parts, how to sharpen the cutting edges, and the correct speeds and feeds for drilling various metals in order to use them most efficiently and prolong their life.

OBJECTIVES

After completing this unit you will be able to:
1. Identify the parts of a twist drill
2. Identify four systems of drill sizes and know where each is used
3. Grind the proper angles and clearances on a twist drill

TWIST DRILL PARTS

Most twist drills used in machine shop work today are made of high-speed steel. High-speed steel drills have replaced carbon-steel drills since they can be operated at double the cutting speed and the cutting edge lasts longer. High-speed steel drills are always stamped with the letters ''H.S.'' or ''H.S.S.'' Since the introduction of *carbide-tipped drills*, speeds for production drilling have increased up to 300 percent over high-speed steel drills. Carbide drills have made it possible to drill certain materials that would not be possible with high-speed steels.

A drill (Fig. 34-1) may be divided into three main parts: *shank, body,* and *point*.

SHANK

Generally drills up to ½ in. or 12.7 mm in diameter have straight shanks, while those over this diameter usually have tapered shanks. *Straight-shank* drills (Fig. 34-2A) are held in a drill chuck; *tapered-shank* drills (Fig. 34-2B) fit into the internal taper of the drill press spindle. A tang is pro-

vided on the end of tapered-shank drills to prevent the drill from slipping while it is cutting and to allow the drill to be removed from the spindle or socket without the shank being damaged.

BODY

The *body* is the portion of the drill between the shank and the point. It consists of a number of parts important to the efficiency of the cutting action.

1. The *flutes* are two or more helical grooves cut around the body of the drill. They form the cutting edges, admit cutting fluid, and allow the chips to escape from the hole.
2. The *margin* is the narrow, raised section on the body of the drill. It is immediately next to the flutes and extends along the entire length of the flutes. Its purpose is to provide a full size to the drill body and cutting edges.
3. The *body clearance* is the undercut portion of the body between the margin and the flutes. It is made smaller to reduce friction between the drill and the hole during the drilling operation.

Fig. 34-1 The main parts of twist drill. (Courtesy The Cleveland Twist Drill Company.)

A

B

Fig. 34-2 Types of drill shanks. (A) Straight; (B) Tapered. (Courtesy National Twist Drill & Tool Company.)

4. The *web* (Fig. 34-3) is the thin partition in the center of the drill which extends the full length of the flutes. This part forms the chisel edge at the cutting end of the drill (Fig. 34-4). The web gradually increases in thickness toward the shank to give the drill strength.

POINT

The *point* of a twist drill (Fig. 34-4) consists of the chisel edge, lips, lip clearance, and heel. The *chisel edge* is the chisel-shaped portion of the drill point. The *lips* (cutting

Fig. 34-3 The web is the tapered metal column which separates the flutes. (Courtesy The Cleveland Twist Drill Company.)

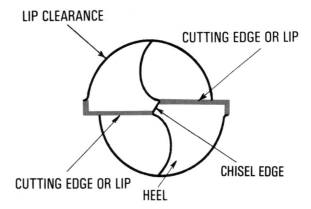

Fig. 34-4 The point of the twist drill.

Fig. 34-5 The lip clearance angle of the cutting edge should be 8 to 12°.

edges) are formed by the intersection of the flutes. The lips must be of equal length and have the same angle so that the drill will run true and will not cut a hole larger than the size of the drill.

The *lip clearance* is the relief which is ground on the point of the drill extending from the cutting lips back to the *heel* (Fig. 34-5). The average lip clearance is from 8 to 12°, depending upon the hardness or softness of the material to be drilled.

DRILL POINT CHARACTERISTICS

Efficient drilling of the wide variety of materials used by industry requires a great variety of drill points. The most important factors determining the size of the drilled hole are the characteristics of the drill point.

A drill is generally considered a roughing tool capable of removing metal quickly. It is not expected to finish a hole to the accuracy possible with a reamer. However, a drill can often be made to cut more accurately and efficiently by proper drill point grinding. The use of various point angles and lip clearances, in conjunction with the thinning of the drill web, will:

1. Control the size, quality, and straightness of the drilled hole
2. Control the size, shape, and formation of the chip
3. Control the chip flow up the flutes
4. Increase the strength of the drill's cutting edges
5. Reduce the rate of wear at the cutting edges
6. Reduce the amount of drilling pressure required
7. Control the amount of burr produced during drilling
8. Reduce the amount of heat generated
9. Permit the use of various speeds and feeds for more efficient drilling

DRILL POINT ANGLES AND CLEARANCES

Drill point angles and clearances are varied to suit the wide variety of material which must be drilled. Three general drill points are commonly used to drill various materials; however, there may be variations of these to suit various drilling conditions.

Conventional point (118°) (Fig. 34-6A). This is the most commonly used drill point and gives satisfactory results for most general-purpose drilling. The 118° point angle should be ground with 8 to 12° lip clearance for best results.

Too much lip clearance weakens the cutting edge and causes the drill to chip and break easily. Too little lip clearance results in the use of heavy drilling pressure; this pressure causes the cutting edges to wear quickly because of the excessive heat generated and also places undue strain on the drill and equipment.

Long angle point (60 to 90°) (Fig. 34-6B). The long angle drill point is commonly used on low helix drills for the drilling of nonferrous metals, soft cast irons, plastics, fibers. wood, etc. The lip clearance on long angle point drills is generally from 12 to 15°. On standard drills, a flat may be ground on the face of the lips to prevent the drill from drawing itself into the soft material.

Flat angle point (135 to 150°) (Fig. 34-6C). The flat angle drill point is generally used to drill hard and tough materials. The lip clearance on flat angle point drills is generally only 6 to 8° to provide as much support as possible for the cutting edges. The shorter cutting edge tends to reduce the friction and heat generated during drilling.

Fig. 34-6 (A) The drill point angle of 118° is suitable for most general work; (B) a drill point angle of 60 to 90° is used for soft materials; (C) a drill point angle of 135 to 150° is best for hard materials.

Fig. 34-7 Checking a drill for size using a drill gage.

SYSTEMS OF DRILL SIZES

Drill sizes are designated under four systems: fractional, number, letter, and millimeter (metric) sizes.

The *fractional* size drills range from 1/64 to 3 1/4 in. varying in steps of 1/64 in. from one size to the next.

The *number* size drills range from #1, measuring 0.228 in. to #97 which measures 0.0059 in.

The *letter* size drills range from A to Z. Letter-A drill is the smallest in the set (0.234 in.) and Z is the largest (0.413 in.).

Millimeter (metric) drills are produced in a wide variety of sizes. Miniature metric drills range from 0.04 to 0.09 mm in steps of 0.01 mm. Straight-shank standard metric drills are available in sizes from 0.5 to 20 mm. Taper-shank metric drills are manufactured in sizes from 8 up to 80 mm.

Drill sizes may be checked by using a drill gage (Fig. 34-7). These gages are available in fractional, letter, number, and millimeter sizes. The size of a drill may also be checked by measuring the drill, *over the margins,* with a micrometer (Fig. 34-8).

SPECIAL DRILLS

The most commonly used drill is the two-fluted twist drill; however, for high-production and special purposes, various types of drills are used. A few of the more common types of special drills and their uses are listed.

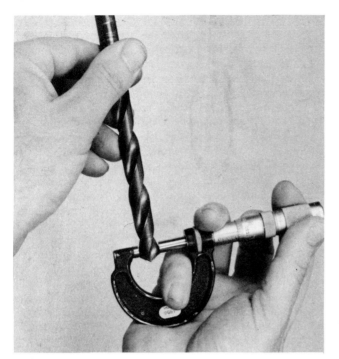

Fig. 34-8 Checking the size of a twist drill using a micrometer.

Fig. 34-9 A high helix drill. (Courtesy The Cleveland Twist Drill Company.)

Fig. 34-10 A core drill.

Fig. 34-11 An oil hole drill.

Fig. 34-12 A straight-fluted drill.

Fig. 34-13 Deep hole or gun drill. (Courtesy The Cleveland Twist Drill Company.)

Fig. 34-14 A spade drill. (Courtesy DoAll Company.)

High helix drills (Fig. 34-9) are designed for drilling deep holes in aluminum, copper, die-cast material, and other metals where the chips have a tendency to jam in a hole. The high helix angle (35 to 40°) and the wider flutes of these drills assist in clearing chips from the hole.

A core drill (Fig. 34-10), designed with three or four flutes, is used primarily to enlarge cored, drilled, or punched holes. This drill has advantages over the two-fluted drills in productivity and finish. In some cases, a core drill may be used in place of a reamer for finishing a hole. Core drills are produced in sizes from ¼ to 3 in. (6.35 to 76.2 mm) in diameter.

Oil hole drills (Fig. 34-11) have one or two oil holes running from the shank to the cutting point through which compressed air, oil, or cutting fluid can be forced when deep holes are being drilled. These drills are generally used on turret lathes and screw machines. The cutting fluid flowing through the oil holes cool the drill's cutting edges and flushes the chips out of the hole.

Straight-fluted drills (Fig. 34-12) are recommended for drilling operations on soft materials such as brass, bronze, copper, and various types of plastic. The straight flute prevents the drill from drawing itself into the material (digging in) while cutting.

If a straight-fluted drill is not available, a conventional drill can be modified by grinding a small flat approximately ¹⁄₁₆ in. (1.58 mm) wide on the face of both cutting edges of the drill.

NOTE: The flat must be ground parallel to the axis of the drill.

Deep hole or gun drills (Fig. 34-13) are used for producing holes from approximately ⅜ to 3 in. (9.52 to 76.2 mm) in diameter and as deep as 20 ft (6.09 m). The most common gun drill consists of a round, tubular stem, on the end of which is fastened a flat, two-fluted drilling insert. Cutting fluid is forced through the center of the stem to flush the chips from the hole. When the drilling insert becomes dull, it can be replaced quickly by loosening one screw which holds it to the tubular stem.

Spade drills are similar to gun drills in that the cutting end is a flat blade with two cutting lips. Spade drills are usually clamped in a holder (Fig. 34-14) and are easily replaced or sharpened. Spade drills are available in a wide range of sizes from very small microdrills to drills up to 12 in. in diameter. Some of the smaller spade drills have replaceable carbide inserts.

Fig. 34-15   A hard-steel drill.   (Courtesy DoAll Company.)

Fig. 34-16   A step drill.   (Courtesy National Twist Drill & Tool Company.)

Fig. 34-17   A saw-type hole cutter.   (Courtesy L. S. Starrett Company.)

A rather unique drill is the *hard-steel drill* (Fig. 34-15) used for drilling hardened steel. These drills are made from a heat-resistant alloy. As the drill is brought into contact with the workpiece, the fluted, triangular point softens the metal by friction and then removes the softened metal ahead of it, in chip form.

Step drills (Fig. 34-16) are used to drill and countersink or drill and counterbore different sizes of holes in one operation. The drill may have two or more diameters ground on it. Each size or step may be separated by a square or angular shoulder, depending on the purpose of the hole. For example, holes to be tapped should have a slight chamfer at the top of the hole to ease the starting of the tap, protect the thread, and leave the tapped hole free of burrs caused by the tap.

A saw-type hole cutter (Fig. 34-17) is a cylindrical diameter cutter with a twist drill in the center to provide a guide for cutting teeth on the hole cutter. This type of cutter is made in various diameters and is generally used for drilling holes in thin materials. It is especially valuable for drilling holes in pipe and sheet metal as little burr is produced and the cutter does not have a tendency to jam as it is breaking through.

DRILLING FACTS AND PROBLEMS

The most common drill problems encountered are illustrated in Fig. 34-18A and B. It is wise to study these various problems to ensure that the amount of drill breakage, regrinding, and downtime is kept to a minimum.

EXCESSIVE SPEED WILL CAUSE WEAR AT OUTER CORNERS OF DRILL. THIS PERMITS FEWER REGRINDS OF DRILL DUE TO AMOUNT OF STOCK TO BE REMOVED IN RECONDITIONING. DISCOLORATION IS WARNING SIGN OF EXCESSIVE SPEED.

EXCESSIVE CLEARANCE RESULTS IN LACK OF SUPPORT BEHIND CUTTING EDGE WITH QUICK DULLING AND POOR TOOL LIFE DESPITE INITIAL FREE CUTTING ACTION. CLEARANCE ANGLE BEHIND CUTTING LIP FOR GENERAL PURPOSES 8 TO 12°

LIP CLEARANCE ANGLE

A — CUTTING LIP LINE
B — HEEL LINE

A
B

AMOUNT OF GRINDING NECESSARY TO REPOINT.
THE USE OF MACHINE-POINT GRINDING IS RECOMMENDED.

INSUFFICIENT CLEARANCE CAUSES THE DRILL TO RUB BEHIND THE CUTTING EDGE. IT WILL MAKE THE DRILL WORK HARD, GENERATE HEAT AND INCREASE END THRUST. THIS RESULTS IN POOR HOLES AND DRILL BREAKAGE.

EXCESSIVE FEED SETS UP ABNORMAL END THRUST WHICH CAUSES BREAKDOWN OF CHISEL POINT AND CUTTING LIPS. FAILURE INDUCED BY THIS CAUSE WILL BE BROKEN OR SPLIT DRILL.

LIP CLEARANCE ANGLE

A — CUTTING LIP LINE
B — HEEL LINE

A
B

Fig. 34-18A   Facts about drills and drilling.   (Courtesy Greenfield Tap & Die Company.)

DRILL GRINDING

The cutting efficiency of a drill is determined by the characteristics and condition of the point of the drill. Most new drills are provided with a general-purpose point (118° point angle and an 8 to 12° lip clearance). As a drill is used, the cutting edges may wear and become chipped or the drill may break. Drills are generally resharpened by hand. However, small drill point grinders or drill-sharpening attachments are inexpensive, readily available, and provide more consistent quality than hand grinding.

To ensure that a drill will perform properly, it is rec-

IMPROPER WEB THINNING IS THE RESULT OF TAKING MORE STOCK FROM ONE CUTTING EDGE THAN FROM THE OTHER, THEREBY DESTROYING THE CONCENTRICITY OF THE WEB AND OUTSIDE DIAMETER.

THE WEB IS THE TAPERED CENTRAL PORTION OF THE BODY THAT JOINS THE LANDS.

WRONG

RIGHT

CUTTING LIPS WITH UNEQUAL ANGLES WILL CAUSE ONE CUTTING EDGE TO WORK HARDER THAN THE OTHER. THIS CAUSES TORSION STRAIN, BELLMOUTH HOLES, RAPID DULLING, POOR TOOL LIFE.

64°
54°

CUTTING LIPS UNEQUAL IN LENGTH CAUSE CHISEL POINT TO BE OFF CENTRE WITH AXIS AND WILL DRILL HOLES OVERSIZE BY APPROXIMATELY TWICE THE AMOUNT OF ECCENTRICITY.

FAST SPIRAL DRILL

REGULAR DRILL

LOADING AND GALLING IS CAUSED BY POOR CHIP REMOVAL WITH INSUFFICIENT DISSIPATION OF HEAT SO THAT MATERIAL ANNEALS ITSELF TO THE CUTTING EDGE AND FLUTE. THIS CONDITION FREQUENTLY RESULTS FROM USING WRONG DRILLS FOR THE JOB OR INADEQUATE CUTTING FLUID APPLICATION.

Fig. 34-18B   Facts about drills and drilling.   (Courtesy Greenfield Tap & Die Company.)

Fig. 34-19 (A) An incorrect point with lips of unequal lengths; (B) lips with unequal angles produce oversize holes. (Courtesy The Cleveland Twist Drill Company.)

Fig. 34-20 The point angle for a general-purpose drill is 118°. (Courtesy The Cleveland Twist Drill Company.)

ommended that the drill point be examined carefully before the drill is mounted in the drill press. A properly ground drill should have the following characteristics:

☐ The length of both cutting lips should be the same.
— Lips of unequal length will force the drill point off center, causing one lip to do more cutting than the other, and produce an oversize hole (Fig. 34-19A).
☐ The angle of both lips should be the same.
— If the angles are unequal, the drill will cut an oversize hole because one lip will do more cutting than the other (Fig. 34-19B).
☐ The lips should be free from nicks or wear.
☐ There should be no sign of wear on the margin.

If the drill does not meet all of these requirements, it should be resharpened. If the drill is not resharpened, it will give poor service, produce inaccurate holes, and may break due to excessive drilling strain.

While a drill is being used, there will be signs to indicate that the drill is not cutting properly and should be resharpened. If the drill is not sharpened at the first sign of dullness, it will require extra power to force the slightly dulled drill into the work. This causes more heat to be generated at the cutting lips and results in a faster rate of wear. When any of the following conditions arise while a drill is in use, it should be examined and reground.

☐ The color and shape of the chips change.
☐ More drilling pressure is required to force the drill into the work.
☐ The drill turns blue due to the excessive heat generated while drilling.
☐ The top of the hole is out of round.
☐ A poor finish is produced in the hole.
☐ The drill chatters when it contacts the metal.
☐ The drill squeals and may jam in the hole.
☐ There is an excessive burr left around the drilled hole.

CAUSES OF DRILL FAILURE

Drills should not be allowed to become so dull that they cannot cut. Overdulling of any metal-cutting tool generally results in poor production rates, inaccurate work, and the shortening of the tool life. Premature dulling of a drill may be caused by any one of a number of factors.

☐ The drill speed may be too high for the hardness of the material being cut.

☐ The feed may be too heavy and overload the cutting lips.
☐ The feed may be too light and cause the lips to scrape rather than cut.
☐ There may be hard spots or scale on the work surface.
☐ The work or drill may not be supported properly, resulting in springing and chatter.
☐ The drill point is incorrect for the material being drilled.
☐ The finish on the lips is poor.

To Grind a Drill

A general-purpose drill has an included point angle of 118° and a lip clearance of from 8 to 12° (Fig. 34-20A and B).

1. Be sure to wear approved safety glasses.
2. Check the grinding wheel and dress it, if necessary, to sharpen and/or straighten the wheel face.
3. Adjust the grinder tool rest so that it is within 1/16 in. (1.59 mm) of the wheel face.
4. Examine the drill point and the margins for wear. If there is any wear on the margins, it will be necessary to grind the point of the drill back until all margin wear has been removed.
5. Hold the drill near the point with one hand, and with the other hand hold the shank of the drill slightly lower than the point (Fig. 34-21).
6. Move the drill so that it is approximately 59° to the face of the grinding wheel (Fig. 34-22).

NOTE: A line scribed on the toolrest at 59° to the wheel-face will assist in holding the drill at the proper angle.

7. Hold the lip or cutting edge of the drill parallel to the grinder toolrest.
8. Bring the lip of the drill against the grinding wheel and slowly lower the drill shank. DO NOT TWIST THE DRILL.
9. Remove the drill from the wheel without moving the position of the body or hands, rotate the drill one-half turn, and grind the other cutting edge.
10. Check the angle of the drill point and length of the lips with a drill point gage (Fig. 34-23).
11. Repeat operations 6 to 10 until the cutting edges are sharp and the lands are free from wear and nicks.

WEB THINNING

Most drills are manufactured with webs that gradually increase in thickness toward the shank to give the drill strength. As the drill becomes shorter, the web becomes

Fig. 34-21 To provide lip clearance, lower the shank of the drill when grinding.

LINE SCRIBED AT 59°

Fig. 34-22 The drill is held at 59° to the face of the grinding wheel.

Fig. 34-23 Checking the drill point angle with a drill point gage.

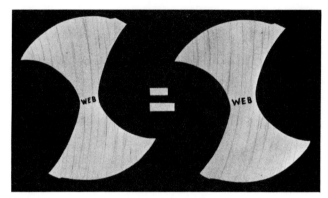

Fig. 34-24 Notice the difference in web thickness between the drill cross section near the shank (right) and the point (left). (Courtesy The Cleveland Twist Drill Company.)

thicker (Fig. 34-24) and more pressure is required in order to cut. This increase in pressure results in more heat, which shortens the drill life. To reduce the amount of drilling pressure and resultant heat, the web of a drill is generally thinned. Webs can be thinned on a special web-thinning grinder, on a tool and cutter grinder, or freehand on a conventional grinder. It is important when thinning a web to grind equal amounts off each edge; otherwise the drill point will be off center (Fig. 34-19A).

REVIEW QUESTIONS

TWIST DRILLS

1. Name three materials used to manufacture drills and state the advantage of any two.
2. Define the body, web, and point of a twist drill.
3. State the purpose or purposes of each of the following parts.
   - (a) tang
   - (b) flutes
   - (c) margin
   - (d) body clearance

DRILL POINT CHARACTERISTICS

4. Why are various drill points and clearances used for drilling operations?

5. Describe and state the purpose of the following drill points:
   - (a) conventional
   - (b) long angle
   - (c) flat angle

6. Why is it necessary to thin the web of a drill?

DRILL SIZES

7. List four systems of drill sizes and give the range of each.
8. State the purpose of the
   - (a) high helix drill
   - (b) core drill
   - (c) oil hole drill
   - (d) straight-fluted drill
   - (e) gun drill
   - (f) hard-steel drill

DRILLING FACTS AND PROBLEMS

9. What problems generally result from the use of excessive speed and excessive feed?
10. Discuss excessive lip clearance and insufficient lip clearance.
11. What is the effect of:
    - (a) drills with unequal angles on the cutting lips?
    - (b) drills with cutting lips of unequal length?

DRILL GRINDING

12. What are the characteristics of a properly ground drill?
13. List the main steps in grinding a general-purpose drill.

CUTTING SPEEDS AND FEEDS
UNIT 35

The most important factors which the operator must consider when selecting the proper speeds and feeds are the diameter and material of the cutting tool, and the type of material being cut. These factors will determine what speeds and feeds should be used and therefore will affect the amount of time it takes to perform the operation. Production time will be wasted if the speed and/or feed are set too low, while the cutting tool will show premature wear if the speed and/or feed are set too high. The ideal speed and feed for any job is where a combination of the best production rate and the best tool life is attained.

OBJECTIVES

After completing this unit you will be able to:
1. Calculate the revolutions per minute (r/min) for inch and metric size drills
2. Select the feed to be used for various operations
3. Calculate the revolutions per minute for the reaming operation

CUTTING SPEED

The speed of a twist drill is generally referred to as *cutting speed, surface speed,* or *peripheral speed*. It is the distance that a point on the circumference of a drill will travel in 1 min.

A wide range of drills and drill sizes is used to cut various metals; an equally wide range of speeds is required for the drill to cut efficiently. For every job, there is the problem of choosing the drill speed which will result in the best production rates and the least amount of downtime for regrinding the drill. The recommended cutting speeds for drilling various types of materials may be found in Table 35-1. The most economical drilling speed depends upon many variables such as:

1. The type and hardness of the material
2. The diameter and material of the drill
3. The depth of the hole
4. The type and condition of the drill press
5. The efficiency of the cutting fluid employed
6. The accuracy and quality of the hole required
7. The rigidity of the work setup

Although all these factors are important in the selection of economical drilling speeds, the type of work material and the diameter of the drill are the most important.

When reference is made to the speed at which a drill should revolve, the cutting speed of the material in *surface feet per minute* (sf/min) or *meters per minute* (m/min) is implied unless otherwise stated. The number of revolutions of the drill necessary to attain the proper cutting speed for the metal being machined is called the *revolutions per minute* (r/min). A small drill operating at the same r/min as a larger drill will travel fewer feet per minute; it naturally would cut more efficiently at a higher number of r/min.

REVOLUTIONS PER MINUTE

To determine the correct number of r/min of a drill press spindle for a given size drill, the following should be known.

1. The type of material to be drilled
2. The recommended cutting speed of the material
3. The type of material from which the drill is made

Formula (Inch)

$$r/min = \frac{CS \text{ (feet per minute)} \times 12}{\pi D \text{ (drill circumference in inches)}}$$

CS = the recommended cutting speed in *feet per minute* for the material being drilled.

NOTE: This would vary, depending on the material from which the drill is made.

D = the diameter of the drill being used.

Since not all machines can be set to the exact calculated speed, π (3.1416) is divided into 12 to arrive at a simplified formula which is accurate enough for most drilling operations.

$$r/min = \frac{CS \times 4}{D}$$

EXAMPLE

Calculate the r/min required to drill a ½-in. hole in cast iron (CS 80) with a high-speed steel drill.

$$r/min = \frac{CS \times 4}{D}$$
$$= \frac{80 \times 4}{\frac{1}{2}}$$
$$= \frac{320}{\frac{1}{2}}$$
$$= 640$$

Formula (Metric)

$$r/min = \frac{CS\ (m)}{\pi D\ (mm)}$$

It is necessary to convert the meters in the numerator to millimeters so that both parts of the equation are in the same unit. To accomplish this, multiply the CS in meters per minute by 1000 to bring it to millimeters per minute.

$$r/min = \frac{CS \times 1000}{\pi \times D}$$

Not all machines have a variable-speed drive and therefore cannot be set to the exact calculated speed. By dividing π (3.1416) into 1000, a simplified formula is derived which is accurate enough for most drilling operations.

$$r/min = \frac{CS \times 320}{D}$$

EXAMPLE

Calculate the r/min required to drill a 15-mm hole in tool steel (CS 18) using a high-speed steel drill.

$$r/min = \frac{CS \times 320}{D}$$
$$= \frac{18 \times 320}{15}$$
$$= \frac{5760}{15}$$
$$= 384$$

FEED

Feed is the distance that a drill advances into the work for each revolution. Drill feeds may be expressed in decimals, fractions of an inch, or millimeters. Since the feed rate is a determining factor in the rate of production and the life of the drill, it should be carefully chosen for each job. The rate of feed is generally governed by:

1. The diameter of the drill
2. The material of the workpiece
3. The condition of the drilling machine

A general rule of thumb is the feed rate increases as the drill size increases. For example, a ¼-in. (6.35-mm) drill should have a feed of only 0.002 to 0.004 in. (0.05 to 0.10 mm), while a 1-in. (25.4-mm) drill will have a feed of 0.010 to 0.025 in. (0.25 to 0.63 mm) per revolution. Too

TABLE 35-1	Speeds for High-Speed Steel Drills									
		STEEL CASTING		TOOL STEEL		CAST IRON		MACHINE STEEL		BRASS AND ALUMINUM
DRILL SIZE		CUTTING SPEEDS IN FEET PER MINUTE OR METERS PER MINUTE								
IN.	MM	40 FT/MIN	12 M/MIN	60 FT/MIN	18 M/MIN	80 FT/MIN	24 M/MIN	100 FT/MIN	30 M/MIN	200 FT/MIN	60 M/MIN
¹⁄₁₆	2	2445	1910	3665	2865	4890	3820	6110	4775	12225	9550
⅛	3	1220	1275	1835	1910	2445	2545	3055	3185	6110	6365
³⁄₁₆	4	815	955	1220	1430	1630	1910	2035	2385	4075	4775
¼	5	610	765	915	1145	1220	1530	1530	1910	3055	3820
⁵⁄₁₆	6	490	635	735	955	980	1275	1220	1590	2445	3180
⅜	7	405	545	610	820	815	1090	1020	1365	2035	2730
⁷⁄₁₆	8	350	475	525	715	700	955	875	1195	1745	2390
½	9	305	425	460	635	610	850	765	1060	1530	2120
⅝	10	245	350	365	520	490	695	610	870	1220	1735
¾	15	205	255	305	380	405	510	510	635	1020	1275
⅞	20	175	190	260	285	350	380	435	475	875	955
1	25	155	150	230	230	305	305	380	380	765	765

Note: There is no direct relationship between the metric and inch drill sizes.

TABLE 35-2   Drill Feeds

DRILL SIZE		FEED PER REVOLUTION	
IN.	MM	IN.	MM
⅛ and smaller	3 and smaller	0.001 to 0.002	0.02 to 0.05
⅛ to ¼	3 to 6	0.002 to 0.004	0.05 to 0.10
¼ to ½	6 to 13	0.004 to 0.007	0.10 to 0.18
½ to 1	13 to 25	0.007 to 0.015	0.18 to 0.38
1 to 1½	25 to 38	0.015 to 0.025	0.38 to 0.63

coarse a feed may chip the cutting edges or break the drill. Too light a feed causes a chattering or scraping noise which quickly dulls the cutting edges of the drill.

The drill feeds listed in Table 35-2 are recommended for general-purpose work. When drilling alloy or hard steels, use a somewhat slower feed. Softer metals such as aluminum, brass, or cast iron can usually be drilled with a faster feed. Whenever chips coming from the hole turn blue, it is wise to stop the machine and examine the drill. Blue chips indicate too much heat at the cutting edge. This heat is caused by either a dull cutting edge or too high a speed.

CUTTING FLUIDS

Drilling work at the recommended cutting speeds and feeds causes considerable heat to be generated at the drill point. This heat must be dissipated as quickly as possible; otherwise, it will cause a drill to dull rapidly.

The purpose of a *cutting fluid* is to provide both cooling and lubrication. For a liquid to be most effective in dissipating heat, it must be able to absorb heat rapidly, have a good resistance to evaporation, and have a high thermal conductivity. Unfortunately, oil has poor cooling qualities. Water is the best coolant; however, it is rarely used by itself because it promotes rust and has no lubrication value. Basically, a good cutting fluid should:

1. Cool the workpiece and tool
2. Reduce friction
3. Improve the cutting action
4. Protect the work against rusting
5. Provide antiweld properties
6. Wash away the chips

See Unit 85, Table 85-1, for the recommended cutting fluids for various metals.

REAMING SPEEDS AND FEEDS

SPEEDS

The selection of the most efficient speed for machine reaming is dependent upon the following factors.

1. The type of material being reamed
2. The rigidity of the setup

3. The tolerance and finish required in the hole

Generally, reaming speeds should be from one-half to two-thirds the speed used for drilling the same material.

Higher reaming speeds can be used when the setup is rigid; slower speeds should be used when the setup is less rigid. A hole requiring close tolerances and a fine finish should be reamed at slower speeds. The use of coolants improves the surface finish, produces a more accurate hole, and allows higher speeds to be used.

Reamers do not work well when they chatter; the speed selected should always be low enough to eliminate chatter. Table 35-3 gives the recommended reaming speeds for high-speed steel reamers. Carbide reamers may be operated at higher speeds.

FEEDS

The feed used for reaming is usually two to three times greater than that used for drilling. The feed rate will vary with the material being reamed; however, it should be approximately 0.001 to 0.004 in. (0.02 to 0.10 mm) per flute per revolution. Feeds which are too low generally result in glazing, excessive reamer wear, and sometimes chatter. Too high a feed tends to reduce the hole accuracy and sometimes results in poor surface finish. Generally, feeds should be the highest possible which will still produce the hole accuracy and finish required.

An exception to these feed rates occurs when tapered holes are being reamed. As tapered reamers cut along their entire length, a light feed is necessary. The reamer should be removed occasionally and the flutes cleaned of chips.

TABLE 35-3   Recommended Reaming Speeds for High-Speed-Steel Reamers

MATERIAL	SPEED	
	FT/MIN	M/MIN
Aluminum	130–200	39–60
Brass	130–180	39–55
Bronze	50–100	15–30
Cast iron	50–80	15–24
Machine steel	50–70	15–21
Steel alloys	30–40	9–12
Stainless steel	40–50	12–15
Magnesium	170–270	52–82

REVIEW QUESTIONS

DRILLING SPEEDS

1. Why is it important that a drill be operated at the correct speed?
2. Explain the difference between cutting speed and r/min.
3. What factors determine the most economical drilling speed?
4. Calculate the r/min which would be required to drill:
    (a) a ⅜-in.-diameter hole in tool steel
    (b) a 1-in.-diameter hole in aluminum
    (c) a 9-mm hole in a steel casting
    (d) a 20-mm hole in cast iron

CUTTING FLUIDS

5. What is the purpose of a cutting fluid?
6. Name four important qualities that a cutting fluid should have.

REAMING SPEEDS AND FEEDS

7. What is the recommended speed and feed for reaming?
8. Discuss the effects of too high or too low reaming speeds and feeds.

DRILLING HOLES
UNIT 36

Early humans used an arrow wrapped around a bow string to drill holes in bone and wood (Fig. 36-1). Although the technique for drilling holes is much different today, the principle remains the same—pressing a rotating cutting tool into the workpiece. Some of the important factors in drilling are securely fastening the workpiece, observing proper safety precautions, and setting the proper speeds and feeds. The pressure applied to the feed handle and the appearance of the chips are good indicators as to how effective the drill operation is being performed.

Fig. 36-1   Primitive man used a bow drill to produce holes.

After completing this unit you will be able to:
1. Measure the size of inch and metric drills
2. Drill the correct size center holes in workpieces
3. Drill small and large holes to an accurate location

DRILL PRESS SAFETY

Before starting with any drill press operation, it is good practice to observe some basic safety precautions. These precautions will not only ensure the safety of the operator, but also prevent damage to the machine, cutting tool, and the workpiece.

1. Always wear approved safety glasses to protect your eyes.
2. Never attempt to hold the work by hand; a table stop or clamp should be used to prevent the work from spinning (Fig. 36-2).
3. Never set speeds or adjust the work unless the machine is stopped.
4. Keep your head well back from revolving parts of a drill press to prevent your hair from being caught.
5. As the drill begins to break through the work, ease up on the drilling pressure and allow the drill to break through gradually.
6. Always remove burrs from a drilled hole with a file or scraper.
7. Never leave a chuck key in the drill chuck *at any time*.
8. Never attempt to grab work which may have caught in the drill. *Stop the machine*.
9. Always keep the floor around a drill press clean and free of tools, chips, and oil (Fig. 36-3). They can cause serious accidents.

DRILLING HINTS

The following hints should help prevent many problems which could affect the accuracy of the hole and the efficiency of the drilling operation.

1. Always examine the condition of the drill point before use, and if necessary, resharpen it.
2. Set the correct revolutions per minute (r/min) for the size of drill being used. Too high speed quickly dulls a drill, and too low speed causes a small drill to break.
3. Arrange the work setup so that the drill will not cut into the table, parallels, or drill vise as it breaks through the workpiece.
4. The work should always be clamped securely for the drilling operation. For small-diameter holes, a clamp or stop fastened to the left-hand side of the table will prevent the work from spinning (Fig. 36-2).
5. The end of the workpiece farthest from the hole should be placed on the left-hand side of the table so that if the work catches, it will not swing toward the operator.
6. Always clean a tapered drill shank, sleeve, and the machine spindle before inserting a drill.

7. It is a good practice to start each hole with a center drill. The small point of the center drill will pick up a center-punch mark accurately; the center-drilled hole will provide a guide for the drill to follow.
8. Thin workpieces, such as sheet metal, should be clamped to a hardwood block for drilling. This prevents the work from catching and also steadies the drill point as it breaks through the workpiece.
9. A drill squeak usually indicates a dull drill. Stop the machine and examine the condition of the drill point; regrind the drill if necessary.
10. When increased pressure must be applied during a drilling operation, the reason is generally a dull drill or a chip caught in the hole between the drill and the work; correct these conditions before proceeding.

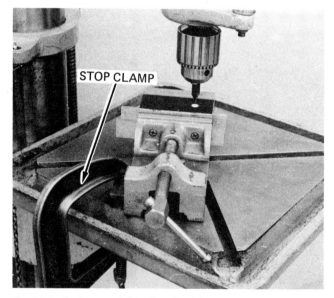

Fig. 36-2 A clamp or table stop should be fastened to the left side of the table.

Fig. 36-3 Poor housekeeping can lead to accidents.

MEASURING THE SIZE OF A DRILL

In order to produce a hole to size, it is important that the correct size drill is used to drill the hole. It is good practice to always check a drill for size before using it to drill a hole. Drills may be checked for size with a drill gage (Fig. 36-4A) and with a micrometer (Fig. 36-4B). Although the size of most drills is stamped on the drill shank, the micrometer is still the most accurate way of measuring the exact size of a drill. When a drill is being checked for size with a micrometer, always be sure that the measurement is taken across the margin of the drill.

LATHE CENTER HOLES

Work to be turned between the centers on a lathe must have a hole drilled in each end so that the work may be supported by the lathe centers. A *combination drill and countersink* (Fig. 36-5), more commonly called a *center drill,* is used for this operation.

To ensure an adequate bearing surface for the work on the lathe center, it is important that center holes be drilled to the correct size and depth (Fig. 36-6A).

A center hole which is too shallow is illustrated in Fig. 36-6B. This results in poor support for the work and possible damage to both the lathe center and the work.

Figure 36-6C shows a center hole which has been drilled too deep. The taper on the lathe center cannot contact the taper of the center hole; the result is poor support for the work.

Fig. 36-5 Two types of center drills. (A) Regular type; (B) Bell type. (Courtesy Cleveland Twist Drill Company.)

Fig. 36-4 (A) Checking a drill for size using a drill gage. (B) Checking the size of a twist drill using a micrometer. (Courtesy Kostel Enterprises Limited.)

Fig. 36-6 (A) A center hole drilled to the proper depth; (B) a center hole drilled too shallow; (C) a center hole drilled too deep.

SIZE		WORK DIAMETER		DIAMETER OF COUNTERSINK C	DRILL POINT DIAMETER	BODY SIZE
REGULAR TYPE	BELL TYPE	INCHES	MILLIMETERS			
1	11	3/16–5/16	3–8	3/32	3/64	1/8
2	12	3/8–1/2	9.5–12.5	9/64	5/64	3/16
3	13	5/8–3/4	15–20	3/16	7/64	1/4
4	14	1–1 1/2	25–40	15/64	1/8	5/16
5	15	2–3	50–75	21/64	3/16	7/16
6	16	3–4	75–100	3/8	7/32	1/2
7	17	4–5	100–125	15/32	1/4	5/8
8	18	6 and over	150 and over	9/16	5/16	3/4

TABLE 36-1 Center Drill Sizes

DRILLING LATHE CENTER HOLES

1. Select the proper size center drill to suit the diameter of the work (See Table 36-1).
2. Fasten the center drill in the drill chuck, having it extend beyond the chuck only about 1/2 in. (12.7 mm).
3. Place the work to be center drilled in the drill vise as shown in Fig. 36-7.
4. Set the drill press at the proper speed and start the machine.
5. Locate the center-punch mark in the work directly below the center drill point.
6. Carefully feed the center drill into the center punch mark in the work for about 1/16 in. (1.58 mm).
7. Raise the center drill, apply a few drops of cutting fluid, and continue drilling.
8. Frequently remove the drill from the hole to apply cutting fluid, remove the chips, and measure the diameter of the top of the center hole.
9. Continue drilling until the hole is the proper size.

SPOTTING A HOLE LOCATION WITH A CENTER DRILL

The chisel edge at the end of the web is generally larger than the center-punch mark on the work and therefore it is difficult to start a drill at the exact location. To prevent a drill from wandering off center, it is considered good practice to first spot every center-punch mark with a center drill. The small point on the center drill will accurately follow the center-punch mark and provide a guide for the larger drill which will be used.

1. Mount a small-size center drill in the drill chuck.
2. Mount the work in a vise or set it on the drill press table. *Do not clamp the work or the vise.*
3. Set the drill press speed to about 1500 *r/min*.
4. Bring the point of the center drill into the center-punch

Fig. 36-7 Work set up for center-hole drilling.

mark and allow the work to center itself with the drill point.
5. Continue drilling until about one-third of the tapered section of the center drill has entered the work (Fig. 36-8).
6. Spot all the holes which are to be drilled.

DRILLING WORK HELD IN A VISE

The most common method of holding small workpieces is by means of a vise which may be held by hand against a table stop or clamped to the table. When drilling holes larger than 1/2 in. (12.7 mm) in diameter, the vise should be clamped to the table.

1. Spot the hole location with a center drill.
2. Mount the correct size drill in the drill chuck.
3. Set the drill press to the proper speed for the size of drill and the type of material to be drilled.
4. Fasten a clamp or stop on the left side of the table (Fig. 36-2).

Fig. 36-8 Spotting a hole location with a center drill.

Fig. 36-9 Tighten the table clamp while the drill is revolving in the hole. (Courtesy Kostel Enterprises Limited.)

5. Mount the work on parallels in a drill vise and tighten it securely.
6. With the vise against the table stop, locate the spotted hole under the center of the drill.
7. Start the drill press spindle and begin to drill the hole.
   a. For holes up to ½ in. (12.7 mm) in diameter, hold the vise against the table

   OR

   stop by hand (Fig. 36-2).
   b. For holes over ½ in. (12.7 mm) in diameter
   • Lightly clamp the vise to the table with a clamp.
   • Drill until the full drill point is into the work.
   • With the drill revolving, keep the drill point in the work and tighten the clamp holding the vise securely (Fig. 36-9).
8. Raise the drill occasionally and apply cutting fluid during the drilling operation.
9. Ease up on the drilling pressure as the drill starts to break through the workpiece.

DRILLING TO AN ACCURATE LAYOUT

If a hole must be drilled to an exact location, the position of the hole must be accurately laid out as shown in Fig. 36-10A. During the drilling operation it may be necessary to draw the drill point over so that it is concentric with the layout (Fig. 36-10B).

1. Clean and coat the surface of the work with layout dye.
2. Locate the position of the hole from *two machined edges* of the workpiece and scribe the lines as shown in Fig. 36-10A.
3. Lightly prick-punch where the two lines intersect.
4. Check the accuracy of the punch mark with a magnifying glass and correct if necessary.
5. With a pair of dividers, scribe a circle to indicate the diameter of the hole required (Fig. 36-10B).
6. Scribe a test circle ⅟₁₆ in. (1.58 mm) smaller than the hole size.
7. Punch four witness marks on circles up to ¾ in. (19.05 mm) in diameter, and eight witness marks on larger circles (Fig. 36-10C).
8. Deepen the center of the hole location with a center punch to provide a larger indentation for the drill to follow.
9. Center drill the work to *just beyond the depth of the drill point*.
10. Mount the proper size drill in the machine and drill a hole to a depth equal to one-half to two-thirds of the drill diameter.

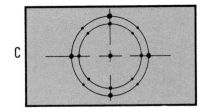

Fig. 36-10 Layout for a hole to be drilled.

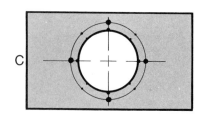

Fig. 36-11   Drawing the drill point to the layout.

PILOT HOLE

Fig. 36-12   Drilling a pilot or lead hole helps the larger drill cut easily and accurately.

11. Examine the drill indentation; it should be concentric with the inner proof circle (Fig. 36-11A).
12. If the spotting is off center, cut shallow V-grooves with a cape or diamond-point chisel on the side toward which the drill must be moved (Fig. 36-11B).
13. Start the drill in the spotted and grooved hole. The drill will be drawn toward the direction of the grooves.
14. Continue cutting grooves into the spotted hole until the drill point is drawn to the center of the scribed circles as shown in Fig. 36-11C.

NOTE: *The drill point must be drawn to the center of the scribed circle before the drill has cut or spotted to the drill's diameter.*

15. Continue to drill the hole to the desired depth.

DRILLING LARGE HOLES

As drills increase in size, the thickness of the web also increases to give the drill added strength. The thicker the web, the thicker will be the point or chisel edge of the drill.

As the chisel edge becomes larger, poorer cutting action results and more pressure must be applied for the drilling operation. A thick web will not follow the center-punch mark accurately and the hole may not be drilled in the proper location. Two methods are generally employed to overcome the poor cutting action of a thick web on large drills.

1. The web is thinned.
2. A lead, or pilot, hole is drilled.

The usual procedure for drilling large holes is that first a pilot, or lead, hole (Fig. 36-12), the diameter of which is slightly larger than the thickness of the web, is drilled. *Care must be taken that the pilot hole is drilled on center.* The pilot hole is then followed with a larger drill. This method may also be used to drill average size holes when the drill press is small and does not have sufficient power to drive the drill through the solid metal.

NEVER drill a pilot hole any bigger than necessary; otherwise, the larger drill may:

1. Cause chattering
2. Drill the hole out-of-round
3. Spoil the top (mouth) of the hole

PROCEDURE

1. Check the print and select the proper drill for the hole required.
2. Measure the thickness of the web at the point. Select a pilot drill with a diameter slightly larger than the web thickness (Fig. 36-13).
3. Mount the workpiece on the table.
4. Adjust the height and position of the table so that the drill chuck can be removed and the larger drill placed in the spindle without having to lower the table after the pilot hole is drilled. Lock the table securely in this position.
5. Place a center drill in the drill chuck, set the proper spindle speed, and accurately drill a center hole.

NOTE: The center drill should be used first since it is short, rigid, and more likely to follow the center-punch mark.

6. Using the proper size pilot drill and correct spindle speed, drill the pilot hole to the required depth (Fig. 36-12). The work may be lightly clamped at this time.
7. Shut off the machine leaving the pilot drill in the hole.
8. Clamp the work securely to the table.
9. Raise the drill spindle and remove the drill and drill chuck.

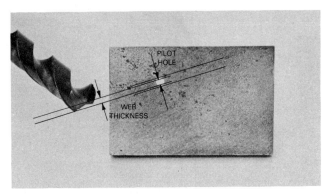

Fig. 36-13 The size of the pilot drill should be a little larger than the thickness of the drill web.

10. Clean the taper shank of the drill and the drill press spindle hole. Remove any burrs on the drill shank with an oilstone.
11. Mount the large drill in the spindle.
12. Set the proper spindle speed and feed and drill the hole to the required depth. If hand feed is used, the feed pressure should be eased as the drill breaks through the work.

REVIEW QUESTIONS

SAFETY

1. Select three of the most important safety suggestions and explain why they should be observed.

DRILLING HINTS

2. What result might be expected from
   (a) too high drill speed?
   (b) too low drill speed?

3. Why is it good practice to start each hole with a center drill?

LATHE CENTER HOLES

4. State three reasons why a center drill should be removed from the work frequently.
5. Why should center holes not be drilled
   (a) too shallow?
   (b) too deep?

SPOTTING A HOLE

6. What is the purpose of spotting a hole before drilling?
7. How deep should each hole be spotted?

DRILLING WORK IN A VISE

8. What is the purpose of fastening a clamp or table stop to the left side of the drill press table?
9. When should drilling pressure be eased?

DRILLING TO AN ACCURATE LAYOUT

10. List the procedure for laying out a hole before drilling to a layout.
11. How deep should the hole be drilled before the drill point indentation is examined?
12. Explain how a drill may be drawn over to a layout.

DRILLING LARGE HOLES

13. List three disadvantages of a thick web found on large drills.
14. What are pilot holes and why are they necessary?
15. Why should pilot holes not be drilled any larger than necessary?
16. How high should the drill press table be adjusted when drilling large holes?

REAMING

UNIT 37

Each component in a product must be made to exact standards in order for that product to function properly. Since it is impossible to produce holes which are round, smooth, and accurate to size by drilling, the reaming operation is very important. Reamers are used to enlarge and finish a hole which was previously formed by drilling or boring. Speed, feed, and reaming allowances are the three main factors which will affect the accuracy and finish of the hole and the life of the reamer.

OBJECTIVES

After completing this unit you should be able to:
1. Identify and state the purpose of hand reamers and machine reamers
2. Know the advantages of carbide-tipped reamers
3. Calculate the reaming allowance required for each reamer
4. Ream a hole by hand in a drill press
5. Machine ream a hole

REAMERS

A reamer is a rotary cutting tool with several straight or helical cutting edges along its body. It is used to accurately size and smooth a hole which has been previously drilled or bored. Some reamers are operated by hand (hand reamers), while others may be used under power in any type of machine tool (machine reamers).

REAMER PARTS

Reamers generally consist of three main parts: shank, body, and angle of chamfer (Fig. 37-1).

The shank, which may be straight or tapered, is used to drive the reamer. The shank of machine reamers may be round or tapered, while hand reamers have a square on the end to accommodate a tap wrench.

The body of a reamer contains several straight or helical grooves, or flutes, and lands (the portion between the flutes). A margin (the top of each tooth) runs from the *angle of chamfer* to the end of the flute. The *body clearance angle* is a relief or clearance behind the margin which reduces friction while the reamer is cutting. The *rake angle* is the angle formed by the face of the tooth when a line is drawn from a point on the front marginal edge through the center of the reamer (Fig. 37-1). If there is no angle on the face of the tooth, the reamer is said to have *radial land*.

The angle of chamfer is the part of the reamer which actually does the cutting. It is ground on the end of each tooth and there is clearance behind each chamfered cutting edge. On rose reamers, the angle of chamfer is ground on the end only and the cutting action occurs at this point. On fluted reamers, each tooth is relieved and most of the cutting is done by the reamer teeth.

TYPES OF REAMERS

Reamers are available in a variety of designs and sizes; however, they all fall into two general classifications: *hand* and *machine* reamers.

Hand Reamers

Hand reamers (Fig. 37-2) are finishing tools used when a hole must be finished to a high degree of accuracy and finish. Holes to be hand-reamed should be bored to within 0.003 to 0.005 in. (0.07 to 0.12 mm) of the finish size. NEVER attempt to ream more than 0.005 in. (0.12 mm) with a hand reamer.

A square on the shank end allows a wrench to be used for turning the reamer into the hole. The teeth on the end of the reamer are tapered slightly for a distance equal to the reamer diameter so that it can enter the hole to be reamed.

A hand reamer should never be used under mechanical power. When using a hand reamer, it is important to keep it true and straight with the hole. The dead center in a lathe or a stub center in a drill press will help keep the reamer aligned during the hand reaming operation.

Taper hand reamers (Fig. 37-3), both roughing and finishing, are available for all standard size tapers. Because chips do not fall out readily, a taper reamer should be removed from the hole and the flutes cleaned frequently.

MACHINE REAMERS

Machine reamers may be used in any machine tool for both roughing and finishing a hole. They are also called *chucking reamers* because of the method used to hold them for the reaming operation. Machine reamers are available in a wide variety of types and styles. Only some of the more common types will be discussed.

Fig. 37-1 The main parts of a reamer.

Fig. 37-2 Straight and helical-fluted hand reamers. (Courtesy The Cleveland Twist Drill Company.)

Fig. 37-3 Taper hand reamers. (A) Roughing; (B) finishing. (Courtesy The Cleveland Twist Drill Company.)

Fig. 37-4 (A) A rose reamer cuts on the end angle only; (B) the fluted reamer has more teeth than the rose reamer and cuts on the sides and end.

Fig. 37-5 A carbide-tipped reamer.

Fig. 37-6 Shell reamers are economical for reaming large holes.

Fig. 37-7 An adjustable reamer with inserted blades.

Fig. 37-8 An expansion-type reamer can be expanded slightly.

Rose reamers (Fig. 37-4A) can be purchased with straight or tapered shanks and with straight or helical flutes. The teeth on the end have a 45° chamfer which is backed off to produce the cutting edge. The lands are nearly as wide as the flutes and are not backed off. Rose reamers cut on the end only and can be used to remove material quickly and bring the hole fairly close to the size required. Rose reamers are usually made 0.003 to 0.005 in. (0.07 to 0.12 mm) under the nominal size.

Fluted reamers (Fig. 37-4B) have more teeth than those reamers for a comparable diameter. The lands are relieved for the entire length, and fluted reamers therefore cut along the side as well as at the chamfer on the end. These reamers are considered finishing tools and are used to bring a hole to size.

Carbide-tipped reamers (Fig. 37-5) were developed to meet the ever-increasing demand for higher production rates. They are similar to rose or fluted reamers, except that carbide tips have been brazed to their cutting edges. Because of the hardness of the carbide tips, these reamers resist abrasion and maintain sharp cutting edges even at high temperatures. Carbide-tipped reamers outlast high-speed steel reamers, especially on castings where hard scale or sand is a problem. Because carbide-tipped reamers can be run at higher speeds and still maintain their size, they are used extensively for long production runs.

Shell reamers (Fig. 37-6) are reamer heads mounted on a driving arbor. The shank of the driving arbor may be straight or tapered, depending on the size and type of shell reamer used. Two slots in the end of this reamer fit into lugs on the driving arbor. Sometimes a locking screw in the end of the arbor holds the shell reamer in place.

The advantages of shell reamers are:

1. They are economical for larger holes.
2. Various head sizes can be easily interchanged on one arbor.
3. When a reamer becomes worn, it may be thrown away while the driving arbor can be used with other reamers.

Adjustable reamers (Fig. 37-7) have inserted blades which can be adjusted approximately at 0.015 in. (0.38 mm) over or under the nominal reamer size. The threaded body has a series of tapered grooves cut lengthwise into which blades are fitted. Adjusting nuts on either end can be used to increase or decrease the diameter of the reamer.

Hand- or machine-adjustable reamers can be readily sharpened and are available with either high-speed steel or carbide inserts.

Expansion reamers (Fig. 37-8) are similar to adjustable reamers; however, the amount they can be expanded is very limited. The body of this reamer is slotted, and a tapered, threaded plug is fitted into the end. Turning this plug will allow a 1-in. (25.4-mm) reamer to expand up to 0.005 in. (0.12 mm). Expansion reamers are meant not to be oversize reamers, but to give longer life to finishing reamers.

Emergency reamers, drills whose corners (at the lip and land) have been slightly rounded and honed, may be used with fairly good results if a reamer of a particular size is not available. First drill the hole as close as possible to the required size. Then run the reaming drill at a fairly high speed and feed it into the hole slowly.

REAMER CARE

The accuracy and surface finish of a hole, as well as the life of a reamer, depend greatly on the care a reamer receives. It is wise to remember that a reamer is a finishing tool and should be handled carefully.

1. Never turn a reamer backward; this will ruin the cutting edges.
2. Always store reamers in separate containers to prevent the cutting edges from being nicked or burred. Plastic or cardboard tubes make excellent reamer containers.
3. Never roll or drop reamers on metal surfaces, such as bench tops, machines, plates, etc.
4. When not in use, a reamer should be oiled, especially on the cutting edges, to prevent rusting.
5. A fine, free-cutting grinding wheel should be used for resharpening reamers. Burring of the cutting edges destroys the life of the reamer, while a rough cutting edge produces a rough hole and the reamer dulls quickly.

REAMING ALLOWANCES

The amount of material left in a hole for the reaming operation depends on a number of factors. If a hole has been punched, rough-drilled, or bored, it requires more metal for reaming than a hole which has already been reamed with a roughing reamer. The type of machining operation prior to reaming must be considered as well as the material being reamed.

General rules for the amount of material which should be left in a hole for machine reaming are as follows.

TABLE 37-1 Recommended Stock Allowances for Reaming			
HOLE SIZE		ALLOWANCE	
IN.	MM	IN.	MM
¼	6.35	0.010	0.25
½	12.7	0.015	0.38
¾	19.05	0.018	0.45
1	25.4	0.020	0.50
1¼	31.75	0.022	0.55
1½	38.1	0.025	0.63
2	50.8	0.030	0.76
3	76.2	0.045	1.14

1. For holes up to ½ in. diameter, allow 1/64 in. for reaming.
2. For holes over ½ in. diameter, allow 1/32 in. for reaming.

NOTE: Never leave more than 0.003 to 0.005 in. in a hole for hand reamers up to ½ in. diameter. On larger holes, a proportional allowance should be left to make a good finish possible.

For metric size reamers allow 0.10 mm for holes up to 12 mm in diameter. For holes over 12 mm, allow 0.20 mm for reaming. See Table 37-1 for the recommended allowances for various size holes.

REAMING SPEEDS AND FEEDS

SPEEDS

The selection of the most efficient speed for machine reaming is dependent upon the following factors.

1. The type of material being reamed
2. The rigidity of the setup
3. The tolerance and finish required in the hole

Generally, reaming speeds should be from 1/2 to 2/3 the speed used for drilling the same material.

Higher reaming speeds can be used when the setup is rigid; slower speeds should be used when the setup is less rigid. A hole requiring close tolerances and a fine finish should be reamed at slower speeds. The use of coolants improves the surface finish and allows higher speeds to be used.

Reamers do not work well when they chatter; the speed selected should always be low enough to eliminate chatter.

Table 37-2 gives the recommended reaming speeds for high-speed steel reamers. Carbide reamers may be operated at higher speeds.

FEEDS

The feed used for reaming is usually two to three times greater than that used for drilling. The feed rate will vary with the material being reamed; however, it should be

TABLE 37-2. Recommended Reaming Speeds for High Speed Steed Reamers		
MATERIAL	SPEED	
	FT./MIN	M/MIN.
Aluminum	130–200	39–60
Brass	130–180	39–55
Bronze	50–100	15–30
Cast iron	50–80	15–24
Machine steel	50–70	15–21
Steel alloys	30–40	9–12
Stainless steel	40–50	12–15
Magnesium	170–270	52–82

approximately 0.001 to 0.004 in. (0.02 to 0.10 mm) per flute per revolution. Feeds which are too low generally result in glazing, excessive reamer wear, and sometimes chatter. Too high a feed tends to reduce the hole accuracy and sometimes results in poor surface finish. Generally, feeds should be the highest possible which will still produce the hole accuracy and finish required.

An exception to these feed rates occurs when tapered holes are being reamed. As tapered reamers cut along their entire length, a light feed is necessary. The reamer should be removed occasionally and the flutes cleaned.

REAMING HINTS

1. Examine a reamer and remove all burrs from the cutting edges with a hone so that good surface finishes will be produced.

2. Cutting fluid should be used in the reaming operation to improve the hole finish and prolong the life of the reamer.

3. Helical-fluted reamers should always be used when long holes and those with keyways or oil grooves are reamed.

4. Straight-fluted reamers are generally used when extreme accuracy is required.

5. To obtain hole accuracy and good surface finish, use a roughing reamer first and then a finishing reamer. An old reamer which is slightly undersize may be used as a roughing reamer.

6. Never, *under any circumstances,* turn a reamer backward.

7. Never attempt to start a reamer on an uneven surface; the reamer will go toward the point of least resistance and will not produce a straight, round hole.

8. If chatter occurs, stop the machine, reduce the speed, and increase the feed. To overcome the chatter marks, it may be necessary to restart the reamer slowly by pulling the drill press belt by hand.

9. To avoid chatter, select a reamer with an incremental cut (unequally spaced teeth).

10. When hand reaming in a drill press, always use a stub center in the drill press spindle to keep the reamer aligned.

REAMING A STRAIGHT HOLE

There are two types of reamers used in machine shop work—hand reamers and machine reamers. Hand reamers have a square on one end and are used to remove no more than 0.005 in. (0.12 mm) from a hole. Machine reamers may have straight shanks which are held and driven by a drill chuck, or tapered shanks which fit directly into the drill press spindle. They are generally used to remove from 1/64 in. (0.39 mm) to 1/32 in. (0.79 mm) of metal from a hole, depending upon the hole diameter.

HAND REAMING A STRAIGHT HOLE

1. Mount the work on parallels in a vise and clamp securely to the table (Fig. 37-9).

Fig. 37-9   Work clamped to the table correctly.

Fig. 37-10   Using a stub center to keep the reamer aligned.

2. Drill the hole to the proper size leaving an allowance for the hand reamer to be used.

NOTE: Reaming allowance should be no more than 0.005 in. (0.12 mm) for a 1-in. (25.4-mm) diameter reamer.

3. Do not move the location of the work or the table; remove the drill and mount a stub center in the drill chuck (Fig. 37-10).

4. Start the end of the reamer in the drilled hole.

5. Fasten a tap wrench on the reamer.

6. Engage the stub center in the center hole on the end of the reamer.

7. With the downfeed lever, apply slight pressure while turning the reamer clockwise by hand.

8. Apply cutting fluid and ream the hole.

9. When removing the reamer, turn it clockwise, never counterclockwise.

MACHINE REAMING A STRAIGHT HOLE

A hole which must be finished to size should be reamed immediately after it has been drilled and while the hole is still aligned with the drill press spindle. This will ensure that the reamer follows the same location as the drill.

1. Mount the work on parallels in a vise and fasten it securely to the table.
2. Select the proper size drill for the reaming allowance required, and drill the hole.

NOTE: *Do not move the work or drill press table at this time.*

3. Mount the proper reamer in the drill press.
4. Adjust the spindle speed to suit the reamer and the work material.
5. Start the drill press and carefully lower the spindle until the chamfer on the reamer starts to cut (Fig. 37-11).
6. Apply cutting fluid and feed the reamer by applying enough pressure to keep the reamer cutting.
7. Remove the reamer from the hole by raising the down-feed handle.
8. Shut off the machine and remove the burr from the edge of the hole.

Fig. 37-11   Reaming a hole in the drill press.

REAMING A TAPERED HOLE

Before reaming a tapered hole in a workpiece, it is necessary to step-drill the hole prior to reaming. The recommended drills to be used can be determined by referring to Fig. 37-12.

1. Mount the work on parallels in a vise or on the drill table and fasten it securely.
2. Align the hole center under the drill point.
3. Drill a hole 1/64 in. (0.39 mm) smaller than the smallest diameter of the tapered hole (Fig. 37-12).
4. Obtain a drill 1/64 in. (0.39 mm) smaller than the size of the finished hole measured at a point B (one-half the length of the tapered section).
5. Drill the hole 1/16 in. (1.59 mm) less than C dimension (Fig. 37-12).
6. Mount a roughing tapered reamer in the drill spindle.
7. Adjust the spindle speed to about one-half the speed used to ream a straight hole.
8. Rough-ream the hole to about 0.005 in. (0.12 mm) undersize while applying cutting fluid.
9. Mount the finish reamer, apply cutting fluid, and finish-ream the hole to size.

A.   (SMALL DRILL SIZE) SHOULD BE THE SMALLEST DIAMETER OF THE TAPERED HOLE MINUS 1/64 in.

B.   (LARGE DRILL SIZE) SHOULD BE THE DIAMETER AT MIDDLE OF TAPERED HOLE MINUS 1/64 in.

C.   (DEPTH OF LARGE HOLE) ONE-HALF THE LENGTH OF THE TAPER MINUS 1/16 in.

Fig. 37-12   Drill sizes for tapered holes.

REVIEW QUESTIONS

REAMERS

1. What is the purpose of a reamer?
2. Define the following reamer parts:
   (a) body
   (b) angle of chamfer
   (c) body clearance angle
   (d) radial land
3. How may a hand reamer be recognized, and for what purpose is it used?
4. Compare a rose reamer and a fluted reamer.

5. State the advantages of carbide-tipped reamers.
6. Describe briefly the following types of reamers and state their purpose.
   (a) shell
   (b) adjustable

REAMER CARE

7. List three important points which should be observed in the care of reamers.

REAMING ALLOWANCES

8. What is the general rule for the amount of material left in a hole for machine reaming?

9. How much should be left in the following holes for reaming?
  (a) 1¼ in.
  (b) 19.05 mm

REAMING HINTS

10. List seven of the most important reaming hints.

11. How much material should be left in a hole for hand reaming?
12. How can the reamer be kept in alignment when hand reaming on a drill press?
13. Why should the work or drill press table not be moved before a hole is reamed?
14. Make a neat sketch to show the procedure for drilling in preparation for reaming a tapered hole.

DRILL PRESS OPERATIONS
UNIT 38

The drill press is a versatile machine tool and can be used to perform a variety of operations other than the drilling of holes. The variety of cutting and finishing tools available allow operations such as tapping, counterboring, countersinking, boring, spot-finishing, and buffing to be performed on a drill press.

OBJECTIVES

After completing this unit you should be able to:
1. Counterbore and countersink holes
2. Select and use the proper tap to thread a hole in a drill press
3. Use three methods to transfer hole locations
4. Finish metal surfaces by buffing

COUNTERBORING

Counterboring is the operation of enlarging the end of a hole which has been drilled previously. A hole is generally counterbored to a depth slightly greater than the head of the bolt, cap screw, or pin which it is to accommodate.

Counterbores (Fig. 38-1) are supplied in a variety of styles, each having a pilot in the end to keep the tool in line with the hole being counterbored. Some counterbores are available with interchangeable pilots to suit a variety of hole sizes.

TO COUNTERBORE A HOLE

1. Set up and fasten the work securely.
2. Drill the proper size hole in the workpiece to suit the body of the pin or screw.
3. Mount the correct size counterbore in the drill press (Fig. 38-2).
4. Set the drill press speed to approximately one-quarter of that used for drilling.

Fig. 38-1 A set of counterbores.

Fig. 38-2 A counterbore is used to enlarge the end of a hole.

5. Bring the counterbore close to the work and see that the pilot turns freely in the drilled hole.
6. Start the machine, apply cutting fluid, and counterbore to the required depth.

COUNTERSINKING

Countersinking is the process of enlarging the top end of a hole to the shape of a cone to accommodate the conical-shaped heads of fasteners so that the head will be flush with or below the surface of the part. These cutting tools, called countersinks (Fig. 38-3), are available with various included angles such as 60, 82, 90, 100, 110, and 120°.

An 82° countersink is used to enlarge the top of a hole so that it will accommodate a flat-head machine screw (Fig. 38-4). The hole is countersunk until the head of the machine screw is flush with or slightly below the top of the work surface (Fig. 38-4). All holes that are to be threaded should be countersunk slightly larger than the tap diameter to protect the start of the thread.

The speed recommended for countersinking is approximately one-quarter of the drilling speed.

TO COUNTERSINK A HOLE FOR A MACHINE SCREW

1. Mount an 82° countersink in the drill chuck.
2. Adjust the spindle speed to about one-half that used for drilling.
3. Place the workpiece on the drill table.
4. With the spindle stopped, lower the countersink into the hole. Clamp the work if necessary. If a pilot-type countersink is used, the pilot should be a slip fit in the drilled hole. The pilot will center the cutting tool and the work.
5. Raise the countersink slightly, start the machine and feed the countersink by hand until the proper depth is reached. The diameter may be checked by placing an inverted screw in the countersunk hole (Fig. 38-4).

Fig. 38-3 Countersinking produces a tapered hole to fit a flat-head machine screw.   Courtesy Kostel Enterprises Limited.)

Fig. 38-4 Countersink until the top of the hole is slightly larger than the diameter of the screw head.

6. If several holes are to be countersunk, set the depth stop so that all of the holes will be the same depth. The gage should be set when the spindle is stationary and the countersink is in the hole.
7. Countersink all of the holes to the depth set on the gage.

TAPPING

Tapping in a drill press may be performed either by hand or under power with the use of a tapping attachment. The advantage of using a drill press for tapping a hole is that

Fig. 38-5 A set of hand taps. (A) Taper; (B) plug; (C) bottoming.

Fig. 38-6 Types of machine taps. (A) Gun; (B) stub flute; (C) spiral flute.

Fig. 38-7 The characteristics of a fluteless tap. (A) Fluteless tap; (B) lobes of the tap.

the tap can be started squarely and maintained that way through the entire length of the hole.

Machine tapping involves the use of a tapping attachment mounted in the drill press spindle. The tapping operation should be done immediately after the drilling operation to obtain the best accuracy and to avoid duplication of the setup. This sequence is especially important when tapping by hand.

The most common taps used for tapping holes in a drill press are hand taps and machine taps. Hand taps (Fig. 38-5) are available in sets containing the taper, plug, and bottoming taps. When using hand taps in a drill press, it is important that the tap be guided by holding it in a drill chuck or supporting it with a stub center.

Machine taps are designed to withstand the torque required to thread the hole and to clear the chips out of the hole quickly. The most common machine taps are the gun taps, stub-flute taps, and spiral-fluted taps (Fig. 38-6).

The *fluteless tap* (Fig. 38-7) is actually a forming tool used to produce internal threads in ductile materials such as copper, brass, aluminum, and leaded steels.

TO TAP A HOLE BY HAND IN A DRILL PRESS

1. Mount the work on parallels with the center-punch mark on the work in line with the spindle, and clamp the work securely to the drill press table.
2. Adjust the drill press table height so that the drill may be removed after the hole has been drilled without moving the table or work.
3. Center drill the hole location.
4. Drill the hole to the correct *tap drill size* for the tap to be used.

NOTE: The work or table *must not* be moved after the drilling.

5. Mount a stub center in the drill chuck (Fig. 38-8).
OR
Remove the drill chuck and mount a special center in the drill press spindle.
6. Fasten a suitable tap wrench on the end of the tap.

Fig. 38-8 Guiding a tap into the workpiece using a stub center held in a drill chuck.

7. Place the tap in the drilled hole, and lower the drill press spindle until the center fits into the center hole in the tap shank.

8. Turn the tap wrench clockwise to start the tap into the hole, and at the same time keep the center in light contact with the tap.

9. Continue to tap the hole in the usual manner; keep the tap aligned by applying light pressure on the drill press downfeed lever.

A *tapping attachment* (Fig. 38-9) may be mounted in a drill press spindle to rotate the tap by power. It has a built-in friction clutch which drives the tap clockwise when the drill press spindle is fed downwards. If there is excessive pressure against the tap because it is stuck or jammed in a hole, the clutch slips before the tap will break. The tapping attachment has a reversing mechanism, engaged by the drill press spindle being raised, to back a tap out of the hole.

Two- or three-fluted machine or gun taps are used for tapping under power because of their ability to clear the chips. Tapping speed for most materials ranges from 60 to 100 r/min.

BORING

Boring is the operation of enlarging a drilled or cored hole by means of a single- or double-edged cutting tool held in a boring bar. The hole produced should be concentric, parallel, and perpendicular to the work surface.

Although boring is generally not done on a drill press, sometimes, because of the nature of the workpiece, it may be necessary to use this machine for a boring operation.

Many large drill presses have a hole in the center of the table into which a bushing can be inserted. A pilot on the end of the boring bar fits into the table bushing and keeps the boring bar rigid and aligned during the machining operation.

TO BORE IN A DRILL PRESS

1. Mount as large a boring bar as possible into the drill press spindle.

2. Swing the drill press table to allow the boring bar pilot to enter the table bushing hole.

3. Lock the drill press table in position.

4. Mount the work on parallels to allow the chips to clear the hole and also to prevent cutting into the table (Fig. 38-10).

5. Centralize the hole to be machined to the boring bar, and clamp the work securely in position.

6. Set the drill press to the correct revolutions per minute (r/min), and feed for the material being cut. Roughing feeds may be as high as 0.015 in. (0.38 mm), while finishing feeds are approximately 0.001 to 0.005 in. (0.02 to 0.12 mm).

7. Set the toolbit and take a light trial cut approximately ¼ in. (6.35 mm) deep.

8. Stop the machine and measure the size of the hole with inside calipers or telescoping gages.

9. Reset the cutting tool for the depth of cut desired.

10. Continue to bore the hole to the required size by setting the cutting tool for each cut.

Fig. 38-9   A tapping attachment held in a drill press.

Fig. 38-10   The work and drill press setup for boring a hole.

NOTE: In order to produce an accurate hole with a good surface finish, keep the drill press spindle adjusted snugly, use a sharp cutting tool, and apply cutting fluid.

TRANSFERRING HOLE LOCATIONS

During the construction of dies, jigs, fixtures, and machine parts, it is often necessary to transfer the location of holes accurately from one part to another. Three common methods of transferring hole locations are:

1. Spotting with a twist drill
2. Using transfer punches
3. Using transfer screws

Regardless of which method is used, the holes in the existing part are used as a master or guide to transfer the hole locations to another part.

TO SPOT WITH A TWIST DRILL

1. Remove the burrs from the mating surfaces on both parts.

Fig. 38-11   Transferring hole locations by spotting with a drill.

Fig. 38-12   Transferring hole locations using a transfer punch.

2. Align both parts accurately and clamp them together.
3. Mount a drill, the same diameter as the hole to be transferred, in the drill press spindle.
4. Start the drill into the hole of the guide part and spot-drill the second part (Fig. 38-11).

NOTE: Never spot-drill deeper than the diameter of the drill.

5. Spot-drill all the holes that are to be transferred.
6. Remove the original part.
7. Drill the spotted holes to the required diameter.

TO USE TRANSFER PUNCHES

1. Remove the burrs from the mating surfaces on both parts.
2. Align both parts accurately and clamp them together.
3. Secure a transfer punch (Fig. 38-12) of the same diameter as the hole to be transferred.
4. Place the punch in the hole and *lightly* strike it with a hammer to mark the hole location.
5. Use the correct size of transfer punch on all the holes that are to be transferred.
6. Remove the original part.
7. Use a divider to lay out proof circles for the holes to be drilled.
8. With a center punch, deepen the existing transfer-punch marks.
9. Use the method outlined in ''Drilling to a Layout'' to drill the holes to location accurately (Unit 36).

TO USE TRANSFER SCREWS

Many times it is necessary to transfer the location of threaded holes. This may be easily accomplished by the use of transfer screws (Fig. 38-13) which have been hardened and sharpened to a point. Two flats are ground on the point to allow the screws to be threaded into a hole with a small wrench or a pair of needle-nose pliers.

1. Remove all burrs from the mating surfaces.
2. Thread transfer screws into the holes to be transferred, allowing the points to extend beyond the work surface approximately 1/32 in. (0.78 mm).

Fig. 38-13   Transfer screws are used to transfer threaded hole locations.

3. Align both parts accurately and then sharply strike one part with a hammer.
4. Remove the original part, and deepen the marks left by the transfer screws with a center punch.
5. Drill all holes to the required size.

BUFFING

Buffing is the operation of producing a highly polished finish on a round or flat surface. To remove the machine marks on the surface of a workpiece, first use coarse and then fine abrasive cloth. This surface should then be brought into contact with a buffing wheel which has been coated with jeweler's rouge or polishing compound.

BUFFING THE SURFACE OF A WORKPIECE

1. Polish the surface with fine abrasive cloth until all scratches and machine marks are removed (Fig. 38-14).
2. Place a buffing wheel on an arbor and tighten it in a drill chuck.
3. Set the drill spindle to its highest speed.
4. Start the machine and hold the polishing compound against the edge of the revolving wheel until the periphery is coated.

NOTE: Do not apply too much compound as the excess compound will be thrown off the buffing wheel and wasted.

5. Grip the work firmly in both hands and press it against the edge of the buffing wheel (Fig. 38-15).

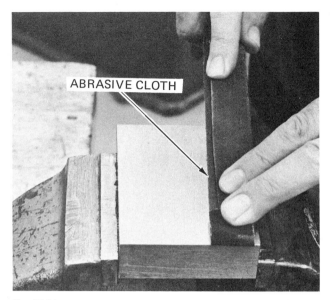

Fig. 38-14 A workpiece being polished with fine abrasive cloth.

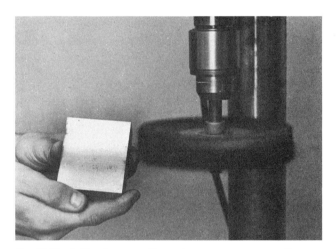

Fig. 38-15 Buffing improves the surface appearance. (Courtesy Kostel Enterprises Limited.)

CAUTION: Do not hold the workpiece with a cloth as it may get caught in the wheel and cause injury to the operator or damage the workpiece, or both.

6. Slowly turn the surface to be polished while keeping it in contact with the rotating buffing wheel.
7. Continue buffing the work surface until a glossy finish is produced.
8. With a clean, dry, soft cloth remove any excess compound from the surface of the workpiece.

DRILL JIGS

A drill jig is used whenever it is necessary to drill holes to an exact location in a large number of identical parts. Drill jigs are used to save layout time, to avoid incorrectly located holes, and to produce holes accurately and economically. The advantages of using a drill jig are:

Fig. 38-16 A drill jig positions any number of identical parts so that a hole or holes may be drilled to an accurate location. (Courtesy The Cleveland Twist Drill Company.)

1. Since it is not necessary to lay out the hole locations, layout time is eliminated.
2. Each part is quickly and accurately aligned.
3. The part is held in position by a clamping mechanism.
4. The drill jig bushings provide a guide for the drill.
5. The hole locations in each part will be exactly the same; therefore the parts produced are interchangeable.
6. Unskilled labor can be used.

A drill jig (Fig. 38-16) is designed so that the part to be drilled may be fastened into it and drilled immediately. Hardened drill jig bushings, used to guide and keep the drill in location, are located in the drill jig wherever holes must be drilled. When two or more different size holes are to be drilled in the same part, it is preferable to have a gang or multispindle drill press set up. A different size drill is mounted in each spindle, and the drill jig is passed from one spindle to the next for each hole.

DRILLING SPECIAL-SHAPED HOLES

Square, hexagon, octagon, and special-shaped holes can be quickly and accurately cut in a drill press. A special floating chuck, angular drill, guide plate, and slip bushing are required for this operation (Fig. 38-17A–D).

The principle of drilling holes with angular sides is the same as that used for drilling round holes. It is an accepted fact that a circle is composed of a series of minute chords (degrees) of which there are 360 to a full circle.

Figure 38-18 illustrates the various positions of the cut-

Fig. 38-17 Components required to produce special-shaped holes. (A) Floating chuck; (B) angular drill; (C) guide plate; (D) slip bushing.

ting lips of a square drill as it rotates in the guide plate. The center of the drill does not follow a circle, but a series of minute alternate cycloidlike curves whose chords are parallel to the sides of the hole being drilled. The full-floating chuck takes up the driving and floating action and allows the drill to operate as freely as an ordinary drill. Special-shaped holes can be drilled on a drill press, lathe, or any spindle-driven machine if the tools illustrated in Fig. 38-17 are used.

TO DRILL SPECIAL-SHAPED HOLES

1. Align and clamp the work to be drilled under the center of the drill press spindle.
2. Locate and fasten the guide plate to the workpiece in the required position.
3. Insert the slip bushing into the guide plate, and drill the lead hole. The lead hole allows the chips to clear and gives the drill freedom to turn the corners. Lead holes should be approximately 25 percent smaller than the distance across the flats of the hole to be drilled.
4. Insert a drill of the correct size into the full-floating chuck. The setscrew in the chuck should be tightened securely on the flat of the drill shank.
5. Mount the chuck into the drill press spindle.
6. With the drill press spindle in its uppermost position, raise the table until the workpiece is close to the drill.

Fig. 38-18 Drilling a square hole in a plate.

7. Snug up the drill press spindle sleeve bearing to prevent chatter.
8. Lower the drill into the guide plate and turn the machine spindle by hand to see that the drill turns freely.
9. Set the drill press to approximately one-quarter the speed and feed used for regular drilling.
10. Start the machine, lubricate the drill, and apply steady cutting pressure to avoid chatter while drilling to depth.

REVIEW QUESTIONS

COUNTERBORING AND COUNTERSINKING

1. List the procedure for counterboring a hole.
2. Why should holes that are to be tapped be countersunk?

TAPPING

3. What is the advantage of tapping a hole by hand in a drill press?
4. Describe the procedure for tapping a hole by hand in a drill press.
5. Explain how a tapping attachment operates.

BORING

6. Explain how a drill press may be set up for boring a hole.
7. What requirements are necessary for an accurate hole with a good surface finish to be produced?

TRANSFERRING HOLE LOCATIONS

8. Name three methods of transferring the location of holes from one part to another.
9. Explain the procedure for spotting with a twist drill.
10. What are transfer punches and how are they used?
11. Describe transfer screws and explain how they are used.

DRILLING SPECIAL-SHAPED HOLES

12. What tools are required for drilling special-shaped holes?
13. Explain the procedure for drilling special-shaped holes on a standard drill press.

THE LATHE
SECTION 11

Historically, the lathe is the forerunner of all machine tools. The first application of the lathe principle was probably the potter's wheel used in prehistoric times. This machine rotated a mass of clay which enabled the clay to be formed into a cylindrical shape.

The modern lathe operates on the same basic principle. The work is held and rotated on its axis while the cutting tool is advanced along the lines of a desired cut (Fig. 1). The lathe is one of the most versatile machine tools used in industry. With suitable attachments, the lathe may be used for turning, tapering, form turning, screw cutting, facing, drilling, boring, spinning, grinding, and polishing operations. Cutting operations are performed with a cutting tool fed either parallel or at right angles to the axis of the work. The cutting tool may also be fed at an angle relative to the axis of the work for machining tapers and angles.

Modern production has led to the development of many special types of lathes, such as the *engine, turret, single-* and *multiple-spindle automatic, tracer,* and *numerically controlled* lathes.

The *engine lathe* (Fig. 2), basically not a production lathe, is found in jobbing shops, school shops, and toolrooms.

When many duplicate parts are required, the *turret lathe* (Fig. 3)

Fig. 1 The cutting action of a lathe. (Courtesy Standard-Modern Tool Company.)

Fig. 2   The engine lathe is the most common lathe found in a machine shop. (Courtesy South Bend Lathe Company.)

Fig. 3   A turret lathe is used to mass produce parts.

may be used. This lathe is equipped with a multisided toolpost called a turret to which several different cutting tools may be mounted. In use, different cutting tools are employed in a given sequence to perform a series of operations on each part. This same sequence may be repeated on many parts without having to change or reset the cutting tools.

When hundreds or thousands of identical small parts are required, they may be produced on *single-* and *multiple-spindle automatic lathes* (Fig. 4). On these machines, six or eight different operations may be performed on as many parts at the same time. Once set up, the machine will produce the parts for as long as required.

Tracer lathes (Fig. 5) are used where a few duplicate parts are required. A hydraulically operated cross-slide (and cutting tool) is controlled by a stylus bearing against a round or flat template.

Fig. 4   A single-spindle automatic lathe.

Fig. 5   A hydraulic tracer attachment mounted on a lathe. (Courtesy Cincinnati Milacron Incorporated.)

Tracer attachments are available for converting most engine lathes into tracer lathes.

Numerically controlled lathes (Fig. 6) have come into widespread use in the past several years. With these machines, the cutting tool movements are controlled by a programmed, punched tape to perform a sequence of operations automatically on the workpiece once the machine has been set up.

A very useful and popular device which may be added to an engine lathe (or any other machine) is the *digital readout system*. Engine lathes equipped with these devices (Fig. 7) can produce duplicate parts to within 0.001 in. (0.02 mm) or less on the diameters and lengths of a workpiece.

Only the engine lathe, which is basic to all lathes, will be discussed in detail in the following units.

Fig. 6   A numerically controlled lathe.

Fig. 7   A lathe equipped with a digital readout system.   (Courtesy Bendix Corporation.)

UNIT 39

The majority of work turned on a lathe, other than in production shops, will be turned on an engine lathe. This lathe is an accurate and versatile machine on which many operations such as turning, tapering, form turning, threading, facing, drilling, boring, grinding, and polishing may be performed. Three common engine lathes are the tool-room, heavy-duty, and gap-bed lathes.

OBJECTIVES

After completing this unit you will be able to:
1. Identify and know the purpose of the main operative parts of the lathe
2. Set the lathe to run at any required speed
3. Set the proper feed for the cut required

LATHE SIZE AND CAPACITY

Lathe size is designated by the *largest work diameter* which can be swung over the lathe ways and generally the *maximum distance between centers* (Fig. 39-1). Some manufacturers designate the lathe size by the largest work diameter which can be swung over the ways and the overall bed length.

Lathes are manufactured in a wide range of sizes, the most common being from 9- to 30-in. swing, with a capacity of 16 in. to 12 ft between centers. A typical lathe may have a 13-in. swing, a 6-ft-long bed, and a capacity to turn work 36 in. long between centers (Fig. 39-1).

The average metric lathe used in school shops may have a 230- to 330-mm swing and have a bed length of from 500 to 3000 mm in length.

PARTS OF THE LATHE

The main parts of a lathe are the bed (and ways), headstock, quick-change gearbox, carriage, and tailstock (Fig. 39-2).

BED

The *bed* is a heavy, rugged casting made to support the working parts of the lathe. On its top section are machined ways that guide and align the major parts of the lathe.

HEADSTOCK

The *headstock* (Fig. 39-3) is clamped on the left-hand side of the bed. The *headstock spindle*, a hollow cylindrical shaft supported by bearings, provides a drive from the motor to work-holding devices. A live center, a faceplate

or a chuck can be fitted to the spindle nose to hold and drive the work. The live center has a 60° point which provides a bearing for the work to turn between centers.

Headstock spindles can be driven either by a stepped pulley and a belt, or by transmission gears in the headstock (Fig. 39-3). The lathe with a stepped pulley drive is generally called a belt-driven lathe; the gear-driven lathe is referred to as a geared-head lathe. Some light- and medium-duty lathes are equipped with a variable speed drive. On this type of drive, the speed is changed while the headstock spindle is revolving. However, when the lathe is shifted from a high- to low-speed range or vice versa, the lathe spindle must be stopped before the speed change is made.

A — SWING
B — DISTANCE BETWEEN CENTRES
C — LENGTH OF BED
D — RADIUS (ONE-HALF OF SWING)

Fig. 39-1 Lathe size is indicated by the swing and the length of the bed.

Fig. 39-2 The parts of an engine lathe. (Courtesy The R. K. LeBlond Machine Tool Co.)

Fig. 39-3 A gear-drive headstock.

The *feed reverse lever,* mounted on the headstock, reverses the rotation of the feed rod and lead screw.

QUICK-CHANGE GEARBOX

The *quick-change gearbox* (Fig. 39-4) containing a number of different sized gears, provides the *feed rod* and *lead screw* with various speeds for turning and thread-cutting operations. The feed rod advances the carriage for turning operations when the *automatic feed lever* is engaged. The lead screw advances the carriage for thread-cutting operations when the *split-nut lever* is engaged.

CARRIAGE

The *carriage* (Fig. 39-5), consisting of three main parts, the *saddle, cross-slide,* and *apron,* is used to move the cutting tool along the lathe bed.

Fig. 39-4 (A) A quick-change gearbox. (Courtesy Standard-Modern Tool Company.) (B) A cutaway section of a quick-change gear box. (Courtesy Cincinnati Lathe & Tool Company.)

Fig. 39-5 The main parts of an engine lathe carriage.

Fig. 39-6 The tailstock assembly. (Courtesy Cincinnati Milacron Incorporated.)

The *saddle,* an H-shaped casting mounted on the top of the lathe ways, provides a means of mounting the cross-slide and the apron.

The *cross-slide,* mounted on top of the saddle, provides a manual or automatic cross movement for the cutting tool. The *compound rest,* fitted on top of the cross-slide, is used to support the cutting tool. It can be swiveled to any angle for taper-turning operations, and is moved manually. The cross-slide and compound rest both have graduated collars which ensure accurate cutting tool settings in thousandths of an inch or hundredths of a millimeter. The apron, fastened to the saddle, houses the gears and mechanism required to move the carriage or cross-slide automatically. A locking-off lever inside the apron prevents engaging the split-nut lever and the automatic feed lever at the same time.

The *apron handwheel* can be turned manually to move the carriage along the lathe bed. This handwheel is connected to a gear which meshes in a rack fastened to the lathe bed.

The *automatic feed lever* engages a clutch which provides automatic feed to the carriage. The *feed-change lever* can be set for longitudinal feed or for cross feed. When in the neutral position, the feed-change lever permits the split-nut lever to be engaged for thread cutting. For thread-cutting operations, the carriage is moved automatically when the split-nut lever is engaged. This causes the threads of the split-nut to engage into the threads of the revolving lead screw and move the carriage at a predetermined rate.

TAILSTOCK

The *tailstock* (Fig. 39-6), consisting of the upper and lower tailstock castings, can be adjusted for taper or parallel turning by two screws set in the base. The tailstock can be locked in any position along the bed of the lathe by the *tailstock clamp.* The *tailstock spindle* has an internal taper to receive the *dead center,* which provides support for the right-hand end of the work. Other standard tapered shank tools, such as reamers and drills, can be held in the tailstock

spindle. A *spindle clamp* is used to hold the tailstock spindle in a fixed position. The *tailstock handwheel* moves the tailstock spindle in or out of the tailstock casting. It can also be used to provide a hand feed for drilling and reaming operations.

SETTING SPEEDS ON A LATHE

Engine lathes are designed to operate at various spindle speeds for machining of different materials. These speeds are measured in revolutions per minute (r/min) and are changed by the stepped pulleys or gear levers.

On a *belt-driven lathe,* various speeds are obtained by changing the flat belt and the back gear drive.

On the *geared-head lathe* (Fig. 39-3) speeds are changed by moving the speed levers into proper positions according to the revolutions per minute chart fastened to the headstock. While shifting the lever positions, place one hand on the faceplate or chuck, and turn the lathe spindle slowly by hand. This will enable the levers to engage the gear teeth without clashing.

NOTE: Never change speeds when the lathe is running. On lathes equipped with *variable-speed drives,* the speed is changed by turning a dial or handle *while the machine is running.*

SETTING FEEDS

The feed of an engine lathe, or the distance the carriage will travel in one revolution of the spindle, depends on the speed of the feed rod or lead screw. This is controlled by the change gears in the quick-change gearbox (Fig. 39-7A). This quick-change gearbox obtains its drive from the headstock spindle through the end gear train (Fig. 39-4B). A chart mounted on the front of the quick-change gearbox indicates the various feeds and metric pitches or threads

A

LEVERS		ENGLISH—THREADS PER INCH					METRIC—PITCH IN M/M			
		SLIDING FEEDS IN THOUSANDS					SURFACING $\frac{1}{4}$		SLIDING	
D	B	60	56	52	48	44	40	38	36	32
		.5 mm	.005	.005	.006	.006	.75 mm	.007	.008	.009
C	B	30	28	26	24	22	20	19	18	16
		1 mm	.010	.011	1.25 mm	.013	1.5 mm	.015	.016	.017
D	A	15	14	13	12	11	10	$9\frac{1}{2}$	9	8
		2 mm	.020	.021	2.5 mm	.025	3 mm	.029	.031	0.34
C	A	$7\frac{1}{2}$	7	$6\frac{1}{2}$	6	$5\frac{1}{2}$	5	$4\frac{3}{4}$	$4\frac{1}{2}$	4
		4 mm	.039	.042	5 mm	.050	6 mm	.058	.061	.068

B

Fig. 39-7   (A) A quick-change gearbox permits fast setting of feeds. (B) A quick-change gearbox chart. Note the settings for metric threads.   (Courtesy Colchester Lathe Company.)

SHEAR PIN

A

B

Fig. 39-8   (A) A shear pin in the end gear train prevents damage to the gears in case of an overload; (B) a spring-ball clutch will slip when too much strain is applied to the feed rod.   (Courtesy Colchester Lathe Company.)

per inch which may be obtained by setting levers to the positions indicated (Fig. 39-7B).

TO SET THE FEED FOR THE APRON (CARRIAGE DRIVE)

1. Select the desired feed on the chart.
2. Move tumbler lever #4 (Fig. 39-7A) into the hole directly below the selected feed.
3. Follow the row in which the selected feed is found to the left, and set the *feed-change levers* (#1 and #2) to the letters indicated.
4. Set lever #3 to disengage the lead screw.

NOTE: Before turning on the lathe, be sure all levers are fully engaged by turning the headstock spindle by hand, and see that the feed rod turns.

SHEAR PINS AND SLIP CLUTCHES

To prevent damage to the feed mechanism due to overload or sudden torque, some lathes are equipped with either *shear pins* or *slip clutches* (Fig. 39-8A, B). Shear pins, usually made of brass, may be found on the feed rod, lead screw, and the end gear train. Spring-loaded slip clutches are found only on feed rods. When the feed mechanism is overloaded, either the shear pin will break or the slip clutch will slip, causing the automatic feed to stop. This prevents damage to the gears or shafts of the feed mechanism.

REVIEW QUESTIONS

1. List the operations which can be performed on a lathe.
2. How is the size of a lathe designated?

PARTS OF THE ENGINE LATHE

3. Name the four *main* units of a lathe.
4. State the purpose of the following:
   (a) headstock spindle
   (b) lead screw and feed rod
   (c) quick-change gearbox
   (d) split-nut lever
   (e) feed-change lever
   (f) cross-slide
   (g) compound rest

SETTING SPEEDS AND FEEDS

5. List three types of lathe drives.
6. Explain how speeds on a geared-head lathe are changed.
7. List the steps to set a feed of 0.010 in. (0.25 mm).
8. What is the purpose of:
   (a) a shear pin?
   (b) a slip clutch?

Many lathe accessories are available to increase the versatility of the lathe and the variety of work which can be machined. Lathe accessories may be divided into two categories:
1. Work-holding, -supporting, and -driving devices
2. Cutting-tool-holding devices

Work-holding, -supporting, and -driving devices include lathe centers, chucks, faceplates, mandrels, steady and follower rests, lathe dogs, and drive plates. Cutting-tool-holding devices include various types of straight and offset toolholders, threading toolholders, boring bars, turret-type toolposts, and quick-change toolpost assemblies.

OBJECTIVES

After completing this unit you will be able to:
1. Identify and know the purpose of the common work-holding and -driving accessories
2. Identify and know the purpose of the common cutting-tool-holding accessories

WORK-HOLDING DEVICES

LATHE CENTERS

Most turning operations can be performed between centers on a lathe. Work to be turned between centers must have a center hole drilled in each end (usually 60°) to provide a bearing surface which allows the work to turn on the centers. The centers merely support the work while the cutting operations are performed. A lathe dog, fitted into a driving plate, provides a drive for the work (Fig. 40-1).

A variety of lathe centers are used to suit various operations or workpieces. Probably the most commonly used centers in school shops are the solid 60° centers with a Morse taper shank (Fig. 40-2). These are generally made from high speed steel or a good grade of machine steel with carbide inserts or tips. Care must be taken when using these centers to adjust and lubricate them occasionally as the work heats up and expands. If this is not done, both the center and the workpiece may be damaged. The damage to the workpiece will be the loss of concentricity which will prevent future operations from being performed using the center holes. The lathe center must also be reground to remove the damaged section before it can be used.

The revolving *dead center* (Fig. 40-3), sometimes called a live dead center, is commonly used to support long work held in a chuck. This center usually contains anti-friction bearings which allow the center to revolve with the work.

No lubrication is required between the center and the workpiece, and the center tension is not affected by workpiece expansion during the cutting action.

A *microset adjustable center* (Fig. 40-4) fits into the tailstock spindle and provides a means of aligning lathe centers or producing slight tapers on work being machined between centers. An eccentric, or sometimes a dovetail slide, allows this type of center to be adjusted a limited amount to each side of center. Lathe centers are quickly and easily aligned with this type of center.

Fig. 40-1 Work mounted between centers is usually driven by a lathe dog.

CARBIDE
INSERT

Fig. 40-2 A 60° lathe center. (Courtesy DoAll Company.)

REMOVABLE POINT
ROTATING COVER
ROLLER AND CAGE ASSEMBLY
SPRING RING
BODY
SPINDLE
LOCKING SCREW
NEEDLE ROLLER BEARING
SCREWED RETAINER
O-RING
END CAP

Fig. 40-3 A revolving dead center.
(Courtesy Enco Manufacturing Company.)

0.006 in.
0.006 in.

Fig. 40-4 A microset adjustable dead center.

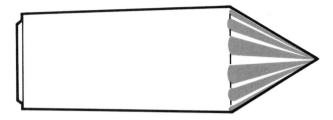

Fig. 40-5 A self-driving live center.

Fig. 40-6 A three-jaw universal geared scroll chuck.

Fig. 40-7 A four-jaw independent chuck.

The *self-driving center* (Fig. 40-5), mounted in the headstock spindle, is used when machining the entire length of a workpiece in one operation and when a chuck or lathe dog could not be used to drive the work. Grooves ground around the circumference of the lathe center point provide the drive for the workpiece. The work (usually a soft material such as aluminum) is forced onto the driving center; a revolving dead center is used to support the work and keep it against the grooves of the driving center.

CHUCKS

Some workpieces, because of their size and shape, cannot be held and machined between lathe centers. Lathe chucks are used extensively for holding work for machining operations. The most commonly used lathe chucks are: the three-jaw universal, four-jaw independent, and the collet chuck.

The three-jaw universal chuck (Fig. 40-6) is used to hold round and hexagonal work. It grasps the work quickly and within a few thousandths of an inch or hundredths of a millimeter of accuracy, because the three jaws move simultaneously when adjusted by the chuck wrench. This simultaneous motion is caused by a scroll plate into which all three jaws fit. Three-jaw chucks are made in various sizes from 4 to 16 in. (101.6 to 406.4 mm) in diameter. They are usually provided with two sets of jaws, one for outside chucking and the other for inside chucking.

The four-jaw independent chuck (Fig. 40-7) has four jaws, each of which can be adjusted independently by a chuck wrench. They are used to hold round, square, hexagonal, and irregular-shaped workpieces. The jaws can be reversed to hold work by the inside diameter.

Universal and independent chucks can be fitted to the three types of headstock spindles. Figure 40-8 shows a threaded spindle nose, Fig. 40-9 a tapered spindle nose, Fig. 40-10 a cam-lock spindle nose. The threaded type screws on in a clockwise direction; the tapered type is held by a lock nut that tightens on the chuck. The cam lock is held by tightening the cam locks using a T-wrench. On the taper and cam-lock types, the chuck is aligned by the taper on the spindle nose.

The collet chuck (Fig. 40-11) is the most accurate chuck and is used for high-precision work. Spring collets are available to hold round, square, or hexagon-shaped workpieces. Each collet has a range of only a few thousandths of an inch or hundredths of a millimeter over or under the size stamped on the collet.

A special adaptor is fitted into the taper of the headstock spindle, and a hollow draw bar having an internal thread is inserted in the opposite end of the headstock spindle. As the handwheel (and draw bar) is revolved, it draws the collet into the tapered adapter, causing the collet to tighten on the workpiece. This type of chuck is also referred to as a *spring-collet chuck*. Another form of spring-collet chuck uses a chuck wrench to tighten the collet on the workpiece. This type is mounted on the spindle nose in the same manner as standard chucks and can hold larger work than the draw-in type.

The Jacobs collet chuck (Fig. 40-12) has a wider range than the spring-collet chuck. Instead of a draw bar, it

Fig. 40-10   The parts of a cam-lock spindle nose.
1. Registration lines on spindle nose
2. Registration lines on cam-lock
3. Cam-locks
4. Cam-lock mating stud on chuck or faceplate
5. Hollow-head retaining screw

Fig. 40-8   A threaded spindle nose.

Fig. 40-9   An American standard lathe spindle nose.

Fig. 40-11   A cross-sectional view of a headstock showing the construction of a draw-in collet assembly.

incorporates an impact-tightening handwheel to close the collets on the workpiece. A set of 11 Rubber-Flex collets, each having an adjustment range of almost ⅛ in. (3.2 mm), makes it possible to hold a wide range of work diameters. When the handwheel is turned clockwise, the rubber-flex collet is forced into a taper, causing it to tighten on the workpiece. When the handwheel is turned counterclockwise, the collet opens and releases the workpiece.

Magnetic chucks (Fig. 40-13) are used to hold iron or steel parts that are too thin or that may be damaged if held in a conventional chuck. These chucks are fitted to an adaptor mounted on the headstock spindle. Work is held lightly for aligning purposes by turning the chuck wrench approximately one-quarter turn. After the work has been trued, the chuck is turned to the full-on position to hold the work securely. This type of chuck is used only for light cuts and for special grinding applications.

Faceplates are used to hold work that is too large or of such a shape that it cannot be held in a chuck or between centers. Faceplates are usually equipped with several slots to permit the use of bolts to secure the work or

Fig. 40-12 The Jacobs collet chuck has a wider range than other collet chucks. (Courtesy The Jacobs Manufacturing Company.)

Fig. 40-13 Work may be held on a magnetic chuck for turning operations.

Fig. 40-14 An angle plate fastened to a faceplate is used to hold the work for machining. (Courtesy Colchester Lathe & Tool Co.)

Fig. 40-15 A steadyrest is often used to support long or slender work during a machining operation.

Fig. 40-16 A follower rest mounted on the saddle may be used to support long, slender work when machining.

Fig. 40-17 A plain mandrel. (Courtesy Ash Precision Equipment Inc.)

angle plate (Fig. 40-14) so that the axis of the workpiece may be aligned with the lathe centers. When work is mounted off center, a counterbalance (Fig. 40-14) should be fastened to the faceplate to prevent imbalance and the resultant vibrations when the lathe is in operation.

A *steadyrest* (Fig. 40-15) is used to support long work held in a chuck or between lathe centers. It is located on, and aligned by the ways of the lathe and may be positioned at any point along the lathe bed provided it clears the carriage travel. The three jaws, tipped with plastic, bronze, or rollers, may be adjusted to support any work diameter within the steadyrest capacity. During machining operations performed on or near the end of a workpiece, a steadyrest supports the end of work held in a chuck, when the work cannot be supported by the tailstock center. A steadyrest also supports the center of long work to prevent springing when the work is machined between centers.

A *follower rest* (Fig. 40-16), mounted on the saddle, travels with the carriage to prevent work from springing up and away from the cutting tool. The cutting tool is generally positioned just ahead of the follower rest to provide a smooth bearing surface for the two jaws of the follower rest.

A *mandrel* holds an internally machined workpiece between centers so that further machining operations are concentric with the bore. There are several types of mandrels, the most common being the *plain mandrel* (Fig. 40-17), the *expanding mandrel, gang mandrel,* and *stub mandrel.*

LATHE DOGS

When work is machined between centers, it is generally driven by a lathe dog. The lathe dog has an opening to receive the work and a setscrew to fasten the dog to the work. The tail of the dog fits into a slot on the driveplate and provides the drive to the workpiece.

Lathe dogs are made in a variety of sizes and types to suit various workpieces:

The *standard bent-tail lathe dog* (Fig. 40-18A) is the most commonly used dog for round workpieces. These dogs are available with square-head setscrews or headless setscrews which are safer since there is no head protruding.

Fig. 40-18 Common types of lathe dogs. (A) Standard bent tail; (B) straight tail; (C) safety clamp; (D) clamp type.

The *straight-tail dog* (Fig. 40-18B) is driven by a stud in the driveplate. Since this is a more balanced type of dog than the bent-tail dog, it is used in precision turning where centrifugal force of a bent-tail dog may cause inaccuracies in the work.

The *safety clamp lathe dog* (Fig. 40-18C) may be used to hold a variety of work since it has a wide range of adjustment. It is particularly useful on finished work where the setscrew of a standard lathe dog may damage the finish.

The *clamp lathe dog* (Fig. 40-18D) has a wider range than the other types and may be used on round, square, rectangular, and odd-shaped workpieces.

CUTTING-TOOL-HOLDING DEVICES

Most toolbits used in lathe-turning operations in school shops are square and are generally held in a standard toolholder (Fig. 40-19). These toolholders are made in various styles and sizes to suit different machining operations. Toolholders for turning operations are available in three styles: left-hand offset, right-hand offset, and straight.

Each of these toolholders has a square hole to accommodate the square toolbit which is held in place by a setscrew. The hole in the toolholder is at an angle of approximately 15 to 20° to the base of the toolholder (Fig. 40-19C). When the cutting tool is set on center, this angle provides the proper amount of back rake in relation to the workpiece.

The *left-hand offset* toolholder (Fig. 40-19A) is designed for machining work close to the chuck or faceplate and for cutting from right to left. This type of toolholder is designated by the letter L to indicate the direction of cut.

The *right-hand offset* toolholder (Fig. 40-19B) is designed for machining work close to the tailstock, for cutting

Fig. 40-19 Common lathe toolholders. (A) left-hand offset; (B) right-hand offset; (C) straight; (D) carbide. (Courtesy J. H. Williams and Company.)

from left to right, and for facing operations. This type of toolholder is designated by the letter R.

The *straight* toolholder (Fig. 40-19C) is a general-purpose type. It can be used for taking cuts in either direction and for general machining operations. This type of toolholder is designated by the letter S.

The *carbide* toolholder (Fig. 40-19D) has a square hole, parallel to the base of the toolholder, to accommodate carbide-tipped toolbits. When using carbide-tipped toolbits, it is important that the toolbit be held so that there is little or no back rake (Fig. 40-19D). Toolholders of this type are designated by the letter C. The correct methods of using this and other types of carbide toolholders are fully explained in Unit 83.

Toolholders for throwaway carbide inserts are shown in Fig. 40-20. The insert in Fig. 40-20A is held in the holder by a cam action. The insert in Fig. 40-20B is secured by means of a clamp. These types of toolholders are available for use with conventional, turret-type, and heavy-duty toolposts.

Cutting-off or *parting tools* are used when the work must be grooved or parted off. The long, thin, cutting-off blade is locked securely in the toolholder by means of either a cam lock or a locking nut. Three types of parting toolholders are shown in Fig. 40-21A–C.

A *threading toolholder* (Fig. 40-22) is designed to hold a special form-relieved thread-cutting tool. Although most machinists grind their own thread-cutting tools, this tool is convenient because it has an accurately ground 60° angle. This angle is maintained throughout the life of the tool because it is only sharpened on the top of the cutting surface when it becomes dull.

Boring toolholders are made in several styles. A *light boring toolholder* (Fig. 40-23A) is held in a standard toolpost and is used for small holes and light cuts. The *medium boring toolholder* (Fig. 40-23B) is suitable for heavier cuts and is also held in a standard toolpost. The cutting tool may be held at 45° or 90° to the axis of the bar.

The *heavy-duty boring bar holder* (Fig. 40-23C) is mounted on the compound rest of the lathe. It has three bars of different diameters to suit the diameter of the hole being bored. With this type of boring bar, use the largest bar possible to get the maximum rigidity and to avoid chatter. The toolbit may be held at 45 or 90° to the axis.

TOOLPOSTS

The *standard* or *round* toolpost (Fig. 40-24) is generally supplied with the engine lathe. This toolpost fits into the T-slot of the compound rest. It provides a means of holding

A

B

A

C

Fig. 40-20 (A) Carbide inserts are held securely in a holder by means of a cam action. (B) Carbide inserts are held securely by a clamp.

Fig. 40-21 Types of solid cutting-off toolholders. (A) Left-hand; (B) straight; (C) right-hand.   (Courtesy J. H. Williams and Company.)

and adjusting the toolholder or cutting tool. The concave ring and the wedge or rocker provides for the adjustment of the cutting-tool height.

The *turret-type* toolpost (Fig. 40-25) provides a convenient means of holding toolbits when several operations, such as turning, threading, grooving, and parting, must be performed on a workpiece. This type of toolholder is suitable for carbide-tipped tools, since no provision is made for back rake. Four different types of toolbits may be mounted in this toolholder and each may be brought into position by loosening the locking handle and rotating the turret head until the desired toolbit is in the cutting position. The handle is then turned to lock the head in place.

Quick-change toolposts and holders provide a quick and accurate means of setting and changing tools for different operations. The dovetailed toolholder fits onto a dovetailed

Fig. 40-22   A preformed thread-cutting tool and toolholder.

A

B

C

Fig. 40-23   (A) A light boring toolholder. (B) a medium boring toolholder; (C) a heavy-duty boring tool.

Fig. 40-24   A standard toolpost.

Fig. 40-25   A turret-type toolpost permits the use of up to four different cutting tools.

231

Fig. 40-26 With a dovetailed toolpost, cutting tools can be rapidly changed and accurately set.

Fig. 40-28 A heavy-duty toolpost is used for taking heavy cuts. (Courtesy Cincinnati Milacron Incorporated.)

Fig. 40-27 Various types of quick-change toolholders. (Courtesy DoAll Company.)

toolpost which is mounted on the compound rest (Fig. 40-26). There are several different holders to accommodate the various types of cutting tools (Fig. 40-27). The cutting tool, which is held in the holder by means of setscrews, is generally sharpened in the holder and preset (in the holder) in the toolroom. After a cutting tool becomes dull, the unit can be quickly replaced with another sharpened and preset unit. This procedure ensures accuracy of size, since each preset toolbit will be in exactly the same position as the previous cutting tool. Each unit may be adjusted vertically on the toolpost by means of the knurled adjusting nut, then locked in position by means of a clamp.

The *heavy-duty toolpost* (Fig. 40-28) is also mounted in the slot of the compound rest. It is designed basically for use with carbide cutting tools and holds a single cutting tool or toolholder. This type of toolpost is used when very heavy cuts are required.

REVIEW QUESTIONS

WORK-HOLDING DEVICES

1. Name the three types of lathe centers and state the purpose of each.

2. What precautions must be taken when turning work supported by solid centers?
3. Describe and state the purpose of the following:
   (a) three-jaw universal chuck
   (b) four-jaw independent chuck
   (c) collet chuck
   (d) magnetic chuck
4. What advantage has a Jacobs collet chuck over a spring-collet chuck?
5. What is the purpose of
   (a) a steadyrest?   (b) a follower rest?
6. Name three types of lathe dogs.
7. What is the disadvantage of the square-head setscrew in a lathe dog?
8. What is the advantage of a straight-tail dog?

CUTTING-TOOL-HOLDING DEVICES

9. Describe three types of standard toolholders and state the purpose of each.
10. How does a carbide toolholder differ from a standard toolholder?
11. State two methods by which throwaway carbide inserts are held in a toolholder.
12. What procedure should be followed when using a heavy-duty boring bar set?
13. Name and state the purpose of four types of toolposts.

UNIT 41

One of the most important components in the machining process is the cutting tool. Its performance will determine the efficiency of the operation. Consequently, much thought should be given not only to the selection of the cutting-tool material but also to the cutting-tool angles required to machine the workpiece material properly.

OBJECTIVES

After completing this unit you will be able to:
1. Know the various types of cutting tools and their application
2. Identify the various rake and clearance angles on a lathe toolbit
3. Grind a general-purpose lathe cutting tool

CUTTING-TOOL MATERIALS

Lathe toolbits are generally made of four materials: high-speed steel, cast alloys (such as stellite), cemented carbides, and ceramics. More exotic cutting tools such as cubic boron nitride (Borazon) and diamonds are used considerably in industry for special applications. Borazon is used to machine hardened alloy steels and tough superalloys. Diamond cutting tools are used to machine nonferrous and nonmetallic materials which require close tolerances and a high surface finish.

The properties possessed by each of these materials are different and the application of each depends on the material being machined and the condition of the machine.

Lathe toolbits should possess the following properties.

1. They should be hard.
2. They should be wear-resistant.
3. The should be capable of standing up to high temperatures developed during the cutting operation.
4. They should be able to withstand shock during the cutting operation.

HIGH-SPEED STEEL TOOLBITS

Probably the toolbit most commonly used in schools for lathe operations is the high-speed steel toolbit. High-speed steels may contain combinations of tungsten, chromium, vanadium, molybdenum, and cobalt. They are capable of taking heavy cuts, withstanding shock, and maintaining a sharp cutting edge under red heat.

High-speed steel toolbits are generally of two types: molybdenum-base (Group M) and tungsten-base (Group T). The most widely used tungsten-base toolbit is known as T_1, which is sometimes called 18-4-1 since it contains about 18 percent tungsten, 4 percent chromium, and 1 percent vanadium.

A general-purpose molybdenum-base high-speed steel toolbit is known as M_1 or 8-2-1. This alloy contains about 8 percent molybdenum, 2 percent tungsten, 1 percent vanadium, and 4 percent chromium.

These two types are general-purpose tools; if more red hardness is desired, one containing more cobalt should be selected. Since there are many different grades of high-speed steel toolbits, it is advisable to refer to the manufacturer's recommendations for a toolbit for a specific job. Table 86-1 indicates the properties imparted to a high-speed steel toolbit by the various alloying elements.

CAST ALLOY TOOLBITS

Cast alloy (stellite) toolbits usually contain 25 to 35 percent chromium, 4 to 25 percent tungsten, 1 to 3 percent carbon; the remainder is cobalt. These toolbits have high hardness, high resistance to wear, and excellent red-hardness qualities. Since they are cast, they are weaker and more brittle than high-speed steel toolbits. Stellite toolbits are capable of high speeds and feeds on deep uninterrupted cuts. They may be operated at about two to two-and-a-half times the speed of a high-speed steel toolbit.

NOTE: When grinding stellite toolbits, light pressure only should be used and the toolbit should not be quenched in water.

CEMENTED CARBIDES

Cemented carbide toolbits for lathe work may be of two types: those brazed to a steel shank and the throwaway type of insert (see Unit 83). Carbide toolbits have low toughness but high hardness and excellent red-hardness qualities. Carbides are available in different grades to suit the material being machined. Tungsten carbide toolbits are generally used for cutting cast iron and nonferrous metals, whereas titanium and tantalum carbides are used for machining various types of steel. Cemented carbide toolbits are capable of speeds three to four times greater than high-speed steel toolbits.

CERAMIC AND DIAMOND CUTTING TOOLS

The most recent development in lathe cutting tools is the application of ceramic and diamond cutting tools. These tools are capable of much higher cutting speeds than those of any other cutting tools. They are discussed in detail in Unit 84.

CUTTING TOOLS NOMENCLATURE

Cutting tools used on a lathe are generally single-pointed, and although the shape of the tool is changed for various applications, the same nomenclature applies to all cutting tools (Fig. 41-1).

The base is the bottom surface of the tool shank.
The cutting edge is the leading edge of the toolbit that does the cutting.
The face is the surface against which the chip bears as it is separated from the work.
The flank is the surface of the tool which is adjacent to and below the cutting edge.
The nose is the tip of the cutting tool formed by the junction of the cutting edge and the front face.
The nose radius is the radius to which the nose is ground. The size of the radius will affect the finish. For rough turning, a small nose radius [about 1/64 in. (0.38 mm)] is used. A larger radius [about 1/16 to 1/8 in. (1.58 to 3.17 mm)] is used for finish cuts.

The point is the end of the tool that has been ground for cutting purposes.
The shank is the body of the toolbit or the part held in the toolholder.

LATHE TOOLBIT ANGLES AND CLEARANCES

Proper performance of a toolbit depends on the clearance and rake angles which must be ground on the toolbit. Although these angles vary for different materials, the nomenclature is the same for all toolbits (see Table 41-1).

Side Cutting Edge Angle: The angle which the cutting edge forms with the side of the tool shank (Fig. 41-2). Side cutting angles for a general-purpose lathe cutting tool may be from 10 to 20°, depending on the material being cut. If this angle is too large (over 30°), the tool will tend to chatter.

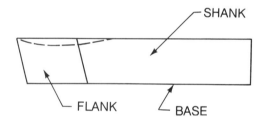

Fig. 41-1 Angles and clearances of lathe cutting tools.

MATERIAL	SIDE RELIEF	FRONT RELIEF	SIDE RAKE	BACK RAKE
Aluminum	12	8	15	35
Brass	10	8	5 to −4	0
Bronze	10	8	5 to −4	0
Cast iron	10	8	12	5
Copper	12	10	20	16
Machine steel	10 to 12	8	12 to 18	8 to 15
Tool steel	10	8	12	8
Stainless steel	10	8	15 to 20	8

TABLE 41-1 Rake and Relief Angles in Degrees for High-Speed Steel Lathe Tools

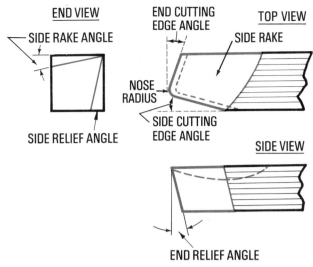

Fig. 41-2 Nomenclature of a general-purpose lathe toolbit.

Fig. 41-3 Lathe cutting-tool angles and clearances.

Fig. 41-4 The shape and dimensions of a general-purpose toolbit.

End Cutting Edge Angle: The angle formed by the end cutting edge and a line at right angles to the center line of the toolbit (Fig. 41-2). This angle may vary from 5 to 30° depending on the type of cut and finish desired. An angle of 5 to 15° is satisfactory for roughing cuts; angles between 15 and 30° are used for general purpose turning tools. The larger angle permits the cutting tool to be swivelled to the left when taking light cuts close to the dog or chuck, or when turning to a shoulder.

Side Relief (Clearance) Angle: The angle ground on the flank of the tool below the cutting edge (Figs. 41-2 and 3). This angle is generally 6 to 10°. The side clearance on a toolbit permits the cutting tool to advance lengthwise into the rotating work and prevents the flank from rubbing against the workpiece.

End Relief (Clearance) Angle: The angle ground below the nose of the toolbit which permits the cutting tool to be fed into the work. It is generally 10 to 15° for general purpose tools (Figs. 41-2 and 3). This angle must be measured when the toolbit is held in the toolholder. The end relief angle varies with the hardness and type of material and the type of cut being taken. The end relief angle is smaller for harder materials, to provide support under the cutting edge.

Side Rake Angle: The angle at which the face is ground away from the cutting edge. For general purpose toolbits, the side rake is generally 14° (Figs. 41-2 and 3). Side rake creates a keener cutting edge and allows the chips to flow away quickly. For softer materials, the side rake angle is generally increased.

Angle of Keenness: The included angle produced by grinding side rake and side clearance on a toolbit. This angle may be altered depending on the type of material being machined.

Back (Top) Rake: The backward slope the tool face away from the nose. The back rake angle is generally about 20° and is provided for in the toolholder (Fig. 41-3). Back rake permits the chips to flow away from the point of the cutting tool.

GRINDING LATHE CUTTING TOOLS

Because of the number of turning operations which can be performed on a lathe, a wide variety of cutting tools are used. In order for these cutting tools to perform effectively, they must possess certain angles and clearances for the material being cut (Table 41-1). All lathe tools cut if they have front and side clearance. The addition of side and top rake enables the chips to escape quickly from the cutting edge, making the tool cut better. All lathe cutting tools must have certain angles and clearances regardless of shape; therefore, only the grinding of a general-purpose cutting tool (Fig. 41-4) will be explained in detail.

TO GRIND A GENERAL-PURPOSE TOOLBIT

1. Grip the toolbit firmly while supporting the hands on the grinder toolrest (Fig. 41-5).
2. Hold the toolbit at the proper angle to grind the cutting edge angle. At the same time, tilt the bottom of the toolbit in toward the wheel and grind the 10° side relief or clearance angle on the cutting edge.

NOTE: The cutting edge should be approximately ½ in. (12.7 mm) long and should be over about one-quarter the width of the toolbit (Fig. 41-5).

3. While grinding, move the toolbit back and forth across

Fig. 41-5 Grinding the side cutting edge and side relief angles on a toolbit.

½ | ¼ OF TOOLBIT WIDTH

70° TO 80° POINT ANGLE

Fig. 41-6 Grinding the end relief angle on a lathe toolbit.

the face of the wheel. This accelerates grinding and prevents grooving the wheel.

4. The toolbit must be cooled frequently during the grinding operation. NEVER OVERHEAT A TOOLBIT.

NOTE: Stellite or cemented carbide tools should never be quenched, and *never* grind carbides with an aluminum oxide wheel.

FRONT CLEARANCE — TOOL ANGLE FOR STEEL
SIDE CLEARANCE
TOOL ANGLE FOR CAST IRON

Fig. 41-7 Checking the end relief angle of a toolbit while it is in a toolholder.

SIDE RAKE-GROUND THE LENGTH OF THE CUTTING EDGE

Fig. 41-8 Grinding the side rake on a lathe toolbit. (Courtesy Kostel Enterprises Limited.)

5. Grind the end cutting edge so that it forms an angle a little less than 90° with the side cutting edge (Fig. 41-6). Hold the tool so that the end cutting edge angle and end relief angle of 15° are ground at the same time.

6. Using a toolbit grinding gage, check the amount of end relief when the toolbit is in the toolholder (Fig. 41-7).

7. Hold the top of the toolbit at approximately 45° to the axis of the wheel and grind the side rake to approximately 14° (Fig. 41-8).

NOTE: When grinding the side rake, *be sure that the top of the cutting edge is not ground below the top of the toolbit*. If a step is ground in the top of the toolbit, a chip trap is formed which greatly reduces the efficiency of the cutting tool.

8. Grind a slight radius on the point of the cutting tool, *being sure to maintain the same front and side clearance angle*.

9. With an oilstone, hone the cutting edge of the toolbit slightly. This will lengthen the life of the toolbit and enable it to produce a better surface finish.

REVIEW QUESTIONS

CUTTING-TOOL MATERIALS

1. What properties should a cutting tool possess?
2. What elements are found in high-speed steel toolbits?
3. State the precaution which should be taken when grinding stellite toolbits.
4. Name two types of carbide toolbits.
5. List three qualities of a carbide tool.
6. Name three types of carbides used in toolbits.

CUTTING-TOOL NOMENCLATURE

7. Make neat labeled sketches of a single-pointed cutting tool and indicate the following parts:
   (a) face
   (b) cutting edge
   (c) nose
   (d) point
   (e) flank
   (f) shank

LATHE TOOLBIT ANGLES AND CLEARANCES

8. State the purpose of the following:
   (a) side cutting edge angle
   (b) side relief angle
   (c) end relief angle
   (d) side rake angle
   (e) back rake
   (f) angle of keenness
9. Draw sketches to illustrate the angles in question 8.

GRINDING LATHE CUTTING TOOLS

10. State the procedure for grinding a general-purpose toolbit.

CUTTING SPEEDS, FEEDS, AND DEPTH OF CUT

UNIT 42

In order to operate a lathe efficiently, the machinist must consider the importance of cutting speeds and feeds. Much time may be lost if the lathe is not set at the proper spindle speed and if the proper feed rate is not selected.

OBJECTIVES

After completing this unit you will be able to:
1. Calculate the speed at which any given workpiece should revolve
2. Determine the proper feed for roughing and finishing cuts
3. Estimate the time required to machine a part in a lathe

CUTTING SPEEDS

Lathe work *cutting speed* (CS) may be defined as the rate at which a point on the work circumference travels past the cutting tool. For instance, if a metal has a CS of 90 ft/min, the spindle speed must be set so that 90 ft of circumference will pass the cutting tool in 1 min. Cutting speed is always expressed in feet per minute (ft/min) or in meters per minute (m/min). Remember that the CS of a metal must not be confused with the number of turns the workpiece will make in 1 min (r/min).

Industry demands that machining operations be performed as quickly as possible; therefore correct CS must be used for the type of material being cut. If a CS is too high, the cutting tool edge breaks down rapidly, resulting in time lost to recondition the tool. With too slow a CS,

| MATERIAL | TURNING AND BORING | | | | THREADING | |
| | ROUGH CUT | | FINISH CUT | | | |
	FT/MIN	M/MIN	FT/MIN	M/MIN	FT/MIN	M/MIN
Machine steel	90	27	100	30	35	11
Tool steel	70	21	90	27	30	9
Cast iron	60	18	80	24	25	8
Bronze	90	27	100	30	25	8
Aluminum	200	61	300	93	60	18

time will be lost for the machining operation, resulting in low production rates. Based on research and testing by steel and cutting-tool manufacturers, the CS for high-speed-steel tools listed in Table 42-1 are recommended for efficient metal removal rates. These speeds may be varied slightly to suit factors such as the condition of the machine, the type of work material, and sand or hard spots in the metal. The CS for cemented-carbide and ceramic cutting tools may be found in Units 83 and 84, Tables 83-4 and 84-2.

To calculate the lathe spindle speed in revolutions per minute (r/min), the CS of the metal and the diameter of the work must be known. The proper spindle speed can then be set by dividing the CS (in inches per minute) by the circumference of the work (in inches). The calculation for determining the spindle speed (r/min) is as follows:

INCH MEASUREMENTS

$$\text{Formula:} \quad \text{r/min} = \frac{\text{CS} \times 12}{\pi D}$$

$$\text{where} \quad \text{CS} = \text{cutting speed}$$

$$D = \text{diameter of work to be turned}$$

However, because most lathes provide only a limited number of speed settings, a simpler formula is usually used:

$$\text{r/min} = \frac{\text{CS} \times 4}{D}$$

Thus, to calculate the r/min required to rough-turn a 2-in.-diameter piece of machine steel (CS 90)

$$\text{r/min} = \frac{\text{CS} \times 4}{D}$$
$$= \frac{90 \times 4}{2}$$
$$= 180$$

NOTE: The recommended speeds and feeds for carbide cutting tools are found in Unit 83. The same formula for calculating spindle speed will be used for these tools.

METRIC MEASUREMENTS

$$\text{Formula:} \quad \text{r/min} = \frac{\text{CS} \times 320}{D}$$

EXAMPLE

Calculate the r/min required to turn a 45-mm-diameter piece of machine steel (CS 40 m/min).

$$\text{r/min} = \frac{\text{CS} \times 320}{D}$$
$$= \frac{40 \times 320}{45}$$
$$= \frac{12,800}{45}$$
$$= 284$$

LATHE FEED

The feed of a lathe is defined as the distance the cutting tool advances along the length of the work for every revolution of the spindle. For example, if the lathe is set for a 0.015-in. (0.40-mm) feed, the cutting tool will travel along the length of the work 0.015 in. (0.40 mm) for every complete turn that the work makes. The feed of an engine lathe is dependent upon the speed of the lead screw or feed rod. The speed is controlled by the change gears in the *quick-change gearbox* (Fig. 39-4A).

Whenever possible, only two cuts should be taken to bring a diameter to size: a roughing cut and a finishing cut. Since the purpose of a roughing cut is to remove excess material quickly and surface finish is not too important, a coarse feed should be used. The finishing cut is used to bring the diameter to size and produce a good surface finish and therefore a fine feed should be used.

For general-purpose machining, a 0.010- to 0.015-in. (0.25- to 0.40-mm) feed for roughing and a 0.003- to 0.005-in. (0.07- to 0.12-mm) feed for finishing is recom-

TABLE 42-2 Feeds for Various Materials (Using a High-Speed Cutting Tool)

MATERIAL	ROUGH CUTS		FINISH CUTS	
	INCHES	MILLIMETERS	INCHES	MILLIMETERS
Machine steel	0.010–0.020	0.25–0.50	0.003–0.010	0.07–0.25
Tool steel	0.010–0.020	0.25–0.50	0.003–0.010	0.07–0.25
Cast iron	0.015–0.025	0.40–0.65	0.005–0.012	0.13–0.30
Bronze	0.015–0.025	0.40–0.65	0.003–0.010	0.07–0.25
Aluminum	0.015–0.030	0.40–0.75	0.005–0.010	0.13–0.25

mended. Table 42-2 lists the recommended feeds for cutting various materials when using a high-speed-steel cutting tool.

DEPTH OF CUT

The depth of cut may be defined as the depth of the chip taken by the cutting tool and is one-half of the total amount removed from the workpiece in one cut. Figure 42-1 shows a ⅛-in. depth of cut being taken on a 2-in.-diameter workpiece. Note that the diameter has been reduced by ¼ in. to 1¾ in. diameter. When machining a workpiece, it is advisable to take one roughing and one finishing cut if possible. If much material must be removed, the roughing cut should be as deep as possible to reduce the diameter to within 0.030 to 0.040 in. (0.76 to 1.02 mm) of the size required. When rough turning, the depth of cut will depend on the following factors:

- ☐ The condition of the machine
- ☐ The type and shape of the cutting tool used
- ☐ The rigidity of the workpiece
- ☐ The rate of feed

When finish turning, the depth of cut will depend on the type of work and the finish required. In any case, it should not be less than 0.005 in. (0.13 mm).

GRADUATED MICROMETER COLLARS

When the diameter of a workpiece must be turned to an accurate size, the graduated micrometer collars should be used.

Graduated micrometer collars are sleeves or bushings that are mounted on the compound rest and crossfeed screws (Fig. 42-2). They assist the lathe operator to set the cutting tool accurately to remove the required amount of material from the workpiece. The collars on lathes using the inch system of measurement are usually graduated in thousandths of an inch (0.001). The micrometer collars on

Fig. 42-1 The depth of cut on a lathe.

Fig. 42-2 Micrometer collars on the compound rest and cross-feed handles permit machining the work to an accurate size. The thumbscrew A is used to lock the collar in place.

lathes using the metric system of measurement are usually graduated in steps of two hundredths of a millimeter (0.02 mm). The circumference of the crossfeed and compound rest screw collars on lathes using inch systems of measurement are usually divided into 100 or 125 equal divisions, each having a value of 0.001 in. Therefore, if the crossfeed screw is turned *clockwise* 10 graduations, the cutting tool will be moved 0.010 in. toward the work. Because the work in a lathe revolves, a 0.010 in. depth of cut will be taken from the entire work circumference, thereby reducing the diameter 0.020 in. (2 × 0.010 in.) (Fig. 42-3). On some lathes however, the graduations are such that when the graduated collar is moved 10 divisions, the cutting tool will move only 0.005 in. and the diameter will be reduced by 0.010 in. or by an amount equal to the reading on the collar.

NOTE: Some industrial machines may have collars with 250 or even 500 graduations which make a much finer infeed possible.

Machine tools equipped with graduated collars generally fall into two classes.

1. *Machines where the work revolves.* These include lathes, vertical boring mills, cylindrical grinders, etc.

When you machine the *circumference of work* on machines where the work revolves, it is wise to remember that since material is removed from the entire circumference, the cutting tool should be moved in only half the amount of material to be removed.

2. *Machines where the work does not revolve.* These include shapers, milling machines, surface grinders, etc.

On machines where the work does not revolve, the material removed from a workpiece is equal to the amount set on the graduated collar, because the machining is taking place only on one surface. Therefore, if a 0.010 in. depth of cut is set, 0.010 in. will be removed from the work (Fig. 42-4).

HINTS ON GRADUATED COLLAR USE

1. If the graduated collar has a locking screw, make sure the collar is secure before setting a depth of cut (Fig. 42-2).

2. All depths of cut must be made by feeding the cutting tool *toward the workpiece*.

3. If the graduated collar is turned past the desired setting, it must be turned backward a half-turn and then fed into the proper setting to remove the backlash (the play between the feed screw and the nut).

4. Never hold a graduated collar when setting a depth of cut. Graduated collars with friction devices can be moved easily if held when setting a depth of cut.

5. The graduated collar on the *compound rest* can be used for accurately setting the depth of cut for the following operations.

☐ *Shoulder turning*—When a series of shoulders must be spaced accurately along a piece of work, the compound

TOOL FEED ⟶ 0.010
MATERIAL REMOVED 0.020

Fig. 42-3   On machines where the work revolves, the cutting tool should be set in for only half the amount to be removed from the diameter.

TOOL FEED .010

.010 REMOVED

Fig. 42-4   On machines where the work does not revolve, the cutting should be set in for the amount of material to be removed.

rest should be set at 90° to the cross-slide. With the carriage locked in position, the graduated collar of the compound rest can be used for the spacing of shoulders to within 0.001 in. (or 0.02 mm on metric collars) accuracy.

☐ *Facing*—The graduated collars may also be used to face a workpiece to length. When the compound rest is swung to 30°, the amount removed from the length of the work when facing will be one-half the amount of feed on the graduated collar.

☐ *Machining accurate diameters*—When accurate diameters must be machined or ground in a lathe, the compound rest should be set to 84°16' to the cross-slide. On inch-calibrated compound rests, a 0.001 in. movement would result in a 0.0001 in. infeed movement of the tool (Fig. 42-5). Similarly, on metric lathes, a 0.02 mm movement of the compound rest results in a 0.002 mm infeed movement of the cutting tool.

Fig. 42-5   The compound rest is set at 84°16' for making fine settings.

CALCULATING MACHINING TIME

Any machinist should be able to estimate the time required to machine a workpiece. Factors such as spindle speed, feed and depth of cut must be considered.

The following formula may be applied to calculate the time requird to machine a workpiece.

$$\text{Time} = \frac{\text{distance}}{\text{rate}}$$

$$\text{Distance} = \text{length of cut}$$
$$\text{Rate} = \text{feed} \times \text{r/min}$$

EXAMPLE

Calculate the time required to machine a 2-in.-diameter machine steel shaft 16 in. long to 1.850-in.-diameter finish size.

SOLUTION

Roughing cut—See Table 42–1.

$$r/min = \frac{CS \times 4}{D}$$
$$= \frac{90 \times 4}{2}$$
$$= 180$$

Roughing feed—See Table 42–2.

$$\text{Feed} = 0.020$$

$$\text{Roughing cut time} = \frac{\text{length of cut}}{\text{feed} \times \text{r/min}}$$
$$= \frac{16}{0.020 \times 180}$$
$$= 4.4 \text{ min}$$

Finishing cut time

$$r/min = \frac{100 \times 4}{1.850}$$
$$= 216$$

$$\text{Finish feed} = 0.003$$

$$\text{Finish cut time} = \frac{16}{0.003 \times 216}$$

$$= 24.7 \text{ min}$$
$$\text{Total machine time} = 4.4 + 24.7 = 29.1 \text{ min}$$

REVIEW QUESTIONS

CUTTING SPEEDS AND FEEDS

1. Define CS and state how it is expressed.
2. Why is proper CS important?
3. At what r/min should the lathe revolve to rough-turn a piece of cast iron 101 mm in diameter when using a high-speed steel toolbit?
4. Calculate the r/min to turn a piece of 3¾-in.-diameter machine steel using a high-speed steel toolbit.
5. Define lathe feed.

DEPTH OF CUT

6. Define depth of cut.
7. How deep should a roughing cut be?
8. What factors will determine the depth of a roughing cut?
9. A 2½-in.-diameter workpiece must be machined to 2.375 finished diameter. What should be the depth of:
   (a) the roughing cut?
   (b) the finishing cut?

GRADUATED MICROMETER COLLARS

10. Name two classes of machines equipped with graduated collars and state the differences between them.
11. What precautions should be taken when setting a depth of cut?
12. What is the value of one graduation on a metric graduated micrometer collar?
13. What size will a 75-mm-diameter workpiece be after a 6.25-mm-deep cut is taken from the work?

MACHINING TIME

14. Calculate the time required to machine a 3⅛-in.-diameter piece of tool steel 14 in. long to a diameter of 3.000 in.

UNIT 43

Any good worker will be aware of the safety requirements in any area of a shop and will always attempt to observe the safety rules. Failure to do so may result in serious injury with the resultant loss of time and pay and in the loss of production to the company.

OBJECTIVES

After completing this unit you will be:
1. Aware of the importance of good safety standards in a shop
2. Familiar with the necessary safety precautions required to run a lathe
3. Able to point out any safety infractions by other workers

SAFETY PRECAUTIONS

The lathe, like most other machine tools, can be hazardous if not operated properly. A good lathe operator is a safe operator who realizes the importance of keeping the machine and the surrounding area clean and tidy. Accidents on any machine do not just happen; they are usually caused by carelessness and could generally be avoided. To minimize the chance of accidents when you are operating a lathe, the following precautions should be observed.

1. Always wear approved safety glasses. When operating a lathe, chips fly and it is important to protect your eyes.

2. Roll up sleeves, remove ties, and tuck in loose clothing. Short sleeves are preferable because loose clothing can get caught by revolving lathe dogs, chucks, and rotating parts of the lathe. You can be drawn in causing a serious injury.

3. Never wear a ring or watch (Fig. 43-1).

    a. Rings or watches can be caught in revolving work or lathe parts and cause serious injuries.

    b. A metal object falling on the hand would bend or break a ring causing much pain and suffering until the ring could be removed.

4. Do not operate a lathe until you fully understand its controls.

    a. Not knowing what might happen when levers or switches are turned on can be very dangerous.

    b. Be sure that you can stop the machine quickly in case something unexpected happens.

5. Never operate a machine if safety guards are removed or not closed properly.

    a. Safety guards are installed by the manufacturer to cover revolving gears, belts, or shafts.

    b. Loose clothing or a hand can be drawn into the revolving parts if guards are not replaced or closed properly.

6. Stop the lathe before you measure the work, clean, oil, or adjust the machine. Measuring revolving parts can result in broken tools or personal injury to the operator.

7. Do not use a rag to clean the work or the machine when the lathe is in operation. The rag may get caught and be drawn in along with your hand.

8. Never attempt to stop a lathe chuck or drive plate with your hand. Your hand can be injured or fingers broken if they get caught in the slots and projections of the drive plate or chuck.

9. Be sure that the chuck or faceplate is mounted securely before starting the lathe.

    a. If the lathe is started with a loose spindle accessory, the rotation would loosen the accessory and cause it to be thrown from the lathe.

Fig. 43-1 Wearing rings and watches in a machine shop can be dangerous.

Fig. 43-2 Never leave a chuck wrench in a chuck.

Fig. 43-3 Good housekeeping can prevent tripping accidents.

b. A heavy accessory, with the velocity created by the revolving spindle, can turn into a dangerous missile.

10. Always remove the chuck wrench after use (Fig. 43-2). *Never leave it in the chuck at any time!* If the lathe is started with a chuck wrench in the chuck, the following could result:

   a. The wrench could fly and injure someone.

   b. The wrench could be jammed against the lathe bed damaging the wrench, lathe bed, chuck, and lathe spindle.

11. Move the carriage to the furthest position of the cut and revolve the lathe spindle one complete turn by hand before you start the lathe.

   a. This will ensure that all parts will clear without jamming.

   b. It will also prevent an accident and damage to the lathe.

12. Keep the floor around the machine free from grease, oil, metal cuttings, tools, and workpieces (Fig. 43-3).

   a. Oil and grease can cause falls which can result in painful injuries.

   b. Objects on the floor are hazards which are the cause of tripping accidents.

13. Avoid horseplay at all times, especially when operating any machine tool. Horseplay can result in tripping accidents or being pushed into the revolving spindle or workpiece.

14. Always clean the chips with a brush and never with your hand or a cloth (Fig. 43-4). Steel chips are sharp and can cause cuts if handled manually or with a cloth that has chips embedded.

15. Whenever you are polishing, filing, cleaning, or making adjustments to the workpiece or machine, always remove the sharp toolbit from the toolholder to prevent serious cuts to arms or hands.

Fig. 43-4 Always remove chips with a brush or hook; *never by hand.*

REVIEW QUESTIONS

1. State three possible results of not observing the safety rules in a shop.

2. What is generally the most important cause of an accident?

3. Why are the following points important when operating a lathe?

   (a) wearing safety glasses

   (b) wearing loose clothing

   (c) wearing watches and rings

   (d) removing a chuck wrench

   (e) keeping the machine and surrounding floor area clean

   (f) stopping the lathe to measure work or clean the machine

4. Why should chips not be cleaned from a lathe with a cloth?

MOUNTING, REMOVING, AND ALIGNING LATHE CENTERS

UNIT 44

Any work machined between lathe centers is generally turned for some portion of its length, then reversed, and the other end finished. It is important, when machining work between centers, that the live center run absolutely true. If the live center does not run true, when the work is reversed to machine the opposite end, the two turned diameters will not be concentric with each other and the part may have to be scrapped.

OBJECTIVES

After completing this unit you will be able to:
1. Properly mount and/or remove the lathe centers
2. Align the lathe centers by visual, trial cut, and dial indicator methods

TO MOUNT LATHE CENTERS

1. Remove any burrs from the lathe spindle, centers, or spindle sleeves (Fig. 44-1).
2. Thoroughly clean the tapers on the lathe centers and in the headstock and tailstock spindles (Fig. 44-2).

NOTE: Never attempt to clean the taper in the headstock spindle while the lathe is running.

3. Partially insert the cleaned center in the lathe spindle.
4. With a quick snap, force the center into the spindle. When mounting a tailstock center, observe the same procedure.

After a center has been mounted in the headstock spindle, it should be checked for trueness. Start the lathe and observe if the center runs true. Whenever accuracy is required, it is wise to check the trueness of the center with a dial indicator (Fig. 44-3). If the center is not running true and has been mounted properly, it should be ground while mounted in the lathe spindle.

TO REMOVE LATHE CENTERS

The *live center* may be removed by using a *knockout bar* that is pushed through the headstock spindle (Fig. 44-4). A slight tap is required to remove the center. When removing the live center with a knockout bar, place a cloth over the center and hold it with one hand to prevent an accident or damage to the center (Fig. 44-5).

Fig. 44-1 Remove burrs from the spindle sleeve before inserting it in a lathe.

Fig. 44-2 Clean the headstock spindle before inserting a center.

Fig. 44-3 Using a dial indicator to check the trueness of the live center.

Fig. 44-4 Removing the live center with a knockout bar.

The dead center can be removed by turning the tailstock handwheel to draw the spindle back into the tailstock. The end of the screw contacts the end of the dead center, forcing it out of the spindle.

ALIGNMENT OF LATHE CENTERS

To produce a parallel diameter when machining work between centers, it is important that the lathe centers be aligned; that is, the two lathe centers must be in line with each other and true with the centerline of the lathe. If the centers are not aligned, the work being machined will be tapered.

Three common methods are used to align lathe centers.

1. By aligning the centerlines on the back of the tailstock with each other (Fig. 44-6). This is only a visual check and therefore not too accurate.
2. The trial cut method (Fig. 44-7), where a small cut is taken from each end of the work and the diameters are measured with a micrometer.
3. By using a parallel test bar and dial indicator (Fig. 44-8). This is the fastest and most accurate method of aligning lathe centers.

TO ALIGN CENTERS BY ADJUSTING THE TAILSTOCK

1. Loosen the tailstock clamp nut or lever.
2. Loosen one of the adjusting screws G or F (Fig. 44-6), depending upon the direction the tailstock must be moved.

Fig. 44-5 Hold a cloth over the point of the lathe center to prevent hand injury when removing. (Courtesy Kostel Enterprises Limited.)

Fig. 44-6 The lines on the tailstock must be in line to produce a parallel diameter.

Tighten the other adjusting screw until the line on the top half of the tailstock aligns exactly with the line on the bottom half.
3. Tighten the loosened adjusting screw to lock both halves of the tailstock in place.
4. Check that the tailstock lines are still aligned and adjust if necessary.
5. Lock the tailstock clamp nut or lever.

TO ALIGN CENTERS BY THE TRIAL CUT METHOD

1. Take a light cut [approximately 0.005 in. (0.12 mm)] to true the diameter, from Section A at the tailstock end for ¼ in. (6.35 mm) long.
2. Stop the feed and note the reading on the graduated collar of the crossfeed handle.
3. Move the cutting tool away from the work with the crossfeed handle.
4. Bring the cutting tool close to the headstock end.
5. Return the cutting tool to the same graduated collar setting as at Section A.

Fig. 44-7   A trial cut at each end of the work is used to check the lathe center alignment.

Fig. 44-8   Tailstock alignment may be accurately checked with a parallel test bar and dial indicator.

6. Cut a ½-in. (12.7-mm) length at Section B and then stop the lathe.

7. Measure both diameters with a micrometer (Fig. 44-7).

8. If both diameters are not the same size, adjust the tailstock either toward or away from the cutting tool one-half the difference of the two readings.

9. Take another light cut at A and B at the same crossfeed graduated collar setting. Measure these diameters and adjust the tailstock, if required.

To Align Centers Using a Dial Indicator and Test Bar

1. Clean the lathe and work centers and mount the test bar.

2. Adjust the test bar snugly between centers and tighten the tailstock spindle clamp.

3. Mount a dial indicator on the toolpost or lathe carriage. Be sure that the indicator plunger is parallel to the lathe bed and that the contact point is set on center.

4. Adjust the cross-slide so that the indicator registers approximately 0.025 in. or 0.65 mm at the tailstock end.

5. Move the carriage by hand so that the indicator registers on the diameter at the headstock end (Fig. 44-8) and note the indicator reading.

6. If both indicator readings are not the same, adjust the tailstock by the adjusting screws until the indicator registers the same reading at both ends.

7. Tighten the adjusting screw which was loosened.

8. Tighten the tailstock clamp nut.

9. Adjust the tailstock spindle until the test bar is snug between the lathe centers.

10. Recheck the indicator readings at both ends and adjust the tailstock, if necessary.

REVIEW QUESTIONS

1. Briefly describe the proper procedure for mounting a lathe center.

2. How may the live center be checked for trueness after it has been mounted in the spindle?

3. What precautions should be observed when removing a live center?

4. Briefly describe how to align the lathe centers by the trial cut method.

5. Why must the test bar be snug between centers when aligning centers with a dial indicator?

FACING BETWEEN CENTERS
UNIT 45

Work mounted between centers can be machined, removed, or set up for additional machining and still maintain the same degree of accuracy. Facing on a lathe is one of the most important machining operations in a machine shop. It is very important that the cutting tool and work be properly set up, or damage to the machine, work, and lathe centers will result.

OBJECTIVES

After completing this unit you will be able to:
1. Set up a workpiece for machining between centers
2. Set up a workpiece to face the ends
3. Face a workpiece to an exact length

TO SET UP A CUTTING TOOL FOR MACHINING

1. Move the toolpost to the *left-hand side* of the T-slot in the compound rest.
2. Mount a toolholder in the toolpost so that the setscrew in the toolholder is approximately 1 in. (25.4 mm) beyond the toolpost (Fig. 45-1).

When taking heavy cuts, it is advisable to set the toolholder at right angles to the work (Fig. 45-2). If the toolholder should move under pressure of the cut, the cutting tool would swing away from the work and make the diameter larger.

Fig. 45-1 The toolholder and toolbit being set up for a machining operation.

If the toolholder were set as in Fig. 45-3 and it moved under pressure of the cut, the toolbit would swing into the work, causing the diameter to be cut undersize. A toolholder can be set as in Fig. 45-3 for light finishing cuts.

3. Insert the proper cutting tool into the toolholder, having the toolbit extend ½ in. (12.7 mm) beyond the toolholder, and never more than twice its thickness.
4. Tighten the toolholder setscrew only two-finger pressure on the wrench to hold the toolbit in the toolholder.

NOTE: If the toolholder setscrew is tightened too tightly, it will break the toolbit, which is very hard and brittle.

5. Set the cutting-tool point to center height. Check it against the lathe center point (Fig. 45-1).
6. Tighten the toolpost securely to prevent it from moving during a cut.

TO MOUNT WORK BETWEEN LATHE CENTERS

1. Check that the live center is running true. If it is not running true, remove the center, clean all surfaces, and replace the center. Check again for trueness.
2. Clean the lathe center points and the center holes in the workpiece.
3. Adjust the tailstock spindle until it projects about 2½ to 3 in. (63.5 to 76.2 mm) beyond the tailstock (Fig. 45-4).
4. Loosen the tailstock clamp nut or lever.
5. Place a lathe dog on the left-hand end of the work and leave it loose.
6. Apply a suitable center lubricant to the right-hand end of the work.
7. Place the end of the work with the lathe dog on the

TOOL SWINGS CLEAR OF WORK

CUTTING TOOL INCLINED TOWARD TAILSTOCK

Fig. 45-2 The toolholder set to prevent the toolbit from digging into the work.

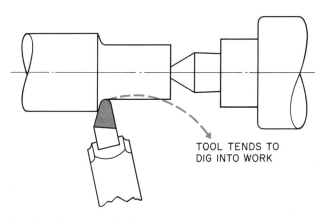

TOOL TENDS TO DIG INTO WORK

Fig. 45-3 The toolholder improperly set for taking a heavy cut.

Fig. 45-4   The tailstock spindle should not extend more than 2½ to 3 in. beyond the tailstock.

Fig. 45-5   Hold one end of the work on the live center while sliding the tailstock up to support the other end.

Fig. 45-6   Tighten the tailstock clamp nut securely.

Fig. 45-7   Snug up the work between centers while holding the tail of the dog to the top of the drive plate slot.

Fig. 45-8   Reverse the tailstock handwheel until the tail of the dog drops in the slot.

Fig. 45-9   Hold the handwheel in position while tightening the tailstock spindle clamp.

live center (Fig. 45-5) and slide the tailstock up until it supports the other end of the work.

8. Tighten the tailstock clamp nut or lever (Fig. 45-6).

9. Adjust the tail of the dog in the drive plate slot, making sure it does not bind, and tighten the lathe dog screw.

10. Turn the drive plate by hand until the tail of the lathe dog is parallel to the bed of the lathe.

11. With the tail of the dog *up* in the slot lightly tighten the tailstock handwheel (Fig. 45-7).

12. Carefully turn the tailstock handwheel backward until the tail of the dog drops in the drive plate slot (Fig. 45-8).

13. Hold the tailstock handwheel in this position, and, with the other hand, tighten the tailstock spindle clamp (Fig. 45-9).

14. Check the center tension. The tail of the dog should drop by gravity into the slot, and there should be no end play between the centers (Fig. 45-10).

Fig. 45-10 The center tension is correct when the tail of the dog drops of its own weight and there is no end play.

15. Move the carriage to the furthest position of the cut and revolve the lathe by hand to make sure that the lathe dog clears the compound rest.

FACING BETWEEN CENTERS

Workpieces to be machined are generally cut a little longer than required, and then end-faced to the proper length. Facing is an operation of machining the ends of a workpiece square with its axis. To produce a flat, square surface when facing between centers, the lathe centers must be in line. Work is often held in a chuck, faced to length, and center-drilled in one setup. This operation is covered in Unit 51.

The purposes of facing are:

1. To provide a true, flat surface, square with the axis of the work
2. To provide an accurate surface from which to take measurements
3. To cut the work to the required length

TO FACE WORK BETWEEN CENTERS

1. Move the toolpost to the *left-hand side* of the compound rest, and set the right-hand facing toolbit to the height of the lathe center point (Fig. 45-11).
2. Clean the lathe and work centers, and mount the work between centers.

NOTE: Use a half-center in the tailstock if it is available (Fig. 45-12).

3. Set the facing toolbit pointing left, as shown in Fig. 45-12.

NOTE: The point of the toolbit must be closest to the work and a space must be left along the side.

4. Set the lathe to the correct speed for the diameter and type of material being cut.
5. Start the lathe and bring the toolbit as close to the center of the work as possible.

Fig. 45-11 The toolholder should be gripped short and the toolbit set to center. (Courtesy Kostel Enterprises Limited.)

SPACE

Fig. 45-12 A half-center allows the entire surface to be faced.

6. Move the carriage to the left, using the apron hand-wheel, until a small cut is started.
7. Feed the cutting tool out by turning the crossfeed handle and cut from the center outward. If the automatic crossfeed is used for feeding the cutting tool, the carriage should be locked in position (Fig. 45-13).

Fig. 45-13 Lock the carriage to produce a flat surface.

8. Repeat operations 5, 6, and 7 until the work is cut to the correct length. (It is advisable, before facing, to mark the correct length with center-punch marks and then face until the punch marks are cut in half.)

NOTE: When facing, finishing cuts should begin at the center of the workpiece and be fed toward the outside.

FACING THE WORKPIECE TO AN ACCURATE LENGTH

When work is to be machined accurately to length, the compound rest graduated micrometer collars may be used. When facing, the compound rest may be set to 30° (to the cross-slide). The side movement of the cutting tool is always half the amount of the compound feed. For example, if the compound rest is fed in 0.010 in. (0.25 mm), the side movement of the tool, or the amount of material removed from the end of the workpiece, will be 0.005 in. (0.12 mm) (Fig. 45-14).

Fig. 45-14 The compound rest may be set at 30° for accurate end facing.

REVIEW QUESTIONS

1. Explain the procedure for setting up a cutting tool for turning.
2. List the main steps for mounting work between centers.

FACING BETWEEN CENTERS

3. State three purposes of facing a workpiece.
4. Explain how a facing toolbit is set up.

MACHINING BETWEEN CENTERS
UNIT 46

In school shops or training programs, where the length of each session is fixed, much of the work machined on a lathe is done between centers. When both the headstock and tailstock are aligned, work can be machined, removed from the lathe at the end of the work period, and replaced for additional machining with the assurance that any machined diameter will run true (concentric) with other diameters. In training programs, it is often necessary to remove and replace work in a lathe many times before it is completed; therefore, machining between centers saves much valuable time in setting up work accurately in comparison to other work-holding methods. The most common operations performed on work mounted between centers are facing, rough and finish turning, shoulder turning, filing, and polishing.

OBJECTIVES

After completing this unit you should be able to:
1. Set up the cutting tool for turning operations
2. Turn parallel diameters to within ±0.002-in. (0.05-mm) accuracy for size
3. Produce a good surface finish by filing and polishing
4. Machine square, filleted, and beveled shoulders to within 1/64-in. (0.30-mm) accuracy

SETTING UP A CUTTING TOOL

When machining between centers, it is very important that the workpiece and cutting tool be set up properly, otherwise the workpiece could be ruined or the lathe centers damaged. Carelessness in setting up work properly could also result in work being thrown out of a lathe and causing injury to the operator.

1. Move the toolpost to the *left-hand side* of the T-slot in the compound rest.
2. Mount a toolholder in the toolpost so that the setscrew in the toolholder is approximately 1 in. (25.4 mm) beyond the toolpost (Fig. 46-1).

When taking heavy cuts, it is advisable to set the toolholder at right angles to the work (Fig. 46-2A). If the toolholder should move under pressure of the cut, the cutting tool would swing away from the work and make the diameter larger.

If the toolholder were set as in Fig. 46-2B and it moved under pressure of the cut, the toolbit would swing into the work, causing the diameter to be cut undersize. A toolholder can be set as in Fig. 46-2B for light finishing cuts.
3. Insert the proper cutting tool into the toolholder, having the tool extend 1/2 in. (12.7 mm) beyond the toolholder and never more than twice its thickness.

TOOL SWINGS CLEAR OF WORK

CUTTING TOOL INCLINED TOWARD TAILSTOCK

A

TOOL TENDS TO DIG INTO WORK

B

Fig. 46-2 (A) The toolholder set to prevent the toolbit from digging into the work; (B) the toolholder improperly set for taking a heavy cut.

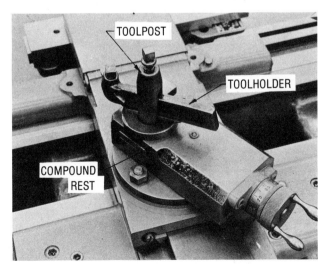

TOOLPOST

TOOLHOLDER

COMPOUND REST

Fig. 46-1 Grip the toolholder short in the toolpost.

251

Fig. 46-3 Set the point of the toolbit even with the center.

4. Set the cutting-tool point to center height. Check it against the lathe center point (Fig. 46-3).

5. Tighten the toolpost *securely* to prevent it from moving during a cut.

MOUNTING WORK BETWEEN CENTERS

Since the procedure for mounting the work between lathe centers for matching is the same as for facing, this operation will not be repeated. See Unit 45 for the proper procedure for mounting work between centers.

PARALLEL TURNING

Work is generally machined on a lathe for two reasons: to cut it to size and to produce a true diameter. Work that must be cut to size and have the same diameter along the entire length of the workpiece involves the operation of parallel turning. Many factors determine the amount of material which can be removed on a lathe at one time. Whenever possible, a diameter should be cut to size in two cuts: a roughing cut and a finishing cut (Fig. 46-4).

NOTE: To remove metal from a cylindrical piece of work and have the same diameter at each end, *the lathe centers must be in line*. (See Unit 44 for the methods of aligning centers.)

Before either the rough or finish cut is taken, the cutting tool must be set accurately for the depth of cut desired.

TO SET AN ACCURATE DEPTH OF CUT

1. Move the toolpost to the left-hand side of the compound rest and set the toolbit height to center.

2. Tighten the toolpost screw securely.

3. Start the lathe and take a light cut [approximately 0.005 in. (0.12 mm) or until a true diameter is produced] ¼ in. (6.35 mm) long at the right end of the work (Fig. 46-5).

4. Stop the lathe, but *DO NOT* move the crossfeed screw handle.

Fig. 46-4 A rough and finish cut.   (Courtesy Kostel Enterprises Limited.)

Fig. 46-5 Take a light trial cut about ¼ in. long to clean up the diameter.

Fig. 46-6 With the cutting tool clear of the work, measure the diameter.

5. Move the cutting tool past the end of the work by turning the carriage handwheel.

6. Measure the work and calculate the amount of material to be removed (Fig. 46-6).

7. Advance the graduated collar clockwise half the amount of material to be removed. For example, if 0.040 in. (1.00 mm) must be removed, the graduated collar should be turned in 0.020 in. (0.50 mm), since the cut is taken off the circumference of the work.

ROUGH TURNING

The operation of rough turning is used to remove as much metal as possible in the shortest length of time. Accuracy and surface finish are not important in this operation; therefore a 0.020- to 0.030-in. (0.50- to 0.76-mm) feed is recommended. Work is generally rough-turned to within 1/32 in. (0.78 mm) of the finished size in as few cuts as possible (preferably one cut).

TO ROUGH-TURN WORK

1. Set the lathe to the correct speed for the type and size of material being cut (Table 42-1).
2. Adjust the quick-change gearbox for a 0.010- to .030-in. (0.25- to 0.76-mm) feed, depending on the depth of cut and condition of the machine.
3. Move the toolholder to the left-hand side of the compound rest and set the toolbit height to center.
4. Tighten the toolpost *securely* to prevent the toolholder from moving during the machining operation.
5. Take a light trial cut at the right-hand end of the work for 1/4-in. (6.35-mm) length (Fig. 46-5).
6. Measure the work and adjust the toolbit for the proper depth of cut.
7. Cut along for 1/4 in. (6.35 mm), stop the lathe, and check the diameter for size. The diameter should be approximately 1/32 in. (0.78 mm) over the finish size (Fig. 46-6).
8. Readjust the depth of cut, if necessary.

FINISH TURNING

Finish turning, which follows rough turning, produces a smooth surface finish and cuts the work to an accurate size. Factors such as the condition of the cutting tool, the rigidity of the machine and work, and the lathe speeds and feeds affect the type of surface finish produced.

TO FINISH-TURN WORK

1. See that the cutting edge of the toolbit is free from nicks, burrs, etc. It is good practice to hone the cutting edge before you take a finish cut.
2. Set the toolbit on center; check it against the lathe center point.
3. Set the lathe to the recommended speed and feed. The feed used depends upon the surface finish required.
4. Take a light trial cut 1/4 in. (6.35 mm) long at the right-hand end of the work to
   a. Produce a true diameter
   b. Set the cutting tool to the diameter
   c. Set the graduated collar to the diameter
5. Stop the lathe and measure the diameter.
6. Set the depth of cut for half the amount of material to be removed.
7. Cut along for 1/4 in. (6.35 mm), stop the lathe, and check the diameter.
8. Readjust the depth of cut, if necessary, and finish turn the diameter.

Fig. 46-7 Move the carriage to the extreme right and cover the ways of the lathe with paper.

NOTE: In order to produce the truest diameter possible, finish turn work to the required size. Should it be necessary to finish a diameter by filing or polishing, *never* leave more than 0.002 to 0.003 in. (0.05 to 0.07 mm) for this operation.

FILING IN A LATHE

Work should be filed in a lathe only to remove a small amount of stock, to remove burrs, or round off sharp corners. Work should always be turned to within 0.002 to 0.003 in. (0.05 to 0.07 mm) of size, if the surface is to be filed. When larger amounts must be removed, the work should be machined, since excessive filing will produce work which is out of round and not parallel. The National Safety Council recommends filing with the left hand, so that the arms and hands can be kept clear of the revolving chuck or drive plate. It is wise to always remove the toolbit from the toolholder before filing unless the machining operation does not permit it. In this case, wind the carriage so that the toolbit is as far as possible from the area being filed.

NOTE: Before attempting to file or polish in a lathe, it is good practice to cover the lathe bed with a piece of paper to prevent filings from getting into the slides and causing excessive wear and damage to the lathe (Fig. 46-7). Cloth is not suitable for this purpose because it tends to get caught in the revolving work or the lathe.

TO FILE IN A LATHE

1. Set the spindle speed to approximately twice that used for turning.
2. Mount the work between centers, lubricate, and carefully adjust the dead center in the workpiece. Use a revolving dead center if one is available.
3. Move the carriage as far to the right as possible and remove the toolpost.
4. Disengage the lead screw and feed rod.
5. Select a 10- or 12-in. (250 or 300 mm) *mill file* or a *long-angle lathe file*.

NOTE: Be sure that the file handle is properly secured on the tang of the file.

6. Start the lathe.

7. Grasp the file handle in the left hand and support the file point with the right-hand fingers (Fig. 46-8).

8. Apply light pressure and push the file forward to its full length. Release pressure on the return stroke.

9. Move the file about half the width of the file for each stroke and continue filing, using 30 to 40 strokes per minute, until the surface is finished.

10. When filing in a lathe, the following points should be observed.

   a. Roll up sleeves above the elbows—short sleeves are preferable.

   b. Remove watches and rings.

   c. Never use a file without a properly fitted handle.

   d. Never apply too much pressure to the file. Excessive pressure produces out-of-roundness and causes the file teeth to clog and damage the work surface.

   e. Clean the file frequently with a file brush. Rub a little chalk into the file teeth to prevent clogging and facilitate cleaning.

POLISHING IN A LATHE

After the work surface has been filed, the finish may be improved by polishing with abrasive cloth.

TO POLISH IN A LATHE

1. Select the correct type and grade of abrasive cloth for the finish desired: use a piece about 6 to 8 in. (150 to 200 mm) long and 1 in. (25 mm) wide. For ferrous metals, use aluminum oxide abrasive cloth. Silicon carbide abrasive cloth should be used for nonferrous metals.

2. Set the lathe to run at high speed.

3. Disengage the feed rod and lead screw.

4. Remove the toolpost and toolholder.

5. Lubricate and adjust the dead center. Use a revolving dead center if one is available.

6. Roll up sleeves above the elbows and tuck in any loose clothing.

7. Start the lathe.

8. Hold the abrasive cloth on the work (Fig. 46-9).

9. With the right hand, press the cloth firmly on the work while *tightly* holding the other end of the abrasive cloth with the left hand. (*Caution:* Do not let the short end of the abrasive cloth wrap around the work.)

10. Move the cloth slowly back and forth along the work.

NOTE: For normal finishes, 80- to 100-grit abrasive cloth should be used. For better finishes, use a finer grit abrasive cloth.

TURNING TO A SHOULDER

When turning more than one diameter on a piece of work, the change in diameters, or step, is known as a shoulder. Three common types of shoulders are illustrated in Fig. 46-10.

Fig. 46-8 Always hold the file handle in the left hand to avoid injury.

Fig. 46-9 A high surface finish can be produced with abrasive cloth.

TO TURN A SQUARE SHOULDER

1. With the work mounted in a lathe, lay out the shoulder position from the finished end of the work. In case of filleted shoulders, allow sufficient length to permit the proper radius to be formed on the finished shoulder.

2. Place the point of the toolbit at this mark and cut a *small* groove around the circumference to mark off the length.

3. With a turning tool, rough- and finish-turn the work to within $\frac{1}{16}$ in. (1.58 mm) of the required length (Fig. 46-11).

4. Set up an end-facing tool, chalk the small diameter of the work, and bring the cutting tool up until it just removes the chalk mark (Fig. 46-12).

5. Note the reading on the graduated collar of the crossfeed handle.

Fig. 46-10   Types of shoulders.

Fig. 46-11   Turn the small diameter to within ¹⁄₁₆ in. of the finished length.

Fig. 46-12   Note the reading on the graduated collar when the toolbit just touches the small diameter.

6. Face (square) the shoulder, cutting to the line *using hand feed*.

7. For successive cuts, return the crossfeed handle to the same graduated collar setting.

TO MACHINE A FILLETED SHOULDER

Fillets are used at a shoulder to overcome the sharpness of a corner and to strengthen the part at this point. If a filleted corner is required, a toolbit having the same radius is used to finish the shoulder.

1. Lay out the length of the shoulder with a center-punch mark or by cutting a light groove (Fig. 46-13).

2. Rough- and finish-turn the small diameter to the correct length *minus the radius to be cut*. For example, a 3-in. (76.2-mm) length with a ⅛-in. (3.17 mm) radius should be turned 2⅞ in. (73 mm) long.

3. Mount the correct radius toolbit and set it to center.

4. Set the lathe for one-half the turning speed.

5. Coat the small diameter, near the shoulder, with chalk or layout dye.

6. Start the lathe and feed the cutting tool in until it lightly marks the small diameter near the shoulder (Fig. 46-14).

7. Slowly feed the cutting tool sideways with the carriage handwheel until the shoulder is cut to the correct length.

Fig. 46-13   The length of the shoulder is indicated by a center-punch mark.

MACHINING A BEVELED (ANGULAR) SHOULDER

Beveled or angular shoulders are used to eliminate sharp corners and edges, to make parts easier to handle, and to improve the appearance of the part. They are sometimes used to strengthen a part by eliminating the sharp corner found on a square shoulder. Shoulders are beveled at angles

Fig. 46-14   Setting the radius toolbit to the small diameter.

Fig. 46-15   Using a protractor to set the toolbit side cutting edge to an angle.

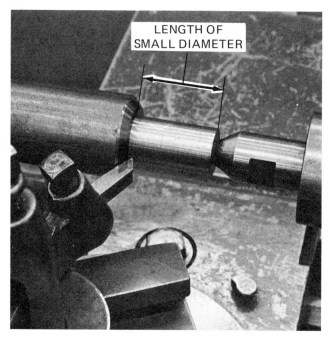

LENGTH OF SMALL DIAMETER

Fig. 46-16   Machining a beveled shoulder with the side of a cutting tool.

Fig. 46-17   The compound rest swiveled to cut a large beveled shoulder.

ranging from 30 to 60°; however, the most common is the 45° bevel.

1. Turn the large diameter to size.
2. Lay out the position of the shoulder along the length of the workpiece.
3. Rough- and finish-turn the small diameter to size.
4. Mount a side cutting tool in the toolholder and set it to center.
5. Use a protractor and set the side cutting edge of the toolbit to the desired angle (Fig. 46-15).
6. Apply chalk or layout dye to the small diameter as close as possible to the shoulder location.
7. Set the lathe spindle to approximately one-half of the turning speed.
8. Bring the point of the toolbit in until it just removes the chalk or layout dye.
9. Turn the carriage handwheel by hand to feed the cutting tool into the shoulder (Fig. 46-16).
10. Apply cutting fluid to assist the cutting action and to produce a good surface finish.
11. Machine the beveled shoulder until it is the required size.

If the size of the shoulder is large and chatter occurs when cutting with the side of the toolbit, it may be necessary to cut the beveled shoulder using the compound rest (Fig. 46-17).

1. Set the compound rest to the desired angle.
2. Adjust the toolbit so that only the point does the cutting.
3. Machine the bevel by feeding the compound rest by hand.

SETTING UP A CUTTING TOOL

1. How should the toolholder and cutting tool be set for machining between centers?

2. What might happen if the toolholder were set to the left and moved under the pressure of the cut?

PARALLEL TURNING

3. What precaution must be taken before parallel turning?

4. Explain the procedure for setting an accurate depth of cut.

5. State the purpose of rough and finish turning.

6. How many cuts should be taken to turn a diameter to size?

7. What is the purpose of a light trial cut at the right-hand end of the work?

FILING AND POLISHING

8. How much should be left on a diameter for filing to size?

9. How should the file be held for filing in a lathe?

10. List two of the most important points to be observed when filing in a lathe.

11. List the main steps for polishing a diameter in a lathe.

SHOULDER TURNING

12. What type of cutting tool should be used to machine a square shoulder?

13. How close to finish length should a diameter which requires a filleted shoulder be cut?

14. Name two methods of cutting angular shoulders on a lathe.

KNURLING, GROOVING, AND FORM TURNING

UNIT 47

Operations such as knurling, grooving, and form turning are used to alter either the shape or the finish of a round workpiece. These operations are more commonly performed on work mounted in a chuck; however, they can also be performed on work mounted between lathe centers if certain precautions are observed. Knurling is used to improve the surface finish on the work and provide a hand grip on the diameter. Grooving is used to provide a relief at the end of a thread or a seat for snap or O-rings. Form turning produces a concave or convex form on internal or external surfaces of a workpiece.

OBJECTIVES

After completing this unit you should be able to:
1. Set up and use knurling tools to produce diamond-shaped or straight patterns on diameters
2. Cut square, round, and V-shaped grooves on work between centers or in a chuck
3. Machine convex or concave forms on diameters freehand

257

Fig. 47-2   A knurling tool with one set of rolls in a self-centering head.   (Courtesy J. H. Williams and Company.)

Fig. 47-1   Fine, medium, and coarse diamond- and straight-pattern knurling rolls.   (Courtesy J. H. Williams and Company.)

Fig. 47-3   A knurling tool with three sets of rolls in a revolving head.   (Courtesy J. H. Williams and Company.)

KNURLING

Knurling is a process of impressing a diamond-shaped or straight-line pattern into the surface of the workpiece to improve its appearance or to provide a better gripping surface. Straight knurling is often used to increase the workpiece diameter when a press fit is required.

Diamond- and straight-pattern rolls are available in three styles: fine, medium, and coarse (Fig. 47-1).

The knurling tool (Fig. 47-2) is a toolpost-type toolholder on which a pair of hardened-steel rolls are mounted. These rolls may be obtained in diamond and straight-line patterns, and in coarse, medium, and fine pitches. Some knurling tools are made with the three various pitched rollers on one holder (Fig. 47-3).

Fig. 47-4   Setting a knurling tool to center.

TO KNURL IN A LATHE

1. Mount the work between centers and mark the required length to be knurled.

NOTE: If the work is held in a chuck for knurling, the right end of the work should be supported with a revolving tailstock center.

2. Set the lathe to run at *one-quarter* the speed required for turning.

3. Set the carriage feed to *0.015 to 0.030 in.* (0.38 mm to 0.76 mm).

4. Set the center of the floating head of the knurling tool even with the dead-center point (Fig. 47-4).

5. Set the knurling tool at right angles to the workpiece and tighten it securely in this position (Fig. 47-5).

6. Start the machine and lightly touch the rolls against the work to see that they are tracking properly (Fig. 47-6). Adjust if necessary.

7. Move the knurling tool to the end of the work so that only half the roll face bears against the work. If the knurl does not extend to the end of the workpiece, the knurling tool is set at the correct limit of the section to be knurled.

8. Force the knurling tool into the work approximately 0.025 in. (0.63 mm) and start the lathe.

OR

Start the lathe and then force the knurling tool into the work until the diamond pattern comes to a point.

9. Stop the lathe and examine the pattern. If necessary, reset the knurling tool.

   a. If the pattern is incorrect (Fig. 47-6) it is usually because the knurling tool is not set on center.

   b. If the knurling tool is on center and the pattern is not

Fig. 47-5 The knurling tool set at 90° and moved near the end of the work.

Fig. 47-6 Correct and incorrect knurling patterns.

correct, it is generally due to worn knurling rolls. In this case it will be necessary to set the knurling tool off square slightly so that the corner of the knurling rolls can start the pattern.

10. Once the pattern is correct, engage the automatic carriage feed and apply cutting fluid to the knurling rolls.

11. Knurl to the proper length.

NOTE: *Do not disengage the feed until the full length has been knurled, otherwise rings will be formed on the knurled pattern* (Fig. 47-7).

12. If the knurling pattern is not to a point after the length has been knurled, reverse the lathe feed and take another pass across the work.

GROOVING

Grooving, commonly called *recessing, undercutting,* or *necking,* is often done at the end of a thread to permit full

Fig. 47-7 Disengaging the automatic feed will damage the knurling pattern.

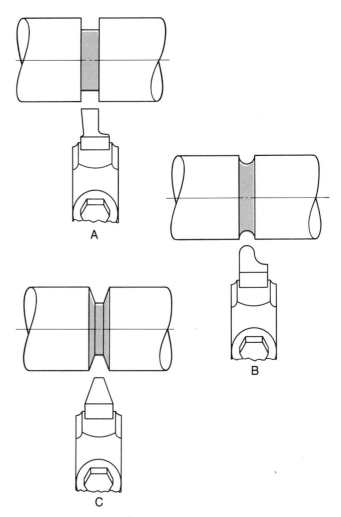

Fig. 47-8 Three types of common grooves.

travel of the nut up to a shoulder, or at the edge of a shoulder to ensure a proper fit of mating parts. Grooves are generally square, round, or V-shaped (Fig. 47-8).

Rounded grooves are usually used where there is a strain on the part and where a square corner would lead to fracturing of the metal at this point.

TO CUT A GROOVE

1. Grind a toolbit to the desired size and shape of the groove required. If a parting tool is used to cut a groove, never grind the width of the tool.
2. Lay out the location of the groove.
3. Set the lathe to half the speed for turning.
4. Mount the workpiece in the lathe.
5. Set the toolbit to center height (Fig. 47-9).
6. Locate the toolbit on the work at the position where the groove is to be cut.
7. Start the lathe and feed the cutting tool toward the work using the crossfeed handle until the toolbit lightly marks the work.
8. Hold the crossfeed handle in position and then set the graduated collar to zero (Fig. 47-10).
9. Calculate how far the crossfeed screw must be turned to cut the groove to the proper depth.
10. Feed the toolbit into the work *slowly* using the cross-feed handle.
11. Apply cutting fluid to the point of the cutting tool. To ensure that the cutting tool will not bind in the groove, move the carriage slightly to the left and to the right while grooving. Should chatter develop, the spindle speed of the lathe should be reduced.
12. Stop the lathe and check the depth of groove with outside calipers or knife-edge verniers.

NOTE: Safety goggles should always be worn when grooving on a lathe.

FORM TURNING ON A LATHE

It is often necessary to form irregular shapes or contours on a workpiece. Form turning may be done on a lathe by three methods.

1. Freehand
2. Form-turning tool
3. Hydraulic tracer attachment

TURNING A FREEHAND FORM OR RADIUS

Freehand form turning probably presents the greatest problem to the beginning lathe operator. In this operation, coordination of both hands is required and practice is important in mastering this skill.

To Turn a ½-in. or 12.7-mm Radius on the End of a Workpiece

1. Mount the workpiece in a chuck and face the end.
2. With the work revolving, mark a line ½ in. (or 12.7 mm) from the end using a pencil (Fig. 47-11).
3. Mount a round-nose turning tool on center.
4. Start the lathe and adjust the toolbit in until it touches the diameter about ¼ in. (or 6.35 mm) from the end.

Fig. 47-9 A grooving tool set to center.

Fig. 47-10 The graduated collar should be set to zero when the tool just touches the work diameter. (Courtesy Kostel Enterprises Limited.)

5. Place one hand on the crossfeed handle and the other on the carriage handwheel.
6. Turn the carriage *handwheel* (*not the handle*) to feed the toolbit slowly toward the end of the work; at the same time, turn the crossfeed handle to move the tool into the work.

NOTE: It will take practice to coordinate the movement of the carriage in relation to the crossfeed. For the first ¼ in. (or 6.35 mm) of the radius, the carriage must be moved faster than the crossfeed handle. However, for the second ¼ in. (or 6.35 mm), the crossfeed handle must be moved faster than the carriage.

7. Back out the toolbit and move the carriage to the left.
8. Take successive cuts as in step 6 until the toolbit starts to cut close to the ½-in. (or 12.7-mm) line.
9. Test the radius with a ½-in. (or 12.7-mm) gage.
10. If the radius is not correct, it may have to be recut. It is often possible to finish the cut to the required shape by filing.

The same procedure as step 6 is followed when cutting internal radii (Fig. 47-12). It is always advisable to start at the large diameter, feeding along and in until the proper radius and diameter are obtained.

Fig. 47-11   Turning a ½-in. or 12.7-mm radius on the end of the workpiece.

Fig. 47-12   Turning a concave and convex radius on a workpiece.

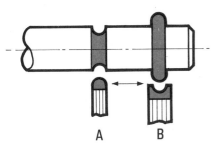

A   B

Fig. 47-13   Toolbits ground to cut (A) a concave radius; (B) a convex radius.

Fig. 47-14   A follower on the cross-slide follows the template in order to produce a special form.

Fig. 47-15   The cross-slide must be disconnected in order to follow the template.

FORM TURNING TOOLS

Smaller radii and contours are conveniently formed on a workpiece by a form turning tool. The lathe toolbit is ground to the desired radius and used to form the contour on the workpiece. Toolbits may also be ground to produce a concave radius (Fig. 47-13A).

This method of forming radii and contours eliminates the need for checking with a gage or template once the toolbit is ground to the desired shape. Duplicate contours may also be formed on several workpieces when the same toolbit is used.

When producing a convex radius, it is necessary to leave a collar of the desired size on the workpiece (Fig. 47-13B).

To produce a good finish by this method, the work should be revolved slowly. The tool should be fed into the work slowly while cutting oil is applied. To eliminate chatter during the cutting operation, the cutting tool should be moved slightly back and forth (longitudinally).

FORM TURNING USING A TEMPLATE AND FOLLOWER

When only a few pieces of a special form are required, they can be accurately produced by using a template and a follower fastened to the cross-slide of the lathe. The ac-curacy of the template, which must be made, determines the accuracy of the form produced.

Cutting a Special Form Using a Template and Follower

1. Make an accurate template to the form desired.
2. Mount the template on a bracket fastened to the back of the lathe (Fig. 47-14).
3. Position the template lengthwise in relation to the workpiece.
4. Mount a round-nose cutting tool in the toolpost.
5. Fasten a follower, the face of which must have the same form as the point of the cutting tool, on the cross-slide of the lathe.
6. Rough-out the form on the workpiece freehand by keeping the follower close to the template by manually operating the carriage and cross-slide controls. While taking roughing cuts, keep the distance between the follower and the template fairly constant. For the final roughing cut, the follower should be kept within approximately ⅟₃₂ in. (0.79 mm) of the template at all times.
7. Disconnect the cross-slide from the crossfeed screw (Fig. 47-15).

Fig. 47-16 A hydraulic tracer attachment being used to machine an intricate form. (Courtesy Retor Developments Ltd.)

HYDRAULIC LINES
STYLUS ADJUSTMENT
TOOL SLIDE
TURRET TOOL POST
CUTTING TOOL
WORKPIECE
FEED
CONTROL VALVE
TEMPLATE SUPPORT

8. Apply light hand pressure on the cross-slide to keep the follower in contact with the template.

9. Engage the automatic carriage feed and take a finish cut from the workpiece while the follower is being kept in contact with the template.

HYDRAULIC TRACER ATTACHMENT

When many duplicate parts having several radii or contours which may be difficult to produce are required, they may be easily made on a hydraulic tracer lathe or on a lathe equipped with a hydraulic tracer attachment (Fig. 47-16).

Hydraulic tracer lathes incorporate a means of moving the cross-slide by controlled oil pressure supplied by a hydraulic pump. A flat template of the desired contour of the finished piece, or a circular template identical to the finished piece, is mounted in an attachment on the lathe. Automatic control of the tool slide and duplication of the part is achieved by a stylus which bears against the template surface. As the carriage is fed along automatically, the stylus follows the contour of the template. The stylus arm actuates a control valve regulating the flow of oil into a cylinder incorporated in the tool slide base. A piston connected to the tool slide is moved in or out by the flow of oil to the cylinder. This movement causes the tool slide (and toolbit) to move in or out as the carriage moves along, duplicating the profile of the template on the workpiece.

ADVANTAGES OF A TRACER ATTACHMENT

1. Intricate forms, difficult to produce by other means, can be readily produced.
2. Various forms, tapers, and shoulders can be produced in one cut.
3. Duplicate parts can be produced rapidly and accurately.
4. Accuracy and finish of the part do not depend on the skill of the operator.

Hints on the Use of a Tracer Attachment

1. The toolbit point and stylus should have the same form and radius.
2. The radius on the toolbit should be smaller than the smallest radius on the template.
3. The stylus must be set to the point on the template giving the smallest diameter of the work.
4. The centerline of the template must be parallel to the ways of the lathe.
5. The form of the template must be smooth.
6. No angle larger than 30°, or the equivalent radius, should be incorporated in the form of the template.
7. Duplicate parts produced between centers must be the same length and have the center holes drilled to the same depth.
8. Duplicate parts held in a chuck should project the same distance from the chuck jaws.
9. The included angle of the tool point should be less than the smallest angle on the template.

REVIEW QUESTIONS

KNURLING

1. Define the process of knurling.
2. Explain how the knurling tool should be set up.
3. Why is it important not to disengage the feed during the knurling operation?

GROOVING

4. For what purposes are grooves used?
5. How can the depth of the cut be gaged when grooving?
6. What should be done to prevent the cutting tool from binding in a deep groove?

7. Name three methods by which form turning may be done on a lathe.

8. Briefly describe the procedure for turning a ½-in. (12.7-mm) radius on the end of a workpiece.

9. What is a template?
10. What types of templates may be used with a tracer lathe?
11. List three advantages of a tracer lathe or tracer attachment.
12. List six points to observe when using a tracer lathe.

TAPERS AND TAPER TURNING
UNIT 48

A *taper* may be defined as a uniform change in the diameter of a work-piece measured along its axis. Tapers in the inch system are expressed in taper per foot, taper per inch, or in degrees. Metric tapers are expressed as a ratio of 1 mm per unit of length; for example 1:20 taper would have a 1-mm change in diameter in 20 mm of length. A taper provides a rapid and accurate method of aligning machine parts and an easy method of holding tools such as twist drills, lathe centers, and reamers.

Machine tapers (those used on machines and tools) are now classified by the American Standards Association as *self-holding tapers* and *steep* or *self-releasing tapers*.

OBJECTIVES

After completing this unit you should be able to:
1. Identify and know the purpose of self-holding and self-releasing tapers
2. Cut short, steep tapers using the compound rest
3. Calculate and cut tapers on work between centers by offsetting the tailstock
4. Calculate and machine tapers with taper attachment

SELF-HOLDING TAPERS

Self-holding tapers, when seated properly, remain in position because of the wedging action of the small taper angle. The most common forms of self-holding tapers are the Morse, the Brown and Sharpe, and the ¾-in.-per-foot machine taper. See Table 48-1.

The smaller sizes of self-holding tapered shanks are provided with a tang to help drive the cutting tool. Larger sizes employ a tang drive with the shank held in by a key or a key drive with the shank held in with a draw bolt.

STEEP TAPERS

Steep tapers (self-releasing) have a 3½ in. taper per foot (tpf). This was formerly called the standard milling ma-chine taper. It is used mainly for alignment of milling machine arbors and accessories. A steep taper has a key drive and uses a draw-in bolt to hold it securely in the milling machine spindle.

STANDARD TAPERS

Although many of the tapers referred to in Table 48-1 are taken from the Morse and Brown and Sharpe taper series, those not listed in this table are classified as nonstandard machine tapers.

The *Morse taper*, which has approximately ⅝-in. tpf, is used for most drills, reamers, and lathe center shanks. Morse tapers are available in eight sizes ranging from #0 to #7.

The *Brown and Sharpe taper*, available in sizes from #4

TABLE 48-1 Basic Dimensions of Self-holding Tapers

NUMBER OF TAPER	TAPER PER FOOT	DIAMETER AT GAGE LINE (A)	DIAMETER AT SMALL END (D)	LENGTH (P)	SERIES ORIGIN
0.239	0.502	0.2392	0.200	¹⁵/₁₆	Brown and Sharpe taper series
0.299	0.502	0.2997	0.250	1³/₁₆	
0.375	0.502	0.3752	0.3125	1½	
*0	0.624	0.3561	0.252	2	Morse taper series
1	0.5986	0.475	0.369	2⅛	
2	0.5994	0.700	0.572	2⁹/₁₆	
3	0.6023	0.938	0.778	3³/₁₆	
4	0.6233	1.231	1.020	4¹/₁₆	
4½	0.624	1.500	1.266	4½	
5	0.6315	1.748	1.475	5³/₁₆	
6	0.6256	2.494	2.116	7¼	
7	0.624	3.270	2.750	10	
200	0.750	2.000	1.703	4¾	¾-in.-tpf series
250	0.750	2.500	2.156	5½	
300	0.750	3.000	2.609	6¼	
350	0.750	3.500	3.063	7	
400	0.750	4.000	3.516	7¾	
450	0.750	4.500	3.969	8½	
500	0.750	5.000	4.422	9¼	
600	0.750	6.000	5.328	10¾	
800	0.750	8.000	7.141	13¾	
1000	0.750	10.000	8.953	16¾	
1200	0.750	12.000	10.766	19¾	

*Taper #0 is not a part of the self-holding taper series. It has been added to complete the Morse taper series.

to #12, has approximately 0.502-in. tpf, except #10 which has a taper of 0.516-in./ft. This self-holding taper is used on Brown and Sharpe machines and drive shanks.

The *Jarno taper*, 0.600-in. tpf, was used on some lathe and drill spindles in sizes from #2 to #20. The taper number indicates the large diameter in eighths of an inch and the small diameter in tenths of an inch. The taper length is indicated by the taper number divided by two.

The *standard taper pins* used for positioning and holding parts together have ¼-in. tpf. Standard sizes in these pins range from #6/0 to #10.

LATHE SPINDLE NOSE TAPERS

There are two types of tapers used on lathe spindle noses.

The *Type D-1* has a very short tapered section (3-in. tpf) and is used on cam-lock spindles (Fig. 48-1).

The *Type L* lathe spindle nose has a taper of 3½-in./ft and has a considerably longer taper than the Type D-1. The chuck or drive plate is held on by a threaded lock ring fitted on the spindle behind the taper nose. A key drive is employed in this type of taper (Fig. 48-2).

TAPER CALCULATIONS

To machine a taper, particularly by the tailstock offset method, it is often necessary to make calculations to ensure accurate results. Since tapers are often expressed in *taper per foot*, *taper per inch*, or in *degrees*, it may be necessary to calculate either of these dimensions.

TO CALCULATE THE TPF

To calculate the tpf it is necessary to know the large diameter, the small diameter, and the length of taper. The tpf can be calculated by applying the following formula.

$$\text{tpf} = \frac{(D - d)}{\text{length of taper}} \times 12$$

To calculate the tpf for the workpiece in Fig. 48.3.

$$\text{tpf} = \frac{(1¼ - 1)}{3} \times 12$$

$$= \frac{1}{4} \times \frac{1}{3} \times 12$$

$$= 1$$

Fig. 48-1 Tapered lathe spindle nose Type D-1.

Fig. 48-2 Tapered lathe spindle nose Type L.

Fig. 48-3 The main part of an inch taper.

Fig. 48-4 Dimensions of a workpiece having a taper.

TO CALCULATE THE TAILSTOCK OFFSET

When calculating the tailstock offset, the tpf and the total length of the work must be known (Fig. 48-4).

1. Tailstock offset $= \dfrac{\text{tpf} \times \text{length of work}}{24}$

$$= \frac{(1\frac{1}{8} - 1)}{3} \times 12$$

$$= \frac{1}{8} \times \frac{1}{3} \times 12$$

$$= 1/2 \text{ in.}$$

2. Tailstock offset $= \dfrac{1/2 \times 6}{24}$

$$= \frac{1}{2} \times \frac{1}{24} \times 6$$

$$= 1/8 \text{ in.}$$

A simplified formula can be used to calculate the tailstock offset if the taper per inch is known.

Taper per inch $= \dfrac{\text{taper per foot}}{12}$

Tailstock offset $= \dfrac{\text{taper per inch} \times \text{OL}}{2}$

In cases where it is not necessary to find the tpf, the following simplified formula can be used to calculate the amount of tailstock offset.

Tailstock offset $= \dfrac{\text{OL}}{\text{TL}} \times \dfrac{(D - d)}{2}$

OL = overall length of work
TL = length of the tapered section
D = diameter at the large end
d = diameter at the small end

For example, to find the tailstock offset required to cut the taper for the work in Fig. 48-4.

Tailstock offset $= \dfrac{6}{3} \times \dfrac{1}{8} \times \dfrac{1}{2}$

$$= 1/8 \text{ in.}$$

INCH TAPER ATTACHMENT OFFSET CALCULATIONS

Most tapers cut on a lathe with the taper attachment are expressed in tpf. If the tpf of the taper on the workpiece is not given, it may be calculated by using the following formula:

$$\text{tpf} = \frac{(D - d) \times 12}{\text{TL}}$$

EXAMPLE

Calculate the tpf for a taper with the following dimensions: large diameter (D), 1⅜ in.; small diameter (d), ¹⁵⁄₁₆ in.; length of tapered section (TL), 7 in.

$$\text{tpf} = \frac{(1\ 3/8 - 15/16) \times 12}{7}$$

$$= \frac{7/16 \times 12}{7}$$

$$= 3/4 \text{ in.}$$

METRIC TAPERS

Metric tapers are expressed as a ratio of 1 mm per unit of length. In Fig. 48-5 the work would taper 1 mm in a distance of 20 mm. This taper would then be expressed as a ratio of 1:20 and would be indicated on a drawing as "taper = 1:20."

Since the work tapers 1 mm in 20 mm of length, the diameter at a point 20 mm from the small diameter (d) will be 1 mm larger ($d + 1$).

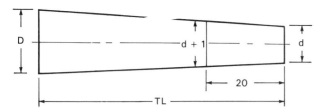

Fig. 48-5 The characteristics of a metric taper.

Fig. 48-6 Dimensions of a metric taper.

Fig. 48-7 Metric taper turning by the tailstock offset method.

Some common metric tapers are:

Milling machine spindle   -1:3.429
Morse taper shank   -(approx.) 1:20
Tapered pins and pipe threads -1:50

Metric Taper Calculations

If the small diameter (d), the unit length of taper (k), and the total length of taper (l) are known, the large diameter (D) may be calculated.

In Fig. 48-6, the large diameter (D) will be equal to the small diameter plus the amount of taper. The amount of taper for the unit length (k) is $(d + l) - (d)$ or 1 mm. Therefore the amount of taper per millimeter of unit length $= 1/k$.

The *total amount of taper* will be the taper per millimeter ($1/k$) multiplied by the total length of taper (l).

$$\text{Total taper} = \frac{1}{k} \times l \text{ or } \frac{l}{k}$$

$$D = d + \text{total amount of taper}$$

$$D = d + \frac{l}{k}$$

EXAMPLE

Calculate the large diameter D for a 1:30 taper having a small diameter of 10 mm and a length of 60 mm.

SOLUTION

Since taper is 1:30, $k = 30$.

$$\begin{aligned}
D &= d + \frac{l}{k} \\
&= 10 + \frac{60}{30} \\
&= 10 + 2 \\
&= 12 \text{ mm}
\end{aligned}$$

Metric Tailstock Offset Calculations

If the taper is to be turned by offsetting the tailstock, the amount of offset is calculated as follows. See Fig. 48-7.

$$\text{Offset} = \frac{D - d}{2 \times l} \times L$$

D = large diameter
d = small diameter
l = length of taper
L = length of work

EXAMPLE

Calculate the tailstock offset required to turn a 1:30 taper × 60 mm long on a workpiece 300 mm long. The small diameter of the tapered section is 20 mm.

SOLUTION

$$\begin{aligned}
D &= d + \frac{l}{k} \\
&= 20 + \frac{60}{30} \\
&= 20 + 2 \\
&= 22 \text{ mm}
\end{aligned}$$

$$\begin{aligned}
\text{Tailstock offset} &= \frac{D - d}{2 \times l} \times L \\
&= \frac{22 - 20}{2 \times 60} \times 300 \\
&= \frac{2}{120} \times 300 \\
&= 5 \text{ mm}
\end{aligned}$$

Metric Taper Attachment Offset Calculations

When the taper attachment is used to turn a taper, the amount the guide bar is set over may be determined as follows:

1. If the angle of taper is given on the blueprint, set the guide bar to one-half the angle (Fig. 48-8).
2. If the angle of taper is not given on the blueprint, use the following formula to find the amount of guide bar setover.

Fig. 48-8 Metric taper turning by the taper attachment method.

Fig. 48-9 Measuring the amount of offset with a rule.

$$\text{Guide bar setover} = \frac{D - d}{2} \times \frac{L}{l}$$

D = large diameter of taper
d = small diameter of taper
l = length of taper
L = length of taper attachment guide bar

EXAMPLE

Calculate the amount of setover for a 500-mm-long guide bar to turn a 1:50 × 250-mm-long taper on a workpiece. The small diameter of the taper is 25 mm.

$$D = d + \frac{l}{k}$$

$$= 25 + \frac{250}{50}$$

$$= 30 \text{ mm}$$

$$\text{Guide bar setover} = \frac{D - d}{2} \times \frac{L}{l}$$

$$= \frac{30 - 25}{2} \times \frac{500}{250}$$

$$= \frac{5}{2} \times 2$$

$$= 5 \text{ mm}$$

TAPER TURNING

Taper turning on a lathe can be performed on work held between centers or in a lathe chuck. There are three methods of producing a taper.

1. By offsetting the tailstock
2. By means of the taper attachment which has been set to the proper tpf or the proper taper angle of the workpiece; on inch tapers by means of the taper attachment which has been set to the tpf or taper angle of the workpiece; or on metric tapers by calculating the guide bar offset
3. By adjusting the compound rest to the angle of the taper

The method used to machine any taper depends on the work length, taper length, taper angle, and number of pieces to be machined.

THE TAILSTOCK OFFSET METHOD

The tailstock offset method is generally used to cut a taper when no taper attachment is available. This involves moving the tailstock center out of line with the headstock center. However, the amount that the tailstock may be offset is limited. This method will not permit steep tapers to be turned or standard tapers to be turned on the end of a long piece of work.

METHODS OF OFFSETTING THE TAILSTOCK

The tailstock may be offset by three methods.

1. By using the graduations on the end of the tailstock (visual method)
2. By means of the graduated collar and feeler gage
3. By means of a dial indicator

To Offset the Tailstock by the Visual Method

1. Loosen the tailstock clamp nut.
2. Offset the upper part of the tailstock by loosening one setscrew and tightening the other until the required amount is indicated on the graduated scale at the end of the tailstock (Fig. 48-9).

NOTE: Before machining the work be sure that both setscrews are snugged up to prevent any lateral movement of the tailstock.

To Offset the Tailstock Accurately

The tailstock may be accurately offset by using a dial indicator (Fig. 48-10).

1. Adjust the tailstock spindle to the distance it will be used in the machining setup and lock the tailstock spindle clamp.

Fig. 48-10 Offsetting a tailstock using a dial indicator. (Courtesy Kostel Enterprises Limited.)

Fig. 48-11 Offsetting the tailstock using the crossfeed graduated collar and a feeler.

2. Mount a dial indicator in the toolpost with the plunger in a horizontal position and on center.

3. Using the crossfeed handle, move the indicator so that it registers approximately 0.020 in. (0.50 mm) on the work, and set the indicator and crossfeed graduated collars to zero.

4. Adjust the tailstock until the required offset is shown on the indicator.

5. Move the cross-slide out until the required offset is shown on the graduated collar. *Be sure to remove the backlash.*

6. Adjust the tailstock until the dial indicator reads zero after both offset screws have been snugged up.

The tailstock may also be offset fairly accurately by using a feeler gage between the toolpost and the tailstock spindle in conjunction with the crossfeed graduated collar (Fig. 48-11).

To Turn a Taper by the Tailstock Offset Method

1. Loosen the tailstock clamp nut.
2. Offset the tailstock the required amount.
3. Set up the cutting tool as for parallel turning.

NOTE: The cutting tool *must* be on center.

4. Starting at the small diameter, take successive cuts until the taper is 0.050- to 0.060-in. (1.27- to 1.52-mm) over-size.

5. Check the taper for accuracy using a taper ring gage, if required. (See Taper Ring Gages, Unit 14.)

6. Finish-turn the taper to the size and fit required.

TAPER TURNING USING THE TAPER ATTACHMENT

The use of a taper attachment provides many advantages when taper turning.

1. The lathe centers remain in alignment preventing the distortion of centers on the workpiece.

2. The setup is simple and permits changing from taper to parallel turning with no time lost to align the centers.

3. The length of the workpiece does not matter, since duplicate tapers may be turned on any length of work.

4. Tapers may be produced on work held between centers, in a chuck, or in a collet.

5. Internal tapers can be produced by this method.

6. Metric taper attachments are graduated in millimeters and degrees, while inch attachments are graduated in both degrees and inches of tpf. This eliminates the need of lengthy calculations and set up.

7. A wider range of tapers may be produced.

There are two types of taper attachments.

1. *The plain taper attachment* (Fig. 48-12)
2. *The telescopic taper attachment* (Fig. 48-13)

When using the plain taper attachment, remove the binding screw which holds the cross-slide to the crossfeed screw nut. The binding screw is then used to connect the sliding block to the slide of the taper attachment. With the plain taper attachment, the depth of cut is made by using the compound rest feed handle.

When a telescopic taper attachment is used, the crossfeed screw is not disengaged and the depth of cut can be set by the crossfeed handle.

To Cut a Taper Using a Telescopic Taper Attachment

1. Clean and oil the guide bar B (Fig. 48-13).
2. Loosen the lock screws D^1 and D^2 and offset the end of the guide bar the required amount or, for inch attachments, set the bar to the required taper in degrees or tpf.
3. Tighten the lock screws.
4. Set up the cutting tool on center.
5. Set the workpiece in the lathe and mark the length of taper.
6. Tighten the connecting screw G on the sliding block E.

NOTE: If a plain taper attachment is being used, remove the binding screw in the cross-slide and use it to connect

Fig. 48-12 The parts of a plain taper attachment.

Fig. 48-13 The telescopic taper attachment.

the sliding block and the connecting slide. The compound rest must also be set at right angles to the lathe bed.

7. Move the carriage until the center of the attachment is opposite the length to be tapered.

8. Lock the anchor bracket A to the lathe bed.

9. Take a cut ¹⁄₁₆ in. (1.58 mm) long, stop the lathe, and check the end of the taper for size.

10. Set the depth of the roughing cut to 0.050 to 0.060 in. (1.27 to 1.52 mm) oversize, and machine the taper.

NOTE: Start the feed about ½ in. (12.7 mm) before the start of the cut to remove any play in the taper attachment.

11. Readjust the taper attachment if necessary, take a light cut, and recheck the taper fit.

12. Finish-turn and fit the taper to a gage.

When standard tapers must be produced on a piece of work, a taper plug gage may be mounted between centers and the taper attachment adjusted to this angle by using a dial indicator mounted on center in the toolpost.

When an *internal taper* is cut, the same procedure is followed, except that the guide bar is set to the side of the centerline opposite to that used when turning an external taper.

When mating external and internal tapers must be cut, it is advisable first to machine the internal taper to a plug gage. The external taper is then fitted to the internal taper.

TAPER TURNING USING THE COMPOUND REST

To produce short or steep tapers stated in degrees, the compound rest method is used. The tool must be fed in by hand, using the compound rest feed handle.

To Cut a Taper with the Compound Rest

1. Refer to the drawing for the amount of taper required in degrees.

If the angle on the drawing is not given in degrees, calculate the compound rest setting as follows:

$$\text{Tan Angle} = \frac{\text{tpf}}{24}$$

EXAMPLE

For the workpiece illustrated in Fig. 48-3, the calculations would be:

$$\text{tpf} = \frac{1}{4} \times \frac{12}{3}$$

$$= 1 \text{ in.}$$

$$\text{Tan angle} = \frac{1}{24}$$

$$= 0.04166$$

By referring to the trigonometry tables in any handbook, it is found that the angle of this taper (and the compound rest setting) is 2°23′.

If a set of trigonometry tables is not available, use the simplified formula to calculate the angle of the taper (and the compound rest setting).

$$\text{Tan angle} = \text{tpf} \times 2°23′$$

2. Loosen the compound rest lock screws.

3. Swivel the compound rest as follows:
   a. where included angles are given on a drawing, swivel the compound rest to one-half the angle (Fig. 48-14 top)
   b. where angles are given on one side only, (Fig. 48-14 center), swivel the compound rest to that angle

4. Tighten the compound rest lock screws using only two-finger pressure on the wrench to avoid stripping the lock screw threads.

5. Set the cutting tool to center with the toolholder at right angles to the taper to be cut.

6. Tighten the toolpost securely.

7. Back off the top slide of the compound rest so that there will be enough travel to machine the length of the taper.

8. Move the carriage to position the cutting tool near the start of the taper and then lock the carriage.

9. Rough turn the taper by feeding the cutting tool using the *compound rest feed handle* (Fig. 48-15).

10. Check the taper for accuracy and readjust the compound rest setting if necessary.

11. Finish turn and check the taper for size and fit.

CHECKING A TAPER

Inch tapers can be checked by scribing two lines exactly 1 in. apart on the taper and carefully measuring the taper at these points with a micrometer (Fig. 48-16). The difference in readings will indicate the tpi of the workpiece. Tapers may be more accurately checked by using a sine bar (see Unit 13).

Fig. 48-14 Direction to swing the compound rest for cutting various angles.

Fig. 48-15   Cutting a short taper using the compound rest.

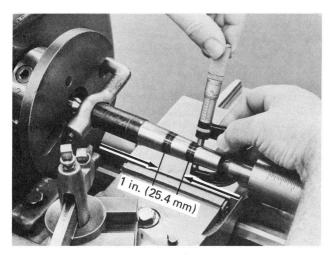

Fig. 48-16   Checking the accuracy of a taper with a micrometer.

Fig. 48-17   Using a taper micrometer for checking an external taper.   (Courtesy Taper Micrometer Corporation.)

To obtain a more accurate taper, a taper ring gage is used to check external tapers. A taper plug gage is used to check internal tapers (see Unit 14).

The *taper micrometer* (Fig. 48-17), quickly and accurately measures tapers while the workpiece is still in the machine. This instrument incorporates an adjustable anvil and a 1 in. sine bar attached to the frame, which is adjustable by the micrometer thimble. The micrometer reading indicates the tpi, which can be readily converted to tpf or angles. The anvil can be adjusted to accommodate a wide range of work sizes.

Taper micrometers are available in various models for measuring internal tapers and dovetails (Fig. 48-18), and in bench models incorporating two indicators for quickly checking the accuracy of tapered parts.

The advantages of taper micrometers are:

1. The taper accuracy can be checked while the workpiece is still in the machine.
2. They provide a quick and accurate means of checking tapers.
3. They are simple to operate.

Fig. 48-18   A taper micrometer for measuring internal tapers. (Courtesy Taper Micrometer Corporation.)

Fig. 48-19 Checking the accuracy of a taper using chalk lines.

4. The need for costly gauging equipment is eliminated.
5. They can be used for measuring external tapers, internal tapers, and dovetails.

TO FIT AN EXTERNAL TAPER

1. Make three equally spaced lines with chalk or mechanics blue along the taper (see Unit 14, Taper Ring Gages).
2. Insert the taper into the ring gage and turn *counterclockwise* for one-half turn. (See Fig. 48-19.)

CAUTION: Do not force the taper into the ring gage.

3. Remove the workpiece and examine the chalk marks. If the chalk has spread along the whole length of the taper, the taper is correct. If the chalk lines are rubbed from only one end, the taper setup must be adjusted.
4. Make a slight adjustment to the taper attachment and taking trial cuts, machine the taper until the fit is correct.

TO CHECK A METRIC TAPER

1. Check the drawing for the taper required.
2. Clean the tapered section of the work and apply layout dye.
3. Lay out two lines on the taper which are the same distance apart as the second number in the taper ratio. For example, if the taper was 1:20, the lines would be 20 mm apart.

NOTE: If the work is long enough, lay out the lines at double or triple the length of the tapered section and increase the difference in diameters by the appropriate amount. For instance, on a 1:20 taper the lines may be laid out 60 mm apart or three times the unit length of the taper. Therefore the difference in diameters would then be 3 × 1 or 3 mm. This will give a more accurate check of the taper.

4. Measure the diameters carefully with a metric micrometer at the two lines. The difference between these two diameters should be 1 mm for each unit of length.

5. If necessary, adjust the taper attachment setting to correct the taper.

REVIEW QUESTIONS

TAPERS

1. Define a taper.
2. Explain the difference between self-holding and steep tapers.
3. State the tpf for the following tapers:
   (a) Morse
   (b) Brown and Sharpe
   (c) Jarno
   (d) Standard taper pin
4. Describe the type D-1 and type L spindle nose and state where each is used.

TAPER CALCULATIONS

5. Calculate the tpf and tailstock offset for the following work:
   (a) D = 1.625 in., d = 1.425 in., TL = 3 in., OL = 10 in.
   (b) D = ⅞ in., d = ⁷⁄₁₆ in., TL = 6 in., OL = 9 in.
6. Calculate the tailstock offset for the following using the simplified tailstock offset formula:
   (a) D = ¾ in., d = ¹⁷⁄₃₂ in., TL = 6 in., OL = 18 in.
   (b) D = ⅞ in., d = ²⁵⁄₃₂ in., TL = 3½ in., OL = 10½ in.
7. Explain what is meant by a metric taper of 1:50.
8. Calculate the large diameter of a 1:50 taper having a small diameter of 15 mm and a length of 75 mm.
9. Calculate the tailstock offset required to turn a 1:40 taper × 100 mm long on a workpiece 450 mm long. The small diameter is 25 mm.

TAPER TURNING

10. Name three methods of offsetting the tailstock for taper turning.
11. List the advantages of a taper attachment.
12. List the main steps required to cut an external taper using the taper attachment.
13. Describe a taper micrometer and state its advantages.
14. Explain in point form how to fit an external taper.
15. Calculate the amount of setover for a 480-mm-long guide bar to turn a 1:40 taper × 320 mm long on a workpiece. The small diameter of the taper is 37.5 mm.
16. At what angle should the compound rest be set to machine a workpiece with a large diameter of 1¼ in., small diameter of ¾ in., length of taper of 1 in.?

UNIT 49

Threads have been used for many hundreds of years for holding parts together, making adjustments to tools and instruments, and for transmitting power and motion. A thread is basically an inclined plane or wedge that spirals around a bolt or nut. Threads have progressed throughout the ages from the early screws, which were filed by hand, to the highly accurate ball screws used on the precision machine tools of today. Although the purpose of a thread is basically the same as when the early Romans developed it, the art of producing threads has continually improved. Today threads are mass-produced by taps, dies, thread rolling, thread milling, and grinding to the very high standards so necessary for accuracy and quality control. Thread cutting is a skill that every machinist should possess since it is still necessary to cut threads on an engine lathe, especially if a special size or form of thread is required.

OBJECTIVES

After completing this unit you should be able to:
1. Recognize and know the purpose of six common thread forms
2. Set up a lathe to cut inch external Unified threads
3. Set up an inch lathe to cut metric threads
4. Set up a lathe and cut internal threads
5. Set up a lathe and cut external Acme threads

THREADS

A thread may be defined as a helical ridge of uniform section formed on the inside or outside of a cylinder or cone. Threads are used for several purposes.

1. For fastening devices such as screws, bolts, studs, and nuts.
2. To provide accurate measurement as in a micrometer.
3. To transmit motion. The threaded lead screw on the lathe causes the carriage to move along when threading.
4. To increase torque. Heavy work can be raised with a screw jack.

THREAD TERMINOLOGY

To understand and calculate thread parts and sizes, the following definitions relating to screw threads should be known. (Fig. 49-1).

A screw thread is a helical ridge of uniform section formed on the inside or outside of a cylinder or cone.
An external thread is cut on an external surface or cone such as on a cap screw or a wood screw.

An internal thread is produced on the inside of a cylinder or cone, such as the thread on the inside of a nut.
The major diameter is the largest diameter of an external or internal thread.
The minor diameter is the smallest diameter of an external or internal thread. This was formerly known as the root diameter.
The pitch diameter is the diameter of an imaginary cylinder which passes through the thread at a point where the groove and thread widths are equal.

The pitch diameter is equal to the major diameter minus a single depth of thread. The tolerance and allowances on threads are given at the pitch diameter line. The pitch diameter is also used to determine the outside diameter for rolled threads. The diameter of the blank is always equal to the pitch diameter of the thread to be rolled. Thread rolling is a displacement operation and the amount of metal displaced is forced up to form the thread above the pitch line.

NOTE: Pitch diameter is not used as a basis for determining International Organization for Standardization (ISO) metric thread dimensions.

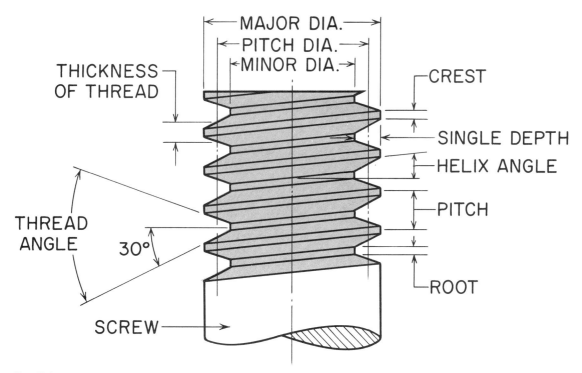

Fig. 49-1 Parts of a screw thread.

The number of threads is the number of crests or roots per inch of threaded section. This term does not apply to metric threads.

Pitch is the distance from a point on one thread to a corresponding point on the next thread, measured parallel to the axis. Pitch is expressed in millimeters for metric threads.

Lead is the distance a screw thread advances axially in one revolution. On a single start thread, the lead and the pitch are equal.

Root is the bottom surface joining the sides of two adjacent threads. The root of an external thread is on its minor diameter. The root of an internal thread is on its major diameter.

Crest is the top surface joining two sides of a thread. The crest of an external thread is on the major diameter while the crest of an internal thread is on the minor diameter.

A *flank* (side) is a thread surface which connects the crest with the root.

Depth of thread is the distance between the crest and root measured perpendicular to the axis.

Angle of thread is the included angle between the sides of a thread measured in an axial plane.

The helix angle (lead angle) is the angle which the thread makes with a plane perpendicular to the thread axis.

A *right-hand thread* is a helical ridge of uniform cross section onto which a nut is threaded in a clockwise direction.

When the thread is held in a horizontal position with its axis pointing from right to left, a right-hand thread will slope *down* and to the right (Fig. 49-2A). When a right-hand thread is cut on a lathe, the toolbit advances from right to left.

Fig. 49-2 Screw threads, (A) Right hand; (B) Left hand.

A *left-hand thread* is a helical ridge of uniform cross section onto which a nut is threaded in a counterclockwise direction.

When the thread is held in a horizontal position with its axis pointing from right to left, the thread will slope *down* and to the left (Fig. 49-2B). When a left-hand thread is cut on a lathe the toolbit advances from left to right.

THREAD FORMS

Over the past several decades, one of the world's major industrial problems has been the lack of an international thread standard whereby the thread standard used in any country could be interchanged with that of another country. In April 1975, the ISO drew up an agreement covering a standard metric thread profile, the sizes and pitches for the various threads in the new ISO Metric Thread Standard. The new series has only 25 thread sizes ranging in diameter from 1.6 to 100 mm. Countries throughout the world have been encouraged to adopt the ISO series (See Table 49-1).

TABLE 49-1 ISO Metric Pitch and Diameter Combinations			
NOMINAL DIA., mm	THREAD PITCH, mm	NOMINAL DIA., mm	THREAD PITCH, mm
1.6	0.35	20	2.5
2.0	0.40	24	3.0
2.5	0.45	30	3.5
3.0	0.50	36	4.0
3.5	0.60	42	4.5
4.0	0.70	48	5.0
5.0	0.80	56	5.5
6.3	1.00	64	6.0
8.0	1.25	72	6.0
10	1.50	80	6.0
12	1.75	90	6.0
14	2.00	100	6.0
16	2.00		

These metric threads are identified by the letter M, the nominal diameter, and the pitch. For example, a metric thread with an outside diameter of 5 mm and a pitch of 0.8 mm would be identified as follows: M 5 × 0.8.

The ISO series will not only simplify thread design but will generally produce stronger threads for a given diameter and pitch and will reduce the large inventory of fasteners now required by industry.

The *ISO metric thread* (Fig. 49-3) has a 60° included angle and a crest equal to 0.125 times the pitch, similar to the National Form thread. The main difference, however, is the depth of thread (D), which is *0.6134* times the pitch. Because of these dimensions, the flat on the root of the thread (FR) is wider than the crest (FC). The root of the ISO metric thread is one-fourth of the pitch (0.250P).

The American National Standard thread (Fig. 49-4) is divided into four main series, all having the same shape and proportions. National Coarse (NC), National Fine (NF), National Special (NS), and National Pipe (NPT). This thread has a 60° angle with a root and crest truncated to one-eighth the pitch. This thread is used in fabrication, machine construction and assembly, and for components where easy assembly is desired.

The formula for calculating the depth of a 100 percent thread is 0.866/N. However, since this thread would be very difficult to cut (especially internally) the following formula, which gives 75 percent of a thread, is generally a standard in the industry.

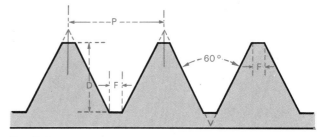

Fig. 49-4   American National thread.

$$D = 0.6495 \times P \quad \text{or} \quad \frac{0.6495}{N}$$

$$F = 0.125 \times P \quad \text{or} \quad \frac{0.125}{N}$$

The British Standard Whitworth thread (BSW) (Fig. 49-5) has a 55° V-form with rounded crests and roots.

Fig. 49-3   ISO metric thread.

$$D \text{ (external)} = 0.54127 \times P$$
$$FC = 0.125 \times P$$
$$FR = 0.250 \times P$$

The most commonly used thread forms in North America at the present time are:

This thread application is the same as for the American National form thread.

Fig. 49-5   Whitworth thread.

$$D = 0.6403 \times P \quad \text{or} \quad \frac{0.6403}{N}$$

$$R = 0.1373 \times P \quad \text{or} \quad \frac{0.1373}{N}$$

The Unified thread (Fig. 49-6) was developed by the United States, Britain, and Canada, so that equipment produced by these countries would have a standardized thread system. Until this thread was developed, many problems were created by the non-interchangeability of threaded parts being used in these countries. The Unified thread is a combination of the British Standard Whitworth and the American National form thread. This thread has a 60° angle with a rounded root, and the crest may be rounded or flat.

Fig. 49-6   Unified thread.

$$D\text{—(external thread)} = 0.6134 \times P \text{ or } \frac{0.6134}{N}$$

$$\text{—(internal thread)} = 0.5413 \times P \text{ or } \frac{0.5413}{N}$$

$$F\text{—(external thread)} = 0.125 \times P \text{ or } \frac{0.125}{N}$$

$$\text{—(internal thread)} = 0.250 \times P \text{ or } \frac{0.250}{N}$$

The American National Acme thread (Fig. 49-7) is replacing the square thread in many cases. It has a 29° angle and is used for feed screws, jacks, and vises.

Fig. 49-7   Acme thread.

$$\begin{aligned}
D &= \text{minimum } 0.500P \\
  &= \text{maximum } 0.500P + 0.010 \\
F &= 0.3707P \\
C &= 0.3707P - 0.0052 \\
  &\quad \text{(for maximum depth)}
\end{aligned}$$

The Brown and Sharp worm thread (Fig. 49-8) has a 29° included angle as the Acme thread; however, the depth is greater and the widths of the crest and root are different. This thread is used to mesh with worm gears and transmit motion between two shafts at right angles to each other but not in the same plane. The self-locking feature makes it adaptable to winches and steering mechanisms.

Fig. 49-8   Brown and Sharpe worm thread.

$$\begin{aligned}
D &= 0.6866P \\
F &= 0.335P \\
C &= 0.310P
\end{aligned}$$

The square thread (Fig. 49-9) is frequently being replaced by the Acme thread due to the difficulty of cutting it, particularly with taps and dies. Square threads were often found on vises and jack screws.

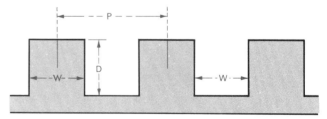

Fig. 49-9   Square thread.

$$\begin{aligned}
D &= 0.500P \\
F &= 0.500P \\
C &= 0.500P + 0.002
\end{aligned}$$

The international metric thread (Fig. 49-10) is a standardized thread used in Europe. This thread has a 60° included angle with a crest and root truncated to one-eighth the depth. Although this thread is used extensively throughout Europe, its use in North America has been confined mainly to spark plugs and the manufacture of instruments.

Fig. 49-10 International metric thread.

$$D = 0.7035P \text{ (maximum)}$$
$$0.6855P \text{ (minimum)}$$
$$F = 0.125P$$
$$R = 0.0633P \text{ (maximum)}$$
$$0.054P \text{ (minimum)}$$

THREAD FITS AND CLASSIFICATIONS

Certain terminology is used when referring to thread classifications and fits. To understand any thread system properly, the terminology relating to thread fits should be understood.

Fit is the relationship between two mating parts which is determined by the amount of clearance or interference when they are assembled.

Allowance is the intentional difference in size of the mating parts or the minimum clearance between mating parts (Fig. 49-11). With threads, the allowance is the permissible difference between the largest external thread and the smallest internal thread. This produces the tightest fit acceptable for any given classification.

The allowance for a 1 in.—8 UNC (Unified National Coarse) Class 2A and 2B fit is:

Minimum pitch diameter of the internal thread (2B)	= 0.9188 in.
Maximum pitch diameter of the external thread (2A)	= 0.9168 in.
Allowance or intentional difference	= 0.002 in.

Tolerance is the variation permitted in part size. The tolerance may be expressed as plus, minus, or both. The total tolerance is the sum of the plus and minus tolerances. For example, if a size is 1.000 ± 0.002, the total tolerance is 0.004. In the Unified and National systems, the tolerance is plus on external threads and minus on internal threads. Thus, when a thread varies from the basic or nominal size, it will ensure a freer rather than a tighter fit.

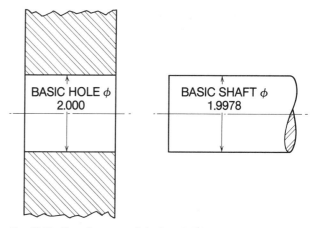

Fig. 49-11 The allowance (intentional difference) between the shaft and hole is 0.0022 in.

The tolerance for a 1 in.—8 UNC Class 2A thread is:

Maximum pitch diameter of the external thread (2A)	= 0.9168 in.
Minimum pitch diameter of the external thread	= 0.9100 in.
Tolerance or permitted variation	= 0.0068 in.

Limits are the maximum and minimum dimensions of a part.

The limits for a 1 in.—8 UNC Class 2A thread are:

Maximum pitch diameter of the external thread (2A)	= 0.9168 in.
Minimum pitch diameter of the external thread (2A)	= 0.9100 in.

The pitch diameter of this thread must be between 0.9168 in. (upper limit) and 0.9100 in. (lower limit).

Nominal size is the designation used to identify the size of the part. For example, in the designation of 1 in.—8 UNC, the figure 1 indicates a 1-in.-diameter thread.

Actual size is the measured size of a thread or part. The basic size is the size from which tolerances are set. Although the basic major diameter of a 1 in.—8 UNC Class 2A thread is 1.000 in., the actual size may vary from 0.998 to 0.983 in.

Classification of Thread Fits

With wide use of threads, it became necesary to establish certain limits and tolerances to properly identify classes of fit.

ISO Metric Tolerances and Allowances The ISO metric screw thread tolerance system provides for allowances and tolerances defined by tolerance grades, tolerance positions, and tolerance classes.

Tolerance Grades A medium tolerance, used on a general-purpose thread, is indicated by the number 6. Any number below 6 indicates a finer tolerance and any number above 6 indicates a greater tolerance. The tolerance for the thread at the pitch line and for the major diameter may be shown on the drawing.

Allowance Symbols are used to indicate the allowance for external threads:

  e indicates a large allowance
  g indicates a small allowance
  h indicates no allowance

For internal threads:

  G indicates a small allowance
  H indicates no allowance

EXAMPLE

An external metric thread may be designated as follows:

Metric	Nominal Size	Pitch	Pitch Diam. Tolerance	Outside Diam. Tolerance
↓	↓	↓	↓	↓
M	6 ×	0.75 −	5g	6g

The thread fit between mating parts is indicated by the internal thread designation followed by the external thread tolerance:

$$M\ 20 \times 2 - 6H/5g\ 6g$$

Unified thread fits have been divided into three categories and the applications of each have been defined by the Screw Thread Committee. External threads are classified as 1A, 2A, and 3A, and internal threads as 1B, 2B, and 3B.

Classes 1A and 1B include those threads for work which must be readily assembled. They have the loosest fit, with no possibility of interference between the mating external and internal threads when the threads are dirty or bruised.

Classes 2A and 2B are used for most of commercial fasteners. These threads provide a medium or free fit and permit power wrenching with minimum galling and seizure.

Classes 3A and 3B are used where a more accurate fit and lead are required. No allowance is provided and the tolerances are 75 percent of those used for 2A and 2B fits.

By classifying the tolerances of threads, the cost of threaded parts is reduced, since manufacturers may use any combination of mating threads that suit their needs. With the former system of identifying classes of tolerances (Classes 1, 2, 3, 4), it was felt that a Class 3 internal thread should be used with a Class 3 external thread.

With reference to the Unified system it should be noted that "Class" refers to tolerance or tolerance and allowance, and does not refer to fit. The fit between the mating parts is determined by the selected combination used for a specific application. For example, if a closer-than-normal fit is required, a Class 3B nut may be used on a Class 2A bolt. The basic dimensions, tolerances, and allowances for these threads may be found in any machine handbook.

THREAD CALCULATIONS

To cut a correct thread on a lathe, it is necessary first to make calculations so that the thread will have proper di-

mensions. The following formulas will be helpful when calculating thread dimensions. The symbols used in these diagrams and formulas are:

  D = single depth of thread
  P = pitch

EXAMPLE

Calculate the pitch, depth, minor diameter, and width of flat for a ¾—10 NC thread.

$$P = \frac{1}{\text{Number of tpi}}$$
$$= \frac{1}{10}$$
$$= 0.100 \text{ in.}$$

$$D = 0.6495 \times P$$
$$= 0.6495 \times \frac{1}{10}$$
$$= 0.065 \text{ in.}$$

Minor diameter = major diameter − $(D + D)$
  = 0.750 − (0.065 + 0.065)
  = 0.620 in.

Width of flat $= \dfrac{P}{8}$
$$= \frac{1}{8} \times \frac{1}{10}$$
$$= 0.0125 \text{ in.}$$

EXAMPLE

What is the pitch, depth, minor diameter, width of crest, and width of root for a M 6.3 × 1.0 thread?

$$P = 1 \text{ mm}$$
$$D = 0.54127 \times 1$$
$$= 0.54 \text{ mm}$$

Minor diameter = major diameter − $(D + D)$
  = 6.3 − (0.54 + 0.54)
  = 5.22 mm

Width of crest = 0.125 × P
  = 0.125 × 1
  = 0.125 mm

Width of root = 0.250 × P
  = 0.250 × 1
  = 0.250 mm

To Set the Quick-Change Gearbox for Threading

The quick-change gearbox provides a means of quickly setting the lathe for the desired pitch of the thread in number of threads per inch on inch-system lathes or millimeters on metric lathes. This unit contains a number of different size gears which vary the ratio between headstock spindle revolutions and the rate of carriage travel when thread cutting.

1. Check the blueprint for the pitch of the thread required.
2. From the chart on the quick-change gearbox, find the *whole number* which represents the pitch in threads per inch or in millimeters.
3. With the lathe stopped, engage the *tumbler lever* into the hole which is in line with the pitch (tpi or in millimeters) (Fig. 49-12).
4. Set the *top lever* into the proper position as indicated on the chart.
5. Engage the *sliding gear* in or out as required.

NOTE: Some lathes have two levers on the gearbox which take the place of the top lever and sliding gear, and these should be set as indicated on the chart.

Fig. 49-12 Quick-change gear mechanism is used for setting the number of tpi.

6. Turn the lathe spindle by hand to ensure that the lead screw revolves.
7. Recheck the lever settings to avoid errors.

THREAD-CHASING DIAL

To cut a thread on a lathe, the lathe spindle and the lead-screw must be in the same relative position for each successive cut. Most lathes have a thread-chasing dial either built into, or attached to, the carriage for this purpose. The chasing dial indicates when the split nut should be engaged with the lead screw in order for the cutting tool to follow the previously cut groove.

The thread-chasing dial is connected to a worm gear which meshes with the threads of the lead screw (Fig. 49-13). The dial is graduated into eight divisions, four numbered and four unnumbered, and it revolves as the lead screw turns. Fig. 49-14 indicates when the split-nut lever should be engaged for cutting various numbers of threads per inch (tpi).

☐ Even threads use any division.
☐ Odd threads must stay on either numbered or unnumbered lines; cannot use both.

THREAD CUTTING

Thread cutting on a lathe is a process that produces a helical ridge of uniform section on a workpiece. This is performed by taking successive cuts with a threading toolbit of the

Fig. 49-13 Thread-chasing dial mechanism.

THREADS PER INCH TO BE CUT	WHEN TO ENGAGE SPLIT NUT	READING ON DIAL
EVEN NUMBER OF THREADS	ENGAGE AT ANY GRADUATION ON THE DIAL	1 1 ½ 2 2 ½ 3 3 ½ 4 4 ½
ODD NUMBER OF THREADS	ENGAGE AT ANY MAIN DIVISION	1 2 3 4
FRACTIONAL NUMBER OF THREADS	1/2 THREADS, E.G.,11 1/2 ENGAGE AT EVERY OTHER MAIN DIVISION 1 & 3, OR 2 & 4 OTHER FRACTIONAL THREADS ENGAGE AT SAME DIVISION EVERY TIME	
THREADS WHICH ARE A MULTIPLE OF THE NUMBER OF THREADS PER INCH IN THE LEAD SCREW	ENGAGE AT ANY TIME THAT SPLIT NUT MESHES	USE OF DIAL UNNECESSARY

Fig. 49-14 Split-nut engagement rules for thread cutting.

same shape as the thread form required. Work to be threaded may be held between centers or in a chuck. If work is held in a chuck, it should be turned to size and threaded before the work is removed.

TO SET UP A LATHE FOR THREADING (60° THREAD)

1. Set the lathe speed to about one-quarter the speed used for turning.
2. Set the quick-change gearbox for the required pitch in threads per inch or in millimeters.
3. Engage the lead screw.
4. Secure a 60° threading toolbit and check the angle using a thread center gage.
5. Set the compound rest at 29° to the right (Fig. 49-15); set it to the left for a left-hand thread.
6. Set the cutting tool to the height of the lathe center point.
7. Mount the work between centers. Make sure the lathe dog is tight on the work. If the work is mounted in a chuck, it must be held tightly.
8. Set the toolbit at right angles to the work, using a thread center gage (Fig. 49-16).

NOTE: Never jam a toolbit into a thread center gage. This can be avoided by aligning only the cutting edge (leading side) of the toolbit with the gage. A piece of paper on the cross-slide under the gage and toolbit makes it easier to check tool alignment.

9. Arrange the apron controls to allow the split-nut lever to be engaged.

THREAD-CUTTING OPERATION

Thread cutting is one of the more interesting operations which can be performed on a lathe. It involves manipulation of the lathe parts, coordination of the hands, and strict attention to the operation. Before proceeding to cut a thread for the first time on any lathe, it is wise to take several trial passes, without cutting, in order to get the feel of the machine.

To Cut a 60° Thread

1. Check the major diameter of the work for size. It is good practice to have the diameter 0.002 in. (or 0.05 mm) undersize.
2. Start the lathe and chamfer the end of the workpiece with the side of the threading tool to just below the minor diameter of the thread.
3. Mark the length to be threaded by cutting a light groove at this point with the threading tool while the lathe is revolving.
4. Move the carriage until the point of the threading tool is near the right-hand end of the work.
5. Turn the *crossfeed handle* until the threading tool is close to the diameter, but stop when the handle is at the 3 o'clock position (Fig. 49-17).

Fig. 49-15   Compound rest set at 29° for thread cutting.

Fig. 49-16   Setting a threading tool square with a thread center gage.

Fig. 49-17   The handle of the crossfeed screw set at approximately 3 o'clock position for thread cutting.

Fig. 49-18 Checking the number of threads per inch using a screw pitch gage.

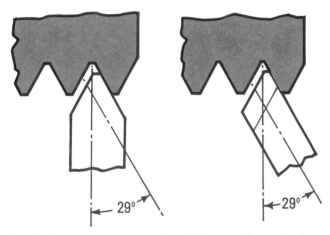

Fig. 49-19 When the tool is fed in at 29° most of the cutting is done by the leading edge of the toolbit.

6. Hold the crossfeed handle in this position and set the graduated collar to zero.

7. Turn the compound rest handle until the threading tool *lightly marks the work.*

8. Move the carriage to the right until the toolbit clears the end of the work.

9. Feed the compound rest *clockwise* about 0.003 in. (0.08 mm).

10. Engage the split-nut lever on the correct line of the thread-chasing dial (Fig. 49-14) and take a trial cut along the length to be threaded.

11. At the end of the cut, turn the crossfeed handle *counterclockwise* to move the toolbit away from the work and disengage the split-nut lever.

12. Stop the lathe and check the number of tpi with a thread pitch gage, rule, or center gage (Fig. 49-18). If the pitch (tpi or millimeters) produced by the trial cut is not correct, recheck the quick-change gearbox setting.

13. After each cut, turn the carriage handwheel to bring the toolbit to the start of the thread and return the crossfeed handle to zero.

14. *Set the depth of all threading cuts with the compound rest handle* (Fig. 49-19). For National Form threads, use Table 49-2; for ISO metric threads, see Table 49-3.

15. Apply cutting fluid and take successive cuts until the

TABLE 49-2 Depth Settings when Cutting 60° National Form Threads

tpi	COMPOUND REST SETTING		
	0°	30°	29°
24	0.027	0.031	0.0308
20	0.0325	0.0375	0.037
18	0.036	0.0417	0.041
16	0.0405	0.0468	0.046
14	0.0465	0.0537	0.0525
13	0.050	0.0577	0.057
11	0.059	0.068	0.0674
10	0.065	0.075	0.074
9	0.072	0.083	0.082
8	0.081	0.0935	0.092
7	0.093	0.1074	0.106
6	0.108	0.1247	0.1235
4	0.1625	0.1876	0.1858

Note: When using this table for cutting National Form threads, the correct width of flat (0.125P) must be ground on the toolbit, otherwise the thread will not be the correct width.

top (crest) and the bottom (root) of the thread are the same width.

16. Remove the burrs from the top of the thread with a file.

17. Check the thread with a master nut and take further cuts, if necessary, until the nut fits the thread freely with no end play (Fig. 49-20).

There are six ways to check threads depending on the accuracy required.

1. Master nut
2. Thread micrometer
3. Three wires
4. Thread roll or snap gage
5. Thread ring or plug gage
6. Optical comparator

To Reset a Threading Tool

A threading tool must be reset whenever it is necessary to remove partly threaded work from the lathe and finish it at

Fig. 49-20 Checking the thread with a master nut.

a later time, if the threading tool is removed for regrinding, or if the work slips under the lathe dog.

1. Set up the lathe and work for thread cutting.
2. Start the lathe, and with the toolbit clear of the work, engage the split-nut lever on the correct line.
3. Allow the carriage to travel until the toolbit is opposite any portion of the unfinished thread (Fig. 49-21).
4. Stop the lathe, *leaving the split-nut lever engaged.*
5. Feed the toolbit into the thread groove using ONLY the compound rest and crossfeed handles until the right-hand edge of the toolbit touches the rear side of the thread (Fig. 49-22).

NOTE: Do not have the cutting edge of the toolbit contacting the thread at this time.

6. Set the crossfeed graduated collar to zero.
7. Back out the threading tool using the crossfeed handle, disengage the split-nut lever, and move the carriage until the toolbit clears the start of the thread.
8. Set the crossfeed handle back to zero and take a trial cut without setting the compound rest.
9. Set the depth of cut using the compound rest handle and finish the thread to the required depth.

To Convert an Inch-Designed Lathe to Metric Threading

Metric threads may be cut on a standard quick-change gear lathe by using a pair of change gears having 50 and 127 teeth, respectively. Since the lead screw has inch dimensions and is designed to cut threads per inch, it is necessary to convert the pitch in millimeters to centimeters and then into threads per inch. To do this, it is first necessary to understand the relationship between inches and centimeters.

$$1 \text{ in.} = 2.54 \text{ cm}$$

Therefore the ratio of inches to centimeters is 1:2.54 or 1/2.54.

To cut a metric thread on an inch lathe, it is necessary to install certain gears in the gear train which will produce a ratio of 1/2.54. These gears are:

Fig. 49-21 With the split-nut lever engaged, stop the machine when the threading tool is over the thread.

Fig. 49-22 Resetting the threading tool in a partially cut groove using *only* the crossfeed and compound rest handles.

$$\frac{1}{2.54} \times \frac{50}{50} = \frac{50}{127} \quad \frac{\text{teeth}}{\text{teeth}}$$

In order to cut metric threads, two gears having 50 and 127 teeth must be placed in the gear train of the lathe. The 50-tooth gear is used as the spindle or drive gear and the 127-tooth gear is placed on the lead screw.

To Cut a 2-mm Metric Thread on a Standard Quick-Change Gear Lathe

1. Mount the 127-tooth gear on the lead screw.
2. Mount the 50-tooth gear on the spindle.
3. Convert the 2-mm pitch to threads per centimeter.

$$10 \text{ mm} = 1 \text{ cm}$$

$$P = \frac{10}{2} = 5 \text{ threads/cm}$$

4. Set the quick-change gearbox to 5 tpi. By means of the 50- and 127-tooth gears, the lathe will now cut 5 threads/cm or 2-mm pitch.
5. Set up the lathe for thread cutting. See *To Set Up a Lathe for Threading (60° Thread)*.
6. Take a light trial cut. At the end of the cut, back out the cutting tool and stop the machine but *do not disengage the split nut.*
7. Reverse the spindle rotation until the cutting tool has just cleared the end of the threaded section.
8. Check the thread with a metric screw pitch gage.
9. Cut the thread to the required depth (Table 49-3).

NOTE: Never disengage the split nut until the thread has been cut to depth.

To Cut a Left-hand Thread (60°)

A left-hand thread is used to replace a right-hand thread on certain applications where the nut may loosen due to the rotation of a spindle.

The procedure for cutting left-hand threads is basically the same as for right-hand threads with a few exceptions.

TABLE 49-3 Depth Setting when Cutting 60° ISO Metric Threads			
PITCH, mm	COMPOUND REST SETTING, mm		
	0°	30°	29°
0.35	0.19	0.21	0.21
0.4	0.21	0.25	0.24
0.45	0.24	0.28	0.27
0.5	0.27	0.31	0.31
0.6	0.32	0.37	0.37
0.7	0.37	0.43	0.43
0.8	0.43	0.50	0.49
1.0	0.54	0.62	0.62
1.25	0.67	0.78	0.77
1.5	0.81	0.93	0.93
1.75	0.94	1.09	1.08
2.0	1.08	1.25	1.24
2.5	1.35	1.56	1.55
3.0	1.62	1.87	1.85
3.5	1.89	2.19	2.16
4.0	2.16	2.50	2.47
4.5	2.44	2.81	2.78
5.0	2.71	3.13	3.09
5.5	2.98	3.44	3.40
6.0	3.25	3.75	3.71

Fig. 49-23 Set the compound rest at 29° to the left for left-hand threads.

1. Set the lathe speed and the quick-change gearbox for the pitch of the thread to be cut.
2. Engage the feed-direction lever so that the lead screw will revolve in the *opposite* direction to that for a right-hand thread.
3. Set the compound rest to 29° to the *LEFT* (Fig. 49-23).
4. Set up the left-hand threading tool and square it with the work.
5. Cut a groove at the left end of the section to be threaded. This gives the cutting tool a starting point.
6. Proceed to cut the thread to the same dimensions as for a right-hand thread.

Cutting a Thread on a Tapered Section

When a tapered thread, such as a pipe thread, is required on the end of a workpiece, either the taper attachment or the offset tailstock may be used for cutting the taper. The same setup is then used as for regular thread cutting. When setting up the threading tool, it is most important that it be set at 90° to the axis of the work and not square with the tapered surface (Fig. 49-24).

THREAD MEASUREMENT

Interchangeable manufacture demands that all parts be made to certain standards in order that, on assembly, they will fit the intended component properly. This is especially important for threaded components, and therefore the measurement and inspection of threads is important.

Threads may be measured by a variety of methods; the most common are:

Fig. 49-24 The toolbit must be set square with the axis of the work when cutting a thread on a tapered section.

1. A thread ring gage
2. A thread plug gage
3. A thread snap gage
4. A screw thread micrometer
5. A thread comparator micrometer
6. An optical comparator
7. The three-wire method

The description and use of the ring gage, plug gage, snap gage, screw thread micrometer, thread comparator micrometer, and optical comparator for checking threads are fully described in Section 5.

Three-Wire Method of Measuring Threads

The three-wire method of measuring threads is recommended by the Bureau of Standards and the National Screw Thread Commission. It is recognized as one of the best methods of checking the pitch diameter because the results are least affected by an error which may be present in the

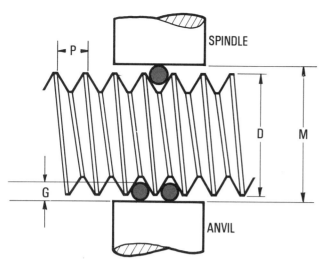

Fig. 49-25   The three-wire method of measuring 60° threads.

For the most accurate thread measurement, the best size wire (0.57735P) should be used, since this wire contacts the thread at the pitch diameter.

EXAMPLE

To find M (measurement over the wires) for a ¾—10 NC thread.

1. Calculate G (wire size).

$$G = \frac{0.57735}{10}$$

$$= 0.0577$$

2. Calculate M (measurement over the wires).

$$M = D + 3G - \frac{1.5155}{N}$$

$$= 0.750 + (3 \times 0.0577) - \frac{1.5155}{10}$$

$$= 0.750 + 0.1731 - 0.1516$$

$$= 0.9231 - 0.1516$$

$$= 0.7715$$

included thread angle. For threads which require an accuracy of 0.001 in. or 0.02 mm, a micrometer can be used to measure the distance over the wires. An electronic comparator should be used to measure the distance over the wires for threads requiring greater accuracy.

Three wires of equal diameter are placed in the thread; two on one side and one on the other side (Fig. 49-25). The wires used should be hardened and lapped to three times the accuracy of the thread to be inspected. A standard micrometer may then be used to measure the distance over the wires (M). Different sizes and pitches of threads require different size wires. For the greatest accuracy, the *best wire size* should be used. This is one that will contact the thread at the pitch diameter (middle of the sloping sides). If the best size wire is used, the pitch diameter of the thread can be calculated by subtracting the wire constant (found in any handbook) from the measurement over the wires.

To Calculate the Measurement Over the Wires

The measurement over the wires for American National Threads (60°) can be calculated by applying the following formula.

$$M = D + 3G - \frac{1.5155}{N}$$

where M = measurement over the wires
     D = major diameter of the thread
     G = diameter of the wire used
     N = number of tpi

Any of the following formulas can be used to calculate G.

1. Largest size wire $= \dfrac{1.010}{N}$ or 1.010P

2. Best size wire $= \dfrac{0.57735}{N}$ or 0.57735P

3. Smallest size wire $= \dfrac{0.505}{N}$ or 0.505P

MULTIPLE THREADS

Multiple threads are used when it is necessary to obtain an increase in lead and a deep, coarse thread cannot be cut. Multiple threads may be double, triple, or quadruple, depending on the number of starts around the periphery of the workpiece (Fig. 49-26).

The *pitch* of a thread is always the distance from a point on one thread to the corresponding point on the next thread. The *lead* is the distance a nut advances lengthwise in one complete revolution. On a single-start thread, the pitch and lead are equal. On a double-start thread, the lead will be twice the pitch. On triple-start threads the lead will be three times the pitch.

Multiple-start threads are not as deep as single-start threads and therefore have a more pleasing appearance. For example, a double-start thread having the same lead as a single-start thread would be cut only half as deep.

Multiple threads may be cut on a lathe by:

1. Using an accurately slotted drive plate or faceplate
2. Disengaging the intermediate gear of the end gear train and rotating the spindle the desired amount
3. Using the thread-chasing dial (only for double-start threads with an odd number lead).

To Cut an 8-tpi Double Thread

1. Set up the lathe and cutting tool as for cutting a single-start thread.
2. Set the quick-change gearbox to 4 tpi. (The lead of this thread is ¼ in.).
3. Cut the first thread to half the depth required for 4 tpi.
4. Leave the crossfeed handle set to the depth of the thread and *note the reading on the compound rest graduated collar.*

Fig. 49-26 The relationship between the pitch and the lead for a single-start and multiple-start threads. (A) Single thread. (B) Double thread. (C) Triple thread.

5. Withdraw the threading tool from the work using the *compound rest handle*.

6. Revolve the work exactly one-half turn by either of the following methods.

    a. **(1)** Remove the work from the lathe with the lathe dog attached.

    (2) Replace the work in the lathe with the tail of the dog in the slot exactly opposite the one used for the first thread.

NOTE: An accurately slotted drive plate or faceplate must be used for this method of indexing. A special indexing plate may also be used for this purpose.

OR

    b. **(1)** Turn the lathe by hand until a tooth of the spindle gear is exactly between two teeth of the intermediate gear.

    (2) With chalk, mark both the spindle tooth and the space in the intermediate gear (Fig. 49-27).

    (3) Disengage the intermediate gear from the spindle gear.

    (4) Starting with the tooth *next* to the marked tooth on the spindle gear, count the number required for a half-revolution of the spindle. For example, if the spindle gear has 24 teeth, count 12 teeth and mark this one with chalk.

    (5) Revolve the lathe spindle by hand one-half turn to bring the marked tooth in line with the chalk mark on the intermediate gear.

    (6) Reengage the intermediate gear.

7. Reset the crossfeed handle to the same position as when cutting the first thread.

8. Cut the second thread, feeding the compound rest handle until the graduated collar is at the same setting as for the first thread.

The Thread-Chasing Dial Method of Cutting Multiple Threads

Double-start threads with an odd-numbered lead (example ⅕, ⅐, etc.) may be cut using the thread-chasing dial.

1. Take one cut on the thread by engaging the split nut at a numbered line on the chasing dial.

2. Without changing the depth of cut, take another cut at

Fig. 49-27 Marking the spindle and intermediate gear before indexing the work exactly one-half turn for cutting a double thread.

an unnumbered line on the chasing dial. The second thread will be exactly in the middle of the first thread.

3. Continue cutting the thread to depth, taking two passes (one on a numbered line, the other on an unnumbered line) for every depth-of-cut setting.

SQUARE THREADS

Square threads were often found in vise screws, jacks, and other devices where maximum power transmission was required. Because of the difficulty of cutting this thread with taps and dies, it is being replaced by Acme thread. With care, square threads can be readily cut on a lathe.

The Shape of a Square Threading Tool

The square threading tool looks like a short cutting-off tool. It differs from it in that both sides of the square threading tool must be ground at an angle to conform to the helix angle of the thread (Fig. 49-28).

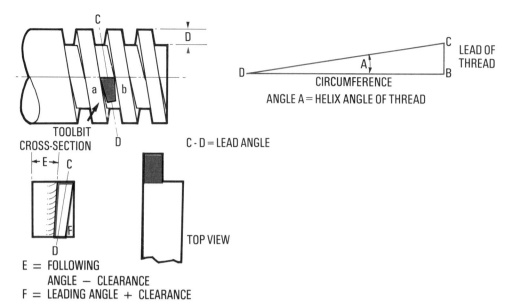

TOOLBIT
CROSS-SECTION

CIRCUMFERENCE
ANGLE A = HELIX ANGLE OF THREAD

C - D = LEAD ANGLE

TOP VIEW

E = FOLLOWING
ANGLE − CLEARANCE
F = LEADING ANGLE + CLEARANCE

Fig. 49-28 The shape of a square threading tool.

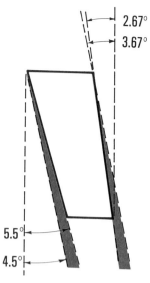

Fig. 49-29 Helix angles of the thread and the clearance angles necessary for a square threading tool.

The helix angle of a thread, and therefore the angle of the square threading tool, depends upon two factors.

1. The helix angle changes for each *different lead* on a given diameter. The greater the lead of the thread, the greater will be the helix angle.

2. The helix angle changes for each *different diameter* of thread for a given lead. The larger the diameter, the smaller will be the helix angle.

The helix angle of either the leading or following side of a square thread can be represented by a right-angle triangle (Fig. 49-28). The side opposite equals the *lead* of the thread, while the side adjacent equals the circumference of either the major or minor diameter of the thread. The angle between the hypotenuse and the side adjacent represents the helix angle of the thread.

To Calculate the Helix Angles of the Leading and Following Sides of a Square Thread

$$\text{Tan leading angle} = \frac{\text{lead of thread}}{\text{circumference of minor diameter}}$$

$$\text{Tan following angle} = \frac{\text{lead of thread}}{\text{circumference of major diameter}}$$

Clearance

If a square toolbit is ground to the same helix angles as the leading and following sides of the thread, it would have no clearance and the sides would rub. To prevent the tool from rubbing, it must be provided with approximately 1° clearance on each side, making it thinner at the bottom (Fig. 49-29). For the leading side of the tool, *add* 1° to the calculated helix angle. On the following side, *subtract* 1° from the calculated angle.

EXAMPLE

To find the leading and following angles of a threading tool to cut a 1¼ in.—4 square thread.

SOLUTION

$$\text{Lead} = 0.250 \text{ in.}$$

$$\text{Single depth} = \frac{0.500}{4}$$

$$= 0.125 \text{ in.}$$

$$\text{Double depth} = 2 \times 0.125$$
$$= 0.250 \text{ in.}$$

$$\text{Minor diameter} = 1.250 - 0.250$$
$$= 1.000 \text{ in.}$$

$$\text{Tan leading angle} = \frac{\text{lead}}{\text{minor dia. circumference}}$$

$$= \frac{0.250}{1.000 \times \pi}$$

$$= \frac{0.250}{3.1416}$$

$$= 0.0795 \text{ in.}$$

$$\therefore \text{ The angle of the thread} = 4°33'$$

$$\text{The toolbit angle} = 4°33' \text{ plus } 1° \text{ clearance}$$
$$= 5°33'$$

$$\text{Tan following angle} = \frac{\text{lead}}{\text{major dia. circumference}}$$

$$= \frac{0.250}{1.250\pi}$$

$$= \frac{0.250}{3.927}$$

$$= 0.0636 \text{ in.}$$

$\therefore$ The angle of the thread = 3°38′

The toolbit angle = 3°38′ minus 1° clearance
$$= 2°38′$$

To Cut a Square Thread

1. Grind a threading tool to the proper leading and following angles. The width of the tool should be approximately 0.002 in. (0.05 mm) wider than the thread groove. This will allow the completed screw to fit the nut readily. Depending on the size of the thread, it may be wise to grind two tools; a roughing tool 0.015 in. (0.38 mm) undersize, and a finishing tool 0.002 in. (0.05 mm) oversize.

2. Align the lathe centers and mount the work.

3. Set the quick-change gearbox for the required number of tpi.

4. Set the compound rest at 30° to the right. This will provide side movement if it becomes necessary to reset the cutting tool.

5. Set the threading tool square with the work and on center.

6. Cut the right-hand end of the work to the minor diameter for approximately 1/16 in. (1.58 mm) long. This will indicate when the thread is cut to the full depth.

7. If the work permits, cut a recess at the end of the thread to the minor diameter. This will provide room for the cutting tool to "run out" at the end of the thread.

8. Calculate the single depth of the thread

$$\left(\frac{0.500}{N} \right)$$

9. Start the lathe and just touch the tool to the work diameter.

10. Set the *crossfeed graduated collar* to zero.

11. Set a 0.003-in. (0.08-mm) depth of cut with the *crossfeed screw* and take a trial cut.

12. Check the thread with a thread pitch gage.

13. Apply cutting fluid and cut the thread to depth, moving the *crossfeed* in from 0.002 to 0.010 in. (0.05 to 0.25 mm) for each cut. The depth of the cut will depend on the thread size and the nature of the workpiece.

NOTE: Since the thread sides are square, *all cuts* must be set using the *crossfeed screw*.

ACME THREAD

The *Acme thread* is gradually replacing the square thread because it is stronger and easier to cut with taps and dies. It is used extensively for lead screws because the 29° angle formed by its sides allows the split nut to be engaged readily during thread cutting.

The Acme thread is provided with 0.010-in. clearance for both the crest and root on all sizes of threads. The hole for an internal Acme thread is cut 0.020 in. larger than the minor diameter of the screw, and the major diameter of a

Fig. 49-30 The toolbit is squared with the work by an Acme thread gage.

tap or internal thread is 0.020 in. larger than the major diameter of the screw. This provides 0.010 in.-clearance between the screw and nut on both the top and bottom.

To Cut an Acme Thread

1. Grind a toolbit to fit the end of the Acme thread gage (Fig. 49-30). Be sure to provide sufficient side clearance so that the tool will not rub while cutting the thread.

2. Grind the point of the tool flat until it fits into the slots of the gage indicating the number of threads per inch to be cut.

NOTE: If a gage is not available, the width of the toolbit point may be calculated:

$$\text{Width of point} = \frac{0.3707}{N} - 0.0052 \text{ in.}$$

3. Set the quick-change gearbox to the required number of threads per inch.

4. Set the compound rest 14½° to the right (half the included thread angle).

5. Set the Acme threading tool on center and square it with the work using the gage shown in Fig. 49-30.

6. At the right-hand end of the work, cut a section 1/16 in. long to the minor diameter. This will indicate when the thread is to the full depth.

7. Cut the thread to the proper depth by feeding the cutting tool, using the *compound rest*.

Measuring Acme Threads

For most purposes, the *one-wire method* of measuring Acme threads is accurate enough. A single wire or pin of the correct diameter is placed in the thread groove (Fig. 49-31) and measured with a micrometer. The thread is the correct size when the micrometer reading over the wire is the same as the major diameter of the thread and the *wire is tight in the thread*.

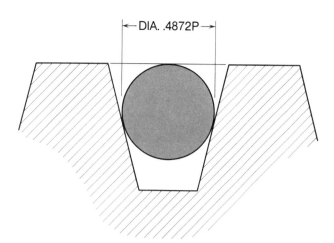

Fig. 49-31 Using one wire to measure the accuracy of an Acme thread.

NOTE: It is important that the burrs be removed from the diameter before using the one-wire method.

The diameter of the wire to be used can be calculated as follows:

$$\text{Wire diameter} = 0.4872 \times \text{pitch}$$

For example, if 6 threads per inch are being cut, the wire diameter should be:

$$\text{Wire size} = 0.4872 \times \frac{1}{6}$$

$$= 0.081 \text{ in.}$$

INTERNAL THREADS

Most internal threads are cut with taps; however, there are times when a tap of a specific size is not available and the thread must be cut on a lathe. Internal threading, or cutting threads in a hole, is an operation performed on work held in a chuck, collet, or mounted on a faceplate. The threading tool is similar to a boring toolbit, except the shape is ground to the form of the thread to be cut.

To Cut a 1⅜ in.—6 NC Internal Thread

1. Calculate the tap drill size of the thread.

$$\text{Tap drill size} = \text{major diameter} - \frac{1}{N}$$

$$= 1.375 - \frac{1}{6}$$

$$= 1.375 - 0.166$$

$$= 1.209 \text{ in.}$$

2. Mount the work to be threaded in a chuck, collet, or on a faceplate.

3. Drill a hole approximately 1/16 in. smaller than the tap drill size in the workpiece. For this thread it would be 1.209 − 0.062 = 1.147 or a 1 5/32-in. hole.

Fig. 49-32 The compound rest is set at 29° to the left for cutting right-hand internal threads.

4. Mount a boring tool in the lathe and bore the hole to the tap drill size (1.209 in.). The boring bar should be as large as possible and held short. The boring operation cuts the hole to size and makes it true.

5. Recess the start of the hole to the major diameter of the thread (1.375 in.) for 1/16 in. length. During the thread-cutting operation, this will indicate when the thread is cut to depth.

6. If the thread does not go through the workpiece, a recess should be cut at the end of the thread to the major diameter (Fig. 49-32). This recess should be wide enough to allow the threading tool to "run out" and permit time to disengage the split-nut lever.

7. Set the compound rest at 29° to the left (Fig. 49-32); to the right for left-hand threads.

8. Mount a threading toolbit into the boring bar and set it to center.

9. Square the threading tool with a thread center gage (Fig. 49-33).

10. Place a mark on the boring bar, measuring from the threading tool, to indicate the length of hole to be threaded. This will show when the split-nut lever should be disengaged.

Fig. 49-33 Using a thread center gage to square the threading tool with the work.

11. Start the lathe and turn the crossfeed handle *out* until the threading tool just scratches the internal diameter.

12. Set the crossfeed graduated collar to zero.

13. Set a 0.003-in. depth of cut by feeding the compound rest *out* and take a trial cut.

14. At the end of *each* cut on an internal thread, disengage the split-nut lever and feed the crossfeed handle *in* to clear the thread.

15. Clear the threading tool from the hole and check the pitch of the thread.

16. Return the crossfeed handle back to zero and set the depth of cut by turning the compound rest *out* the desired amount.

17. Cut the thread to depth; check the fit with a screw or threaded plug gage.

NOTE: The last few cuts should not be deeper than 0.001 in. each to eliminate the spring of the boring bar.

REVIEW QUESTIONS

THREADS

1. Define a thread.
2. List four purposes of threads.

THREAD TERMINOLOGY

3. Define: pitch diameter, pitch, lead, root, crest.
4. How is the pitch desginated for:
   (a) UNC threads?
   (b) metric threads?
5. Why is the diameter of the blank for a rolled thread equal to the pitch diameter?
6. How may a right-hand thread be distinguished from a left-hand thread?

THREAD FITS AND CLASSIFICATIONS

7. Define: fit, allowance, tolerance, limits.
8. How are external UNC threads classified?
9. Name and describe three classifications of UNC fits.
10. How are thread fits designated for the ISO threads?
11. Why was the ISO metric system of threads adopted?
12. Describe a thread designated as M 56 × 5.5.
13. Name five thread forms used in North America and state the included angle of each.
14. What are the principal differences between the American and Unified threads?

THREAD CALCULATIONS

15. For an M 20 × 2.5 thread, calculate:
    (a) pitch
    (b) depth
    (c) minor diameter
    (d) width of crest
    (e) width of root
16. Sketch a UNC thread and show the dimensions of the parts.
17. How does a Brown and Sharpe worm thread differ from an Acme thread?
18. For a 1 in.—8 NC thread, calculate:
    (a) minor diameter
    (b) width of toolbit point
    (c) amount of compound rest feed
19. What is the purpose of the quick-change gearbox?
20. Describe a thread-chasing dial and state its purpose.
21. The lead screw on a lathe has 6 tpi; at what point or points on the thread-chasing dial may the split-nut lever be engaged to cut the following threads: 8, 9, 11½, 12?

THREAD CUTTING

22. List the main steps required to set up the lathe for cutting a 60° thread.
23. List the main steps required to cut a 60° thread.
24. It has been necessary to remove a partially threaded piece of work from the lathe and to finish it at a later time. Describe how to reset the threading tool to "pick up" the thread.

METRIC THREADS

25. (a) What two change gears are required to cut a metric thread on a standard lathe?
    (b) Where are these gears mounted?

26. Describe how a lathe (having a quick-change gearbox) is set up to cut a 2.5-mm thread.
27. What precaution should be taken when cutting a metric thread?

THREAD MEASUREMENT

28. Make a sketch to illustrate how a thread is measured using the three-wire method.

29. Calculate the best wire size and the measurement over the wires for the following threads:
- (a) ¼ in.—20 NC
- (b) ⅝ in.—11 NC
- (c) 1¼ in.—7 NC

MULTIPLE THREADS

30. What is the purpose of a multiple thread?
31. If the pitch of a multiple thread is ⅛ in., what will be the lead for a double-start thread? A triple-start thread?
32. List three methods by which multiple threads may be cut.

SQUARE THREADS

33. Name two factors that affect the helix angle of a thread.

34. Calculate the leading and following angles of a square threading toolbit required to cut a 1½ in.—6 square thread.
35. List the main steps required to cut a square thread.

ACME THREAD

36. If the width of the root of an Acme thread is $0.3707P$ at minimum depth, why is the Acme threading tool ground to $0.3707P - 0.0052$?
37. Describe how an Acme thread may be measured.

INTERNAL THREADS

38. List the steps required to cut a 1¼ in.—7 UNC thread 2 in. deep in a block of steel 3 in. × 3 in. × 3 in. Show all necessary calculations.

STEADY RESTS, FOLLOWER RESTS, MANDRELS

UNIT 50

Various accessories make it possible to machine different lengths and shapes of workpieces in a lathe. Steady and follower rests are used to support long or slender workpieces and prevent them from springing during a machining operation either between lathe centers or in a chuck. Steady rests are attached to the lathe bed and are generally set in about the middle of a workpiece. Follower rests are attached to the lathe carriage and are set up immediately to the right of the cutting tool. Since the relationship of the attachment and the cutting tool remains the same, the workpiece is supported throughout the cut.

Mandrels generally fit the hole of a thin part such as gears, flanges, and pulleys, and they allow the outside diameter of the work to be machined between centers or in a chuck.

OBJECTIVES

After completing this unit you should be able to:
1. Set up and use a steady rest for machining a long shaft
2. Set up and use a follower rest when machining a long shaft
3. Mount and machine work on a mandrel
4. Lay out and machine an eccentric on work held between centers

STEADY REST

A steady rest (Fig. 50-1) is used to support long, slender work and prevent it from springing while being machined between centers. A steady rest may also be used when it is necessary to perform a machining operation on the end of a workpiece which is held in a chuck. The steady rest is fastened to the lathe bed and its three jaws are adjusted to the surface of the work to provide a supporting bearing. The jaws on a steady rest are generally made of soft ma-

Fig. 50-1 Using a steady rest to support the end of long work held in a chuck. (Courtesy South Bend Lathe, Incorporated.)

Fig. 50-2 The cathead can be adjusted to provide a true bearing surface for the steady rest, even if square work is being machined.

Fig. 50-3 A 60° spotting tool set up to recut a damaged center.

terial, such as fiber or brass, to prevent damaging the work surface. Other steady rests have rollers attached to the jaws to provide good support for the work.

TO SET UP A STEADY REST

1. Mount the work between centers.
OR
Set up and true the work in a chuck.
2. a. If the work diameter is not round, turn a true spot on the diameter (slightly wider than the steady rest jaws) at the point where the steady rest will be supporting the work. Long work in a chuck should be first supported by the tailstock center. If the diameter is rough, turn a section for the steady rest and one near the chuck to the same diameter.
b. If it is impossible to turn a true diameter (due to the shape of the workpiece), mount and adjust a *cathead* (Fig. 50-2) on the work.
3. Move the carriage to the tailstock end of the lathe.
4. Place the steady rest on the lathe bed at the desired position. If work diameter is turned and held in a chuck, slide the steady rest up close to the chuck.
5. Adjust the lower two jaws to the work diameter, using a paper feeler to provide clearance between the jaws and the work.
6. Slide the steady rest to the desired position and fasten it in place.
7. Close the top section of the steady rest and adjust the top jaw, using a paper feeler.
8. Apply white or red lead to the diameter at the steady rest jaws.
9. Start the lathe and carefully adjust each jaw until it just touches the diameter.

NOTE: The white or red lead will smear just as the jaw contacts the work.

10. Tighten the lock screw on each jaw and then apply a suitable lubricant.
11. Before machining, it is wise to indicate the top and front of the turned diameter at the chuck and at the steady rest to check for alignment. If the indicator reading varies, adjust the steady rest until it is correct.

TO TRUE A DAMAGED CENTER HOLE

1. Mount and true the work in a chuck and steady rest, if necessary.
2. Grind a 60° spotting tool (Fig. 50-3) and mount it on center in the tool holder.
3. Start the lathe and gradually bring the spotting tool into the damaged center hole.
4. Recut the center hole until the damaged section is removed.
5. Remove the workpiece, mount it between centers, and turn the diameter as required.

FOLLOWER REST

A follower rest, mounted on the saddle, moves along with the carriage to prevent work from springing up and away from the cutting tool. The follower rest, positioned immediately behind the cutting tool, can be used to support long work for successive operations such as thread cutting (Fig. 50-4).

TO SET UP A FOLLOWER REST

1. Mount the work between centers.
2. Fasten the follower rest to the saddle of the lathe.

Fig. 50-4 A follower rest used to support long, slender work between centers while thread cutting.

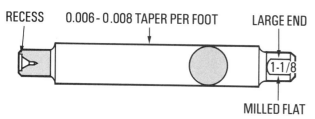

Fig. 50-5 A standard solid mandrel.

Fig. 50-6 An expansion mandrel.

Fig. 50-7 A gang mandrel.

3. Position the cutting tool in the toolpost so that it is just to the left of the follower rest jaws.

4. Turn the work diameter, for approximately 1½ in. (38.1 mm) long, to the desired size.

5. Adjust both jaws of the follower rest until they lightly contact the turned diameter.

6. Tighten the lock screw on each jaw.

7. Lubricate the work and the follower rest jaws to prevent marring the finished diameter.

8. If successive cuts are required when reducing the diameter of a workpiece, readjust the follower rest jaws as in steps 4 to 7.

MANDRELS

A *mandrel* (Fig. 50-5) is a precision tool which, when pressed into the hole of a workpiece, provides centers for a machining operation. They are especially valuable for thin work such as flanges, pulleys, gears, etc., where the outside diameter must run true with the inside diameter and it would be difficult to hold the work in a chuck.

CHARACTERISTICS OF A STANDARD MANDREL

1. Mandrels are usually hardened and ground and tapered 0.006 to 0.008 in./ft (0.50 to 0.66 mm/m).

2. The nominal size is near the middle and the small end is usually 0.001 in. (0.02 mm) under; the large end is 0.004 in. (0.10 mm) over the nominal size.

3. Both ends are turned smaller than the body and provided with a flat so that the lathe dog does not damage the accuracy of the mandrel.

4. The size of the mandrel is stamped on the large end.

5. The center holes, which are recessed slightly, are large enough to provide a good bearing surface and to withstand the strain caused by machining a workpiece.

TYPES OF MANDRELS

Many types of mandrels are used to suit various types of workpieces or machining operations. Descriptions and purposes of some of the more common types follow.

The solid mandrel (Fig. 50-5) is available for most of the standard hole sizes. It is a general-purpose mandrel which may be used for a variety of workpieces.

The expansion mandrel (Fig. 50-6) consists of a sleeve with four or more grooves cut lengthwise fitted over a solid mandrel. A taper pin fits into the sleeve to expand it to hold work that does not have a standard size hole. Another form of expansion mandrel has a slotted bushing fitting over a tapered mandrel. Bushings of various sizes can be used with this mandrel, increasing its range.

The gang mandrel (Fig. 50-7) is used to hold a number of identical parts for a machining operation. The body of the mandrel is parallel (no taper) and has a shoulder or flange on one end. The other end is threaded for a locking nut.

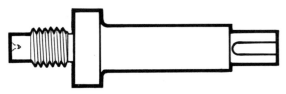

Fig. 50-8 A threaded mandrel.

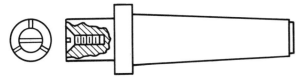

Fig. 50-9 A taper-shank mandrel.

1. CLEAN & OIL
THESE SURFACES

2. REMOVE BURRS
FROM EDGE OF HOLE

3. START SMALL END
OF MANDREL

Fig. 50-10 Preparing the mandrel and workpiece before mounting.

The threaded mandrel (Fig. 50-8) is used for holding workpieces having a threaded hole. An undercut at the shoulder ensures that the workpiece will seat squarely and is not canted on the threads.

The taper-shank mandrel (Fig. 50-9) may be fitted to the tapered hole in the headstock spindle. The projecting portion may be machined to any desired form to suit the workpiece. This type of mandrel is often used for small workpieces or those which have blind holes.

TO MOUNT WORK ON A PLAIN MANDREL

1. Secure a mandrel to fit the hole in the workpiece.
2. Thoroughly clean the mandrel and apply a thin film of oil on the diameter.
3. Clean and remove any burrs from the hole in the workpiece (Fig. 50-10).
4. Start the small end of the mandrel into the hole (the large end has the size stamped on it) by hand.
5. Place the work on an arbor press with a machined surface down so that the hole is at right angles (Fig. 50-11).
6. Press the mandrel firmly into the workpiece.

TO TURN WORK ON A MANDREL

Work pressed on a mandrel is held in position by friction; therefore cutting operations should be toward the large end

Fig. 50-11 Using an arbor press to force a mandrel into a workpiece.

Fig. 50-12 On an eccentric the axes are parallel but not in line.

of the mandrel. This will tend to keep the work tight on the mandrel.

1. Fasten the lathe dog on the *large end* of the mandrel (where the size is stamped).
2. Clean the lathe and mandrel centers and then mount the work.
3. If the entire side of the work must be faced, it is good practice to use a paper feeler between the toolbit point and mandrel for setting the toolbit. This will prevent marring or scoring the surface of the mandrel.
4. When turning the outside diameter of work, always cut toward the large end of the mandrel.
5. On large diameter work, it is advisable to take light cuts to prevent the work from slipping on the mandrel or chattering.

ECCENTRICS

An *eccentric* (Fig. 50-12) is a shaft which may have two or more turned diameters parallel to each other but not concentric with the normal axis of the work. Eccentrics are used in locking devices, in the feed mechanism on some shapers, and in the crank shaft of an automobile, etc., where it is necessary to *convert rotary motion into reciprocating motion,* or vice versa.

The amount of eccentricity, or *throw* of an eccentric, is the distance that a set of center holes has been offset from the normal work axis. If the center holes were offset ¼ in. (6.35 mm) from the work axis, the amount of throw would be ¼ in. (6.35 mm), but the total travel of the eccentric would be ½ in. (12.7 mm).

There are three types of eccentrics which are generally cut on a lathe in the following ways.

1. When the throw enables all centers to be located on the ends of the workpiece.
2. When the throw is too small to allow all centers to be located on the workpiece at the same time.
3. When the throw is so great that all centers cannot be located on the workpiece.

Fig. 50-13   Laying out the centers of an eccentric.

Fig. 50-14   The setup required for turning an eccentric with a large throw.

TO TURN AN ECCENTRIC WITH A 0.375-in. OR 10-mm THROW

1. Place the work in a chuck and face it to length. If the center holes are to be removed later, leave the work ¾ in. (19.05 mm) longer.

2. Place the work in a V-block on a surface plate and apply layout dye to both ends of the work.

3. Set a vernier height gage to the top of the work and note the vernier reading.

4. Subtract half the work diameter from the reading and set the gage to this dimension.

5. Scribe a centerline on both ends of the work.

6. Rotate the work 90° and scribe another centerline on both ends at the same height-gage setting (Fig. 50-13).

7. Lower or raise the height-gage setting 0.375 in. (or 10 mm) and scribe the lines for the offset centers on both ends.

8. Carefully center-punch the four scribed centers, and drill the center holes in each end.

9. Mount the work in a lathe and turn the diameter with the true centers.

10. Set the work on the offset centers and turn the eccentric (center section) to the required diameter.

TO CUT AN ECCENTRIC WITH A SMALL THROW

This procedure should be followed when the centers are too close to be located on the workpiece at the same time.

1. Cut the work ¾ in. (19.05 mm) longer than required.

2. Face the ends and drill one set of center holes in the lathe.

3. Mount the work between centers and turn the large diameter to size.

4. Cut off the ends to remove the center holes.

5. Lay out and drill a new set of center holes, offsetting them from center the required throw.

6. Turn the eccentric diameter to size.

TO TURN AN ECCENTRIC WITH A LARGE THROW

1. Set the work on the normal centers and turn both ends to size.

2. Secure or make a set of support blocks as shown in Fig. 50-14. The hole in the support block should fit the turned ends of the work snugly. A setscrew in each block is used to securely fasten the support blocks to the work. The number of centers required should be laid out and drilled in the support blocks (Fig. 50-14).

3. Align both support blocks parallel on the work and lock them in position.

4. Counterbalance the lathe to prevent undue vibration.

5. Turn the various diameters as required.

REVIEW QUESTIONS

FOLLOWER RESTS AND STEADY RESTS

1. State the purpose of a follower rest.

2. Explain how to set up a steady rest for turning a long shaft held between centers.

3. Describe a cathead and state when it is used.

4. Explain how a damaged center hole may be trued.

MANDRELS

5. State the purpose of a mandrel.

6. Draw a 1-in. (25.4-mm) standard mandrel and include all specifications.

7. Name and describe four types of mandrels and state their purpose.

8. List the precautions that must be taken when using a mandrel (include mounting and turning).

ECCENTRICS

9. Define an eccentric and state its purpose.

10. Explain the difference between the throw and total travel of an eccentric.

11. A crankshaft 6 in. (152.4 mm) long having equal-length journals is required to produce a travel of 1½ in. (38.1 mm) on a piston. The journal (finished shaft) size is 1 in. (25.4 mm).

   (a) Describe how to lay out this eccentric.

   (b) What size material would be required if ⅛ in. (3.17 mm) is allowed for "cleaning up"? This can be calculated easily with the aid of a sketch.

12. What precautions must be taken when turning an eccentric having a large throw?

MACHINING IN A CHUCK

UNIT 51

Spindle-mounted accessories, such as chucks, centers, drive plates, face-plates, etc., are fitted to and driven by the headstock spindle. The most versatile and commonly used spindle accessories are the lathe chucks. The jaws of a chuck are adjustable and therefore workpieces which might be difficult to hold by another method can be held securely. Three-jaw chucks, whose jaws move simultaneously, are generally used to hold round finished diameters. Four-jaw chucks, whose jaws move independently, are generally used to hold odd-shaped pieces.

All the machining operations which can be performed on work between centers, such as turning, knurling, threading, etc., can be performed on work held in a chuck. Most chuck work is generally fairly short; however, if workpieces longer than three times their diameter must be machined in a chuck, the end of the work should be supported to prevent it from springing and being thrown out of the lathe.

OBJECTIVES

After completing this unit, you should be able to:
1. Mount and machine work in a three-jaw chuck
2. Mount and machine work in a four-jaw chuck
3. Face, groove, and cut off work held in a chuck

MOUNTING AND REMOVING LATHE CHUCKS

The proper procedure for mounting and removing chucks must be carefully followed in order to not damage the lathe spindle and/or the chuck and to preserve the accuracy of the lathe. There are three types of spindle noses found on engine lathes on which lathe chucks may be mounted. They are the threaded spindle nose, the tapered spindle nose, and the cam-lock spindle nose. The procedure for mounting a chuck on each type of spindle nose follows.

TO MOUNT A CHUCK

1. Set the lathe to the slowest speed. *SHUT OFF THE ELECTRICAL SWITCH*.
2. Remove the drive plate and live center.
3. Clean all surfaces of the spindle nose and the mating parts of the chuck.

NOTE: Steel chips or any dirt will destroy the accuracy of the spindle nose and the mating taper in the chuck.

4. Place a cradle block on the lathe bed in front of the spindle and place the chuck on the cradle (Fig. 51-1).

CHUCK CRADLE

Fig. 51-1 A properly fitted cradle block makes mounting and removal of chucks easy and safe. (Courtesy Kostel Enterprises Limited.)

Fig. 51-2 Align the keyway in the chuck with the key on the spindle.

Fig. 51-3 Strike the spanner wrench sharply to tighten the lock ring.

5. Slide the cradle close to the lathe spindle nose and mount the chuck.

a. Threaded Spindle Nose Chucks
(1) Revolve the lathe spindle *by hand* in a counterclockwise direction and bring the chuck up to the spindle. *NEVER START THE MACHINE.*
(2) If the chuck and spindle are correctly aligned, the chuck should easily thread onto the lathe spindle.
(3) When the chuck adapter plate is within ¹⁄₁₆ in. (1.58 mm) of the spindle shoulder, give the chuck a quick turn to seat it against the spindle shoulder.
(4) Do not jam a chuck against the shoulder too tightly, it may damage the threads and make the chuck difficult to remove.

b. Taper Spindle Nose Chucks
(1) Revolve the lathe spindle by hand until the key on the spindle nose aligns with the keyway in the tapered hole of the chuck (Fig. 51-2).
(2) Slide the chuck onto the lathe spindle.
(3) Turn the lock ring in a counterclockwise direction until it is hand-tight.
(4) Tighten the lock ring securely with a spanner wrench by striking it sharply downward (Fig. 51-3).

c. Cam-Lock Spindle Nose Chucks
(1) Align the registration line of each cam lock with the registration line on the lathe spindle nose.
(2) Revolve the lathe spindle by hand until the holes in the spindle align with the cam-lock studs of the chuck (Fig. 51-4).
(3) Slide the chuck onto the spindle.
(4) Tighten each cam lock in a *clockwise* direction (Fig. 51-5).

Fig. 51-4 Mounting a chuck on a cam-lock spindle nose.

TO REMOVE A CHUCK

The following procedures for removing a lathe chuck also apply to other spindle-mounted accessories such as drive plates and faceplates.

1. Set the lathe in the slowest speed. *STOP THE MOTOR.*

Fig. 51-5 Tighten the cam locks in a clockwise direction.

2. Place a cradle block under the chuck (Fig. 51-1).
3. Remove the chuck by the following methods.

a. Threaded Spindle Nose Chucks
(1) Turn the chuck until a wrench hole is in the top position.

Fig. 51-6   A hardwood block can be used to remove a chuck from a *threaded* spindle nose.

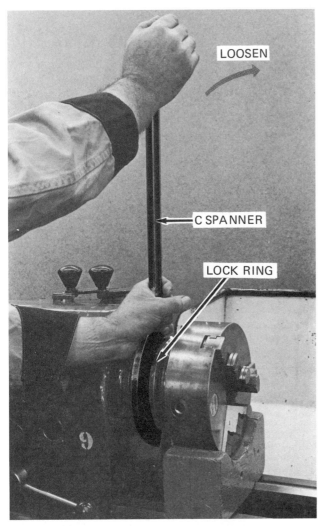

Fig. 51-7   A C-spanner wrench is used to loosen the locking ring which holds the chuck on the taper nose spindle.

Fig. 51-8   Loosen each cam lock until the registration lines on the lock and spindle match.   (Courtesy Kostel Enterprises Limited.)

(2) Insert the chuck wrench into the hole and pull it *sharply* toward the front of the lathe.

OR

(1) Place a block or short stick under the chuck jaw as shown in Fig. 51-6.

(2) Revolve the lathe spindle by hand in a *clockwise* direction until the chuck is loosened on the spindle.

(3) Remove the chuck from the spindle and store it where it will not be damaged.

b. *Taper Spindle Nose Chucks*

(1) Secure the proper C-spanner wrench.

(2) Place it around the lock ring of the spindle with the handle in an upright position.

(3) Place one hand on the curve of the spanner wrench to prevent it from slipping off the lock ring (Fig. 51-7).

(4) With the palm of the other hand, *sharply* strike the handle of the wrench in a clockwise direction.

(5) Hold the chuck with one hand, and with the other hand remove the lock ring from the chuck.

NOTE: The lock ring may turn a few turns and then become tight. It may be necessary to use the spanner wrench again to loosen the taper contact between the chuck and the spindle nose.

(6) Remove the chuck from the spindle and store it with the jaws in the up position.

c. *Cam-Lock Spindle Nose Chucks*

(1) With the chuck wrench, turn each cam lock in a counterclockwise direction until its registration line coincides with the registration line on the lathe spindle nose (Fig. 51-8).

(2) Place one hand on the chuck face, and with the palm of the other hand *sharply* strike the top of the chuck. This is necessary to break the taper contact between the chuck and the lathe spindle (Fig. 51-9).

NOTE: Sometimes it may be necessary to use a soft-faced hammer for this operation.

4. Slide the chuck clear of the spindle and place it carefully in a storage compartment.

Fig. 51-9   Strike the top of the chuck sharply with a hand to break the taper contact.

MOUNTING WORK IN A CHUCK

In order that workpieces be held securely for machining operations in either a three- or four-jaw chuck, certain procedures should be followed. Since the construction and purpose of three- and four-jaw chucks differ, the method of mounting work in each chuck also differs in certain aspects.

TO MOUNT WORK IN A THREE-JAW CHUCK

1. Clean the chuck jaws and the surfaces of the workpiece.
2. Use the proper size chuck wrench and open the chuck jaws slightly more than the diameter of the work.
3. Place the work in the chuck leaving no more than three times the diameter extending beyond the chuck jaws.
4. Tighten the chuck jaws with the wrench in the left hand while the right hand slowly rotates the workpiece.
5. Tighten the chuck jaws securely by *using the chuck wrench only.*

ASSEMBLING JAWS IN A THREE-JAW UNIVERSAL CHUCK

Three-jaw chucks are supplied with two sets of chuck jaws: one set for outside gripping and one set for inside gripping (Fig. 51-10). The jaws supplied with each chuck are marked with the same serial number and should *never be used with another chuck.* When it is necessary to change the chuck jaws, it is very important that the proper sequence be followed, otherwise the work held in the jaws will not run true.

1. Thoroughly clean the jaws and the jaw slides in the chuck.
2. Turn the chuck wrench clockwise until the start of the scroll thread is almost showing at the back edge of slide 1.
3. Insert jaw 1 (in slot 1) and press down with one hand while turning the chuck wrench clockwise with the other (Fig. 51-11).

Fig. 51-10   A three-jaw universal chuck with reversible jaws.

Fig. 51-11   Inserting the first jaw into the start of the scroll plate.

4. After the scroll thread has engaged in the jaw, continue turning the chuck wrench clockwise until the start of the scroll is near the back edge of groove 2.
5. Insert jaw 2 and repeat steps 3 and 4 (Fig. 51-12).
6. Insert the third jaw in the same manner (Fig. 51-13).

Some chucks are equipped with one set of top jaws which are fastened on by allen screws. In order to reverse these jaws, it is necessary to remove the screws, clean the mating parts, and replace the jaws in the reverse direction. The screws should be tightened uniformly and securely so that the jaws are not distorted.

TO MOUNT WORK IN A FOUR-JAW CHUCK

Work which must run absolutely true must be mounted in a four-jaw chuck because each jaw can be adjusted independently. The chuck jaws, which are reversible, can hold round, square, or irregular shaped workpieces securely. The work can be adjusted either concentric or off center.

1. Measure the diameter of the work to be chucked.
2. With a chuck wrench adjust the jaws to the approximate

Fig. 51-12 The scroll plate turned to start the second jaw.

Fig. 51-13 Inserting the third jaw into a three-jaw chuck.

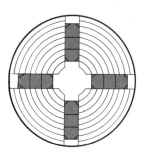

Fig. 51-14 The lines on a four-jaw chuck can be used as a guide when truing work.

size according to the ring marks on the face of the chuck (Fig. 51-14).

3. Set the work in the chuck and set the jaws snugly against the work surface.

4. True the workpiece by any of the following methods:

a. Chalk Method

(1) Start the lathe, and with a piece of chalk held in hand, lightly mark the high spot on the diameter (Fig. 51-15).

(2) Stop the lathe and check the chalk mark. If it is an even, lightly marked line around the work, the work is true.

(3) If there is only one mark, loosen the jaw opposite the chalk mark and tighten the jaw next to the chalk mark.

(4) Continue this operation until the chalk marks lightly around the work or leaves two marks opposite each other.

b. Surface Gage Method

(1) Place a *surface gage* on the lathe bed and adjust the point of the scriber so that it is close to the work surface (Fig. 51-16).

(2) Revolve the lathe by hand to find the low spot on the work.

(3) Loosen the jaw *nearest* the low spot and tighten the jaw *opposite* the low spot to adjust the work closer to center.

(4) Repeat steps (2) and (3) until the work is running true.

TO TRUE WORK IN A FOUR-JAW CHUCK USING A DIAL INDICATOR

A dial indicator should be used whenever a machined diameter must be aligned to within a few thousandths of an inch or hundredths of a millimeter.

1. Mount the work and true it approximately, using either the chalk or surface gage method.

2. Mount an indicator, with a range of at least 0.100 in. (2.54 mm) in the toolpost of the lathe (Fig. 51-17).

Fig. 51-15 Using chalk to indicate the high spot on the work. (Courtesy Kostel Enterprises Ltd.)

Fig. 51-16 Using a surface gage to true work in a four-jaw chuck.

Fig. 51-17 Truing work in a chuck using a dial indicator.

3. Set the indicator spindle in a *horizontal position* with the contact point set to center height.

4. Bring the indicator point against the work diameter so that it registers approximately 0.020 in. (0.50 mm) and revolve the lathe *by hand*.

5. Note the highest and the lowest reading on the dial indicator.

6. Slightly loosen the chuck jaw at the lowest reading, and tighten the jaw at the high reading until the work is moved half the difference between the two indicator readings.

7. Continue to adjust *only these two opposite jaws* until the indicator registers the same at both jaws.

NOTE: Disregard the indicator readings on the work between these two jaws.

8. Adjust the other set of opposite jaws in the same manner until the indicator registers the same at any point on the work circumference.

9. Tighten all jaws evenly to secure the workpiece firmly.

10. Rotate the lathe spindle by hand and recheck the indicator reading.

TO FACE WORK HELD IN A CHUCK

The purpose of facing work in a chuck is the same as that of facing between centers: to obtain a true, flat surface and to cut the work to length.

1. True up the work in a chuck using the chalk or dial indicator method. (This is not necessary if a three-jaw universal chuck is used.)

2. Have at least an amount equal to the diameter of the work projecting from the chuck jaws.

3. Swivel the compound rest at 90° (right angles) to the cross-slide when facing a series of shoulders (Fig. 51-18).

OR

Swivel the compound rest 30° to the right if only one surface on the work must be faced (Fig. 51-19).

4. Set up the facing toolbit to the height of the dead center and pointing slightly to the left.

5. Lock the carriage in position (Fig. 51-18).

6. Set the depth of cut by using the graduated collar on the compound rest screw.

   a. Twice the amount to be removed if the compound rest is set at 30°.

   b. The same as the amount to be removed if the compound rest is set at 90° to the cross-slide.

7. Face the work to length.

TO ROUGH- AND FINISH-TURN WORK IN A CHUCK

Although much work is turned in a lathe in school shops and training programs because of the ease of resetting the work accurately, most work machined in industry is held in a chuck. Since most work held in a chuck is relatively

Fig. 51-18   The compound rest set at 90° for facing accurate lengths.

Fig. 51-19   Feed the compound rest twice the amount to be faced off a surface.

short, in many cases there is no need to support the end of the work because it is held securely enough by the lathe chuck. Workpieces that extend more than three times their diameter must be supported to prevent the work from springing during the machining operation. The most common methods of supporting the end of long workpieces is by the use of a revolving tailstock center or with a steady-rest.

The operations of rough and finish turning in a chuck are the same as turning between centers (Unit 46) and will not be detailed here. Whenever possible, a diameter should be machined to size in two cuts: one roughing cut and one finishing cut.

ROUGH AND FINISH TURNING

1. Mount the work securely in a chuck having no more than three times the diameter extending beyond the chuck jaws.

2. Move the toolpost to the left of the compound rest and grip the toolholder short (Fig. 51-20).

Fig. 51-20 The toolpost set on the left side of the compound rest and the toolholder held short.

Fig. 51-21 Measuring the diameter of a trial cut.

Fig. 51-22 Inserted blade cutoff or parting tools. (A) Left-hand offset. (B) Straight. (C) Right-hand offset.

3. Fasten a general-purpose toolbit in the toolholder and set the point to center.

4. Tighten the toolpost screw securely.

5. Set the lathe to the correct speed and feed for rough turning.

6. Take a light trial cut at the end of the work and measure the diameter (Fig. 51-21).

7. Adjust the crossfeed graduated collar in one-half of the amount of metal to be removed.

8. Rough-turn the diameter to the correct length.

9. Set the lathe speed and feed for finish turning.

10. Take a light trial cut at the end of the workpiece.

11. Set the graduated collar in one-half the amount of metal to be removed.

12. Take another trial cut, measure the diameter, and if it is correct, finish-turn the diameter.

CUTTING OFF WORK IN A CHUCK

Cutoff tools, often called parting tools, are used for cutting off work projecting from a chuck, for grooving, and for under-cutting. The inserted blade-type parting tool is most commonly used, and it is provided in three holders (Fig. 51-22).

TO CUT OFF WORK IN A CHUCK

1. Mount the work in the chuck with the part to be cut off as close to the jaws as possible.

2. Mount the cutoff tool on the left-hand side of the compound rest with the cutting edge set on center (Fig. 51-23).

3. Place the holder as close to the toolpost as possible to prevent vibration and chatter.

Fig. 51-23 The tip of the cutoff blade being set to center height.

Fig. 51-24 The cutoff blade being moved to the correct position. (Courtesy Kostel Enterprises Limited.)

4. Extend the cutting blade beyond the holder half the diameter of the work to be cut, plus ⅛ in. (or 3 mm) for clearance.

5. Set the lathe to approximately one-half the turning speed.

6. Move the cutting tool into position (Fig. 51-24).

7. Start the lathe and feed the cutoff tool into the work by hand, keeping a steady feed during the operation. Cut brass and cast iron dry, but use cutting fluid for steel.

8. When grooving or cutting off deeper than ¼ in. (6.35 mm), it is good practice to move the parting tool sideways slightly. This may be accomplished by moving the carriage handwheel back and forth a few thousandths of an inch or hundredths of a millimeter during the cutting operation. This side motion cuts a little wider groove and prevents the tool from jamming.

9. Before the cut is completed, remove the burrs from each side of the groove with a file.

NOTE: To avoid chatter, keep the tool cutting and apply cutting fluid constantly during the operation. Feed slowly when the part is almost cut off.

REVIEW QUESTIONS

MOUNTING AND REMOVING CHUCKS

1. What safety precautions should be observed before attempting to mount or remove chucks?

2. Why is it necessary that the taper on the lathe spindle nose and the mating taper in the chuck be thoroughly cleaned before mounting?

3. Explain the procedure for mounting a chuck on a taper spindle nose.

4. Explain the procedure for removing a chuck from a cam-lock spindle nose.

MOUNTING WORK IN A CHUCK

5. How far should work extend beyond the chuck jaws before it must be supported?

6. Briefly list the procedure for assembling jaws in a three-jaw universal chuck.

7. For what purpose are four-jaw chucks used? Explain why.

8. Briefly explain how to true a piece of work to within 0.001 in. (0.02 mm) accuracy in a four-jaw chuck.

FACING IN A CHUCK

9. How should the cutting tool be set for facing?

10. How much should the compound rest graduated collar be turned in for facing:

    (a) 0.050 in. with the compound rest set at 90° to the cross-slide?

    (b) 0.025 in. when the compound rest is set at 30°?

ROUGH AND FINISH TURNING

11. How should the toolholder and the cutting tool be set for machining work in a chuck?

12. What is the purpose of a light trial cut at the end of a workpiece?

CUTTING OFF

13. How far should the blade extend beyond the holder when cutting off?

14. What procedure should be used when cutting grooves deeper than ¼ in. (6.35 mm)?

DRILLING, BORING, REAMING, AND TAPPING

UNIT 52

Internal operations such as drilling, boring, reaming, and tapping can be performed on work being held in a chuck. Boring tools are mounted in the toolpost, while drills, reamers, and taps may be held either in a drill chuck mounted in the tailstock spindle or directly in the tailstock spindle. Since the work held in a chuck is generally machined true, these operations are usually machined concentric with the outside diameter of the workpiece.

OBJECTIVES

After completing this unit you should be able to:
1. Drill small- and large-diameter holes in a lathe
2. Ream holes to an accurate size and good surface finish
3. Bore a hole to within 0.001-in. (0.02-mm) accuracy
4. Use a tap to produce an internal thread which is concentric with the outside diameter

TO SPOT AND DRILL WORK IN A CHUCK

Spotting ensures that a drill will start in the center of the work. A spotting tool is used to make a shallow, V-shaped hole in the center of the work, which provides a guide for the drill to follow. In most cases, a hole can be spotted quickly and fairly accurately by using a center drill (Fig. 52-1). Where extreme accuracy in spotting is necessary, a spotting toolbit should be used.

1. Mount work true in a chuck.

2. Set the lathe to the proper speed for the type of material to be drilled.
3. Check the tailstock and make sure it is in line.
4. Spot the hole with a center drill or spotting tool.
5. Mount the twist drill in the tailstock spindle, in a drill chuck, or in a drill holder (Figs. 52-2 to 52-4).

NOTE:
   a. When a tapered-shank drill is mounted directly in the tailstock spindle, a dog should be used to stop the drill from turning and scoring the tailstock spindle taper (Fig. 52-2).

Fig. 52-1   Using a center drill to spot a hole.

Fig. 52-2   A tapered-shank drill mounted in the spindle is prevented from turning by a lathe dog.

Fig. 52-3 Supporting the end of a drill will prevent it from wobbling when starting a hole.

Fig. 52-4 Drilling a hole with a drill mounted in a drill holder and supported on the dead center.

b. The end of the drill may be supported with the end of a toolholder so that the drill will start on center (Fig. 52-3).

c. Tapered-shank drills are often mounted in a drill holder. The point of the drill is positioned in the hole, while the end of the holder is supported by the dead center. The handle of the holder rests on the toolholder and against the toolpost to prevent the drill from turning and from pulling into the work (Fig. 52-4).

6. Start the lathe and drill to the desired depth, applying cutting fluid frequently.

7. To gage the depth of the hole, use the graduations on the tailstock spindle, or measure the depth with a steel rule (Fig. 52-5).

8. Withdraw the drill frequently to remove the chips and measure the depth of the hole.

CAUTION: Always ease the force on the feed as the drill breaks through the work.

BORING

Boring is the operation of enlarging and truing a drilled or cored hole with a single-point cutting tool. Special-diameter holes, for which no drills are available, can be produced by boring.

Holes may be drilled in a lathe; however, such holes are generally not considered accurate although the drill may have started straight. During the drilling process, the drill may become dull or hit a hard spot or blowhole in the metal, which will cause the drill to wander or run off center. If such a hole is reamed, the reamer will follow the drilled hole, and, as a result, the hole will not be straight. Therefore, if it is important that a reamed hole be straight and true, the hole should first be drilled, then bored and reamed.

Fig. 52-5 Measuring the depth of a drilled hole using a rule.

TO BORE WORK IN A CHUCK

1. Mount the work in a chuck, face, spot, and drill the hole approximately ¹⁄₁₆ in. (1.58 mm) undersize.

2. Select a boring bar as large as possible and have it extend beyond the holder only enough to clear the depth of the hole to be bored.

3. Mount the boring bar holder in the toolpost on the left-hand side of the compound rest.

4. Set the boring toolbit to center (Fig. 52-6).

5. Set the lathe to the proper speed and select a medium feed.

6. Start the lathe and bring the boring tool into contact with the inside diameter of the hole.

7. Take a light trial cut [approximately 0.005 in. (0.12 mm) or until a true diameter is produced] ¼ in. (6.35 mm) long at the right-hand end of the work.

8. Stop the lathe and measure the hole diameter with a telescopic gage or inside micrometer.

9. Determine the amount of material to be removed from the hole.

NOTE: Leave approximately 0.010 to 0.020 in. (0.25 to 0.50 mm) for a finish cut.

10. Set the depth of cut for half the amount of metal to be removed.
11. Start the lathe and take the roughing cut.

NOTE: If chatter or vibration occurs during machining, slow the lathe speed and increase the feed until it is eliminated.

12. Stop the lathe and bring the boring tool out of the hole without moving the crossfeed handle.
13. Set the depth of the finish cut and bore the hole to size. For a good surface finish, a fine feed is recommended.

REAMING

Reaming may be performed in a lathe to quickly obtain an accurately sized hole and to produce a good surface finish. Reaming may be performed after a hole has been drilled or bored. If a true, accurate hole is required, it should be bored before the reaming operation.

TO REAM WORK IN A LATHE

1. Mount the work in a chuck; face, spot, and drill the hole to size.
   Holes under ½ in. (12.7 mm) in diameter, drill ¹⁄₆₄ in. (0.39 mm) undersize; holes over 1½ in. (28.7 mm) in diameter drill ¹⁄₃₂ in. (0.79 mm) undersize. If the holes must be true, they should be bored 0.010 in. (0.25 mm) undersize.
2. Mount the reamer in a drill chuck or drill holder (Fig. 52-7). When reaming holes ⅝ in. (15.87 mm) in diameter and larger, fasten a lathe dog near the reamer shank and support the tail on the compound rest to prevent the reamer from turning.
3. Set the lathe to approximately half the drilling speed.
4. Bring the reamer close to the hole and lock the tailstock in position.
5. Start the lathe, apply cutting fluid to the reamer, and slowly feed it into the drilled or bored hole with the tailstock handwheel.
6. Occasionally remove the reamer from the hole to clear chips from the flutes and apply cutting fluid.
7. Once the hole has been reamed, stop the lathe and remove the reamer from the hole.

CAREFUL: *Never turn the lathe spindle or reamer backward for any reason. This will damage a reamer.*

8. Clean the reamer and store it carefully to prevent it from being nicked or damaged.

TAPPING

Tapping is one method of producing an internal thread on a lathe. The tap is aligned by placing the point of the lathe

Fig. 52-6 The boring bar should be held short and the toolbit set on center to machine inside diameters. (Courtesy R.K. LeBlond Machine Tool Company.)

Fig. 52-7 The setup for reaming in a lathe.

Fig. 52-8 Using tap to cut internal threads.

dead center in the shank end of the tap to guide it while the tap is turned by a tap wrench. A standard tap may be used for this operation; however, a gun tap is preferred because the chips are cleared ahead of the tap. When tapping a hole in a lathe, the spindle is locked and the tap is turned by hand (Fig. 52-8).

TO TAP A HOLE IN A LATHE

1. Mount the work in the chuck, face, and center-drill.
2. Select the proper tap drill for the tap to be used.
3. Set the lathe to the proper speed.
4. Drill with the tap drill to the required depth. Use cutting fluid if required.
5. Chamfer the edge of the hole slightly larger than the tap diameter.
6. Stop the lathe and lock the spindle, or put the lathe in its lowest speed.
7. Place a taper tap in the hole and support the shank with the tailstock center.
8. With a suitable wrench, turn the tap, keeping the dead center snug into the shank of the tap by turning the tailstock handwheel.
9. Apply cutting fluid while tapping the hole (Fig. 52-8).
10. Back off the tap frequently to break the chip.
11. Remove the taper tap and finish tapping the hole with a plug or bottoming tap.

GRINDING ON A LATHE

Cylindrical and internal grinding may be done on a lathe if a proper grinding machine is not available. A toolpost grinder, mounted on a lathe, may be used for cylindrical and taper grinding as well as angular grinding of lathe centers. An internal grinding attachment for the toolpost grinder permits the grinding of straight and tapered holes. Grinding should be done on a lathe only when no other machine is available, or when the cost of performing a small grinding operation on a part would not warrant setting up a regular grinding machine. Since the work should rotate in an opposite direction to the grinding wheel, the lathe must be equipped with a reversing switch.

TO GRIND A LATHE CENTER

1. Remove the chuck or drive plate from the lathe spindle.
2. Mount the lathe center to be ground in the headstock spindle.
3. Set a slow spindle speed.
4. Swing the compound rest to 30° (Fig. 52-9) with the centerline of the lathe.
5. Protect the ways of the lathe with cloth or canvas and place a pan of water below the lathe center.
6. Mount the toolpost grinder and adjust the center of the grinding spindle to center height.
7. Mount the proper grinding wheel, true, and dress.
8. Start the lathe, with the spindle revolving in reverse.
9. Start the grinder and adjust the grinding wheel until it sparks lightly against the revolving center.
10. Lock the carriage in this position.
11. Feed the grinding wheel in 0.001 in. (0.02 mm), using the crossfeed handle.
12. Move the grinder along the face of the center using the compound rest feed.
13. Check the angle of the center using a center gage, and adjust the compound rest if necessary.

Fig. 52-9 The compound rest set at 30° to grind a lathe center.

14. Finish-grind the center.

NOTE: If a high finish is desired, polish the center with abrasive cloth at a high spindle speed.

REVIEW QUESTIONS

DRILLING

1. Why is spotting important before a hole is drilled?
2. Name three methods of holding various size drills in a tailstock.
3. How can the depth of a drilled hole be gaged?

BORING

4. Define the boring process.
5. Why should a hole be bored before it is reamed?
6. How should the boring bar and cutting tool be set up for boring?
7. Briefly explain how to bore a hole to 1.750 in. (44.45 mm).

REAMING

8. What is the purpose of reaming in a lathe?
9. How much material should be left in a hole for reaming a
   (a) ⁷⁄₁₆-in. hole?
   (b) 1-in. hole?
10. Why should a reamer never be turned backward?

TAPPING

11. How is the tap started and guided so that the thread will be true to the bored hole?

SHAPERS
SECTION 12

Shapers are used primarily to produce flat surfaces by means of single or multiple cutting tools. Shapers use a reciprocating motion and produce a straight-line cut on the work during the forward stroke. On the backward stroke, the tool returns to the starting position and the table (and work) is advanced either by automatic or hand feed the desired amount for the next cut. The cutting action of a shaper is illustrated below.

Since the shaper cuts only on the forward stroke and does not cut on the return stroke, it is a relatively slow machine in relation to the rate of metal removal. Industry has long recognized that shapers are not very economical machine tools, and they have been replacing shapers with more efficient machines such as vertical milling machines and machining centers. Most machinery manufacturers have recognized that the day of the shaper has passed and have discontinued their manufacture. The basic operations on a shaper have been included in this book because they will still be found in some school shops and jobbing shops, and it is important that a person beginning in the trade be familiar with their operation.

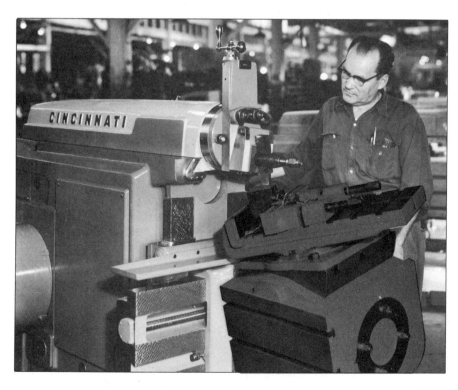

SHAPER CONSTRUCTION, SPEEDS, FEEDS

UNIT 53

The most common type of shaper was the horizontal type and it was available with either the crank-type or hydraulic driving mechanism. The size of a shaper is determined by the largest cube which it can machine. For example, a 14-in. shaper can machine a block 14 in. × 14 in. × 14 in.; a 300-mm shaper can machine a block 300 mm × 300 mm × 300 mm.

Since an apprentice may still come across these machines, it is wise to be familiar with their main operative parts and operation. As with other machine tools, the proper use of cutting tools and the setting of speeds and feeds are important to performing work satisfactorily.

OBJECTIVES

After completing this unit you should be able to:
1. Identify and know the purpose of the main operative parts
2. Calculate and set the proper speeds and feeds for a machining operation

PRINCIPLE OF THE CRANK-TYPE SHAPER

There are several mechanical methods of driving the reciprocating shaper ram, the crank-type being the most commonly used. Figure 53-1 shows the complete cycle of the crank-type mechanism, which is contained inside the column of the shaper (Fig. 53-2). The quick return of the ram on the back or noncutting stroke is obtained when the crank pin is traveling through 140° of the cycle. The forward or cutting stroke of the ram is made when the crank pin is in the upper portion of the vibrating (rocker) arm slot.

The crank pin passes through 220° of the cycle during the cutting stroke. As the crank pin passes center and moves further down the vibrating arm slot, it begins the return movement of the arm and the ram. At this time, the crank pin is in the lower part of the slot and in the lower 140° arc of the circle. Therefore, the movement of the upper end of the vibrating arm to which the ram is attached is more rapid but less powerful than on the forward stroke, even though the large gear and its crank pin are moving at a steady rate throughout the cycle.

PRINCIPLE OF THE HYDRAULIC SHAPER

Hydraulically driven shapers (Fig. 53-3) are similar in nearly all respects to the crank type; only the drive principle

Fig. 53-1 The cycle of the crank-type shaper.

DOWN-FEED HANDLE

RAM POSITIONING LOCK

STROKE INDICATOR

RAM

TOOLHEAD

TOOLPOST

FEED SELECTOR LEVER

STROKE REGULATOR LOCK NUT

GEAR CHANGE LEVERS

STROKE REGULATOR SHAFT

TABLE

CROSS RAIL

CROSS-FEED DIRECTION LEVER

CROSS-FEED TRAVERSE CRANK

VERTICAL TRAVERSE SHAFT

Fig. 53-2   A crank-type shaper.

FEED SCREW DIAL AND CRANK

ELECTRIC CLUTCH AND BRAKE CONTROL

RAM

START AND STOP BUTTONS

SWIVEL HEAD

OIL PRESSURE GAUGE

TOOL LIFTER

TOOLPOST

GEAR SHIFTER LEVER

BACK GEAR SELECTOR LEVER

POWER CROSS-FEED SELECTOR

RAM ADJUSTING SHAFT

VISE

STROKE INDICATOR DIAL

TABLE

STROKE ADJUSTING SHAFT

TABLE SUPPORT

POWER RAPID TRAVERSE LEVER

RAIL CLAMP CONTROL

CROSS-FEED ENGAGEMENT LEVER

APRON

RAIL ELEVATING MANUAL CONTROL

CROSS-FEED MANUAL CONTROL

CROSS RAIL

Fig. 53-3   A hydraulic shaper.

is different (Fig. 53-4). Oil pressure is provided by a pump through the *reversing valve* to the right-hand end of the operating cylinder where it moves the piston which drives the ram forward for the cutting stroke. The return stroke is obtained by the reversing valve directing the flow of oil to the left-hand end of the operating cylinder. The length of the stroke can be controlled by setting the position of the reversing stops on the ram. The table feed is operated by a table feed cylinder and piston. The speed and feed changes are made by control valves which allow the setting of variable speeds and feeds while the machine is operating.

The main advantages of a hydraulic shaper are:

SHAPER CONSTRUCTION, SPEEDS, FEEDS

Fig. 53-4 Hydraulic shaper principle. (Courtesy Mobil Oil Corporation.)

Fig. 53-5 Parts of a shaper toolhead.

1. It provides a uniform positive motion.
2. A wide range of speeds and feeds is available.
3. Speeds and feeds are independent and constant during the cutting stroke.
4. The surface finish obtained is free from gear chatter marks.
5. The speeds and feeds can be readily adjusted without stopping the shaper.
6. Cutting speeds and feeds are shown on a dial.

SHAPER PARTS

Shaper parts are the same for all horizontal types. The *ram* is a semicylindrical form of heavy construction on which the toolhead is mounted. The *driving mechanism* which provides a reciprocating action to the ram may be of the crank or hydraulic type. Fastened to the *shaper toolhead* (Fig. 53-5) is the *apron,* which consists of the *clapper box, clapper block, hinge pin,* and the *toolpost.* The hinged clapper block prevents the cutting tool from dragging on the return stroke.

SHAPER SPEEDS

To operate the shaper efficiently, it is necessary to calculate the speed at which the ram should operate. The speed depends on three factors: the type of material being cut, its length, and the type of cutting tool used. *Cutting speeds for materials are expressed in either feet per minute (ft/ min) or meters per minute (m/min).* A list of recommended cutting speeds for various metals is shown in Table 53-1. Shaper speeds, however, are calculated in strokes per minute or meters per minute. To calculate the proper shaper speed, one should understand the theory of the cutting and return stroke.

CUTTING AND RETURN STROKE

Since the length of the cutting stroke is the same as that of the return stroke, the pin and crosshead fastened to the gear pass through 220° on the cutting stroke and only 140° on the return stroke; therefore, the ratio is 220:140 or about a 3:2 ratio (Fig. 53-1). That is, if 3 seconds are required for the cutting stroke, 2 seconds are needed for the return stroke. Thus the return stroke is called the *quick return.* Since the sum of these figures is 5, the cutting stroke

TABLE 53-1	Shaper Speeds and Feeds																
	MACHINE STEEL				TOOL STEEL				CAST IRON				BRASS			
CUTTING TOOL	SPEED PER MINUTE		FEED		SPEED PER MINUTE		FEED		SPEED PER MINUTE		FEED		SPEED PER MINUTE		FEED	
	ft	m	in.	mm	ft	m	in.	mm	ft	m	in.	mm	ft	m	in.	mm
High-speed steel	80	24	0.010	0.25	50	15	0.015	0.38	60	18	0.020	0.51	160	48	0.010	0.25
Carbide	150	46	0.010	0.25	150	46	0.012	0.30	100	30	0.012	0.30	300	92	0.015	0.38

requires ⅗ of the time of a complete cycle, while the return stroke requires ⅖.

FORMULA FOR CALCULATING STROKES PER MINUTE

Inch Speed Calculations

When the length of the stroke in inches and the number of strokes per minute are given, the product is equal to the total inches cut in 1 min. of the machine operating time. Since the recommended cutting speeds are stated in feet per minute and the length of the cutting stroke in inches, it is necessary to change the inches into feet by multiplying by $\frac{1}{12}$. As the cutting stroke takes up ⅗ of the cycle, it is necessary to divide the above calculation by ⅗ (or multiply by $\frac{5}{3}$) to arrive at the cutting strokes per minute. Instead of multiplying by $\frac{1}{12}$ and then by $\frac{5}{3}$ for every calculation, it is easier to use 0.14 or $\frac{1}{7}$, which is the result of $\frac{1}{12} \times \frac{5}{3}$.

Therefore, to calculate strokes per minute, divide the cutting speed of the metal by the length of the stroke $\times$ 0.14 or $\frac{1}{7}$

$$N = \frac{CS}{L \times 0.14}$$

OR

$$N = \frac{CS}{L \times \frac{1}{7}}$$

OR

$$N = \frac{CS \times 7}{L}$$

N = Number of strokes per minute
CS = Cutting speed of the material in feet per minute
L = Length of stroke in inches

NOTE: The length of the stroke is always the length of the work plus 1 in. for tool clearance.

EXAMPLE

Find the number of strokes per minute for a shaper to machine a piece of 11-in.-long cast iron at 60 ft/min.

$$N = \frac{CS \times 7}{L}$$

$$N = \frac{60 \times 7}{12}$$

$$= 35 \text{ strokes per minute}$$

Metric Speed Calculations

When shaping a workpiece with metric dimensions, set the length of stroke to 25 mm longer than the workpiece to allow for clearance at each end of the stroke. The number of strokes per minute (N) is calculated as follows:

$$\text{Strokes per minute } N = \frac{CS \text{ (meters)}}{L \text{ (meters)}} \times \frac{3}{5} \text{ or } 0.6$$

EXAMPLE

How many strokes per minute will be required to machine a piece of tool steel 330 mm long (CS 15)?

Length of stroke (L) = 330 + 25 mm clearance

$$N = \frac{CS}{L} \times 0.6$$

$$= \frac{15}{0.36} \times 0.6$$

$$= 25 \text{ strokes per minute}$$

SHAPER FEEDS

Shaper feed is defined as the distance the work is moved sideways toward the cutting tool for each forward stroke of the ram. Feed rate is an important factor in determining the time required for the machining operation and the surface finish produced. In order for metals to be machined efficiently in a shaper, the following factors must be considered when speeds and feeds are selected.

☐ The type of material being cut
☐ The condition of the shaper
☐ The type of cut (roughing or finishing)

☐ The type of cutting tool used
☐ The surface finish required

The feed must be set so that the table and the workpiece are moved on the return stroke of the ram. If the feed is set to operate on the cutting (forward) stroke, the cutting tool will be damaged and the work surface will be rough and irregular.

As in most machining operations, it is important that metal be removed as quickly as possible during the roughing cut. Coarse feeds and light cuts are not as efficient as fine feeds and deep cuts because a thick chip will not curl as readily and requires more power to drive the tool. Coarse feeds cause the metal to tear more readily, resulting in a rougher surface. It is good practice to set the feed to produce the desired surface finish and then gradually increase the depth of cut to the capacity of the machine. For finish cuts, a finer feed with a light cut is recommended. The purpose of roughing cuts is to remove excess material as quickly as possible; finishing cuts are used to produce the desired accuracy and the proper surface finish.

In crank-type shapers, the amount and direction of feed are obtained by the principle illustrated in Fig. 53-6. Gear A is fastened to the crank-gear shaft. When the crank gear turns to produce the return stroke of the ram, gear A turns 140°. This turns gear B 140° and moves the connecting rod in the direction of the arrows. This in turn causes the pawl to turn the sprocket wheel, and since the sprocket wheel is keyed to the table feed screw shaft, this shaft also turns, moving the table on the return stroke. To vary the rate of feed, set the connecting rod closer to the center of the T-slot on the feed driving mechanism for finer feeds and away from the center for coarser feeds (Fig. 53-6). To disengage the feed, raise the pawl knob and turn it 90° to prevent the pawl from engaging the teeth in the sprocket wheel.

On hydraulic shapers, the feed direction and the amount of feed are regulated by control and reversing valves. The feed is driven by a gear and a rack which are controlled by a hydraulic system. Feeds can be increased or decreased while the machine is running by adjustment of the valves.

VERTICAL FEED

Vertical feed on a shaper may be used for machining vertical or angular surfaces. Vertical feed (Fig. 53-7) may be operated manually by the downfeed screw on the toolhead or automatically by the power downfeed. The power feed is attached to the downfeed screw by a bevel gear drive, and on the return stroke of the ram, the feed mechanism indexes the amount of vertical feed per cut. Feeds used for flat or angular surfaces should always be operated on the return stroke. When a vertical surface is to be machined, the top of the clapper box should be swiveled away from the surface to be machined (Fig. 53-7). This will prevent the cutting tool from binding against the work on the return stroke.

CUTTING TOOLS

Shaping operations vary from producing a flat surface to producing contours, grooves, and various intricate shapes.

Fig. 53-6 Shaper table feed mechanism.

Fig. 53-7 Shaper vertical feed mechanism. Note that the clapper box is swung for machining a vertical surface.   (Courtesy The Cincinnati Shaper Co.)

To produce these surfaces, a wide variety of cutting tools ground at different shapes and angles is required. The shape of the cutting tool varies with the work, type of cut, and the condition of the machine. Figure 53-8 illustrates cutting tools with the recommended clearance angles for cutting steel.

The shaper toolbits must have cutting clearances as do lathe toolbits, even though the shaper tool is held in a vertical position and the lathe toolbit is held horizontally (Fig. 53-9). Since the shaper tool does not feed sideways into the workpiece on the cutting stroke, about 2 to 4° *side clearance* is sufficient. The *front clearance* is usually about 4°. Too much front clearance will cause the toolbit to dull rapidly because of insufficient support behind the cutting edge. When insufficient front clearance is ground on the toolbit, the toolbit rubs and causes poor cutting action, leaving a rough surface finish.

Side clearance and side rake influence the angle of keenness. The side rake for steel is generally from 10 to 20°; the side clearance of 2 to 8° depends upon the tool used and the hardness of the metal being machined. Cutting tools

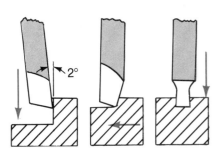

Fig. 53-8   Common shaper toolbits.

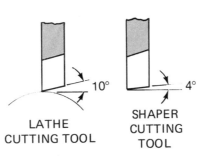

Fig. 53-9   A comparison of the end relief angles of lathe and shaper toolbits.

FINISHING TOOLBIT FOR CAST IRON

FINISHING TOOLBIT FOR STEEL

Fig. 53-10   Finishing shaper tools for various materials. (A) Cast iron; (B) steel.

A

B

C

Fig. 53-11   Types of shaper toolholders. (A) Straight; (B) offset; (Courtesy J.H. Williams and Company) (C) multipositional. (Courtesy Armstrong Brothers Tool Company.)

Fig. 53-12   A tool lifter raises the cutting tool on the return stroke. (Courtesy The Cincinnati Shaper Co.)

can be made from the solid shank type or from toolbit blanks. They are ground for left-hand or right-hand cutting, depending upon the direction of cut. The left-hand cutting tool is perhaps the most commonly used. The square-nose tool (Fig. 53-10) may be used for finishing cuts on cast iron and steel. Generally 0.010 to 0.015 in. (0.25 to 0.38 mm) for steel and 0.005 to 0.010 in. (0.12 to 0.25 mm) for cast iron should be left for finishing with this type of tool.

Toolholders used for shaping may be straight, offset, or multipositional (Fig. 53-11).

NOTE: The hole that accommodates the cutting tool in a shaper toolholder is parallel to the base, not inclined as in a lathe toolholder.

CARBIDE TOOLS

Carbide tools can be used efficiently only if the shaper can exceed a cutting speed of 100 ft/min (or 30 m/min). Greater efficiency may be obtained with carbide cutting tools on a hydraulic shaper because of the constant speed and feed rate which can be maintained. When using carbide cutting tools, the shaper should be equipped with a tool lifter (Fig. 53-12) to lift the tool clear of the work on the return stroke. This is necessary to prevent excessive tool wear and chipping of the cutting edge. Without these factors, it is not economical to use carbides instead of high-speed steel cutting tools.

Most materials are machined at cutting speeds two or three times faster with carbides, and the feeds are generally about the same as used with high-speed steel cutting tools. When machining with carbide cutting tools, it is more efficient to increase the cutting speed than the feed. A deep cut with a light feed is better than a light cut and a heavy feed because the chip pressure is distributed over a longer cutting edge. As the depth of cut is increased, the speed is decreased. Power consumed in shaping is directly propor-

313

Fig. 53-13   Brazed carbide shaper tools.

Fig. 53-14   Replaceable carbide inserts and toolholder.   (Courtesy Armstrong Brothers Tool Company.)

Fig. 53-15   Shaper carbide tool nomenclature.

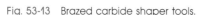

tional to the cutting speed.

Carbide cutting tools can be supplied in two common designs. *Brazed-on tips* (Fig. 53-13) are generally used for fine cuts and light-duty work. *Replacement inserts* (Fig. 53-14) are used for deep cuts, interrupted cuts, and large-area cuts. The latter is preferred to the brazed tool because it can he held more firmly and is easily replaced.

Carbide Tool Nomenclature

A—side rake angle
B—side clearance angle
C—front clearance
D—back rake angle
E—side cutting angle
F—nose radius
G—land width
H—point on the cutting edge

Shaper carbide tools have specified angles and design (Fig. 53-15) to eliminate tool damage and maintain tool life and surface finish. *Side clearance angle B* is generally held to a minimum to provide the maximum support for the cutting edge without rubbing. Also, the *front clearance C* is approximately 4°, which provides a strong support to the cutting edge. The *back rake angle D,* used on most roughing tools or machining cast iron and steels, may vary between 0 and −20°. This allows the cutting to start above the edge H, which is the weakest part of the tool, and protects the cutting edge against shock of starting and of intermittent cuts. On finishing tools, a 10 to 15° positive angle is often used so that tool chatter is avoided during light finishing cuts. A large cutting angle of 30 to 40° should be used with as large a radius at F as possible, particularly on roughing tools. Machining with carbides can be economical if proper machining and tool practices are followed.

To Maintain Proper Tool Finishes Free from Chatter Marks

1. Check the cutting tool for proper clearance angles.
2. Inspect the cutting edge for dullness, radii, and fractured edge.
3. See that the work is fastened securely.
4. Reduce the speed and increase the feed.
5. Adjust the gibs on the table, ram, toolhead, and clapper block tapered pin to eliminate any play.

6. Avoid too much toolhead overhand and have the cutting tool as close to the toolpost screw as possible.

REVIEW QUESTIONS

SHAPERS

1. Explain briefly the cutting action of the shaper.
2. Why is the return stroke of the crank-type shaper faster than the cutting stroke?
3. How can the length of the stroke be controlled on a hydraulic shaper?
4. List the advantages of a hydraulic shaper.

SPEEDS AND FEEDS

5. List the factors that determine the speed at which the shaper should operate.
6. (a) Calculate the strokes per minute using a high-speed steel cutting tool for the following:
   (i) bronze 7 in. long
   (ii) cast iron 14 in. long
   (iii) machine steel 10 in. long
   (b) How many strokes per minute should be used to provide a cutting speed of 18 m/min if the work is 250 mm long?
7. What four main factors govern the selection of the feed?
8. Define: cutting speed, feed.
9. Explain why the feed should operate on the return stroke.
10. How is the feed regulated on a hydraulic shaper?

CUTTING TOOLS

11. Sketch a shaper cutting tool, show the clearance angles, and explain why it differs from a lathe toolbit.
12. Why is it necessary to have a tool lifter on a shaper when using carbide cutting tools?
13. State four machine practices that help to prevent tool chatter.

SHAPER SETUPS AND OPERATIONS

UNIT 54

Although shapers are being replaced by more efficient machine tools, it is still possible for a beginner to the trade to come across these machines. Since these machines still provide a work station in school shops, it is wise to be familiar with the basic setups and operations common to shapers. As with all machine tools, a good worker is one that takes the necessary safety precautions to avoid personal injury, prevent damage to the equipment, and avoid spoiling workpieces.

OBJECTIVES

After completing this unit you should be able to:
1. Set the length and position of a shaper stroke
2. Align the shaper vise parallel and at 90° to the ram travel
3. Perform basic machining operations on a shaper

SAFETY PRECAUTIONS

It is very important that the operator be familiar with the various areas of a shaper which could be hazardous. By following a few basic safety precautions and using common sense, most accidents can be avoided.

1. Always wear safety glasses to protect your eyes from flying chips (Fig. 54-1).
2. Never operate any machine until you have been properly instructed.
3. *Before starting* a shaper, make sure that the work, vise, tool, and ram are securely fastened.

4. Check that the tool and toolhead will clear the work and also the column on the return stroke.
5. Always prevent chips from flying by using a metal shield or wire screen. This could prevent injury to others.
6. *Always* stand parallel to the cutting stroke and *not* in front of it.
7. Never attempt to remove chips or reach across the table while the ram is in motion.
8. Keep the area around a machine neat and tidy (Fig. 54-2).
9. Clean oil and grease from the floor immediately to prevent dangerous falls.
10. Never attempt to adjust a machine while it is in motion.

Fig. 54-1 Always wear safety glasses when operating a shaper.

Fig. 54-2 Poor housekeeping can lead to accidents.

315

Fig. 54-3 Aligning the shaper vise at 90° to the ram travel.

Fig. 54-4 Use the crossfeed screw to move the vise jaw past the indicator.

ALIGNING THE SHAPER VISE

Most work machined in a shaper is held in a vise bolted to a table. The body of the vise can be swiveled on a base plate to any angular setting required. This method is especially useful for cutting angular surfaces. When work must be shaped parallel or at right angles to the direction of the cut, the vise jaw must be aligned. To align the vise accurately, a dial indicator is attached to the toolhead of the shaper and is passed over the entire length of the solid jaw or a parallel held in the vise. With this method, a vise may be aligned to within 0.001-in. (0.02-mm) accuracy.

TO ALIGN THE SHAPER VISE AT 90° TO THE RAM TRAVEL

1. Adjust the vise until the lines on the vise base and swivel plate are in line.
2. Tighten the clamping nuts lightly.
3. Set the dial indicator in the toolholder, and position it

Fig. 54-5 Aligning the vise parallel to the ram travel.

against the solid vise jaw or against a parallel held in the vise (Fig. 54-3).
4. Move the shaper ram by hand until the dial indicator registers approximately one-half revolution against the solid jaw of the vise or the parallel.
5. Set the indicator bezel to zero.
6. Move the indicator along the length of the solid jaw or parallel by turning the crossfeed screw, and note the indicator reading (Fig. 54-4).
7. Adjust the vise to one-half the amount registered on the indicator.

NOTE: *Always* adjust the vise *away* from the indicator.

8. Tighten the clamping nuts, recheck the accuracy of the vise, and adjust if necessary.

TO ALIGN THE SHAPER VISE PARALLEL TO THE RAM TRAVEL

1. Adjust the vise parallel to the ram travel by the lines on the vise base and swivel plate.
2. Tighten the clamping nuts lightly.
3. Adjust the length of the shaper stroke so that it is slightly shorter than the length of the shaper vise.
4. Set the dial indicator in the toolholder and position it on the solid jaw or on a parallel held in the vise (Fig. 54-5).
5. Move the shaper table until the dial indicator registers approximately one-half revolution against the solid jaw or the parallel.
6. Set the indicator bezel to zero.
7. By hand, move the shaper ram along the length of the parallel or solid jaw and note the indicator reading.

CAUTION: The indicator should not travel beyond the vise jaw or parallel.

8. Adjust the vise to one-half the amount registered on the indicator (Fig. 54-6).

NOTE: Always adjust the vise *away* from the indicator.

9. Tighten the clamping nuts, recheck the accuracy of the vise, and adjust if necessary.

Fig. 54-6 Always tap the vise away from and not toward the indicator.

Fig. 54-8 Testing for parallelism along the second parallel.

Fig. 54-7 Testing for parallelism along one parallel.

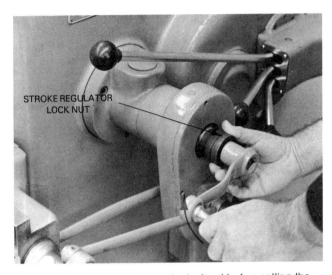

STROKE REGULATOR LOCK NUT

Fig. 54-9 Loosen the stroke regular lock nut before setting the shaper stroke.

TO TEST THE WORK SEAT FOR PARALLELISM

1. Set up the indicator as shown in Fig. 54-7.
2. Place a set of parallels on the bottom of the vise.
3. Adjust the toolhead until the dial indicator registers approximately one-quarter of a revolution on one parallel.
4. Move the ram forward by hand for the length of the parallel and note the indicator reading.
5. With the crossfeed screw handle, move the vise to obtain a reading across the second parallel (Fig. 54-8).
6. To correct any errors in alignment, remove the vise from the table, clean it, and remove burrs from all clamping surfaces on the vise and table.
7. Remount the vise on the table and test for accuracy.

NOTE: If the shaper is equipped with a swivel or universal table, the table surface should be checked for alignment before the vise is removed.

SETTING THE SHAPER STROKE

In order to cut work as efficiently as the shaper permits, it is good practice to not have the shaper stroke any longer

than is necessary. The length of the stroke should be about 1 in. (25.4 mm) longer than the work surface to be cut. This allows the cutting tool to clear each end of the work by ½ in. (12.7 mm). A stroke which is any longer would only waste time and take longer to machine the workpiece.

TO SET THE SHAPER STROKE LENGTH

1. Measure the length of the piece to be machined and add 1 in. (or 25 mm) to the length to allow clearance for the cutting tool.
2. Start the shaper and stop it when the ram is at the back end of the stroke.
3. Loosen the stroke regulator lock nut (Fig. 54-9).
4. Adjust to the length required by turning the stroke regulator screw.
5. Tighten the stroke regulator lock nut.

NOTE: In Fig. 54-10 the *crosshead* that fits into the *vibrating arm* is the part moved by the stroke regulator screw away from the center to lengthen the stroke and toward the center to shorten it.

Fig. 54-10 The shaper stroke-adjusting mechanism.

TO SET THE SHAPER STROKE POSITION

To set the position of the shaper stroke, the length of stroke must be set first. The ram must then be adjusted so that its travel will machine the full length of the piece. For example, if a workpiece 8½ in. long is to be machined, the stroke is set at 9½ in., allowing ½ in. for cutting tool clearance at each end.

1. Loosen the ram lock or clamp, with the ram at the back end of the stroke.
2. Adjust the ram by turning the ram adjusting screw.
3. Tighten the ram clamp to lock the ram in a fixed position.
4. Start the shaper and check the position of the ram to see if the cutting tool clears each end of the work.

SHAPER OPERATIONS

TO SHAPE A FLAT SURFACE

In Fig. 54-11, the toolhead is set up ready to shape a flat surface.

1. Clean the vise and remove all burrs from the workpiece.
2. Place parallels under the work if necessary, so that the machined surface will be approximately ¼ in. (6.35 mm) above the vise jaws.
3. Mount the work in the center of the vise.
4. At each corner of the work, place a short paper feeler between the work and parallels.
5. Tighten the vise securely and tap the work lightly with a lead hammer until all the paper feelers are tight (Fig. 54-12).
6. Swing the toolhead apron to the right (Fig. 54-11) to allow the cutting edge to swing free of the work on the return stroke.
7. Set the length and position of the shaper stroke.

Fig. 54-11 The correct method of clamping a shaper tool.

Fig. 54-12 Tap the work down until all feelers are tight.

8. Set the shaper speed to the required number of strokes per minute.
9. Adjust the toolbit to the work surface by using a paper feeler (Fig. 54-13).
10. Move the cutting tool clear of the work.
11. Set the depth of cut required by using the graduated downfeed screw collar.
12. Engage the automatic table feed, making sure that it feeds on the return stroke of the ram.

MACHINING A BLOCK SQUARE AND PARALLEL

To machine the four sides of a piece of work so that its sides are square and parallel, it is important that each side be machined in a definite order (Fig. 54-14). It is very important that dirt and burrs be removed from the work,

Fig. 54-13   Setting the cutting tool to the surface of the work.

Fig. 54-14   The sequence for machining the sides of a rectangular piece square and parallel.

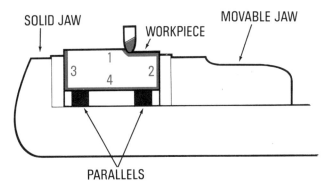

Fig. 54-15   Setup for shaping the first side. The stroke is set to cut the length rather than the width.

Fig. 54-16   Setup for shaping the second side. Notice that the first side machined is placed against the solid jaw.

Fig. 54-17   Setup for shaping the third side. The width should be the correct size after this operation.

Fig. 54-18   Setup for shaping the fourth side. The thickness should be correct after this operation.

vise, and parallels since any one of these can cause inaccurate work.

To Machine a Block Square and Parallel

To machine work in a shaper or milling machine vise so that it will be square and parallel, proceed as follows.

1. Set the work in the vise, with one of the largest sides up, and machine side 1 (Fig. 54-15).

NOTE: Remove all burrs after machining each side.

2. Place side 1 against the solid jaw and a round bar between side 4 and the movable jaw so the work will be held flat against the solid jaw, and machine side 2 at right angles to 1 (Fig. 54-16).

NOTE: *The round bar should be in the center of the amount of work being held inside the vise.*

3. Place side 1 against the solid jaw and 2 at the bottom, with a round bar between side 4 and the movable jaw, and machine side 3 (Fig. 54-17).

4. Place side 4 up as in Fig. 54-18 and machine it flat.

NOTE: With three finished surfaces, the round bar is not required in machining side 4.

5. Use a square (Fig. 54-19) to set up the work for machining the ends at right angles to the sides.
6. Check all surfaces with a square.

SHAPING WITH A HOLD-DOWN

Hold-downs are flat, wedge-shaped pieces of hardened steel generally used to hold thin pieces in a vise. They have a wide edge ground at an angle of approximately 8° and a narrow edge which may be ground straight or have a slight radius. The wide edge is placed against the vise jaw, and it forces or tips the narrow edge downward when the vise is tightened. Figure 54-20 shows a hold-down being used to hold a thin workpiece for a machining operation.

When the top surface of a workpiece must be machined parallel to the bottom surface, two hold-downs are used, one against each jaw and directly opposite each other. The action of the hold-down is to force the work against the parallels under the work or to the base of the vise.

Procedure for Machining with Hold-Downs

1. Set the work in the vise on parallels with a machined side down.

NOTE: Use thin paper strips between each corner of the work and the parallels.

2. Place the machined edge against the solid jaw.
3. Hold the wide edge of the hold-down against the movable jaw with the thin edge against the edge of the work surface.

NOTE: The hold-down is generally placed at about the center of the workpiece. The angle ground on the wide edge of the hold-down will force the work onto the parallel when the vise is tightened.

4. Tap the work *lightly* and retighten the work in the vise.
5. If the paper strips are not tight, remove the work and check all surfaces for burrs or cuttings.
6. Replace the work in the vise and again check that the work is seated firmly on the parallels.

NOTE: When thin workpieces are to be machined, it may be necessary to use two hold-downs, one against each vise jaw.

To Shape a Vertical Surface

When it is necessary to shape the ends of a long workpiece square with the sides, the work should be set up as in Fig. 54-21.

1. Align vise at right angles to the ram.
2. Place a square on the bottom of the vise and align the toolhead (vertically) with the blade (Fig. 54-22).
3. Place the work in the vise with the edge to be machined extending approximately ½ in. (12.7 mm) beyond the vise jaws.

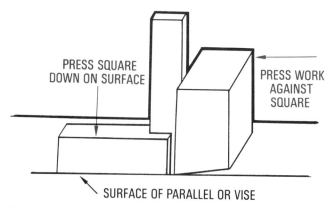
Fig. 54-19   Method of setting up work for machining the ends.

Fig. 54-20   Work set up in a vise using hold-downs.

Fig. 54-21   The clapper box must be swiveled for making vertical cuts.

4. Swing the top of the apron away from the surface being machined (Fig. 54-21). This will permit the cutting tool to swing away from the work so that it will not bind on the return stroke.
5. Offset the toolholder enough to permit the cutting tool to machine the entire edge in one pass.
6. Set the length and position of the stroke.
7. Check to see that the toolhead clears the frame or work at the end of the vertical travel.

Fig. 54-22 Aligning the toolhead with a square.

Fig. 54-23 Work set on an angle with the layout line parallel to the top of the vise.

Fig. 54-24 A vise swiveled to cut an angle.

Fig. 54-25 The toolhead set to machine an angle.

8. Set the depth of cut, using the crossfeed traverse crank.
9. Start the machine and feed the cutting tool down, on the return stroke, using the downfeed handle.

MACHINING ANGULAR SURFACES

Most angular surfaces can be machined by setting the work on an angle in the vise (Fig. 54-23), swiveling the universal vise (Fig. 54-24), or by setting the toolhead to an angle and feeding the toolbit in the plane of the toolhead (Fig. 54-25).

To Swivel the Toolhead for Machining Angular Surfaces

To machine an angular surface, the toolhead must be swiveled from the vertical position to allow the downfeed screw to move the cutting tool on an angular plane (Fig. 54-25). Whenever the toolhead is swiveled, it is important to extend it to the full depth of cut so that the toolhead clears the shaper column on the return stroke. To produce an angle

of 60° to the base, the toolhead must be adjusted to 30° from the vertical position. The clapper box should also be swiveled to the right (Fig. 54-25) to allow the tool to swing clear of the work on the return stroke of the ram. This prevents the tool from dragging, reduces damage to the cutting edge, and produces a better surface finish.

SERRATING A FLAT SURFACE

Serrating is an operation performed on a surface to form a series of equally spaced grooves. Serrations are often used on vise jaws, plates, washers, and other parts that are intended to provide a good grip (Fig. 54-26).

1. Mount the work in the center of the vise with the surface to be machined far enough above the vise jaws to allow clearance for the depth of the serrations.
2. Swivel the vise to the required angle of the serrations.
3. Set the cutting tool to the surface of the work using a paper feeler.
4. Set the graduated collar on the downfeed screw to *zero* without moving the position of the handle.
5. Turn the crossfeed traverse crank until the cutting tool is at the edge of the work.
6. Set the graduated collar on the crossfeed traverse screw to *zero* without moving the position of the handle.
7. Move the table with the crossfeed screw using the

Fig. 54-26   Serrations being cut on a shaper.   (Courtesy Kostel Enterprises Limited.)

graduated collar to obtain the spacing pitch of the serrations required.

8. Start the shaper and feed the cutting tool down about 0.005 in. (0.12 mm) on each return stroke of the ram.

9. When the groove being cut is the correct depth, stop the shaper.

10. Raise the cutting tool until it just clears the top of the work.

11. With the crossfeed traverse crank, move the table the correct distance for the next groove or serration.

12. Repeat steps 7 to 11 until all the grooves are cut in one direction.

13. Swivel the vise or turn the work in the vise and cut the grooves in the other direction if required.

REVIEW QUESTIONS

ALIGNMENT OF A VISE

1. In point form, list the steps necessary to align the vise jaws:

(a) parallel to the shaper stroke
(b) at 90° to the shaper stroke

2. Explain how the work seat may be tested for parallelism.

TO SET THE LENGTH OF THE SHAPER STROKE

3. What information is required before setting the length of the stroke?

4. List the steps required to set the shaper stroke to machine a piece of work 7 in. (177.8 mm) long.

TO SET THE POSITION OF THE SHAPER STROKE

5. What parts of a shaper are used to set the position of the stroke?

6. Why is it important to set the position of stroke after the length has been set?

7. Why must the ram lock always be tightened before the shaper is started?

TO SHAPE A FLAT SURFACE

8. Explain why all cuttings and burrs must be removed before setting up work in a vise.

9. What size parallels should be selected when setting up work in a vise?

10. Why should paper feelers be used between the parallels and the work?

11. On what stroke should the automatic feed operate when shaping?

TO MACHINE A BLOCK SQUARE AND PARALLEL

12. How is the first surface selected?

13. Why is a round rod used between the work and the movable jaw?

14. In what position should the round bar be set?

15. Make neat diagrams of the setups required to machine the four sides of a block square and parallel.

TO SHAPE A VERTICAL SURFACE

16. Why must the vise be aligned before shaping a vertical surface?

17. Explain how the toolhead, apron, and cutting tool should be set for shaping a vertical surface.

MILLING MACHINES
SECTION 13

Milling machines are machine tools used to produce one or more machined surfaces accurately on a piece of material, *the workpiece;* this is done by one or more rotary milling cutters having single or multiple cutting edges. The workpiece is held securely on the *work table* of the machine or in a holding device clamped to the table. It is then brought into contact with a revolving cutter.

The milling machine is a versatile machine tool which can handle a variety of operations normally performed by other machine tools. It is used not only for the milling of flat and irregular-shaped surfaces but also for gear and thread cutting, drilling, boring, reaming, and slotting operations. Its versatility makes it one of the most important machine tools used in machine shop work.

MILLING MACHINES AND ACCESSORIES

UNIT 55

A wide variety of milling machines are required by industry to suit the job requirements of the many parts which must be machined. To make milling machines more versatile, a large variety of accessories and attachments are available so that each machine can perform more operations on each workpiece.

OBJECTIVES

After completing this unit you should be able to:
1. Recognize and know the purpose of four milling machines
2. Know the purpose of the main operational parts of a horizontal and vertical milling machine
3. Recognize and know the purpose of four milling machine accessories and attachments

HORIZONTAL MILLING MACHINES

In order to meet many different industrial requirements, milling machines are made in a wide variety of types and sizes. They are classified under the following headings:

1. *Manufacturing type,* in which the cutter height is controlled by vertical movement of the headstock.
2. *Special type,* designed for specific milling operations.
3. *Knee-and-column type,* in which the relationship between the cutter height and the work is controlled by vertical movement of the table.

MANUFACTURING-TYPE MILLING MACHINES

Manufacturing-type milling machines are used primarily for quantity production of identical parts. This type of machine may be either semiautomatic or fully automatic and is of simple but sturdy construction. Fixtures clamped to the table hold the workpiece for a variety of milling operations, depending upon the type of cutters or special spindle attachments used. Some of the distinctive features of manufacturing-type machines are the *automatic cycle* of cutter and work approach, the *rapid movement* during the noncutting part of the cycle, and the *automatic spindle stop.* After this machine has been set up, the operator is required only to load and unload the machine and start the automatic cycle controlled by cams and preset *trip dogs.*

Some of the more common manufacturing-type milling operations are:

The plain manufacturing type (Fig. 55-1) has one horizontal spindle and one headstock. This machine is sometimes equipped with a reciprocating table cycle which permits feeding and rapid traversing in both directions. On this type of machine, *reciprocal milling* is possible. Two identical fixtures may be mounted on opposite ends of the table. While work is being machined at one end, a new piece is being loaded into the fixture at the other end.

The duplex manufacturing type (Fig. 55-2) is similar to the plain type, except that it has two horizontal spindles mounted in two independently adjustable headstocks. It can be used to perform two identical or two different milling operations on one or more pieces at the same time.

The small plain automatic knee-and-column type (Fig. 55-3) is similar to the plain horizontal mill. It is used for milling small or medium size parts in different quantities. The table is operated by power and controlled automatically by trip dogs mounted on the front of the table.

The unit-type automatic fixed-bed milling machine (Fig. 55-4) consists of a group of small size manufacturing milling machine units mounted on a common base. As many as four units can be mounted on one base for producing small parts at a high rate of speed.

The small plain automatic fixed-bed type (Fig. 55-5) has complete automatic cycle control of the table by a cycle

Fig. 55-1 A plain manufacturing-type milling machine.

Fig. 55-2 A duplex manufacturing-type milling machine. (Courtesy Cincinnati Milacron Incorporated.)

Fig. 55-3 A small plain automatic knee-and-column-type milling machine. (Courtesy Cincinnati Milacron Incorporated.)

Fig. 55-4 A unit-type automatic fixed-bed milling machine. (Courtesy Cincinnati Milacron Incorporated.)

model. This machine is used for efficient, accurate reproduction of curved or irregular surfaces. If the tracer is disengaged it can be used for standard milling operations.

SPECIAL-TYPE MILLING MACHINES

These machines are designed for individual milling operations and are used for only one particular job. They may be completely automatic and are used for production purposes when hundreds or thousands of similar pieces are to be machined.

selector and trip dogs. The machine may also be equipped with *automatic rise and fall* of the spindle carrier, permitting quick and economical milling of surfaces on different levels or between obstructions.

The tracer-controlled milling machine (Fig. 55-6) has a hydraulic or electrical circuit designed to automatically control the relative positions of the cutter and the workpiece by a tracer stylus riding on a cam, template, or

MACHINING CENTERS

A special type of machine which is finding wide applications in the industrial world is the horizontal numerically-controlled machining center (Fig. 55-7). This machine is capable of handling a wide variety of work such as straight and contour milling, drilling, reaming, tapping, and boring all in one setup. Because of the rugged construction, reli-

Fig. 55-6   A tracer-controlled milling machine.   (Courtesy Cincinnati Milacron Incorporated.)

Fig. 55-5   A small plain automatic fixed-bed milling machine. (Courtesy Cincinnati Milacron Incorporated.)

Fig. 55-7   A numerically controlled machining center.   (Courtesy Cincinnati Milacron Incorporated.)

able controls, and antibacklash ball screws, they are capable of high production rates while still maintaining a high degree of accuracy. This machine and its contribution to manufacturing will be discussed in greater detail in Section 15, Unit 70, "Computer-Age Machining."

KNEE-AND-COLUMN-TYPE MILLING MACHINES

Machines in this class fall into three categories.

1. Plain horizontal milling machines
2. Universal horizontal milling machines
3. Vertical milling machines

UNIVERSAL HORIZONTAL MILLING MACHINES

The universal horizontal milling machine is essential for advanced machine shop work, and the difference between this machine and the plain horizontal mill will be dealt with in this unit.

Figure 55-8 shows the parts of a universal horizontal mill. The only difference between this mill and the plain horizontal machine is the addition of a *table swivel housing* between the table and the saddle. This housing permits the table to be swiveled 45° in either direction in a horizontal plane for such operations as the milling of helical grooves in twist drills, milling cutters, and gears.

PARTS OF THE MILLING MACHINE

The *base* gives support and rigidity to the machine and also acts as a reservoir for the cutting fluids.

The *column face* is a precision-machined and scraped section used to support and guide the knee when it is moved vertically.

The *knee* is attached to the column face and may be moved vertically on the column face either manually or automatically. It houses the feed mechanism (Fig. 55-9).

Fig. 55-8 Universal horizontal milling machine.

COLUMN

OVERARM

SPINDLE

CINCINNATI

BASE

ELEVATING SCREW

TABLE

SWIVEL HOUSING

SADDLE

KNEE

Fig. 55-9   The main parts of a milling machine.   (Courtesy Cincinnati Milacron Incorporated.)

The *saddle* is fitted on top of the knee and may be moved in or out manually by means of the crossfeed handwheel or automatically by the crossfeed engaging lever (Fig. 55-9).

The *swivel table housing*, fastened to the saddle on a universal milling machine, enables the table to be swiveled 45° to either side of the centerline.

The *table* rests on guideways in the saddle and travels longitudinally in a horizontal plane. It supports the vise and the work (Fig. 55-9).

The *crossfeed handwheel* is used to move the table toward or away from the column.

The *table handwheel* is used to move the table horizontally back and forth in front of the column.

The *feed dial* is used to regulate the table feeds.

The *spindle* provides the drive for arbors, cutters, and attachments used on a milling machine.

The *overarm* provides for correct alignment and support of the arbor and various attachments. It can be adjusted and locked in various positions depending on the length of the arbor and the position of the cutter.

The *arbor support* is fitted to the overarm and can be clamped at any location on the overarm. Its purpose is to align and support various arbors and attachments.

The *elevating screw* is controlled by hand or an automatic feed. It gives an upward or downward movement to the knee and the table.

The *spindle speed dial* is set by a crank that is turned to regulate the spindle speed. On some milling machines the spindle speed changes are made by means of two levers. When making speed changes, always check to see whether the change can be made when the machine is running or if it must be stopped.

BACKLASH ELIMINATOR

A feature on most milling machines is the addition of a *backlash eliminator*. This device, when engaged, eliminates the backlash (play) between the nut and the table lead screw, permitting the operation of *climb* (down) milling. Figure 55-10 shows a diagrammatic sketch of the Cincinnati backlash eliminator.

The backlash eliminator works as follows. Two independent nuts are mounted on the lead screw. These nuts engage a common crown gear which in turn meshes with a rack. Axial movement of the rack is controlled by the backlash eliminator engaging knob on the front of the saddle. By turning the knob in, the nuts are forced to move along the lead screw in opposite directions, removing all backlash. The nuts-gear-rack arrangement is shown in Fig. 55-10.

MILLING MACHINE ACCESSORIES

A wide variety of accessories, which greatly increase its versatility and productivity, are available for the milling machine. These accessories may be classed as *fixtures* or as *attachments*.

FIXTURES

A fixture (Fig. 55-11) is a work-holding device fastened to the table of a machine or to a machine accessory, such as a rotary table. It is designed to hold workpieces that cannot be readily held in a vise or in production work when large quantities are to be machined. The fixture must be designed so that identical parts, when held in the fixture, will be positioned exactly and held securely. Fixtures may be constructed to hold one or several parts at one time and should permit the quick changing of workpieces. The work may be positioned by stops, such as pins, strips, or setscrews, and held in place by clamps, cam-lock levers, or setscrews.

To produce uniform workpieces, clean the chips and cuttings from a fixture before mounting a new workpiece.

Fig. 55-10   Cross section of a Cincinnati Milacron backlash eliminator.   (Courtesy Cincinnati Milacron Incorporated.)

Fig. 55-11   Milling a workpiece held in a fixture.   (Courtesy Cincinnati Milacron Incorporated.)

MILLING MACHINE ATTACHMENTS

Milling machine attachments may be divided into three classes.

1. Those designed to hold special attachments; these are attached to the spindle and column of the machine. They are the *vertical, high-speed, universal, rack milling,* and *slotting* attachments. These attachments are designed to increase the versatility of the machine.
2. *Arbors, collets,* and *adaptors* which are designed to hold standard cutters.
3. Those designed to hold the workpiece, such as a *vise, rotary table,* and *indexing or dividing head.*

Vertical Milling Attachment

The *vertical milling attachment* (Fig. 55-12) which may be mounted on the face of the column or the overarm, enables a plain or universal milling machine to be used as a vertical milling machine. Angular surfaces may be machined by swiveling the head, parallel to the face of the column, to any angle up to 45° on either side of the vertical position. On some models the head may be swiveled as much as 90° to either side. Vertical attachments enable the horizontal milling machine to be used for such operations as face milling, end milling, drilling, boring, and T-slot milling.

A modification of the vertical milling attachment is the *universal milling attachment* which may be swiveled in two planes, parallel to the column and at right angles to it, for the cutting of compound angles. The vertical and universal attachments are also manufactured in *high-speed* models which permit the efficient use of small and medium size end mills and cutters for such operations as die sinking and key seating.

Rack Milling Attachment

The *rack milling attachment* (Fig. 55-13A) and the *rack indexing attachment* (Fig. 55-13B) are used to mill longer

Fig. 55-12 A vertical milling attachment. (Courtesy Cincinnati Milacron Incorporated.)

gear racks (flat gears) than could be cut with the standard horizontal milling machine. These attachments will be discussed later in the section with gear-cutting accessories.

Slotting Attachments

The *slotting attachment* (Fig. 55-14) converts the rotary motion of the spindle into reciprocating motion for cutting keyways, splines, templates, and irregularly shaped surfaces. The length of the stroke is controlled by an adjustable crank. The tool-slide may be swiveled to any angle in a plane parallel to the face of the column, making the slotting attachment especially valuable in die work.

Arbors, Collets, and Adapters

Arbors, used for mounting the milling cutter, are inserted and held in the main spindle by a draw bolt or a special quick-change adapter (Fig. 55-15).

Shell-end mill arbors may fit into the main spindle or the spindle of the vertical attachment. These devices permit face milling to be done either horizontally or vertically.

Collet adapters are used for mounting drills or other tapered-shank tools in the main spindle of the machine or the vertical milling attachment.

A *quick-change adapter,* mounted in the spindle, permits such operations as drilling, boring, and milling without a change in the setup of the workpiece.

Vises

Milling machine vises are the most widely used workholding devices for milling; they are manufactured in three styles.

The *plain vise* (Fig. 55-16A) may be bolted to the table so that its jaws are parallel or at right angles to the axis of the spindle. The vise is positioned quickly and accurately

Fig. 55-13   (A) A rack milling attachment; (B) a rack indexing attachment.   (Courtesy Cincinnati Milacron Incorporated.)

Fig. 55-14 A slotting attachment. (Courtesy Cincinnati Milacron Incorporated.)

Fig. 55-15 Arbors, collets, and adapters.

A

B

by keys on the bottom which fit into the T-slots on the table.

The *swivel base vise* (Fig. 55-16B) is similar to the plain vise, except that it has a swivel base which enables the vise to be swiveled through 360° in a horizontal plane.

The *universal vise* (Fig. 55-16C) may be swiveled through 360° in a horizontal plane and may be tilted from 0 to 90° in a vertical plane. It is used chiefly by toolmakers, moldmakers, and diemakers, since it permits the setting of compound angles for milling.

Indexing or Dividing Head

This very useful accessory permits the cutting of bolt heads, gear teeth, and ratchets, etc. When connected to the lead screw of the milling machine, it will revolve the work as required to cut helical gears and flutes in drills and reamers, etc. The dividing head will be fully discussed later in the section.

REVIEW QUESTIONS

HORIZONTAL MILLING MACHINE

1. Name six operations that can be performed on a milling machine.

MANUFACTURING-TYPE MILLING MACHINE

2. List four features of a manufacturing-type milling machine.

UNIVERSAL HORIZONTAL MILLING MACHINE

3. What is the difference between a plain horizontal and a universal horizontal milling machine?
4. What is the purpose of the backlash eliminator?

MILLING MACHINE ACCESSORIES AND ATTACHMENTS

5. What is the purpose of a fixture?

C

Fig. 55-16 (A) A plain vise (Courtesy Cincinnati Milacron Incorporated.); (B) a swivel base vise; (C) a universal vise.

6. List three milling machine attachments and give examples of each.
7. Describe the purpose of the following devices:
   (a) vertical milling attachment
   (b) rack milling attachment
   (c) slotting attachment
8. Name three methods of holding cutters on a milling machine.
9. What are the features of the:
   (a) plain vise
   (b) swivel base vise
   (c) universal vise

MILLING CUTTERS
UNIT 56

The proper selection, use, and care of milling cutters must be practiced if the best results are to be achieved with the milling machine. The operator or apprentice must be able not only to determine the proper spindle speed for any cutter but should also constantly observe how the milling machine performs with different cutters and clearances.

Milling cutters are manufactured in many types and sizes. Only the most commonly used cutters will be discussed.

OBJECTIVES

After completing this unit you should be able to:
1. Identify and state the purpose of six standard milling cutters
2. Identify and state the purpose of four special-purpose cutters
3. Use high-speed steel and carbide cutters for proper applications

MILLING CUTTER MATERIALS

In the milling process, as with most metal-cutting operations, the cutting tool must possess certain qualities in order to function satisfactorily. Cutters must be harder than the metal being machined and strong enough to withstand pressures developed during the cutting operation. They must also be tough to resist the shock resulting from the contact of the tooth with the work. To maintain keen cutting edges, they must be able to resist the heat and abrasion of the cutting process.

Most milling cutters today are made of high-speed steel or tungsten carbide. Special-purpose cutters, made in the plant for a special job, may be made from plain carbon steel.

High-speed steel, consisting of iron with various amounts of carbon, tungsten, chromium, molybdenum, and vanadium, is used for most solid milling cutters since it possesses all the qualities required for a milling cutter. In this steel, carbon is the hardening agent, while tungsten and molybdenum enable the steel to retain its hardness up to

red heat. Vanadium increases the tensile strength and chromium increases the toughness and wear resistance.

When a higher rate of production is desired and when harder metals are being machined, cemented carbides replace high-speed steel cutters. Cemented-carbide cutters, (Fig. 56-1A), although more expensive, may be operated from 3 to 10 times faster than high-speed steel cutters. Cemented-carbide tips may be either brazed to a steel body or held in place by means of a locking or clamping device (Fig. 56-1B).

When cemented-carbide cutters are to be used, care must be taken to select the proper type of carbide for the job. Straight tungsten carbide is used for machining cast iron, machine steel, most nonferrous alloys, and plastics. Tantalum carbide is used to machine low- and medium-carbon steels, and tungsten-titanium carbide is used for high-carbon steels.

Although cemented carbides have many advantages as cutting tools, several disadvantages limit extensive use.

1. Cemented-carbide cutters are more costly to buy, maintain, and sharpen.

Fig. 56-1   (A) A cemented-carbide milling cutter; (B) cemented-carbide tips are held in place by a locking device.

Fig. 56-2   Plain milling cutters. (A) Light duty; (B) light duty helical; (C) heavy duty; (D) high helix.

2. In order to use these cutters efficiently, machines must be rigid and have greater horsepower and speed than is required for high-speed cutters.

3. Carbide cutters are brittle, and therefore the edges break easily if misused.

4. Special grinders with silicon carbide and diamond wheels are required to sharpen carbide cutters properly.

PLAIN MILLING CUTTERS

Probably the most widely used milling cutter is the plain milling cutter, which is a cylinder made of tool steel with teeth cut on the periphery; it is used to produce a *flat surface*. These cutters may be of several types as shown in Fig. 56-2.

Light-duty plain milling cutters (Fig. 56-2A) which are less than ¾ in. (19.05 mm) wide will usually have straight teeth; those over ¾ in. (19.05 mm) wide have a helix angle of about 25° (Fig. 56-2B). This type of cutter is used only for light milling operations since it has too many teeth to permit the chip clearance required for heavier cuts.

Heavy-duty plain milling cutters (Fig. 56-2C) have fewer teeth than the light-duty type, which provide for better chip clearance. The helix angle varies up to 45°. This helix angle on the teeth produces a smoother surface because of the shearing action and reduces chatter. Less power is required with this type of cutter than with straight-tooth and small-helix-angle cutters.

High-helix plain milling cutters (Fig. 56-2D) have helix angles from 45 to over 60°. They are particularly suited to the milling of wide and intermittent surfaces on contour and profile milling. Although this type of cutter is usually mounted on the milling machine arbor, it is sometimes shank-mounted with a pilot on the end and used for milling elongated slots.

333

STANDARD SHANK-TYPE HELICAL MILLING CUTTERS

Standard shank-type helical milling cutters (Fig. 56-3), also called *arbor-type cutters,* are used for milling forms from solid metal; for example, they are used when making yokes or forks. Shank-type milling cutters are also used for removing inner sections from solids. These cutters are inserted through a previously drilled hole and supported at the outer end with type A arbor supports. Special spindle adapters are used to hold these cutters.

SIDE MILLING CUTTERS

Side milling cutters (Fig. 56-4) are comparatively narrow cylindrical milling cutters with teeth on each side as well as on the periphery. They are used for cutting slots and for face and straddle milling operations. Side milling cutters may have straight teeth (Fig. 56-4A) or staggered teeth (Fig. 56-4B). Staggered-tooth cutters have each tooth set alternately to the right and left with an alternately opposite helix angle on the periphery. These cutters have free cutting action at high speeds and feeds. They are particiularly suited for milling deep, narrow slots.

Half-side milling cutters (Fig. 56-4C) are used when only one side of the cutter is required, as in end facing. These cutters are also made with interlocking faces so that two cutters may be placed side by side for slot milling. The interlocking type is more suited for slot cutting than the solid-type staggered-tooth cutter since the amount ground from the side of the cutter during regrinding may be compensated by a washer between the cutters. Half-side milling cutters have considerable rake and, therefore, are able to take heavy cuts.

FACE MILLING CUTTERS

Face milling cutters (Fig. 56-5) are generally over 6 in. (152.4 mm) in diameter and have *inserted teeth* held in place by a wedging device. The teeth may be of high-speed steel, cast tool steel, or they may be tipped with sintered-carbide cutting edges. The corners of this type of cutter are bevelled; most of the cutting action occurs at these points and the periphery of the cutter. The face of the tooth removes a small amount of stock left by the spring of the work or cutter. To prevent chatter, only a small portion of the tooth face near the periphery is in contact with the work; the remainder is ground with a suitable clearance (8 to 10°).

This type of cutter is often used as a combination cutter, making the roughing and finishing cut in one pass. The

Fig. 56-5   A face milling cutter.   (Courtesy The BUtterfield Division, Union Twist Drill Company.)

Fig. 56-3   Standard shank-type helical milling cutter.   (Courtesy The BUtterfield Division, Union Twist Drill Company.)

A   B

C

Fig. 56-4   Side milling cutters (A) Plain; (B) staggered tooth; (C) half side.

Fig. 56-6 Shell end mill and adapter.

Fig. 56-7 (A) Single-angle milling cutter. (B) Double-angle milling cutter. (Courtesy The Butterfield Division, Union Twist Drill Co.)

roughing and finishing blades are mounted on the same body, with a limited number of finishing blades being set to a smaller diameter and extending slightly farther from the face than the roughing blades. The finishing blades have a slightly wider cutting face surface which creates a better surface finish.

Face milling cutters under 6 in. (152.4 mm) are called *shell end mills* (Fig. 56-6). They are solid, multiple-tooth cutters with teeth on the face and the periphery. They are usually held on a stub arbor which may be threaded or employ a key in the shank to drive the cutter. Shell end mills are more economical than large solid end mills because they are cheaper to replace when broken or worn out.

ANGULAR CUTTERS

Angular cutters have teeth that are neither parallel nor perpendicular to the cutting axis. They are used for milling angular surfaces, such as grooves, serrations, chamfers, reamer teeth, etc. They may be divided into two groups:

1. *Single-angle milling cutters* (Fig. 56-7A) have teeth on the conical surface and may or may not have teeth on the flat side. The included angle between the flat face and the conical face designates the cutters, such as 45 or 60° angular cutter.
2. *Double-angle milling cutters* (Fig. 56-7B) have two intersecting conical surfaces with cutting teeth on both. When

these cutters have equal angles on both sides of the line at a right angle to the axis (symmetrical), they are designated by the size of the included angle. When the angles formed with this line are not the same (unsymmetrical), the cutters are designated by specifying the angle on either side of the plane or line, such as 12 to 48° double-angle milling cutter.

FORMED CUTTERS

Formed cutters (Fig. 56-8) incorporate the exact shape of the part to be produced, permitting exact duplication of irregularly shaped parts more economically than most other means. Formed cutters are particularly useful for the production of small parts. Each tooth of a formed cutter is identical in shape, and the clearance is machined on the full thickness of each tooth by the form or master tool in a cam-controlled relieving machine. Examples of *form-relieved cutters* are concave, convex, and gear cutters.

Formed cutters are sharpened by grinding the tooth face. Tooth faces are radial and may have positive, zero, or negative rake, depending on the cutter application. *It is important* that the original rake on the tooth be maintained so that the profiles of the tooth and of the work are not changed. If the tooth face rake is maintained exactly, the cutter may be resharpened until the teeth are too thin for use; thus the exact, original shape of the tooth can be maintained. These cutters are sometimes produced with an angular face which causes a shearing action and reduces chatter during the cutting process. They are, however, more difficult to sharpen.

METAL SAWS

Metal-slitting saws (Fig. 56-9) are basically thin plain milling cutters with sides relieved or "dished" to prevent rubbing or binding when in use. Slitting saws are made in widths from 1/32 to 3/16 in. (0.8 mm to 5 mm). Because of

Fig. 56-8   Types of formed cutters. (A) Concave; (B) convex; (C) gear tooth.

Fig. 56-9   Metal-slitting saws.   (Courtesy The Butterfield Division, Union Twist Drill Company.)

their thin cross section, they should be operated at approximately one-quarter to one-eighth of the feed per tooth used for other cutters. For nonferrous metals, their speed can be increased. Unless a special driving flange is used for slitting saws, it is *not* advisable to key the saw to the milling arbor. The arbor nut should be pulled up as tightly as possible *by hand only*. Since slitting saws are so easily broken, some operators find it desirable to ''climb'' or ''down'' mill when sawing. However, to overcome the play between the lead screw and nut, the backlash eliminator should be engaged.

END MILLS

End mills have cutting teeth on the end as well as on the periphery and are fitted to the spindle by a suitable adapter.

They are of two types, the *solid end mill* in which the shank and the cutter are integral, and the *shell end mill*, which, as previously stated, employs a separate shank.

Solid end mills, generally smaller than shell end mills, may have either straight or helical flutes. They are manufactured with straight and taper shanks and with two or more flutes. The two-flute type (Fig. 56-10A) has flutes which meet at the cutting end, forming two cutting lips across the bottom. These lips are of different lengths, one extending beyond the center axis of the cutter. This arrangement eliminates the center and permits the two-flute end mill to be used in a milling machine for drilling a hole to start a slot that does not extend to the edge of the metal. When a slot is being cut with a two-flute end mill, the depth of cut should not exceed one-half the diameter of the cutter. When the four-flute end mill (Fig. 56-10B) is used for slot cutting, it is usually started at the edge of the metal.

A

B

Fig. 56-10A (A) A two-flute end mill; (B) a four-flute end mill.

T-SLOT CUTTER

The *T-slot cutter* (Fig. 56-11A) is used to cut the wide horizontal groove at the bottom of a T-slot after the narrow vertical groove has been machined with an end mill or a side milling cutter. It consists of a small side milling cutter with teeth on both sides and an integral shank for mounting, similar to an end mill.

DOVETAIL CUTTER

The *dovetail cutter* (Fig. 56-11B) is similar to a single-angle milling cutter with an integral shank. Some dovetail cutters are manufactured with an internal thread and are mounted on a special threaded shank. They are used to form the sides of a dovetail after the tongue or the groove has been machined with another suitable cutter, usually a side milling cutter. Dovetail cutters may be obtained with 45, 50, 55, or 60° angles.

WOODRUFF KEYSEAT CUTTER

The *Woodruff keyseat cutter* (Fig. 56-11C) is similar in design to plain and side milling cutters. Smaller sizes up to approximately 2 in. (50.8 mm) in diameter are made with a solid shank and straight teeth; larger sizes are mounted on an arbor and have staggered teeth on both the sides and the periphery. They are used for milling semi-cylindrical keyseats in shafts.

Woodruff cutters are designated by a number system. The right-hand two digits give the nominal diameter in eighths of an inch, while the preceding digits give the width of the cutter in thirty-seconds of an inch. For example, a #406 cutter would have dimensions as follows:

Diameter   06 × ⅛ = ¾ in.
Width   4 × 1/32 = ⅛ in.

FLYCUTTERS

The *flycutter* (Fig. 56-12) is a single-pointed cutting tool with the cutting end ground to the desired shape. It is mounted in a special adapter or arbor. Since all the cutting is done with one tool, a fine feed must be used. Flycutters are used in experimental work and where the high cost of a special-shaped cutter would not be warranted.

A

B

C

Fig. 56-11A (A) A T-slot cutter. (Courtesy The Butterfield Division, Union Twist Drill Company.); (B) a dovetail cutter; (C) a Woodruff keyseat cutter.

Fig. 56-12 A flycutter.

REVIEW QUESTIONS

1. Name two materials which are used to make milling cutters.
2. Name five elements that may be added to iron or steel to produce high-speed steel, and state the purpose of each additive.
3. Discuss the advantages and disadvantages of cemented-carbide tools.
4. Describe three types of plain milling cutters and explain their uses.
5. Describe three types of side milling cutters, and state where each is used.
6. List two ways in which formed cutters differ from plain milling cutters.
7. What advantage does a two-flute end mill have.

CUTTING SPEEDS, FEEDS, AND DEPTH OF CUT
UNIT 57

The most important factors affecting the efficiency of a milling operation are cutter speed, feed, and depth of cut. If the cutter is run too slowly, valuable time will be wasted, while excessive speed results in loss of time in replacing and regrinding cutters. Somewhere between these two extremes is the efficient *cutting speed* for the material being machined.

The rate at which the work is fed into the revolving cutter is important. If the work is fed too slowly, time will be wasted and cutter chatter may occur, which shortens the life of the cutter. If work is fed too fast, the cutter teeth can be broken. Much time will be wasted if several shallow cuts are taken instead of one deep or roughing cut. Therefore *speed, feed,* and *depth of cut* are three important factors in any milling operation.

OBJECTIVES

After completing this unit you should be able to:
1. Select cutting speeds and calculate the r/min. for various cutters and materials
2. Select and calculate the proper feeds for various cutters and materials
3. Follow the correct procedure for taking roughing and finishing cuts

CUTTING SPEEDS

One of the most important factors affecting the efficiency of a milling operation is cutter speed. The cutting speed of a metal may be defined as *the speed, in surface feet per minute* (sf/min) *or meters per minute* (m/min) *at which the metal may be machined efficiently.* When the work is machined in a lathe, it must be turned at a specific number of revolutions per minute (r/min), *depending on its diameter,* to achieve the proper cutting speed. When work is machined in a milling machine, *the cutter* must be revolved at a specified number of r/min, *depending upon its diameter,* to achieve the proper cutting speed.

Since different types of metals vary in hardness, structure, and machinability, different cutting speeds must be used for each type of metal and for various cutter materials. Several factors must be considered when determining the proper r/min at which to machine a metal. The most important are:

☐ The type of work material
☐ The cutter material
☐ The diameter of the cutter
☐ The surface finish required
☐ The depth of cut being taken
☐ The rigidity of the machine and work setup

The cutting speeds for the more common metals are shown in Table 57-1.

INCH CALCULATIONS

To get optimum use from a cutter, the proper speed at which the cutter should be revolved must be determined. When machining machine steel, a high-speed steel cutter would have to achieve a surface speed of about 90 ft/min (27.43 m/min). Since the diameter of the cutter affects this speed, it is necessary to consider its diameter in the calculation. The following example illustrates how the formula is developed.

EXAMPLE

Calculate the speed required to revolve a 3-in.-diameter high-speed steel milling cutter when cutting machine steel (90 sf/min).

1. First, determine the circumference of the cutter or the distance a point on the cutter would travel in one revolution.
Circumference of cutter = 3 in. × 3.1416
2. To determine the proper cutter speed or *r/min*, it is necessary only to divide the cutting speed (CS) by the circumference of the cutter.

TABLE 57-1 Milling Machine Cutting Speeds

MATERIAL	HIGH-SPEED STEEL CUTTER		CARBIDE CUTTER	
	ft/min	m/min	ft/min	m/min
Machine steel	70-100	21-30	150-250	45-75
Tool steel	60-70	18-20	125-200	40-60
Cast iron	50-80	15-25	125-200	40-60
Bronze	65-120	20-35	200-400	60-120
Aluminum	500-1000	150-300	1000-2000	150-300

$$r/min = \frac{CS \text{ (ft)}}{\text{circumference (in.)}}$$

$$= \frac{90}{3 \times 3.1416}$$

Since the numerator is in feet and the denominator in inches, the numerator must be changed to inches. Therefore;

$$r/min = \frac{12 \times CS}{3 \times 3.1416}$$

Because it is usually impossible to set a machine to the exact r/min, it is permissible to consider that 3.1416 will divide into 12 approximately 4 times. The formula now becomes

$$r/min = \frac{4 \times CS}{D}$$

Using this formula and Table 57-1, it is possible to calculate the proper cutter or spindle speed for any material and cutter diameter.

EXAMPLE

At what speed should a 2-in.-diameter carbide cutter revolve to mill a piece of cast iron (CS 150)?

SOLUTION

$$r/min = \frac{4 \times 150}{2}$$

$$= 300$$

METRIC CALCULATIONS

The r/min at which the milling machine should be set when using metric measurements is as follows:

$$r/min = \frac{CS \text{ (m)} \times 1000}{\pi \times D \text{ (mm)}}$$

$$= \frac{CS \times 1000}{3.1416 \times D}$$

Since only a few machines are equipped with variable-speed drives which allow them to be set to the exact cal-culated speed, a simplified formula can be used to calculate r/min. The π (3.1416) on the bottom line of the formula will divide into the 1000 of the top line approximately 320 times. This results in a simplified formula which is close enough for most milling operations.

$$r/min = \frac{CS \text{ (m)} \times 320}{D \text{ (mm)}}$$

EXAMPLE

Calculate the r/min required for a 75-mm-diameter high-speed steel milling cutter when cutting machine steel (*CS* 30 m/min).

$$r/min = \frac{30 \times 320}{75}$$

$$= \frac{9600}{75}$$

$$= 128$$

Although these formulas are helpful in calculating the cutter (spindle) speed, it should be remembered that they are approximate only, and the speed may have to be altered because of the hardness of the metal and/or the machine condition. Best results may be obtained if the following rules are observed.

1. For longer cutter life, use the lower CS in the recommended range.
2. Know the hardness of the material to be machined.
3. When starting a new job, use the lower range of the CS and gradually increase to the higher range if conditions permit.
4. If a fine finish is required, reduce the feed rather than increase the cutter speed.
5. The use of coolant, properly applied, will generally produce a better finish and lengthen the life of the cutter because it absorbs heat, acts as a lubricant, and washes chips away.

MILLING FEEDS AND DEPTH OF CUT

The two other factors which affect the efficiency of a milling operation are the *milling feed,* or the rate at which the work is fed into the milling cutter, and the *depth of cut* taken on each pass.

FEED

Milling machine feed may be defined as the distance in inches (or millimeters) per minute that the work moves into the cutter. On most milling machines, the feed is regulated in inches (or millimeters) per minute and is independent of the spindle speed. This arrangement permits faster feeds for larger, slowly rotating cutters.

Feed is the rate at which the work moves into the revolving cutter, and it is measured either in inches per minute or in millimeters per minute. The *milling feed* is determined by multiplying the chip size (chip per tooth) desired, the number of teeth in the cutter, and the r/min of the cutter.

Chip per tooth (cpt) is the amount of material which should be removed by each tooth of the cutter as it revolves and advances into the work. See Tables 57-2 and 57-3 for the recommended cpt for some of the more common metals.

The feed rate used on a milling machine depends upon a variety of factors, such as,

1. The depth and width of cut
2. The design or type of cutter
3. The sharpness of the cutter
4. The workpiece material
5. The strength and uniformity of the workpiece
6. The type of finish and accuracy required
7. The power and rigidity of the machine

As the work advances into the cutter, each successive tooth advances into the work an equal amount, producing chips of equal thickness. It is this thickness of the chips or the *feed per tooth,* along with the number of teeth in the cutter, which form the basis for determining the rate of feed. The ideal rate of feed may be determined as follows.

$$\text{Feed} = \text{no. of cutter teeth} \times \text{feed/tooth} \times \text{cutter r/min}$$

Inch Calculations

The formula used to find the work feed in inches per minute is:

$$\text{Feed (in/min)} = N \times \text{cpt} \times \text{r/min}$$

N = number of teeth in the milling cutter
cpt = chip per tooth for a particular cutter and metal as given in Tables 57-2 and 57-3
r/min = revolutions per minute of the milling cutter

EXAMPLE

Find the feed in inches per minute using a 3.5-in-diameter 12-tooth helical cutter to cut machine steel (CS 80). It

TABLE 57-2 Recommended Feed per Tooth (High-Speed Steel Cutters)													
MATERIAL	**FACE MILLS**		**HELICAL MILLS**		**SLOTTING AND SIDE MILLS**		**END MILLS**		**FORM-RELIEVED CUTTERS**		**CIRCULAR SAWS**	
	in.	mm	in.	mm	in.	mm	in.	mm	in.	mm	in.	mm
Aluminum	0.022	0.55	0.018	0.45	0.013	0.33	0.011	0.28	0.007	0.18	0.005	0.13
Brass and bronze (medium)	0.014	0.35	0.011	0.28	0.008	0.20	0.007	0.18	0.004	0.10	0.003	0.08
Cast iron (medium)	0.013	0.33	0.010	0.25	0.007	0.18	0.007	0.18	0.004	0.10	0.003	0.08
Machine steel	0.012	0.30	0.010	0.25	0.007	0.18	0.006	0.15	0.004	0.10	0.003	0.08
Tool steel (medium)	0.010	0.25	0.008	0.20	0.006	0.15	0.005	0.13	0.003	0.08	0.003	0.08
Stainless steel	0.006	0.15	0.005	0.13	0.004	0.10	0.003	0.08	0.002	0.05	0.002	0.05

Table 57-3 gives the same information for cemented-carbide-tipped cutters. Because of the machine condition and the often limited power of the machine, it is sometimes necessary to reduce the feed per tooth.

TABLE 57-3 Recommended Feed per Tooth (Cemented-Carbide-Tipped Cutters)													
MATERIAL	**FACE MILLS**		**HELICAL MILLS**		**SLOTTING AND SIDE MILLS**		**END MILLS**		**FORM-RELIEVED CUTTERS**		**CIRCULAR SAWS**	
	in.	mm	in.	mm	in.	mm	in.	mm	in.	mm	in.	mm
Aluminum	0.020	0.50	0.016	0.40	0.012	0.30	0.010	0.25	0.006	0.15	0.005	0.13
Brass and bronze (medium)	0.012	0.30	0.010	0.25	0.007	0.18	0.006	0.15	0.004	0.10	0.003	0.08
Cast iron (medium)	0.016	0.40	0.013	0.33	0.010	0.25	0.008	0.20	0.005	0.13	0.004	0.10
Machine steel	0.016	0.40	0.013	0.33	0.009	0.23	0.008	0.20	0.005	0.13	0.004	0.10
Tool steel (medium)	0.014	0.35	0.011	0.28	0.008	0.20	0.007	0.18	0.004	0.10	0.004	0.10
Stainless steel	0.010	0.25	0.008	0.20	0.006	0.15	0.005	0.13	0.003	0.08	0.003	0.08

would first be necessary to calculate the proper r/min for the cutter.

$$r/min = \frac{4 \times CS}{D}$$

$$= \frac{4 \times 80}{3.5}$$

$$= 91$$

$$Feed\ (in/min) = N \times cpt \times r/min$$

$$= 12 \times 0.010 \times 91$$

$$= 10.9\ or\ 11$$

Metric Calculations

The formula used to find work feed in millimeters per minute is the same as the formula used to find the feed in inches per minute, except that mm/min is substituted for in./min.

EXAMPLE

Calculate the feed in millimeters per minute for a 75-mm-diameter, six-tooth helical milling cutter when machining a cast iron workpiece (CS 60).

First calculate the *r/min* of the cutter.

$$r/min = \frac{CS \times 320}{D}$$

$$= \frac{60 \times 320}{75}$$

$$= \frac{19,200}{75}$$

$$= 256$$

$$Feed\ (mm/min) = N \times cpt \times r/min$$

$$= 6 \times 0.18 \times 256$$

$$= 276.4$$

$$= 276\ mm/min$$

The calculated feeds would be possible only under ideal conditions. Under average operating conditions, it is suggested that the milling machine feed be set to approximately one-third or one-half the amount calculated. The feed can then be gradually increased to the capacity of the machine and the finish desired.

Tables 57-2 and 57-3 give suggested feed per tooth for various types of milling cutters for roughing cuts under average conditions. For finish cuts, the feed per tooth would be reduced to one-half or even one-third of the value shown.

One final consideration should be made concerning feed, that is, the direction in which the work is fed into the cutter.

The most commonly used method of feeding is to feed the work against the rotation direction of the cutter (*conventional or up milling*) (Fig. 57-1). However, if the machine is equipped with a backlash eliminator certain types of work can best be milled by *climb milling* (Fig. 57-2).

Although climb milling is not as widely used as conventional milling, it has certain advantages and disadvantages.

Advantages

1. It is particularly suited for machining thin and hard-to-hold parts since the workpiece is forced against the table or holding device by the cutter.
2. Work need not be clamped as tightly.
3. Consistent parallelism and size may be maintained, particularly on thin parts.
4. It may be used where break-out at the edge of the workpiece could not be tolerated.
5. It requires up to 20 percent less power to cut by this method.
6. It may be used when cutting off stock or when milling deep, thin slots.

Disadvantages

1. *This method cannot be used unless the machine has a backlash eliminator* and the table gibs have been tightened.

Fig. 57-1 Conventional or up milling.

Fig. 57-2 Climb or down milling.

CUTTING SPEEDS, FEEDS, AND DEPTH OF CUT

2. It cannot be used for machining castings or hot-rolled steel since the hard outer scale will damage the cutter.

DEPTH OF CUT

Where a smooth accurate finish is desired, it is considered good milling practice to take a roughing and finishing cut. Roughing cuts should be deep, with a feed as heavy as the work and the machine will permit. Heavier cuts may be taken with helical cutters having fewer teeth since they are stronger and have a greater chip clearance than cutters with more teeth.

Finishing cuts should be light, with a finer feed than is used for roughing cuts. The depth of the cut should be at least 1/64 in. (0.38 mm). Lighter cuts and extremely fine feeds are not advisable since the chip taken by each tooth will be thin and the cutter will often rub on the surface of the work, rather than bite into it, thus dulling the cutter. When a fine finish is required, the feed should be reduced rather than the cutter speeded up; more cutters are dulled by high speeds than by high feeds.

NOTE: To prevent damage to the finished surface, *never* stop the feed when the cutter is revolving over the workpiece. For the same reason, stop the cutter before returning the work to the starting position upon completion of the cut.

MILLING CUTTER FAILURE

The problem of obtaining maximum efficiency in the use of milling cutters is one which faces every user. The solution to this problem lies in constant awareness of factors that contribute to tool failure or poor performance other than the style or type of end mill being used.

EXCESSIVE HEAT

Excessive heat (Fig. 57-3) is one of the main causes of total cutting edge failure or shortened tool life. Heat, which is present in all milling operations, is caused by cutting edges rubbing on the workpiece and by chips sliding along the faces of teeth when they are separated from the part being milled. When this heat becomes excessive, the hardness of the cutting edges is affected, resulting in less resistance to wear and subsequently rapid dulling.

The heat problem is one of an ever-expanding cycle. As tool-wear lands increase, cutting temperatures rise because of increased friction. As cutting edges dull, more force is needed to part chips and the sliding pressure of chips on the faces of teeth increases, resulting in still more heat. As more heat is generated, the coolant, which is used to reduce temperature, becomes less efficient, temperatures rise higher, the hardness and strength of the cutting edges is further affected, and the cycle is repeated. Heat cannot be totally eliminated but it can be minimized by using properly designed and sharpened tools, by operating at the speeds and feeds recommended for the workpiece material, and by efficiently applying a suitable coolant.

Fig. 57-3   Excessive heat shortens the cutter life.   (Courtesy The Weldon Tool Company.)

Fig. 57-4   Abrasion causes the cutting edges to dull.   (Courtesy The Weldon Tool Company.)

ABRASION

Abrasion (Fig. 57-4) is an abrading or wearing-away action caused by the metallurgy of the workpiece. It dulls cutting edges and causes "wear lands" to develop around the periphery of a cutter. As dulling increases and wear lands grow wider, friction increases and greater force horsepower is required to keep the cutter cutting. The rapid rise in friction, heat, and rotational forces resulting from the abrasive nature of workpiece material can reach a point where the cutter ceases to function effectively or is totally destroyed. Since heat and abrasion are related, the suggestions for minimizing heat will also apply to abrasion. Because some materials are much more abrasive than others, it is

CHIPPING OR CRUMBLING OF CUTTING EDGES

When cutting forces impose a greater load on cutting edges than their strength can withstand, small fractures occur and small areas of the cutting edges chip out (Fig. 57-5). The material that is left uncut by these chipped-out portions imposes a still greater cutting load on the following cutter tooth, aggravating the problem. This condition is progressive, and once started will lead to total cutter failure since chipped edges are dull edges that increase friction, heat, and horsepower requirements.

The major causes of chipping and fracturing of cutting edges are:

☐ Excessive feed per tooth (fpt)
☐ Poor cutter design
☐ Brittleness of the end mill due to improper heat treatment
☐ Running cutters backward
☐ Chattering due to a nonrigid condition in either the fixture, workpiece, or the machine
☐ Inefficient chip washout which permits chips to be recut or compressed between workpiece surfaces and cutting edges
☐ Built-up edge break-away

Fig. 57-5 Too heavy a load on the cutting edge will cause it to chip. (Courtesy The Weldon Tool Company.)

CLOGGING

Some workpiece materials have a "gummy" composition that causes chips to be long, stringy, and compressible (Fig. 57-6). Chips from still other materials may have a tendency to cold-weld or gall to cutting edges and/or the faces of teeth. When milling these materials, very often the chips clog or jam into the flute area resulting in cutter breakage. This condition can be minimized by reducing the depth or width of cut, by reducing the fpt, by using tools with fewer teeth creating more chip space, and by more effectively applying a coolant under pressure that lubricates the tooth face and is directed so that it flushes out the flute area, keeping it free of chips. It might be necessary at times to use two coolant nozzles to accomplish this.

Fig. 57-6 Clogging reduces the chip space and can cause a cutter to break. (Courtesy The Weldon Tool Company.)

BUILT-UP EDGES

Built-up edges (Fig. 57-7) occur when particles of the material being machined cold-weld, gall, or otherwise adhere to the faces of teeth adjacent to the cutting edges. This process will continue until the build-up functions as the cutting edge. When this happens, more power is required and a poor surface finish is usually produced on the workpiece. Periodically, the built-up material will break away from the tooth face and go with either the chip or with the workpiece. This intermittent break-away often takes with it a portion of the cutting edge, causing a complete tool breakdown with all of its attendant problems. This condition is quite prevalent on a tool such as a milling cutter because as each tooth engages the cut in an intermittent manner, the possibility of built-up material break-away and edge chipping is increased.

Fig. 57-7 Built-up edges will create a poor cutting action. (Courtesy The Weldon Tool Company.)

343

The built-up edge problem can often be moderated by reducing feed and/or depth of cut. However, the most effective solution to the problem is usually found in a forceful application of a good cutting fluid that gets into the area where the chip is being formed. Optimum results are gained when the coolant coats the cutting edges with a thin fluid or oxide layer that forms a cushion between the chip and the tool to prevent a built-up edge from forming.

WORK HARDENING OF THE WORKPIECE

Work hardening of the workpiece (Fig. 57-8) can cause milling cutter failure. This condition, sometimes called strain hardening, cold working, or glazing, is the result of the action of cutting edges deforming or compressing the surface of the workpiece causing a change in the work material structure that increases its hardness. Usually, this increase in hardness is evident by a smooth, highly glazed surface that resists the wetting action of coolants and offers extreme resistance to the penetration of cutting edges. Fortunately, not all materials are subject to work hardening, but this condition can occur when machining most of the high-temperature and high-strength superalloys, all of the austenitic stainless steels, and many of the highly alloyed carbon tool steels.

Since it is basically the rubbing action of cutting edges that causes work hardening on the surface of these materials, it is extremely important to use sharp tools operating at generous power feeds to keep rubbing contact between tool and work at a minimum. The use of proper CS is important, and climb cutting with a generous application of activated cutting oil is highly recommended. Avoid the use of dull milling cutters and light finishing cuts, and never permit a tool to dwell or rotate in contact with the workpiece without feeding.

Machines and work-holding fixtures should be massive and rigid, and tool overhang as short as possible in order to keep deflection during cut at a minimum. Should a surface that is to be milled be already glazed or work-hardened from some previous cutting operation, it is very beneficial to break up the glaze by vapor honing or abrading the surface with coarse emery cloth. This will reduce the slipperiness of the surface and make it easier for tool cutting edges to bite in. Also, breaking the glaze on the surface will permit the coolant to give better wetting action which is very helpful in prolonging tool life.

CRATERING

Cratering (Fig. 57-9) is caused by chips sliding on the tooth face adjacent to the cutting edge. This is the area of high heat and extreme abrasion due to high chip pressures. The sliding and curling of the chips erodes a narrow hollow or a groove into the tooth face. Once this cratering starts it can get progressively worse until it results in total tool failure. This problem can be minimized by efficiently applying a coolant that provides a high-pressure fluid film or a chemical oxide film on the tool to prevent metal-to-metal contact between chips and tooth face. Also beneficial are

Fig. 57-8   Work hardening of the workpiece can cause cutter failure.   (Courtesy The Weldon Tool Company.)

Fig. 57-9   Cratering can be reduced with the use of cutting fluid. (Courtesy The Weldon Tool Company.)

tool surface treatments properly applied that impart a high-abrasion-resisting superficial hardness to faces of the teeth.

REVIEW QUESTIONS

1. Name three factors that affect the efficiency of a milling operation.

2. List six factors which must be considered when selecting the proper r/min for milling.
3. At what maximum speed should a 3½-in. cemented-carbide milling cutter revolve to machine cast iron?
4. At what r/min should a 115-mm high-speed steel cutter revolve to machine a piece of tool steel?
5. What rules should be observed to obtain best results when using milling cutters?

FEED

6. Name three factors which determine milling feed.

7. Define chip per tooth.

8. Calculate the feed (in./min) for an eight-tooth, 3-in.-diameter cemented-carbide face milling cutter to cut cast iron.

9. Calculate the feed (m/min) for a twelve-tooth 90-mm-diameter high-speed steel helical milling cutter to machine aluminum.

DEPTH OF CUT

10. What is considered to be a proper roughing cut?

11. Why is a very light cut not desirable as a finish cut?

MILLING CUTTER FAILURE

12. List seven main causes of milling cutter failure and state how each can be minimized.

MILLING MACHINE SETUPS
UNIT 58

Before any operation is performed on a milling machine, it is important that the machine be properly set up. Proper setup of the machine will prolong the life of the machine and its accessories and produce accurate work. The operator should also follow safety procedures to prevent injury to the worker and to avoid damaging the machine and spoiling the workpiece.

OBJECTIVES

After completing this unit you should be able to:
1. Mount and remove a milling machine arbor
2. Mount and remove a milling cutter
3. Align the milling machine table and vise

MILLING MACHINE SAFETY

The milling machine, like any other machine, demands the total attention of the operator and a thorough understanding of the hazards associated with its operation. The following points should be observed when operating the milling machine.

1. Be sure that the work and the cutter are mounted securely before taking a cut.

2. Always wear safety glasses.

3. When mounting or removing milling cutters, always hold them with a cloth to avoid being cut.

4. When setting up work, move the table as far as possible from the cutter to avoid cutting your hands.

5. Be sure that the cutter and machine parts clear the work (Fig. 58-1).

6. *Never* attempt to mount, measure, or adjust work until the cutter is *completely* stopped.

7. Keep hands, brushes, and rags away from a revolving milling cutter *at all times*.

Fig. 58-1   Before taking a cut be sure the arbor and arbor support will clear the work.

MILLING MACHINE SETUPS

Fig. 58-2 Poor housekeeping can cause tripping accidents.

Fig. 58-3 Place tools on a piece of masonite or plywood to protect the machine table.

8. When using milling cutters, do not use an excessively heavy cut or feed. This can cause a cutter to break and the flying pieces could cause serious injury.

9. Always use a *brush,* not a rag, to remove the cuttings after the cutter has stopped revolving.

10. Never reach over or near a revolving cutter; keep hands at least 12 in. or 300 mm from a revolving cutter.

11. Keep the floor around the machine free of chips, oil, and cutting fluid (Fig. 58-2).

MILLING MACHINE SETUPS

To prolong the life of a milling machine and its accessories and to produce accurate work, the following points should be carefully observed when making milling machine setups.

1. Prior to mounting any accessory or attachment, check that both the machine surface and the accessory are free of dirt and chips.

2. Do not place the tools, cutters, or parts on the milling machine table. Place them on a piece of masonite, board, or a bench kept for this purpose to prevent damaging the table or machined surfaces (Fig. 58-3).

3. When mounting cutters, be sure to use keys on all but slitting saws.

4. Check that the arbor spacers and bushings are clean and free of burrs.

5. When tightening the arbor nut take care to make it only hand tight with a wrench. NEVER USE A HAMMER ON THE NUT OR HOLDING DEVICE ON ANY MACHINE. The use of a hammer and wrench to tighten nuts will strip the threads and bend or damage the accessory or part.

6. When work is mounted in a vise, tighten the vise securely by hand and tap it into place with a LEAD or SOFT-FACED HAMMER.

MOUNTING AND REMOVING A MILLING MACHINE ARBOR

The milling arbor is used to hold the cutter during the machine operation. When mounting or removing an arbor,

Fig. 58-4 The milling machine arbor holds and drives the milling cutter.

Fig. 58-5 The draw-in bar holds the arbor firmly in place in the spindle.

the proper procedure should be followed in order to preserve the accuracy of the machine. An improperly mounted arbor may damage the taper surfaces of the arbor or machine spindle, cause the arbor to bend, or make the cutter run out of true.

THE ARBOR ASSEMBLY

The milling cutter is driven by a key which fits into the keyways on the arbor and cutter (Fig. 58-4). This prevents the cutter from turning on the arbor. Spacer and bearing bushings hold the cutter in position on the arbor after the nut has been tightened. The tapered end of the arbor is held securely in the machine spindle by a draw-in bar (Fig. 58-5). The outer end of the arbor assembly is supported by the bearing bushing and the arbor support.

TO MOUNT AN ARBOR

1. Clean the tapered hole in the spindle and the taper on the arbor using a clean cloth.

Fig. 58-6 Removing a milling machine arbor.

Fig. 58-7 Locking the arbor into the spindle with a draw-in bar.

2. Check that there are no cuttings or burrs in the taper which would prevent the arbor from running true.

3. Check the bearing bushing and remove any burrs using a honing stone.

4. Place the tapered end of the arbor in the spindle and align the spindle driving lugs with the arbor slots.

5. Place the right hand on the draw-in bar (Fig. 58-6) and turn the thread into the arbor approximately 1 in. (25.4 mm).

6. Tighten the draw-in bar lock nut securely against the back of the spindle (Fig. 58-7).

TO REMOVE AN ARBOR

1. Remove the milling machine cutter.

2. Loosen the lock nut on the draw-in bar approximately two turns.

3. With a soft-faced hammer, sharply strike the end of the draw-in bar until the arbor taper is free (Fig. 58-8).

4. With one hand hold the arbor and unscrew the draw-in bar from the arbor with the other hand (Fig. 58-6).

5. Carefully remove the arbor from the tapered spindle so as not to damage the spindle or arbor tapers.

6. Leave the draw-in bar in the spindle for further use.

7. Store the arbor in a suitable rack to prevent it from being damaged or bent (Fig. 58-9).

Fig. 58-8 Removing the arbor from the spindle.

MOUNTING AND REMOVING MILLING CUTTERS

Milling cutters must be changed frequently to perform various operations, so it is important that the following sequence be followed in order not to damage the cutter, the machine, or the arbor.

TO MOUNT A MILLING CUTTER

1. Remove the arbor nut and collar and place them on a piece of masonite (Fig. 58-3).

2. Clean all spacing collar surfaces of cuttings and burrs.

3. Set the machine to the slowest spindle speed.

4. Check the direction of the arbor rotation by starting and stopping the machine spindle.

Fig. 58-9 Storing the arbor in a suitable rack.

5. Slide the spacing collars on the arbor to the position desired for the cutter (Fig. 58-10).

6. Fit a key into the arbor keyway at the position where the cutter is to be located.

7. Hold the cutter with a cloth and mount it on the arbor, being sure that the cutter teeth point in the direction of the arbor rotation (Fig. 58-10).

8. Slide the arbor support in place and be sure that it is on a bearing bushing on the arbor.

9. Put on additional spacers leaving room for the arbor nut. TIGHTEN THE NUT BY HAND.

10. Lock the arbor support in position. (Fig. 58-11).

347

Fig. 58-10 Use spacer collars to set the cutter in position.

Fig. 58-11 Tightening the arbor support.

11. Tighten the arbor nut firmly with a wrench using only hand pressure.
12. Lubricate the bearing collar in the arbor support.
13. Make sure that the arbor and arbor support will clear the work (Fig. 58-1).

REMOVING A MILLING CUTTER

1. Be sure that the arbor support is in place and supporting the arbor on a bearing bushing before using a wrench on the arbor nut. This will prevent bending the arbor.
2. Clean all cuttings from the arbor and cutter.
3. Set the machine to the lowest spindle speed.
4. Loosen the arbor nut with a properly fitting wrench.

NOTE: Most threads on an arbor are left-hand and therefore loosen in a clockwise direction.

5. Loosen the arbor support and remove it from the over-arm.
6. Remove the nut, spacers and cutter. Place them on a board (Fig. 58-12) and not on the table surface.
7. Clean the spacer and nut surfaces and replace them on the arbor. *Do not use a wrench to tighten the arbor nut at this time.*
8. Store the cutter in a proper place to prevent the cutting edges from being damaged.

Fig. 58-12 A piece of masonite is used to protect the milling machine table when changing a cutter.

MAGNETIC BASE

Fig. 58-13 Aligning universal milling machine table.

ALIGNING THE TABLE ON A UNIVERSAL MILLING MACHINE

If the workpiece is to be machined accurately to a layout or have cuts made square or parallel to a surface, it is always good practice to align the table of a universal milling machine prior to aligning the vise or fixture.

NOTE: If a long keyway is to be milled in a shaft, it is of utmost importance to align the table. If it is not aligned properly, the keyway will not be milled parallel to the axis of the shaft.

PROCEDURE

1. Clean the table and the face of the column thoroughly.
2. Mount a dial indicator *on the table* by means of a magnetic base or any suitable mounting device (Fig. 58-13).
3. Move the table toward the column until the dial indicator registers approximately one-quarter of a revolution.
4. Set the indicator bezel to zero by turning the dial.
5. Using the table feed handwheel, move the table along the width of the column.
6. Note the reading on the dial indicator and compare it to the zero reading at the other end of the column.
7. If there is any movement of the indicator hand, loosen the locking nuts on the *swivel table housing*.

Fig. 58-14 A dial indicator mounted on the arbor for alignment purposes.

Fig. 58-15 Accurately aligning the vise parallel to the table travel.

PARALLELS

Fig. 58-16 (A) Aligning an angle plate parallel to the column; (B) and (C) aligning an angle plate at right angles to the column.

8. Adjust the table for half the difference of the needle movement and lock the table housing.

9. *Recheck* the table for alignment and adjust if necessary until there is no movement in the indicator as it is moved along the column.

NOTE: *Always indicate from the table to the face of the column, never from the column to the table.*

ALIGNING THE MILLING MACHINE VISE

Whenever accuracy is required on the workpiece, it is necessary to align the device which holds the workpiece. This may be a vise, angle plate, or a special fixture. Since most work is held in a vise, the alignment of this accessory will be outlined.

TO ALIGN THE VISE PARALLEL TO THE TABLE TRAVEL

1. Clean the surface of the table and the base of the vise.

2. Mount and fasten the vise on the table.

3. Swivel the vise until the solid jaw is approximately parallel with the table slots.

4. Mount an indicator to the arbor or cutter (Fig. 58-14).

5. Make sure the solid jaw is clean and free of burrs.

6. Adjust the table until the indicator registers about one-quarter of a revolution against a parallel held between the jaws of the vise.

7. Set the bezel to zero (Fig. 58-15A).

8. Move the table along for the length of the parallel and note the reading of the indicator (Fig. 58-15B). Compare it to the zero reading at the other end of the parallel.

9. Loosen the nuts on the upper or swivel part of the vise.

10. Adjust the vise to half the difference of the indicator readings by tapping it by hand or with a *soft* hammer in the appropriate direction.

NOTE: *Never tap the vise so that the parallel moves against the indicator plunger. This will damage the indicator.*

11. *Recheck* the vise for alignment and adjust if necessary until there is no movement in the indicator as it is moved along the parallel.

Methods of aligning an angle plate are shown in Fig. 58-16.

1. Why is the proper setup of the machine important?

SAFETY

2. What can happen if an excessively heavy cut is taken or too fast a feed used?
3. How far should your hands be kept from a revolving cutter?

MOUNT AND REMOVE ARBORS

4. What is the purpose of an arbor?
5. How is the arbor held securely in the machine spindle?
6. How is the outer end of the arbor supported?

7. How should the arbor be stored when it is not in the machine?
8. In which direction should the cutter teeth point when it is mounted on an arbor?
9. Why should the arbor support be in place before attempting to remove a cutter from the arbor?

ALIGNING THE TABLE AND VISE

10. Why is the table alignment so important to most milling operations?
11. Why must the indicator be mounted on the table and not the column or the arbor when aligning a universal table?
12. By means of suitable sketches, show how the vise may be aligned:
   (a) parallel to the line of longitudinal table travel
   (b) at right angles to the milling machine column

MILLING OPERATIONS
UNIT 59

The milling machine is one of the most versatile machine tools found in industry. The wide variety of operations which can be performed depends upon the type of machine used, the cutter employed, and the attachments and accessories available for the milling machine.
1. *Plain milling*, where the surface cut is parallel to the periphery of the cutter. These surfaces may either be flat or formed (Fig. 59-1).
2. *Face milling*, where the surface cut is at right angles to the axis of the cutter (Fig. 59-2).

OBJECTIVES

After completing this unit you will be able to:
1. Set a cutter to the proper depth
2. Mill a flat surface on a workpiece
3. Know how to perform operations such as face, side, straddle, and gang milling

SETTING THE CUTTER TO THE WORK SURFACE

Before setting a depth of cut the operator should check that the work and the cutter are properly mounted and that the cutter is revolving in the proper direction.

TO SET THE CUTTER TO THE WORK SURFACE

1. Raise the work to within ¼ in. (6.35 mm) of the cutter and directly under it.

2. Hold a long piece of thin paper on the surface to the work (Fig. 59-3).

NOTE: *Have the paper long enough to prevent the fingers from coming in contact with the revolving cutter.*

3. Start the cutter rotating.
4. With the left hand on the elevating screw handle, move the work up slowly until the cutter grips the paper.
5. Stop the spindle.
6. Move the machine table so that the cutter just clears the end of the workpiece (Fig. 59-4).
7. Raise the knee 0.002 in. (0.05 mm) for paper thickness.

Fig. 59-1 Milling the surface of a workpiece held in a fixture.

Fig. 59-2 Face milling on a manufacturing-type milling machine. (Courtesy Cincinnati Milacron Incorporated.)

Fig. 59-3 Setting the cutter to the work.

Fig. 59-4 Be sure that the cutter clears the end of the work before setting the depth of cut.

Fig. 59-5 Remove all burrs before setting work in a vise.

8. Set the graduated collar on the elevating screw handle to zero. Do not move the elevating screw handle.
9. Move the work clear of the cutter and raise the table to the desired depth of cut.

NOTE: *If the knee is moved up beyond the desired amount, turn the handle backward one-half turn and come up to the required line. This will take up the backlash in the thread movement.*

This method can also be used when setting the edge of a cutter to the side of a piece of work. In this case, the paper will be placed between the side of the cutter and the side of the workpiece.

MILLING A FLAT SURFACE

One of the most common operations performed on a milling machine is that of machining a flat surface. Flat surfaces are generally machined on a workpiece with a helical mill-

ing cutter. The work may be held in a vise or clamped to a table.

1. Remove all burrs from all edges of the work with a mill file (Fig. 59-5).
2. Clean the vise and workpiece.

Fig. 59-6 Tap the work down with a soft-faced hammer until all paper feelers are tight.

Fig. 59-7 Most work is machined by conventional milling.

3. Align the vise to the column face of the milling machine using a dial indicator.

4. Set the work in the vise using parallels and paper feelers under each corner to make sure that the work is seated on the parallels.

5. Tighten the vise securely by hand. DO NOT USE A HAMMER ON THE VISE WRENCH.

6. Tap the work *lightly* at the four corners with a soft hammer until the paper feelers are tight between the work and the parallels (Fig. 59-6).

7. Select a plain helical cutter wider than the work to be machined.

8. Mount the cutter on the arbor for conventional milling (Fig. 59-7).

9. Set the proper speed for the size of cutter and the type of work material.

10. Set the feed to approximately 0.003 to 0.005 (0.08 to 0.13 mm) chip per tooth (cpt).

11. Start the cutter and raise the work, using a paper feeler between the cutter and the work (Fig. 59-3).

12. Stop the spindle when the cutter just cuts the paper.

13. Raise the knee 0.002 in. (0.05 mm) for paper thickness.

14. Set the graduated collar on the elevating screw handwheel to zero (Fig. 59-8).

15. Move the work clear of the cutter and set the depth of cut using the graduated collar.

16. For roughing cuts use a depth of not less than 1/8 in. (3.17 mm) and 0.010 to 0.025 in. (0.25 to 0.63 mm) for finish cuts.

17. Set the table dogs for the length of cut.

18. Engage the feed and cut side 1.

19. Set up and cut the remaining sides as required.

NOTE: See Unit 65, "Machining a Block Square and Parallel," for the setup procedures if the four sides of a block must be machined.

Fig. 59-8 After setting the cutter to the work, set the graduated collar to zero.

Fig. 59-9 Face milling produces flat, vertical surfaces.

FACE MILLING

Face milling is the process of producing a flat vertical surface at right angles to the cutter axis (Fig. 59-9). The cutters used for face milling are generally inserted-tooth cutters or shell end mills. *Face milling cutters* are made in sizes of 6 in. (152.4 mm) diameter and over; cutters less than 6 in. (152.4 mm) diameter are usually called *shell end mills.*

1. When face milling a large surface, use an inserted-tooth cutter and mount it into the spindle of the machine.

2. When milling smaller surfaces, the face milling cutter should be about 1 in. (25.4 mm) larger than the width of the workpiece.

Fig. 59-10 Side milling a vertical surface.

EQUAL DISTANCES

Fig. 59-11 Locating the cutter over the center of a round shaft.

3. Set the speeds and feeds for the type of cutter and the material being cut.

4. Set up the work on the milling machine being sure that the work-holding clamps do not interfere with the cutting action.

5. Use cutting fluid if the cutter material or work will allow.

6. Take as large a roughing cut as possible to bring the surface close to within 1/32 to 1/16 in. (0.79 to 1.58 mm) of the finish size.

7. Set the depth of the finish cut and machine the surface to size.

8. After the operation is complete, clean and store the cutter and clamping equipment in their proper places.

SIDE MILLING

Side milling is often used to machine a vertical surface on the sides or the ends of a workpiece (Fig. 59-10).

1. Set up the work in a vise and on parallels.

NOTE: Be sure that the layout line on the surface to be cut extends about 1/2 in. (12.7 mm) beyond the edge of the vise and the parallels to prevent the cutter, vise, or parallels from being damaged.

Fig. 59-12 Straddle milling.

2. Tighten the vise securely by hand. DO NOT USE A HAMMER ON THE VISE WRENCH.

3. Tap the work *lightly* at the four corners with a soft hammer until the paper feelers are tight between the work and the parallels.

4. Mount a side milling cutter as close to the spindle bearing as possible to provide maximum rigidity when milling.

5. Set the proper speed and feed for the cutter being used.

6. Start the machine and move the table until the top corner of the work just touches the revolving cutter.

7. Set the crossfeed graduated collar to zero.

8. With the table handwheel move the work clear of the cutter.

9. Set the required depth of cut with the crossfeed handle.

10. Lock the saddle to prevent movement during the cut.

11. Take the cut across the surface.

CENTERING A CUTTER TO MILL A SLOT

1. Locate the cutter as close to the center of the work as possible.

2. Using a steel square and rule (Fig. 59-11) or a gage block, adjust the work to the center by using the crossfeed screw dial.

3. Lock the saddle to prevent movement during the cut.

4. Move the work clear of the cutter and set the depth of cut.

5. Proceed to cut the slot using the same methods as for milling a flat surface.

STRADDLE MILLING

Straddle milling (Fig. 59-12) involves the use of two side milling cutters to machine the opposite sides of a workpiece parallel in one cut. The cutters are separated on the arbor

Fig. 59-13 Gang milling.

by a spacer or spacers of the required length so that the distance between the inside faces of the cutters is equal to the desired size. An adjustable micrometer arbor spacer may be used to vary the distance between the two cutters and also to compensate for the wear or the regrinding of the side milling cutters. Applications of straddle milling are the milling of square and hexagonal heads on bolts.

The following points should be observed when straddle milling.

1. Select two sharp side milling cutters, preferably staggered tooth, of the suitable sizes.
2. Mount the cutters, with suitable arbor spacers, as close to the column as the work will permit.
3. Mount the arbor support as close to the cutters as possible to provide rigidity for the cutters and arbor.
4. Center the cutters on the workpiece in the proper location.
5. Tighten the saddle lock to prevent any movement during a cut.
6. Set the cutter to the work surface.
7. Move the table so that the cutter clears the end of the workpiece.
8. Set the depth of cut required and tighten the knee clamp.
9. Set the proper speeds and feeds for the cutter size and the type of work material.
10. Use a good supply of cutting fluid and do the straddle milling operation in one cut.

GANG MILLING

Gang milling (Fig. 59-13) is a fast method of milling used a great deal in production work. It is performed by using two or more cutters on the arbor to produce the desired shape. The cutters may be a combination of plain and side milling cutters. If more than one helical milling cutter is used, a right- and left-hand helix cutter should be used to

Fig. 59-14 Sawing and slitting. (Courtesy Kostel Enterprises Limited.)

offset the thrust created by this type of cutter and to minimize the possibility of chatter. If several cutters are used and the slots are to be the same size, it is important that the diameter and width of each cutter be the same.

The following points should be observed to avoid problems during a gang milling operation.

1. Select cutters which are as close as possible to the same size and the same number of teeth. This will allow maximum speeds and feeds to be used.
2. Mount the correct cutters on the arbor as close to the machine column as possible.
3. Fasten the arbor support as close to the cutters as the work will permit.
4. Set the spindle speed to suit the largest-diameter cutter.
5. Be sure that the work is fastened securely and that the workholding devices will not come in contact with the cutters.
6. Use a good flow of cutting fluid to assist the cutting action and produce a good surface finish.

SAWING AND SLITTING

Metal-slitting saws may be used for milling narrow slots and for cutting off work (Fig. 59-14). Plain slitting saws, because of their thin cross section, are rather fragile cutting tools, and if they are not used carefully they will break easily.

To get the maximum life from a slitting saw take the following precautions.

1. Never key a slitting saw on the arbor unless it is mounted in a special flanged mounting collar. If the slitting saw is

keyed and jams in the work, it will rotate and cut through the key, and it will be difficult to remove the broken saw.

NOTE: The arbor nut must be drawn up as tightly as possible by *hand* only.

2. When selecting a slitting saw, choose one with the smallest diameter that will permit adequate clearance between the arbor collars or supports and the clamping bolts, holding device, or workpiece.

3. Mount the saw close to the column face and have the outer arbor support as close as possible to the saw.

4. Always use a sharp cutter.

5. Be sure that the table gibs are drawn up to eliminate any play between the table and the saddle.

6. Operate saws at approximately one-quarter to one-eighth the feed per tooth used for side milling cutters.

7. When sawing or slitting fairly long or deep slots, it is advisable to climb-mill to prevent the cutter from crowding sideways and breaking. The table should be fed carefully by hand and the backlash eliminator engaged.

REVIEW QUESTIONS

SETTING CUTTER TO WORK

1. Why should a long, thin piece of paper be used when setting a cutter to the work surface?

2. What should be done when the knee is moved beyond the desired amount?

MILLING A FLAT SURFACE

3. List the steps required to set work in a vise for milling.

4. Explain how the cutter should be set to the surface of the work.

FACE MILLING

5. Describe the process of face milling.

6. How large should a face milling cutter be in relation to the width of the work?

SIDE MILLING

7. For what purpose is side milling often used?

8. How should the work be set up in a vise for side milling?

9. Why should the table saddle be locked before machining?

10. Explain how a cutter can be aligned with the center of a round workpiece.

STRADDLE AND GANG MILLING

11. What is the difference between straddle and gang milling?

12. Why should the arbor support be mounted as close to the cutters as possible?

SAWING AND SLITTING

13. List five of the most important precautions to be observed when sawing or slitting.

THE INDEXING OR DIVIDING HEAD

UNIT 60

The *indexing or dividing head* is one of the most important attachments for the milling machine. It is used to divide the circumference of a workpiece into equally spaced divisions when milling such items as gears, splines, squares, hexagons, etc. It may also be used to rotate the workpiece at a predetermined ratio to the table feed rate to produce cams and helical grooves on gears, drills, reamers, etc.

OBJECTIVES

After completing this unit you will be able to:
1. Calculate and mill flats by simple and direct indexing
2. Calculate the indexing necessary with a wide-range divider
3. Calculate the indexing necessary for angular and differential indexing

INDEX HEAD PARTS

The universal dividing head set consists of the *headstock* with *index plates*, headstock *change gears* and *quadrant*, *univeral chuck*, *footstock*, and the *center rest* (Fig. 60-1). A *swiveling block*, mounted in the *base*, enables the headstock to be tilted from 5° below the horizontal position to 10° beyond the vertical position. The side of the base and the block are graduated to indicate the angle of the setting. Mounted in the swiveling block is a *spindle*, with a *40-tooth worm wheel* attached, which meshes with a worm (Fig. 60-2). The worm, at right angles to the spindle, is connected to the *index crank*, the pin of which engages in the *index plate* (Fig. 60-3). A *direct indexing plate* is attached to the front of the spindle.

A 60° center may be inserted into the front of the spindle, and a *universal chuck* may be threaded onto the end of the spindle.

The *footstock* is used in conjunction with the headstock to support work held between centers or the end of work held in a chuck. The footstock center may be adjusted longitudinally to accommodate various lengths of work and may be raised or lowered off center. It may also be tilted out of parallel with the base when cuts are being made on tapered work.

Long, slender work held between centers is prevented from bending by the *adjustable center rest*.

METHODS OF INDEXING

The main purpose of the indexing or dividing head is to divide the workpiece circumference accurately into any number of divisions. This may be accomplished by the following indexing methods; *direct, simple, angular,* and *differential*.

DIRECT INDEXING

Direct indexing is the simplest form of indexing. It is performed by disengaging the worm shaft from the worm wheel by means of an eccentric device in the dividing head. Some direct dividing heads do not have a worm and worm wheel but rotate on bearings. The index plates contain slots which are numbered and a spring-loaded tongue lock is used to engage in the proper slot. Direct indexing is used for quick indexing of the workpiece when cutting flutes, hexagons, squares, etc.

The work is rotated the required amount and held in place by a pin which engages into a hole or slot in the *direct indexing plate* mounted on the end of the dividing head spindle. The direct indexing plate usually contains three sets of hole circles or slots: 24, 30, and 36. The number of divisions it is possible to index is limited to numbers which are factors of either 24, 30, or 36. The common divisions that can be obtained by direct indexing are listed in Table 60-1.

EXAMPLE

What direct indexing is necessary to mill eight flutes on a reamer blank?

As the 24-hole circle is the only one divisible by 8 (the required number of divisions), it is the only circle which can be used in this case.

$$\text{Indexing} = \frac{24}{8} = 3 \text{ holes on a 24-hole circle}$$

NOTE: *NEVER* count the hole or slot in which the index pin is engaged.

Fig. 60-1   A universal dividing head set.

Fig. 60-2 Section through dividing head showing the worm wheel and worm shaft.

Fig. 60-3 Section through dividing head showing the spindle and index plate.

TABLE 60-1	Divisions Possible with Direct Indexing														
PLATE HOLE CIRCLES															
24	2	3	4	—	6	8	—	—	12	—	—	—	24	—	—
30	2	3	—	5	6	—	—	10	—	15	—	—	30	—	
36	2	3	4	—	6	—	9	—	12	—	18	—	—	36	

To Mill a Square by Direct Indexing

1. Disengage the worm and worm shaft by turning the worm disengaging shaft lever if the dividing head is so equipped.

2. Adjust the plunger behind the index plate into the 24-hole circle (Fig. 60-4).

3. Mount the workpiece in the dividing head chuck or between centers.

4. Adjust the cutter height and cut the first side.

5. Remove the plunger pin by turning the plunger pin lever (Fig. 60-5).

Fig. 60-4 Adjusting the plunger pin to fit into the proper hole circle.

Fig. 60-5 The plunger pin and the direct indexing plate are used for indexing a limited number of divisions.

6. Turn the plate, attached to the dividing head spindle, one-half turn (12 holes) and engage the plunger pin.

7. Take the second cut.

8. Measure the work across the flats and adjust the work height if required.

9. Cut the remaining sides by indexing every six holes until all surfaces are cut.

357

SIMPLE INDEXING

In *simple indexing,* the work is positioned by means of the crank, index plate, and sector arms. The worm, attached to the crank must be engaged with the worm wheel on the dividing head spindle. Since there are 40 teeth on the worm wheel, one complete turn of the index crank will cause the spindle and the work to rotate one-fortieth of a turn. Similarly, 40 turns of the crank will revolve the spindle and work one turn. Thus there is a ratio of 40:1 between the turns of the index crank and the dividing head spindle.

To calculate the indexing or the number of turns of the crank for most divisions, it is necessary only to divide 40 by the number of divisions (*N*) to be cut.

$$\text{Indexing} = \frac{40}{N}$$

EXAMPLE

The indexing required to cut eight flutes would be:

$$\frac{40}{8} = 5 \text{ full turns of the index crank}$$

If, however, it was necessary to cut seven flutes, the indexing would be

$$\frac{40}{7} = 5\frac{5}{7} \text{ turns}$$

Five complete turns are easily made; however, the five-sevenths of a turn involves the use of the index plate and sector arms.

INDEX PLATE AND SECTOR ARMS

The *index plate* is a circular plate provided with a series of equally spaced holes into which the index crank pin engages. The *sector arms* fit on the front of this plate and may be set to any portion of a complete turn.

To get five-sevenths of a turn, choose any hole circle (Table 60-2) which is divisible by the denominator 7, such as 21, then take five-sevenths of 21 = 15 holes on a 21-hole circle. Therefore, the indexing for seven flutes would be 40/7 = 5⁵⁄₇ turns or 5 complete turns plus 15 holes on the 21-hole circle. When extreme accuracy is required for indexing, choose the circle with the most holes.

The procedure for cutting seven flutes would be as follows:

1. Mount the proper index plate on the dividing head.
2. Loosen the index crank nut and set the index pin into a hole on the 21-hole circle.
3. Tighten the index crank nut and check to see that the pin enters the hole easily.
4. Loosen the setscrew on the sector arm.
5. Place the narrow edge of the left arm against the index pin.
6. Count 15 holes on the 21-hole circle. *Do not include the hole in which the index crank pin is engaged.*

TABLE 60-2 Index-Plate Hole Circles	
BROWN AND SHARPE	
Plate 1	15-16-17-18-19-20
Plate 2	21-23-27-29-31-33
Plate 3	37-39-41-43-47-49
CINCINNATI STANDARD PLATE	
One side	24-25-28-30-34-37-38-39-41-42-43
Other side	46-47-49-51-53-54-57-58-59-62-66

Fig. 60-6   Setting the cutter to the top of the work using a paper feeler.   (Courtesy Kostel Enterprises Limited.)

7. Move the right sector arm slightly beyond the 15th hole and tighten the sector arm setscrew.
8. Align the cutter with the workpiece.
9. Start the machine and set the cutter to the top of the work by using a paper feeler (Fig. 60-6).
10. Move the table so that the cutter clears the end of the work.
11. Tighten the friction lock on the dividing head before making each cut and loosen the lock when indexing for spaces.
12. Set the depth of cut and take the first cut.
13. After the first flute has been cut, return the table to the original starting position.
14. Withdraw the index pin and turn the crank clockwise five turns plus the 15 holes indicated by the right sector arm. Release the index pin between the 14th and 15th holes, and gently tap it until it drops into the 15th hole.
15. Turn the sector arm *furthest from the pin* clockwise until it is against the index pin.

NOTE: It is important that the arm *furthest from the pin* be held and turned. If the arm *next* to the pin were held and turned, the spacing between both sector arms could be increased when the other arm hit the pin. This could result in an indexing error which would not be noticeable until the work was completed.

16. Lock the dividing head; then continue machining and indexing for the remaining flutes. Whenever the crank pin

is moved past the required hole, *remove the backlash* between the worm and worm wheel by turning the crank *counterclockwise* approximately one-half turn and then carefully *clockwise* until the pin engages into the proper hole.

ANGULAR INDEXING

When the angular distance between divisions is given instead of the number of divisions, the setup for simple indexing may be used; however, the method of calculating the indexing is changed.

One complete turn of the index crank turns the work one-fortieth of a turn, or one-fortieth of 360° which equals 9°.

When the angular dimension is given in *degrees,* the indexing is then calculated as follows:

$$\text{Indexing in degrees} = \frac{\text{no. of degrees required}}{9}$$

EXAMPLE

Calculate the indexing for 45°.

$$\text{Indexing} = \frac{45}{9}$$
$$= 5 \text{ complete turns}$$

EXAMPLE

Calculate the indexing for 60°.

$$\text{Indexing} = \frac{60}{9}$$
$$= 6\frac{2}{3}$$
$$= 6 \text{ complete turns, 12 holes}$$
$$\text{on an 18-hole circle}$$

If the dimensions are given in *degrees* and *minutes,* it will be necessary to convert the degrees into minutes (number of degrees × 60′) and add this sum to the minutes required.

The indexing in minutes is calculated as follows:

$$\text{Indexing in minutes} = \frac{\text{no. of minutes required}}{540}$$

EXAMPLE

Calculate the indexing necessary for 24′.

$$\text{Indexing} = \frac{24}{540}$$
$$= \frac{4}{90}$$
$$= \frac{1}{22.5}$$

The indexing for 24′ would be 1 hole on the 22.5-hole circle. As the 23-hole circle is the nearest hole circle, the indexing would be one hole on the 23-hole circle. Since in this case there is a slight error (approximately one-half minute) in indexing, it is advisable to use this method only for a few divisions if extreme accuracy is required.

EXAMPLE

Calculate the indexing for 24°30′.

Convert 24° into minutes
$$(24 \times 60) = 1440'$$
$$\text{Add } 30' = \underline{30'}$$
$$\text{Total} = 1470'$$

$$\text{Indexing} = \frac{1470}{540}$$
$$= 2\text{-}13/18 \text{ turns}$$
$$= 2 \text{ complete turns plus 13 holes}$$
$$\text{on an 18-hole circle}$$

When indexing for degrees and half degrees (30′), use the 18-hole circle (Brown and Sharpe).

$$1/2° (30') = 1 \text{ hole on the 18-hole circle}$$
$$1° = 2 \text{ holes on the 18-hole circle}$$

When indexing for 1/3° (20′) and 2/3° (40′), the 27-hole circle should be used (Brown and Sharpe).

$$1/3° (20') = 1 \text{ hole on the 27-hole circle}$$
$$2/3° (40') = 2 \text{ holes on the 27-hole circle}$$

When indexing for minutes using a Cincinnati dividing head, note that one space on the 54-hole circle will rotate the work 10′ (1/54 × 540).

DIFFERENTIAL INDEXING

When it is impossible to calculate the required indexing by the simple indexing method, that is, when the fraction 40/N cannot be reduced to a factor of one of the available hole circles, it is necessary to use *differential indexing.*

With this method of indexing, the index plate must be revolved either forward or backward a part of a turn while the index crank is turned to attain the proper spacing or indexing.

In differential indexing, as in simple indexing, the index crank rotates the dividing head spindle. The spindle rotates the index plate, after the locking pin has been disengaged, by means of change gears connecting the dividing head spindle and the worm shaft (Fig. 60-7). The rotation of the plate may be either in the same direction (positive) or in the opposite direction (negative) of the index crank. This change of rotation is effected by an idler gear or gears in the gear train.

When it is necessary to calculate the indexing for a required number of divisions by the differential method, a number is chosen close to the required divisions which can be indexed by simple indexing.

To illustrate the principle of differential indexing, assume that the index crank has to be rotated one-ninth of a turn and there is only an 8-hole circle available.

If the crank is moved one-ninth of a turn, the index pin will contact the plate at a spot before the first hole on the 8-hole circle. The exact position of this spot would be the difference between one-eighth and one-ninth of a revolution of the crank. This would be

$$ \tfrac{1}{8} - \tfrac{1}{9} = \frac{9 - 8}{72} = \tfrac{1}{72} \text{ of a turn } less $$

than one-eighth of a turn, or one-seventy-second of a turn short of the first hole. Since there is no hole at this point into which the pin could engage, it is necessary to cause the plate to rotate backwards by means of change gears one-seventy-second of a turn in order that the pin will engage in a hole. At this point the index crank will be locked at exactly one-ninth of a turn.

The method of calculating the change gears (Fig. 60-7) required to rotate the plate the proper amount is as follows:

$$ \text{Change gear ratio} = (A - N) \times \frac{40}{A} $$

$$ = \frac{\text{driver (spindle) gear}}{\text{driven (worm) gear}} $$

$$ A = \text{approximate number of divisions} $$

$$ N = \text{required number of divisions} $$

When the approximate number of divisions is larger than the required number, the resulting fraction is plus and the index plate must move in the same direction as the crank (clockwise). This *positive rotation* is accomplished by using an idler gear. However, if the approximate number is smaller than the required number, the resulting fraction is minus and the index plate must move in a counter-clockwise direction. This *negative rotation* requires the use of two idler gears. The numerator of the fraction represents the driving (spindle) gear or gears, while the denominator represents the driven (worm) gear or gears. The gearing may be either simple or compound and the rotation is as follows.

Simple gearing: One idler for a positive rotation of the index plate
Two idlers for a negative rotation of the index plate
Compound gearing: One idler for a negative rotation of the index plate
Two idlers for a positive rotation of the index plate

EXAMPLE

Calculate the indexing and change gears required for 57 divisions.

The change gears supplied with the dividing head are as follows: 24, 24, 28, 32, 40, 44, 48, 56, 64, 72, 86, 100.

Fig. 60-7 Headstock geared for differential indexing.

The available index plate hole circles are as follows:

Plate 1: 15, 16, 17, 18, 19, 20
Plate 2: 21, 23, 27, 29, 31, 33
Plate 3: 37, 39, 41, 43, 47, 49

PROCEDURE

1. Indexing $= \dfrac{40}{N} = \dfrac{40}{57}$

Since there is no 57-hole circle and since it is impossible to reduce this fraction to suit any hole circle, it is necessary to select an approximate number close to 57, for which simple indexing may be calculated.

2. Let the approximate number of divisions equal 56.

3. Indexing for 56 divisions $= \dfrac{40}{56} = \dfrac{5}{7}$

or 15 holes on the 21-hole circle.

4.
$$ \text{Gear ratio} = (A - N) \times \frac{40}{A} $$
$$ = (56 - 57) \times \frac{40}{56} $$
$$ = -1 \times \frac{40}{56} $$
$$ = -\frac{5}{7} $$
$$ \text{Change gears} = -\frac{5}{7} \times \frac{8}{8} $$
$$ = -\frac{40 \text{ (spindle gear)}}{56 \text{ (worm gear)}} $$

Therefore, for indexing 57 divisions, a 40-tooth gear is mounted on the dividing head spindle, and a 56-tooth gear is mounted on the worm shaft. Since the fraction is a negative quantity and simple gearing is to be used, the index plate rotation is negative, or counter-clockwise, and two idlers must be used. After the proper gears are installed, the simple indexing procedure for 56 divisions should be followed.

THE WIDE-RANGE DIVIDING HEAD

Although simple or differential indexing is satisfactory for most indexing problems, there may be certain divisions which cannot be indexed by either of these methods. Cincinnati Milacron, Inc., manufactures a *wide-range divider* that may be applied to a Cincinnati universal dividing head. With this attachment, it is possible to obtain from 2 up to 400,000 divisions.

The wide-range divider consists of a large index plate *A* (Fig. 60-8), sector arms, and crank *B* which engages in the plate *A*. This large plate contains 11-hole circles on each side. Mounted in front of the large index plate is a small index plate *C* containing a 54-hole and a 100-hole circle. The crank *D* operates through a reduction of gears having a ratio of 100:1. These gears are mounted in the housing *G*. The ratio between the worm (and the crank *B*) and the dividing head spindle is 40:1.

Fig. 60-8   A wide-range dividing head.   (Courtesy Cincinnati Milacron Incorporated.)

INDEXING FOR DIVISIONS

The ratio of the large index crank to the dividing head is 40:1, as in simple indexing. The ratio of the small index crank which drives the large crank by planetary gearing is 100:1. Therefore *one turn of the small crank* drives the index head spindle 1/100 of 1/40, or 1/4000 of a turn. One hole on the 100-hole circle of the small index plate *C* = 1/100 × 1/4000, or 1/400,000 of a turn. Therefore the formula for indexing *divisions* with a wide-range divider is 400,000/*N* and is applied as follows.

As the ratio of the large index crank is 40:1, any number which divide into 40 (the two numbers to the left of the short vertical line) represent full turns of the large index crank. If a 100-hole circle is used with the large crank, *one hole* on this circle will produce 1/100 or 1/40, or 1/4000 of a turn. Therefore, any numbers which divided into 4000 (the two numbers to the left of the long vertical line) are indexed on the 100-hole circle of the large plate. The numbers to the right of the long vertical line are indexed on the 100-hole circle of the small plate. Thus for the 1250 divisions, the indexing would be: 3 holes on the 100-hole circle of the large plate plus 20 holes on the 100-hole circle of the small plate.

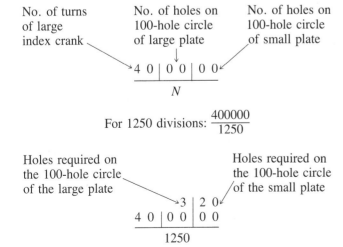

No. of turns of large index crank   No. of holes on 100-hole circle of large plate   No. of holes on 100-hole circle of small plate

$$4\ 0\ |\ 0\ 0\ |\ 0\ 0$$
$$N$$

For 1250 divisions: $\dfrac{400000}{1250}$

Holes required on the 100-hole circle of the large plate   Holes required on the 100-hole circle of the small plate

$$3\ |\ 2\ 0$$
$$4\ 0\ |\ 0\ 0\ |\ 0\ 0$$
$$1250$$

ANGULAR INDEXING WITH THE WIDE-RANGE DIVIDER

The wide-range divider is especially suited for accurate angular indexing. Indexing in degrees, minutes, and seconds is easily accomplished without the complicated calculations necessary with standard dividing heads.

For angular indexing, both the large and small index cranks are set on the 54-hole circle of each plate. Each space on the 54-hole circle of the large plate will cause the dividing head spindle to rotate 10'. Each space on the 54-hole circle of the small plate will cause the work to rotate 6''. Therefore, for indexing angles with a wide-range divider, the following formulas are used.

$$\text{Degrees} = \frac{N}{9} \text{ (indexed on the large plate)}$$

$$\text{Minutes} = \frac{N}{10} \text{ (indexed on the large plate)}$$

$$\text{Seconds} = \frac{N}{6} \text{ (indexed on the small plate)}$$

EXAMPLE

To index for an angle of 17°36'18''

1. Degrees $= \dfrac{17}{9}$

   $= 1\frac{8}{9}$ turns

   OR

One turn plus 48 holes on the 54-hole circle of the large index plate.

Fig. 60-9 Linear graduating.

Fig. 60-10 Gearing required for linear graduating.

2. Minutes $= \dfrac{36}{10}$

   = 3 holes on a 54-hole circle of the large index plate, leaving a remainder of 6′

3. Convert the 6′ into seconds (6 × 60 = 360″) and add it to the 18″ still required.

4. Seconds $= \dfrac{378}{6}$

   = 63 holes on a 54-hole circle of the small plate

OR

One turn and 9 holes on the 54-hole circle. Therefore, to index for 17°36′18″ would require one turn and 51 holes (48 + 3) on the 54-hole circle of the large plate plus one turn and 9 holes on the 54-hole circle of the small plate.

LINEAR GRADUATING

The operation of producing accurate spaces on a piece of flat stock, or that of *linear graduating,* is easily accomplished on the horizontal milling machine (Fig. 60-9).

In this process the work may be clamped to the table or held in a vise, depending on the shape and size of the part. Care must be taken to align the workpiece parallel with the table travel.

To produce an *accurate* longitudinal movement of the table, the dividing head spindle is geared to the lead screw of the milling machine (Fig. 60-10).

If the dividing head spindle and the lead screw were connected with gears with equal number of teeth and the index crank turned one revolution, the spindle and lead screw on an inch milling machine would revolve one-fortieth of a revolution. This rotation of the lead screw

[having 4 threads per inch (tpi)] would cause the table to move $\frac{1}{40} \times \frac{1}{4}$ (one turn of the lead screw) = $\frac{1}{160}$ = 0.00625 in. (0.15 mm). Thus five turns of the index crank would move the table 5 × 0.00625, or $\frac{1}{32}$ in. (0.78 mm).

The formula for calculating the indexing for linear graduations in thousandths of an inch is

$$\frac{N}{0.00625}$$

Very small movements of the table, such as 0.001, may be obtained by applying the formula

$$\frac{0.001}{0.00625} = \frac{1}{6\frac{1}{4}} \text{ turns}$$

(4/25 turn), or 4 holes on the 25-hole circle.

If the lead screw of a metric milling machine had a pitch of 5 mm, one turn of the index crank would move the table one-fortieth of 5 mm, or 0.125 mm. Therefore it would require four complete turns on the index crank to move the table 0.5 mm.

The formula for calculating the indexing for linear graduations in millimeters is

$$\frac{N}{0.125}$$

For a small movement of the table, such as 0.025 mm, apply the formula:

$$\frac{0.025}{0.125} = 1/5 \text{ turn, or 5 holes on a 25-hole circle.}$$

Other suitable table movements may be obtained by using the appropriate hole circle and/or different change gears.

The point of the toolbit used for graduating is generally ground to a V-shape, although other special forms may be desired. The tool is mounted vertically in a suitable arbor which is of sufficient length to extend the toolbit over the workpiece (Fig. 60-9).

The uniformity of the length of the lines is controlled by the *accurate* movement of the crossfeed handwheel or by stops suitably mounted on the ways of the knee.

When graduating is done, the starting point on the workpiece is positioned under the point of the *stationary,* vertical toolbit. The work is moved clear of the tool by the crossfeed handwheel and the proper depth is set by means of the vertical feed crank. The table is then locked in place. For a uniform width of lines to be maintained, the work must be held absolutely flat and the table height must never be adjusted.

REVIEW QUESTIONS

INDEXING OR DIVIDING HEAD

1. Name four parts of a dividing head set.
2. Name four methods of indexing that may be performed using the dividing head.
3. For what purpose is direct indexing used?

SIMPLE INDEXING

4. Explain how the ratio of 40:1 is determined on a standard dividing head.
5. Calculate the simple indexing, using a Brown and Sharpe dividing head, for the following divisions: 37, 41, 22, 34, and 120.
6. What procedure should be followed in order to set the sector arms for 12 holes on an 18-hole circle?

ANGULAR INDEXING

7. Explain the principle of angular indexing.
8. Calculate the indexing, using a Cincinnati dividing head, for the following angles: 21°, 37°, 21°30', and 37°40'.

DIFFERENTIAL INDEXING

9. For what purpose is differential indexing used?
10. What is meant by positive rotation and negative rotation of the index plate?
11. Using a Brown and Sharpe dividing head, calculate the indexing and change gears for the following divisions: 53, 59, 101, and 175. A standard set of change gears, having the following numbers of teeth, is supplied: 24, 24, 28, 32, 40, 44, 48, 56, 64, 72, 86, 100.

WIDE-RANGE DIVIDING HEAD

12. How does the wide-range dividing head differ from a standard dividing head?
13. What two ratios are found in a wide-range dividing head?
14. Calculate the indexing for (a) 1000 and (b) 1200 divisions, using the wide-range dividing head.
15. Calculate the angular indexing for the following, using a wide-range dividing head: 20°45', 25°15'32".

LINEAR GRADUATING

16. How are the dividing head and the milling machine geared for linear graduating?
17. What indexing would be required to move the table 0.003 in. when using equal gearing on the dividing head and the lead screw?

GEARS

UNIT 61

When it is required to transmit rotary motion from one shaft to another, several methods may be used, such as belts, pulleys, and gears. If the shafts are parallel to each other and quite a distance apart, a flat belt and large pulleys may be used to drive the second shaft, the speed of which may be controlled by the size of the pulleys.

When the shafts are closer together, such as in the case of the sensitive drill press, a V-belt, which tends to reduce the excessive slippage of a flat belt, may be used. Here the speed of the driven shaft may be controlled by means of stepped or variable-speed pulleys. When the shafts are close together and parallel, some power may be transmitted by two rollers in contact, with one roller mounted on each shaft. Slippage is the main problem here, and the desired speed of the driven shaft could not be maintained.

The methods outlined are means by which power may be transmitted from one shaft to another, but the speed of the driven shaft may not be accurate in all cases due to slippage between the driving and driven members (belts, pulleys, or rollers). In order to eliminate slippage and produce a positive drive, gears are used.

OBJECTIVES

After completing this unit you should be able to:
1. Identify and know the purpose of six types of gears used in industry
2. Apply various formulas for calculating gear-tooth dimensions

GEARS AND GEARING

Gears are used to transmit power positively from one shaft to another by means of successively engaging teeth (in two gears). They are used in place of belt drives and other forms of friction drives when exact speed ratios and power transmission must be maintained. Gears may also be used to increase or decrease the speed of the driven shaft, thus decreasing or increasing the *torque* of the driven member.

Shafts in a gear drive or train are generally parallel. They may, however, be driven at any angle by means of suitably designed gears.

TYPES OF GEARS

Spur gears (Fig. 61-1) are generally used to transmit power between two parallel shafts. The teeth on these gears are straight and parallel to the shafts to which they are attached. When two gears of different sizes are in mesh, the larger is called the *gear* while the smaller is called the *pinion*. Spur gears are used where slow to moderate-speed drives are required.

Internal gears (Fig. 61-2) are used where the shafts are parallel and the centers must be closer together than could be achieved with spur or helical gearing. This arrangement provides for a stronger drive since there is a greater area of contact than with the conventional gear drive. It also provides speed reductions with a minimum space requirement. Internal gears are used on heavy-duty tractors where much torque is required.

Helical gears (Fig. 61-3) may be used to connect parallel shafts or shafts which are at an angle. Because of the progressive rather than intermittent action of the teeth, helical gears run more smoothly and quietly than spur gears. Since there is more than one tooth in engagement at any

Fig. 61-1 A spur gear and pinion are used for slower speeds.

Fig. 61-2 Internal gears provide speed reductions with a minimum space requirement.

Fig. 61-3 Helical gears. (A) For drives which are parallel to each other; (B) for drives which are at right angles to each other.

Fig. 61-4 Herringbone gears eliminate end thrust on shafts.

Fig. 61-5 (A) Bevel gears transmit power at 90°; (B) the driver and driven miter gears are the same size; (C) angular bevel gears are used for shafts which are not at right angles; (D) hypoid gears are used in automotive drives.

one time, helical gears are stronger than spur gears of the same size and pitch. However, special bearings (thrust bearings) are often required on shafts to overcome the end thrust produced by these gears as they turn.

On most installations where it is necesary to overcome end thrust, *herringbone gears* (Fig. 61-4) are used. This type of gear resembles two helical gears placed side by side, with one-half having a left-hand helix and the other half a right-hand helix. These gears have a smooth continuous action and eliminate the need for thrust bearings.

When two shafts are located at an angle with their axial lines intersecting at 90°, power is generally transmitted by means of *bevel gears* (Fig. 61-5A). When the shafts are at right angles and the gears are of the same size, they are called *miter gears* (Fig. 61-5B). However, it is not necessary that the shafts be only at right angles in order to transmit power. If the axes of the shafts intersect at any angle other than 90°, the gears are known as *angular bevel gears* (Fig. 61-5C). Bevel gears have straight teeth very similar to spur gears. Modified bevel gears having helical teeth are known as *hypoid gears* (Fig. 61-5D). The shafts of these gears, although at right angles, are not in the same plane and, therefore, do not intersect. Hypoid gears are used in automobile drives.

Fig. 61-6   A worm and worm gear is used for speed reduction.

Fig. 61-7   A rack and pinion converts rotary motion to linear motion.

When shafts are at right angles and considerable reduction in speed is required, a *worm and worm gear* (Fig. 61-6) may be used. The worm which meshes with the worm gear may be a single- or multiple-start thread. A worm with a double-start thread will revolve the worm gear twice as fast as a worm with a single-start thread and the same pitch.

When it is necesary to convert rotary motion to linear motion, a *rack and pinion* (Fig. 61-7) may be used. The rack, which is actually a straight or flat gear, may have straight teeth to mesh with a spur gear, or angular teeth to mesh with a helical gear.

GEAR TERMINOLOGY

A knowledge of the more common gear terms is desirable to understand gearing and to make the calculations neccessary to cut a gear (Fig. 61-8). Most of these terms are applicable to either inch or metric gearing, although the method of calculating dimensions may differ. These methods are explained as applicable to the inch and metric gear-cutting sections.

Addendum is the radial distance between the pitch circle and the outside diameter or the height or the tooth above the pitch circle.
Center distance is the shortest distance between the axes of two mating gears or the distance equal to one-half the sum of the pitch diameters.
Chordal addendum is the radial distance measured from the top of the tooth to a point where the chordal thickness and the pitch circle intersect the edge of the tooth.
Chordal thickness is the thickness of the tooth measured

at the pitch circle or the length of the chord which subtends the arc of the pitch circle.
Circular pitch is the distance from a point on one tooth to a corresponding point on the next tooth measured on the pitch circle.
Circular thickness is the thickness of the tooth measured on the pitch circle; it is also known as the *arc thickness*.
Clearance is the radial distance between the top of one tooth and the bottom of the mating tooth space.
Dedendum is the radial distance from the pitch circle to the bottom of the tooth space. The dedendum is equal to the addendum plus the clearance.
Diametral pitch (inch gears) is the ratio of the number of teeth for each inch of pitch diameter of the gear. For example, a gear of 10 diametral pitch and a 3-in. *pitch diameter* would have 10×3 or 30 teeth.
Involute is the curved line produced by a point of a stretched string when it is unwrapped from a given cylinder (Fig. 61-9).
Linear pitch is the distance from a point on one tooth to the corresponding point on the next tooth of a gear rack.
Module (metric gears) is the pitch diameter of a gear divided by the number of teeth. It is an actual dimension, unlike diametral pitch, which is a ratio of the number of teeth to the pitch diameter.
Outside diameter is the overall diameter of the gear, which is the pitch circle plus two addendums.
Pitch circle is a circle which has the radius of half the pitch diameter with its center at the axis of the gear.
Pitch circumference is the circumference of the pitch circle.
Pitch diameter is the diameter of the pitch circle which is equal to the outside diameter minus two addendums.

Pressure angle is the angle formed by a line through the point of contact of two mating teeth and tangent to the two *base circles* and a line at right angles to the centerline of the gears.

Root circle is the circle formed by the bottoms of the tooth spaces.

Root diameter is the diameter of the root circle.

Tooth thickness is the thickness of the tooth measured on the pitch circle.

Whole depth is the full depth of the tooth or the distance equal to the addendum plus the dedendum.

Working depth is the distance that a gear tooth extends into the tooth space of a mating gear which is equal to two addendums.

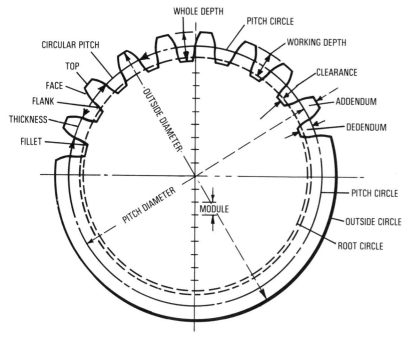

Fig. 61-8 Parts of a gear.

Fig. 61-9 Methods of generating and involute.

TABLE 61-1 Rules and Formulas for Spur Gears			
TO OBTAIN	**KNOWING**	**RULE**	**FORMULA**
Addendum (A)	Circular pitch	Multiply the circular pitch by 0.3183.	$A = CP \times 0.3183$
Addendum	Diametral pitch	Divide 1 by the diametral pitch.	$A = \dfrac{1}{DP}$
Center distance (CD)	Circular pitch	Multiply the number of teeth in both gears by the circular pitch, and divide the product by 6.2832.	$CD = \dfrac{(N + n) \times CP}{6.2832}$
Center distance	Diametral pitch	Divide the total number of teeth in both gears by twice the diametral pitch.	$CD = \dfrac{N + n}{2 \times DP}$
Chordal (corrected) addendum (CA)	Pitch diameter addendum, and number of teeth	Subtract from 1 the cosine of the result of 90° divided by the number of teeth. Multiply this result by half the pitch diameter. To this product, add the addendum.	$CA = \left[\left(1 - \cos\dfrac{90}{N}\right)\dfrac{PD}{2}\right] + A$
Chordal thickness (CT)	Pitch diameter and number of teeth	Divide 90 by the number of teeth; find the sine of this result and multiply by the pitch diameters.	$CT = \sin\dfrac{90}{N} \times PD$
Circular pitch (CP)	Center-to-center distance	Multiply center to center distance by 6.2832, and divide the product by the total number of teeth in both gears.	$CP = \dfrac{CD \times 6.2832}{N + n}$
			(Cont. on next page)

TABLE 61-1 (Continued)

TO OBTAIN	KNOWING	RULE	FORMULA
Circular pitch	Diametral pitch	Divide 3.1416 by the diametral pitch.	$CP = \dfrac{3.1416}{DP}$
Circular pitch	Pitch diameter and number of teeth	Multiply pitch diameter by 3.1416 and divide by the number of teeth.	$CP = \dfrac{PD \times 3.1416}{N}$
Clearance (CL)	Circular pitch	Divide circular pitch by 20.	$CL = \dfrac{CP}{20}$
Clearance	Diametral pitch	Divide 0.157 by the diametral pitch.	$CL = \dfrac{0.157}{DP}$
Dedendum (D)	Circular pitch	Multiply the circular pitch by 0.3683.	$D = CP \times 0.3683$
Dedendum	Diametral pitch	Divide 1.157 by the diametral pitch.	$D = \dfrac{1.157}{DP}$
Diametral pitch (DP)	Circular pitch	Divide 3.1416 by the circular pitch.	$DP = \dfrac{3.1416}{CP}$
Diametral pitch	Number of teeth and outside diameter	Add 2 to the number of teeth and divide the sum by the outside diameter.	$DP = \dfrac{N + 2}{OD}$
Diametral pitch	Number of teeth and pitch diameter	Divide the number of teeth by the pitch diameter.	$DP = \dfrac{N}{PD}$
Number of teeth (N)	Outside diameter and diametral pitch	Multiply the outside diameter by the diametral pitch and subtract 2.	$N = OD \times DP - 2$
Number of teeth	Pitch diameter and circular pitch	Multiply the pitch diameter by 3.1416 and divide by the circular pitch.	$N = \dfrac{PD \times 3.1416}{CP}$
Number of teeth	Pitch diameter and diametral pitch	Multiply the pitch diameter by the diametral pitch.	$N = PD \times DP$
Outside diameter (OD)	Number of teeth and circular pitch	Add 2 to the number of teeth and multiply the sum by the circular pitch. Divide this product by 3.1416.	$OD = \dfrac{(N + 2) \times CP}{3.1416}$
Outside diameter	Number of teeth and diametral pitch	Add 2 to the number of teeth and divide the sum by the diametral pitch.	$OD = \dfrac{N + 2}{DP}$
Outside diameter	Pitch diameter and diametral pitch	To the pitch diameter add 2 and divide by the diametral pitch.	$OD = PD + \dfrac{2}{DP}$
Pitch diameter (PD)	Number of teeth and circular pitch	Multiply the number of teeth by the circular pitch and divide by 3.1416.	$PD = \dfrac{N \times CP}{3.1416}$
Pitch diameter	Number of teeth and diametral pitch	Divide the number of teeth by the diametral pitch.	$PD = \dfrac{N}{DP}$
Pitch diameter	Outside diameter and number of teeth	Multiply the number of teeth by the outside diameter, and divide the product by the number of teeth plus 2.	$PD = \dfrac{N \times OD}{N + 2}$

TABLE 61-1 (Continued)

TO OBTAIN	KNOWING	RULE	FORMULA
Tooth thickness *(T)*	Circular pitch	Divide the circular pitch by 2.	$T = \dfrac{CP}{2}$
Tooth thickness	Circular pitch	Multiply the circular pitch by 0.5.	$T = CP \times 0.5$
Tooth thickness	Diametral pitch	Divide 1.5708 by the diametral pitch.	$T = \dfrac{1.5708}{DP}$
Whole depth (WD)	Circular pitch	Multiply the circular pitch by 0.6866.	$WD = CP \times 0.6866$
Whole depth	Diametral pitch	Divide 2.157 by the diametral pitch.	$WD = \dfrac{2.157}{DP}$

Note: There are three commonly used gear-tooth forms which have pressure angles of 14½, 20, and 25°. The 20 and 25° tooth forms are now replacing the 14½° tooth form because the base of the tooth is wider and stronger. For the formula regarding the 20 and 25° tooth forms, see *Machinery's* or the *American Machinist's* handbooks.

REVIEW QUESTIONS

1. List six types of gears, and state where each may be used.
2. Define the following gear terms, and state the formula used to determine each. Use the formulas involving diametral pitch where applicable, *not* circular pitch.
   (a) pitch diameter
   (b) diametral pitch
   (c) addendum
   (d) dedendum
   (e) clearance
   (f) outside diameter
   (g) number of teeth
3. Calculate the pitch diameter, outside diameter, and whole depth of tooth for the following gears:
   (a) 8 DP having 36 teeth
   (b) 12 DP having 81 teeth
   (c) 16 DP having 100 teeth
   (d) 6 DP having 23 teeth
   (e) 4 DP having 54 teeth

GEAR CUTTING
UNIT 62

Most gears cut on a milling machine are generally used to repair or replace a gear which has been broken, lost, or no longer carried in inventory. Industry generally mass-produces gears on special machines which have been designed for this purpose. The most common types of *gear-generating machines are the gear-shaping machines and the gear-hobbing machines.* It is generally more economical to buy gears from a firm which specializes in gear manufacture unless it is very important that the machine be back in operation soon. At this time, it is quite possible that a machinist would be called on to cut a gear in order to repair the machine quickly and get it back into production.

After completing this unit you should be able to:
1. Select the proper cutter for any gear to be cut
2. Calculate gear tooth dimensions for inch gears
3. Calculate gear tooth dimensions for metric gears
4. Set up and cut a spur gear

INVOLUTE GEAR CUTTERS

Gear cutters are an example of formed cutters. This type of cutter is sharpened on the face and ensures exact duplication of the shape of the teeth, regardless of how far back the face of the tooth has been ground.

Gear cutters are available in many sizes ranging from 1 to 48 diametral pitch (DP). Cutters for teeth smaller than 48 DP are available as special cutters. Comparative sizes of teeth ranging from 4 to 16 DP are shown in Fig. 62-1.

When gear teeth are cut on any gear, a cutter must be chosen to suit both the DP and the number of teeth (*N*). The tooth space for a small pinion cannot be of the same shape as the tooth space for a large mating gear. The teeth on smaller gears must be more "curved" to prevent binding of meshing gear teeth. Therefore, sets of gear cutters are made in a series of slightly different shapes to permit the cutting of any desired number of teeth in a gear, with the assurance that the teeth will mesh properly with those of another gear of the *same* DP.

These cutters are generally made in sets of eight and are numbered from 1 to 8 (Fig. 62-2). Notice the gradual change in shape from the #1 cutter, which has almost straight sides, to the much more curved sides of the #8 cutter. As shown in Table 62-1, the #1 cutter is used for cutting any number of teeth in a gear from 135 teeth to a rack, while the #8 cutter will cut only 12 and 13 teeth. It should be noted that in order for gears to mesh, they must be of the same DP; the cutter number permits *only* a more accurate meshing of the teeth.

Some gear cutter manufacturers have augmented the set of eight cutters with seven additional cutters in half sizes, making a total 15 cutters in the set, numbered 1, 1½, 2, 2½, etc. In the half series, a #1½ cutter would be used to cut from 80 to 134 teeth, while a 7½ cutter would cut 13 teeth only (Table 62-1).

EXAMPLE

A 10-DP gear and a pinion in mesh have 100 teeth and 24 teeth respectively. What cutters should be used to cut these gears?

Cutter Selection

Since the gears are in mesh, both must be cut with a 10-DP cutter.

A #2 cutter should be used to cut the teeth on the gear, since it will cut from 55 to 134 teeth.

A #5 cutter should be used to cut the pinion, since it cuts from 21 to 25 teeth.

TO CUT A SPUR GEAR

The procedure for machining a spur gear is outlined in the following example.

EXAMPLE

A 52-tooth gear with an 8 DP is required.

PROCEDURE

1. Calculate all the necessary gear data. (See Unit 61)

   a. Outside diameter $= \dfrac{N + 2}{DP}$

   $= \dfrac{54}{8}$

   $= 6.750$ in.

   b. Whole depth of tooth $= \dfrac{2.157}{DP}$

   $= \dfrac{2.157}{8}$

   $= 0.2697$ in.

   c. Cutter number $= 3$ (35 to 54 teeth)

Fig. 62-1 Comparative gear tooth sizes 4 to 16 DP.

16 DP
14 DP
12 D
10 DP
8 DP
7 DP
6 DP
5 DP
4 DP

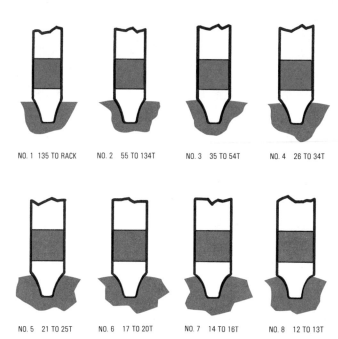

NO. 1 135 TO RACK   NO. 2 55 TO 134T   NO. 3 35 TO 54T   NO. 4 26 TO 34T

NO. 5 21 TO 25T   NO. 6 17 TO 20T   NO. 7 14 TO 16T   NO. 8 12 TO 13T

Fig. 62-2 Involute profiles for a set of gear cutters.

| TABLE 62-1 Involute Gear Cutters | |
CUTTER NUMBER	RANGE
1	135 teeth to a rack
1½	80 to 134 teeth
2	55 to 134 "
2½	42 to 54 "
3	35 to 54 "
3½	30 to 34 "
4	26 to 34 "
4½	23 to 25 "
5	21 to 25 "
5½	19 and 20 "
6	17 to 20 "
6½	15 and 16 "
7	14 to 16 "
7½	13 teeth
8	12 to 13 teeth

d. Indexing (using Cincinnati standard plate)

$$= \frac{40}{N}$$

$$= \frac{40}{52}$$

$$= \frac{10}{13} \times \frac{3}{3} = 30 \text{ holes on the 39-hole circle}$$

2. Turn the gear blank to the proper outside diameter (6.750 in.)

3. Press the gear blank firmly onto the mandrel.

NOTE: If the blank was turned on a mandrel, be sure that

Fig. 62-3 Checking the alignment of the index centers with a dial indicator.

it is tight because the heat caused by turning might have expanded the blank slightly.

4. Mount the index head and footstock, and check the alignment of the index centers (Fig. 62-3).

5. Set the dividing head so that the index pin fits into a hole on the 39-hole circle and the sector arms are set for 30 holes.

NOTE: Do not count the hole in which the pin is engaged.

6. Mount the mandrel (and workpiece), with the large end toward the indexing head, between the index centers.

NOTE:
(a.) The footstock center should be adjusted up tightly into the mandrel and locked in position.
(b.) The dog should be tightened properly on the mandrel and the tail of the dog should not bind in the slot.
(c.) The tail of the dog should then be locked in the driving fork of the dividing head by means of the setscrews. This will ensure that there will be no play between the dividing head and the mandrel.
(d.) The dog should be far enough from the gear blank to ensure that the cutter will not hit the dog when the gear is being cut.

7. Move the table close to the column to keep the setup as rigid as possible.

8. Mount an 8 DP#3 cutter on the milling machine arbor over the approximate center of the gear. Be sure to have the cutter rotating in the direction of the indexing head.

9. Center the gear blank with the cutter by either of the following methods:
(a.) Place a square against the outside diameter of the gear (Fig. 62-4). With a pair of inside calipers or a rule, check the distance between the square and the side of the cutter. Adjust the table until the distances from both sides of the gear blank to the sides of the cutter are the same.

Fig. 62-4   Centering a gear cutter and the workpiece.

Fig. 62-5   Setting the gear cutter to the diameter of the work.

Fig. 62-6   Notching all gear teeth eliminates errors.

(b.) A more accurate method of centralizing the cutter is to use gage blocks instead of the inside calipers or rule.

10. *LOCK THE CROSS SLIDE.*

11. Start the milling cutter and run the work under the cutter.

12. Raise the table until the cutter *just* touches the work. This can be done by using a chalk mark on the gear blank or a piece of paper between the gear blank and the cutter to indicate when the cutter is just touching the work (Fig. 62-5).

13. Set the graduated feed collar on the vertical feed to zero.

14. Move the work clear of the cutter by means of the longitudinal feed handle and raise the table to about two-thirds the depth of the tooth (0.180 in.); then *lock the knee clamp.*

NOTE: A special stocking cutter is sometimes used to rough out the teeth.

15. Slightly notch all gear teeth on the end of the work to check for correct indexing (Fig. 62-6).

16. Rough out the first tooth and set the automatic feed trip dog after the cutter is clear of the work.

17. Return the table to the starting position.

NOTE: Clear the end of the work with a cutter.

18. Cut the remaining teeth and return the table to the starting position.

19. Loosen the knee clamp, raise the table to the proper depth of 0.270 in., and *lock the knee clamp.*

NOTE: It is advisable to remove the crank from the knee-elevating shaft so that it will not be moved accidentally and change the setting.

20. Finish-cut all teeth.

NOTE: After each tooth has been cut, the cutter should be stopped before the table is returned to prevent marring the finish on the gear teeth.

METRIC GEARS AND GEAR CUTTING

Countries which have been using a metric system of measurement usually use the *module* system of gearing. The *module (M)* of a gear equals the pitch diameter (PD) divided by the number of teeth *(N)* ($M = PD/N$), whereas the DP of a gear is the ratio of *N* to the PD (DP = *N*/PD). The DP of a gear is the *ratio* of the number of teeth per inch of diameter whereas *M is an actual dimension*. Most of the terms used in DP gears remain the same for module gears; however, the method of calculating the dimensions has changed in some instances. Table 62-2 gives the necessary rules and formulas for metric spur gears.

TABLE 62-2 Rules and Formulas for Metric Module Spur Gears

TO OBTAIN	KNOWING	RULE	FORUMLA
Addendum *(A)*	Normal Module	Addendum equals module	$A = M$
Circular pitch (CP)	Module	Multiply module by π	$CP = M \times 3.1416$
	Pitch diameter and number of teeth	Multiply pitch diameter by π and divide by number of teeth.	$CP = \dfrac{PD \times 3.1416}{N}$
	Outside diameter and number of teeth	Multiply outside diameter by π and divide by number of teeth minus 2.	$CP = \dfrac{OD \times 3.1416}{N - 2}$
Chordal thickness (CT)	Module and outside diameter	Divide 90° by the number of teeth. Find the sine of this angle and multiply by the pitch diameter.	$CT = PD \times \sin \dfrac{90}{N}$
	Module	Multiply module by π and divide by 2.	$CT = \dfrac{M \times 3.1416}{2}$
	Circular pitch	Divide circular pitch by 2.	$CT = \dfrac{CP}{2}$
Clearance (CL)	Module	Multiply module by 0.166 mm.	$CL = M \times 0.166$
Dedendum (D)	Module	Multiply module by 1.166 mm.	$D = M \times 1.166$
Module *(M)*	Pitch diameter and number of teeth	Divide pitch diameter by the number of teeth.	$M = \dfrac{PD}{N}$
	Circular pitch	Divide circular pitch by π.	$M = \dfrac{CP}{3.1416}$
	Outside diameter and number of teeth	Divide outside diameter by number of teeth + 2.	$M = \dfrac{OD}{N + 2}$
Number of teeth *(N)*	Pitch diameter and module	Divide pitch diameter by the module.	$N = \dfrac{PD}{M}$
	Pitch diameter and circular pitch	Multiply pitch diameter by π and divide product by circular pitch.	$N = \dfrac{PD \times 3.1416}{CP}$
Outside diameter (OD)	Number of teeth and module	Add 2 to the number of teeth and multiply sum of module.	$OD = (N + 2) \times M$
	Pitch diameter and module	Add 2 modules to pitch diameter.	$OD = PD + 2M$
Pitch diameter (PD)	Module and number of teeth	Multiply module by number of teeth.	$PD = M \times N$
	Outside diameter and module	Subtract 2 modules from outside diameter.	$OD = PD + 2M$
Pitch diameter	Number of teeth and outside diameter	Multiply number of teeth by outside diameter and divide product by number of teeth + 2.	$PD = \dfrac{N \times OD}{N + 2}$
Whole depth (WD)	Module	Multiply module by 2.166 mm.	$WD = M \times 2.166$
Center-to-center distance (CD)	Pitch diameters	Divide the sum of the pitch diameters by 2.	$CD = \dfrac{PD_1 + PD_2}{2}$

METRIC MODULE GEAR CUTTERS

The most common metric gear cutters are available in modules ranging from 0.5 to 10 mm (Table 62-3). However, metric module gear cutters are available up to 75 mm. Any metric module size is available in a set of eight cutters, numbered from #1 to #8. The range of each cutter is reversed to those of DP cutters. For instance, a #1 metric module cutter will cut from 12 to 13 teeth; a #8 DP cutter will cut from 135 teeth to a rack. Table 62-3 shows the cutters available and the range of each cutter in the set.

EXAMPLE

A spur gear has PD of 60 mm and 20 teeth. Calculate:

1. Module
2. Circular pitch
3. Addendum
4. Outside diameter
5. Dedendum
6. Whole depth
7. Cutter number

1. $M = \dfrac{PD}{N}$

   $= \dfrac{60}{20}$

   $= 3$ mm

2. $CP = M \times \pi$

   $= 3 \times 3.1416$

   $= 9.425$ mm

3. $A = M$

   $= 3$ mm

4. $OD = (N + 2) \times M$

   $= 22 \times 3$

   $= 66$ mm

5. $D = M \times 1.666$

   $= 3 \times 1.666$

   $= 4.998$ mm

6. $WD = M \times 2.166$

   $= 3 \times 2.166$

   $= 6.498$ mm

7. Cutter number (see Table 62-3) $= 3$

EXAMPLE

Two identical gears in mesh have a CD of 120 mm. Each gear has 24 teeth. Calculate:

1. pitch diameter
2. module
3. outside diameter
4. whole depth
5. circular pitch
6. chordal thickness

1. $PD = \dfrac{2 \times CD}{2}$ (equal gears)

   $= \dfrac{2 \times 120}{2}$

   $= \dfrac{240}{2}$

   $= 120$ mm

2. $M = \dfrac{PD}{N}$

   $= \dfrac{120}{24}$

   $= 5$

3. $OD = (N + 2) \times M$

   $= 26 \times 5$

   $= 130$ mm

4. $WD = M \times 2.166$

   $= 5 \times 2.166$

   $= 10.83$ mm

5. $CP = M \times \pi$

   $= 5 \times 3.1416$

   $= 15.708$ mm

6. $CT = \dfrac{M \times \pi}{2}$

   $= \dfrac{5 \times 3.1416}{2}$

   $= 7.85$ mm

GEAR TOOTH MEASUREMENT

To ensure that the gear teeth are of the proper dimensions, they should be measured with a gear tooth vernier caliper. The caliper should be set to the corrected addendum, a dimension which may be found in most handbooks.

Gear sizes may also be accurately checked by measuring over wires or pins of a specific diameter, which have been placed in two diametrically opposite tooth spaces of the gear (Fig. 62-7A). For gears having an odd number of teeth, the wires are placed as nearly opposite as possible (Fig. 62-7B). A measurement taken over these wires is checked against tables found in most handbooks. These tables indicate the measurement over the wires for any gear having a given number of teeth and a specific pressure angle. Since these tables are far too extensive to be printed in this book, the reader is asked to refer to any handbook of them.

In order to measure inch gears accurately, the diametral pitch and number of teeth of the gear must be known. In order to measure metric gears, module must be known. The wire or pin size to be used is determined as follows:

1. For external inch spur gears the wire or pin size is equal to 1.728 divided by the diametral pitch of the gear.
2. For internal inch spur gears, the wire size is equal to 1.44 divided by the diametral pitch of the gear.
3. Metric module gears are measured using a wire size equal to 1.728 multiplied by module of the gear. The mea-

TABLE 62-3 Metric Module Gear Cutters			
MODULE SIZE, mm		MILLING CUTTER NUMBERS	
0.50	3.50	CUTTER NO.	FOR CUTTING
0.75	3.75	1	12-13 teeth
1.00	4.00	2	14-16 teeth
1.25	4.50	3	17-20 teeth
1.50	5.00	4	21-25 teeth
1.75	5.50	5	26-34 tooth
2.00	6.00	6	35-54 teeth
2.25	6.50	7	55-134 teeth
2.50	7.00	8	135 teeth to rack
2.75	8.00		
3.00	9.00		
3.25	10.00		

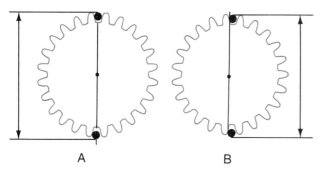

Fig. 62-7 Wires or pins are used to check gear sizes accurately. (Courtesy Kostel Enterprises Limited.)

surement over the wires should equal the value shown in the handbook tables multiplied by module.

EXAMPLE (inch)

Determine the wire size and the measurement over the wires for a 10-DP external gear having 28 teeth and a 14.50° pressure angle.

Wire size $= \dfrac{1.728}{10}$

$\qquad = 0.1728$ in.

By referring to handbook tables, the size over the wires for a gear having 28 teeth and a 14.50° pressure angle should be 30.4374 in. divided by the DP. Therefore the measurement over the wires should be

$$\dfrac{30.4374}{10} = 3.0437 \text{ in.}$$

If the measurement is larger than this size, the PD is too large and the depth of cut will have to be increased. If it is less than the determined size, the gear is undersize. Gears having an odd number of teeth are calculated in a similar manner but using the proper tables of these gears.

REVIEW QUESTIONS

1. What cutter numbers should be used to cut the following gears?
   (a) 8 DP–36 teeth
   (b) 12 DP–81 teeth
   (c) 16 DP–100 teeth
   (d) 6 DP–23 teeth
2. Describe two methods of centering the gear blank with the cutter when machining a spur gear.
3. What precautions should be observed when mounting the gear blank between the index head and centers?
4. What is the purpose of notching before gear teeth are cut?
5. Compare the terms *module* and *pitch diameter*.
6. How does the metric module numbering system for gear cutters differ from that for diametral pitch systems?
7. For a 40-tooth spur gear, 240 mm pitch diameter, calculate:
   (a) module
   (b) circular pitch
   (c) outside diameter
   (d) addendum
   (e) dedendum
   (f) whole depth
   (g) cutter number

GEAR TOOTH MEASUREMENT

8. State two methods of measuring gear teeth.
9. How is the wire size determined for:
   (a) external gears?
   (b) internal gears?

The process of milling helical grooves, such as flutes in a drill, teeth in helical gears, or the worm thread on a shaft, is known as *helical milling*. It is performed on the universal milling machine by gearing the dividing head through the worm shaft to the lead screw of the milling machine.

OBJECTIVES

After completing this unit you will be able to:
1. Calculate the lead and helix angle of a helical gear
2. Set up a milling machine to machine a helix
3. Make the calculations and setup for milling a helical gear

HELICAL TERMS

The term *spiral* is often used incorrectly in place of a *helix*.

A *helix* is a theoretical line or path generated on a *cylindrical* surface by a cutting tool which is fed lengthwise at a uniform rate, while the cylinder is also rotated at a uniform rate (Fig. 63-1). The flutes on a drill or the threads on a bolt are examples of helices.

A *spiral* is the path generated by a point moving at a fixed rate of advance along the surface of a *rotating cone or plane* (Fig. 63-2). Threads on a wood screw and pipe threads are examples of conical spirals, while watch springs and scroll threads on a universal lathe chuck are examples of plane or flat spirals.

In order to cut either an inch or metric helix, any two of the following must be known:

1. *The lead* of a helix is the longitudinal distance the helix advances axially in one complete revolution of the work.
2. *The angle of the helix* is formed by the intersection of the helix with the axis of the workpiece.

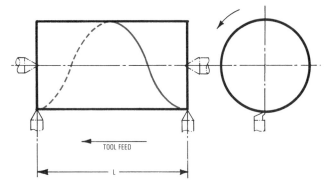

Fig. 63-1 A helix will be generated if the work is turned as the tool is moved along uniformly.

3. *The diameter (and circumference) of the workpiece.*

In comparing two different helices, it will be noticed that the greater the angle with the centerline, the shorter will be the lead. However, if the diameter is increased but the helix angle remains the same, the greater will be the lead. Thus it is evident that the lead of a helix varies with:

1. the diameter of the work
2. the angle of the helix

The relationship between the diameter (and circumference), the helix angle, and the lead is shown in Fig. 63-3. Note that if the surface of the cylinder could be unwound to produce a flat surface, the helix would form the hypotenuse of a right-angled triangle, with the circumference forming the side opposite and the lead the side adjacent.

CUTTING A HELIX

To cut a helix on a cylinder, the following steps are necessary:

1. Swing the table in the proper direction to the angle of the helix to ensure that a groove of the same contour as the cutter is produced.
2. The work must rotate one turn while the table travels lengthwise the distance equal to the lead. This is achieved by installing the proper change gears between the worm shaft on the dividing head and on the milling machine lead screw.

DETERMINING THE HELIX ANGLE

To ensure that a groove of the same contour as the cutter is produced, the table must be swung to the angle of the

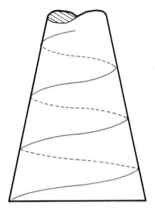

Fig. 63-2 A spiral is produced on a conical surface.

Fig. 63-3 Relationship of lead circumference and helix angle.

Fig. 63-5 The grooves of a right-hand helical cutter slope down and to the right.

A

DIRECTION OF TABLE FEED

B

Fig. 63-4 (A) When the table is swiveled to the helix angle, the exact profile of the cutter will be generated; (B) an incorrect angle produces an incorrect profile.

helix (Fig. 63-4A). The importance of this is shown in Fig. 63-4B.

Note that when the table is not swung (Fig. 63-4B), a helix having the proper lead but an improper contour will be generated. By referring to Fig. 63-3, it can easily be seen that the angle may be calculated as follows:

$$\text{Tangent of the helix angle} = \frac{\text{circumference of work}}{\text{lead of helix}}$$

$$= \frac{3.1416 \times \text{diameter } (D)}{\text{lead of helix}}$$

EXAMPLE (inch)

To what angle must the milling machine table be swiveled to cut a helix having a lead of 10.882 in. on a piece of work 2 in. in diameter?

$$\text{Tangent of helix angle} = \frac{3.1416 \times D}{\text{lead of helix}}$$

$$= \frac{3.1416 \times 2}{10.882}$$

$$= \frac{6.2832}{10.882}$$

$$= 0.57739$$

$$\text{Helix angle} = 30°$$

After the helix angle has been calculated, it is necessary to determine the *direction* in which to swivel the table to produce the proper hand of helix (that is, right or left hand).

EXAMPLE (metric)

To what angle must a milling machine table be swiveled to cut a helix having a lead of 450 mm on a workpiece 40 mm in diameter?

$$\text{Tangent of helix angle} = \frac{3.1416 \times D \text{ (mm)}}{\text{Lead of helix (mm)}}$$

$$= \frac{3.1416 \times 40}{450}$$

$$= 0.2792$$

$$= 15°36'$$

DETERMINING THE DIRECTION TO SWING THE TABLE

In order to determine the hand of a helix, hold the cylinder on which the helix is cut in a horizontal plane with its axis running in a right-left direction.

If the helix slopes *down* and to the right, it is a right-hand helix (Fig. 63-5). A left-hand helix slopes *down* and

Fig. 63-6   Worm shaft and the lead screw are connected for helical milling.

to the left. When a *left-hand helix* is to be cut, the table of the milling machine must be swiveled in a clockwise direction (operator standing in front of the machine). A right-hand helix may be produced similarly by moving the right end of the table in toward the column or by moving it in a counterclockwise direction.

CALCULATING THE CHANGE GEARS TO PRODUCE THE REQUIRED LEAD

To cut a helix, it is necessary to have the work move lengthwise and rotate at the same time. The amount the work (and table) travels lengthwise as the work revolves one complete revolution is the *lead*. The rotation of the work is caused by gearing the worm shaft of the dividing head to the lead screw of the machine (Fig. 63-6).

INCH CALCULATIONS

To cut a helix on an inch milling machine, it is necessary first to understand how to calculate the required change gears for any desired lead. Assume that the dividing head worm shaft is geared to the table lead screw with equal gears (for example, both having 24-tooth gears). The dividing head ratio is 40:1, while a standard milling machine lead screw has 4 threads per inch (tpi). The lead screw, as it revolves one turn, would revolve the dividing head spindle one-fortieth of a revolution. In order for the dividing head spindle to revolve one turn, it would be necessary for the lead screw to revolve 40 times. Thus the table would travel $40 \times \frac{1}{4}$ in. or 10 in. while the work revolves one turn. Therefore, the lead of a milling machine is said to be 10 in. when the lead screw (4 tpi) is connected to the dividing head (40:1 ratio) with equal gears.

In calculating the change gears required to cut any lead, the following formula may be used:

$$\frac{\text{Lead of helix}}{\text{Lead of machine (10 in.)}} = \frac{\text{product of driven gears}}{\text{product of driver gears}}$$

The ratio of gears required to produce any lead on a milling machine having a lead screw with 4 tpi is always equal to a fraction having the lead of the helix for the numerator and 10 for the denominator.

NOTE: The preceding formula may be inverted if preferred.

$$\frac{\text{Lead of the machine}}{\text{Lead of the helix}} = \frac{\text{product of driver gears}}{\text{product of driven gears}}$$

EXAMPLE

Calculate the change gears required to produce a helix having a lead of 25 in. on a piece of work. The available change gears have the following number of teeth: 24, 24, 28, 32, 40, 44, 48, 56, 64, 72, 86, 100.

SOLUTION

$$\text{Gear ratio} = \frac{\text{lead of helix (driven gears)}}{\text{lead of machine (driver gears)}}$$
$$= \frac{25}{10}$$

Since 10- and 25-tooth gears are not supplied with standard dividing heads, it is necessary to multiply the 25:10 ratio by any number that will suit the change gears available.

$$\text{Gear ratio} = \frac{25}{10} \times \frac{4}{4}$$
$$= \frac{100 \text{ (driven gear)}}{40 \text{ (driver gear)}}$$

As both 100-tooth and 40-tooth gears are available, simple gearing may be used.

EXAMPLE

Calculate the change gears required to produce a helix having a lead of 27 in. The available change gears are as in the above example.

SOLUTION

$$\text{Gear ratio} = \frac{\text{lead of helix (driven gears)}}{\text{lead of machine (driver gears)}}$$
$$= \frac{27}{10}$$

Since there are no gears in the set which are multiples of both 27 and 10, it is impossible to use simple gearing. Compounding gearing must therefore be used, and it becomes necessary to factor the fraction 27/10 as follows:

$$\text{Gear ratio} = \frac{27}{10}$$

$$= \frac{3}{2} \times \frac{9 \text{ (driven)}}{5 \text{ (driver)}}$$

It is now necessary to multiply both the numerator and denominator of each fraction by the same number in order to bring the ratio into the range of the gears available.

NOTE: This does not change the value of the fraction.

$$\frac{3 \times 16}{2 \times 16} = \frac{48}{32}$$

$$\frac{9 \times 8}{5 \times 8} = \frac{72}{40}$$

$$\text{The gear ratio} = \frac{48 \times 72 \text{ (driven gears)}}{32 \times 40 \text{ (driver gears)}}$$

∴The driven gears are 48 and 72 and the driver gears are 32 and 40.

The gears would be placed in the train as follows (Fig. 63-6):

Gear on worm	72	(driven)
First gear on stud	32	(driver)
Second gear on stud	48	(driven)
Gear on lead screw	40	(driver)

The preceding order is not absolutely necessary; the two driven gears may be interchanged and/or the two driver gears may be interchanged, *provided a driver is not interchanged with a driven gear.*

METRIC CALCULATIONS

The pitch of the lead screw on a metric milling machine is stated in millimeters. Most milling machine lead screws have a 5-mm pitch and the dividing head has a ratio of 40:1. As the lead screw revolves one turn, it would revolve the dividing head spindle one-fortieth of a revolution. In order for the dividing head spindle (and work) to revolve one full turn, the lead screw must make 40 complete revolutions. Therefore the lead of the machine would be 40 times the pitch of the lead screw.

For metric calculations, the change gears required are calculated as follows:

$$\frac{\text{Lead of helix (mm)}}{\text{Lead of machine (mm)}} = \frac{\text{product of driven gears}}{\text{product of driver gears}}$$

The normal change gears in a set are 24, 24, 28, 32, 36, 40, 44, 48, 56, 64, 72, 86, 100.

EXAMPLE

Calculate the change gears required to cut a helix having a lead of 500 mm on a workpiece using a standard set of gears. The milling machine lead screw has a pitch of 5 mm.

$$\frac{\text{Driven gears}}{\text{Driver gears}} = \frac{\text{lead of helix}}{\text{pitch of lead screw} \times 40}$$

$$= \frac{500}{5 \times 40}$$

Fig. 63-7 A second idler reverses the direction of rotation.

$$= \frac{500}{200}$$

$$= \frac{5}{2} \times \frac{20}{20}$$

$$= \frac{100}{40}$$

Driven gear = 100
Driven gear = 40

DIRECTION OF SPINDLE ROTATION

Figure 63-6 illustrates the setup required to cut a left-hand helix on most machines. Note that the gear on the lead screw and the worm gear revolve in the same direction. To cut a right-hand helix, the spindle must revolve in the opposite direction, and therefore another idler must be inserted as in Fig. 63-7. The idler in this case acts neither as a driven nor a driver gear and is not considered in the calculation of the gear train. It acts merely as a means of changing the direction of rotation of the dividing head spindle. It should also be noted that the direction of spindle rotation for simple gearing will be opposite to that for compound gearing.

CUTTING SHORT LEAD HELICES

When it is necessary to cut leads smaller than those shown in most machinery handbooks, it is advisable to disengage the dividing head worm and wormwheel and connect the change gears directly from the table lead screw to the dividing head spindle rather than to the worm shaft. This method permits machining leads to one-fortieth of the leads shown in the handbook tables. Thus, if the machine is geared to cut a lead of 4.000 in., by connecting the worm shaft and the lead screw, the same gearing would produce a lead of 1/40 × 4.000 in., or 0.100 in. when geared directly to the dividing head spindle.

PROBLEM

A plain helical milling cutter is required to the following specifications:

Diameter: 4 in.
Number of teeth: 9
Helix: right hand
Helix angle: 25°
Rake angle: 10° positive radial rake
Angle of flute: 55°
Depth of flute: ½
Length: 4 in.
Material: tool steel

PROCEDURE

1. Turn blank to sizes indicated (Fig. 63-8).
2. Apply layout die to the end of the blank, and lay out as in Fig. 63-9.
3. Lay out a line on periphery to indicate direction of the *right-hand* helix (Fig. 63-10).
4. Press the cutter blank firmly on the mandrel. If a threaded mandrel is used, be sure to tighten the nut securely.
5. Mount the dividing head and footstock.
6. Calculate the indexing for nine divisions.

$$\text{Indexing} = \frac{40}{9}$$
$$= 4\frac{4}{9}$$
$$= 4 \text{ turns, 8 holes on an 18-hole circle}$$

7. Set the sector arms to 8 holes on the 18-hole circle.

NOTE: Do not count the hole in which the pin is engaged.

8. Disengage the index plate locking device.
9. Calculate the lead of the helix.

$$\text{Lead} = \frac{3.1416 \times D}{\tan \text{ helix angle}}$$
$$= 3.1416 \times D \cot \text{ helix angle}$$
$$\left(\text{since } \frac{1}{\tan} = \cot\right)$$
$$= 3.1416 \times 4 \times 2.1445$$
$$= 26.949 \text{ in.}$$

10. Consult any handbook for the change gears to cut the lead closest to 26.949 in. Obviously this is 27.
11. If a handbook is not available, change gears can be calculated for the closest lead, which is 27 in.
12. Change gears required for 27-in. lead.

$$\frac{\text{Required lead}}{\text{Lead of machine}} = \frac{27}{10}$$
$$= \frac{9}{5} \times \frac{3}{2}$$

$$\frac{9 \times 8}{5 \times 8} = \frac{72}{40} \qquad \frac{3 \times 16}{2 \times 16} = \frac{48}{32}$$

$$\text{Change gears} = \frac{72 \times 48 \text{ (driven gears)}}{40 \times 32 \text{ (driver gears)}}$$

13. Mount the change gears, allowing a slight clearance between mating teeth.
14. Mount the work between the centers with the large end of the mandrel against the dividing head.
15. Swivel the table 25° in counterclockwise direction.
16. Adjust the crossfeed handwheel until the table is about 1 in. from the face of the column. This is to ensure that the table clears the column when machining the cutter.
17. Swing the table back to 0.
18. Mount a 55° double-angle cutter so that it revolves toward the dividing head, and center it approximately over the flute layout.
19. Rotate the blank until the flute layout is aligned with

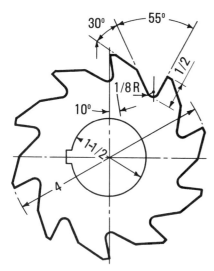

Fig. 63-8 Dimensions of a helical milling cutter.

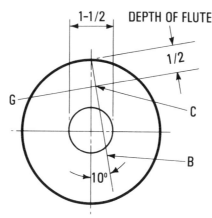

Fig. 63-9 Locating the first tooth on the cutter. (Courtesy Cincinnati Milacron Incorporated.)

Fig. 63-10 Laying out the direction of the flute. (Courtesy Cincinnati Milacron Incorporated.)

the cutter edge. This may be checked with a rule or straight-edge (Fig. 63-11).

20. Move the blank over, using the crossfeed for the distance of M (Fig. 63-11) or until the point C (Fig. 63-9) is in line with the centerline of the cutter.

21. With the work clear of the cutter, set the depth to 0.500 in.

22. Rotate the table 25° and lock securely (right end in toward the column).

23. Carefully cut the first tooth space, checking the accuracy of the location and the depth.

24. Index for and cut the remaining flutes.

25. Remove the fluting cutter and mount a plain helical milling cutter.

26. Rotate the work (using the index crank) until a line at 30° to the side of the flute is parallel to the table (Fig. 63-9). This may be checked by means of a surface gage. The blank, however, may be rotated by indexing an amount equal to

$$90 - (30 + \frac{55}{2})$$

$$= 90 - 57.5$$
$$= 32.5° \ (32°30')$$

Indexing for 32°30':

$$32° \times 60' = 1920'$$
$$30' = 30'$$
$$32°30' = \overline{1950'}$$

$$= \frac{1950}{540}$$

$$= \frac{330}{540}$$

$$= 3 \ \frac{11}{18}$$

$$= 3 \text{ turns} + 11 \text{ holes on} \\ \text{the 18-hole circle}$$

27. Adjust the workpiece under the cutter.

28. With the cutter rotating, raise the table until the width of the land on the workpiece is about 1/32 in. wide.

29. Cut the secondary clearance (30° angle) on all teeth of the workpiece.

HELICAL MILLING CALCULATIONS (METRIC)

For any helical milling calculation is it necessary to determine:

1. The angle at which the table must be swiveled to produce the proper helix angle
2. The change gears required to revolve the work one turn as the work travels the distance of the lead

For metric helices, these calculations are as follows:

1. Helix angle or angle to which to swivel the table

$$\text{Tangent } \angle = \frac{\text{circumference of workpiece}}{\text{lead of helix}}$$

2. Change gears required

$$\frac{\text{Driven gears}}{\text{Driver gears}} = \frac{\text{lead of helix}}{\text{pitch of lead screw} \times 40}$$

The normal change gears in a set are: 24, 24, 28, 32, 36, 40, 44, 48, 56, 64, 72, 86, 100.

EXAMPLE

A lead of 480 mm is to be cut on a workpiece 40 mm in diameter. The lead screw of the milling machine has a pitch of 5 mm. The dividing head has a ratio of 40:1 (40 turns of the crank are required to revolve the work one turn). Calculate the angle at which to set the table and the change gears necessary to produce the required lead:

1. Helix angle

$$\text{Tan helix angle} = \frac{\text{circumference of work}}{\text{lead of helix}}$$

$$= \frac{40 \times 3.1416}{480}$$

$$= 0.26180$$

$$\text{Helix angle} = 14°40'$$

2. Change gears

$$\frac{\text{Driven gears}}{\text{Driver gears}} = \frac{\text{lead of helix}}{\text{pitch of lead screw} \times 40}$$

$$= \frac{480}{5 \times 40}$$

$$= \frac{480}{200}$$

$$= \frac{12}{5} \ \frac{(6 \times 2)}{(5 \times 1)}$$

$$\frac{6}{5} \times \frac{8}{8} = \frac{48}{40} \qquad \frac{2}{1} \times \frac{28}{28} = \frac{56}{28}$$

$$\text{Gears} = \frac{48}{40} \times \frac{56}{28}$$

$$\text{Driven gears} = 48 \text{ and } 56$$
$$\text{Driver gears} = 40 \text{ and } 28$$

HELICAL GEARING

Helical gearing requires a thorough knowledge of all the aspects of spur gearing, as well as the formulas and pro-

Fig. 63-12 Relationship between the circular pitch and the normal circular pitch.

cedures used in helical milling. In helical gear cutting two new terms are encountered: *normal circular pitch (NCP)* and *normal diametrical pitch (NDP)*.

The NCP is the distance from a point on one tooth to a corresponding point on the next tooth, measured on the pitch circle at right angles to the face of the tooth. The *circular pitch (CP)* is measured on the face of the gear in a plane at right angles to the axis of the gear. Figure 63-12 illustrates the relationship between the CP and the NCP. It will be noted that a right-angle triangle is formed, with a line representing the CP as the hypotenuse and a line representing the NCP as the side opposite. By simple geometry, it may be proved that the angle a is equal to the angle a_1, which is the angle of the helix. Therefore, in the triangle illustrated in Fig. 63-12,

$$\text{Angle } a = \frac{NCP}{CP}$$

Therefore, the relationship between NCP and CP is exactly proportional to the length of these lines. The number of teeth in a helical and a spur gear of the same size and pitch will also be proportional to the length of these lines.

It is obvious that the CP will increase as the helix angle increases; therefore, the greater the helix angle, the fewer the teeth in a helical gear as compared to a spur gear of the same pitch diameter and diametral pitch (DP).

Since most gearing is calculated on the DP system, it will be necessary to convert the CP into DP terms. Referring to the spur gearing formulas,

$$DP = \frac{3.1416}{CP} \quad \text{or} \quad CP = \frac{3.1416}{DP}$$

Since *normal diametral pitch* (NDP) bears the same relation to NCP as DP does to CP, the following formulas will apply:

$$NDP = \frac{3.1416}{NCP}$$

$$NCP = \frac{3.1416}{NDP}$$

Referring to Fig. 63-12, note that

$$\text{Cos helix angle} = \frac{NCP}{CP}$$

This may be converted to DP as follows:

$$\text{Cos helix angle} = \frac{NCP}{CP}$$

$$= \frac{(3.1416/NDP)}{(3.1416/DP)}$$

$$= \frac{3.1416}{NDP} \times \frac{DP}{3.1416}$$

$$= \frac{DP}{NDP}$$

$$\therefore \text{ Cos helix angle} = \frac{DP}{NDP}$$

Other formulas may be derived as follows:

$$NDP = \frac{DP}{\cos \text{ helix angle}}$$

$$DP = NDP \times \cos \text{ helix angle}$$

CUTTER SELECTION

The type of cutter used for cutting helical gears is the same as for spur gears. However, the DP of a spur gear and the cutter now becomes the NDP of the helical gear. The thickness of the cutter at the pitch line for cutting helical gears should be equal to one-half the NCP.

When a spur gear is being cut, the number of the cutter used is dependent on the number of teeth being cut (Table 62-1). However, when a helical gear is being cut, this table does not apply since the shape of the tooth would be hanged because of the helix angle of the gear teeth. To determine the proper cutter for machining a helical gear, it is necesary to consider the number of teeth being cut and the helix angle of the tooth. The cutter number may be determined by dividing the number of teeth (N) by the cube of the cosine of the angle. Thus

$$\text{Cutter number} = \frac{N}{(\cos a)^3}$$

EXAMPLE

Determine the cutter number required to cut 38 teeth on a helical gear having an angle of 45°.

$$\begin{aligned}
\text{Number of teeth for} \\
\text{which to select the cutter}
\end{aligned} = \frac{N}{(\cos a)^3}$$

$$= \frac{38}{(0.7071)^3}$$

$$= \frac{38}{0.3534}$$

$$= 108$$

Therefore, a #2 cutter would be used since it cuts from 55 to 134 teeth.

HELICAL GEAR CALCULATIONS AND FORMULAS

Many of the calculations and formulas that apply to spur gears also apply to helical gears. Others may be calculated from Table 63-1 as required. It is well to remember the following when making helical gears:

1. The DP of a spur gear is the NDP of a helical gear.
2. The NDP of the helical gear is the DP of the cutter
3. The cutter number must be calculated using the formula:

Number of teeth for which to select a cutter $= \dfrac{N}{(\cos a)^3}$

The alignment chart (Table 63-2) will be found convenient when it is necessary to determine cutter numbers for helical gears.

By means of a straightedge, align the actual number of teeth in the left-hand column with the helix angle in the right-hand column. Read the cutter number from the center column where the straightedge crosses the line.

TABLE 63-1 Rules for Calculating Inch Helical Gears

NO. OF RULE		TO FIND RULE	FORMULA
1	Addendum (A)	Divide by the normal diametral pitch.	$A = \dfrac{1}{NDP}$
2	Cutter number (CN)	Divide the number of teeth by the cube of the cosine of the helix angle.	$CN = \dfrac{N}{(\cos \angle)^3}$
3	Helix angle of teeth ($\angle$)	Divide the normal circular pitch by the circular pitch. The quotient equals the cosine of the helix angle.	$\cos \angle = \dfrac{NCP}{CP}$
4	Helix angle of teeth	Divide the diametral pitch by the normal diametral pitch. The quotient equals the cosine of the helix angle.	$\cos \angle = \dfrac{DP}{NDP}$
5	Helix angle of teeth	Divide the number of teeth by the product of the normal diametral pitch and the pitch diameter. The quotient equals the cosine of the helix angle.	$\cos \angle = \dfrac{N}{NDP \times PD}$
6	Lead of the helix ($\angle$)	Divide the product of the number of teeth and the circular pitch by the tangent of the helix angle.	$Lead = \dfrac{N \times CP}{\tan \angle}$
7	Lead of the helix	Multiply the pitch diameter by 3.1416 times the cotangent of the helix angle.	$L = 3.1416 \times PD \cot \angle$
8	Normal diameter pitch (NDP)	Divide the number of teeth by the cosine of the helix angle, add 2, and divide the sum by the outside diameter.	$NDP = \left(\dfrac{N}{\cos \angle} + 2\right) \div OD$
9	Normal circular pitch (NCP)	Multiply the circular pitch by the cosine of the helix angle.	$NCP = CP\cos \angle$
10	Outside diameter of gear blank (OD)	Add 0.6366 times the normal circular pitch to the pitch diameter.	$OD = 0.6366 \times NCP + PD$
11	Pitch diameter	Multiply the product of the circular pitch and the number of teeth by 0.3183 and divide the product by the cosine of the helix angle.	$PD = \dfrac{0.3183 \times CP \times N}{\cos \angle}$
12	Pitch diameter	Divide the number of teeth by the product of the normal diameter pitch and the cosine of the helix angle.	$PD = \dfrac{N}{NDP \cos \angle}$
13	Thickness of tooth at pitch line (T)	Divide 1.571 by the normal diametral pitch.	$T = \dfrac{1.571}{NDP}$
14	Whole depth of tooth (WD)	Divide 2.157 by the normal diametral pitch.	$WD = \dfrac{2.157}{NDP}$

TABLE 63-2 Helical Gear Cutter (Inch) Nomograph

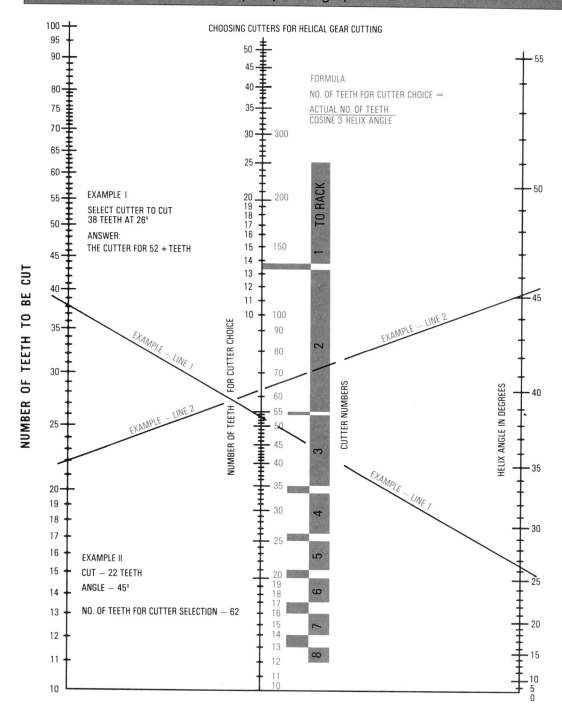

CHOOSING CUTTERS FOR HELICAL GEAR CUTTING

FORMULA

NO. OF TEETH FOR CUTTER CHOICE =

$$\frac{\text{ACTUAL NO. OF TEETH}}{\text{COSINE 3 HELIX ANGLE}}$$

EXAMPLE I

SELECT CUTTER TO CUT
38 TEETH AT 26°

ANSWER:
THE CUTTER FOR 52 + TEETH

EXAMPLE II

CUT — 22 TEETH

ANGLE — 45°

NO. OF TEETH FOR CUTTER SELECTION — 62

Metric Helical Gear Calculations

Most metric module helical gear calculations are similar to those for DP helical gears and module spur gears. There are, however, certain changes which must be understood before any calculations are attempted. Refer to Fig. 63-12 to more clearly understand the following terms.

The *circular pitch* (CP) is the distance from one tooth to a corresponding point on the next tooth, measured on the

pitch circle. The *real (pitch) module (M)* is calculated on the pitch circle.

The *normal circular pitch* (NCP) is the distance from one tooth to a corresponding point on the next tooth measured at right angles to the tooth face. The *normal (pitch) module* (nM) is calculated on this distance.

For all other calculations, see Unit 62 on metric spur gears and helical milling. Table 63-3 covers the rules and formulas for calculating metric module helical gears.

TABLE 63-3 Rules and Formulas for Module Helical Gears

TO OBTAIN	KNOWING	RULE	FORMULA
Addendum (A)	Normal module	Addendum equals normal module	$A = nM$
Dedendum (D)	Normal module	Multiply normal module by 1.166	$D = nM \times 1.166$
Circular pitch (CP)	Module	Multiply the module by π	$CP = M \times 3.1416$
	Normal circular pitch and helix angle	Divide the normal circular pitch by the cosine of the helix angle	$CP = \dfrac{NCP}{\text{Cos } \angle}$
	Normal module and helix angle	Multiply the normal module by π and divide the product by the cosine of the helix angle.	$CP = \dfrac{nM \times 3.1416}{\text{Cos } \angle}$
Normal Circular Pitch (NCP)	Normal module	Multiply normal module by π	$NCP = nM \times 3.1416$
	Circular pitch and helix angle	Multiply the circular pitch by the cosine of the helix angle.	$NCP = CP \times \text{Cos } \angle$
	Pitch diameter, helix angle, and number of teeth	Multiply the pitch circumference by the cosine of the helix angle and divide the product by the number of teeth.	$NCP = \dfrac{PD \times 3.1416 \times \text{Cos } \angle}{N}$
Helix angle of Teeth ($\angle$)	Normal module and module	Normal module divided by the module equals the cosine of the helix angle.	$\text{Cos } \angle = \dfrac{nM}{M}$
	Number of teeth, normal module, and pitch diameter	The number of teeth multiplied by the normal module and divided by the pitch diameter equals the cosine of the helix angle.	$\text{Cos } \angle = \dfrac{N \times nM}{PD}$
Lead of helix (L)	Pitch diameter and helix angle	Multiply the circumference of the pitch circle by the cotangent of the helix angle.	$L = PD \times 3.1416 \times \text{Cot } \angle$
	Number of teeth, module and helix angle	Multiply the number of teeth by the module times 3.1416. Multiply the product by the cotangent of the helix angle.	$L = N \times M \times 3.1416 \times \text{Cot } \angle$
Normal module (nM)	Module and helix angle	Multiply the module by the cosine of the helix angle.	$nM = M \times \text{Cos } \angle$
	Normal circular pitch	Divide the normal circular pitch by π.	$nM = \dfrac{NCP}{3.1416}$
	Circular pitch and helix angle	Multiply the circular pitch by the cosine of the helix angle and divide the product by π.	$nM = \dfrac{CP \times \text{Cos } \angle}{3.1416}$
Module (M)	Normal module and helix angle	Divide normal module by the cosine of the helix angle.	$M = \dfrac{nM}{\text{Cos } \angle}$
	Circular pitch	Divide the circular pitch by π.	$M = \dfrac{CP}{3.1416}$
	Normal circular pitch and helix angle	Divide the normal circular pitch by π times the cosine of the helix angle.	$M = \dfrac{NCP}{3.1416 \times \text{Cos } \angle}$

(Cont. on next page)

	TABLE 63-3 (Continued)		
TO OBTAIN	**KNOWING**	**RULE**	**FORMULA**
Outside diameter (OD)	Pitch diameter and normal module	Add 2 normal modules to the pitch diameter.	$OD = PD + 2nM$
	Normal module, number of teeth, and helix angle	Divide the number of teeth by the cosine of the helix angle and add 2 to the quotient. Multiply this number by the normal module.	$OD = nM \left(\dfrac{N}{\cos \angle} + 2 \right)$
Pitch diameter (PD)	Normal module and number of teeth	Multiply the number of teeth by the normal module and divide the product by the cosine of the helix angle.	$PD = \dfrac{N \times nM}{\cos \angle}$
Number of teeth (N)	Pitch diameter and circular pitch	Multiply the pitch diameter by π and divide the product by the circular pitch.	$N = \dfrac{PD \times 3.1416}{CP}$
	Pitch diameter and module	Divide the pitch diameter by the module.	$N = \dfrac{PD}{M}$
	Pitch diameter, normal module, and helix angle	Multiply the pitch diameter by the cosine of the helix angle and divide the product by the normal module.	$N = \dfrac{PD \times \cos \angle}{nM}$
Whole depth of tooth (WD)	Normal module	Multiply the normal module by 2.166	$WD = nM \times 2.166$

EXAMPLE

It is required to cut a helical metric 5 (normal) module gear having 38 teeth. The pitch diameter (PD) of the gear is 200 mm. Calculate:

1. Real module (*M*)
2. Normal module (nM)
3. Circular pitch (CP)
4. Normal circular pitch (NCP)
5. Addendum (*A*)
6. Outside diameter (OD)
7. Depth of tooth (WD)
8. Helix angle ($\angle$)
9. Lead (*L*)
10. Indexing

SOLUTION

1. $\quad M = \dfrac{200}{38}$
   $\quad\quad = 5.26$ mm

2. $\quad nM = 5$

3. $\quad CP = M \times 3.1416$
   $\quad\quad = 5.26 \times 3.1416$
   $\quad\quad = 16.52$ mm

4. $\quad nCP = nM \times 3.1416$
   $\quad\quad = 5 \times 3.1416$
   $\quad\quad = 15.70$ mm

5. $\quad A = nM$
   $\quad\quad = 5$ mm

6. $\quad OD = PD + 2nM$
   $\quad\quad = 200 + 10$
   $\quad\quad = 210$ mm

7. $\quad WD = nM \times 2.166$
   $\quad\quad = 10.83$ mm

8. $\quad \cos \angle = \dfrac{nM}{M}$
   $\quad\quad = \dfrac{5}{5.2632}$
   $\quad\quad = 0.950$
   $\quad\quad = 18°11'$

9. $\quad L = PD \times 3.1416 \times \cot \angle$
   $\quad\quad = 200 \times 3.1416 \times 3.0445$
   $\quad\quad = 1913$ mm

10. Indexing $= \dfrac{40}{N}$
    $\quad\quad = \dfrac{40}{38}$

$$= 1\text{-}2/38$$
$$= 1\text{-}1/19$$
= 1 turn of crank + 1 hole in the 19-hole circle

TO MILL AN INCH HELICAL GEAR

The following problem will serve as a guide in making helical gear calculations. If the steps are followed in sequence, little difficulty should be encountered.

PROBLEM

It is required to cut two mating helical gears having a 3:2 ratio, 12 DP pitch, and 3.000-in. center-to-center distance.

Calculations

Pitch Diameters The sum of the PDs is 3.000 in. $\times$ 2 = 6.000 in. Since the total of the ratios is 3 + 2 = 5, the large gear will occupy three-fifths of the total distance of the pitch diameters. (6.000 in.), while the pinion will be two-fifths of the total distance.

$$\text{PD of gear} = 3/5 \times 6.0 = 3.600$$
$$\underline{\text{PD of pinion} = 2/5 \times 6.0 = 2.400}$$
$$6.000 \text{ in.}$$
(proof)

Number of Teeth

$$N = DP \times PD$$
$$N \text{ in gear} = 3.6 \times 12$$
$$= 43.2$$
$$N \text{ in pinion} = 2.4 \times 12$$
$$= 28.8$$

Since it is impossible to cut 43.2 or 28.8 teeth, it is necessary to change the number of teeth in each gear to two whole numbers which have a 3:2 ratio. The teeth of the gear are cut at an angle, and there will be fewer teeth in a helical gear than in a corresponding spur gear. It is therefore necessary to select two whole numbers which are *less* than 43.2 and 28.8 and which have a ratio of 3:2.

The largest numbers under 43.2 and 28.8 having this ratio are 42 and 28.

Therefore:

$$N \text{ in gear} = 42$$
$$N \text{ in pinion} = 28$$

Diametral Pitch

$$DP = \frac{N}{PD}$$

$$DP \text{ of gear} = \frac{42}{3.6}$$
$$= 11.666 \text{ teeth}$$

OR

$$DP \text{ of pinion} = \frac{28}{2.4}$$
$$= 11.666 \text{ teeth (proof)}$$

It has now been established that the DP, or the number of teeth per inch of PD of each gear, is 11.666. It is, however, required to use a 12-DP cutter which will cut 12 teeth per inch of PD. This will necessitate swinging the table to the proper helix angle to cut the gears.

Helix Angle

$$\text{Cos helix angle} = \frac{DP}{NDP}$$
$$= \frac{11.666}{12}$$
$$= 0.97216$$
$$\therefore \text{ helix angle} = 13°33'$$

Outside Diameter

$$OD = PD + \frac{2}{DP}$$

$$OD \text{ of gear} = 3.600 + \frac{2}{12}$$
$$= 3.766 \text{ in.}$$

$$OD \text{ of pinion} = 2.400 + \frac{2}{12}$$
$$= 2.566 \text{ in.}$$

Leads

$$\text{Lead} = PD \times 3.1416 \cot \text{ helix angle}$$

$$\text{Lead of gear} = 3.6 \times 3.1416 \cot 13°33'$$
$$= 3.6 \times 3.1416 \times 4.1493$$
$$= 46.926 \text{ in.}$$

$$\text{Lead of pinion} = 2.4 \times 3.1416 \times 4.1493$$
$$= 31.285 \text{ in.}$$

Change Gears The change gears may be determined from the tables supplied with the dividing head or most machinery handbooks. To determine the change gears for the gear having a lead of 46.926 in., check a handbook for the closest lead which is 46.880 in. The gears for this lead are as follows:

$$\frac{\text{Driven gears}}{\text{Driver gears}} = \frac{100 \times 72}{32 \times 48}$$

The error in this lead is 46.926 − 46.880 = 0.046 in. in approximately 47 in. Change gears for pinion having a lead of 31.285 in.; the closest lead is 31.270 in.

$$\frac{\text{Driven gears}}{\text{Driver gears}} = \frac{86 \times 64}{40 \times 44}$$

The error in this lead is 31.285 − 31.270 = 0.015 in. in approximately 31 in.

NOTE: If a chart or handbook is not available, the change gears may be calculated by *continued fractions* which are dealt with later in this Unit where the examples for this pinion are shown.

$$\text{Number of teeth in gear for which to select a cutter} = \frac{\text{number of teeth in gear}}{(\cos \text{ helix angle})^3}$$

$$= \frac{42}{(0.97216)^3}$$

$$= 45.71 \text{ teeth}$$

A #3 cutter would be used since it will cut from 35 to 54 teeth.

$$\text{Number of teeth in pinion for which to select a cutter} = \frac{28}{(0.97216)^3}$$

$$= 30.47 \text{ teeth}$$

A #4 cutter would be selected since it will cut from 26 to 34 teeth.

The alignment chart (Table 63-2) may be used to determine the proper cutter number. This will eliminate the need to calculate the required number of the cutter.

CUTTING THE GEARS

The setup for cutting the gears is exactly the same as for helical milling. Having made all the necessary calculations for the gear and the pinion, list all the pertinent information, including the depth of tooth and indexing for both gears, on a separate sheet of paper as follows.

PROCEDURE

1. Mount the dividing head at the end of the milling machine table.
2. Mount the work between centers.
3. Set the indexing crank and the sector arms.
4. Disengage the index plate locking device.
5. Mount the change in gears as required.
6. Swing the table to the helix angle and in the proper direction.

NOTE: When cutting mating helical gears for parallel shafts, it is always necessary to cut one gear with a left-hand helix and the other with a right-hand helix so that the gears will mesh.

7. Move the table, using the crossfeed handle, to within 1 in. (25.4 mm) of the column to ensure clearance between the table and column.
8. Swing the table back to zero.
9. Mount the cutter on the arbor over the approximate center of the work.
10. Center the work with the cutter.
11. Raise the table until the cutter just touches the work.
12. Set the vertical feed screw collar to zero.
13. With the work clear of the cutter, raise the table to the proper depth of the tooth.
14. Swing the table in the proper direction.

NOTE: When the table is swung in the other direction to cut the opposite helix, it will be necessary to add an idler

to the gear train in order to reverse the direction of rotation of the work.

15. Nick each tooth to ensure the proper indexing.
16. Cut all teeth.

NOTE: Lower the table one full turn of the handwheel before returning it to the starting position to overcome backlash and so that the cutter will not damage the finish on the teeth.

CONTINUED FRACTIONS

Most handbooks contain a table of inch change gears for milling various inch leads. If there are no tables available or if the lead is not within the range shown in the tables, the change gears may be calculated by means of *continued fractions* or *successive quotients*.

To calculate the change gears by this method, it is necessary to form a fraction using the required lead and the lead of the machine.

PROBLEM

Calculate the gears required to produce a lead of 5.189 in. on a helical gear. The lead of the milling machine is 10 in. The available change gears are as follows: 24, 24, 28, 32, 40, 44, 48, 56, 64, 72, 86, 100.

PROCEDURE

1. $\text{Gear ratio} = \dfrac{\text{required lead}}{\text{lead of machine}}$

$$= \frac{\text{product driven}}{\text{product drivers}}$$

$$= \frac{5.189}{10}$$

$$= \frac{5189}{10000}$$

2. Since this fraction cannot be factored, the required gears may be obtained by dividing the smaller number into the larger number to obtain the successive quotients. Continue to divide the remainder into the previous divisor until the remainder is zero (Fig. 63-13).

3. Arrange the successive quotients on a chart (Fig. 63-14). In the left-hand columns, place the figures 1 and 0 as in Fig. 63-15. The numbers are arranged in this way only when the gear ratio is a proper fraction, that is 5189/10,000, or when the lead of the helix is less than the lead of the machine.

If the ratio is an improper fraction, such as that produced by the lead required for the larger helical gear, which has a lead greater than the lead of the machine, as in the foregoing problem 15,566/10,000, the numbers must be arranged as in Fig. 63-16.

4. After the numbers have been arranged on the chart, the convergents are found by multiplying each quotient *(A)* by

Helical Gear Data		
	GEAR	PINION
DP	12	12
PD	3.600 in.	2.400 in.
OD	3.766 in.	2.566 in.
N	42	28
Cutter number	3	4
Indexing (Cincinnati plate)	20 holes on 21-hole circle	1 turn 9 holes on 21-hole circle
Depth of tooth	0.1797 in.	0.1797 in.
Helix angle	13°33′ (right-hand)	13°33′ (left-hand)
Direction to swing table	counterclockwise	clockwise
Lead	46.926 in.	31.285 in.
Change gears—Worm gear	100	86
First stud gear	32	40
Second stud gear	72	64
Lead screw gear	48	44

Fig. 63-13

Fig. 63-14

Fig. 63-15

Fig. 63-16

the number in the box *(B)*, the upper row of convergents, which is located below and one square to the left of the quotient. To this product add the number in the next box to the left *(C)* (Fig. 63-17). The result is then placed in the convergent box below the quotient used *(A)*. For example,

$$A \times B + C \text{ (in the upper section)} = 1 \times 0 + 1$$
$$= 1$$

5. Move to the next quotient in the upper row and follow the same procedure.

NOTE: Always multiply the number in the upper row by the number in the box below and to the left. To this product, add the value of the next box (second row) to the left. For example,

$$12 \times 1 + 1 = 13$$

which would be placed below the 12 (Fig. 63-18). Continue this process until the original number (5189) is arrived at (Fig. 63-19).

6. Follow the same procedure for the lower set of convergents which should end with the original number (10,000). The calculation is proved correct when the original number appears in the runs of convergents. Examination of Fig. 63-19 reveals several sets of fractions in the convergent squares. These are really ratios which become more accurate as they move to the right until the original numbers finally appear.

7. Starting at the right, examine each ratio until one is found that will factor into a set of numbers which can be fitted into the range of change gears available. In this case (Fig. 63-19), the ratio or fraction is 14/27.

8. Factor this fraction as follows:

$$\frac{14}{27} = \frac{7 \times 2}{9 \times 3}$$

9. These fractions may then be multiplied by a fraction having the value of 1 to suit the change gears available.

$$\frac{7}{9} \times \frac{8}{8} = \frac{56}{72}$$

$$\frac{2}{3} \times \frac{16}{16} = \frac{32}{48}$$

Therefore, the gears used to cut a lead reasonably close to 5.189 in. are:

$$\frac{56 \times 32}{72 \times 48} = \frac{\text{driven gears}}{\text{driver gears}}$$

It should be noted that the further to the right that the factorable fraction lies, the more accurate will be the lead produced by the gears.

CALCULATING THE ERROR IN THE LEAD

To calculate the error in the lead produced by these gears, proceed as follows.

$$\text{Lead produced} = \frac{56}{72} \times \frac{32}{48}$$
$$= \frac{14}{27} \times 10 \text{ (lead of machine)}$$
$$= 5.1895 \text{ in.}$$
$$\text{Desired lead} = 5.189$$
$$\text{Error} = 5.189 - 5.185$$
$$= 0.004 \text{ in.}$$

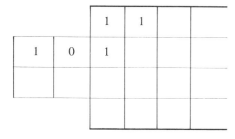

Fig. 63-17

Fig. 63-18

	1	1	12	1	2	1	2	34	QUOTIENTS
1	0	1	1	13	14	41	55	151	5189	CONVERGENTS
0	1	1	2	25	27	79	106	291	10000	CONVERGENTS
	1	1	12	1	2	1	2	34	QUOTIENTS

Fig. 63-19

REVIEW QUESTIONS

HELICAL MILLING

1. Define:
   - (a) helix
   - (b) spiral
   - (c) lead
   - (d) angle of helix
2. Make a sketch to illustrate the relationship between the lead, circumference, and helix angles.
3. List two factors which affect the lead of a helix.
4. To what angle must the table be swung to cut the following helices?
   - (a) lead 10.290 in., diameter of workpiece 3¼ in.
   - (b) lead 12.000 in., diameter of workpiece 2¾ in.
   - (c) lead = 600 mm, work diameter = 100 mm
   - (d) lead = 232 mm, work diameter = 25 mm
5. How may a right- and left-hand helix be recognized, and in which direction should the table be swiveled for each?
6. (a) Calculate the change gears to cut the following leads:
   1. 6.000 in.
   2. 7.500 in.
   3. 9.600 in.

   (b) The lead screw of a milling machine has a pitch of 5 mm. The available change gears are 24, 24, 28, 32, 36, 40, 44, 48, 56, 64, 72, 86, 100. Calculate the change gears required for the following leads: 800 mm, 560 mm.
7. It is required to make a helical milling cutter having the following specifications:
   Diameter: 3.475 in.
   Helix: left-hand
   Rake angle: 5° positive
   Depth of flute: 0.5 in.
   Material: tool steel
   Number of teeth: 7
   Helix angle: 20°
   Angle of flute: 55°
   Length: 3 in.
   Calculate:
   - (a) indexing
   - (b) lead
   - (c) change gears required to cut this lead

HELICAL GEARING

8. Compare circular pitch to normal circular pitch. Illustrate by means of a suitable sketch.
9. (a) The following information applies to helical gears having parallel shafts.

	Pitch	Ratio	Center distance
(1)	3	3:1	10.5 in.
(2)	10	4:1	5.5 in.
(3)	6	3:1	7.5 in.
(4)	8	12:5	6.0625 in.

For the above gears and pinions calculate:
   - (1) pitch diameter
   - (2) number of teeth
   - (3) diametral pitch
   - (4) helix angle
   - (5) outside diameter
   - (6) lead
   - (7) change gears
   - (8) cutter number to be used

NOTE: A standard set of change gears is supplied.

   (b) It is required to cut a 34-tooth helical metric 6 module gear with a 600-mm lead. The pitch diameter of the gear is 180 mm and the pitch of the lead screw is 5 mm. Calculate:
   - (1) module
   - (2) circular pitch
   - (3) normal circular pitch
   - (4) addendum
   - (5) outside diameter
   - (6) depth of tooth
   - (7) helix angle
   - (8) indexing

CONTINUED FRACTIONS

10. Using continued fractions, calculate the change gears required to produce the following leads, when a standard set of change gears is supplied:
   - (a) 20.570 in.
   - (b) 31.360 in.
   - (c) 9.778 in.
   - (d) 6.667 in.

CAM, RACK, WORM, AND CLUTCH MILLING

UNIT 64

The variety of attachments available make the milling machine a very versatile machine tool. Besides the standard operations generally performed on a milling machine, with the proper setups and attachments, it is possible to cut cams, racks, worms, and clutches. Although a machinist may not be called on too frequently, it is wise to be familiar with these operations so that you will be able to cut the form required.

OBJECTIVES

After completing this unit you will be able to:
1. Calculate and cut a uniform-motion cam
2. Set up the machine and cut a rack
3. Understand how a worm is cut
4. Set up the machine and cut a clutch

CAMS AND CAM MILLING

A *cam* is a device generally applied to a machine to change rotary motion into straight-line or reciprocating motion and to transmit this motion to other parts of the machine through a follower. The cam shaft on an automobile engine incorporates several cams which control the opening and closing of the intake and exhaust valves. Many machine operations, especially on automatic machines, are controlled by cams which transmit the desired motion to the cutting tool through a follower and some type of push rod.

Cams are also used to transform linear motion into a reciprocating motion of the follower. Cams of this type are called *plate*, or *bar*, cams or are often referred to as *templates*. Templates are often used on tracer-type milling machines and lathes where parts must be produced to the profile of the template.

Cams may also be used as locking devices. Extensive applications are found in jig and fixture design and in quick-locking clamps.

CAMS USED TO IMPART MOTION

Cams of this type are generally found on machines and may be of two types, the *positive* and the *nonpositive*.

Positive-type cams, such as the cylindrical and grooved plate (Fig. 64-1), control the follower at all times. That is, the follower remains engaged in the groove on the face or the periphery of the cam and uses no other means to maintain this engagement.

Example of the *non-positive-type cams* are the plate, toe and wiper, and crown (Fig. 64-2). In the nonpositive types, the cam pushes the follower in a given direction and depends on some external force, such as gravity or springs, to keep the follower bearing against the cam surface.

Followers may be of several types.

The roller type (Fig. 64-3A) has the least frictional drag and requires little or no lubrication.
The tapered roller type (Fig. 64-3B) is used with grooved plate or cylindrical cams.
The flat, or plunger, type (Fig. 64-3C) is used to transmit large forces and required lubrication.
The knife-edge, or pointed, type (Fig. 64-3D) is used on more intricate cams as it permits sharp contours to be followed more readily than with a roller cam.

CAM MOTIONS

There are three standard types of motions imparted by cams to followers and machine parts. These are:

1. Uniform motion
2. Harmonic motion
3. Uniformly accelerated and decelerated motion

The *uniform-motion* cam moves the follower at the same rate from the beginning to the end of the stroke. Since the movement starts from zero to full speed and ends in the same abrupt way, there is a distinct shock at the beginning and the end of the stroke. Machines using this type of cam must be rigid and sturdy enough to withstand this constant shock.

Fig. 64-1 (A) Cylindrical or drum cam with a taper roller follower; (B) grooved-plate cam with a roller follower.

Fig. 64-2 (A) Plate cam with a knife-edge follower; (B) toe and wiper cam; (C) crown cam.

Fig. 64-3 Types of cam followers. (A) Roller; (B) tapered roller; (C) flat; (D) knife edge or pointed.

The *harmonic-motion* cam provides a smooth start and stop to the cycle. It is used when uniformity of motion is not essential and where high speeds are required.

The *uniformly accelerated and decelerated* cam moves the follower slowly at first, then accelerates or decelerates at a uniform rate. It then gradually decreases in speed, permitting the follower to come to a slow stop before reversal takes place. This type is considered the smoothest of the three motions and is used on high-speed machines.

RADIAL CAM TERMS

A lobe is a projecting part of the cam which imparts a reciprocal motion to the follower. Cams may have one or several lobes, depending on the application to the machine (Figs. 64-4 and 64-5).

Rise is the distance one lobe will raise or lower the follower as the cam revolves.

Lead is the total travel which would be imparted to the

Fig. 64-4 Single-lobe uni-form-rise cam.

Fig. 64-5 Double-lobe uni-form-rise cam.

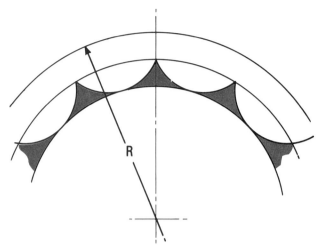

Fig. 64-6 Ridges left by incremental cuts.

follower in one revolution of a uniform-rise cam, having only one lobe in 360°. In Fig. 64-5, the lead for a double-lobe cam is twice the lead of a single-lobe cam having the same rise. It is the *lead* of the cam and not the rise that controls the gear selection in cam milling.

Uniform rise is the rise generated on a cam which moves inward at an even rate around the cam, assuming the shape of an Archimedes spiral. This is caused by uniform feed and rotation of the work when a cam is being machined.

CAM MILLING

In the majority of plate cams which do not have a uniform rise, the cam must be laid out and machined by incremental cuts. By this method, the blank is rotated through an angular increment and the cut is taken to the layout line or a predetermined point. This process is repeated until the outline of the cam is produced as closely as possible. The ridges left between each successive cut are then removed by filing and polishing (Fig. 64-6).

Uniform-rise cams may be produced in the milling machine with a vertical head, by the combined *uniform rotation* of the cam blank, held in the spindle of a dividing head and the *uniform feed* of the table.

When a cam is machined by this method, the work and the vertical head are usually swung at an angle so that the axis of the work and the axis of the mill are parallel (Fig. 64-7).

If the work and the vertical milling attachment are maintained in a vertical position, only a cam having the same lead for which the machine is geared can be cut. When the work and the vertical milling attachment are inclined, any desired lead may be produced, providing that the desired lead is *less* than the lead for which the machine is geared. In other words, the required lead to be cut on the cam must always be less than the forward feed of the table during one revolution of the work.

The principle involved in swinging the head is as follows. If a milling machine is set up to cut a cam and has equal gears on the dividing head and the lead screw, with the work and the vertical attachment in a vertical position (Fig. 64-8), the table would advance 10 in. while the work revolved one turn. An Archimedes spiral would be generated, and the cam would have a lead of 10 in. which is also a rise of 10 in. in 360°.

Fig. 64-7 Machine set up for cam milling using a short lead milling attachment.

Rotate both the work and cutter so that they are parallel to the table and reading zero (Fig. 64-9). If a cut is taken, the work will move along the length of the end mill (which in this case would have to be 10 in.), and a circle having no lead would be generated.

From these two examples, it can be seen that it is possible to mill any lead between zero and that for which the machine is geared, if the cutter and the work are inclined to any given angle between 0 and 90°. If this method were not used, it would be necessary to have a different change gear combination for every lead to be cut. This would be impossible because of the large number of change gears required and the time required to change the gears for each different lead.

The calculations required to set the machine for cam cutting are as follows.

From the drawing determine the *lead* of the lobe or lobes of the cam; that is, determine the amount of rise of each lobe if it were continued for the full circumference of the cam.

Fig. 64-8 Vertical head and work set at 90°.

Fig. 64-9 Vertical attachment and work at zero.

Fig. 64-10 Relation between cam lead and table lead. (Courtesy Cincinnati Milacron Incorporated.)

If the space occupied by the lobe is indicated in degrees on the drawing, the lead would be calculated as follows:

$$\text{Lead} = \frac{\text{rise of lobe in inches} \times 360}{\text{number of degrees of circumference occupied by the lobe}}$$

However, if the circumference is divided into 100 equal parts, it will be necessary to calculate the lead as follows:

$$\text{Lead} = \frac{\text{rise of lobe in inches} \times 100}{\text{percent of the circumference occupied by the lobe}}$$

EXAMPLE

It is required to cut a uniform-rise cam having a rise of 0.375 in. in 360°. Calculate the required lead, the inclination of the work, and the vertical head.

PROCEDURE

$$\text{Lead of cam} = 0.375 \text{ in.}$$

The machine and the dividing head should then be geared to 0.375 in. This is impossible, since the shortest lead which can be cut with regular change gears on a milling machine is generally 0.670 in.

NOTE: Any handbook will contain these milling tables.

To cut a lead of 0.375 in., it will be necessary to gear the machine to something more than 0.375 in.; in this case it will be 0.670 in. By consulting a handbook, it will be noted that the gears required for a 0.670-in. lead are 24, 86, 24, 100. It will also be necessary to swing the work and the vertical head to a definite angle. Figure 64-10 illustrates how this is calculated.

In the diagram,

$L = $ lead to which the machine is geared
$H = $ *lead* of the cam

$i = $ angle of inclination of the dividing head spindle in degrees

$$\text{Sin } i = \frac{H}{L}$$

Therefore,

$$\text{Sine angle of inclination} = \frac{\text{lead of cam}}{\text{lead of machine}}$$

$$\text{Sine angle} = \frac{0.375}{0.670}$$

$$= 0.55970$$

$$\text{Angle} = 34°2'$$

Figure 64-10 illustrates that when the work travels along the distance L (0.670), it will rotate one turn and reach a point on the cutter which is 0.375 in. higher than the starting point. This will produce a rise and a lead of 0.375 in. on the cam.

EXAMPLE

It is required to cut a uniform-rise cam having three lobes, each lobe occupying 120° and each having a rise of 0.150 in.

PROCEDURE

1. $\text{Lead of cam} = \dfrac{0.150 \times 360}{120}$

$$= 0.450 \text{ in.}$$

2. The smallest lead over 0.450 in. to which a machine can be geared is 0.670 in. (handbook).
3. Change gears required to produce a lead of 0.670 in.

395

$$= \frac{24}{86} \times \frac{24 \text{ (driven)}}{100 \text{ (drivers)}}$$

4. Mount the dividing head, and connect the worm shaft and the lead screw with the above gears. *Disengage the index plate locking device.*
5. Scribe three marks 120° apart on the periphery of the cam.
6. Mount the work in the dividing head chuck.
7. Mount an end mill of sufficient length in the vertical attachment.
8. Calculate the offset of the work and the vertical head.

$$\text{Sine} = \frac{0.450}{0.670}$$
$$= 0.67164$$
$$\therefore \text{Angle} = 42°12'$$

9. Swivel the dividing head to 42°12′.
10. Swivel the vertical milling attachment to 90° − 42°12′ = 47°48′.
11. Centralize the work and the cutter.
12. Rotate work with the index crank until one scribed mark on the cam blank is exactly at the bottom dead center.
13. Adjust the table until the cutter is touching the lower side of the work and the center of the cutter is in line with the mark.

NOTE: When the cutter is below the work, the setup will be more rigid and there will be no chance of chips obscuring any layout lines.

14. Set the vertical feed collar to zero.
15. Start the machine.
16. Using the index head crank, rotate the work one-third turn or until the second scribed line is in line with the forward edge of the cutter.

NOTE: If the machine is geared to a lead over 2½ in., the automatic feed may be used. If it is geared to a lead of less than 2½ in., a short lead attachment should be used or the table should be fed along manually by means of the index head crank.

17. Lower the table slightly and disengage the gear train, or disengage the dividing head worm.
18. Return the table to the starting position.
19. Rotate the work until the next line on the circumference is in line with the center of the cutter.
20. Reengage the gear train or the dividing head worm.
21. Cut the second lobe.
22. Repeat steps 17, 18, 19, and 20, and cut the third lobe.

NOTE: When calculating the offset as in step 8, the lead of the machine is often set to the whole number nearest to *twice* the cam lead. When the lead of the machine is exactly twice the lead of the cam, the offset will always equal 30° because the lead of the machine, which is the hypotenuse (Fig. 64-10), is twice the lead of the cam. Thus the sine of the angle of inclination is equal to 1/2 or 0.500, which then makes the angle equal to 30°.

RACK MILLING

A *rack,* in conjunction with a gear (pinion), is used to convert rotary motion into longitudinal motion. Racks are found on lathes, drill presses, and many other machines in a shop. A rack may be considered as a spur gear which has been straightened out so that the teeth are all in one plane. The circumference of the pitch circle of this gear would now become a straight line which would just touch the pitch circle of a gear meshing with the rack. Thus the pitch line of a rack is the distance of one addendum (1/DP) below the top of the tooth.

The pitch of a rack is measured in linear (circular) pitch, which is obtained by dividing 3.1416 by the diametral pitch:

$$\frac{3.1416}{DP}$$

The method used to cut a rack will depend generally on the length of the rack. If the rack is reasonably short [10 in. (254 mm) or less], it may be held in the milling machine vise in a position parallel to the cutter arbor. On short racks, the teeth may be cut by accurately moving the cross-slide of the machine an amount equal to the circular pitch of the gear and then moving the table longitudinally to cut each tooth. If the rack is longer than the cross travel of the milling machine table, it must be held longitudinally on the table and is generally held in a special fixture.

The milling cutter is held in a *rack milling attachment* (Fig. 64-11). When cutting a straight tooth, the cutter is held at 90° to the position used when cutting a spur gear.

It is possible to mount slotting or narrow side milling cutters on the rack milling attachment for milling operations that can be handled more easily by using the machine crossfeed.

RACK INDEXING ATTACHMENT

When cutting a rack using the rack milling attachment, the table is often moved (indexed) for each tooth by means of the rack indexing attachment (Fig. 64-11). This consists of an indexing plate with two diametrically opposed notches and a locking pin. Two change gears selected from a set of 14 are mounted as shown in Fig. 64-11. Different combinations of change gears permit the machine table to be moved accurately in increments, corresponding to the linear (circular) pitch of the rack, by making either a half-turn or one complete turn of the plate. For indexing requiring one complete turn only, provision is made to close off one of the slots, thus preventing any error in indexing.

This attachment permits the indexing of all diametrical pitches from 4 to 32, as well as all circular pitches from 1/8 in. to 3/4 in., varying by 16ths. The following table movements can also be produced: 1/7, 1/6, 1/5, 2/7, 1/3, and 2/5 in.

WORMS AND WORM GEARS

Worms and worm gears are used when a great ratio reduction is required between the driving and driven shafts. A worm is a cylinder on which is cut a single- or multiple-

Fig. 64-11 Cutting the teeth on a helical rack using the rack milling and indexing attachments. (Courtesy Cincinnati Milacron Incorporated.)

Fig. 64-12 Milling machine setup for machining a worm. (Courtesy Cincinnati Milacron Incorporated.)

Fig. 64-13 Types of clutch teeth. (A) Straight tooth, (B) inclined tooth, (C) saw tooth.

start Acme-type thread. The angle of this thread ranges from a 14.5 to 30° pressure angle. As the lead angle of the worm increases, the greater the pressure angle should be on the side of the thread. The teeth on a worm gear are machined on a peripheral groove which has a radius equal to half the root diameter of the worm. The drive ratio between a worm and worm gear assembly is one to the number of teeth in the worm gear. Thus if a worm gear had 50 teeth, the ratio would be 50:1, providing the worm had a single-start thread. If it had a double-start thread, the ratio would be 50:2, or 25:1.

TO MILL A WORM

Worms are often cut on a milling machine with a rack milling attachment and a thread milling cutter (Fig. 64-12). The setup of the cutter is similar to that for rack milling. The work is held between index centers and is rotated by suitable gears between the worm shaft and the lead screw of the milling machine. This is similar to the setup for helical milling. A short lead attachment is generally used when a worm is being milled because the thread usually has a short lead. If a short lead attachment is not available, the work may be rotated and the table moved lengthwise by means of the index crank on the dividing head.

PROCEDURE

1. Calculate all dimensions of the thread, that is, lead, pitch, depth, and angle of thread.

NOTE: The angle of the thread is calculated using the pitch diameter.

2. Mount the worm blank between the dividing head centers located at the end of the milling machine table.
3. Determine the proper gears for the lead, and mount them so that they connect the worm shaft and the lead screw.
4. Disengage the index plate locking device.
5. Mount the proper thread milling cutter on the rack milling attachment.

6. Swing the rack milling attachment to the required helix angle of the worm thread and in the proper direction for the lead of the worm.
7. Center the work under the cutter.
8. Raise the work up to the cutter.
9. Move the work clear of the cutter, and raise the table to the required depth of thread.
10. Cut the thread using the automatic feed or by turning the index crank handle to feed the table manually.

CLUTCHES

Positive-drive clutches are used extensively to drive or disconnect gears and shafts in machine gearboxes. The headstocks on most lathes use clutches, machined on the hubs of gears, to engage or disengage gears to provide different spindle speeds. The positive drive on this type of clutch is produced by means of interlocking teeth or projections on the driving and driven parts and does not rely on friction drive as in the case of friction-type clutches.

Three forms of positive-drive clutches are shown in Fig. 64-13. The *straight-tooth clutch* (Fig. 64-13A) permits rotation in either direction. This type is more difficult to engage since the mating teeth and grooves must be in perfect alignment before engagement is possible.

The *inclined-tooth clutch* (Fig. 64-13B) provides an easier means of engaging or disengaging the driving and driven members because of an 8 or 9° angle machined on the faces of the teeth. Since this type of clutch tends to disengage more rapidly, it must be provided with a positive means of locking it in engagement. Clutches of this type permit the shafts to run in either direction without backlash.

The *sawtooth clutch* (Fig. 64-13C) permits drive in only one direction but is more easily engaged than the other two types of clutches. The angle of the teeth in this type is generally 60°.

TO MACHINE A STRAIGHT-TOOTH CLUTCH HAVING THREE TEETH

NOTE: This method applies to all clutches having an odd number of teeth.

1. Mount the dividing head on the milling machine table.
2. Mount a three-jaw chuck in the dividing head.
3. Mount the workpiece in the chuck.

NOTE: The spindle and the chuck of the dividing head may be positioned either horizontally or vertically. For this operation, it is assumed to be in a vertical position.

4. Set the sector arms to the proper indexing, that is, 40/3 = 13⅓ turns or 13 turns and 13 holes on a 39-hole circle.
5. Mount a side milling cutter on the milling machine arbor. The cutter should be no wider than the narrowest or innermost part of the groove.
6. Set the proper spindle speed and table feed.
7. Start the cutter and adjust the work until the edge nearest the front of the machine *just* touches the inner side of the cutter (Fig. 64-14). Set the cross-feed graduated feed collar to zero.
8. Move the table longitudinally until the work is clear of the cutter.
9. Move the table laterally half the diameter of the work plus about 0.001 in. (0.02 mm) for clearance. Lock the saddle in this position.
10. Set the depth of cut, and lock the knee clamps.
11. Take a cut across the full width of the workpiece as in Fig. 64-15.
12. Return the table to starting position.
13. Index for the next tooth and take the second cut as shown in Fig. 64-16.
14. Return the table to starting position.
15. Index for the next tooth.
16. Take the third cut as shown in Fig. 64-17.

TO MACHINE A STRAIGHT-TOOTH CLUTCH HAVING FOUR TEETH

When machining a straight- or inclined-tooth clutch having an even number of teeth, it is necessary to machine one side of each tooth first and then machine the second side of each. This obviously requires more time; therefore, it is desirable that clutches be designed with an odd number of teeth to reduce machining time and the chance of error.

Fig. 64-14 Adjusting the work to the cutter.

Fig. 64-15 Cutting the first tooth.

Fig. 64-16 Cutting the second tooth.

Fig. 64-17 Cutting the third tooth.

1. Mount the work as in the previous example and set the proper indexing, if required.
2. Start the cutter and adjust the workpiece until the edge nearest the front of the machine just touches the inner edge of the cutter (Fig. 64-14).
3. With the work clear of the cutter, move the saddle over half the diameter of the work plus the thickness of the cutter minus 0.001 in. (or 0.02 mm) for clearance.
4. Adjust the work until the cutter is over the center hole of the clutch. It is advisable to cut from the center to the outside of the clutch in order to minimize the vibration.
5. Set the depth and lock the knee clamp.
6. Take the first cut.

NOTE: When cutting an even number of clutch teeth, it is absolutely necessary that the cut be made through one wall only.

7. Index for and cut the remaining teeth on the one side. (Indexing = 10 turns for each tooth.)
8. Revolve the work one-eighth of a turn (five turns of the index crank).
9. Touch the opposite side of the work to the other side of the cutter, that is, the outside of the cutter to the inner edge of the work.

10. Repeat operations 3, 4, 5, 6, and 7 until all the teeth are cut. If any pie-shaped pieces of metal remain in the tooth spaces, they must be removed with an additional cut through the center of the space.

REVIEW QUESTIONS

CAMS AND CAM MILLING

1. Define a cam.

2. Name four types of cams.

3. Name four types of followers, and state where each is used.

4. List three cam motions, and describe the type of motion which is imparted to the follower in one revolution of the cam.

5. In a single-lobe cam and a double-lobe cam, what is the relationship of the lead to the rise?

6. Calculate:

(a) lead of cam

(b) the change gears required to produce this lead

(c) the angle at which to swing the dividing head

(d) the angle at which to swing the vertical head for each of the following examples of uniform rise:

(1) a single-lobe cam having a rise of 0.125 in. in 360°

(2) a two-lobed cam, each lobe having a rise of 0.187 in. in 180°

(3) a three-lobed cam, each lobe having a rise of 0.200 in. in 120°

RACK MILLING

7. Define a rack and state its purpose.

8. A 10-DP gear is in mesh with a rack. The gear has 42 teeth. The rack is 1 in. thick from the top of the tooth to the bottom of the rack. Calculate the distance from the center of the gear to the bottom of the rack.

9. Calculate the linear pitch of a 5-pitch rack, an 8-pitch rack, a 14-pitch rack.

WORMS AND WORM GEARS

10. Define a worm and worm gear, and describe their use.

11. Describe briefly how a worm is cut on a milling machine.

CLUTCHES

12. List three types of clutches, and state the application of each.

13. Why are clutches with an odd number of teeth preferred to those with an even number of teeth?

14. After the outside surface of a 3-in.-diameter (or 76-mm) clutch blank has been touched up to a 1/2-in.-wide (or 12-mm) side cutter, how far must the table be moved over when cutting

(a) a clutch having five teeth?

(b) a clutch having six teeth?

THE VERTICAL MILLING MACHINE—CONSTRUCTION AND OPERATION

UNIT 65

Much of the work on a milling machine is best done with a vertical milling attachment. The time involved in setting up a vertical attachment prevents the milling machine from being used for other milling operations at the same time; therefore, the vertical milling machine has become popular in industry. This machine offers versatility not found in any other machine. Some of the operations which can conveniently be carried out on these machines are face milling, end milling, keyway cutting, dovetail cutting, T-slot and circular slot cutting, gear cutting, drilling, boring, and jig boring. Because of the machine construction (vertical spindle), many of the facing operations can be done with a fly cutter, which reduces the cost of cutters

399

considerably. Also, since most cutters are much smaller than for the horizontal mill, the cost of the cutters for the same job is usually much less for a vertical milling machine.

OBJECTIVES

After you have completed this unit you will be able to:
1. Name and know the purpose of the main operative parts
2. Align the vertical head and the vise to within an accuracy of ± 0.001 in. (0.02 mm)
3. Mill flat and angular surfaces
4. Mill a block square and parallel
5. Mill keyways in a shaft

The standard vertical milling machine (Fig. 65-1) has all the construction features of a plain horizontal milling machine except that the cutter spindle is mounted in a vertical position. The spindle head on most vertical milling machines may be swiveled, which readily permits the machining of angular surfaces. The spindles on most vertical milling machines have a short travel, which facilitates step milling and the drilling and boring of holes.

The cutters used are of the end mill or shell end mill type. This type of machine is particularly suited to the use of the rotary table, permitting the machining of circular grooves and the positioning of holes which have been laid out with angular measurements.

The *ram-type* vertical milling machine (Fig. 65-2) is lighter than the standard vertical machine. Because of its simplicity and the ease of setup, it has become increasingly popular. Machines of this type are generally used for lighter types of milling machine work.

PARTS OF THE RAM-TYPE VERTICAL MILL

The base is made of ribbed cast iron. It may contain a coolant reservoir.

The column is often cast integrally with the base. The machined face of the column provides the ways for vertical movement of the knee. The upper part of the column is machined to receive a *turret* on which the overarm is mounted.

The overarm is round, or of the ram type, as illustrated in Fig. 65-2. It may be adjusted toward or away from the column to increase the capacity of the machine.

The head is attached to the end of the ram. Provision is made to swivel the head in one plane. On universal-type machines, the head may be swiveled in two planes. Mounted on top of the head is the *motor* which provides drive to the *spindle*, usually through V-belts. Spindle speed changes are effected by means of a variable-speed pulley and crank or by belt changes and a reduction gear. The spindle may be fed by means of a hand lever, a handwheel, or by automatic power feed. Most machines are equipped with a micrometer quill stop for precision drilling and boring to depth.

The knee moves up and down on the face of the column and supports the saddle and the table. The knee in this type of machine does *not* contain the gears for the au-

tomatic feed as in the horizontal milling machine and the standard vertical milling machine. The automatic feed on most of these types of machines is not a standard feature and is usually added as an accessory. It is an external device and controls only the longitudinal feeds of the table. Most cutting on the vertical milling machine is done by end mill cutters; therefore, it is not necessary to swing the table even when cutting a helix. As a result, vertical milling machines are equipped with plain tables only.

ALIGNING THE VERTICAL HEAD

Proper alignment of the head is of the utmost importance when machining holes or when face milling. If the head is not at an angle of 90° to the table, the holes will not be square with the work surface when the cutting tool is fed

Fig. 65-1 A standard vertical milling machine.

HEAD

QUILL FEED HAND
LEVER

QUILL FINE FEED
HANDWHEEL

RAM

SPINDLE

TABLE

COLUMN

TABLE
TRAVERSE
HANDWHEEL

CROSSFEED
HANDWHEEL

VERTICAL TRAVERSE
LEVER

KNEE

BASE

Fig. 65-2 A ram-type vertical milling machine.

by hand or automatically. When face milling, the machined surface will be stepped if the head is not square with the table. Although all heads are graduated in degrees and some have vernier devices used for setting the head, it is well to check the spindle alignment as follows.

PROCEDURE

1. Mount a dial indicator on a suitable rod, bent at 90° and held in the spindle (Fig. 65-3).
2. Position the indicator over the front of the table.

NOTE: If a ground cylindrical plate is available, place it on the clean table and position the indicator near the outer edge.

Fig. 65-3 An indicator assembly fastened to a rod in a chuck.

Fig. 65-4 Setting the indicator to zero on the right side of the table.

Fig. 65-5 The spindle rotated 180°.

3. Carefully lower the spindle until the indicator button touches the table and the dial indicator registers about one-half revolution, then set the bezel to zero (Fig. 65-4). Lock the spindle in place.

4. Carefully rotate the vertical mill spindle 180° by hand until the button bears on the opposite side of the table (Fig. 65-5). Compare the readings.

5. If there is any discrepancy in the readings, loosen the locking nuts on the swivel mounting and adjust the head until the indicator registers one-half the difference between the two readings. Tighten the locking nuts.

6. Recheck the accuracy of the head and adjust if necessary.

7. Rotate the vertical mill spindle 90°, and set the dial indicator as in step 3.

8. Rotate the machine spindle 180°, and check the reading at the other end of the table.

9. If the two readings do not coincide, repeat step 5 until the readings are the same.

10. Tighten the locking nuts on the swivel mount.

11. Recheck the readings and adjust if necessary.

Fig. 65-6 The indicator assembly for checking the vise for alignment.

NOTE: When readings are taken, it is important that the indicator button does not catch in the T-slots of the table. To prevent this, it is advisable to work from the high reading first and then rotate to the low reading. It should be apparent that the longer the rod used, the more accurate the setting will be.

ALIGNING THE VISE

When the vise is aligned on a vertical milling machine, the dial indicator may be attached to the quill or the head by any convenient means, such as clamps (Fig. 65-6) or magnetic base. The same method of alignment should be followed as was outlined in this unit for aligning the vise on a horizontal milling machine.

TO MACHINE A FLAT SURFACE

1. Clean the vise and mount the work securely in the vise, on parallels if necessary (Fig. 65-7).

Fig. 65-7 Tap the work down with a soft-faced hammer until all feelers are tight.

2. Check that the vertical head is square with the table.

3. If possible, select a cutter which will just overlap the edges of the work. This will then require only one cut to be taken to machine the surface. If the surface to be machined is fairly narrow, an end mill slightly larger in diameter than the width of the work should be used. If the surface is large and requires several passes, a shell end mill or a suitable fly cutter should be used.

Fig. 65-8   Measuring the work after a light trial cut.

Fig. 65-9   Side 1 or the surface with the largest area should be facing up.

NOTE: It is not advisable to use a facing cutter which is too wide since the head may be thrown out of alignment if the cutter should jam.

4. Set the proper spindle speed for the size and type of cutter and the material being machined.
5. Tighten the quill clamps.
6. Start the machine, and adjust the table until the end of the work is under the edge of the cutter.
7. Raise the table until the work surface just touches the cutter. Move the work clear of the cutter.
8. Raise the table about 1/32 in. (0.78 mm) and take a trial cut for approximately 1/4 in. (6.35 mm).
9. Move the work clear of the cutter, *stop the cutter,* and measure the work (Fig. 65-8).
10. Raise the table the desired amount, and lock the knee clamp.
11. Mill the surface to size using the automatic feed if the machine is so equipped.

Fig. 65-10   The finished side (1) should be placed against the cleaned solid jaw for machining side 2.

MACHINING A BLOCK SQUARE AND PARALLEL

In order to mill the four sides of a piece of work so that its sides are square and parallel, it is important that each side be machined in a definite order. It is very important that dirt and burrs be removed from the work, vise, and parallels, since they can cause inaccurate work.

MACHINING SIDE 1

1. Clean the vise thoroughly and remove all burrs from the workpiece.
2. Set the work on parallels in the center of the vise with the largest surface (side 1) facing up (Fig. 65-9).
3. Place short paper feelers under each corner between the parallels and the work.
4. Tighten the vise securely.
5. With a soft-faced hammer, tap the workpiece down until all paper feelers are tight.
6. Mount a flycutter in the milling machine spindle.
7. Set the machine for the proper speed for the size of the cutter and the material to be machined.

8. Start the machine and raise the table until the cutter just touches near the right-hand end of side 1.
9. Move work clear of the cutter.
10. Raise the table about 0.030 in. (0.76 mm) and machine side 1 using a steady feed rate (Fig. 65-9).
11. Take the work out of the vise and remove all burrs from the edges with a file.

MACHINING SIDE 2

12. Clean the vise, work, and parallels thoroughly.
13. Place the work on parallels, if necessary, with side 1 against the solid jaw and side 2 up (Fig. 65-10).
14. Place short paper feelers under each corner between the parallels and the work.
15. Place a round bar between side 4 and the movable jaw.

NOTE: The round bar must be in the *center* of the amount of work being held inside the vise jaws.

16. Tighten the vise securely and tap the work down until the paper feelers are tight.
17. Follow steps 8 to 11 and machine side 2.

Fig. 65-11   Place side 1 against the solid jaw, with side 2 down for machining side 3.

Fig. 65-12   Measuring the trial cut on side 3.

MACHINING SIDE 3

18. Clean the vise, work, and parallels thoroughly.
19. Place side 1 against the solid vise jaw with side 2 resting on parallels if necessary (Fig. 65-11).
20. Push the parallel to the left so that the right edge of the work extends about ¼ in. (6.35 mm) beyond the parallel.
21. Place short paper feelers under each corner between the parallels and the work.
22. Place a round bar between side 4 and the movable jaw being sure that the round bar is in the center of the amount of work held *inside the vise*.
23. Tighten the vise securely and tap the work down until the paper feelers are tight.
24. Start the machine and raise the table until the cutter just touches near the right-hand end of side 3.
25. Move the work clear of the cutter and raise the table about 0.010 in. (0.25 mm).
26. Take a trial cut about ¼ in. (6.35 mm) long, stop the machine, and measure the width of the work (Fig. 65-12).

Fig. 65-13   The work is set up properly for machining side 4. Note the paper feelers between the parallel and the work.

27. Raise the table the required amount and machine side 3 to the correct width.
28. Remove the work and file off all burrs.

MACHINING SIDE 4

29. Clean the vise, work, and parallels thoroughly.
30. Place side 1 down on the parallels with side 4 up and tighten the vise securely (Fig. 65-13).

NOTE: With three finished surfaces, the round bar is not required when machining side 4.

31. Place short paper feelers under each corner between the parallels and work (Fig. 65-13).
32. Tighten the vise securely.
33. Tap the work down until the paper feelers are tight.
34. Follow steps 24 to 27 and machine side 4 to the correct thickness.

MACHINING THE ENDS SQUARE

Two common methods are used to square the ends of workpieces in a vertical mill. Short pieces are generally held vertically in the vise and are machined with an end mill or flycutter (Fig. 65-14). Long pieces are generally held flat in the vise with one end extending past the end of the vise. The end surface is cut square with the body of an end mill.

PROCEDURE FOR SHORT WORK

  1. Set the work in the center of the vise with one of the ends up and tighten the vise lightly.
  2. Hold a square down firmly on top of the vise jaws and bring the blade into light contact with the side of the work.
  3. Tap the work until its edge is aligned with the blade of the square (Fig. 65-14).

Fig. 65-14 Setting the edge of the work square for machining the ends.

Fig. 65-15 One end machined square with the sides of the workpiece.

Fig. 65-16 Measuring the length of the workpiece with a depth micrometer.

4. Tighten the vise securely and recheck the squareness of the side.

5. Take about 0.030-in.-deep (0.76-mm) cut and machine the end square (Fig. 65-15).

6. Remove the burrs from the end of the machined surface.

7. Clean the vise and set the machined end on paper feelers in the bottom of the vise.

8. Tighten the vise securely and tap the work down until the paper feelers are tight.

9. Take a trial cut from the end until the surface cleans up.

10. Measure the length of the workpiece with a depth micrometer (Fig. 65-16).

11. Raise the table the required amount and machine the work to length.

TO MACHINE AN ANGULAR SURFACE

1. Lay out the angular surface.

2. Clean the vise.

Fig. 65-17 The head swiveled to machine an angle.

3. Align the vise with the direction of the feed. This is of the utmost importance.

4. Mount the work on parallels in the vise.

5. Swivel the vertical head to the required angle (Fig. 65-17).

6. Tighten the quill clamp.

7. Start the machine and raise the table until the cutter touches the work. Carefully raise the table until the cut is of the desired depth.

8. Take a trial cut for about 1/2 in. (12.7 mm).
9. Check the angle with a protractor.
10. If the angle is correct, continue the cut.

NOTE: It is always advisable to feed the work into the rotation of the cutter, rather than with the rotation of the cutter which may draw the work into the cutter and cause damage to the work, the cutter, or both.

11. Machine to the required depth, taking several cuts if necessary.

Alternate Method

Angles may sometimes be milled by leaving the head in a vertical position and setting the work on an angle in the vise (Fig. 65-18). This will depend on the shape and size of the workpiece.

PROCEDURE

1. Check that the vertical head is square with the table.
2. Clean the vise.
3. Lock the quill clamp.
4. Set the workpiece in the vise with the layout line parallel to the top of the vise jaws and about 1/4 in. (6.35 mm) above them.
5. Adjust the work under the cutter so that the cut will start at the narrow side of the taper and progress into the thicker metal.
6. Take successive cuts of about 0.125 to 0.150 in. (3.17 to 3.81 mm), or until the cut is about 1/32 in. (0.78 mm) above the layout line.
7. Check to see that the cut and the layout line are parallel.
8. Raise the table until the cutter just touches the layout line.
9. Clamp the knee at this setting.
10. Take the finishing cut.

Fig. 65-18 Machining an angle by adjusting the work.

Fig. 65-19 Layout of a keyseat on a shaft.

CUTTING SLOTS AND KEYSEATS

Slots and keyseats with one or two blind ends may be cut in shafts more easily on a vertical milling machine, using a two- or three-fluted end mill, than with a horizontal mill and a side facing cutter.

PROCEDURE

1. Lay out the position of the keyseat on the shaft, and scribe reference lines on the end of the shaft (Fig. 65-19).
2. Secure the workpiece in a vise on a parallel. If the shaft is long, it may be clamped directly to the table by placing it in one of the table slots or in V-blocks.
3. Using the layout lines on the end of the shaft, set up the shaft so that the keyseat layout is in the proper position on top of the shaft.

4. Mount a two- or three-fluted end mill of a diameter equal to the width of the keyseat, in the milling machine spindle.

NOTE: If the keyseat has two blind ends, a two- or three-lip end mill must be used, since they may be used as a drill to start the slot. If the slot is at one end of the shaft (one blind end only), a four-fluted end mill may be used, although a two- or three-lip end mill will give better chip clearance.

5. Center the workpiece by carefully touching the cutter up to one side of the shaft (Fig. 65-20). This may also be done by placing a piece of thin paper between the shaft and the cutter.

NOTE: Paper may be made to adhere to shafts or work surfaces by wetting it with coolant or oil before applying it to the surface. This eliminates the necessity of holding

Fig. 65-20 Setting the cutter to the side of the work.

Fig. 65-21 The cutter centered with the layout.

Fig. 65-22 The keyseat milled to depth and length.

paper between the cutter and the work, thus making it a safer operation.

6. Lower the table until the cutter clears the workpiece.

7. Move the table over an amount equal to half the diameter of the shaft plus half the diameter of the cutter plus the thickness of the paper (Fig. 65-21). For example, if a 1/4-in. (6.35-mm) slot is required in a 2-in (50.8-mm) shaft and the thickness of the paper used is 0.002 in. (0.05 mm), the table would be moved over $1.000 + 0.125 + 0.002 = 1.127$ in. $(25.4 + 3.17 + 0.05 = 28.62$ mm).

8. If the keyseat being cut has two blind ends, adjust the work until the end of the keyseat is aligned with the edge of the cutter.

9. Feed the cutter down (or the table up) until the cutter *just* cuts to its full diameter. If the keyseat has one blind end only, the work is adjusted so that this cut is taken at the end of the work. The work would now be moved clear of the cutter.

10. Adjust the depth of cut to one-half the thickness of the key, and machine the keyseat to the proper length (Fig. 65-22).

MEASURING THE DEPTH OF KEYSEATS

If the keyseat is at the end of a shaft, the proper depth of the keyseat is checked by measuring diametrically from the bottom of the keyseat to the opposite side of the shaft.

This distance may be calculated as follows (Fig. 65-23):

$$M = D - \left(\frac{W}{2} + h\right)$$

$$h = 0.5 \ (D - \sqrt{D^2 - W^2})$$

M = the distance from the bottom of the keyseat to the opposite side of the shaft
D = diameter of the shaft
W = width of the keyseat
h = height of the segment above the width of the keyseat

PROBLEM

Calculate the measurement M if the shaft is 50 mm in diameter and a 6-mm key is to be used.

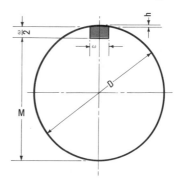

Fig. 65-23 Keyseat calculations.

$$\begin{aligned}
h &= (D - \sqrt{D^2 - W^2}) \\
&= 0.5(50 - \sqrt{2500 - 36}) \\
&= 0.5(50 \ \sqrt{2464}) \\
&= 0.5(50 - 49.64) \\
&= 0.5(0.36) \\
&= 0.18 \text{ mm} \\
M &= D - \left(\frac{W}{2} + h\right) \\
&= 50 - \left(\frac{6}{2} + 0.18\right) \\
&= 50 - (3 + 0.18) \\
&= 50 - 3.18 \\
&= 46.82 \text{ mm}
\end{aligned}$$

WOODRUFF KEYS

Woodruff keys are used when keying shafts and mating parts (Fig. 65-24A). Woodruff keyseats are more quickly cut than are square keyseats, and the key should not require any fitting after the keyseat has been cut. These keys are semicircular in shape and can be purchased in standard sizes which are designated by E numbers. They can be conveniently made from round bar stock of the required diameter.

Woodruff keyseat cutters (Fig. 65-24B) have shank diameters of ½ in. for cutters up to 1½ in. in diameter. The

407

Fig. 65-24 Woodruff key and keyseat cutters.

shank is undercut adjacent to the cutter to permit the cutter to go into the proper depth. The sides of the cutter are slightly tapered toward the center to permit clearance while cutting. Cutters over 2 in. in diameter are mounted on an arbor.

The size of the cutter is stamped on the shank. The last two digits of the number indicate the nominal diameter in eighths of an inch. The digit or digits preceding the last two numbers indicate the nominal width of the cutter in thirty-seconds of an inch. Thus a cutter marked 608 would be 8 × ⅛, or 1 in. in diameter, and 6 × 1/32, or 3/16 in. wide. The key would be a semicircular cross section to fit the groove exactly.

To Cut a Woodruff Keyseat

1. Align the spindle of the vertical milling machine 90° to the table.
2. Lay out the position of the keyseat.
3. Set the shaft in the vise of the milling machine or on V-blocks. Be sure that the shaft is level (parallel to the table).
4. Mount the cutter of the proper size in the spindle.
5. Start the cutter, and touch the bottom of the cutter to the top of the workpiece. Set the vertical graduated feed collar to zero.
6. Move the work clear of the cutter. Raise the table half the diameter of the work plus half the thickness of the cutter. Lock the knee at this setting.
7. Position the center of the slot with the center of the cutter. Lock the table in this position.
8. Touch the revolving cutter to the work. Use a strip of paper between the cutter and the work if desired. Set the crossfeed screw collar to zero.
9. Cut keyseat to the proper depth.

NOTE: Keyseat proportions may be found in any handbook.

REVIEW QUESTIONS

1. Why has the vertical milling machine been so readily accepted by industry?
2. Name six operations which can be performed in a vertical mill.

3. Describe:
   (a) a standard vertical mill
   (b) a ram type vertical mill
4. State the purpose of the following parts:
   (a) column
   (b) overarm
   (c) head
   (d) knee

ALIGNING THE VERTICAL HEAD AND VISE

5. Why is it necessary to align the vertical head square with the table?
6. Describe briefly how the vertical head may be aligned with the table surface.
7. Why is it important that the vise be aligned with the table travel?
8. Describe one method of aligning the vise parallel to the table travel.

MACHINING A FLAT SURFACE

9. What precautions should be observed before machining a flat surface?
10. What type of cutter should be used when machining a large surface?
11. How can the edge of a flat piece be machined square with a flat finished surface?

MACHINING A BLOCK SQUARE AND PARALLEL

12. How should the work be set in a vise to machine side 1?
13. How should the work be set in a vise to machine side 2?
14. Draw a sketch of the work setup required for machining side 3 and side 4.

MACHINING ANGULAR SURFACES

15. Describe briefly two methods of machining angular surfaces.

CUTTING SLOTS AND KEYWAYS

16. What type of cutter should be used to cut slots or keyways having one or two blind ends?
17. How can round work be held for machining slots and keyways?
18. Explain one method of aligning the end mill with the center of a shaft.
19. How may the depth of a keyway be measured?
20. Calculate the measurement for a ½-in.-wide keyseat from the bottom of a 2-in. diameter shaft to the bottom of the keyseat (½ × ½ key).
21. Describe a Woodruff key and state its purpose.
22. Explain the markings 810 found on a Woodruff key cutter.
23. Explain how to center a Woodruff key cutter with a shaft.

SPECIAL MILLING OPERATIONS

UNIT 66

The versatility of the vertical milling machine is further increased by the use of special-shaped cutters and machine accessories. Operations such as milling of T-slots and dovetails are possible because special milling cutters are manufactured for each purpose. The rotary table accessory permits the milling of radii and circular slots. A boring head mounted in the machine spindle and measuring rods on the table allow the vertical milling machine to be used for the accurate jig boring of holes.

OBJECTIVES

After completing this unit you will be able to:
1. Set up and use the rotary table to mill a circular slot
2. Set up and mill internal and external dovetails
3. Jig bore holes on a vertical mill

THE ROTARY TABLE

The *rotary table* or the *circular milling attachment* (Fig. 66-1) can be used on plain universal vertical milling machines and slotters. It may provide rotary motion to the workpiece in addition to the longitudinal and vertical motion provided by the machine. With this attachment, it is possible to cut radii, circular grooves, and circular sections not possible by other means. The drilling and boring of holes which have been designated by angular measurements, as well as other indexing operations, are easily accomplished with this accessory. This attachment is also suitable for use with the slotting attachment on a milling machine.

Rotary tables may be of two types: those having hand feed and those having power feed. The construction of these is basically the same, the only exception being the automatic feed mechanism.

HAND FEED ROTARY TABLE CONSTRUCTION

The rotary table unit consists of a *base* which is bolted to the milling machine table. Fitting into the base is the *rotary table,* on the bottom of which is mounted a *worm gear* (Fig. 66-1). A *worm shaft* mounted in the base meshes with and drives the worm gear. The worm shaft may be quickly disengaged when rapid rotation of the table is required, as when setting work concentric with the table. A *handwheel* is mounted on the outer end of the worm shaft. The bottom edge of the table is graduated in half degrees.

On most rotary table units, there is a *vernier scale* on the handwheel collar which permits setting to within two minutes of a degree. The table has T-slots cut into the top surface to permit the clamping of work.

A hole in the center of the table accommodates test plugs to permit easy centering of the table with the machine spindle. Work may be centered with the table by means of test plugs or arbors.

Some rotary tables use an *indexing attachment* in place of a handwheel (Fig. 66-2). This attachment is often supplied as an accessory to the standard rotary table. It not

Fig. 66-1 A hand feed rotary table.

Fig. 66-2   A rotary table with an indexing attachment.   (Courtesy Cincinnati Milacron Incorporated.)

Fig. 66-3   A rotary table with power feed.   (Courtesy Cincinnati Milacron Incorporated.)

only serves the same purpose as a handwheel, but also permits the indexing of work with dividing head accuracy. The worm and wormwheel ratio of rotary tables is not necessarily 40:1 as on most dividing heads. The ratio may be 72:1, 80:1, 90:1, 120:1, or any other ratio. Larger ratios are usually found on larger tables. The method of calculating the indexing is the same as for the dividing head except that the number of teeth in the wormwheel is used rather than 40 as in the dividing head calculations.

EXAMPLE

Calculate the indexing for five equally spaced holes on a circular plate using a rotary table with an 80:1 ratio.

$$\text{Indexing} = \frac{\text{number of teeth in wormwheel}}{\text{number of divisions required}}$$

$$= \frac{80}{5}$$

$$= 16 \text{ turns}$$

Accurately spaced holes on circles and segments (such as clutch teeth and teeth in gears too large to be held between index centers) are some of the applications of this type of rotary table. This type may be supplied with a power feed mechanism.

The third type of rotary table has provision for power rotation of the table (Fig. 66-3). Here the worm shaft is connected to the milling machine lead screw drive gear by a special shaft and an end gear train. The rate of rotation is controlled by the feed mechanism of the milling machine. When operated by power, the rotation of the table may be controlled by the operator using a feed lever attached to the unit or by trip dogs located on the periphery of the table. The operation of the table may be controlled by the handwheel or the automatic feed lever as required.

This type of attachment is particularly suited for production work and continuous milling operations when large numbers of small identical parts are required. In this operation, the parts are mounted in suitable fixtures on the table, and the rotary feed moves the parts under the cutter. After the piece has passed beneath the cutter, the finished piece is removed and replaced with an unfinished workpiece.

To Center the Rotary Table with the Vertical Mill Spindle

1. Square the vertical head with the machine table.
2. Mount the rotary table on the milling machine.
3. Place a test plug in the center hole of the rotary table.
4. Mount an indicator with a grasshopper leg in the machine spindle.
5. With the indicator just clearing the top of the test plug, rotate the machine spindle by hand and approximately align the plug with the spindle.
6. Bring the indicator into contact with the diameter of the plug, and rotate the spindle by hand.
7. Adjust the machine table by the longitudinal and cross-feed handles until the dial indicator registers no movement.
8. Lock the machine table and saddle, and recheck the alignment.
9. Readjust if necessary.

To Center a Workpiece with the Rotary Table

Often it is necessary to perform a rotary table operation on several identical workpieces, each having a machined hole in the center. To align each workpiece quickly, a special plug can be made to fit the center hole of the workpiece and the hole in the rotary table. Once the machine spindle has been aligned with the rotary table, each succeeding piece can be aligned quickly and accurately by placing it over the plug.

If there are only a few pieces, which would not justify the manufacture of a special plug, or if the workpiece does not have a hole through its center, the following method can be used to center the workpiece on the rotary table.

1. Align the rotary table with the vertical head spindle.
2. Lightly clamp the workpiece to the rotary table in the approximate center.

Fig. 66-4   Milling a circular slot in a workpiece.

NOTE: *Do not* move the crossfeed or longitudinal feed handles.

3. Disengage the rotary table worm mechanism.
4. Mount an indicator in the machine spindle or on the milling machine table, depending upon the workpiece.
5. Bring the indicator into contact with the surface to be indicated, and revolve the rotary table by hand.
6. With a soft metal bar, tap the work (away from the indicator movement) until no movement is registered on the indicator in a complete revolution of the rotary table.
7. Clamp the workpiece tightly, and recheck the accuracy of the setup.

NOTE: If a center-punch mark must be aligned, a wiggler instead of an indicator is mounted in the milling machine spindle and the punch mark is positioned under the point of the wiggler.

RADIUS MILLING

When it is required to mill the ends on a workpiece to a certain radius or to machine circular slots having a definite radius, a sequence should be followed. Figure 66-4 illustrates a typical setup.

PROCEDURE

1. Align the vertical milling machine spindle at 90° to the table.

2. Mount a circular milling attachment (rotary table) on the milling machine table.
3. Center the rotary table with the machine spindle using a test plug in the table and a dial indicator in the spindle.
4. Set the longitudinal feed dials and the crossfeed dial to zero.
5. Mount the work on the rotary table, aligning the center of the radial cuts with the center of the table. A special arbor may be used for this. Another method is to align the center of the radial cut with a wiggler mounted in the machine spindle.
6. Move either the crossfeed or the longitudinal feed (whichever is more convenient) an amount equal to the radius required.
7. *Lock both the table and the saddle,* and remove the handles if convenient.
8. Mount the proper end mill.
9. Rotate the work, using the rotary table feed handwheel, to the starting point of the cut.
10. Set the depth of cut and machine the slot to the size indicated on the drawing, using hand or power feed.

MILLING A T-SLOT

T-slots are machined in the tops of machine tables and accessories to receive bolts for clamping workpieces. They are machined in two operations.

PROCEDURE

1. Consult a handbook for the T-slot dimensions.
2. Lay out the position of the T-slot.
3. Square the vertical milling machine spindle with the machine table.
4. Mount the work on the milling machine. If the work is to be held in a vise, the vise jaw must be aligned with the table travel. If the work is clamped to the table, the position of the slot must be aligned with the table travel.
5. Mount an end mill having a diameter slightly larger than the diameter of the bolt body. The size of the end mill to be used is shown in the T-slot tables.
6. Machine the center slot to the proper depth of the T-slot, using the end mill.
7. Remove the end mill, and mount the proper T-slot cutter.
8. Set the T-slot cutter depth to the bottom of the slot.
9. Machine the lower part of the slot.

MILLING DOVETAILS

Dovetails are used to permit reciprocating motion between two elements of a machine. They are composed of an external or an internal part and are adjusted by means of a gib. Dovetails may be machined on a vertical milling machine or on a horizontal mill equipped with a vertical milling attachment. A dovetail cutter is a special single-angle end mill type of cutter ground to the angle of the dovetail required.

Milling an Internal Dovetail

1. Refer to a handbook for the method of measuring a dovetail.
2. Check the measurements of the workpiece in which the dovetail is to be cut. Remove all burrs.
3. Lay out the position of the dovetail.
4. Mount the workpiece in a vise, clamp it on a rotary table, or if the work is long, bolt it directly to the machine table.
5. Indicate the side of the workpiece or the slot layout to see that it is parallel to the line of table travel.
6. Mount an end mill of a diameter narrower than the center section of the dovetail (Fig. 66-5A).
7. Start the end mill, and touch up to the side of the work.
8. Set the crossfeed dial to zero.
9. Move the work over until the end mill is in the center of the dovetail. In this case, it will be the distance from the side of the workpiece to the center of the dovetail plus half the diameter of the cutter (plus the thickness of paper, if used).
10. Lock the saddle in this position, and set the crossfeed dial to zero.
11. Touch the edge of the cutter to the top of the work.
12. Move the work clear of the cutter, and set the depth of cut. Lock the knee in this position. Note the depth of this slot should be 0.030 to 0.050 in. (0.76 to 1.27 mm) deeper than the bottom of the dovetail to prevent drag and to provide clearance for dirt and chips between the mating dovetail parts.
13. Mill the channel to the width of the cutter (Fig. 66-5A).
14. Move the work over an amount equal to half the difference between the machined slot size and the size of the dovetail at the top. *Check for backlash.*
15. Take this finish cut along the one side of the work.
16. Check the width of the slot.
17. Move the work over to the finished width of the top of the dovetail. *Check for backlash.*
18. Cut the second side and check the width of the slot.
19. Return the saddle to zero.
20. Mount a dovetail cutter.
21. Set the depth for a roughing cut. This should be about 0.005 to 0.010 in. (0.12 to 0.25 mm) less than the finish depth.
22. Calculate the width of the dovetail at the bottom.
23. Move the work over 0.010 in. (0.25 mm) less than the finished size of this side. This will leave enough for the finish cut. Note the readings on the crossfeed dial.
24. Rough out the angle on the first side.
25. Move the work over to the other side the same amount from the centerline, and rough-cut the other side.
26. Set the cutter to the proper depth.
27. Machine the bottom surface (both sides) of the dovetail to the finished depth.
28. Using two rods, measure the dovetail for size.
29. Move the table over half the difference between the rough dovetail and the finished size. *Check for backlash.*
30. Take the finish cut on one side.
31. Move the table over the required amount, and finish the other side (Fig. 66-5B).
32. Check the finished size of the dovetail.

NOTE: If the work is mounted centrally on a rotary table, it is possible to rotate the work a half-turn (180°) after step 18 and take the same cuts on each side of the block for each successive step.

To Mill an External Dovetail

1. Center the cutter over the dovetail position.
2. Remove as much material as possible from each side of the external dovetail, that is, cut it to the largest size of the dovetail. In this operation it will be necessary to note the readings from the centerline and remove the backlash for each side.
3. Mount a dovetail cutter, and center it with the workpiece.
4. Move the work over one-half the width of the dovetail plus half the diameter of the cutter. Allow 0.010 in. (0.25 mm) for a finish cut.
5. Take this roughing cut.
6. Move the work over an equal amount to the other side of the centerline, taking up the backlash.
7. Rough-cut the second side.
8. Adjust the work over, and take the finish cut on the one side.
9. Measure the dovetail using two rods.
10. Adjust for the finish cut on the second side, and take this cut.
11. Measure the width of the finished dovetail.

NOTE: If the work is mounted centrally on a rotary table, the work may be rotated a half-turn and the same cuts taken on each side.

JIG BORING ON A VERTICAL MILLING MACHINE

If a jig borer is not available, the vertical milling machine may be used for jig boring purposes.

When the vertical milling machine is used for accurate hole location, the same setup, locating, and machining methods used in a jig borer apply. The coordinate system of hole location is used in each case. Since this topic is covered extensively in Unit 68, the reader is advised to refer to this unit before jig boring on the vertical milling machine.

Since the vertical milling machine does not have the same lead screw accuracy as a jig borer, it must have some external measuring system to ensure the accuracy of the table setting. Measuring rods and dial indicators, a vernier scale, or optical measuring devices, such as digital readout boxes, may be used for this purpose.

Inch precision end-measuring rods (Fig. 66-6) are generally supplied in sets of 11 rods, including two micrometer heads capable of measuring from 4 to 5 in. to accuracy of 0.0001. These micrometer heads (Fig. 66-6B) are usually furnished with a red identifying ring on one and a black

Fig. 66-5 Milling a dovetail. (A) Center section roughed out; (B) both sides of the dovetail machined.

Fig. 66-6 Precision end-measuring rods. (A) A set of measuring rods; (B) a micrometer head; (C) a 3-in. measuring rod. (Courtesy The L.S. Starrett Company.)

Fig. 66-7 Using measuring rods to set the table position accurately. (Courtesy The L.S. Starrett Company.)

identifying ring on the other. One of these is used for longitudinal settings and the other for transverse settings. The solid rods are made of hardened and ground tool steel and have several concentric collars (Fig. 66-6C). The nine rods in a standard set are as follows: two in each of the following sizes: lengths of 1, 2, and 6 in. and one 12-in. rod. Other rods are available in lengths of 4, 5, 7, 8, 10, and 15 in. The rod ends (and the micrometer head ends) are hardened, ground, and precision lapped, providing extreme accuracy.

The rods are held in V-shaped troughs, one mounted on the milling machine table and the other on the saddle (Fig. 66-7). A stop rod is mounted at one end of the trough, while at the other end a long-range dial indicator graduated in 0.0001 is mounted.

A *metric precision end-measuring rod* set consists of two micrometer heads capable of measuring to an accuracy of 0.002 mm and a number of solid measuring rods.

A standard measuring rod set consists of two rods in the following lengths: 25, 50, 75, 150, and 300 mm. Other rods are available in lengths of 100, 125, 175, 200, 250, and 375 mm.

The table is positioned accurately from the X and Y coordinates by adding (or subtracting) specified lengths or buildups of various rod combinations and micrometer settings.

TO POSITION THE MILLING MACHINE TABLE USING MEASURING RODS

1. Set the spindle to the edge of the work. See Unit 68 for methods of locating an edge.
2. Clean the trough and ends of the stop rods and indicator rods.
3. Check the indicator for free operation.
4. Place the required number of rods, including the micrometer head, in the trough to take up the space between the stop rod and the indicator rod.
5. Adjust the micrometer until it has extended enough to cause the indicator needle to move a half-turn.
6. Lock the table.
7. Set the indicator bezel to zero.
8. Increase the rod and micrometer buildup by the length of the measurement between the side of the workpiece and the hole location.
9. Move the table along more than this required distance.
10. Insert the rods and the micrometer head.
11. Move the table back until the needle moves the half-turn and registers zero.
12. Lock the table.
13. Recheck the setting and adjust if necessary.

DIGITAL READOUT BOXES

Electronically controlled measuring devices, suitably mounted on the table and the saddle, indicate the table travel to an accuracy of 0.0005 or 0.0001 in. (0.012 or 0.002 mm). A digital readout box (Fig. 66-8) resembles the odometer on an automobile dashboard and indicates the distance traveled through a series of numbers visible through a glass front on the box. This arrangement permits quick and accurate setting of the machine table.

VERNIER SCALE

The milling machine may be equipped with scales on the table and saddle with pointers suitably mounted on the machine. The scales are generally graduated in increments of tenths of an inch or 2.0 mm. A vernier arrangement is mounted adjacent to the feed screw collars which permits the reading in 0.0001 in. or 0.002 mm.

NOTE: Settings on this machine must be made in one direction only in order to remove the backlash.

Vertical milling machines may be used to position holes without the use of any of the aforementioned equipment; however, this method is not too accurate.

The table is positioned by means of the graduated feed collars. This method may be used if the accuracy of the location must not be less than 0.002 in. or 0.05 mm. In order to eliminate backlash, it is important that all settings be made in one direction only.

Fig. 66-8 A digital readout box. (Courtesy Bendix Corporation.)

Fig. 66-9 Cutting an internal keyway with a slotting attachment. (Courtesy Cincinnati Milacron Incorporated.)

VERTICAL MILLING MACHINE ATTACHMENTS

The versatility of the vertical milling machine may be further increased by the use of the following attachments.

The *rack milling attachment* permits the machining of racks and broaches on the vertical mill. The spindle of this attachment is fitted into the spindle of the machine while the housing is clamped to the machine quill. It operates on the same principle as a rack milling attachment on the horizontal milling machine.

The *slotting attachment* (Fig. 66-9) is generally fitted to the back end of the overarm, which may be rotated 180° to permit the use of this device. The slotting attachment

Fig. 66-10   A tracer-controlled vertical milling machine.   (Courtesy Cincinnati Milacron Incorporated.)

operates independently of the machine drive on most vertical milling machines. It transmits a reciprocating motion to a single-pointed tool by means of a motor-driven eccentric. This attachment may be used for cutting keyways and slotting out small blanking dies.

DIE SINKING

One important application of the vertical milling machine is that of *die sinking*. Dies used in drop forging and die casting have impressions or cavities cut in them by means of die sinking. The operation of machining a die cavity on a vertical mill is generally done by hand control of the machine, using various end-mill-type cutters. This is usually followed by considerable filing, scraping, and polishing to produce the highly finished, properly curved and contoured surfaces required on the die.

Complicated shapes and patterns on molds and dies are made more conveniently and accurately on a vertical milling machine equipped with tracer control. In addition to the regular cutting head, the machine is equipped with a tracer head. Electrical discharge machining is finding ever-increasing use in the manufacture of molds and dies. Intricate shapes and forms, which would be difficult or impossible to produce on conventional machines, can be quickly and easily reproduced in even supertough, space-age metals.

On a tracer-controlled machine (Fig. 66-10) the form of a master or pattern is transferred to the workpiece by means of a hydraulic tracer unit actuated by a stylus or tracer finger. This contacts the master and moves the cutter up or down with the vertical or side travel of the stylus.

Since both the work and the master are fastened to the table, they both travel at the same rate. As the pattern moves under the stylus, the workpiece assumes the identical form of the master. Some machines may be equipped with ratio arms or devices, and the master can be made considerably larger than the finished workpiece. If a master is made 10 times larger, the arms on the machine are set at a 10:1 ratio and then the part is cut. An error of 0.010 in. (0.25 mm) on the master will only result in a 0.001 in. (0.025 mm) error on the workpiece.

Only a light pressure on the stylus, in contact with the master or template, is necessary to deflect the tracer arm and actuate the control valve which controls the movement of the cutting head.

REVIEW QUESTIONS

ROTARY TABLE

1. For what purpose may a rotary table be used?
2. Describe briefly the construction of a rotary table.
3. What is the purpose of the hole in the center of a rotary table?
4. What common ratios are found on rotary tables?
5. Describe briefly how a rotary table may be centered with the vertical mill spindle.
6. Explain how a number of identical parts, having a hole in the center, can be quickly aligned on a rotary table.

RADIUS AND T-SLOT MILLING

7. Explain how a large radius may be cut using a rotary table.
8. What purpose do T-slots serve?
9. List the two operations necessary in order to cut a T-slot.

MILLING DOVETAILS

10. What is the purpose of a dovetail?

11. What is the procedure for machining the center section of an internal dovetail?
12. Explain how the first angular side of a dovetail is cut.
13. How can an internal dovetail be measured accurately for size?

JIG BORING ON A VERTICAL MILL

14. Name and describe three types of measuring systems used on vertical mills for jig boring.
15. Name three methods of locating an edge.
16. How may a table be moved exactly 2.6836 in. (68.163 mm) from one location to another by using measuring rods?

VERTICAL MILLING MACHINE ATTACHMENTS

17. What is the purpose of the rack milling and slotting attachments?
18. How can a vertical mill be used for various die-sinking operations
    (a) manually?
    (b) automatically?
19. Explain the purpose of ratio arms or devices in die sinking.

THE JIG BORER AND JIG GRINDER

SECTION 14

The jig borer was developed primarily to overcome the toolmaker's perpetual problem of accurately locating and producing holes. The jig borer is especially useful in the manufacture of jigs and fixtures when there must be an accurate dimensional relationship between the locators aligning the workpiece and the bushing holes which are used to provide accurate hole locations. It is an invaluable machine tool in the manufacture of simple, compound, progressive, and lamination dies which require great accuracy between a variety of locating parts. Holes in the punch pad and die plate, pilot holes, and bushing holes which align the stripper plate are obvious "naturals" for a jig borer. Such holes can be produced quickly and accurately. By interchanging the tools in the spindle of the jig borer, operations such as drilling, boring, reaming, and counterboring can be readily performed.

The jig grinder was developed because of the need for accurately located holes in work material which had to be hardened. Much work produced in toolrooms must be hardened so that the part does not wear too quickly and that it will provide years of service. During the hardening process, material has a tendency to warp and distort, therefore the holes must be jig-ground to reposition the holes accurately.

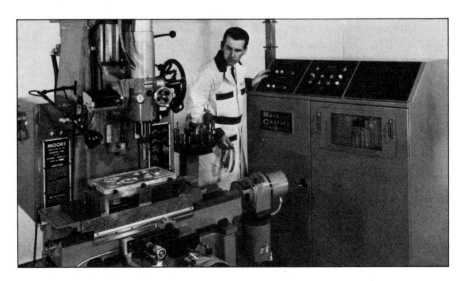

A jig borer (Fig. 67-1), although similar to a vertical milling machine, is much more accurate and built closer to the floor so that the worker can operate it while in a seated position. The precision-ground lead screws controlling the table movements are capable of infinitely fine incremental divisions and permit simultaneous measurement and positioning to within an accuracy of 0.0001 in. (0.002 mm) over the table length. This machine must be not only rugged for the heavy cuts necessary for roughing purposes but also sensitive for the more accurate finishing cuts.

OBJECTIVES

After completing this unit you will be able to:
1. Identify and state the purpose of the main operative parts of a jig borer
2. Use various accessories and work-holding devices for setting up and boring holes

JIG BORER PARTS

The *variable-pitch pulley drive* is operated by pushing a button on the *electric control panel* to provide the spindle with a variable-speed range from 60 to 2250 revolutions per minute (r/min).

The *quill housing* can be raised or lowered to accommodate various sizes of work if first the *quill housing clamp* is loosened and then the *quill housing vertical positioning handle* is turned.

The *brake level* is manually operated to stop the rotation of the spindle. It is especially useful while various tools are being removed or replaced in the spindle.

The *rapid feed handwheel* allows the spindle to be raised or lowered rapidly by hand.

The *friction clutch* may be used to engage or disengage the handfeed of the quill.

The *graduated downfeed dial*, by means of a vernier, reads the distance of vertical spindle travel in hundredths of a millimeter or in thousandths of an inch for inch-designed dials.

The *adjustable stop for hole depths* can be adjusted to allow the spindle to move to a predetermined depth for drilling or boring a hole.

The *spindle* revolves inside the quill and supplies drive for the cutting tools. An internal taper in the spindle allows a variety of tools to be rapidly and accurately interchanged.

The *reference scales* (longitudinal and crossfeed) serve as reference points in moving the table into position. They determine the position of the starting or reference point of the job.

The graduated dials, with microsetting verniers on the *longitudinal* and *crossfeed screw handwheels*, allow the table to be positioned quickly and accurately to within 0.002 mm or to within 0.0001 in. for inch-graduated dials.

TO INSERT SHANKS IN THE SPINDLE

Toolroom work, with a variety of hole sizes, requires frequent changing of tools (Fig. 67-2). The tapered hole in the spindle allows this to be done quickly and accurately if certain precautions are taken.

1. The taper shank on the tool being inserted and the hole in the spindle must be *perfectly clean;* otherwise the tapers will be damaged and inaccuracy will occur.

2. Protect the taper shanks from finger perspiration, especially if the shank will be in the machine for some time. This may cause both the shank and the spindle to rust.

3. Avoid inserting shanks too tightly, especially when the spindle is warmer than the inserted shank. When the shank warms up, it expands and may jam in the spindle.

4. When removing or replacing tools in the spindle, apply the brake firmly with the left hand. The wrench should be held carefully to prevent it from slipping out of the hand and damaging the machine table.

ACCESSORIES AND SMALL TOOLS

A wide variety of accessories enable a jig borer to meet three basic requirements: *accuracy, versatility,* and *productivity.* Only accessories concerned with drilling, boring, and reaming will be dealt with.

VARIABLE PITCH PULLEY DRIVE

FEED SELECTOR LEVER

QUILL HOUSING VERTICAL POSITIONING HANDLE

GRADUATED DIAL DOWNFEED

RAPID FEED HANDWHEEL

BRAKE LEVER

FRICTION CLUTCH POWER FEED

POWER FEED CLUTCH

QUILL HOUSING

SPINDLE

MICROMETER DEPTH STOP

LONGITUDINAL REFERENCE SCALE

ELECTRIC CONTROL PANEL

TABLE CLAMP

CROSS REFERENCE SCALE

LONGITUDINAL LEADSCREW HANDWHEEL

MICRO-SETTING VERNIER

CROSS-SLIDE CLAMP

CROSS LEADSCREW HANDWHEEL

Fig. 67-1   A #3 Moore precision jig borer. (Courtesy Moore Special Tool Company Inc.)

Fig. 67-2   Replacing a tool into the jig borer spindle.   (Courtesy Moore Special Tool Company Inc.)

Fig. 67-3   A set of solid boring bars.

Fig. 67-4   A swivel block boring chuck.
(Courtesy Moore Special Tool Company Inc.)

Fig. 67-5   A rugged, versatile boring chuck designed for heavy cuts.

Fig. 67-6   A DeVlieg microbore boring bar.   (Courtesy DeVlieg Machine Company.)

DRILLING

Key-type and keyless chucks are used to hold smaller-sized straight-shank spotting tools, drills, and reamers. Special collets are used to hold larger straight-shank spotting tools, drills, and reamers. A setscrew in the collet is tightened against a flat on the tool shank, providing a positive grip which eliminates twisting and scuffing of the tool shank.

BORING

Single-point boring, the most accurate method of attaining locational accuracy in jig boring, makes it necessary to have a wide variety of boring tools. The most commonly used boring tools are a solid boring bar, a swivel block boring chuck, an offset boring chuck, and a DeVlieg microbore boring bar.

The solid boring bar (Fig. 67-3) is fitted with an adjusting screw which, when adjusted, advances the toolbit over a relatively short range. Solid boring bars are rigid, making them especially useful in boring deep holes; a number of solid boring bars may be left set at a specific size for repetitive boring.

The swivel block boring chuck (Fig. 67-4) provides a greater range of adjustment than other types in proportion to its diameter and better visibility to the operator while boring. One disadvantage of this type of chuck is that since the tool swings in an arc, the graduations for adjusting the tool travel vary depending upon the length of cutting tool used.

The offset boring chuck (Fig. 67-5) is a versatile tool which permits the cutting tool to be moved outward at 90° to the spindle axis of the machine. This allows the use of a wide variety of cutting tools without altering the value of the adjusting graduations. This chuck makes it possible to perform operations such as boring, counterboring, facing, undercutting, and machining outside diameters.

The DeVlieg microbore boring bar (Fig. 67-6) is equipped with a micrometer vernier adjustment which permits the cutting tool to be accurately adjusted to within a tenth of a thousandth.

Single-Point Boring Tools

Since single-point boring is the most accurate method of generating accurate hole location, a wide variety of cutting tools is available for this operation. The toolbits shown in Fig. 67-7 are generally used for small holes; however, with the necessary chuck attachments, larger holes may be bored. These toolbits are available in high-speed steel and also with brazed, cemented-carbide tips.

Collets and Chucks

An assortment of collets and chucks (Fig. 67-8) is available for a jig borer spindle to hold straight-shank spotting tools, drills, and precision end-cutting reamers.

REAMERS

Two types of reamers, the rose, or fluted, and the precision end cutting, are used in jig boring for bringing a hole to size quickly. The *rose,* or *fluted reamer,* is used after a hole has been bored and provides an accurate method of sizing a hole. If handled carefully and used to remove only about 0.001 to 0.003 in. (0.02 to 0.07 mm), these reamers maintain fairly accurate hole sizes.

Precision end-cutting reamers, with short, sturdy shanks (Fig. 67-8), provide the fastest method of locating and sizing holes to within an accuracy of ±0.0005 in. (0.01 mm). The end-cutting reamer, held rigidly and running true with the spindle, acts like a boring tool and reamer, locating and sizing the hole at the same time. End-cutting reamers eliminate the use of boring tools if the accuracy of the hole diameter and location does not require closer tolerance than ±0.0005 in. (0.01 mm).

WORK-HOLDING DEVICES

A wide variety of work-holding devices is necessary to fasten work for jig boring. Parallel setup blocks, matched parallel setup angle irons, matched box parallels, and extension parallels help to set up and align a variety of work shapes. To prevent machining into the table surface, most work held on a jig borer is mounted on parallels or another suitable device. *Bolts* and *strap clamps* (Fig. 67-9) provide an efficient and convenient method of holding most types of work. The heel rests for these clamps are made of brass to avoid marring the table and will build up to any height from approximately ⅜ to 9 in. (9.52 to 228.6 mm).

The precision vise (Fig. 67-10) is a valuable accessory for holding work too small to be held with bolts and strap clamps. It has stepped jaws which serve as parallels and a V-slot for holding round work. The vise is mounted on a base plate which is ground square and parallel to the stationary jaw, allowing the vise to be aligned quickly and accurately against the straightedge of the machine.

Fig. 67-7   Single-point boring tools.   (Courtesy Moore Special Tool Company Inc.)

Fig. 67-8   A hardened and ground collet holding a precision end-cutting reamer. (Courtesy Moore Special Tool Company Inc.)

Fig. 67-9   Bolts, strap clamps, heel rests.   (Courtesy Moore Special Tool Company Inc.)

Fig. 67-10   A precision vise.   (Courtesy Moore Special Tool Company Inc.)

V-blocks are used to support and align cylindrical work-pieces. If the work is long enough to require the use of two V-blocks (Fig. 67-11), it is important that a matched set be used to ensure parallelism with the table surface. The work must also be aligned parallel with the table travel. Indicate along the diameter of the cylinder and tap the V-blocks to correct any error.

Fig. 67-11 Cylindrical work set up in V-blocks.   (Courtesy Moore Special Tool Company Inc.)

The microsine plate (Fig. 67-12), based on the sine bar principle, is used to hold work for the machining of angular holes. It can be set to any angle from 0 to 90° by the use of the proper gage block buildup. A clamp rod is used to prevent the microsine plate from moving during machining operations. The surface of the sine plate is large enough to allow the rotary table (Fig. 67-12) to be mounted for work requiring compound angular setups and spacings.

The rotary table (Fig. 67-12) can be mounted on the machine table and then used for spacing holes accurately in a circle. The table is accurately graduated around its circumference in half-degree divisions and, by means of a vernier on the handwheel dial, settings accurate to within ± 12″ or less can be made. When holes are required at 90° to the axis of the work, the *Moore Precision Rotary Table* is constructed so that it may be mounted in a vertical position on the table of the jig borer. A lapped bushing in the center of the table may be used for aligning workpieces centrally. On some rotary tables, the handwheel can be removed and replaced with an index plate attachment. This is especially valuable where a large number of holes or graduations are required, since it eliminates errors resulting from calculation of the angles.

REVIEW QUESTIONS

JIG BORER

1. For what purpose were jig borers developed?
2. For what type of work are they especially valuable?
3. Name the operations which can be performed on a jig borer.
4. Explain the difference between a jig borer and a vertical milling machine.

Fig. 67-12 Combining the microsine plate and the rotary table for compound angle setups and indexing spacing.   (Courtesy Moore Special Tool Company Inc.)

5. State the purpose of each of the following: quill housing, vertical positioning handle, brake lever, rapid feed handwheel, adjustable stop for hole depths, spindle.

6. What purpose do reference scales serve in jig boring?

7. Explain why longitudinal and crossfeed settings may be set quickly and accurately to 0.0001 in. (0.002 mm).

8. List three precautions that should be observed while inserting shanks in the machine spindle.

ACCESSORIES AND SMALL TOOLS

9. Name four common boring tools and explain the advantages of each.

10. Name two types of reamers used in jig boring and explain the advantages of each.

WORK-HOLDING DEVICES

11. Explain why a precision vise is a very valuable jig borer accessory.

12. How should a long, cylindrical piece of work be set up and aligned?

13. State the purpose of
  (a) a microsine plate
  (b) a rotary table

14. Explain how the versatility of the rotary table may be increased.

JIG-BORING HOLES

UNIT 68

Since the beginning of the machine age, the problem of locating holes accurately has plagued the toolmaker. Before the development of accurate locating and measuring machines, the toolmaker was faced with a tedious and costly, though reasonably accurate, method of locating holes. The *jig borer,* first developed in 1917, now provides the toolmaker with a means of quickly and accurately locating holes to within an accuracy of 0.000090 in. over 18 in. of length (0.002 mm over 457.2 mm of length). It is used to finish-bore holes in material left soft or to rough-bore holes in work which will later be hardened and jig-ground.

OBJECTIVES

After completing this unit you will be able to:
1. Set up and align workpieces for jig boring
2. Calculate hole locations using the coordinate system
3. Jig-bore and inspect holes

TO SET UP WORK

Several methods of setting up work are used in jig boring. The most common are:

1. Setting the work parallel to the machine table and aligning one edge of the workpiece with the table travel
2. Mounting the work on a sine plate for angular machining
3. Mounting the work on a rotary table for the angular spacing of holes in a circle.

The most common type of workpiece is flat and rectangular and is one of the easiest to set up. The basic requirement for this type of work is that *two edges be accurately ground at right angles to each other*. This greatly assists in aligning the workpiece and also provides an accurate surface for aligning the center of the spindle with an edge.

The work is usually placed on parallels to avoid cutting the table and is fastened by means of bolts and strap clamps. It is important that the supporting parallels be directly under the clamps to prevent the work from being sprung or bowed, as in Fig. 68-1.

The edge of the workpiece may be aligned with the table travel by setting one of the two ground edges against the straightedge of the machine. Should it be desirable to have the work closer to the center of the table, parallel setup blocks or gage blocks may be used between the straightedge and the work (Fig. 68-2). To ensure correct alignment, it is good practice to check the work with a dial indicator.

TO ALIGN THE WORK EDGE WITH AN INDICATOR

On work that does not allow the use of the straightedge, alignment may be accomplished by holding an indicator in the machine spindle (Fig. 68-3).

1. Clamp the work lightly to the table.
2. Hold the indicator shank in a collet or drill chuck in the machine spindle.
3. Bring the indicator against the ground edge of the work and make it register approximately 0.020 in. (0.50 mm).
4. Turn the machine spindle slightly in each direction and stop when the indicator registers the highest reading. This establishes a 90° relationship between the indicator point and the work edge.
5. Set the indicator dial to zero.
6. Turn the jig borer table handwheel to move the workpiece edge past the indicator point. In this way, the indicator will show how much error there is in the work alignment.

Fig. 68-1 The spring caused by improper clamping will result in holes machined out of square. (Courtesy Moore Special Tool Company Inc.)

7. Then gently tap the work until the indicator shows no movement while the work is moved past the indicator.

NOTE: The work should *always be tapped away from the indicator,* not toward it, to prevent damage to the indicator.

8. When no indicator movement is shown, tighten the clamping nuts and *recheck the alignment* with the indicator.

METHODS OF LOCATING AN EDGE

Once the work has been set parallel with the table travel, it is necessary to align the center of the jig borer spindle to some reference or starting point. The reference points vary greatly and may include a hole, pin, boss, scribed line, slot or contour, or an edge. The two most commonly used reference points are a hole or an edge.

BACKLASH

When a reference point is being picked up, table dials set, or the table positioned, *the movement must always be made in the direction of the arrows on the dials* (Fig. 68-9) to eliminate errors as the result of *backlash.* If the crossfeed or longitudinal dial is turned past the required setting, it is necessary to back the dial away from the setting approximately one-quarter turn to eliminate backlash, and then reset it by turning it in the proper direction. When it is necessary to move the table backward (in the opposite direction of the dial arrows), turn the dial past the setting required by at least one-quarter turn, and then make the final setting in the direction of the arrows.

Fig. 68-2 Setup blocks spacing the work from the straightedge and parallel to the table travel. (Courtesy Moore Special Tool Company Inc.)

Fig. 68-3 Aligning the edge of the work parallel with the table travel with an indicator. (Courtesy Moore Special Tool Company Inc.)

Fig. 68-4 Picking up the edge with an edge-finder and indicator. The mirror assists in reading the indicator when it is turned away from the operator. (Courtesy Moore Special Tool Company Inc.)

Fig. 68-5 A spring-loaded edgefinder. (Courtesy Brown & Sharpe Mfg. Co.)

Fig. 68-6 Picking up an edge without an edgefinder. The indicator is set against the work, raised, rotated 180°, and set against a gage block held against the edge. (Courtesy Moore Special Tool Company Inc.)

PICKING UP AN EDGE WITH AN EDGEFINDER

The *edgefinder* (Fig. 68-4) is a valuable accessory for picking up an edge. It is constructed so that the surface of the edgefinder which is held against the edge of the work is exactly in the center of the slot used for indicating purposes.

1. Hold the edgefinder firmly against the work. On some workpieces, it may be possible to hold the edgefinder to the work with the aid of a clamp.
2. Adjust the indicator holder so that the indicator point just touches one edge of the slot.

3. Bring the indicator into contact with one side of the slot, and have it register approximately 0.010 to 0.020 in. (0.25 to 0.50 mm).
4. Make sure that the indicator is at right angles to the edge by *slightly turning* the jig borer spindle backward and forward. Stop when the indicator registers its highest reading and *set the indicator dial to zero*.
5. Turn the machine spindle one-half turn, stopping when the indicator registers its highest reading on the other edge (Fig. 68-4).
6. Turn the table handwheel in the direction of the arrow one-half the difference between the two indicator readings.
7. Repeat steps 4, 5, and 6 until both indicator readings are exactly the same.
8. When both indicator readings are identical, the center of the machine spindle is located on the edge of the work.

Work may be quickly located to within 0.0005 in. (0.01 mm) by means of a spring-loaded edgefinder (Fig. 68-5) mounted in a drill chuck or collet. This device, which has a movable head the same diameter as the shank, moves approximately 1/32 in. (0.79 mm) off center when the edgefinder is rotated. As the edgefinder is brought against the edge of the workpiece, the eccentricity of the head decreases. When the head of the edgefinder runs concentric with the body, the edge of the workpiece will be half the diameter of the head from the centerline of the spindle.

TO PICK UP AN EDGE WITHOUT AN EDGEFINDER

If an edgefinder is not available, it is still possible to pick up an edge fairly accurately by using the method illustrated in Fig. 68-6.

1. Mount the indicator into the machine spindle.
2. Adjust the indicator holder so that the indicator point is as close as possible to the center of the machine spindle.
3. Bring the indicator into contact with the edge of the work, and have it register approximately 0.010 to 0.020 in. (0.25 to 0.50 mm).
4. Make sure the indicator is at right angles to the edge by *slightly turning* the jig borer spindle backward and forward. Stop when the indicator registers its lowest reading and *set the indicator dial to zero*.
5. Raise the machine spindle to clear the top of the work and rotate the spindle 180° (one-half turn).
6. Place a gage block against the work edge (Fig. 68-6), and turn the machine spindle slightly backward and forward. Stop when the indicator registers its lowest reading.
7. Turn the table handwheel in the direction of the arrow, one-half the difference between the two indicator readings.
8. Repeat steps 4, 5, 6, and 7 until both indicator readings are exactly the same.

TO LOCATE WITH A LINEFINDER

A method which is not very accurate but is sometimes used to pick up an edge or a scribed line, is to locate with a *linefinder* or *wiggler*.

1. Mount the wiggler (Fig. 68-7) into a drill chuck held in the machine spindle.

Fig. 68-7 Picking up an edge with a line-finder. (Courtesy Moore Special Tool Company Inc.)

Fig. 68-8 The Moore locating microscope permits optical pickups where conventional means are impractical. (Courtesy Moore Special Tool Company Inc.)

Fig. 68-9 The relationship of the reference scales and dials with the coordinate dimensions. (Courtesy Moore Special Tool Company Inc.)

2. Start the machine and make the wiggler run true by bringing a finger into contact with the wobbling point.
3. Move the table so that the point of the wiggler is as close as possible to the reference line or edge.
4. Use a magnifying glass so the final settings may be made as accurately as possible.

TO LOCATE WITH A MICROSCOPE

There are times when the reference point on the work does not suit the pickup methods of an indicator or linefinder. When small or partial holes, irregular contours, slots, and punch marks are used as reference points, a *locating microscope* (Fig. 68-8) is used. This microscope has a 40 × magnification, great enough to permit 0.0001 in. (0.002 mm) to be seen. The reticle reference on the microscope consists of a number of concentric circles and two pairs of crossed centerlines for picking up a wide variety of reference points.

THE COORDINATE LOCATING SYSTEM

The *coordinate* locating system is the most efficient method of establishing hole locations for jig-boring operations. This system eliminates the tedious and often inaccurate methods of hole location, such as buttoning and layout. It also permits the best sequence of operations, such as spotting, drilling, and rough boring all holes on a workpiece before the finish-boring operation. The coordinate locating system provides the best conditions for maintaining precision hole location in jig boring.

The two general types of coordinates used are:

1. *Rectangular coordinates,* for dimensions given in straight lines
2. *Polar coordinates,* used with rotary table for holes on circles and showing angles and distances from a zero line or center

RECTANGULAR COORDINATES

The rectangular coordinate system consists of establishing the relationship of the work to be jig-bored to a pair of crossed ordinates, or *zero lines*. On the Moore jig borer, the zero lines are at the upper left-hand corner of the table. These crossed ordinates, sometimes referred to as the X and Y axes, represent zero readings on both the crossfeed and longitudinal reference scales of the jig borer. Figure 68-9 shows the relationship of the reference scales and the lead screw dials with the coordinate dimensions. (Most jig borers on the market are capable of providing readouts in both inches and millimeters.)

After the work is fastened to the table and the spindle aligned with *both edges* of the work, the reference scales (longitudinal and crossfeed) should be set to the nearest inch line. For example, if the crossfeed reference scale is set at 4.000 in. and the longitudinal reference scale is set at 6.000 in., the figures 4.000 in. and 6.000 in. are immediately written on the corresponding edge of the drawing in the correct relationship to the table movement.

Figure 68-10 shows a conventionally dimensioned drawing where some of the dimensions are from an edge, while others are from hole to hole. Figure 68-11 shows the same drawing dimensioned using the coordinate system. The 4.000 in. on the upper right-hand corner and the 6.000 in. on the upper left-hand corner correspond to the settings on the crossfeed and longitudinal reference scales. All dimensions are added to these two points and are labeled on the diagram correspondingly (Fig. 68-11). For example, the three holes on top are all 0.500 in. from the edge which now becomes 4.500 in. from the reference point on the scale. The distance between these three and the hole immediately below is 0.5625 in. which now becomes 5.0625 in. (4.000 + 0.500 + 0.5625) from the reference point.

POLAR COORDINATES

This is the conventional dimensioning system used to indicate the location of holes on a circle which are all the same distance from a common center. Such dimensions may be given either in the form of angles between holes, or by the number of equally spaced holes required on the circle. To convert the number of equally spaced holes into an angular value between each hole, divide this number into 360°. Great care must be taken in these calculations to avoid error, especially if the number is not evenly divisible into 360°.

Once polar coordinates are calculated, they lend themselves directly to use with the rotary table. Here also, if many holes that do not divide evenly into 360° are required, great care must be taken in the indexing of the rotary table since a small error between each hole can result in a sizable cumulative error. The example in Fig. 68-12 shows the problems which can be encountered in calculating the angular spacing when the number is not evenly divisble into 360°.

The Woodworth tables (see Appendix) were developed by W. J. Woodworth and J. D. Woodworth to eliminate the problems and inaccuracies encountered in polar coordinates. The tables establish rectangular coordinates for each hole from an upper and left-hand tangent line, enabling holes to be accurately located without the use of a rotary table. A complete set of coordinate tables for up to 100 holes may be found in the book *Holes, Contours and Surfaces,* published by the Moore Special Tool Company, Bridgeport, Connecticut.

To Calculate Rectangular Coordinates Using Woodworth Tables

1. Determine the coordinates of line A (left tangent) and line B (upper tangent). These two coordinates would be taken from the reference scales on the jig borer table.
2. Multiply the A factor by the diameter of the circle for the location of each hole from the left tangent line (Fig. 68-13).
3. Add the result of the calculation to the coordinate of the A line.
4. Multiply the B factor by the diameter of the circle for the location of each hole from the upper tangent line (Fig. 68-13).
5. Add the result of the calculation to the coordinate of the B line.

PREFIGURING COORDINATES

In order to save operators' time and secure the greatest amount of production from a jig borer, many firms are supplying drawings where dimensions are already given in coordinate readings. In this way, after the job is clamped to the table, the operator is required only to set the reference scales to correspond to the starting point indicated on the drawing and proceed with the job. When coordinate di-

Fig. 68-10   A conventional dimensioned drawing.   (Courtesy Moore Special Tool Company Inc.)

Fig. 68-11   The same drawing with rectangular coordinates.   (Courtesy Moore Special Tool Company Inc.)

PROBLEM:
CHART ANGLES OF HOLES
IN 67-HOLE CIRCLE

360° ÷ 67 = 5 DEGREES 22 MINUTES 23-19/67 SECONDS
STARTING AT ZERO, ADD AND TABULATE AS FOLLOWS.

HOLE NO.	DEGREES	MINUTES	SECONDS
0	0		
1	5	22	23-19/67
ADD	5	22	23-19/67
2	10	44	46-38/67
ADD	5	22	23-19/67
3	16	7	9-57/67
ADD	5	22	23-19/67
4	21	29	33-9/67
ADD	5	22	23-19/67
5	26	51	56-28/67
ADD	5	22	23-19/67
6	32	14	19-47/67
	etc.	etc.	etc.

Fig. 68-12   Steps required to calculate angles between holes on a circle. To prevent cumulative error, each small fraction of an angle must be added every time. (Courtesy W.J. & J.D. Woodworth.)

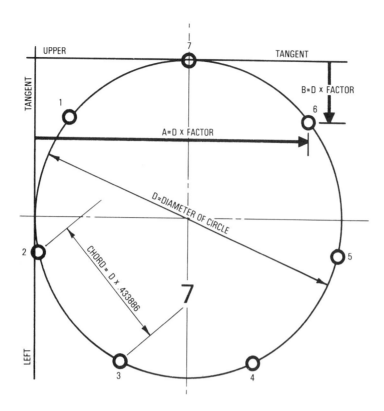

Fig. 68-13   Rectangular coordinate factors and angles for seven evenly spaced holes.   (Courtesy W.J. & J.D. Woodworth.)

→	Factor for A		Factor for B	↓	Angle of Hole		
					Degree	Minute	Second
1	.109084	1	.188255	1	51	25	42-6/7
2	.012536	2	.611261	2	102	51	25-5/7
3	.283058	3	.950484	3	154	17	8-4/7
4	.716942	4	.950484	4	205	42	51-3/7
5	.987464	5	.611261	5	257	8	34-2/7
6	.890916	6	.188255	6	308	34	17-1/7
7	.500000	7	.000000	7	360	0	0

mensions are not supplied, it is good practice for the operator to convert all measurements into coordinate dimensions before proceeding with the job. This procedure will not only save time, but also prevent many errors occurring as a result of alternation between jig-boring operations and locational calculations.

MAKING SETTINGS

Some jig borer manufacturers use measuring rods and indicators, while others, such as the Moore Special Tool Company, use accurate lead screws with vernier readings on the dial to move the table to the required location. Regardless of the system employed, it is important to remember to make *all locational settings* by turning the table handwheels *in one direction only* to avoid errors resulting from backlash. On Moore jig borers, this direction is indicated by an arrow on the crossfeed and longitudinal dials. Whenever it is necessary to move the table in the opposite direction, it is important that the table be moved past the setting by approximately one-quarter turn and then brought to the desired setting by the handwheel being turned in the direction of the arrows. This procedure will eliminate errors as a result of backlash.

Fig. 68-14 A telescopic gauge used to measure hole size. (Courtesy Moore Special Tool Company Inc.)

Fig. 68-15 A plug gage is an accurate means of checking final hole size. (Courtesy Moore Special Tool Company Inc.)

Fig. 68-16 Hole size is read directly from the leaf taper gage at the point of contact with the edges of the hole. (Courtesy Moore Special Tool Company Inc.)

JIG-BORING PROCEDURE

Because of the wide variety of work encountered in jig boring, no standard procedure would always apply. However, the following sequence should be followed whenever possible.

1. Set up and carefully align the work parallel to the table level.
2. Align the center of the machine spindle with the reference point on the work. Set the reference scales of the machine to the nearest major measurement line.
3. Calculate the coordinate location of all holes and mark them on the drawing of the workpiece.
4. Spot the location of all holes lightly with a center drill or spotting tool.
5. Respot all the holes to a depth which will provide a good guide for future drilling operations. This operation is a means of rechecking the initial spotting operation to ensure that errors have not been made in reading the scale or dials.
6. Drill holes over ½ in. (12.7 mm) in diameter to within 1/32 in. (0.79 mm) of finish size. For holes less than ½ in. (12.7 mm) in diameter, drill to within 1/64 in. (0.39 mm) of finish size.

NOTE: When drilling large holes, it is advisable to enlarge the hole in ¼-in. (6.35-mm) steps to avoid undue drilling pressure which may move the work location.

7. Recheck the work alignment, as well as the alignment of the spindle with the reference points, to make sure it has not shifted during the roughing operation. If any error is evident, it will be necessary to repeat steps 1 and 2 before proceeding.
8. Rough-bore all holes to within 0.003 to 0.005 in. (or 0.07 to 0.12 mm) of the size required. Work requiring precise accuracy should be allowed to return to *room temperature* before the finish-boring operation.
9. Finish-bore all holes to the required size.
10. Mount an indicator in the machine spindle and inspect the accuracy of the jig boring before removing the work from the machine.

NOTE: If the accuracy of the work is not required to closer tolerances than ±0.0005 in. (0.01 mm), a precision end-cutting reamer may be used to finish the hole, a procedure which eliminates steps 8 and 9.

MEASUREMENT AND INSPECTION OF HOLES

MEASURING HOLE SIZE

A wide variety of instruments are available for measuring hole sizes to various accuracies.

Telescopic and small hole gages are used in the same manner as inside calipers. However, because they are more rigid, greater measuring accuracy is possible. Figure 68-14 shows a telescopic gage being used to measure the hole diameter.

A plug gage (Fig. 68-15) is more accurate than either calipers or telescopic gages for checking a hole diameter. It has a serious limitation in that it is effective only in checking a hole whose size is exactly that of the gage. Since a plug gage does not relate hole size in measurement units, it is impossible to determine how much material must still be removed from the hole.

Flat leaf taper gages (Fig. 68-16) provide a rapid means of checking hole sizes, especially during the rough-boring operation. The most common set consists of 36 leaf taper gages, allowing holes from 0.095 to 1.005 in. (2.41 to 25.52 mm) to be measured. Each leaf taper gage is 1½ in. (38.1 mm) long, marked off in 0.001-in. (0.025-mm) graduations, and has a range of 0.030 in. (0.76 mm). Because these gages are tapered, they cannot determine whether a hole is bell-mouthed or tapered and cannot measure to within a tenth of a thousandth.

Vernier calipers and various types of *inside micrometers* are instruments capable of measuring hole sizes accurately to within hundredths of a millimeter or thousandths of an inch. They are direct-reading instruments and, therefore, do not require their settings to be transferred to another instrument.

The internal indicator gage (Fig. 68-17) is an extremely accurate instrument capable of measuring hole sizes to 0.0001 in. (0.002 mm). This gage is set for a particular size against either a standard ring gage or a micrometer, and then the indicator dial is revolved until the needle is on the zero line. The gage is then inserted into the hole and the hole size is compared to the indicator setting. Errors, such as taper, bell mouth, or out-of-roundness, are easily checked with this type of instrument.

INSPECTING HOLE LOCATION

After the work has been jig-bored, it is important to verify the accuracy of the hole locations. Many methods of inspecting hole locations may be employed; the method chosen depends upon the accuracy required.

It is good practice to use the *jig borer* to inspect the accuracy of hole locations before removing the work from the machine. This is one of the most accurate and convenient methods of inspection since the work is already set up in the machine. An *indicator* (Fig. 68-18) is mounted in the machine spindle and the original reference point is then picked up. From this reference point, the table is moved to the various hole locations, which are checked for accuracy by having the indicator entered into the hole and the machine spindle revolved. The accuracy of the hole location is shown by the amount the indicator needle varies. A locating microscope, mounted in the jig borer spindle, is especially valuable in inspecting locations given from reference points, such as small or partial holes, contours, or slots.

The advantages of using the jig borer for inspecting hole locations are:

1. The work is already set up, saving time and eliminating errors which could occur in setting up work again.
2. The same coordinate dimensions which were used for jig boring are used for inspection.
3. The machine's measuring system is just as accurate as most measuring standards.
4. The indicator mounted in the machine spindle may be used to check the location, out-of-roundness, bell mouth, or taper of a hole.
5. Work which was bored to polar coordinates can be inspected with the use of rectangular coordinates; the accuracy of rotary table calculations and settings will also be checked.

Vernier calipers (Fig. 68-19) provide a rapid, but not very accurate method of measuring the distance between two holes. The accuracy in using vernier calipers is subject to the following factors.

1. Improper tension on the instrument
2. Mistakes in reading the vernier caliper
3. Errors in angular alignment which are not easily noticed

An outside micrometer may be used to measure the distance between tightly fitting plugs which have been inserted into the holes. The accuracy of this method is subject to the following factors.

1. Holes being out of square with each other

Fig. 68-17 The internal indicator gage provides extremely accurate measurements. (Courtesy Moore Special Tool Company Inc.)

Fig. 68-18 An indicator mounted in the spindle for inspection purposes. (Courtesy Moore Special Tool Company Inc.)

2. Looseness between the plug and the hole
3. Burrs or dirt in the hole or on the plug

A fairly accurate method of inspecting hole location is by means of *gage blocks and a dial indicator* (Fig. 68-20). The work is clamped to an angle plate with the finished edge resting on the surface plate, and the distance from the work edge to the hole surface or inserted plug is calculated. A gage block buildup for the calculated dimension is set up and the dial indicator set to the blocks. The indicator is then passed over the surface of the plug or into the hole, comparing this location with the gage block buildup.

The measuring machine, specially developed for inspecting hole locations to less than 0.0001 in. (0.002 mm), is the most accurate instrument used for inspecting purposes. The Moore measuring machine (Fig. 68-21) incorporates the same basic principles used in jig borers for accurately establishing hole locations.

Fig. 68-19 Measuring the distance between two holes with a vernier caliper. (Courtesy Moore Special Tool Company Inc.)

Fig. 68-20 Measuring the distance from the edge of the work to a plug fitted in a hole. (Courtesy Moore Special Tool Company Inc.)

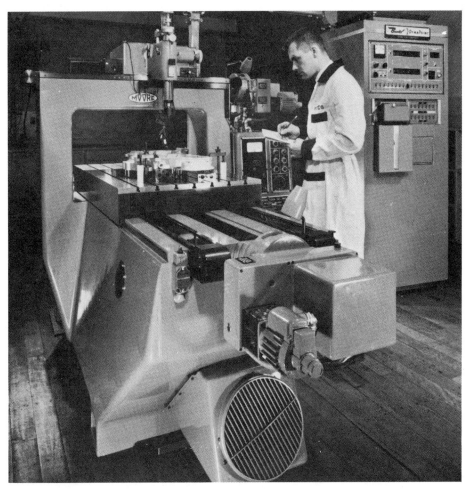

Fig. 68-21 The Moore measuring machine. (Courtesy Moore Special Tool Company Inc.)

REVIEW QUESTIONS

1. Name three methods of setting up work on a jig borer.
2. What requirements are necessary for rectangular work before it is set up in a machine? Explain why these are necessary.
3. What precautions should be observed for clamping?
4. Briefly describe two methods of aligning the edge of a workpiece parallel to the table travel.
5. Why should the work always be tapped away from an indicator, not toward it?

METHODS OF LOCATING AN EDGE

6. What must be done to eliminate errors resulting from backlash?
7. If the dial is turned past the required setting, how is backlash eliminated?
8. Name four methods used to pick up an edge or reference point.
9. Describe a locating microscope and state the purpose for which it is used.

THE COORDINATE LOCATING SYSTEM

10. Define and state the purpose of the two types of coordinates.
11. Explain the principle of the coordinate locating system and how it is used in jig boring.
12. What purpose do the reference scales serve in the coordinate locating system?

MAKING SETTINGS

13. Why should all locational settings be made in the same direction?
14. What must be done when it is necessary to move the table in the direction opposite to that used for locational settings?

JIG-BORING PROCEDURE

15. Why is it advisable to calculate all the coordinate dimensions before proceeding with jig boring?
16. What is the purpose of respotting all holes after the initial spotting operation?
17. What precaution should be observed after the drilling operation to ensure accurate hole locations?
18. When is it advisable to use precision end-cutting reamers?

MEASUREMENT AND INSPECTION OF HOLES

19. List four instruments used to measure hole size.
20. Describe a precision instrument which can accurately measure hole size.
21. State the advantages of using the internal indicator gage for checking holes.
22. List the advantages of using the jig borer for inspection purposes.

THE JIG GRINDER
UNIT 69

The need for accurate hole locations in hardened material led to the development of the jig grinder in 1940. While it was originally developed to position and grind accurately straight or tapered holes, many other uses have been found for the jig grinder over the years. The most important of these has been the grinding of contour forms which may include a combination of radii, tangents, angles, and flats (Fig. 69-1).

The advantages of jig grinding are:
1. Holes distorted during the hardening process can be accurately brought to correct size and position.
2. Holes and contours requiring taper or draft may be ground. Mating parts, such as punches and dies, can be finished to size, eliminating the tedious job of hand fitting.
3. Because more accurate fits and better surface finishes are possible, the service life of the part is greatly prolonged.

4. Many parts requiring contours can be made in a solid form, rather than in sections as was formerly necessary.

OBJECTIVES

After completing this unit you will be able to:
1. Select the wheels and grinding methods required for jig-grinding holes
2. Set up work and jig-grind a straight hole to within a tolerance of ±0.0002 in. (0.005 mm)

JIG GRINDER PARTS

The jig grinder (Fig. 69-2) is similar to a jig borer, both having precision-ground lead screws capable of positioning the table within 0.0001-in. (0.002-mm) accuracy over its entire length. Both are vertical spindle machines and em-

Fig. 69-1   A flanged punch represents an ideal example of jig grinding. (Courtesy Moore Special Tool Company Inc.)

Fig. 69-2   A #3 Moore jig grinder. (Courtesy Moore Special Tool Company Inc.)

ploy the same basic cutting principle encountered in single-point boring. The main difference between these two machines is in the spindles.

The jig grinder is equipped with a high-speed pneumatic turbine grinding spindle for holding and driving the grinding wheel. The spindle construction permits outfeed grinding (Fig. 69-3), and also the grinding of tapered holes (Fig. 69-4).

GRINDING HEAD OUTFEED

A horizontal dovetail slide connects the grinding head to the main spindle of the jig grinder. The grinding head may

MAIN SPINDLE

GRINDING SPINDLE

Fig. 69-3   The grinding spindle may be offset from the main spindle. Lower view shows the planetary path of rotation.

TAPER SETTING SCREWS

PIVOT POINT FOR TAPER SETTING

Fig. 69-4   Main spindle assembly.

433

Fig. 69-5 Assembly for controlling size by outfeed. Dial setting outfeed is graduated in tenths. (Courtesy Moore Special Tool Company Inc.)

be offset from the center of the main spindle to grind various size holes. The amount of eccentricity (offset) of the grinding head can be *accurately* controlled by the internally threaded *outfeed dial* which is mounted on the nonrotating yoke at the top of the jig-grinding spindle (Fig. 69-5). The dial is graduated in steps of 0.0001 in. (0.002 mm) permitting accurate control of the hole size during grinding.

Coarse adjustment of the grinding wheel position is attained by a fine pitch-adjusting screw within the dovetail slide (Fig. 69-5). This coarse-adjusting screw is accessible only when the machine spindle is stopped.

DEPTH-MEASURING DEVICES

The Moore jig grinder has three distinct features for controlling and measuring the depth of holes (Fig. 69-6A and B).

1. The *adjustable positive stop* is on the left-hand end of the pinion shaft. Microadjustment can be made by a limiting screw.
2. The *graduated dial* on the downfeed handwheel indicates the travel of the quill. It can be set to zero at any position and reads the travel depth in steps of 0.001 in. (0.02 mm).
3. The *micrometer stop* (Fig. 69-6), fastened to the column of the grinder, controls hole depth.

Fig. 69-6 Depth-measuring devices. (Courtesy Moore Special Tool Company Inc.)

THE JIG BORER AND JIG GRINDER

DIAMOND DRESSING ARM

Jig grinders must rapidly dress grinding wheels without disturbing the setup and location of the grinding spindle.

The diamond dressing arm (Fig. 69-7) may be quickly swung into the approximate grinding wheel location and then locked into position. The final approach to the grinding wheel is done by a fine-adjusting knurled screw which advances the diamond through the dressing arm.

GRINDING METHODS

The removal of material from a hole with a conventional grinding wheel is carried out by two methods: *outfeed* and *plunge* grinding. Each method has its advantages, and at times both can be used effectively to grind the same hole. Small holes, less than ¼ in. (6.35 mm) in diameter, can be effectively ground by using diamond-charged mandrels. Holes larger in diameter than the normal machine range can be ground effectively if an extension plate (Fig. 69-8) is used between the grinding spindle and the main spindle. With the use of an extension plate, holes up to 9 in. (228.6 mm) in diameter may be ground.

OUTFEED GRINDING

Outfeed grinding is similar to internal grinding where the wheel is fed radially into the work with passes as fine as 0.0001 in. (0.002 mm) at a time. The cutting action takes place with the periphery of the grinding wheel. Outfeed grinding is generally used to remove small amounts of stock when high finish and accurate hole size are required.

PLUNGE GRINDING

Plunge grinding with a grinding wheel can be compared to the cutting action of a boring tool. The grinding wheel is fed radially to the desired diameter and then into the work (Fig. 69-9). Cutting is done with the bottom corner of the wheel only. It is a rapid method of removing excess stock, and if the wheel is properly dressed, it produces satisfactory finishes for some jobs. The sharp cutting action which results from the small contact area of the wheel keeps the work cooler than outfeed grinding.

Diamond-charged mandrels (Fig. 69-10) are used instead of conventional grinding wheels for grinding holes less than ¼ in. (6.35 mm) in diameter. These mandrels should be made of cold-rolled steel which has been turned to the correct size and shape in relation to the hole to be ground. The grinding end of the mandrel is placed in diamond dust

Fig. 69-7   Dressing a wheel with a diamond dressing arm.

Fig. 69-8   Grinding a large hole using an extension plate.   (Courtesy Moore Special Tool Company Inc.)

Fig. 69-9 Hole size may be increased by 1/16 in. (1.59 mm) in one cut by plunge grinding.

DIAMOND CHARGED MANDREL

CEMENT

MOUNTED GRINDING WHEEL

Fig. 69-10 The strength and rigidity of a diamond-charged mandrel exceed those of a mounted grinding wheel. (Courtesy Moore Special Tool Company Inc.)

Fig. 69-11 The hand-held diamond dresser is convenient and effective.

and is tapped sharply with a small hardened hammer to embed the diamond dust in the mandrel surface.

The advantages of this tool over a conventional grinding wheel are:

1. Mandrels have maximum strength and rigidity.
2. Mandrels can be made the ideal diameter and length for each hole.
3. The velocity required for efficient grinding is approximately one-quarter of that for a wheel.
4. The cost per hole is less due to the greater efficiency.

GRINDING WHEELS

SELECTION

Selection of the proper wheel is necessary for satisfactory grinding performance. Since many factors influence the selection of the grinding wheel to be used, a few general principles are outlined.

1. The shank or mandrel of mounted wheels should be as short as possible to assure rigidity.
2. Wherever possible, the grinding wheel diameter should be approximately three-quarters of the diameter of the hole to the ground.
3. Widely spaced abrasive grains in the bond increase the penetrating power of the wheel.
4. When soft, low-tensile-strength materials are being ground, a hard abrasive grain with a fairly strong or hard bond should be used.
5. For grinding high-alloy hardened steels, a hard abrasive grain in a soft or weak bond is recommended.

WHEEL SPEED

The majority of grinding wheels are operated most efficiently at about 6000 surface feet per minute (sf/min) (1828

m/min). Diamond-charged mandrels used for small hole grinding should be operated at approximately 1500 sf/min (457 m/min). The spindle speed can be varied for the different types and diameters of wheels used; three grinding heads are available for the Moore jig grinder. With these heads, a range from 12,000 to 60,000 revolutions per minute (r/min) is possible. The speed of each head may be varied within its range by adjustment of the pressure regulator which controls its air supply.

WHEEL DRESSING

For a grinding wheel to perform efficiently, it is important that it be dressed or trued properly. An improperly dressed wheel will tend to produce the following conditions.

1. Poor surface finish on the hole
2. Surface burns
3. Holes which are out-of-round
4. Taper or bell-mouth holes
5. Locational error

Care in dressing a grinding wheel can prevent many of these undesirable conditions from developing. To develop the best cutting characteristics of the wheel, use the following recommended techniques for dressing a grinding wheel on a jig grinder.

1. While the wheel is running at a reduced rate, dress the top and bottom face with an abrasive stick held in the hand.
2. Dress the diameter of the wheel with a *sharp* diamond (Fig. 69-11).
3. Repeat steps 1 and 2 with the wheel at the proper operating speed.
4. Relieve the upper portion of the diameter (Fig. 69-12) so that approximately 1/4 in. (6.35 mm) of the cutting face remains.
5. The bottom face of the wheel should be concaved slightly with an abrasive stick for grinding to a shoulder or to the bottom of a hole.

436

Fig. 69-12   The width of the wheel face is reduced to avoid excessive side pressure while grinding.   (Courtesy Moore Special Tool Company Inc.)

In *outfeed grinding,* only the diameter of the wheel should be dressed when required. When *plunge grinding,* dress the bottom face of the wheel with an abrasive stick.

GRINDING ALLOWANCES

Many factors determine the amount of material which should be left in a hole for the jig grinding operation. Some of the more common factors are:

1. Type of surface finish in the bored hole
2. Size of the hole
3. Material of the workpiece
4. Distortions which occur during the hardening process

It is difficult to set specific rules on the amount of material which should be left for grinding, because of the many variable factors involved. However, general rules which would apply in most cases are as follows.

1. Holes up to ½ in. (12.7 mm) in diameter should be 0.005 to 0.008 in. (0.12 to 0.20 mm) undersize for the grinding operation.
2. Holes over ½ in. (12.7 mm) in diameter should be 0.010 to 0.015 in. (0.25 to 0.40 mm) undersize for the grinding operation.

SETTING UP WORK

When setting up work for jig grinding, take care to avoid distortion of the workpiece or machine table due to clamping pressures. Keep the following points in mind while mounting the work.

1. When bolts or strap clamps are used, keep the bolts as close as possible to the work.
2. Strap clamps should be placed exactly over the parallels supporting the work. Distortion of the work can occur if a strap clamp is tightened over a part of the work not supported by parallels.
3. Bolts should *not* be tightened any more than is required to hold the workpiece. There is less pressure exerted during jig grinding than during jig boring.
4. Do not clamp work too tightly in the precision vise clamp since this may spring the stationary jaw, dislocating the aligned edge of the work.

5. Set up work on parallels high enough to allow the bottom of the hole being ground to be measured.

TO LOCATE THE WORKPIECE

The same basic methods as in jig boring are used to locate accurately a workpiece on the jig grinder; the straightedge, edge-finder, and indicator are used, for example. Distortion of the workpiece during the heat treating process may necessitate "juggling" during the setting-up process to ensure that all holes will "clean up." The workpiece may be set up parallel to the table travel by three methods.

1. Indicate an edge of the workpiece.
2. Set the work against the table straightedge; then check the alignment with an indicator.
3. On a heat-treated piece, indicate two or more holes and set up the work to suit the average location of a group of holes.

GRINDING SEQUENCE

When a series of holes in a workpiece must be accurately related to each other, consideration must be given to the sequence of grinding operations. The following sequence is suggested when a number of different holes are required, for example, straight, tapered, blind, or holes with shoulders.

1. Rough-grind all holes first. When a high degree of accuracy is required, allow the work to cool to room temperature before proceeding to finish-grind.
2. Finish-grind all holes which can be ground with the same grinding head. This avoids continually changing grinding heads.
3. Holes whose relationship to others is most important should be ground in one continuous period of time.
4. Grind holes with shoulders or steps only once to avoid making accurate depth settings twice.

TO GRIND A TAPERED HOLE

The grinder spindle can be set for grinding tapers in either direction by loosening one adjusting screw and tightening the other. For most taper hole grinding, it is accurate enough to set the grinder spindle to the degrees indicated on the taper-setting plate (Fig. 69-13).

When an *extremely accurate* angular setting is required, the following steps are suggested.

1. Convert the angle into thousandths taper per inch (or into hundredths of a millimeter per 25 mm of length) by mathematical calculations.
2. Mount an indicator in the machine spindle.
3. Set an angle plate or master square on the machine table.
4. Move the indicator through 1 in. (or 25.4 mm) of vertical movement as read on the downfeed dial.
5. Set the adjusting screws until the desired taper is attained.

NOTE: If the adjusting screws are too tight, they will

Fig. 69-13 The taper angle is indicated on the taper-setting plate. (Courtesy Moore Special Tool Company Inc.)

Fig. 69-14 Formula for calculating X at the top of a tapered hole to produced the desired length of section A. (Courtesy Moore Special Tool Company Inc.)

$$X = B - 2A \text{ (TAN angle)}$$

Fig. 69-15 *Left:* Grinding wheel too large and cannot grind a flat surface. *Right:* Wheel is small enough to clear the opposite side of the hole.

bind the vertical movement of the spindle; if too loose, the machine may grind out of round.

By reversing this procedure, one can accurately reset the machine spindle for straight-hole grinding.

Most tapered holes also have a straight section, and the taper is ground to a certain distance from the top. Sometimes it is difficult to observe exactly where the taper begins because of the high finish in the hole and the gradual advance of most tapers. Two methods are generally used to show where the tapered section begins.

1. Apply layout dye to the top portion of the hole with a pipe cleaner. The dye will be removed from the taper portion during the grinding operation, allowing the length of the straight hole to be measured.
2. On holes too small or difficult to see and measure, it is recommended that the taper be ground first to dimension X in Fig. 69-14. This involves the use of the formula in Fig. 69-14 in order to calculate what size the hole would be at dimension X. Once the tapered hole has been correctly ground to size, the straight hole is ground to the proper

diameter. This will automatically produce the proper length of the straight hole A.

TO GRIND SHOULDERED HOLES

Many times it is necessary to grind not only the diameter of a hole, but also the bottom of blind or shouldered holes. Grinding shouldered holes presents a few problems not encountered in straight or taper grinding, and the following suggestions are offered.

1. Select the proper size grinding wheel for the hole size (Fig. 69-15).
2. Make the bottom of the wheel slightly concave with an abrasive stick.
3. Set the depth stop so that the wheel just touches the bottom or shoulder of the hole.
4. Rough-grind the sides and shoulder of the hole at the same time. This eliminates leaving a slight step near the bottom of the hole.
5. Dress the wheel and proceed to finish-grind the hole.

Jig-Grinding Hints

1. Calculate all coordinate hole locations first.
2. Clamp work just enough to hold it in place. Clamping too tightly may cause distortion.
3. Select a grinding wheel which is three-quarters the diameter of the hole to be ground.
4. A wheel with widely spaced grains should be selected for rough grinding.
5. Relieve the wheel diameters so that only ¼ in. (6.35 mm) of the cutting face remains.
6. Never use a glazed wheel for grinding.
7. Rough-grind all holes by plunge grinding.
8. Allow the work to cool before finish grinding.
9. Finish-grind holes with a freshly dressed wheel by outfeed grinding.

REVIEW QUESTIONS

1. Why was the jig grinder developed?
2. State the advantages of jig grinding.

JIG GRINDING PARTS

3. How is the jig grinder spindle constructed to allow for the grinding of taper holes?
4. Explain how the grinding wheel may be positioned to the hole diameter.
5. What three methods may be used for controlling and measuring the depth of a hole?
6. Name two methods of dressing a wheel on a jig grinder.

GRINDING METHODS

7. Compare outfeed and plunge grinding.
8. How may large holes be ground on a jig grinder?
9. State the advantages of diamond-charged mandrels over grinding wheels for small hole grinding.

GRINDING WHEELS

10. List four general principles which should be observed in selecting a grinding wheel.
11. At what speed should the following be operated:
   (a) grinding wheels?
   (b) diamond-charged mandrels?
12. What undesirable conditions will an improperly dressed wheel cause?
13. Explain the procedure for dressing a grinding wheel.

GRINDING ALLOWANCES

14. Name the factors which determine the amount of material that should be left in a hole for jig grinding.
15. State the grinding allowance which would apply in most cases for holes
   (a) under ½ in. (12.7 mm)
   (b) over ½ in. (12.7 mm)

SETTING UP WORK

16. Explain how bolts and clamps should be placed when setting up work.
17. Why is it important that bolts not be tightened too tightly?
18. Name three methods which are used to set up work parallel to the table travel.

GRINDING SEQUENCE

19. List the sequence suggested when grinding a variety of holes.
20. Explain how the grinding head may be set to an accurate angle.
21. Calculate the X dimension for a $\frac{3}{16}$-in.-diameter hole with a 0.200-in. straight section. The included angle of the tapered hole is 2° (see Fig. 69-14).
22. Explain the procedure for grinding shouldered holes.

COMPUTER-AGE MACHINING

SECTION 15

No single invention since the industrial revolution has left such an impact on society as the computer. Today computers can guide and direct spaceships to the moon and outer space and bring them back safely to earth. They route long-distance telephone calls, schedule and control train and plane operations, determine weather forecasts, produce an instantaneous report of our bank balance. In chain stores, the cash register (connected with a central computer) totals bills, posts sales, and updates the inventory with every entry. These are but a few of the applications of the computer in our society.

During the past two to three decades, basic computers were applied to machine tools to program and control the machine operations. These devices have been steadily improved until they are now highly sophisticated units capable of totally controlling the operation of a single machine, a group of machines, or soon even a complete manufacturing plant.

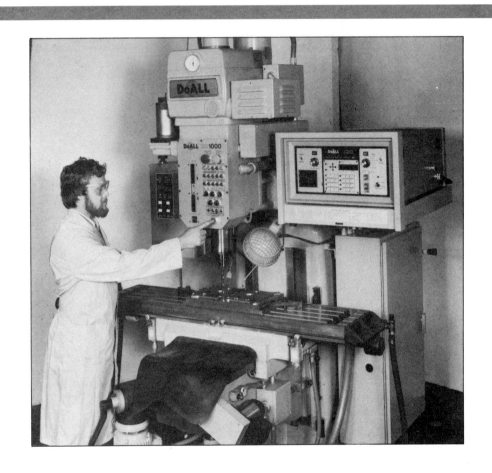

Ever since primitive people became aware of the concept of quantity, people have used some device to count and perform calculations.

Primitive people used their fingers, toes, and stones to count (Fig. 70-1). Around 4000 B.C., the abacus, really the first computer, was developed in the Orient. It uses the principle of moving beads on several wires to make calculations (Fig. 70-2). The abacus is very accurate when properly used. It may still be found in some of the older and smaller Oriental businesses today.

OBJECTIVES

After completing this unit you will be aware of:
1. The development of computers over the ages
2. The effect of computers on everyday life

HISTORY OF THE COMPUTER

In 1642, the first mechanical calculator was constructed by a Frenchman named Blaise Pascal. It consisted of eight wheels or dials each with the numbers 0 to 9 and each wheel representing units, tens, hundreds, thousands, etc. It could however, only add or subtract. Multiplication or division was done by repeated additions or subtractions.

In 1671, a German mathematician added the capability of multiplication and division. However, this advanced machine could only do arithmetical problems.

Charles Babbage, a 19th-century English mathematician, produced a machine called the difference engine that could rapidly and accurately calculate long lists of various functions, including logarithms.

In 1804, a French mechanician, J. M. Jacquard, introduced a punch card system to direct the operations of a weaving loom. In the United States, Herman Hollerith introduced the use of punched cards to record personal information such as age, sex, race, marital status, etc. for the 1890 United States census (Fig. 70-3). The information

was encoded on cards and read and tabulated by electric sensors. This use of punched cards led to the development of the early office machines for the tabulation of data.

In the 1930s, a German named Konrad Zuse built a simple computer which, among other things, was used to calculate wing designs for the German aircraft industry.

A mathematician named George Stibitz produced a similar device in 1939 for the Bell Telephone Laboratories in

Fig. 70-2 The abacus was the first real computer.

Fig. 70-1 Primitive means of counting.

TOES AND FINGERS   STONES   BEADS

Fig. 70-3 Punched cards were our first method of data processing.

Fig. 70-4 Thousands of bits of information may be stored on a tiny silicon chip. (Courtesy Rockwell International.)

the United States. This machine was capable of doing calculations over telephone wires; thus was born the first remote data processing machine.

During World War II, the British built a computer called the Colossus I which helped break the German military codes.

The earliest digital computers used electro-mechanical on-off switches or relays. The first large computer, the Mark I, assembled at Harvard University by IBM, could multiply two 23-digit numbers in about 5 seconds—a very slow feat compared to today's machines.

In 1946, the world's first electronic digital computer, the ENIAC (*e*lectronic *n*umerical *i*ntegrating *a*utomatic *c*omputer) was produced. It contained more than 19,000 vacuum tubes, weighed almost 30 tons, and occupied more than 15,000 ft^2 of floor space. It was a much faster computer—able to add two numbers in $\frac{1}{5000}$ of a second. A machine of this size had many operational problems, particularly with tubes burning out and circuit wiring.

In 1947, the first transistor was produced by the Bell Laboratories. These were used as switches to control the flow of electrons. They were much smaller than vacuum tubes, had fewer failures, gave off less heat, and were much cheaper to make. Computers were then assembled using transistors but still the problem of extensive hand wiring existed. This problem led to the development of the printed circuit.

In the late 1950s, Kilby of Texas Instruments and Noyce of Fairchild discovered that any number of transistors along with the connections between them could be etched on a small piece of silicon (about $\frac{1}{4} \times \frac{1}{4} \times \frac{1}{32}$ in. thick). These chips, called *integrated circuits (ICs),* contained entire sections of the computer such as a logic circuit or a memory register. These chips have been further improved until today thousands of transistors and circuits are crammed into this tiny silicon chip (Fig. 70-4). The only problem with this advanced chip was that the circuits were rigidly fixed and the chips could only do the duties for which they were designed.

In 1971, the Intel Corporation produced the microprocessor—a chip which contained the entire *central processing unit (CPU)* for a simple computer. This single chip could be programmed to do any number of tasks from steering a spacecraft to operating a watch or controlling the new personal computers.

THE ROLE OF THE COMPUTER

Although today's computers seem to amaze us (the older generation particularly), they have become part of our everyday life. They will continue to do so even to a much greater extent in the years to come.

We are amazed that some computers today can perform 1 million calculations per second because of the thousands of transistors and circuits jammed into the tiny chips (ICs). Computer scientists can foresee the day when 1 billion transistors or electronic switches (with the necessary connections) will be crowded into a single chip. A single chip will have a memory large enough to store the text of 200 long novels. Advances of this type will decrease the size of computers considerably.

Some computer scientists feel that by 1990 the prototype of a thinking computer incorporating *artificial intelligence (AI)* will be introduced. They feel that the commercial product will be about 5 years later. This machine will be able to recognize natural speech and written language and will be able to translate and type documents automatically. Once a verbal command is given, the computer will act, unless it does not understand the command. At this point the machine will ask questions until it is able to form its own judgments and act. It will also learn by recalling and studying its errors.

Computers today are being used in larger medical centers to catalog all known diseases with their symptoms, known cures, etc. This knowledge is more than any doctor could remember. Doctors now are able to patch their own computer into the central computer and get an immediate and accurate diagnosis of the patient's problem, thus saving many hours and even days of awaiting the results of routine tests.

Because of the computer, children will learn more at a younger age and, as may be expected, future generations will have a much broader and deeper knowledge than those of past generations. It was said that in the past we doubled our knowledge every 25 years. Now with the computer, the amount of knowledge is said to double every 3 years. With this greater knowledge, the human race will explore and develop new sciences and areas that are unknown to us today (much the same as the computer has done to the older people of this era).

In other areas, the computer has been and will continue to be used to forecast weather; guide and direct planes, spaceships, missiles, and military artillery; and to monitor industrial environment.

In our everyday life, everyone is and will continue to be affected by the computer. Department store computers list and total your purchases, at the same time keeping the inventory up to date and advising the company of peoples' buying habits. Thus the computer permits the company to buy more wisely. Credit bureau computers know how much every adult owes, to whom, and how the debt is being repaid. School computers record students' courses, grades, and other student information. Hospital and medical records

443

Fig. 70-5 CAD systems are invaluable to engineers who research and design products.

the computer along with the speed and direction of each. The information is stored and the future positions of the planes are calculated.

In the manufacturing industry the computer has contributed to the efficient manufacture of all goods. It appears that the impact of the computer will be even greater in the years to come. Computers will continue to improve productivity through *computer-aided design (CAD)* by which the design of a product can be researched, fully developed, and tested before production has started (Fig. 70-5). *Computer-assisted manufacturing (CAM)* results in less scrap, more reliability through the computer control of the machining sequence and the cutting speeds and feeds.

Robots, which are computer-controlled, are being used by industry to an increasing extent. Robots can be programmed to paint cars, weld, feed forges, load and unload machinery, assemble electric motors, and perform dangerous and boring tasks formerly done by humans.

are kept on anyone who has been admitted to a hospital.

Police agencies have access to a national computer which can produce the police records of any known offender. The census bureaus and tax departments of any country have the necessary information on all of its citizens on computer.

On the office front, computers have relieved the accountants of the drudgery of repetitious jobs such as payroll processing. In the future it is expected that many office workers will work from their home on a company computer. This will eliminate the necessity of traveling long distances to work and the need for baby-sitting services required by so many young working families of today.

In air defense control systems, the position and course of all planes from the network of radar stations are fed into

REVIEW QUESTIONS

1. Name three methods of measuring used by primitive people.
2. What was the first computer ever developed?
3. For what purpose were the first punched cards used in the United States?
4. How many transistors and circuits can be found on a silicon chip?
5. How are computers used in medical centers?
6. How are computers affecting the manufacturing industry?

NUMERICAL CONTROL
UNIT 71

Since the latter half of the 1950s, the use of numerical control (NC) for the accurate operation of metal-cutting machines has increased at a tremendous rate throughout the world.

Numerical control may be defined as a method of accurately managing the movement of machine tools by a series of *programmed numerical data* which activate the motors of the machine tool. There is nothing complex or magical about this system. It is based on the simple fundamental that combines automatic measurement of the amount a machine table slide will move with a series of programmed instructions.

The numerically controlled machine is supplied with detailed information regarding the part by means of punched tape. The machine decodes this punched information and electronic devices activate the various motors on the machine tool, causing them to position the work and follow specific machining instructions. The measuring and recording devices incorporated into numerically controlled machine tools assure that the part being manufactured will be accurate. Numerically controlled machines minimize the possibility of human error which existed before their development.

Over recent years, there have been great improvements to NC systems. In many cases, the tape reader on NC machines is bypassed and the machine receives its instructions directly from a computer. The authors, however, felt that the student should be made aware of the basics of NC before being exposed to the more recent developments of computerized numerical control (CNC) and direct numerical control (DNC).

OBJECTIVES

After completing this unit you should
1. Understand the principle of NC
2. Know the difference between open- and closed-loop control
3. Know two methods of positioning and tool path control
4. Be able to read and prepare a simulated tape for a simple block of information

Numerical control is an efficient method of conveying part print information to the motors which control the speeds, feeds, and various motions of the machine tool. The designer's information is transferred from the print to punched tape which is put into the machine tool reader. The punched tape may be compared to the roll of punched paper used to operate the keys of a player piano and produce music. Each hole on the paper roll corresponds to a key (note) on the player piano keyboard. As air passes through the hole, it moves a key on the keyboard, producing a note. On numerically controlled tape-reading devices, beams of light are generally used to decode the tape information. The beams of light are connected to an electrical circuit. Each time a hole in the tape appears beneath the beam of light, a specific circuit is activated, sending signals to start or stop motors (which position the work) and control various functions of the machine tool. Other numerically controlled tape-reading devices commonly used to decode the tape information are brushes, fingers, and magnetic heads.

MEASUREMENT FUNDAMENTALS

Accurate measurement is fundamental to the NC of machine tools. The workpiece or cutting tool must be moved a definite distance in one direction before the machining operation is performed. The distances moved must be accurately measured by some method that the NC system will understand.

There are many methods of measurement, but all consist of comparing the distance to be measured to some known standard. On a milling machine the distance the table is moved could be roughly measured by a rule mounted beside the table or slide. In most cases, however, the table is positioned by turning the lead screw a specific number of turns, depending upon the distance the table is to be moved. As the lead screw has a certain pitch (or number of threads per inch) and was manufactured to a standard, the number of turns is transformed into a specific distance.

In NC systems, accurate measurement is usually achieved by automatically comparing the amount the lead screw has been rotated in accordance with the information on the tape. The degree of accuracy obtained will depend on the type of NC system being used. Some of the most precise NC machines use linear, vernier-like scales with readings made electro-optically and fed back to the control unit.

TYPES OF NUMERICAL CONTROL SYSTEMS

Open loop and *closed loop* are the two main types of control systems used for NC machine tools.

OPEN-LOOP SYSTEM

In the *open-loop system* (Fig. 71-1), the tape is fed into a *tape reader* which decodes the information punched on the tape and stores it briefly until the machine is ready to use it and then converts it into electrical pulses or signals. These signals are sent to the *control unit* which energizes the *servo control units*. The servo control units direct the *servo motors* to perform certain functions according to the information supplied by the tape. The amount each servo motor will move depends upon the number of electrical pulses it receives from the servo control unit. Precision lead screws, usually having 10 threads per inch (tpi) are used on NC machines. If the servo motor connected to the lead screw receives 1000 electrical pulses, the machine slide will move

Fig. 71-1　An open-loop NC system.

Fig. 71-2　A closed-loop NC system showing X axis only.

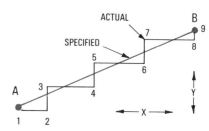

Fig. 71-3　A comparison of point-to-point control and straight cut control.

COMPUTER-AGE MACHINING

1 in. (25.4 mm). Therefore, one pulse will cause the machine slide to move 0.001 in. (0.0254 mm). The open-loop system is fairly simple; however, since there is no means of checking whether the servo motor has performed its function correctly, it is not generally used where an accuracy greater than 0.001 in. (0.025 mm) is required. The open-loop system may be compared to a gun crew that has made all the calculations necessary to hit a distant target, but does not have an observer to confirm the accuracy of the shot.

CLOSED-LOOP SYSTEM

The *closed-loop system* (Fig. 71-2) can be compared to the same gun crew that now has an observer to confirm the accuracy of the shot. The observer relays the information regarding the accuracy of the shot to the gun crew, which then makes the necessary adjustments to hit the target.

The closed-loop system is similar to the open-loop system with the exception that a *feedback unit* (Fig. 71-2) is introduced in the electrical circuit. This feedback unit, often called a *transducer,* compares the amount the machine table has been moved by the servo motor with the signal sent by the control unit. The control unit instructs the servo motor to make whatever adjustments are necessary until both the signal from the control unit and the one from the servo unit are equal. In the closed-loop system, 10,000 electrical pulses are required to move the machine slide 1 in. (25.4 mm). Therefore, on this type of system, one pulse will cause 0.0001-in. (0.0025-mm) movement of the machine slide. Closed-loop NC systems are very accurate because the accuracy of the command signal is recorded and there is an automatic compensation for error. Additional accuracy has now been built into most NC-controlled machines with either open- or closed-loop systems by the use of *ball screws*. These screws typically have an accuracy of ±0.0008 in. over 26 in. (0.020 mm in 660 mm) and a repeatability of ±0.0003 in. (0.007 mm).

TYPES OF CONTROL

There are two types of controls available which position and control the cutting tool and the workpiece during the machining operation. These are *point-to-point* control and the *continuous-path or contouring control*. The type of work to be performed on a machine will determine the type of control that should be purchased.

POINT-TO-POINT POSITIONING

With this type of control, the machine slide is instructed to go from point A to point B. The path taken by the machine slide between these two points does not matter since the operations required must be performed only at points A and B. The tool slide actually moves from point A to B in a series of small steps (Fig. 71-3). As illustrated, there is a difference between the specified and the actual path of the machine slide. The point-to-point commands are ideal for drilling, boring, and tapping operations.

CONTINUOUS-PATH CONTROL

When contours are being machined, the control unit must direct the path of the cutting tool at all times. If a machining operation, for example, a straight-line cut, must be performed between points A and B, additional devices must be added to the system. Much more detailed signals are sent to the machine slide so that each step it takes to get from points A to B is minute. The *continuous path* or *straight-cut system* would follow the specified path AB illustrated in Fig. 71-3 in minute steps at a controlled rate, producing a straight-line cut. A computer is essential for the continuous-path system because of the large number of coordinates required to guide the machine tool on a straight-line path.

By further upgrading the continuous-path system, the path a machine follows can be controlled to form any type of contour. The complex part illustrated in Fig. 71-4 is an example of the types of forms which can be machined easily on a contouring system. This system also requires the use of a computer in programming because the time it would take to calculate all the necessary coordinates would be prohibitive.

INPUT MEDIA

With the development of NC, a variety of input media was used to convey the information from the drawing to the machine. The most common types of input media used were magnetic tape, punched cards, and punched tape.

MAGNETIC TAPE

Usually 1-in.-wide (25.4-mm) tape was used, and on a 4-in. (101.6-mm) length as many as 2000 pieces of information could be included. One major disadvantage of magnetic tape was that it could be affected by nearby electrical transformers or disturbances which could remove some of the information on the tape.

Fig. 71-4   A complex part produced by a contouring system. (Courtesy Cincinnati Milacron Incorporated.)

PUNCHED CARDS

Some NC systems employed 80- or 90-column punched cards which could carry a great deal of information. As many as 120 cards per minute could be decoded by the card reader. The disadvantage of the card system was that cards became bent and would not process through the reader. If not handled with extreme care, the sequence of the cards could be altered.

PUNCHED TAPE

Because each different type of input medium required different types of systems to convey the print information to the machine tool, the Electronics Industries Association decided to standardize the NC input media. A 1-in. (25.4-mm), 8-channel punched tape was selected as a standard since it offered a number of advantages over the magnetic tape and punched cards.

Standard Punched Tape

Since the 1-in.-wide (25.4-mm), 8-channel tape using the *binary coded decimal (BCD) system* has been selected as a standard by the Electronics Industries Association, it will be discussed in greater detail. The following section will give the reader a better understanding of what types are available, how the numerical information from the print is transferred to the tape, and the method of reading tapes.

Types of Tapes

Standard NC tapes which have a continuous row of sprocket holes for feeding purposes (Fig. 71-5) are manufactured from a number of different materials.

1. *Paper:* Tapes made of paper are available in a variety of colors and may be purchased oiled or non-oiled. The oiled tapes are recommended because they help to lubricate the punches in the special tape typewriter which is used to punch the holes in the tape. Paper tape is inexpensive; however, it is not recommended for shop use if more than a few pieces are to be machined, since it tears easily. Paper tapes are usually used for file purposes.
2. *Mylar:* Tapes made of Mylar (a type of plastic) laminated between two strips of paper have provided a sturdy tape which is almost impossible to tear. Although it is more expensive than the plain paper tapes, it is highly recommended for shop use because of its sturdy quality. Aluminized Mylar tapes are recommended for light-source tape readers; however, they are slightly more expensive than the paper Mylar type. They also create considerable wear on the punches used to produce the tape.
3. *Foil:* Foil tapes are sturdy but not generally recommended because they are very hard on the punches of the tape-preparation equipment.

Tape Coding

The Electronics Industries Association adopted the *BCD system* for standardizing information punched on tapes used for NC. The standard coding system used for the 1-in.-wide (25.4-mm), 8-channel tape is illustrated in Fig. 71-5.

Fig. 71-5   Electronics Industries Association standard coding for a 1-in.-wide, 8-channel tape.

The numbers 1 to 8 on the top of the tape represent only the number of each channel and have no relationship to the holes punched in the tape.

1. *The sprocket holes* between channels 3 and 4 are used to drive the tape through the machine. The sprocket holes are set off center so that it is impossible to reverse the tape in the machine.
2. *Channels 1, 2, 3, and 4* are used for *numerical data,* such as dimensions, speeds, feeds, etc.
3. *Channel 5,* marked CH, is called the *parity check.* An odd number of holes must appear in each row; otherwise the tape reader will stop the machine. If an even number of holes must be punched in one row to get the correct information on the tape, and additional hole must be punched in channel 5. The odd parity check looks for any errors or mechanical failures in the tape preparation.
4. *Channel 6* always represents a zero.
5. *Channel 7,* marked X, is used to select a letter to identify various machine operations and is not used for dimensions. Each letter (a to z) in the left-hand vertical column (Fig. 71-5) represents a certain machine function or operation, such as drilling, boring, reaming, etc. When channels 6 and 7 are punched together in conjunction with channels 1, 2, 3, or 4, operations a to i will be selected. When only channel 7 is punched in conjunction with channels 1, 2, 3, or 4, operations j to r will be selected. If only channel 6 is punched in conjunction with channels 1, 2, 3, or 4, operations s to z will be selected.
6. *Channel 8,* marked EL (or EOB), represents the *end of*

a line or end of a block of information. It generally is at the beginning of a tape and always at the end of each block of information.

The *TAB* code, punched in channels 2, 3, 4, 5, and 6, is used to separate each operation or dimension on some programs. The remaining codes in Fig. 71-5 beginning with a "." (a period) and ending with *lowercase* are all acceptable codes used in NC tapes.

Examples of Punched Information

1. Assume that the numerical value of 1 must be recorded. A hole must be punched in channel 1 because it has a numerical value of 1.

2. Assume that the numerical value of 5 must be recorded. A hole must be punched in channels 3 and 1 because channel 3 has a numerical value of 4 and channel 1 has a numerical value of 1. These two totaled equal a numerical value of 5.

NOTE: Since only two holes are punched across this row, an extra hole must be punched in the parity check channel (5), as the tape reader will not recognize an even number of holes (Fig. 71-5). If a hole is not punched in the parity check channel, the tape reader will stop the machine.

3. Operation g would be recorded by punching holes in channels 1, 2, 3, 6, and 7.

Tape Format

Each block of information must contain five complete words or pieces of information. If five complete words are not included in each block, the tape reader will not recognize the information on the tape, and therefore will not activate the control unit. Figure 71-6 shows a punched tape containing one complete block of information consisting of five complete words. From left to right, the information punched on the tape is as follows:

1. A hole is punched in channel 8, which represents the end or the beginning of a line or block of information.

2. The *first word* of the block represents the number of the operation. H001 represents the first operation on the tape. The *letter H* is recorded by punching holes in channels 4, 6, and 7 (Fig. 71-6).
The *two zeros* are recorded by punching channel 6 in two successive rows.
The *1* is recorded by punching channel 1 which has a numerical value of 1.

3. The *TAB* is used to separate each word into a block of information.

4. The *second word* in the block represents the type of operation to be performed. G81 is a drill cycle on Cincinnati Milacron NC systems.
The *letter G* is recorded by punching channels 1, 2, 3, 6, and 7.
The *8* is recorded by punching channel 4 which has a numerical value of 8.
The *1* is recorded by punching channel 1 which has a numerical value of 1.

5. The *third word* represents the distance the table slide must move from the *X* axis. The information contained in the third word of Fig. 71-6 is X12345. As standard tapes on closed-loop systems program all dimensions to ten-

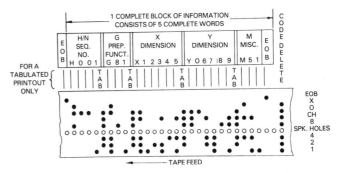

Fig. 71-6 A complete block of information punched on a tape.

thousandths of an inch, the machine will move 1.2345 in from the *X* axis.

6. The *fourth word* represents the distance the table slide must move from the *Y* axis. In Fig. 71-6, Y06789 represents a table movement of 0.6789 in. from the *Y* axis.

7. The *fifth word* represents a miscellaneous machining function. M51 would select the proper cam so that the hole in the workpiece is drilled to the required depth.

Tape Preparation

A special tape data-punching machine, similar to a typewriter, is used to punch information on the NC tape. A programmer reviews the part drawing, determines the sequence of operations required, and records this information on a program sheet. A typist transfers the information contained on the program sheet to the tape using a special tape-punching machine (Fig. 71-7).

On some types of tape-punching machines, the finished tape may be run back through the machine, which makes a duplicate tape and at the same time produces a typed copy of the program. The typed copy of the program can be compared with the original program to determine its accuracy.

In computer-assisted part programming, the tape punch is linked directly with the computer.

NC OPERATION

Although NC systems differ greatly in detail and complexity according to the manufacturer, all have basically the same elements. Regardless of the type of input medium used, all operate a machine tool in a similar manner. A complete sequence of operations beginning with the programmer transferring the information from a print to a program sheet until the finished part is removed from the machine is illustrated in Fig. 71-7.

Numerical control systems require that the part print must be dimensioned by either the *incremental* or *absolute* system. Incremental measuring means that there is no common reference point for all dimensions. Each new point or position originates from the last established point. Absolute measuring is where a reference point is fixed or established and all measurements are given from this point.

1. The programmer reviews the print of the part to be machined, determines the sequence of operations required, and lists the particulars about each operation on a program sheet (Fig. 71-7A).

The example below illustrates how one cycle, or block, of taped information is processed and acted upon by a tape controlled indexer.

(A) Assume the Table is in motion, moving to position 6 as directed by the indexer.

(B) When the Indexer completes both the X and Y counts, it returns a signal to the Cycle Control.

(C) The Cycle Control then energizes the Tool Advance Solenoid.

(D) When the tool advances, the Feedback Switch closes, telling the Tape Reader to read the next block of information.

(E) Tape information goes to the Decoder. In this case:

7   The sequence number is for reference only and performs no function in the program.

TAB The first TAB signal in a block opens the gate to the "X" section of the Command Memory.

—   Tool motion is to operator's left

2   } Stored in Command Memory "X" section

5   Tool is to move 250 steps

0   The second TAB signal in a block closes the gate to the

TAB "X" section of the Command Memory and opens the gate to the "Y" section.

2   Tool motion is 200 steps away from operator

0   } Stored in Command Memory "Y" section

0

EOB Information in this block is complete.

(F) EOB signal is relayed to Cycle Control, which stops Tape Reader before reading next character.

(G) The Feedback Switch opens when the tool is fully retracted, telling Cycle Control the tool cycle is finished.

(H) Cycle Control, having received both "EOB" signal from Decoder and completion signal from the Feedback Switch, tells the Command Memory to transfer to the indexer the motor stepping information stored in Step E.

(J) Indexer starts the motors simultaneously, each stopping when it has completed its required motion to reach position 7. In this case, the "Y" motor will stop first, having moved 200 steps. The "X" motor will continue until it has moved 250 steps.

(K) The cycle is complete, return to Step B.

Fig. 71-7   Numerical control sequence of operation.   (Courtesy The American Superior Electric Company.)

2. A typist transfers the information contained on the program sheet to the tape using a special tape-punching machine (Fig. 71-7B).

3. The punched tape and a copy of the program sheet are then handed to the machine operator. After positioning the part to be machined on the machine table according to the instructions contained on the program sheet, the operator threads the tape into the tape reader (Figs. 71-7C and 71-8).

4. The tape reader, which automatically advances the tape, is then started.

5. As each block of information is decoded by the tape reader, it sends the necessary information to the *cycle control*.

6. After the end of each machining operation, the *feedback switch* informs the cycle control that the previous operation has been completed.

7. The cycle control instructs the *command memory unit* to transfer the tape instructions for the next operation to the *indexer* (Fig. 71-7F).

8. The indexer starts the servo motors, which in turn move the machine table slides the required amount.

ADVANTAGES

Numerical control has been applied to a variety of machines and has gained wide acceptance by industry. Machine tools such as lathes, turret drills, and milling and boring machines are some of the more common types of equipment which employ NC. The following are some of the advantages of NC equipment.

1. The machine has greater flexibility. A machine can now perform multiple operations such as drilling and milling for example.

2. Once the program has been set up, there is over a 20 to 30 percent increase in production.

3. The reliability of the system eliminates most of the human error associated with manual operation, thereby reducing scrap loss.

4. Special jigs and fixtures usually required for positioning are eliminated since the machine can locate positions quickly and accurately.

5. The time required for setting up and locating the workpiece is reduced.

6. Complex operations can be performed with ease.

7. Single parts or limited production runs can be made with minimum effort and cost.

8. The program can be quickly changed by inserting a new tape in the machine.

9. Inspection costs are reduced because of the reliability of the system.

10. Once the program and tooling have been tested, the equipment does not require the traditional high operator skills.

11. Inventory, repair, maintenance, and storage of jigs and fixtures is eliminated. Tapes require little space to store and virtually no maintenance.

DISADVANTAGES

1. The initial cost of NC machines is higher than that of conventional machines.

Fig. 71-8 Tape control reader. (Courtesy The American Superior Electric Company.)

2. Personnel trained in electronics are required to service the equipment.

3. Additional floor space may be required for this equipment.

4. Personnel must be trained in the programming and operation of this equipment.

RECENT ADVANCES IN NUMERICAL CONTROL

Although NC has provided industry with a method of increasing productivity while maintaining a high degree of accuracy, many changes have been made in machining procedures resulting in lower manufacturing costs.

With advancing technology, more efficient use of NC has been developed. The continued developments in microelectronics and computer technology have made this possible by shrinking the size of machine control hardware while providing greater capacity and capabilities. Most machine tools being made today are controlled by computers or other computerlike devices such as programmable controllers.

TYPES OF COMPUTER CONTROL

There are two different types of computer control for NC machines: *computerized numerical control (CNC)* and *direct numerical control (DNC)*.

Computerized Numerical Control

Most NC systems being built are CNC systems. The housings for these systems have tended to look much like those of the older NC systems, and the controls operate and are used similarly. The control unit is mounted on the machine or stands alongside of it (Fig. 71-9).

The CNC system, built around a powerful minicomputer, incorporates much more memory capacity and has many more features to assist in programming, program editing at the machine, setup of the machine, operation of it, and maintenance. Many of these features are sets of machine

Fig. 71-9 A computer-controlled turning center. (Courtesy Cincinnati Milacron Incorporated.)

and control instructions stored in memory, which can be called into use by the part program or by the machine user.

Most CNC systems still have tape readers, with the part programs still being prepared in an office off-line unit and delivered to the machine in the form of punched tape. In some cases, however, the tape is read once, with the part program being stored in memory for repetitive machining.

In some advanced CNC designs now appearing, mini-computers incorporate microprocessors, are complemented by microprocessors, or are replaced by them. Increasing use of microprocessors is tending to divide the control functions among several small boxes mounted on the machine rather than concentrate them in one large control console.

Now emerging are some CNC units that have more intelligence built into the computer. This permits the machine operator to input the data manually that describes the part to be machined. The computer then generates the part program which is held in memory.

Advantages of CNC

☐ It is more flexible in that modifications can be made to the program rather than making a completely new tape as with older conventional (hand-wired) controls.
☐ It can diagnose programs and before the part is produced detect machine and control malfunctions as or before they occur.
☐ It can be integrated with DNC systems in highly sophisticated manufacturing systems.

DIRECT NUMERICAL CONTROL

In this type of system, a number of CNC-equipped machines are controlled remotely from a mainframe computer. Most commonly, the DNC computer handles the scheduling of work and downloads into the machine's CNC memory a complete program when a new part is to be machined. It

bypasses the control's tape reader. If automatic work handling is provided at the machines, the computer can control automatic cycling of each machine. In this case, no machine operator is needed except for initial setup or troubleshooting.

Much less common is the practice of completely running the machines in real time directly from the mainframe central computer. In this case, the control unit at each machine may have no tape reader, no speed and feed override, and none of the features that enable the machine to be operated independently when the need arises.

Advantages of DNC

☐ The often troublesome tape reader on the machine is bypassed.
☐ Stored programs and complicated data are easily accessible and can be revised and edited easily.

REVIEW QUESTIONS

1. Define NC.
2. What methods may be used to decode the punched information?
3. What purpose do transducers serve in a NC system?

TYPES OF NC SYSTEMS

4. Briefly define an open-loop system.
5. In the closed-loop system, how many electrical pulses are required to move the machine slide 1 in. (25.4 mm)?
6. How many electrical pulses would be required to move a machine slide ¼ in. (6.35 mm) in an open-loop system?
7. Explain the difference between an open-loop and a closed-loop system.

8. What is meant by point-to-point and continuous-path commands?

INPUT MEDIA

9. Name three types of input media and state their advantages.
10. State the advantages of paper and Mylar tapes.
11. What feature on standard tape prevents it from being mounted incorrectly in a machine?
12. Explain a parity check and state its purpose.
13. What holes must be punched on a tape to signify operation j9?
14. Explain the purpose of the TAB and EL code.
15. By means of a suitable sketch, show what holes must be punched to indicate the following block of information on a tape.

   (a) operation number: a26

   (b) drilling operation: G81
   (c) x dimension: 2.6783
   (d) y dimension: .250
   (e) drilling depth: 1.060

16. How is tape checked prior to its initial use to ensure its accuracy?

NC OPERATION

17. Briefly list each step in a NC sequence of operations.
18. List six important advantages of NC.

CNC AND DNC

19. List four advantages of CNC.
20. State the difference between the CNC and the DNC systems.

COMPUTER-AIDED DESIGN

UNIT 72

The advent of the computer proved a boon to the design engineer in that it simplified the long and tedious calculations often required for a project. As time went on, it was felt that the computer could be further extended in the design area, and in the 1960s a new system called *computer-aided design (CAD)* was introduced. Computer-aided design allows the designer or engineer to produce finished engineering drawings from simple pencil sketches or from models, and to modify the drawings if they appear not to be functional. From three-drawing orthographic views, it is possible to transform drawings into a three-dimensional view and with appropriate computer software to show the anticipated performance of the part. This has proved extremely valuable to designers, engineers, and drafting people. This new system has streamlined the design and drafting processes and its use is increasing by an estimated 40 percent per year.

OBJECTIVES

After completing this unit you will be able to:
1. Understand the principle and purpose of CAD
2. Name and state the purpose of the main parts of a CAD system
3. State six advantages of CAD

CAD COMPONENTS

Computer-aided design is a television-like system that produces a picture from electronic signals relayed from the computer. Most systems in larger plants use two levels of computers. Smaller desk-top computers, each of which has its own television-like *cathode ray tube (CRT)*, are con-

nected to the large, main, host computer. By adding to this system a keyboard, a light pen, or an electronic tablet and a plotter, the operator can direct the computer to produce any drawing or view required (Fig. 72-1).

The desk-top computer is used by the operator to insert all the required data for the job. Because the memory bank of the desk-top will be much smaller than that of the host

Fig. 72-1 A total Computer Aided Design (CAD) system. (Courtesy Bausch & Lomb.)

computer, only the commonly used data required by the operator will be stored in its memory bank. Additional and more complex information, as well as final part information, will be stored in the memory of the larger host computer. The desk-top computer may call on the host computer to perform tasks beyond its capability. The operator can at any time, through the desk-top computer, recall information stored in the host computer.

DESIGNING THE PART

The operator may start off with a pencil sketch, and using the light pen or an electronic tablet can produce a properly scaled drawing of the part on the screen and also record it in the computer memory. The computer will calculate the coordinates of the ends of the line, the points of intersection, as well as the required arcs, circles, radii, etc., producing a view of the evolving part on the CRT as each feature is added. Basically the operator provides the input and the computer does the calculating.

If any design changes are necessary at this time, the designer makes the changes on the screen with the light pen and these changes are automatically made in the computer memory. In most systems, CRT screens can display multiple orthographic views of a design (front, top, and side) in combination with a three-dimensional isometric view. These views are frequently displayed simultaneously on a split screen and any design changes made on one view are always automatically added to the other views. The designer is able to create and change parts and lines on the CRT with a light pen, an electro-mechanical cursor, or an electronic tablet (Fig. 72-2). It should also be noted that the operator can rotate any drawing to any position desired so that he or she can better study it or compare it to another component.

Most systems have a typewriter-like keyboard for entering the text and commands into the computer. Some systems also provide a list or *menu* of standard commands

Fig. 72-2 An electronic tablet being used to make design changes on a part illustrated on the CRT. (Courtesy Bausch & Lomb.)

containing drawing functions for specifying the size and location of lines, arcs, texts, cross-hatching, standard drafting symbols, and other items (Fig. 72-3). These menus are designed to increase the speed at which drawing data can be entered. Any desired function contained in the menu is entered with the push of a single button. Menus may also be displayed on the screen or on a data tablet and items may be selected from these with the electronic or light pen.

DESIGN DATA AND TESTING

The information stored in the computer memory provides a data base from which the computer can extract and process data for future needs. For instance, it, or other computers linked with it, can calculate the volume, mass, center of gravity, etc. of the part to be produced. It can also perform stress analysis, provide English to metric conversion of the figures, generate a bill of material, and even

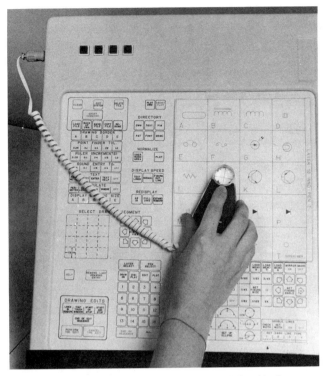

Fig. 72-3 Standard drawing functions are contained in the menu and can be added to a drawing at the push of a button. (Courtesy Bausch & Lomb.)

produce a *numerical control (NC)* tape and the instructions for machining the part.

After the design is completed, an engineer (or designer) can ask the computer questions regarding the anticipated performance of the part. For instance, if the engineer designs a tower which must support a weight of 220,000 lb (100,000 kg or 100 tonnes), he or she can ascertain from the computer if the structure will support this weight. The computer will tell if it will support this weight or if it will break and at what point. If the information displayed shows that the structure will not support this weight, the designer merely makes the required modifications on the screen (using the light pen) and then asks the same question of the computer. This process is repeated until an affirmative answer is received from the computer.

Once the details of a drawing are entered into the system, the operator, engineer, or designer can make changes easily and quickly to any area of the drawing without having to redraw the original. With this equipment, the accuracy of the drawing is greatly improved, which permits a quick check of critical components for clearance and tolerance while the "product" is still in the drawing stage.

TESTING THE FUNCTION OF COMPONENTS

The operator can also assemble the parts electronically by laying the drawing for one component over the other and zoom in on the critical areas to check visually for clearance, interference, etc. By using CAD, engineers are able to make performance analyses on the part. Here they can check the part for stress and strain and determine if it will stand up in use. At this point it is possible to alter the structure and strengthen any weak areas. For example, the part may be thickened at one point or a larger radius added where two surfaces join. Any changes made at this time will also be recorded in the computer memory. At this time it is also possible to estimate the cost of the finished part, run a weight analysis, etc., and even prepare a NC tape for the machining process—all while the part is still in the drawing stage.

After the operator has carefully examined the drawing on the display screen and is satisfied that the detail is accurate, the data may then be directed to the *plotter* to produce the finished drawing. The plotter will hold the drafting paper and produce permanent drawings in ink or ball-point pen. Some systems offer a photoplotter which projects a highly accurate light beam onto photosensitive paper to produce a drawing with sharp, high-contrast lines.

CAD ADVANTAGES

Let us look at another example of the value of this machine. Imagine that an aircraft design engineer is studying the image of an aircraft fuselage on the CRT. By pushing a button he or she could display the landing gear. As the engineer studies the components, some interference may be noted between the extended landing gear and the fuselage. The engineer will then press a button on the keyboard and with the light pen alter the shape of the fuselage, permitting the landing gear to operate freely. By the proper use of CAD equipment, it is possible to correct an error in less than a minute that would have taken months to resolve by the standard procedures formerly used, such as making a prototype or model of the plane.

Following is a list of CAD advantages.

☐ Greater productivity of drafting personnel
☐ Less drawing production time
☐ Better drawing change procedure
☐ Greater drawing and design accuracy
☐ Greater detail in layouts
☐ Superior drawing appearance
☐ Greater parts standardization
☐ Better factory assembly procedures
☐ Less scrap

REVIEW QUESTIONS

1. What is CAD?
2. Name the main components of a CAD system.
3. Briefly explain how CAD can be used to design a part.
4. Define a menu and state its purpose.
5. What design and testing data can be obtained from a CAD system?
6. Why is CAD so important to designing a new product?

CHUCKING AND TURNING CENTERS

UNIT 73

Extensive studies during the mid-1960s showed that about 40 percent of all metal-cutting operations were performed on lathes. Up until this time, most work was done on engine or turret lathes, which were not very efficient by present-day standards. Intensive research led to the development of numerically controlled turning centers and chucking lathes which could produce work of almost any configuration automatically and much more efficiently (Fig. 73-1A). In recent years these have been updated to more powerful computer-controlled units capable of greater precision and higher production rates than their predecessors.

OBJECTIVES

After completing this unit you will be familiar with:
1. The purpose and functions of chucking and turning centers
2. The applications of computerized numerical control (CNC) for turning centers
3. The machining operations which may be performed simultaneously

Fig. 73-1 (A) Turning and chucking centers are versatile and very productive.   (Courtesy Cincinnati Milacron Incorporated.)

Fig. 73-1 (B) A chucking center is designed to increase productivity. (Courtesy Cincinnati Milacron Incorporated.)

Fig. 73-2 Turning and boring operations may be carried on simultaneously. (Courtesy Cincinnati Milacron Incorporated.)

Fig. 73-3 Both turrets being used for internal operations. (Courtesy Cincinnati Milacron Incorporated.)

CNC CHUCKING CENTERS

Chucking centers (Fig. 73-1B), designed to machine most work held in a chuck, are available in a variety of chuck sizes from 8 to 36 in. in diameter. Because there are many similar types of machines made by different manufacturers and all performing similar functions, only one chucking lathe, a four-axis model, will be described.

This chucking center incorporates two turrets, operating independently on separate slides, to machine the workpiece simultaneously. While the seven-tool upper turret is machining the inside diameter (ID), the lower turret, which

also mounts seven different cutting tools, may be machining the outside diameter (OD) (Fig. 73-2). However, if the workpiece requires mainly internal operations, both turrets can work on the inside of the workpiece simultaneously (Fig. 73-3). This type of operation is suitable for large-diameter parts which require boring, chamfering, threading, internal radii, or retaining grooves.

For parts with mostly OD operations, the upper turret can be equipped with turning tools so that both turrets can machine the OD. Figure 73-4 shows a diameter and a chamfer being machined at the same time.

When longer parts must be machined, the right-hand end

Fig. 73-4 Both turrets being used for external turning. (Courtesy Cincinnati Milacron Incorporated.)

Fig. 73-5 A long shaft being supported by a center mounted in the upper turret. (Courtesy Cincinnati Milacron Incorporated.)

Fig. 73-6 Turning and facing operations performed by using the two turrets. (Courtesy Cincinnati Milacron Incorporated.)

Fig. 73-7 Cutting internal and external threads simultaneously. (Courtesy Cincinnati Milacron Incorporated.)

of the shaft may be supported with a center mounted in the upper turret while the lower turret performs the external machining operations (Fig. 73-5). Other operations which may be performed simultaneously are turning and facing (Fig. 73-6) and internal and external threading (Fig. 73-7).

Other chucking centers are chiefly two-axis models. They may have a single disklike turret on which both ID and OD tools are mounted, or they may have two turrets (usually on the same slide). In the latter case, one turret is normally designed for OD tools and the other for ID tools. Whatever the arrangement, the two-axis control will drive only one turret at a time.

CONSTRUCTION

Because of the high spindle speeds [up to 2240 revolutions per minute (r/min)] and the high horsepower requirements,

machines of this type must be rugged to absorb the high cutting forces. The bed and machine frame is a heavy one-piece casting. To permit easy chip removal and easy loading and unloading of the workpieces, the bed is slanted 30° from the vertical plane.

Direct-current servo drives provide accurate positioning and travel of the cutting tools. The servo motors position the turret cross-slides by means of high-precision preloaded ball screws which assures slide positioning repeatability to ±0.0002 in. These turret slides operate at speeds from 0 to 200 in./min (0 to 5080 mm/min) and 400 in./min (10,160 mm/min) on rapid traverse.

TOOLING

Both the upper turret and the lower turret can accommodate seven different types of tools. The toolholders for machin-

Fig. 73-8 A gage for presetting cutting tools for turning centers. (Courtesy Cincinnati Milacron Incorporated.)

Fig. 73-9 A chucking center control panel contains the computer, video display, and machine control functions.   (Courtesy Cincinnati Milacron Incorporated.)

ing ODs are located in the lower turret and are qualified (preset) and only require changing the carbide insert when the tool must be replaced. Tools for machining the ID are mounted in a dovetailed block and preset off the machine by means of a tool-setting gage (Fig. 73-8). The dovetailed

block is then mounted in the upper turret assuring proper tool point positioning every time.

COMPUTER NUMERICAL CONTROL

The control console shown in Fig. 73-9 is the "brains" of the chucking center. It features a minicomputer, a cathode ray tube (CRT) (video display), a tape reader, part program storage, part program edit, and maintenance diagnostic display.

The minicomputer performs control logic, mechanism control and input-output control. The CRT provides a visual display of slide positions, the spindle operating condition, the sequence numbers, preparatory functions, system fault conditions (diagnostics), operator instructions, and keyboard data.

The keyboard input provides the means of communicating with the CNC system. It is used to enter setup and tooling data on new parts programs or for correcting tooling data. The keyboard is also used to run diagnostic checks on the systems with the results (regarding oil pressure, spindle operating conditions, etc.) being displayed on the CRT.

Part program storage can store the data from up to 300 ft (91.4 m) of part program tape. A part program edit feature allows the program to be modified at time of use. Other features include a constant surface speed monitor which checks the part diameter and controls the spindle speed accordingly.

CNC TURNING CENTERS

Computerized numerical control turning centers (Fig. 73-10), while not unlike chucking centers, are designed mainly for machining shaft-type workpieces which are supported by a chuck and a heavy-duty tailstock center.

On four-axis machines two opposed turrets, each capable of holding seven different tools, are mounted on separate cross-slides, one above and one below the centerline of the work. Because the turrets balance the cutting forces applied to the work, extremely heavy cuts can be taken on a workpiece when it is supported by the tailstock. The dual turrets also lend themselves to other operations such as:

☐ Roughing and finishing cuts in one pass
☐ Machining different diameters on a shaft simultaneously (Fig. 73-11)
☐ Finish turning and threading simultaneously
☐ Cutting two different sections of a shaft at the same time

Because the lower turret is designed to accept ID tools, parts may be gripped in the chuck and be machined inside and outside at the same time (Fig. 73-12).

When the turning center is equipped with a steady rest, operations such as end facing, threading, etc. may be performed on the end of a shaft.

A bar-feeding mechanism (Fig. 73-13), supplied as an option, permits the machining of shafts and parts from bar stock providing it is smaller than the diameter of the spindle through-hole.

Fig. 73-10 A CNC turning center is designed for maximum efficiency on shaft-type work-pieces. (Courtesy Cincinnati Milacron Incorporated.)

Fig. 73-11 Both turrets being used to machine a shaft having different diameters. (Courtesy Cincinnati Milacron Incorporated.)

Fig. 73-12 The inside and outside diameters of a part being machined at the same time. (Courtesy Cincinnati Milacron Incorporated.)

Fig. 73-13 A bar-feeding attachment allows parts to be manufactured from bar stock. (Courtesy Cincinnati Milacron Incorporated.)

1. Why were chucking and turning centers developed?
2. Describe the construction of the two turrets of the chucking lathe.

3. What is the purpose of the slanted bed on a chucking or turning center?
4. State the purpose of the following CNC parts.
   (a) minicomputer   (b) CRT   (c) keyboard
5. What is the main purpose of a turning center?
6. Why can extremely heavy cuts be taken on a turning center?

MACHINING CENTERS
UNIT 74

In the 1960s industrial surveys showed that smaller machine components requiring several operations took a long time to complete. The reason was that the part had to be sent to several machines before it was finished and often had to wait a week or more at each machine before being processed. In some cases, parts requiring many different operations spent as many as 20 weeks in the shop before completion. Another startling fact was that during all the time in the shop, the part was only in a machine for 5 percent of the time, and while it was in the machine only 30 percent of the time was spent in machining. Therefore, a part was machined for only 30 percent of 5 percent or 1.5 percent of its time in the shop. In contrast, larger parts like large machine castings took much less time because all the machining was usually done on one machine.

Also in the 1960s, there was much "operator intervention" during the machining process. The operator had to watch the performance of the cutting tool and change spindle speeds and feeds to suit the job and the machine. The operator was frequently changing cutters as they became dull and always had to set depths of roughing and finishing cuts. All of these problems were recognized by machine tool manufacturers, and in the late 1960s and early 1970s they began designing machines which would perform several operations and probably do about 90 percent of the machining on one machine. One of the results of this research was the *machining center* and the later more elaborate version called the *processing center*. The machines can perform the operations of drilling, milling, boring, tapping, and precision profiling very efficiently.

OBJECTIVES
After completing this unit you will be familiar with:
1. The development of the machining center
2. The types and construction of the machining center
3. The operation of the machining center and the processing center

CONSTRUCTION AND OPERATION

There are two main types of machining centers, the *horizontal* and the *vertical spindle* types. Only the horizontal spindle type will be discussed in this unit.

There are two main types of horizontal spindle machining centers. They are:

1. The *traveling-column type* (Fig. 74-1) equipped with one or usually two tables on which the work is mounted. With this type the column (and the cutter) move toward the work and while it is machining the work on one table, the operator is changing the workpiece on the other table.
2. The *fixed-column type* (Fig. 74-2) is equipped with a pallet shuttle. The pallet is a removable table on which the

461

Fig. 74-1 A traveling-column-type machining center. (Courtesy Cincinnati Milacron Incorporated.)

Fig. 74-2 A fixed-column-type machining center. (Courtesy Cincinnati Milacron Incorporated.)

COMPUTER-AGE MACHINING

Fig. 74-3 (A) A new pallet ready for loading while the other pallet is ready for unloading. (B) Pallet being shuttled onto the receiver. (C) The pallet being moved into machining position. (Courtesy Cincinnati Milacron Incorporated.)

workpiece is mounted. With this type, after one workpiece has been machined (Fig. 74-3), the pallet (and workpiece) is forced off the receiver onto the shuttle. The shuttle is then rotated bringing the new pallet into position for the shuttle and the finished work pallet into position for unloading.

In Fig. 74-3B the new pallet is shuttled and clamped onto the receiver, while in Fig. 74-3C, the pallet and workpiece are positioned into the programmed machining position. The cycle for changing and shuttling the pallets off and onto the receiver takes only 20 s.

TOOL CHANGER

All machining centers are equipped for automatic, numerically controlled tool changing, which is much faster and more reliable than manual tool changing. The tools (up to 90 depending on the type of machine) are held in a storage chain (matrix) and each is identified by either the tool number or the storage pocket number. This information is stored in the computer memory. While one operation is being performed on the workpiece, the tool required for the next operation is moved to the pick-up position (Fig. 74-4A) where the tool-change arm removes and holds it.

Immediately on completion of the machining cycle, the tool change unit swings 90° to the tool change position (Fig. 74-4B). The tool-change arm then rotates 90° and removes the cutting tool from the spindle. It then rotates 90° and places the new cutting tool in the spindle after which it returns the old cutting tool to its position in the tool carrier (Fig. 74-4C). This whole operation is completed in about 11 s.

COMPUTER

The computerized numerical control (CNC) is built around a minicomputer and is either mounted on the machine or in a separate unit beside the machine. Most computer controls have similar features such as a *cathode ray tube* (CRT) or display unit and an alphanumeric keyboard for passing additional information to the computer. There will also be storage for 15 to 20 programs for machining different parts as well as program editing capabilities. Most computer controls now are capable of adjusting automatic backlash and lead screw compensation as well as indicating any

malfunction in the machine or any errors in the program functions. This diagnostic display informs the operator of any problems in the machine or control and, by referring to a maintenance manual, the user normally can correct the problem. On some controls, diagnostic units can be tied in by long-distance telephone to a mainframe computer in the machine or control manufacturer's plant. There the problem is assessed immediately and instructions given by phone to the user.

In unmanned machining centers, the computer controls the total program. It programs the parts onto the machine and up to the spindle. It selects the proper program for the part and then carries out the machining program and centers the loading and unloading often performed by robots.

MACHINING CENTER ACCESSORIES

TORQUE CONTROL MACHINING

A feature that is fast becoming popular is that of *torque control machining*, the torque being calculated from measurements at the spindle drive motor (Fig. 74-5). This device will increase productivity by preventing or sensing damage to the cutting tool. The torque is measured when the machine is turning but not cutting, and this value is stored in the computer memory.

As the machining operation begins, the stored value is subtracted from the torque reading at the motor. This will give the net cutting torque, which is compared to the programmed torque or limits stored in the computer [or on numerical control (NC) tape]. If the net cutting torque exceeds the programmed torque limits, the computer will act by reducing the feed rate, turning on the coolant, or even stopping the cycle. The feed rate will be lowered whenever the horsepower requirements exceed the rated motor capacity or the programmed code value.

The system display of three yellow lights advises the operator of the operational conditions in the machine at the time. A left-hand yellow light indicates that the torque control unit is in operation. The middle yellow light indicates that the horsepower limits are being exceeded. The right-hand light comes on when the feed rate drops below

| A | B | C |

Fig. 74-4 (A) The cutting tool required being removed from the storage chain. (B) The tool-change unit swings 90° to change tools in the spindle. (C) The old cutting tool has been removed and replaced with the tool required for the next operation. (Courtesy Cincinnati Milacron Incorporated.)

Fig. 74-5 Torque control raises or lowers the feed rate depending on the depth of cut at any time during the cutting cycle. (Courtesy Cincinnati Milacron Incorporated.)

60 percent of the programmed rate. The meter (Fig. 74-5) indicates the cutting torque (or operational feed rates) as a percent of the programmed feed rate.

As the tool gets dull, the torque will increase and the machine will back off on the feed and ascertain the problem. It could be that there is excessive material on the workpiece or a very dull or broken tool. If the tool is dull, the machine will finish the operation and a new backup tool of the same size will be selected from the storage chain when that operation is to be performed again. If the torque is too great, the machine will stop the operation on the workpiece and program the next piece into position for machining.

PRECISION SURFACE-SENSING PROBE

The *precision probe* (Fig. 74-6) is an extremely sensitive stylus used for setting up and checking workpieces on the machining center. The probe is stored in the tool storage matrix (Fig. 74-7) and is loaded in the spindle when required, to determine if the work is mounted and positioned properly. If there is not enough (or too much) material on the workpiece for proper machining, or if the workpiece is not positioned properly on the pallet, the probe will reject the part and the computer would call up the next workpiece.

The probe may also be used for checking that the proper fixture has been selected, or for checking the amount of core shift in castings before machining. It can also be used to calibrate the accuracy of the machine and to inspect the accuracy of the finished workpiece. One company had previously taken 25 h to inspect a complex part using standard inspection equipment at the bench. The probe performed the same inspection while the part was still mounted on the machine in 12 min. This type of precision surface-sensing probe may also be used for setting up and checking parts which are to be machined in a turning center.

THE PROCESSING CENTER

The foregoing machining center (Cincinnati Milacron T-10) may be equipped with an eight-pallet automatic work

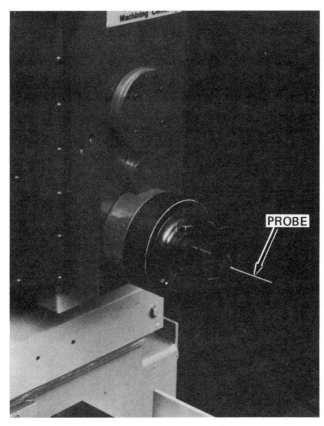

Fig. 74-6 The precision sensing probe is used to quickly check the position of the workpiece. (Courtesy Cincinnati Milacron Incorporated.)

Fig. 74-7 The probe may be loaded into the automatic tool changer for use whenever required. (Courtesy Cincinnati Milacron Incorporated.)

changer (AWC) that includes a manually indexible load-unload work station to permit loading and unloading while the machine is running. With this addition, eight alike or different parts may be loaded on the pallets and presented to the cutting tool(s) for the machining operations required (Fig. 74-8). The operations for each part are programmed into the computer and if the workpiece has been set up properly and accepted by the probe, the computer will select the proper program and the machine will perform the necessary operations on it. This setup permits the machining of eight different parts automatically.

Fig. 74-8 The processing center is a machine designed for the factory of the future. (Courtesy Cincinnati Milacron Incorporated.)

ADVANTAGE OF MACHINING CENTERS

1. Increased productivity.
2. This type of machine can machine to closer tolerances. This means:

☐ Fewer finishing operations required such as grinding or scraping.
☐ Less chance of error because fewer machines are being used.
☐ Less inspection required.
☐ Less paperwork needed.
☐ Lower rejection or scrap rate.

3. Reduction in inventory because there is less floor time between operations.

REVIEW QUESTIONS

1. What factors led to the development of the machining center?

CONSTRUCTION AND OPERATION

2. Name two types of horizontal spindle machining centers and briefly state the principle of each.
3. Describe how the required tool is mounted in the spindle.
4. What is the purpose of
   (a) cathode ray tube?   (b) diagnostic display?

MACHINING CENTER ACCESSORIES

5. Briefly describe the operation of the torque-control device.
6. List three factors that may cause the precision surface-sensing probe to reject a workpiece.
7. State four other uses of the probe.

UNIT 75

Over the past two decades, industry has realized that in order to be competitive in the world market, they had to increase productivity and reduce manufacturing costs. Since the source of skilled workers was dwindling and it was difficult to find people who would perform tasks which were considered to be monotonous, physically difficult, or environmentally unpleasant, industry has found it necessary to automate many manufacturing processes. The development of the computer has made it possible for industry to produce reliable machine tools and robots which are making manufacturing processes more productive and reliable, thereby improving their user's competitive position in the world market.

Over the past 15 years, industry has slowly introduced robots to various aspects of the manufacturing industry. With the need to be competitive in the world market, new control systems, more dependable and lower-priced computers, and more and better robots are being developed in the industrial world. Today's industrial robots are tough, tireless, very accurate, and are helping to make human drudgery a thing of the past in hundreds of factories throughout the world, and at the same time improving productivity and reducing manufacturing costs. The word *robot* is derived from a Czechoslovakian word *robota* meaning "work." The word "robot" has been used for many years in science fiction to mean a "mechanical being." Today's industrial robots still have a long way to go before they are capable of performing like the science-fiction robot, but they are capable of performing tasks which are monotonous, physically difficult, or unsafe for human workers.

OBJECTIVES

After completing this unit you will know:
1. The purpose of the industrial robot in manufacturing
2. The construction and operation of an industrial robot
3. The applications and benefits of robots in manufacturing

THE INDUSTRIAL ROBOT

The industrial robot of today (Fig. 75-1) is basically a single-arm device that can manipulate parts or tools through a sequence of operations or motions as programmed by the computer. These operations or sequences may or may not necessarily be repetitive, because the robot, through the computer, has the ability to make logical decisions. The robot can be applied to many different operations in industry and is capable of being taken from one operation and easily "taught" to do another. This ability is based on the concept of *flexible automation*, where one machine is capable of economically performing many different operations with a minimum amount of special engineering or debugging. This differs from the "hard," or dedicated, system which is designed only to perform one operation or task.

INDUSTRIAL ROBOT REQUIREMENTS

In order for an industrial robot to be most effective, it must meet three criteria: be flexible for many applications, reliable, and easy to teach.

APPLICATION FLEXIBILITY

The ideal industrial robot should be capable of performing many different operations. There are two general categories of applications where robots are being used extensively.

Fig. 75-1 Industrial robots perform back-breaking, boring or repetitive jobs. (Courtesy Cincinnati Milacron Incorporated.)

Fig. 75-2 The robot's arm has six rotary axes of movement. (Courtesy Cincinnati Milacron Incorporated.)

1. *Handling Applications.* Robots used in handling applications must be equipped with some type of device, such as a pair of gripping fingers or vacuum cups, in order to pick up or move the material. They can be used to load or unload machine tools, forging presses, injection-molding machines, and even inspecting or gaging systems. It also includes the handling of materials (moving it from one station to another), retrieving parts from storage areas or conveyor systems, packaging, and palletizing (placing parts on a pallet for transporting to other stations).

2. *Processor Applications.* Robots in this category carry out operations such as spot welding, seam welding, spray painting, metallizing, cleaning, or any other operation where the robot can operate a tool to carry out a manufacturing process.

RELIABILITY

The industrial robot must be very reliable since it will be working in conjunction with some very sensitive machinery which should not be left idle when a robot breaks down. The robot should not only be reliable but also have the ability to diagnose a problem quickly when it occurs and, if possible, take corrective action.

Fig. 75-3 The robot arm is designed to operate over a wide range. (Courtesy Cincinnati Milacron Incorporated.)

EASE OF TEACHING

An industrial robot is taught by example by a human operator who must lead it through the sequence of steps or operations needed for each specific application. This programmed teaching routine is necessary to teach the robot to make logical decisions based on information generated by two-way communications with its working environment.

Robots which are easy to teach do not require that the worker know how to program it or how its control system operates. For simplicity, the human worker should only need to know two things: what the robot has to do to perform an operation, and what the worker has to do to *teach the robot*.

INDUSTRIAL ROBOT DESIGN

A robot system is a means of moving a pair of grippers or some other form of end effector and directing it through a certain sequence of motions. At certain points in the sequence, the system directs some function to occur such as ''close the grippers'' or ''start the machine tool cycle.''

The Tomorrow Tool (T3) manufactured by Cincinnati Milacron Inc. is a hydraulically powered, servo-operated, and computer-controlled industrial robot system which has been designed for maximum flexibility. The arm has six rotary axes of movement such as arm sweep, shoulder swivel, and elbow extension, along with pitch, yaw, and roll of the wrist (Fig. 75-2). The elbow axis is driven by a piston-cylinder arrangement, while all other axes are powered by direct-coupled rotary actuators. This arm provides a mechanical system which has no gears or linear slides and a minimum number of rotating parts. This produces a very reliable system which creates very few maintenance problems.

Because of the construction design, the arm can reach over 150 in. (3800 mm) vertically upward and extend out horizontally almost 100 in. (2500 mm) (Fig. 75-3). It can also operate through 240° of arm sweep. Standard arms are capable of lifting up to a 100-lb (45-kg) load located at 10 in. (250 mm) from the roll axis plate, while heavy-duty

Fig. 75-4 The robot control unit consists of a computer, video display, and control panel. (Courtesy Cincinnati Milacron Incorporated.)

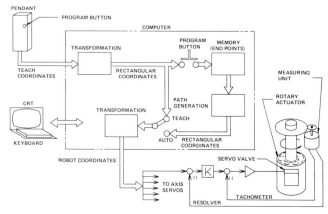

Fig. 75-5 The role of the computer in the control system. (Courtesy Cincinnati Milacron Incorporated.)

arms can lift up to 225 lb (100 kg). The arm can repeat for position anywhere in its range within an accuracy of ±0.050 in. (1.27 mm). However, with the addition of a special system option, the repeatability can be improved to ±0.025 in. (0.64 mm).

THE CONTROL SYSTEM

The purpose of an industrial robot's system is to direct the robot through its movements and sequences of operations and to provide a means for the operator to teach the robot the motions and sequences required to perform various operations.

The control unit used for robots is based on a minicomputer because it is very flexible and allows the operator to teach the robot. The robot's control unit (Fig. 75-4) consists of a minicomputer, a cathode ray tube (CRT) and keyboard, a control panel, axis servo-control electronics, servo test panel, a teach pendant, and terminals for communications with other equipment (Fig. 75-5).

In order to operate effectively, the robot's control system should have the following characteristics:

1. When a robot is operating, it should move from position to position in a straight line with all six axes moving proportionally so that they all start and finish at the same time.
2. The robot should be able to communicate between itself and the control unit in order to start and complete cycles, make logical decisions on its next move, check that an operation has been completed.
3. The robot should be able to diagnose problems as they occur in order to reduce downtime and minimize loss of production.
4. During the teaching mode, the human teacher should be able to guide the tool center point (TCP) along a rectilinear (XYZ) coordinate system or even a cylindrical system (Fig. 75-6).
5. The human teacher should be able to edit a previously taught program by easily adding or deleting points if necessary. Therefore, channels of communication must exist between the teacher and the control.

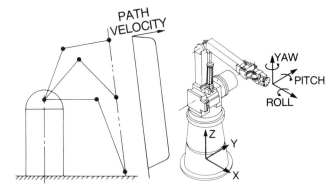

Fig. 75-6 The key operating concepts of a robot. (A) Controlled path—straight line; (B) rectangular and angular coordinates. (Courtesy Cincinnati Milacron Incorporated.)

MODES OF OPERATION FOR ROBOTS

The T3 Cincinnati Milacron robot has three modes of operation: manual, teach, and automatic. The mode desired is selected on the control console.

1. *Manual Mode.* In the manual mode, the human worker can move each axis of the arm independent of all the others with the use of the teach pendant (Fig. 75-7). The purpose of the manual mode is to allow the operator to align the arm to its "home" position. Home is a unique location where the six coordinates of the robot's arm are known to the control system. Before the human operator can program the robot through the teach mode or enter the automatic mode, the arm must be aligned to home. A special display on the control console and arrows on the robot arm can be used for this alignment.
2. *Teach Mode.* In the teach mode, the operator can teach the robot arm to perform a certain task or operation. The axes of the robot arm can be commanded to move in a coordinated fashion. The three coordinate system options available are cylindrical, rectangular, and hand (Fig. 75-8), and the way that coordination occurs is a function of the teach coordinate system chosen.

Movement within a given coordinate system is controlled by the positional pushbuttons on the pendant. For the rec-

Fig. 75-7 The teach pendant allows the robot to be taught easily. (Courtesy Cincinnati Milacron Incorporated.)

Fig. 75-8 The teach coordinate systems. (Courtesy Cincinnati Milacron Incorporated.)

tangular system, two positional buttons will cause the tool center point (TCP) to move left or right; two other buttons cause the TCP to move up and down, and two more buttons cause the TCP to move in and out. The pushbuttons on the teach pendant make it very simple for the human operator to ''teach'' the robot to perform specific tasks using any coordinate system.

3. *The Automatic Mode.* In the automatic mode, the robot performs its taught cycle continuously until either the ''end of cycle stop'' or ''interrupt'' buttons on the computer console are depressed. The end of cycle stop button makes the arm stop the next time it reaches the last point in the cycle. The interrupt button makes the robot arm stop immediately. An error code appears on the control console display whenever something goes wrong with the system or the human worker issues an unacceptable command.

INDUSTRIAL APPLICATIONS FOR ROBOTS

As mentioned previously, the industrial robot is finding many applications where tasks are monotonous, physically difficult, or environmentally unpleasant for the human worker. Because of the flexibility of the robot arm and the ease of teaching it to do new tasks, the robot is continually finding new industrial applications. The most common robot applications are:

1. *Loading and Unloading Machine Tools.* Robots are finding wide use in loading and unloading machine tools, forging presses, and injection-molding machines. (Fig. 75-9). It is not uncommon to find one robot loading and unloading two or more machines at the same time.

Fig. 75-9 A robot being used to load and unload machine tools.

2. *Welding.* Computer-controlled industrial robots can spot-weld car bodies as they are moving on a conveyor. Unique tracking capabilities enable the robot to spot the welds without stopping the line.

Seam welding is another application where robots are finding wide use (Fig. 75-10). The Cincinnati Milacron Milatrac seam-tracking device monitors the weld path ''through the arc'' and adjusts the robot's position to compensate for variations in the start point or seam path to produce consistently accurate weld weaves.

3. *Moving Heavy Parts.* Robots are being used to move heavy parts from an input station, through various machining and gaging stations, to an exit conveyor. They can also bring materials from storage areas, package and palletize, inspect, sort, and assemble (Fig. 75-11).

Fig. 75-10   Robots are used extensively for welding operations. (Courtesy Cincinnati Milacron Incorporated.)

Fig. 75-11   Car bodies are checked for assembly accuracy by robots.   (Courtesy Cincinnati Milacron Incorporated.)

4. *Spray Painting.* The painting of automobile bodies is an example of a job which would be environmentally unpleasant for a human worker. The robot can be "taught" (programmed) to follow the contour of an automobile body or part and apply a consistently even coat of paint to the surface.

5. *Assembly.* Assembly jobs such as stringing a wire harness around a series of pegs is an example of a task which is very monotonous for the human worker. The robot arm is fitted with a spool of wire and a feeding mechanism, and then taught to proceed around pegs and through closely spaced slots to produce a harness.

6. *Machining Operations.* Some repetitive machining operations are ideally suited for the computer-controlled industrial robot. Figure 75-12 illustrates a robot being used to drill airplane parts. As another example, the robot puts vent holes in casting molds.

BENEFITS OF COMPUTER-CONTROLLED INDUSTRIAL ROBOTS

The computer-controlled industrial robot has many advantages and features which make it a very important industrial manufacturing tool.

1. The *controlled path algorithm,* made possible by the computer, controls the position, orientation, velocity, and acceleration of the path of the TCP at all times.

2. The robot can easily be taught because the robot can be moved in different coordinate systems with the teach pendant. The console display provides communication between the teacher and the control.

3. The computer can store system software and taught data in its memory. This allows easy editing of the taught program at any time, especially when there are changes in the application.

4. The memory of the computer can be easily expanded to allow for the inclusion of more taught points or extra functions.

Fig. 75-12   A robot performing a drilling operation on airplane parts.

5. The robot can be interfaced with other equipment and can receive and interpret signals from various sensor devices. It can also provide valuable data to a higher-level computer on manufacturing conditions, quantities, and problems.

6. The system can easily be modified to suit changes in production. The robot computer can be linked to a supervisory computer which can direct the robot to convert to other operations as required.

7. The computer is able to monitor the robot and the equipment it services, and diagnostic messages appear on the console screen to indicate various error conditions. If necessary, the computer may direct the robot to stop operations.

ROBOT SAFETY

Robots, like other pieces of machinery, must be treated with respect in order to prevent accidents. Although robots are taking over hazardous jobs, it is still good practice to observe various safety precautions.

1. The robot working area should be enclosed by some form of barrier to prevent people from entering while the robot is working. Entrance gates, if any, should have controls which automatically stop the robot if anyone enters.
2. Emergency stop buttons, which stop all robot action and cut off the power, should be readily accessible outside the working range of the robot.
3. Robots should be programmed, serviced, and operated only by trained workers who fully understand and can anticipate the robot's action.
4. Care should be exercised during the programming cycle if it is necessary to have people within the robot's working area.
5. The work performed by a robot should not present a health or safety hazard to human workers in nearby areas.
6. Hydraulic and electrical cables should be located so that they cannot be damaged by the operation of the robot or any related equipment.

The six-axis computer-controlled industrial robot has found a variety of uses in industry. As more sophisticated robots are developed, it is quite possible that the robots of the future will play an ever-increasing role in manufacturing. Under development are auditory systems, vision systems, and sensors which will enable a robot to see, hear, and feel.

REVIEW QUESTIONS

1. Why were robots developed?
2. Define an industrial robot.
3. Name three criteria of an industrial robot.
4. Name the six rotary axes of movement of a robot arm.
5. List the components in a robot control unit.
6. What three modes can be used to operate robots?
7. List six applications of industrial robots.
8. Name three important safety precautions which should be observed with robots.

MANUFACTURING SYSTEMS
UNIT 76

A manufacturing system may be defined as a collection of two or more manufacturing units, interconnected with material-handling machinery, under the supervision of one or more dedicated, supervisory (executive) computers.

There are two types of manufacturing systems, the *random* (or flexible) and the *dedicated* systems.

The *random system* consists of standard numerical control (NC) machine tools linked together with appropriate handling devices and total system control. This system is designed for low-volume, random-order machining of parts of any size or shape. This type of system is quite flexible and will permit changes in production of the parts.

The *dedicated system* consists of specialized machines and/or some standard NC machines linked together with less-flexible handling devices and complete system control which may be a computer or a programmable controller. Because the specialized machines are capable of performing only one certain operation and cannot be changed and/or because the handling devices require operations to be done in a fixed sequence, the machine (and the system) is said to be dedicated. Since some of the machines are specialized and do not provide for frequent tool changes, and because the handling system lacks adaptability, the dedicated system requires medium-volume production runs.

After completing this unit you will
1. Know the purpose and function of a flexible manufacturing system
2. Know the purpose and function of a dedicated manufacturing system

FLEXIBLE MANUFACTURING SYSTEM

The fully flexible manufacturing system is in effect a *random system* because it consists of standard NC machine modules linked together with a flexible parts-handling device and total-system computer control.

The flexible manufacturing system shown in Fig. 76-1 consists of two turning centers served by a centrally located robot and a suitable conveyor. Two or more similar NC machines that are grouped to handle workpieces of similar size and shape and are controlled by a central computer are usually called a *cell*. When a robot is used, its computer serves as a system supervisor and is linked to each of the machine computer control units.

OPERATION

The rough part is removed from the conveyor by the programmed *robot* and mounted in the lathe. Proper positioning is assured by the *probe*, which checks surfaces and will later measure outside and inside diameters in process and instruct the computer to make appropriate tool adjustments.

As the machining process continues, the cutting tools will become dull. A *tool cycle time monitor* incorporated in the computer keeps track of tool usage and will signal for a backup tool when the expected life of the tool expires. After the particular machining cycle has been completed, the dull cutting tool is automatically replaced with a sharp one from the tool chain matrix. The old tool is automatically replaced in the tool magazine rack.

Fig. 76-1 A flexible manufacturing system with two turning centers, robot, and conveyor system. (Courtesy Cincinnati Milacron Incorporated.)

To ensure continuous machining with the fewest possible problems, a *machining monitor* constantly measures the torque at the machine spindle. This feature will detect dull or broken tools and other malfunctions in the lathe. Should a problem occur, the monitor will retract the cutting tool and shut off the machine.

After the part has been finished, it will be removed from the lathe by the robot and placed in the storage facility or on an exit conveyor. As an intermediate step, after the part has been machined it may be presented by the robot to be checked by a *post-process gaging control*. Here the final part size is checked by laser for accuracy of size. Should any out-of-tolerance dimensions be detected, the gaging control will send feedback signals through the computer to the machine which automatically adjusts the tool to the correct position.

Flexible machining systems are not limited to turning centers. Figure 76-2 shows a flexible manufacturing system for automating the production of rectangular or box-type parts. Here two traveling-column machining centers automatically machine the parts, which are delivered to the stations by computer-controlled carts. Replacement tooling is also delivered to the machines when required by these carts. Flexible manufacturing systems are usually used on the second and third shifts where it is difficult to obtain skilled help.

DEDICATED MANUFACTURING SYSTEMS

A dedicated system often includes machines which are designed to perform only specialized operations on a part, and additional standard NC machine tools all linked together with appropriate parts handling equipment. The total system is controlled by a computer. Dedicated systems are designed for "medium-volume" production of a part or a restricted family of parts (restricted by shape and size).

The system illustrated in Fig. 76-3 is composed of three machines (stations) which are capable of performing only specialized operations (dedicated). The workpieces, each weighing 2000 lb, are mounted on pallets and are moved automatically from one station to the next for the various operations required.

Another type of dedicated system is shown in Fig. 76-4. Here standard machining centers are combined with special (dedicated) machines for washing and inspecting the parts. The workpieces are shuttled from station to station on pallets. This system is fully automated and produces and inspects several similar and dissimilar parts. As a part is moved into position for machining, the probe and the computer will select the proper machining program for the part. The washing and gaging stations in this system are dedicated.

Fig. 76-2   A flexible manufacturing system automated for the production of boxlike parts.
(Courtesy Cincinnati Milacron Incorporated.)

Fig. 76-3 Dedicated manufacturing systems performed only specialized operations. (Courtesy Cincinnati Milacron Incorporated.)

Fig. 76-4 A dedicated manufacturing system which combines flexible and standard machines. (Courtesy Cincinnati Milacron Incorporated.)

Machines which are grouped to make up dedicated systems will have similar features and controls to those outlined for flexible systems. However, the work moves from station to station in a fixed sequence. Pallets tend to be less universal than handling devices in flexible or random-order systems, and system software is less sophisticated.

REVIEW QUESTIONS

1. Define a manufacturing system.
2. What is the difference between a random and a dedicated system?

FLEXIBLE MANUFACTURING SYSTEM

3. Describe the composition of a flexible manufacturing system.
4. Explain the function of the following components:
   (a) robot
   (b) probe
   (c) tool cycle monitor
   (d) machining monitor

DEDICATED MANUFACTURING SYSTEM

5. How does the dedicated system differ from the flexible manufacturing system?
6. Explain what occurs as a part is moved into position for machining.

FACTORIES OF THE FUTURE
UNIT 77

One of the large problems in the field of manufacturing is that of "batch manufacturing," whereby small quantities of hundreds of different parts are produced. At least three-quarters of all industrial production falls into this category. Large quantities of one item can be produced most efficiently on a transfer (production) line, or by a series of dedicated machines. These processes, however, are not suitable to the manufacture of small numbers of one part (batch). This problem, along with the small percentage of time that machines are being used (often for only one shift), and the lack of skilled personnel for the second and third shifts, has led manufacturers to look to more efficient production methods.

OBJECTIVES

After completing this unit you will:
1. Have an overview of the factory of tomorrow
2. Understand the impact of the computer on our society and manufacturing in particular

VISION OF THE FUTURE

Industry has taken steps in the past few years to overcome production problems. *Computers,* the key to the new technology, have been applied to the automation of equipment to increase machine use, reduce the complexity of batch manufacturing, and improve productivity.

In order to achieve optimum efficiency in manufacturing, machine tools should be run 8760 h per year. Planners are now looking for ways to achieve this goal and some of the means to implement this follow. The factory of the future will be operated at full production with very few people present to produce a great variety of different parts in small batches (Fig. 77-1). Many objects now made of metal will probably be made of synthetic materials and by different processes.

Machines will be grouped together in small clusters or *cells* with *automatic handling* of all materials in and between different cells. Machines will be *computer-controlled,* which will greatly improve the accuracy and consistency of the machine production. The computer memory permits the repetition of a series of complicated operations

Fig. 77-1 A designer's view of the factory of the future. (Courtesy Cincinnati Milacron Incorporated.)

with the push of a button. These programs can be stored in the computer for future use. *Automatic tool changing* (Fig. 77-2) will be common on all machines. With these improvements, it will be possible to remove the operator and employ the person as a monitor or supervisor to control all the production in the cell. The operator will be required to step into the cell only in case of a breakdown. With the addition of sensing devices which will warn of impending machine failures, the cell will be able to perform batch operations continuously without human intervention.

In order to satisfy production needs, several cells will be set up, each connected with the others. These in turn will be connected with areas of the plant such as material storage, transport, and data communications with the executive and mainline or central computers. These several cells will be controlled by one person in the executive computer room.

COMPLETE MANUFACTURING SYSTEMS

In order to make the system complete, all functions in the manufacturing process, that is, *design, planning, inventory control, scheduling,* and *shop floor control,* must be able to communicate with each other automatically. In this way they will all share the common data base and all other current information. Each of the cell systems can take any independent action required, based on this shared information. The mainline or central computer reports to the staff and calls attention to problems which may require attention.

An overview of the factory of the future is shown in Fig. 77-3. There are three general phases of manufacturing: *design, planning,* and *production.*

DESIGN

The design process will be improved by computer-aided design (CAD), whose main purposes are drafting and design. The design engineer may make or modify design drawings by using the *cathode ray tube (CRT)* and the light pen (Fig. 77-4). Color graphics will be added to make three-dimensional shapes easier to comprehend and to reduce the chance of design error. Analysis pertaining to the function of the final product will be carried out by CAD.

PLANNING

All operations relating to the production planning will be computerized. *Capacity Resource Planning (CRP)* will determine equipment and labor requirements. By the use of animation, parts can be moved from one cell to another (on the CRT) to ascertain any problems and determine machine utilization.

PRODUCTION

In order to further increase productivity, *group technology,* whereby parts are classified and coded by similarities in design and processing, will be employed. Computer-aided process planning will be employed. Here, with input from

Fig. 77-2 Automatic tool changing will play an important part in manufacturing systems. (Courtesy Cincinnati Milacron Incorporated.)

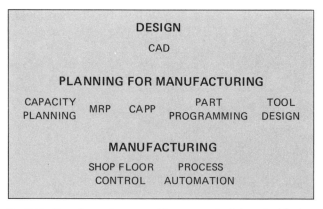

Fig. 77-3 Top-down diagram of the computer-integrated factory. (Courtesy Cincinnati Milacron Incorporated.)

CAD and from the data base along with the use of group technology, the computer will be able to determine the sequence of operations as well as the time standards and cost estimates.

Compuer-aided parts programming will take input from other sources and produce instructions for the individual machine programming. These programs will be stored in general memory and available to shop floor computers whenever required. During the final planning process, cutting tool movements can be simulated by solid modeling on the computer. The movements are recorded in the computer which calculates the numerical description of the path.

TOOL DESIGN

Special tooling and mold design will be improved by the use of CAD, which can compare the proposed design against the design of the part to see if it is functional. On the shop floor, all operations, such as machine and tool scheduling, management, and monitoring, will be totally computer controlled.

Sensing devices, linked to the computer, are essential for machine operation without operators. The *probe* (Fig. 77-5) will be widely used to locate, realign, and inspect

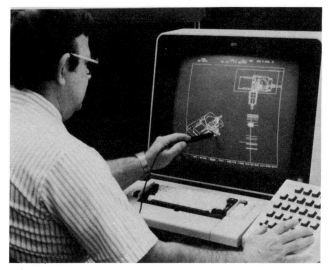

Fig. 77-4 A light pen being used to change the design of a part on a CAD system.

Fig. 77-6 Robots will find ever-increasing roles in the factory of the future. (Courtesy Cincinnati Milacron Incorporated.)

Fig. 77-5 The probe will be widely used to locate, realign, and inspect workpieces. (Courtesy Cincinnati Milacron Incorporated.)

PROBE

Fig. 77-7 Computer-controlled, wire-guided carts will transport tools and materials in the factory of the future. (Courtesy Cincinnati Milacron Incorporated.)

work. Advanced sensors, which will give each tool the sense of touch for positioning work, will no doubt be developed in the near future. Damage to the machine and tools will be controlled by means of *torque-control* devices mounted at the motor drive. These will warn the computer of excessive cutting torque and tool wear. Should other machine problems develop, a *diagnostic system* in the machine computer will warn the executive and central computers of these and they will take appropriate action.

Another area controlled by the computer will be the *material-handling* methods. Computer-controlled *programmable robots* (Fig. 77-6) will play a large part in the factory of the future. Robots will be used to perform dangerous and tedious tasks now performed by people. They may also be used for loading and unloading machines and performing other tasks that fixed automation cannot do.

Material-handling devices, such as wire-guided carts (Fig. 77-7), carousels, and conveyors, will be computerized and integrated into the whole system to supply materials and tools to the machines and remove the finished products.

Inspection and quality control will also be computerized. Constant feedback to the rest of the system will be maintained and any trend which may be developing in part sizes, etc., will be detected early and appropriate computer adjustments will be made to the machine which produced the part.

CONCLUSION

Each of these elements of factory automation is in use in some factories today, but not on a total basis in any one factory. When total integration is obtained, factories will be able to operate 8760 h per year virtually unattended.

REVIEW QUESTIONS

1. Why is the computer so important to the factory of the future?
2. How will the factory of the future be operated?
3. Define a *cell*.
4. What functions must be able to communicate with each other so that a manufacturing system will be complete?
5. State the purpose of:
   (a) capacity resource planning
   (b) computer-aided parts programming
   (c) the probe
   (d) the diagnostic system
   (e) wire-guided carts

GRINDING
SECTION 16

Early humans used abrasive stones to sharpen tools and produce a smooth surface. Over the years, the manufacture of abrasives and grinding tools has developed gradually to produce stronger and better abrasives. The use of abrasives in modern industry has contributed much to our mass production methods. Because of abrasives and modern grinding machines, it has been possible to produce parts and products to the close tolerances and high surface finish required by industry and the machine tool trade.

In order to function properly, an abrasive must have certain characteristics.

1. It must be harder than the material being ground.
2. It must be strong enough to withstand grinding pressures.
3. It must be heat-resistant so that it does not become dull at grinding temperatures.
4. It must be friable (capable of fracturing) so that when the cutting edges become dull, they will break off and present new sharp surfaces to the material being ground.

UNIT 78

Abrasives may be divided into two classes: *natural* and *artificial.*

Natural abrasives, such as sandstone, garnet, flint, emery, quartz, and corundum, were used extensively prior to the early part of the 20th century. However, they have been almost totally replaced by manufactured abrasives with their inherent advantages. One of the best natural abrasives is diamond but, because of the high cost of industrial diamonds (bort), its use in the past was limited mainly to grinding cemented carbides and glass, and sawing concrete, marble, limestone, and granite. However, due to the introduction of synthetic or manufactured diamonds, industrial natural diamonds will become cheaper in cost and will be used on many more grinding applications.

Manufactured abrasives are used extensively because their grain size, shape, and purity can be closely controlled. This uniformity of grain size and shape, which ensures that each grain does its share of work, is not possible with natural abrasives.

There are several types of manufactured abrasives: *aluminum oxide, silicon, carbide, boron carbide, cubic boron nitride,* and *manufactured diamond.*

OBJECTIVES

After completing this unit you will be able to:
1. Describe the manufacture of aluminum oxide and silicon carbide abrasives
2. Select the proper grinding wheel for each type of work material
3. Know the applications of grinding wheels and abrasive products

ALUMINUM OXIDE

Aluminum oxide is probably the most important abrasive, since about 75 percent of the grinding wheels manufactured are made of this material. It is generally used for high-tensile-strength materials including all ferrous metals except cast iron.

Aluminum oxide is manufactured with various degrees of purity for different applications, the hardness and brittleness augmenting with increasing purity. Regular aluminum oxide (Al_2O_3) is about 94.5 percent pure and is a tough abrasive capable of withstanding abuse. It is a grayish color and is used for grinding strong, tough materials, such as steel, malleable and wrought iron, and tough bronzes.

Aluminum oxide of about 97.5 percent purity is more brittle and not as tough as the regular aluminum oxide. This gray abrasive is used in the manufacture of grinding wheels for centerless, cylindrical, and internal grinding of steel and cast iron.

The purest form of aluminum oxide for grinding wheels is a white material which produces a sharp cutting edge when fractured. It is used for grinding the hardest steels and stellite and for die and gage grinding.

MANUFACTURE OF ALUMINUM OXIDE

Bauxite ore, from which aluminum oxide is made, is mined by the open-pit method in Arkansas and in the Guianas of South America. The bauxite ore is usually calcined (reduced to powder form) in a large furnace where most of the water is removed. The calcined bauxite is then loaded into a cylindrical, unlined steel shell about 5 ft (1.52 m) in diameter by 5 ft (1.52 m) deep (Fig. 78-1). This furnace is open at the top and lined with carbon bricks at the bottom. Two or three electrodes of carbon or graphite project into the open top of the furnace. During the operation, the outside of the furnace is cooled by a circumferential spray of water. The furnace is half-filled with a mixture of baux-

Fig. 78-1 Aluminum oxide is produced in an arc-type furnace. (Courtesy The Carborundum Company.)

Fig. 78-2 Silicon carbide is produced in a resistance-type furnace. (Courtesy The Carborundum Company.)

ite, coke screenings, and iron borings in the proper proportions. The coke is used to reduce the impurities in the ore to their metals, which combine with the iron and sink to the bottom of the furnace. The electrodes are then lowered onto the top of the charge, a starting batch of coke screenings is placed between the electrodes, and the current is applied. The coke is rapidly heated to incandescence and the fusion of the bauxite starts. After a small amount of bauxite is fused it becomes the conductor and carries the current. After the fusion of the bauxite has begun, more bauxite and coke screenings are added, and the process continues until the shell is full. During the entire operation, the height of the electrodes is automatically adjusted to maintain a constant rate of power input. When the furnace is full, the power is shut off, and the furnace is allowed to cool for about 12 h. The fused ingot is removed from the shell and allowed to cool for about a week. The ingot is broken up and fed into crushers; the material then is washed, screened, and graded to size.

SILICON CARBIDE

Silicon carbide is suited for grinding materials which have a low tensile strength (aluminum, brass, bronze) and high density, such as cemented carbides, stone, and ceramics. It is also used for cast iron and most nonferrous and nonmetal materials. It is harder and tougher than aluminum oxide. Silicon carbide may vary in color from green to black. Green silicon carbide is used mainly for grinding cemented carbides and other hard materials. Black silicon carbide is used for grinding cast iron and soft nonferrous metals, such as aluminum, brass, and copper. It is also suitable for grinding ceramics.

MANUFACTURE OF SILICON CARBIDE

A mixture of silica sand and high-purity coke is heated in an electric resistance furnace (Fig. 78-2). Sawdust is added to produce porosity in the finished product and to permit the escape of the large volume of gas formed during the operation. Sodium chloride (salt) is added to assist in removing some of the impurities.

The furnace is a brick-lined rectangle about 50 ft (15.24 m) long, 8 ft (2.43 m) wide, and 8 ft (2.43 m) high. It is open at the top. One or more electrodes protrude from each end of the furnace. The mixture of sand, coke, and sawdust is loaded into the furnace to the height of the electrodes. A granular core of coke is placed around the electrodes for the full length of the furnace. The core is covered with more sand, coke, and sawdust mixture and heaped to the top of the furnace. The current is then applied to the furnace and the voltage closely regulated to maintain the desired rate of power input. The time required for this operation is about 36 h.

After the furnace has cooled for about 12 h, the brick sidewalls are removed and the unfused mixture falls to the floor; the silicon carbide ingot can then cool more rapidly. After cooling for several days, the ingot is broken up and the silicon carbide removed. Care must be taken to remove the pure silicon carbide since the outer layer has not been properly fused and is not usable. The inner core surrounding the electrodes is also unusable since it is only graphitized coke.

The resultant silicon carbide is then crushed, treated with acid and alkalis to remove any remaining impurities, screened, and graded to size.

ZIRCONIA-ALUMINUM OXIDE

One of the more recent additions to the abrasive family has been the development of zirconia-aluminum oxide. This material, containing about 40 percent zirconia, is made by fusing zirconium oxide and aluminum oxide at extremely high temperatures [3450°F (1900°C)]. This is the first *alloy abrasive* ever produced. It is used for heavy-duty rough

481

and finish grinding in steel mills, snagging in foundries, and in metal fabrication plants for the rapid rough and finish grinding of welds. The performance of zirconia-aluminum abrasive is superior to that of standard aluminum oxide for rough grinding and snagging operations because the actions of the two grain types are quite different.

During any grinding process, much frictional heat is generated which breaks down the organic bond (rubber, shellac, or phenalic) on the wheel or the coated disk and lets the wheel "self-dress." Standard aluminum oxide abrasive wheels will, at first, penetrate the surface of the workpiece and remove the metal effectively. Soon the abrasive grains will become dull, the wheel will lose its penetrating power and ride on the surface of the work. Frictional heat builds up which softens the organic bond and the entire grain is expelled after about only 25 to 30 percent of its life. With the newer zirconia-alumina abrasive, the grain action is quite different. The cutting action starts in the same manner, but when the abrasive grain starts to dull, microfracturing takes place just below the abrasive grain surface. As the small particles break off, sharp new cutting points are generated while the grain still stays secured in the bond. As the small broken off particles of abrasive are ejected, they take with them much of the heat generated by the grinding action. Due to the cooler cutting action of this abrasive, sharp new cutting edges are produced many times before the grain is finally expelled from the bond and only after about 75 to 80 percent of the abrasive grain has been used. Wheels and disks made with zirconia-alumina base will last from two to five times longer than standard aluminum oxide wheels, depending on the application.

Zirconia-alumina applications offer several advantages over standard abrasives for heavy-duty grinding.

☐ Higher grain strength
☐ Higher impact strength
☐ Longer grain life
☐ Maintains its shape and cutting ability under high pressure and temperature
☐ Higher production per wheel or disk
☐ Less operator time spent changing wheels or disks

BORON CARBIDE

Another of the new abrasives is boron carbide. It is harder than silicon carbide and, next to the diamond, it is the hardest material manufactured. Boron carbide is not suitable for use in grinding wheels and is used only as a loose abrasive and a relatively cheap substitute for diamond dust. Because of its extreme hardness, it is used in the manufacture of precision gages and sand blast nozzles. Crush dressing rolls made of boron carbide have proved superior to tungsten carbide rolls for the dressing of grinding wheels on multiform grinders. Boron carbide is also widely accepted as an abrasive used in ultrasonic machining applications.

MANUFACTURE OF BORON CARBIDE

Boron carbide is produced by dehydrated boric acid being mixed with high-quality coke. The mixture is heated in a horizontal steel cylinder which is completely enclosed except for a hole in each end to accommodate a graphite electrode and vent holes to prevent the escape of the gases formed. The outside of the furnace is sprayed with water to prevent the shell from melting. During the heating process, air must be excluded from the furnace. This is done by dampening the mixture with kerosene, which will volatize and expel the air from the furnace as it is heated. A high current at low voltage is applied for about 24 h, after which the furnace is cooled. The resulting product, boron carbide, is a hard, black, lustrous material.

CUBIC BORON NITRIDE

One of the more recent developments in the abrasive field has been the introduction of cubic boron nitride. This synthetic abrasive has hardness properties between silicon carbide and diamond. The crystal known as Borazon® CBN (cubic boron nitride) was developed by the General Electric Company in 1969. This material is capable of grinding high-speed steel with ease and accuracy and is superior to diamond in many applications.

Cubic boron nitride is about twice as hard as aluminum oxide and is capable of withstanding high grinding temperatures of up to 2500°F (1371°C) before breaking down. CBN is cool-cutting and chemically resistant to all inorganic salts and organic compounds. Because of the extreme hardness of this material, grinding wheels made of CBN are capable of maintaining very close tolerances. These wheels require very little dressing and are capable of removing a constant amount of material across the face of a large work surface without having to compensate for wheel wear. Because of the cool-cutting action of CBN wheels, there is little or no surface damage to the work surface.

Manufacture

Cubic boron nitride is synthesized in crystal form from hexagonal boron nitride ("white graphite") with the aid of a catalyst, heat, and pressure (Fig. 78-3A). The combination of extreme heat [2725°F (1496°C)] and a tremendous pressure (47,540,000 psi) on the hexagonal boron nitride and the catalyst produces a strong, hard, blocky, crystalline structure with sharp corners known as CBN (Fig. 78-3B).

There are two types of CBN:

Borazon CBN is an uncoated abrasive which can be used on plated mandrels and in metal-bonded grinding wheels. This type of wheel is used for general-purpose grinding and for internal grinding on hardened steel.
Borazon Type II CBN is nickel-plated grains of cubic boron nitride used in resin bonds for general-purpose dry and wet grinding of hardened steel. Uses of these wheels range from resurfacing blanking dies to the sharpening of high-speed steel end mills.

MANUFACTURED DIAMONDS

Diamond, the hardest substance known, was primarily used in machine shop work for truing and dressing grinding wheels. Because of the high cost of natural diamonds, industry began to look for cheaper, more reliable sources. In 1954, the General Electric Company, after four years of research, produced Man-Made® diamonds in their laboratory. In 1957, the General Electric Company, after more researching and testing, began the commercial production of these diamonds.

Many forms of carbon were used in experiments to manufacture diamonds. After much experimentation with various materials, the first success came when carbon and iron sulfide in a granite tube closed with tantalum disks were subjected to a pressure of 66,536,750 psi and temperatures between 2550 and 4260°F (1400 and 2350°C). Various diamond configurations are produced by using other metal catalysts such as chromium, manganese, tantalum, cobalt, nickel, or platinum in place of iron. The temperatures used must be high enough to melt the metal saturated with carbon and start the diamond growth.

DIAMOND TYPES

Because the temperature, pressure, and catalyst-solvent can be varied, it is possible to produce diamonds of various sizes, shapes, and crystal structure best suited to a particular need.

Type RVG Diamond

This manufactured diamond is an elongated, friable crystal with rough edges (Fig. 78-4A). The letters RVG indicate that this type may be used with a resinoid or vitrified bond and is used for grinding ultrahard materials such as tungsten carbide, silicon carbide, and space-age alloys. The RVG diamond may be used for wet and dry grinding.

Type MBG-II Diamond

This tough, blocky-shaped crystal is not as friable as the RVG type and is used in metal-bonded grinding (MBG) wheels (Fig. 78-4B). It is used for grinding cemented carbides, sapphires, and ceramics as well as in electrolytic grinding.

Type MBS Diamond

This is a blocky, extremely tough crystal with a smooth, regular surface which is not very friable (Fig. 78-4C). It is used in metal-bonded saws (*MBS*) to cut concrete, marble, tile, granite, stone, and masonry materials.

Diamonds may be coated with nickel or copper to provide a better holding surface in the bond and to prolong the life of the wheel.

	BORON AND NITROGEN AS HEXAGONAL BORON NITRIDE
	HP – HT
	BORAZON CUBIC BORON NITRIDE SINGLE CRYSTAL ABRASIVE
	BORAZON TYPE II

A

Fig. 78-3 (A) The manufacture of Borazon® (cubic boron nitride), (B) crystals of cubic boron nitride. (Courtesy General Electric Company.)

Fig. 78-4 (A) Type RVG diamond is used to grind ultrahard materials. (B) Type MBG-II is a tough diamond crystal used in metal-bonded grinding wheels. (C) Type MBS is a very tough, large crystal used in metal-bonded saws.

ABRASIVE PRODUCTS

After the abrasive has been produced, it is formed into products such as grinding wheels, coated abrasives, polishing and lapping powders, and abrasive sticks, all of which are used extensively in machine shops.

GRINDING WHEELS

Grinding wheels, the most important products made from abrasives, are composed of abrasive material held together with a suitable bond. The basic functions of grinding wheels in a machine shop are:

1. Generation of cylindrical, flat, and curved surfaces
2. Removal of stock
3. Production of highly finished surfaces
4. Cutting-off operations
5. Production of sharp edges and points

For grinding wheels to function properly, they must be hard and tough, and the wheel surface must be capable of gradually breaking down to expose new sharp cutting edges to the material being ground.

The material components of a grinding wheel are the *abrasive grain* and the *bond;* however, there are other physical characteristics, such as *grade* and *structure,* that must be considered in grinding wheel manufacture and selection.

ABRASIVE GRAIN

The abrasive used in most grinding wheels is either *aluminum oxide* or *silicon carbide*. The function of the abrasive is to remove material from the surface of the work being ground. Each abrasive grain on the working surface of a grinding wheel acts as a separate cutting tool and removes a small metal chip as it passes over the surface of the work. As the grain becomes dull, it fractures and presents a new sharp cutting edge to the material. The fracturing action reduces the heat of friction which would be caused if the grain became dull, producing a relatively cool cutting action. As a result of hundreds of thousands of individual grains all working on the surface of a grinding wheel, a smooth surface can be produced on the workpiece.

One important factor to consider in grinding wheel manufacture and selection is the *grain size*. After the abrasive ingot, or pig, is removed from the electric furnace, it is crushed, and the abrasive grains are cleaned and sized by passing them through screens which contain a certain number of meshes, or openings, per inch. A #8 grain size would pass through a screen having eight meshes per linear inch and would be approximately 1/8 in. across. The sizing of the abrasive grain is rather an important operation since undersize grains in a wheel will fail to do their share of the work, while oversize grains will scratch the surface of the work.

Commercial grain sizes are classified as follows:

VERY COARSE	COARSE	MEDIUM
6	14	30
8	16	36
10	20	46
12	24	54
		60

FINE	VERY FINE	FLOUR SIZE
70	150	280
80	180	320
90	220	400
100	240	500
120		600

Relative grain sizes are shown in Fig. 78-5.
The factors affecting the selection of grains sizes are:

1. *The type of finish desired*. Coarse grains are best suited for rapid removal of metal. Fine grains are used for producing smooth and accurate finishes.
2. *The type of material being ground*. Generally coarse grains are used on soft material, while fine grains are used for hard materials.
3. *The amount of material to be removed*. Where a large amount of material is to be removed and surface finish is not important, a coarse-grain wheel should be used. For finish grinding, a fine-grain wheel is recommended.
4. *The area of contact between the wheel and the workpiece*. If the area of contact is wide, a coarse-grain wheel

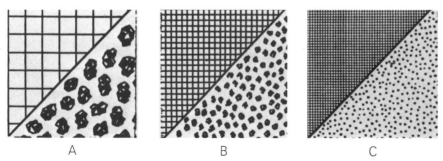

Fig. 78-5   Relative abrasive grain size. (A) 8-grain size; (B) 24-grain size (C) 60-grain size.

is generally used. Fine-grain wheels are used when the area of contact between the wheel and the work is small.

BOND TYPES

The function of the bond is to hold the abrasive grains together in the form of a wheel. There are six common bond types used in grinding wheel manufacture. They are: vitrified, resinoid, rubber, shellac, silicate, and metal.

Vitrified　Vitrified bond is used on most grinding wheels. It is made of clay or feldspar, which fuses at a high temperature and when cooled forms a glassy bond around each grain. Vitrified bonds are strong but break down readily on the wheel surface to expose new grains during the grinding operation. This bond is particularly suited to wheels used for the rapid removal of metal. Vitrified wheels are not affected by water, oil, or acid and may be used in all types of grinding operations. Vitrified wheels should be operated between 6300 and 6500 surface feet per minute (sf/min) (1920 and 1981 m/min).

Resinoid Bond　Synthetic resins are used as bonding agents in resinoid wheels. The majority of resinoid wheels generally operate at 9500 sf/min (2895 m/min); however, the modern trend is toward greater power and faster speeds for faster stock removal. Special resinoid wheels are manufactured to operate at speeds of 12,500 to 22,500 sf/min (3810 to 6858 m/min) for certain applications. These wheels are cool-cutting and remove stock rapidly. They are used for cutting-off operations, snagging, and rough grinding, as well as for roll grinding.

Rubber Bond　Rubber-bonded wheels produce high finishes such as those required on ball bearing races. Because of the strength and flexibility of this wheel, it is used for thin cutoff wheels. Rubber-bonded wheels are also used as regulating wheels on centerless grinders.

Shellac Bond　Shellac-bonded wheels are used for producing high finishes on parts such as cutlery, cam shafts, and paper mill rolls. They are not suitable for rough or heavy grinding.

Silicate Bond　Silicate-bonded wheels are not used to any extent in industry. Silicate bond is used principally for large wheels and for small wheels where it is necessary to keep heat generation to a minimum. The bond (silicate of soda) releases the abrasive grains more rapidly than does the vitrified bond.

Metal Bond　Metal bonds (generally nonferrous) are used on diamond wheels and for electrolytic grinding operations where the current must pass through the wheel.

GRADE

The grade of a grinding wheel may be defined as the degree of strength with which the bond holds the abrasive particles in the bond setting. If the bond posts are very strong (Fig. 78-6C), that is, if they retain the abrasive grains in the wheel during the grinding operation, the wheel is said to be of a hard grade. If the grains are released rapidly during the grinding operation, the wheel is classified as a soft grade (Fig. 78-6A).

The selection of the proper grade of wheel is important. Wheels which are too hard do not release the grains readily; consequently, the grains become dull and do not cut effectively. This is known as glazing. Wheels which are too soft release the grain too quickly, and the wheel will wear rapidly.

It is well to remember that all abrasive grains are hard, and the hardness of a wheel refers to grade (the strength of the bond) and not to the hardness of the grain. Wheel grade symbols are indicated alphabetically ranging from A (softest) to Z (hardest). The grade selected for a particular job depends on the following factors.

1. *Hardness of the material.* A hard wheel is generally used on soft material and soft grades on hard materials.
2. *Area of contact.* Soft wheels are used where the area of contact between the wheel and the workpiece is large. Small areas of contact require harder wheels.
3. *Condition of the machine.* If the machine is rigid, a softer grade of wheel is recommended. Light-duty machines or machines with loose spindle bearings require harder wheels.
4. *The speed of the grinding wheel and the workpiece.* The higher the wheel speed in relation to the workpiece, the softer the wheel should be. Wheels which revolve slowly wear faster; therefore, a harder wheel should be used at slow speeds.

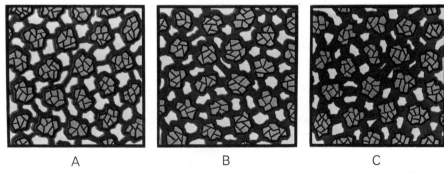

Fig. 78-6　Grades of grinding wheels. (A) Weak bond posts; (B) medium bond posts; (C) strong bond posts.

a   b   c

Fig. 78-7 Structure of a grinding wheel. (A) Dense; (B) medium; (C) open.

Fig. 78-8 An open-structure wheel provides greater chip clearance. (Courtesy The Carborundum Company.)

5. *Rate of feed*. Higher rates of feed require the use of harder wheels since the pressure on the grinding wheel is greater than with slower feeds.

6. *Operator characteristics*. An operator who removes the material quickly requires a harder wheel than one who removes the material more slowly. This is particularly evident in offhand grinding and where piece-work programs are involved.

STRUCTURE

The structure of a grinding wheel is the space relationship of the grain and bonding material to the voids that separate them. In brief, it is the density of the wheel.

If the spacing of the grains is close, the structure is dense (Fig. 78-7A). If the spacing of the grains is relatively wide, the structure is open (Fig. 78-7C).

Selection of the wheel structure depends on the type of work required. Wheels with open structures (Fig. 78-8) provide greater chip clearance than those with dense structures and remove material faster than dense wheels.

The structure of grinding wheels is indicated by numbers ranging from 1 (dense) to 15 (open). Selection of the proper wheel structure is affected by the following factors.

1. *The type of material being ground*. Soft materials will require greater chip clearance; therefore an open wheel should be used.

2. *Area of contact*. The greater the area of contact, the more open should be the structure to provide better chip clearance.

3. *Finish required*. Dense wheels will give a better, more accurate finish.

4. *Method of cooling*. Open structure wheels provide a better supply of coolant for machines using "through the wheel" coolant systems.

In summary, Table 78-1 will serve as a guide to the factors that must be considered in selecting a grinding wheel.

GRINDING WHEEL MANUFACTURE

Most grinding wheels used for machine shop operations are manufactured with vitrified bonds; therefore, the manufacture of only this type of wheel will be discussed. The main operations in the manufacture of vitrified grinding wheels are as follows.

Mixing The correct proportions of abrasive grain and bond are carefully weighed and thoroughly mixed in a rotary power mixing machine (Fig. 78-9). A certain percentage of water is added to moisten the mix.

Molding The proper amount of this mixture is placed in a steel mold of the desired wheel shape and compressed in a hydraulic press (Fig. 78-10) to form a wheel slightly larger than the finished size. The amount of pressure used varies with the size of the wheel and the structure required.

Shaving Although the majority of wheels are molded to shape and size, some machines require special wheel shapes and recesses. These are shaped or shaved to size in the green, or unburned, state on a shaving machine which resembles a potter's wheel.

Firing (Burning) The "green" wheels are carefully stacked on cars and are moved slowly through a long kiln 250 to 300 ft (76 to 91 m) long. The temperature of the kiln is held at approximately 2300°F (1260°C). This operation, which takes about five days, causes the bond to melt and form a glassy case around each grain; the product is a hard wheel.

TABLE 78-1 Factors to Be Considered when Selecting a Grinding Wheel

GRINDING FACTORS	WHEEL CONSIDERATIONS				
	ABRASIVE TYPE	GRAIN SIZE	BOND	GRADE	STRUCTURE
MATERIAL TO BE GROUND (high or low tensile strength, hard, soft)	X	X		X	X
TYPE OF OPERATION (cylindrical, centerless, surface, cutoff, snagging, etc.)			X		
MACHINE CHARACTERISTICS (rugged, light, loose bearings)				X	
WHEEL SPEED (slow, fast)			X	X	
RATE OF FEED (slow, rapid)				X	
AREA OF CONTACT (large, small)		X		X	X
OPERATOR CHARACTERISTICS				X	
AMOUNT OF STOCK TO BE REMOVED (light cut, heavy cut)			X	X	X
FINISH REQUIRED		X	X		X
USE OF COOLANT (wet or dry grinding)			X	X	X

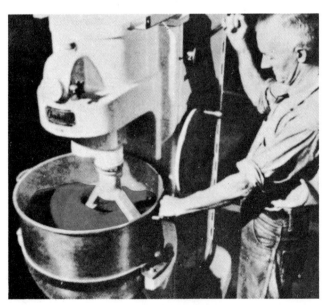

Fig. 78-9 Mixing the abrasive grain and bond. (Courtesy The Carborundum Company.)

Fig. 78-10 Molding grinding wheels using a hydraulic press. (Courtesy Cincinnati Milacron Incorporated.)

Truing   The cured wheels are mounted in a special lathe and turned to the required size and shape by hardened-steel conical cutters, diamond tools, or special grinding wheels.

Bushing   The arbor hole in a grinding wheel is fitted with a lead or plastic-type bushing to fit a specific spindle size. The edges of the bushing are then trimmed to the thickness of the wheel.

Balancing   To remove vibration which may occur while a wheel is revolving, each wheel is balanced. Generally, small, shallow holes are drilled in the "light" side of the wheel and filled with lead to assure proper balance.

Speed Testing   Wheels are rotated in a special, heavy, enclosed case and revolved at speeds at least 50 percent above normal operating speeds. This ensures that the wheel will not break under normal operating speeds and conditions.

STANDARD GRINDING WHEEL SHAPES

Nine standard grinding wheel shapes have been established by the United States Department of Commerce, the Grinding Wheel Manufacturers, and the Grinding Machine Manufacturers. Dimensional sizes for each of the shapes have also been standardized. Each of these nine shapes is identified by a number as shown in Table 78-2.

MOUNTED GRINDING WHEELS

Mounted grinding wheels (Fig. 78-11) are driven by a steel shank mounted in the wheel. They are produced in a variety of shapes for use with jig grinders, internal grinders, portable grinders, toolpost grinders, and flexible shafts. They are manufactured in both aluminum oxide and silicon carbide types.

TABLE 78-2   Common Grinding Wheel Shapes and Applications

SHAPE	NAME	APPLICATIONS
	Straight (Type 1)	Cylindrical, centerless, internal, cutter, surface, and offhand grinding operations
	Cylinder (Type 2)	Surface grinding on horizontal and vertical spindle grinders
	Tapered (both sides) (Type 4)	Snagging operations. The tapered sides lessen the chance of the wheel breaking.
	Recessed (one side) (Type 5)	Cylindrical, centerless, internal, and surface grinders. The recess provides clearance for the mounting flange.
	Straight cup (Type 6)	Cutter and tool grinder and surface grinding on vertical and horizontal spindle machines
	Recessed (both sides) (Type 7)	Cylindrical, centerless, and surface grinders. The recesses provide clearance for mounting flanges.
	Flaring cup (Type 11)	Cutter and tool grinder. Used mainly for sharpening milling cutters and reamers
	Dish (Type 12)	Cutter and tool grinder. Its thin edge permits it to be used in narrow slots
	Saucer (Type 13)	Saw gumming, gashing, milling cutter teeth

GRINDING WHEEL MARKINGS

The standard marking system chart (Fig. 78-12) is used by the manufacturers to identify grinding wheels. This information is found on the blotter of all small- and medium-size grinding wheels. It is stenciled on the side of larger wheels.

The six positions shown in the standard sequence are followed by all manufacturers of grinding wheels. The prefix shown is a manufacturer's symbol and is not always used by all grinding wheel producers.

NOTE: This marking system is used only for aluminum oxide and silicon carbide wheels; it is not used for diamond wheels.

Fig. 78-11 A variety of mounted grinding wheels.   (Courtesy The Carborundum Company.)

SELECTING A GRINDING WHEEL FOR A SPECIFIC JOB

From the foregoing information, the machinist should be able to select the proper wheel for the job required.

EXAMPLE

It is required to rough surface grind a piece of SAE 1045 steel using a straight wheel. Coolant is to be used.

Type of Abrasive Because steel is to be ground, *aluminum oxide* should be used.

Size of Grain Since the surface is not precision-finished, a medium grain can be used—about *46 grit*.

Grade A *medium-grade* wheel which will break down reasonably well should be selected. Use grade *J*.

Structure Since this steel is of medium hardness, the wheel should be of medium density—about 7.

Bond Type Since the operation is standard surface grinding and since coolant is to be used, a *vitrified* bond should be selected.

After the various factors have been considered, an A46-J7-V grinding wheel should be selected to rough-grind SAE 1045 steel.

NOTE: These specifications do not include the manufacturer's prefix or the manufacturer's records.

Fig. 78-12 Straight grinding wheel marking system chart.   (Courtesy Grinding Wheel Institute.)

It is required to finish-grind a high-speed steel milling cutter on the cutter and tool grinder.

Type of Abrasive   Since the cutter is steel, an *aluminum oxide* wheel should be used.

Size of Grain   Since the milling cutter must have a smooth finish, a medium to fine grain should be used. About a *60 grit* is recommended for this type of operation.

Grade   It is important that a cool-cutting wheel be used to prevent burning the cutting edge of the cutter. A wheel which breaks downs reasonably well will permit cool grinding. Use a medium-soft grade such as *J*.

Structure   In order to produce a smooth cut, a medium-dense wheel should be used. For this application, use a #6.

Bond Type   Because most cutter and tool grinders are designed for standard speeds, a *vitrified* bond should be used. When the speed is excessive for the wheel size, a *resinoid* bond should be used.

The wheel selected for this job (disregarding the manufacturer's prefix and records) should be A60-J6-V.

NOTE: If the cutter is chipped or if considerable metal must be removed to resharpen, it is advisable to first rough-grind using a 46-grit wheel.

HANDLING AND STORAGE OF GRINDING WHEELS

Since all grinding wheels are breakable, proper handling and storage is important. Damaged wheels can be dangerous if used; therefore, the following rules should be observed.

1. Do not drop or bump grinding wheels.
2. Always store wheels properly on shelves or in bins provided. Flat and tapered wheels can be stored on edge, while large cup wheels and cylindrical wheels should be stored on the flat sides with a suitable layer of packing between each wheel.
3. Thin organic (resinoid and rubber) bonded wheels should be laid flat on a horizontal surface, away from excessive heat, to prevent warping.
4. Small cup and small internal grinding wheels may be stored separately in boxes, bins, or drawers.

INSPECTION OF WHEELS

After wheels have been received, they should be inspected to see that they have not been damaged in transit.

For further assurance that wheels have not been damaged, they should be suspended and tapped lightly with a

Fig. 78-13   Testing a grinding wheel for cracks.

screwdriver handle for small wheels (Fig. 78-13) or with a wooden mallet for larger wheels. If vitrified or silicate wheels are sound, they give a clear, metallic ring. Organic-bonded wheels give a duller ring, and cracked wheels do not produce a ring. Wheels must be dry and free of sawdust before testing; otherwise the sound will be deadened.

DIAMOND WHEELS

Diamond wheels are used for grinding cemented carbides and hard vitreous materials, such as glass and ceramics.

Diamond wheels are manufactured in a variety of shapes, such as straight, cup, dish, and thin cutoff wheels.

Wheels of ½ in. (12.7 mm) diameter or less have diamond particles throughout the wheel. Wheels larger than ½ in. (12.7 mm) are made with a diamond surface on the grinding face only. The diamonds for this purpose are made in grain sizes ranging from 100 to 400. The proportions of the diamond and bond mixture vary with the application. This diamond concentration is identified by the letters A, B, or C. C concentration will contain four times the number of diamonds of a grinding wheel with an A concentration. This mixture is coated on the grinding face of the wheel in thicknesses ranging from $\frac{1}{32}$ to $\frac{1}{4}$ in. (0.78 to 6.35 mm).

BONDS

There are three types of bonds available for diamond wheels: resinoid, metal, and vitrified.

Resinoid-bonded wheels give a maximum cutting rate and require very little dressing. These wheels remain sharp for a long time and are well suited to grinding carbides.

A recent development in the manufacture of resinoid-bonded diamond wheels has been the coating of the diamond particles with nickel plating by means of an electroplating process. This process is carried out before the diamonds are mixed with the resin. It reduces the tendency of the diamonds to chip and results in cooler-grinding, longer-lasting wheels.

Metal bonds, generally nonferrous, are particularly suited to offhand grinding and cutting-off operations. This type

ABRASIVE		
DIAMOND		
D = Mined		
SD = Man-Made		

GRIT SIZE		
24	120	400S
36	150	500
46	180	500S
60	220	600S
80	240	800S
100	320	1200S
100S	400	1500S

GRADE				
Res.	Metal		Vit.	
H	R	L	Q	H	P
J		N	R	J	R
L		O	T	L	T
N		P		N	

CONCENTRATION	
(Low) =	25
	50
	75
(High) =	100

BOND TYPE
B = Resinoid
M = Metal
V = Vitrified

DEPTH OF DIAMOND SECTION		
1/16	1/8	1/4

BOND MODIFICATION
Numeral to designate special bond modification. Example: Resinoid — 6 and 11. This symbol may be sometimes omitted

NOTE: No grade is shown for hand hones.

* Manufacturer's identification symbol

Fig. 78-14   Diamond grinding wheel marking system chart.   (Courtesy The Norton Company of Canada Ltd.)

of wheel holds its form extremely well and does not wear on radius work or on small areas of contact.

Vitrified-bonded wheels remove stock rapidly but require frequent cleaning with a boron carbide abrasive stick to prevent the wheel from loading. These wheels are particularly suited for offhand and surface grinding of cemented carbides.

DIAMOND WHEEL IDENTIFICATION

The method used to identify diamond wheels differs from that used for other grinding wheels (Fig. 78-14).

COATED ABRASIVES

Coated abrasives (Fig. 78-15) consist of a flexible backing (cloth or paper) to which abrasive grains have been bonded. Garnet, flint, and emery (natural abrasives) are being replaced by aluminum oxide and silicon carbide in the manufacture of coated abrasives. This is due to the greater toughness and the more uniform grain size and shape of manufactured abrasives.

Coated abrasives serve two purposes in the machine shop: metal grinding and polishing.

Metal grinding may be done on a belt or disk grinder and up until the last few years was a rapid, nonprecision method of removing metal. Coarse-grit coated abrasives are used for rapid removal of metal, whereas fine grits are used for polishing.

Emery, a natural abrasive which is black in appearance, is used to manufacture coated abrasives, such as emery cloth and emery paper. Since the grains are not as sharp as artificial abrasives, emery is generally used for polishing metal by hand.

Fig. 78-15   A variety of coated abrasives.

SELECTION OF COATED ABRASIVES

Aluminum Oxide

Aluminum oxide, grey in appearance, is used for high-tensile-strength materials, such as steels, alloy steels, high-carbon steels, and tough bronzes. Aluminum oxide is characterized by the long life of its cutting edges.

For *hand operations,* 60 to 80 grit is used for fast cutting (roughing), while 120 to 180 grit is recommended for finishing operations.

For *machine operations,* such as on belt and disk grinders, 36 to 60 grit is used for roughing, while 80 to 120 grit is recommended for finishing operations.

Silicon Carbide

Silicon carbide, bluish-black in appearance, is used for low-tensile-strength materials, such as cast iron, aluminum, brass, copper, glass, and plastics. The selection of grit size for hand and machine operations is the same as for aluminum oxide–coated abrasives.

COATED ABRASIVE MACHINING

Over the past few years, coated abrasive machining has become widely used in industry. With improved abrasives and bonding materials, better grain structure, more uniform belt splicing, and new polyester belt backing, abrasive belt machining is being used to a much greater extent. Some types of work which had formerly been performed by milling, turning, and abrasive wheel grinding in machine shops, steel mills, steel fabrication plants, and foundries, are now being done more efficiently by belt or disk grinding operations. Coated abrasive operations are now capable of grinding to less than 0.001 in. (0.03 mm) tolerance and with a surface finish of 10 to 20 microinches (0.3 to 0.5 micrometers).

As a result of greatly improved coated abrasive products, heavier and better machines requiring much higher horsepower are being produced in this field. The basic machines have become more automated and now include automatic loaders, feeders, and unloaders. Belt machining has been applied very successfully to the centerless grinding concept where it is used extensively for bar and tube grinding in steel and metalworking plants. In many lumber mills, belt abrasive machining is now used to finish the lumber to size, an operation that previously was done by planing.

REVIEW QUESTIONS

1. What characteristics must an abrasive have in order for it to function properly?

TYPES OF ABRASIVES

2. Name four natural abrasives.
3. What is "bort" and for what purpose is it used?
4. Name three manufactured abrasives and state why they are used extensively today.
5. For what purposes are aluminum oxide and silicon carbide abrasives used?
6. Describe the manufacture of aluminum oxide.
7. Describe the manufacture of silicon carbide.
8. List three uses of zirconia-alumina abrasive.
9. Describe the cutting action of zirconia-alumina abrasive.

10. List four advantages of zirconia-alumina abrasive wheels.
11. How does the manufacture of boron carbide differ from that of the other manufactured abrasives?
12. List the uses for boron carbide.
13. List five advantages of cubic boron nitride grinding wheels.
14. Name two types of Borazon℗ (cubic boron nitride) and state where each is used.
15. List three types of manufactured diamonds and state where each is used.
16. (a) Name two materials used to coat the diamonds.
    (b) What is the purpose of coating the diamonds?

GRINDING WHEELS

17. What are the basic functions of a grinding wheel?
18. Describe the function of each abrasive grain.
19. How is grain size determined and why is it important?
20. What factors affect the selection of the proper grain size?
21. What is the function of a bond and how does it affect the grade of a grinding wheel?
22. Name six types of bonds and state the purpose of each type in the manufacture of grinding wheels.
23. Why is the selection of the grade of wheel important to the grinding operation?
24. What factors should be considered when the grade of wheel is being selected for a particular job?
25. Define the structure of a grinding wheel and state how this is indicated.
26. What factors affect the selection of the proper wheel structure?

GRINDING WHEEL MANUFACTURE

27. Describe briefly the manufacture of a vitrified grinding wheel.

STANDARD GRINDING WHEEL SHAPES

28. Describe the following grinding wheels and state their purpose: Types 1, 5, 6, 11.
29. For what purpose are mounted grinding wheels used?

GRINDING WHEEL MARKINGS

30. Explain the meaning of the following grinding wheel markings: A80-G8-S, C-60-L4-V.
31. What wheel should be selected for grinding
    a. machine steel?
    b. cemented carbide?
    c. cast iron?

INSPECTION OF WHEELS

32. Explain why it is important to inspect a grinding wheel before it is mounted on a grinder.
33. How may grinding wheels be inspected?

34. For what purpose are diamond wheels used?
35. Explain ''diamond concentration'' in a grinding wheel.
36. Name three types of bonds used in diamond wheels and state the purpose of each.
37. Define the following diamond wheel markings: D 120–N 100–B ⅛.

38. Name the three common abrasives used in the manufacture of coated abrasives and state their purpose.
39. What grit size is recommended for
    a. hand operations?
    b. machine operations?

SURFACE GRINDERS AND ACCESSORIES
UNIT 79

Grinding is an important part of the machine trade. Improved grinding machine construction has permitted the production of parts to extremely fine tolerances with improved surface finishes and accuracy. Because of the dimensional accuracy obtained by grinding, interchangeable manufacture has become commonplace in most industries.

Grinding has also, in many cases, eliminated the need for conventional machining. With the development of new abrasives and better machines, the rough part is often finished in one grinding operation, thus eliminating the need for other machining processes. The role of grinding machines has changed over the years; initially they were used on hardened work and for truing hardened parts which had been distorted by heat-treating. Today, grinding is applied extensively to the production of unhardened

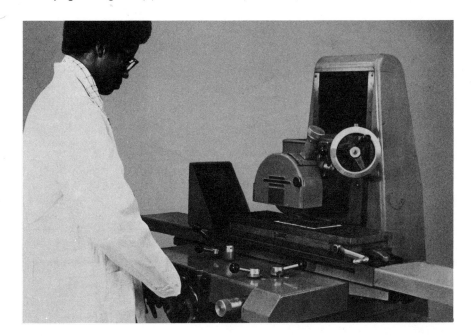

parts where high accuracy and surface finish are required. In many cases, modern grinding machines permit the manufacture of intricate parts faster and more accurately than other machining methods.

OBJECTIVES

After completing this unit you will be able to:
1. Name four methods of surface grinding and state the advantage of each
2. True and dress a grinding wheel
3. Select the proper grinding wheel to be used for each type of work material

THE GRINDING PROCESS

In the grinding process the workpiece is brought into contact with a revolving grinding wheel. Each small abrasive grain on the periphery of the wheel acts as an individual cutting tool and removes a chip of metal (Fig. 79-1). As the abrasive grains become dull, the pressure and heat created between the wheel and the workpiece cause the dull face to break away, leaving new sharp cutting edges.

Regardless of the grinding method used, whether it be cylindrical, centerless, or surface grinding, the grinding process is the same and certain general rules will apply in all cases.

1. Use a silicon carbide wheel for low-tensile-strength materials and an aluminum oxide wheel for high-tensile-strength materials.
2. Use a hard wheel on soft materials and a soft wheel on hard materials.
3. If the wheel is too hard, increase the speed of the work or decrease the speed of the wheel to make it act as a softer wheel.
4. If the wheel appears too soft or wears rapidly, decrease the speed of the work or increase the speed of the wheel, but not above its recommended speed.
5. A glazed wheel will affect the finish, accuracy, and metal removal rate. The main causes of wheel glazing are:
   a. The wheel speed is too fast.

Fig. 79-1 The cutting action of the abrasive grain. (Courtesy The Carborundum Company.)

   b. The work speed is too slow.
   c. The wheel is too hard.
   d. The grain is too small.
   e. The structure is too dense, which causes the wheel to load.
6. If a wheel wears too quickly, the cause may be any of the following.
   a. The wheel is too soft.
   b. The wheel speed is too slow.
   c. The work speed is too fast.
   d. The feed rate is too great.
   e. The face of the wheel is too narrow.
   f. The surface of the work is interrupted by holes or grooves.

SURFACE GRINDING

Surface grinding is a technical term referring to the production of flat, contoured, and irregular surfaces on a piece of work which is passed against a revolving grinding wheel.

TYPES OF SURFACE GRINDERS

There are four distinct types of surface-grinding machines (Fig. 79-2), all of which provide a means of holding the metal and bringing it into contact with the grinding wheel.

The horizontal spindle grinder with a reciprocating table (Fig. 79-2A) is probably the most common surface grinder for toolroom work. The work is reciprocated (moved back and forth) under the grinding wheel, which is fed down to provide the desired depth of cut. Feed is obtained by a transverse movement of the table at the end of each stroke.

The horizontal spindle grinder with a rotary table (Fig. 79-2B) is often found in toolrooms for the grinding of flat circular parts. The surface pattern it produces makes it particularly suitable for grinding parts which must rotate in contact with each other. The work is held on the magnetic chuck of a rotating table and passed under a grinding wheel. Feed is obtained by the transverse movement of the wheelhead. This type of machine permits faster grinding of circular parts since the wheel is always in contact with the workpiece.

The vertical spindle grinder with a rotary table (Fig.

Fig. 79-2 (A) Horizontal spindle grinder with a reciprocating table; (B) horizontal spindle grinder with a rotary table; (C) vertical spindle grinder with a rotary table; (D) vertical spindle grinder with a reciprocating table.   (Courtesy The Carborundum Company.)

79-2C) produces a finished surface by grinding with the face of the wheel rather than the periphery, as in horizontal spindle machines. The surface pattern appears as a series of intersecting arcs. Vertical spindle grinders have a higher metal removal rate than the horizontal-type spindle machines. It is probably the most efficient and accurate form of grinder for the production of flat surfaces.

The vertical spindle grinder with a reciprocating table (Fig. 79-2D) grinds on the face of the wheel while the work is moved back and forth under the wheel. Because of its vertical spindle and greater area of contact between the wheel and the work, this machine is capable of heavy cuts. Material up to ½ in. (12.7 mm) thick may be removed in one pass on larger machines of this type. Provision is made on most of the these grinders to tilt the wheelhead a few degrees from the vertical. This permits greater pressure where the rim of the wheel contacts the workpiece and results in faster metal removal. When the wheelhead is vertical and grinding is done on the face of the wheel, the surface pattern produced is a series of uniform intersecting arcs. If the wheelhead is tilted, it produces a semicircular pattern.

The Horizontal Spindle Reciprocating Table Surface Grinder

The horizontal spindle reciprocating table surface grinder is the most commonly used, and will be discussed in detail.

Machines of this type may be either hand or hydraulically operated (Fig. 79-3).

Parts of a Hydraulic Surface Grinder

The *base* is generally of heavy cast-iron construction. It usually contains the hydraulic reservoir and pump used to operate the table and power feeds. The top of the base has accurately machined ways to receive the saddle.

The *saddle* may be moved in or out across the ways, manually or by automatic feed.

The *table* is mounted on the top of the saddle. The ways for the table are at right angles to those on the base. Thus the table reciprocates across the upper ways on the saddle, while the saddle (and table) moves in or out on the ways of the base.

The *column,* mounted on the back of the frame, contains the ways for the *spindle housing* and *wheelhead.* The wheel feed handwheel provides a means of moving the wheelhead vertically to set the depth of cut.

The reciprocating action of the table may be controlled manually by the *table traverse handwheel* or by the *hydraulic control valve lever.*

The direction of the table is reversed when one of the *stop dogs* mounted on the side of the table strikes the *table traverse reverse lever.*

The table may be fed toward or away from the column manually by means of the *crossfeed handwheel* or automatically by the *power crossfeed control.* This operation moves the work laterally under the wheel.

Fig. 79-3   A hydraulic surface grinder.

Fig. 79-4   A grinding wheel properly mounted on the grinder spindle.

GRINDING WHEEL CARE

In order to ensure the best results in any surface-grinding operation, proper care of the grinding wheel must be taken.

1. When not in use, all grinding wheels should be properly stored.
2. Wheels should be tested for cracks prior to use.
3. Select the proper type of wheel for the job.
4. Grinding wheels should be properly mounted and operated at the recommended speed.

MOUNTING A GRINDING WHEEL

After the correct wheel has been selected for the job, proper mounting of the grinding wheel ensures the best grinding performance.

PROCEDURE

1. Test the wheel to see that it is not cracked by ring testing with the handle of a screwdriver or hammer.
2. Clean the grinding wheel adapter.
3. Mount the adapter through the wheel and tighten the threaded flange (Fig. 79-4).
   a. Be sure that the blotter is on each side of the wheel prior to mounting. A perforated blotter should be used for through-the-wheel coolant. A rubber washer is sometimes used in place of the blotter on some grinders.
   b. The wheel should be a good fit on the adapter or spindle. If it is too tight or too loose, the wheel should not be mounted.

c. To comply with the Wheel Manufacturer's Safety Code, the diameter of the flanges should not be less than one-third the diameter of the wheel.
4. Tighten the wheel adapter flanges only enough to hold the wheel firmly. If it is tightened too much, it may damage the flanges or break the wheel.

BALANCING A GRINDING WHEEL

Proper balance of a mounted grinding wheel is very important since improper balance will greatly affect the surface finish and accuracy of the work. Excessive imbalance creates vibration which will damage the spindle bearings.

There are two methods of balancing a wheel:

1. *Static balancing.* On some grinders, the wheel is balanced off the machine with the use of a balancing stand and arbor. Counterweights in the wheel flange must be correctly positioned in order to balance the grinding wheel.
2. *Dynamic balancing.* Most new grinding machines are equipped with ball-bearing balancing devices which automatically balance a wheel in a matter of seconds while it is revolving on the grinder.

After the wheel has been mounted on the adapter it should be balanced, if provision is made in the adapter for balancing.

To Balance a Grinding Wheel

1. Mount the wheel and adapter on the surface grinder and true the wheel with a diamond dresser.

Fig. 79-5 A grinding wheel balancing stand.

Fig. 79-6 Method of adjusting counterbalances to balance a grinding wheel.

2. Remove the wheel assembly and mount a special tapered balancing arbor in the hole of the adapter.

3. Place the wheel and arbor on a balancing stand (Fig. 79-5) which has been leveled.

4. Allow the wheel to rotate until it stops. This will indicate that the heavy side is at the bottom. Mark this point with chalk.

5. Rotate the wheel and stop it at three positions, one-quarter, one-half, and three-quarters of a turn, to check the balance. If the wheel moves from any of these positions, it is not balanced.

6. Loosen the setscrews in the wheel counterbalances, in the grooved recess of the flange, and move the counterbalances opposite the chalk mark (Fig. 79-6).

7. Check the wheel in the four positions mentioned in steps 4 and 5.

8. Move the counterbalances around the groove an equal amount on each side of the centerline and check for balance again.

9. Continue to move the balances away from the heavy side until the wheel remains stationary at any position.

10. Tighten the counterbalances in place.

TRUING AND DRESSING A GRINDING WHEEL

After mounting a grinding wheel, it is necessary to *true* the wheel to ensure that it will be concentric with the spindle. *Truing* a wheel is the operation of removing any high spots on the wheel, thereby causing it to run concentric with the spindle. Proper grinding practice requires that a wheel be trued before use.

Dressing a wheel is the operation of removing the dull

Fig. 79-7 A diamond dresser used to dress a grinding wheel. (Courtesy Kostel Enterprises Limited.)

grains and metal particles. This operation exposes sharp cutting edges of the abrasive grains to make the wheel cut better. A dull, glazed, or loaded wheel should be dressed for the following reasons.

1. To reduce the heat generated between the work surfaces and the grinding wheel
2. To reduce the strain on the grinding wheel and the machine
3. To improve the surface finish and accuracy of the work
4. To increase the rate of metal removal

An industrial diamond, mounted in a suitable holder on the magnetic chuck, is generally used to true and dress a grinding wheel (Fig. 79-7).

To True and Dress a Grinding Wheel

1. Check the diamond for wear and, if necessary, turn it in the holder to expose a sharp cutting edge to the wheel.

NOTE: Most diamonds are mounted in the holder at an angle of 5 to 15° from the vertical to prevent the possibility

497

Fig. 79-8   Proper positioning of the diamond dresser in relation to the grinding wheel.

TABLE (END VIEW)

Fig. 79-9   Positioning the diamond under the high point of the wheel.

of chatter and of the diamond digging into the wheel. It also permits the diamond to wear on an angle so that a sharp cutting edge may be obtained by merely turning the diamond in the holder.

2. Clean the magnetic chuck thoroughly with a cloth and wipe it with the palm of the hand to remove all grit and dirt.

3. Place a piece of paper, slightly larger than the base of the diamond holder, on the left-hand end of the magnetic chuck. This prevents scratching of the chuck when the diamond holder is being removed.

4. Place the diamond holder on the paper, covering as many magnetic inserts as possible, and energize the chuck. The diamond should be pointing in the same direction as the grinding wheel rotation (Fig. 79-8).

5. Raise the wheel above the height of the diamond.

6. Move the table longitudinally so that the diamond is offset approximately ¼ in. (6.35 mm) to the left of the centerline of the wheel.

7. Adjust the table laterally so that the diamond is positioned under the high point on the face of the wheel (Fig. 79-9). This is important since grinding wheels will wear more quickly on the edges of the wheel, leaving the center of the face higher than the edges.

8. Start the wheel revolving and carefully lower the wheel until the high point touches the diamond.

9. Move the table laterally, using the crossfeed handwheel to feed the diamond across the face of the wheel.

10. Lower the grinding wheel about 0.001 to 0.002 in. (0.02 to 0.05 mm) per pass, and rough-dress the face of the wheel until it is flat and has dressed all around the circumference.

11. Lower the wheel 0.0005 in. (0.01 mm) and take several passes across the face of the wheel. The rate of crossfeed will vary with the structure of the wheel. A rule of thumb is to use a fast crossfeed with coarse wheels and a slow, but regular, crossfeed for fine, closely spaced grains.

The following additional points may be helpful when truing or dressing a grinding wheel.

1. To minimize wear on the diamond, the grinding wheel may be rough-dressed with an abrasive stick.

2. If coolant is to be used during the grinding operation,

it is advisable to use coolant when dressing the wheel. This will protect the diamond and the wheel from excessive heat.

3. A loaded wheel is indicated by a discoloration on the periphery or grinding wheel face. When dressing the wheel, sufficient material should be taken off to remove completely any discoloration on the wheel face.

4. If rapid removal of metal is more important than the surface finish, do not finish-dress the wheel. After the wheel has been rough-dressed, some operators will take a final pass of 0.001 to 0.002 in. (0.02 to 0.05 mm) at a high rate of feed. The rough surface produced by this operation will remove the metal more rapidly than a finish-dressed wheel.

WORK-HOLDING DEVICES

THE MAGNETIC CHUCK

In some surface-grinding operations the work may be held in a vise, on V-blocks, or bolted directly to the table. However, most of the ferrous work ground on a surface grinder is held on a *magnetic chuck* which is clamped to the table of the grinder.

Magnetic chucks may be of two types: the *electromagnetic chuck* and the *permanent magnetic chuck*.

The *electromagnetic chuck* uses electromagnets to provide the holding power. It has the following advantages.

1. The holding power of the chuck may be varied to suit the area of contact and the thickness of the work.

2. A special switch neutralizes the residual magnetism in the chuck, permitting the work to be removed easily.

The *permanent magnetic chuck* provides a convenient means of holding most workpieces to be ground. The holding power is provided by means of permanent magnets. The principle of operation is the same for both electromagnetic and permanent magnetic chucks.

Fig. 79-10   Construction of a permanent magnetic chuck. (Courtesy James Neill & Co. Ltd.)

Fig. 79-11   (A) Magnetic chuck in the on position; (B) magnetic chuck in the off position.   (Courtesy James Neill & Co. Ltd.)

Fig. 79-12   (A) Thin work held on a magnetic chuck top plate attracts fewer magnetic lines of force; (B) Thin work held on an adapter plate.   (Courtesy James Neill & Co. Ltd.)

PERMANENT MAGNETIC CHUCK CONSTRUCTION (Fig. 79-10)

The *base plate* provides a base for the chuck and a means of clamping it to the table of the grinder.

The *grid* or *magnetic pack* houses the *magnets* and the *grid conductor bars*. It is moved longitudinally by a handle when the chuck is placed in the on or off position.

The *case* houses the grid assembly and permits longitudinal movement of the grid. It also provides an oil reservoir for the lubrication of the moving parts.

The *top plate* contains *inserts* or *pole pieces* which are separated magnetically from the surrounding plate by means of white metal. This separation provides the poles necessary to conduct the magnetic lines of flux.

When the work is placed on the face of the chuck (top plate) and the handle moved to the on position (Fig. 79-11A), the grid conductor bars and the inserts in the top plate are in line. This permits the magnetic flux to pass through the work, holding it onto the top plate.

When the handle is rotated 180° to the off position (Fig. 79-11B), the grid assembly moves the grid conductor bars and the inserts out of line. In this position, the magnetic lines of flux enter the top plate and inserts, but not the work.

MAGNETIC CHUCK ACCESSORIES

Often it is not possible to hold all work in the chuck. The size, shape, and type of work will dictate how the work should be held for surface grinding. The holding power of a magnetic chuck is dependent on the size of the workpiece, the area of contact, and the thickness of the workpiece. A highly finished piece will be held better than a poorly machined workpiece.

Very thin workpieces will not be held too securely on the face of a magnetic chuck because there are too few magnetic lines of force entering the workpiece (Fig. 79-12A).

An *adapter plate* (Fig. 79-12B) is used to securely hold thin work [less than ¼ in. (6.35 mm)]. The plate's alternate layers of steel and brass convert the wider pole spacing of the chuck to finer spacing with more but *weaker* flux paths. This method is particularly suited to small, thin pieces and reduces the possibility of distortion when grinding thin work.

Magnetic chuck blocks (Fig. 79-13) provide a means of extending the flux paths to hold workpieces that cannot be held securely on the chuck face. V-blocks may be used to hold round or square stock for *light* grinding.

NOTE: Set the chuck blocks so that the maximum number of magnetic loops pass through the workpiece. They must also be placed so that the laminations are in line with the inner poles or inserts.

Fig. 79-13 Applications of magnetic chuck blocks. (Courtesy James Neill & Co. Ltd.)

If the following points are observed magnetic chuck blocks will last longer.

1. Clean thoroughly before and after use.
2. Store in a covered wooden box.
3. Check frequently for accuracy and burrs.
4. If regrinding is necessary to restore accuracy, take light cuts with a dressed wheel. Use coolant when grinding to prevent the magnetic chuck blocks from heating (even slightly).

Sine chuck. When it is required to grind an angle on a workpiece, the work may be set up with a sine bar and clamped to an angle plate. Often a *sine chuck* (Fig. 79-14), which is a form of a magnetic sine plate, is used to hold the work. The buildup for the angles is the same as for the sine bar. Compound sine chucks, which have two plates hinged at right angles to each other, are available for grinding compound angles.

Magna-vise clamps (Fig. 79-15) may be used when the workpiece does not have a large bearing area on the chuck, or when the work is nonmagnetic. These magnetically actuated clamps consist of comblike bars attached to a solid bar by a piece of spring steel. When work is held with these clamps, the solid bar of one clamp is placed against the backing plate of the magnetic chuck. The work is placed on the chuck surface between the toothed edges of two

Fig. 79-14   A compound sine chuck.   (Courtesy The Taft-Pierce Manufacturing Company.)

clamps as shown. The toothed edges on the bars in contact with the work should be above the magnetic chuck face. When the chuck is energized, the jaws of the clamps are brought down toward the face of the chuck, locking the work in place.

Double-faced tape is often used for holding thin, non-magnetic pieces on the chuck for grinding. The tape, having two adhesive sides, is placed between the chuck and the work, causing the work to be held securely enough for light grinding.

Special fixtures are often used to hold nonmagnetic materials and odd-shaped workpieces, particularly when a large number of workpieces must be ground.

GRINDING FLUIDS

Although work is ground dry in many cases, most machines have provision for applying grinding fluids or coolants.

Grinding fluids serve four purposes.

1. *Reduction of grinding heat,* which affects work accuracy, surface finish, and wheel wear.
2. *Lubrication* of the surface between the workpiece and the grinding wheel, which results in a better surface finish.
3. *Removal of swarf* (small metal chips and abrasive grains) from the cutting area.
4. *Control of grinding dust,* which may present a health hazard.

TYPES OF GRINDING FLUIDS

1. *Soluble oil and water* when mixed form a milky solution which provides excellent cooling, lubricating, and rust-resistant qualities. This solution is generally applied by flooding the surface of the work.

NONMAGNETIC MATERIAL

CLAMP — BACKING PLATE

— SPRING STEEL

IN UNCLAMPED POSITION

B IN CLAMPED POSITION

Fig. 79-15 (A) Nonmagnetic work being held for grinding with magna-vise clamps. (Courtesy Magna Lock Corporation.); (B) magna-vise clamps used to hold work on a magnetic chuck. (Courtesy Brown & Sharpe Mfg. Co.).

Fig. 79-16 Many grinding operations use the flood system to keep the workpiece cool.   (Courtesy DoAll Company.)

DRIP CONTROLS

FLOOD CONTROL

FLOOD COOLANT SPOUT

Fig. 79-17 Through-the-wheel system applies coolant to the point of grinding contact.   (Courtesy DoAll Company.)

2. *Soluble chemical grinding fluids and water* when mixed form a grinding fluid which may be used with flood cooling or "through-the-wheel" cooling systems. The chemical grinding fluid contains rust inhibitors and bactericides to minimize odors and skin irritation.

3. *Straight oil grinding fluids,* generally applied by the flood system, are used where high finish, accuracy, and long wheel life are required. These fluids have better lubricating qualities than the water-soluble fluids but do not have as high heat-dissipating capacity.

METHODS OF APPLYING COOLANTS

The flood system (Fig. 79-16) is probably the commonest form of coolant application. By this method, the coolant is directed onto the workpiece by a nozzle and is recirculated through a system containing a reservoir, a pump, a filter, and a control valve.

Through-the-wheel cooling provides a convenient and efficient method of applying coolant to the area being ground. The fluid is pumped through a tube and discharged into a dovetailed groove in the wheel flange

(Fig. 79-17). Holes through the flange and corresponding holes in the wheel blotter permit the fluid to be discharged into the porous grinding wheel. The centrifugal force, created by the high-speed rotation of the wheel, forces the fluid through the wheel onto the area of contact between the wheel and the work. Some machines have a coolant reservoir above the grinding wheel guard which feeds the coolant into the wheel flange groove by gravity.

The mist cooling system, which supplies coolant in the form of a mist, uses the atomizer principle. Air passes through a line containing a T-connection which leads to the coolant reservoir. The velocity of the air as it passes through the T-connection draws a small amount of coolant from the reservoir and discharges it through a small nozzle in the form of vapor. The nozzle is directed to the point of contact between the work and the wheel. The air and the vapor, as it evaporates, cause the cooling action. The force of the air also blows away the swarf.

SURFACE FINISH

The finish produced by a surface grinder is important, and the factors affecting it should be considered. Some parts that are ground do not require a fine surface finish, and time should not be spent producing fine finishes if not required.

The following factors affect the surface finish.

Material Being Ground Soft material such as brass and aluminum will not permit as high a finish as harder ferrous materials. A much finer finish can be produced on hardened-steel workpieces than can be produced on soft steel or cast iron.

Amount of Material Being Removed If a large amount of material is to be removed, a coarse-grit, open-structure wheel should be used. This will not produce as fine a finish as a fine-grit, dense wheel.

Grinding Wheel Selection A wheel containing abrasive grains which are friable, or fracture easily, will produce a better finish than a wheel made up of tough grains. A fine-grit, dense-structure wheel produces a smoother surface than a coarse-grit, open wheel. A grinding wheel which is too soft releases the abrasive grains too easily, causing them to roll between the wheel and the work, creating deep scratches in the work (Fig. 79-18).

Grinding Wheel Dressing An improperly dressed wheel will leave a pattern of scratches on the work. Care should be taken when finish dressing the wheel to move the diamond slowly across the wheelface. Always dress the wheel sufficiently to expose new abrasive grains and ensure that all glazing or foreign particles have been removed from the periphery of the wheel. New grinding wheels *which have not been properly balanced and trued* will produce a chatter pattern on the surface of the work.

Condition of the Machine A light machine or one with loose spindle bearings will not produce the accuracy and fine surface finish possible in a rigid machine with properly adjusted spindle bearings. Also, to ensure optimum accuracy and surface finish, the machine should be kept clean.

Feed Coarse feeds tend to produce a rough finish. If "feed lines" persist when a fine feed is used, the wheel edges should be rounded slightly with an abrasive stick.

REVIEW QUESTIONS

GRINDING MACHINES AND PROCESS

1. Discuss how the role of grinding has contributed to interchangeable manufacture.
2. How has the role of grinding changed over the years?
3. Outline the action that takes place during a grinding operation.
4. List five important rules which apply to any grinding operation.

SURFACE GRINDING

5. Define surface grinding.

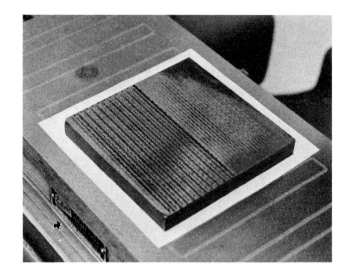

Fig. 79-18 Poor surface finish may be caused by a clogged wheel or by using the wrong grinding wheel.

6. Name four different types of surface grinders and briefly outline the principle of each.

PARTS OF THE HYDRAULIC SURFACE GRINDER

7. Name five *main* parts of the hydraulic surface grinder.
8. Name and state the purpose of five controls found on the hydraulic surface grinder.

GRINDING WHEEL CARE

9. List four points to be observed in grinding wheel care.

TO MOUNT A GRINDING WHEEL

10. List the steps required to mount a grinding wheel.
11. Why is a blotter on each side of the wheel necessary when mounting a grinding wheel?

TO BALANCE A GRINDING WHEEL

12. Why is the proper balance of a grinding wheel essential?
13. Describe briefly the procedure for balancing a grinding wheel.

TO TRUE AND DRESS A GRINDING WHEEL

14. Define truing and dressing.
15. Why are most diamond dressers mounted at an angle of 10 to 15° to the base?
16. Explain how a grinding wheel should be finish-dressed for:
    (a) rough grinding
    (b) finish grinding

WORK-HOLDING DEVICES

17. List the advantages and disadvantages of an electromagnetic chuck.

18. Describe the construction and operation of a permanent magnetic chuck.

MAGNETIC CHUCK ACCESSORIES

19. How are thin workpieces held for surface grinding? Why is this necessary?
20. What precautions must be observed when using magnetic chuck blocks?
21. Describe and state the purpose of magna-vise clamps.

GRINDING FLUIDS

22. State four purposes of grinding fluids.
23. Name three methods of applying coolant and briefly describe the method by which each is applied.

SURFACE FINISH

24. List any five factors that affect the surface finish on the part being ground. How do these factors affect the surface finish?

SURFACE-GRINDING OPERATIONS

UNIT 80

The surface grinder is used primarily for grinding flat surfaces on hardened or unhardened workpieces. Since the workpiece can be held by various methods and the wheel face can be shaped by dressing, it is possible to perform such operations as form, angular, and vertical grinding. Surface grinding brings the work to close tolerances and produces a high surface finish.

Good surface-grinding results depend on several factors, such as the proper mounting of the work and the proper wheel selection for the job. A knowledge of the controls and grinding safety practices are of the utmost importance before anyone attempts to use a surface grinder.

OBJECTIVES

After completing this unit you should be able to:
1. Set up various workpieces for grinding
2. Observe the safety rules to operate the grinder
3. Grind flat vertical and angular surfaces

MOUNTING THE WORKPIECE FOR GRINDING

The size, shape, and type of work will determine the method by which the work should be held for surface grinding.

FLAT WORK OR PLATES

1. Remove all burrs from the surface of the work.
2. Clean the chuck surface with a clean cloth and then wipe the palm of the hand over the surface to remove dirt.
3. Place a piece of paper slightly larger than the workpiece in the center of the magnetic chuck face.
4. Place the work on top of the paper, and be sure to straddle as many magnetic inserts as possible.
5. If the workpiece is warped and rocks on the chuck face, it is advisable to shim the work where necessary to prevent rocking. This will avoid distortion when the work is removed from the magnetic chuck.
6. Turn the handle to the on position.
7. Check the work to see that it is held securely by trying to remove the workpiece.

THIN WORKPIECES

Thin workpieces tend to warp because of the heat created during the grinding operation. To minimize the amount of heat generated, it is advisable to mount the workpiece at an angle of approximately 15 to 30° from the side of the chuck (Fig. 80-1). This reduces the length of time the wheel is in contact with the work, which in turn reduces the amount of heat generated per pass. If an adaptor is available, it should be used, and the work mounted at an angle.

SHORT WORKPIECES

Work which does not straddle three magnetic poles will generally not be held firmly enough for grinding. It is advisable to straddle as many poles as possible and to set parallels or steel pieces around the work to prevent it from moving during the grinding operation (Fig. 80-2). The parallels or steel pieces should be slightly thinner than the workpiece to provide maximum support.

GRINDER SAFETY

When operating any type of grinder, it is important that certain basic, time-tested safety precautions be observed. Generally, the safest grinding practice is also the most efficient.

1. Before mounting a grinding wheel, ring test the wheel to check for defects.
2. Be sure that the grinding wheel is properly mounted on the spindle.
3. See that the wheel guard covers at least one-half the wheel.
4. Make sure that the magnetic chuck has been turned on by trying to remove the work.
5. See that the grinding wheel clears the work before starting a grinder.
6. Be sure that the grinder is operating at the correct speed for the wheel being used.
7. When starting a grinder, always stand to one side of the wheel and make sure no one is in line with the grinding wheel in case it breaks on startup.
8. Never attempt to clean the magnetic chuck or mount and remove work until the wheel has stopped completely.
9. *Always* wear safety glasses when grinding.

GRINDING OPERATIONS

The most common operation performed on a surface grinder is the grinding of flat (horizontal) surfaces. Regardless of the type of grinding operation, it is important that the correct wheel be mounted and the work held securely.

TO GRIND A FLAT (HORIZONTAL) SURFACE

1. Remove all burrs and dirt from the workpieces and the face of the magnetic chuck.

Fig. 80-1 Small, thin work should be set at an angle on the adapter plate to minimize warpage due to grinding heat.

Fig. 80-2 Steel blocks or parallels are placed around short work to prevent it from moving during grinding.   (Courtesy Kostel Enterprises Limited.)

2. Mount the work on the chuck, placing a piece of paper between the chuck and the workpiece.
3. Check to see that the work is held firmly.
4. Set the table reverse dogs so that the center of the grinding wheel clears each end of the work by approximately 1 in. (25.4 mm).
5. Set the crossfeed for the type of grinding operation—roughing cuts, 0.030 to 0.050 in. (0.76 to 1.27 mm); finishing cuts, 0.005 to 0.020 in. (0.12 to 0.50 mm).
6. Bring the work under the grinding wheel by hand, *having about ⅛ in. (3.17 mm) of the wheel edge over the work* (Fig. 80-3).
7. Start the grinder and lower the wheelhead until the wheel just sparks the work.
8. The wheel may have been set on a low spot of the work. It is good practice, therefore, to always raise the wheel about 0.005 in. (0.12 mm).
9. Start the table traveling automatically and feed the entire width of the work under the wheel to check for high spots.
10. Lower the wheel for every cut until the surface is completed—roughing cuts 0.005 to 0.015 in. (0.12 to 0.38 mm), finishing cuts 0.001 to 0.002 in. (0.02 to 0.05 mm).
11. Release the magnet and remove the workpiece, by

Fig. 80-3 The edge of the wheel should overlap the work by about ⅛ in. (3.17 mm). (Courtesy Kostel Enterprises Limited.)

Fig. 80-4 The workpiece may be clamped to an angle plate for grinding the edges square. (Courtesy Kostel Enterprises Limited.)

raising one edge, to break the magnetic attraction. This will prevent scratching the chuck surface.

NOTE: Cutting fluid should be used whenever possible to aid the grinding action and keep the work cool.

GRINDING THE EDGES OF A WORKPIECE

Much work which is machined on a surface grinder must have the edges ground square and parallel so that these edges may be used for further layout or machining operations.

Work which is to be ground all over should be machined to about 0.010 in. (0.25 mm) over the finished size for each surface. The large, flat surfaces are usually ground first which then permits them to be used as reference surfaces for further setups.

When the four edges of a workpiece must be ground, clamp the work to an angle plate so that two adjacent sides may be ground square without moving the workpiece.

Setting Up the Workpiece

1. Clean and remove all burrs from the workpiece, the angle plate, and the magnetic chuck.
2. Place a piece of paper which is slightly larger than the angle plate on the magnetic chuck.
3. Place one end of the angle plate on the paper (Fig. 80-4).
4. Place a flat-ground surface of the workpiece against the angle plate so that the top and one edge of the workpiece project about ½ in. (12.7 mm) beyond the edges of the angle plate (Fig. 80-4).

NOTE: Be sure that the one edge of the work does not project beyond the base of the angle plate. If the work is

Fig. 80-5 A workpiece set up to grind the first edge. (Courtesy Kostel Enterprises Limited.)

smaller than the angle plate, a suitable parallel must be used to bring the top surface beyond the end of the angle plate.

5. Hold the work firmly against the angle plate and turn on the magnetic chuck.
6. Clamp the work securely to the angle plate and set the clamps so that they will not interfere with the grinding operation.

NOTE: Place a thin piece of brass or aluminum between the clamp and the work to prevent marring the finished surface.

7. Turn off the magnetic chuck and carefully place the base of the angle plate on the magnetic chuck (Fig. 80-5).

8. Carefully fasten two more clamps on the end of the workpiece to hold the work securely.

Grinding the Edges of a Workpiece Square and Parallel

After the work has been properly set up on the magnetic chuck, the following procedure should be followed for grinding the four edges of the workpiece.

1. Raise the wheelhead so that it is about ½ in. (12.7 mm) above the top of the work.
2. Set the table reverse dogs so that each end of the work clears the grinding wheel by about 1 in. (25.4 mm).
3. With the work under the center of the wheel, turn the crossfeed handle until *about ⅛ in. (3.17 mm) of the wheel edge overlaps the edge of the work* (Fig. 80-3).
4. Start the grinding wheel and lower the wheelhead until the wheel just sparks the work.
5. Move the work clear of the wheel with the crossfeed handle.
6. Raise the wheel about 0.005 to 0.010 in. (0.12 to 0.25 mm) in case the wheel had been set to a low spot on the work.
7. Check for further high spots by feeding the table by hand so that the entire length of the work passes under the wheel. Raise the wheel if necessary.
8. Engage the table reverse lever and grind the surface until all marks are removed. The depth of cut should be 0.003 to 0.007 in. (0.07 to 0.17 mm) for roughing cuts and 0.0005 to 0.001 in. (0.01 to 0.02 mm) for finishing cuts.
9. Stop the machine and remove the clamps from the right-hand end of the work.
10. Turn off the magnetic chuck and remove the angle plate and workpiece as one unit. Be careful not to jar the work setup.
11. Clean the chuck and the angle plate.
12. Place the angle plate (with the attached workpiece) on its end with the surface to be ground at the top (Fig. 80-6).
13. Fasten two clamps to the right-hand side of the workpiece and the angle plate.
14. Remove the original clamps from the top of the setup.
15. Repeat steps 1 to 8 and grind the second edge.
16. Remove the assembly from the chuck and remove the workpiece from the angle plate.

Grinding the Third and Fourth Edges

When two adjacent sides have been ground, they are then used as reference surfaces to grind the other two sides square and parallel.

1. Clean the workpiece, the angle plate, and the magnetic chuck thoroughly and remove any burrs.
2. Place a clean piece of paper on the magnetic chuck.
3. Place a ground edge of the workpiece on the paper.
   a. If the workpiece is at least 1 in. (25.4 mm) thick and long enough to span three magnetic poles on the chuck, no angle plate is required (Fig. 80-7).

Fig. 80-6 Angle plate and work set for grinding the second edge of the workpiece at 90° to the first edge. (Courtesy Kostel Enterprises Limited.)

Fig. 80-7 Work which has sufficient bearing surface may be set on the chuck for finishing the remaining edges. (Courtesy Kostel Enterprises Limited.)

   b. If the work is less than 1 in. (25.4 mm) thick and does not span three magnetic poles, it should be fastened to an angle plate (Fig. 80-8).
   (1) Place a ground edge on the paper and place an angle plate no higher than the workpiece against the workpiece.

NOTE: A suitable parallel may be required to raise the edge of the work above the edge of the angle plate.

Fig. 80-8 An angle plate may be required to finish the third and fourth edges. (Courtesy Kostel Enterprises Limited.)

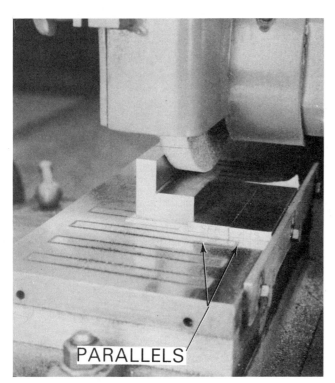

PARALLELS

Fig. 80-9 The setup for vertical grinding.

   (2) Turn on the chuck and carefully clamp the work to the angle plate.
4. Grind the third edge to the required size.
5. Repeat operations 1 to 3 and grind the fourth edge.

TO GRIND A VERTICAL SURFACE

Although most grinding performed on a surface grinder is the grinding of flat horizontal surfaces, it is often necessary

Fig. 80-10 Corner of work is relieved to provide clearance and prevent the corner of the wheel from breaking down. (Courtesy Kostel Enterprises Limited.)

to grind a vertical surface (Fig. 80-9). Extreme care must be taken in the setup of the workpiece when grinding the vertical surface. Before grinding a vertical surface, it is necessary to relieve the corner of the work (Fig. 80-10) to ensure clearance for the edge of the wheel.

PROCEDURE

1. Mount the proper grinding wheel; true, dress, and balance as required.
2. Dress the side of the wheel to give it a slight clearance (Fig. 80-11).
3. Clean the surface of the magnetic chuck and mount the work.
4. With an indicator, align the edge of the work parallel to the table travel.

OR

Place the work against the stop bar of the magnetic chuck which has been aligned. If the work cannot be set against the stop bar, parallels may be used to position the work on the magnetic chuck (Fig. 80-9).

5. Turn on the magnetic chuck and test to see that the work is held securely.
6. Set the reversing dogs, allowing sufficient table travel to permit clearance for the wheel at each end of the stroke.
7. Bring the side of the wheel close to the vertical surface to be ground.
8. Lower the wheel to within 0.002 to 0.005 in. (0.05 to 0.12 mm) of the flat or horizontal surface which has been finish ground.
9. Start the table traveling *slowly* and feed the wheel until it just sparks the vertical surface.
10. Rough-grind the vertical surface to within 0.002 in. (0.05 mm) of size by feeding the table in approximately 0.001 in. (0.02 mm) per pass.
11. Redress the side of the wheel if necessary.
12. Finish-grind the vertical surface by feeding the table approximately 0.0005 in. (0.01 mm) per pass.

Fig. 80-11 (A) Dressing the side of the wheel for grinding a vertical surface; (B) the side of the wheel should be slightly concave for grinding a vertical surface.

Fig. 80-12 Work may be set to a sine bar when an accurate angle must be ground.

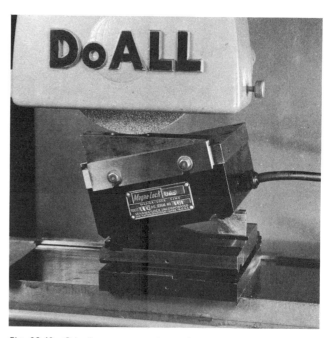

Fig. 80-13 Grinding an accurate angle using a sine chuck. (Courtesy DoAll Company.)

TO GRIND AN ANGULAR SURFACE

When it is necessary to grind an angular surface, the work may be held at an angle by a sine bar and angle plate (Fig. 80-12), a sine chuck (Fig. 80-13), or an adjustable angle vise (Fig. 80-14). When work is held by any of these methods, grinding is done with a flat-dressed wheel.

Angular surfaces may also be ground by holding the work flat and dressing the grinding wheel to the required angle with a sine dresser (Fig. 80-15).

When a sine dresser is not available, a parallel set to the desired angle by means of a sine bar may be clamped to an angle plate. This setup is then placed on a magnetic chuck beneath the grinding wheel (Fig. 80-16).

Fig. 80-14   Work held for grinding in an adjustable angle vise.

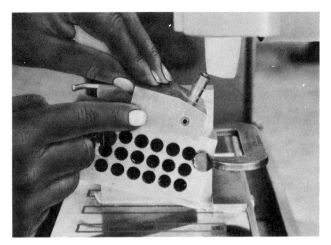

Fig. 80-16   Dressing a wheel using a parallel set to an angle.

Fig. 80-15   Dressing the wheel to an angle using a sine dresser.

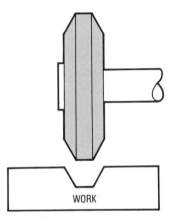

WORK

Fig. 80-17   When form grinding, the wheel is dressed to the reverse profile of the form required.

FORM GRINDING

Form grinding refers to the production of curved and angular surfaces produced by means of a specially dressed wheel.

The reverse form or contour required on the workpiece is dressed on the grinding wheel (Fig. 80-17).

Contours and radii may be produced on the grinding wheel by means of a radius wheel dresser (Fig. 80-18A).

TO DRESS A CONVEX RADIUS ON A GRINDING WHEEL

1. Mount the radius dresser (Fig. 80-18A) squarely on a clean magnetic chuck.

2. Set both stops on the radius dresser so that it can only be rotated one-quarter of a turn. The two stops should be 90° apart.

3. Fasten the diamond height setting bar in the radius dresser. The bottom surface of the height setting bar is the center of the radius dresser.

4. Place a gage block buildup using wear blocks on each side, equal to the radius required on the grinding wheel, between the height-setting bar and the diamond point.

5. Raise the diamond until it just touches the gage blocks (Fig. 80-18B) and then lock it in this position.

NOTE: When dressing a *concave radius,* the diamond point must be set *above* the center of the radius dresser a distance equal to the radius desired.

6. Move the table longitudinally until the diamond is under the center of the grinding wheel (Fig. 80-18A).

7. Lock the table to prevent longitudinal movement.

8. Rotate the arm of the radius dresser one-quarter of a turn, so that the diamond is in a horizontal position.

9. Start the machine and, using the cross-feed handle, bring the diamond in until it just touches the side of the grinding wheel.

10. Lock the table cross-slide in this position.

11. Stop the grinder and raise the wheel until it clears the diamond.

12. Start the grinder, and while slowly rotating the diamond back and forth through the 90° arc, lower the wheel until it just touches the diamond.

13. Feed the wheel down approximately 0.002 to 0.003

Fig. 80-18 (A) A radius wheel dresser mounted on the magnetic chuck; (B) gage blocks being used to set the diamond to the correct height for the radius to be dressed on a grinding wheel; (C) a convex radius being dressed on a grinding wheel.

in. (0.05 to 0.07 mm) for every rotation of the dresser.

14. Continue to dress the radius until the periphery of the wheel just touches the diamond when it is in a vertical position. This indicates that the radius is completely formed (Fig. 80-18C).

15. Stop the grinder, raise the wheel and remove the radius dresser.

Fig. 80-19 Crush-dressing a wheel to grind serrations, with a roll mounted on the machine.

Fig. 80-20 Principle of crush-form grinding.

SPECIAL FORMS

When complex profiles must be ground on long production runs, the wheel may often be crush-formed. A tool-steel or carbide roll, having the desired form or contour of the finished workpiece, is forced into the slowly revolving grinding wheel [200 to 300 sf/min (61 to 91 m/min)] (Fig. 80-19). The grinding wheel assumes the reverse form of the crushing roll. The wheel is then used to grind the form or contour on the workpiece (Fig. 80-20). Grinding wheels may be crush-formed to tolerances as close as ±0.002 in. (0.05 mm), and to radii as small as 0.005 in. (0.12 mm), depending on the grit size and structure of the wheel. As the wheel is used, it will gradually wear out of tolerance and the form must be redressed, using the crush roll. When the crush roll wears, as a result of many redressings, it must be reground to the original tolerance.

Some surface grinders are not designed for crush-form dressing of the wheel. It is not advisable to perform crush-form dressing on any machine equipped with a ball bearing spindle, since the bearings are subjected to a considerable load in crush dressing and may be damaged. Machines equipped with roller bearings have proved quite satisfactory for crush-dressing operations.

CUTTING-OFF OPERATIONS

The surface grinder may be used for cutting off hardened materials by using thin cutoff wheels. The work may be

Fig. 80-21 Cutting off work held in a magnetic V-block on a surface grinder. (Courtesy Kostel Enterprises Limited.)

clamped in a fixture or vise and positioned below the wheel (Fig. 80-21). For thin, short pieces, the wheelhead may be fed straight down to cut off the work. If longer pieces are to be cut, the work is properly mounted and the table is reciprocated as in normal grinding while the wheel is fed down. Diamond wheels may also be mounted for cutting-off operations on carbides.

REVIEW QUESTIONS

TO MOUNT THE WORKPIECE FOR GRINDING

1. Describe briefly the procedure for mounting the following for grinding:
   (a) flat work   (b) short work

TABLE 80-1 Surface-Grinding Problems, Causes, and Remedies		
GRINDING PROBLEM	**CAUSE**	**POSSIBLE REMEDY**
Burning or discoloration	Wheel is too hard	Use a softer, free-cutting wheel. Decrease wheel speed. Increase work speed. Coarse-dress the wheel. Take lighter cuts and dress the wheel frequently. Use coolant directed at point of contact between the wheel and the work.
Burnished work surface (work is highly polished in irregular patches)	Wheel is glazed	Dress the wheel. Use a coarser-grit wheel. Use a softer wheel. Use a more open structure wheel.
Chatter or wavy pattern	Wheel out of balance. Wheel out of round. Spindle bearing too loose. Wheel too hard. Glazing of wheel.	Rebalance. True and dress. Adjust or replace bearings. Use a softer wheel, coarser grit, or more open structure. Increase table speed. Redress.
Scratches on the work surface	Grinding wheel is too soft. (Abrasive grains break off too readily and catch between the wheel and work surface.) Wheel is too coarse. Loose particles of swarf fall onto the work from the wheel guard. Dirty coolant carries dirt particles onto the work surface. Feed lines.	Use a harder wheel. Use a finer-grit wheel. Clean the grinding wheel guard when changing a wheel. Clean the coolant tank and replace the coolant. Slightly round the edges of the wheel.

GRINDER SAFETY

2. How should a wheel be checked for defects?

3. Why is it necessary to see that no one is in line with the grinding wheel before starting the grinder?

4. List five grinder safety rules which you consider to be most important and explain the reason for the selection of each.

TO GRIND A FLAT SURFACE

5. Explain how the grinding wheel should be set to the surface of the work.

6. How should work be removed from a magnetic chuck?

GRINDING THE EDGES OF A WORKPIECE

7. Why are the edges of workpieces often ground square and parallel?

8. How much allowance should be left on a surface for grinding?

9. What precautions should be observed before mounting any workpiece or accessory on the magnetic chuck?

10. What precautions should be observed when mounting a workpiece on an angle plate for grinding two adjacent sides?

11. After the first side of a workpiece has been ground, how is the work set to grind the second side square to the first?

12. What precautions should be observed when setting the grinding wheel to the work surface?

TO GRIND A VERTICAL SURFACE

13. Outline briefly the procedure for grinding a vertical surface.

14. When grinding a vertical surface, why is it necessary first to relieve the corner between the two adjacent surfaces?

15. How is the grinding wheel dressed when grinding a vertical surface? Why is this necessary?

TO GRIND AN ANGULAR SURFACE

16. By what methods may work be held for grinding an angle when using a flat dressed wheel?

17. Name two methods of dressing a wheel to an angle.

FORM GRINDING

18. Name two methods of dressing contours and radii on grinding wheels.

19. Describe the principle of crush-form dressing.

20. On what types of surface grinders should crush-form dressing be performed?

SURFACE GRINDING PROBLEMS, CAUSES, AND REMEDIES

21. List three causes and three remedies for each of the following grinding problems:
   (a) chatter or wavy pattern
   (b) scratches on the work surface

CYLINDRICAL GRINDERS
UNIT 81

When the diameter of a workpiece must be ground accurately to size and to a high surface finish, it may be done on a *cylindrical* grinder. There are two types of machines suitable for cylindrical grinding—the *center type* and the *centerless type*—each with its special applications.

OBJECTIVES

After completing this unit you will be able to:
1. Set up and grind work on a cylindrical grinder
2. Internal-grind on a universal cylindrical grinder
3. Identify and know the principle of three methods of centerless grinding

HEADSTOCK

INTERNAL GRINDING HEAD

WHEELHEAD

FOOTSTOCK

TABLE REVERSE DOG

UPPER TABLE

CROSS-FEED HANDWHEEL

TABLE REVERSE LEVER

LOWER TABLE

TABLE TRAVERSE HANDWHEEL

BASE

Fig. 81-1 Parts of a universal cylindrical grinder.

CENTER-TYPE CYLINDRICAL GRINDERS

Work which is to be finished on a cylindrical grinder is generally held between centers but may also be held in a chuck.

There are two types of cylindrical grinders (center type), the *plain* and the *universal*. The plain-type grinder is generally a manufacturing type of machine. The universal cylindrical grinder (Fig. 81-1) is more versatile, since both the wheelhead and headstock may be swiveled.

PARTS OF THE UNIVERSAL CYLINDRICAL GRINDER

The *base* is of heavy cast-iron construction to provide rigidity. The top of the base is machined to form the *ways* for the table.

The *wheelhead* is mounted on a cross-slide at the back of the machine. The ways on which it is mounted are at right angles to the table ways, permitting the wheelhead to be fed toward the table and the work, either automatically or by hand. On universal machines, the wheelhead may be swiveled to permit the grinding of steep tapers by plunge grinding.

The *table*, mounted on the ways, is driven back and forth by hydraulic or mechanical means. The reversal of the table is controlled by *trip dogs*. The table is composed of the *lower table*, which rests on the ways, and the *upper table*, which may be swiveled for grinding tapers and alignment purposes. The *headstock* and *footstock*, used to support work held between centers, are mounted on the table.

The *headstock* unit is mounted on the left end of the table and contains a motor for rotating the work. A dead center is mounted in the headstock spindle. When work is mounted between centers, it is rotated on *two dead centers* by means of a dog and a drive plate which is attached to, and revolves with, the headstock spindle. The purpose of a dead center in the headstock is to overcome any spindle inaccuracies (looseness, burrs, etc.) which may be transferred to the workpiece. Grinding work on two dead centers results in truer diameters which are concentric with the centerline of the work. Work may also be held in a chuck which is mounted on the spindle nose of the headstock.

The *footstock* supports the right end of the work and is adjustable along the length of the table. The dead center, on which the work is mounted, is spring-loaded to provide the proper center tension on the workpiece.

The *backrest or steadyrest* provides support for long, slender work and prevents it from springing. Outward and downward movement of the workpiece is prevented by means of adjustable supports in front of and below the workpiece. It may be positioned anywhere along the length of the table.

The *center rest,* which resembles a lathe steadyrest, may be mounted at any point on the table. It is used to support the right end of the work when external grinding is confined to the end of the workpiece. Long workpieces on which internal grinding is to be performed are also supported on the end by the center rest.

An *internal grinding attachment* may be mounted on the wheelhead on most machines for internal grinding. It is usually driven by a separate motor.

A *diamond wheel dresser* may be clamped to the table to dress the grinding wheel as required. On some types of grinders, diamond dressers may be permanently mounted on the footstock.

The *coolant system* is built into all cylindrical grinders to provide dust control, temperature control, and a better surface finish on the workpiece.

MACHINE PREPARATION FOR GRINDING

MOUNTING THE WHEEL

All precautions, such as wheel balancing and mounting procedures, used on surface grinders should be observed for cylindrical grinding.

To True and Dress the Wheel

1. Start the grinding wheel to permit the spindle bearings to warm up.
2. Mount the proper diamond in the holder and clamp it to the table. The diamond should be mounted at an angle of 10 to 15° to the wheelface and should be held on or slightly below the centerline of the wheel (Fig. 81-2).
3. Adjust the wheel until the diamond almost touches the high point of the wheel, which is usually in the center of the wheelface.
4. Turn on the coolant if this is to be used for the grinding operation.
5. Feed the wheel into the diamond about 0.001 in. (0.02 mm) per pass, and move it back and forth across the wheelface at a medium rate until the wheelface has been completely dressed.
6. Finish-dress the wheel by using an infeed of 0.0005 in. (0.01 mm) and a slow traverse feed. When a very fine finish is required, the diamond should be traversed slowly across the wheelface for a few passes without any additional infeed.

To Parallel-Grind an Outside Diameter

1. Lubricate the machine as required.
2. Start the grinding wheel to warm up the spindle bearings. This will ensure the utmost accuracy when grinding.

Fig. 81-2 Dressing a grinding wheel on a cylindrical grinder. (Courtesy Cincinnati Milacron Incorporated.)

3. True and dress the grinding wheel if required.
4. Clean the machine centers and the center holes of the work. If the grinder centers are damaged, they must be reground. On hardened-steel workpieces, the center holes should be honed or lapped to ensure the utmost accuracy.
5. Align the headstock and footstock centers with a test bar and indicator.
6. Lubricate the center holes with a suitable lubricant.
7. Set the headstock and footstock for the proper length of work so that the center of the work will be over the center of the table.
8. Mount the work between centers with the dog mounted loosely on the left end of the work.
9. Tighten the dog on the end of the work and engage the drive plate pin in the fork of the dog.
10. Adjust the table dogs so that the wheel will overrun each end of the work by about one-third of the width of the wheelface. If grinding must be done up to a shoulder, the table traverse must be carefully set so that it reverses just before the wheel touches the shoulder.
11. Set the grinder to the proper speed for the wheel being used. Some machines are provided with a means of increasing the wheel speed as the wheel becomes smaller. If this is not done, the wheel will act softer and wear quickly.
12. Set the work speed for the diameter and type of material being ground. Proper work speed is very important. A slow speed causes heating and distortion of the work. High work speeds will cause the wheel to act softer and break down quickly.
13. Set the headstock spindle to rotate the work in an opposite direction to that of the grinding wheel. When grinding, the sparks should be directed down toward the table.
14. If the machine is so equipped, set the automatic infeed for each table reversal. Also set the dwell or "tarry" time, which permits the wheel to clear itself at each end of the stroke.
15. Select and set the desired table traverse or speed. This should be such that the table will move one-half to two-thirds of the wheel width per revolution of the work. Finish grinding is done at a slower rate of table traverse.
16. Start the table and move the wheel up to the workpiece until it just sparks.

17. Engage the wheel feed clutch lever and grind until the work is cleaned up.

18. Check the work for taper and adjust if necessary.

19. Determine the amount of material to be removed and set the feed index for this amount. The infeed of the wheel will stop automatically when the work is at the proper diameter.

20. Stop the machine with the wheel clear of the work and measure the size of the workpiece. If necessary, make a correction on the index setting and grind the piece to size.

TO GRIND A TAPERED WORKPIECE

Tapered work, held between centers, is ground in the same manner as parallel work, except that the table is swiveled to half the included angle of the taper. The taper should be checked for accuracy after the workpiece is cleaned up and the table adjusted if necessary.

Short, steep tapers may be ground on work held in a chuck or between centers by swiveling the wheelhead to the desired angle and plunge grinding the tapered surface with the face of the wheel.

PLUNGE GRINDING

When a short tapered or parallel surface is to be ground on a workpiece, it may be plunge-ground by feeding the wheel into the revolving work, with the table remaining stationary. The grinding wheel may be fed in automatically to the setting on the feed index. It then dwells for a suitable time to permit "spark out" and retracts automatically. In the case of tapered work, the wheelhead must be swung to half the included angle. When plunge grinding, the length of the surface to be ground must be no longer than the width of the grinding wheelface.

INTERNAL GRINDERS

Internal grinding may be defined as the accurate finishing of holes in a workpiece by a grinding wheel. Although internal grinders were originally designed for hardened workpieces, they are now used extensively for finishing holes to size and accuracy in soft material.

Production internal grinding is done on internal grinders designed exclusively for this type of work. The wheel is fed into the work automatically until the hole reaches the required diameter. When the hole is finished to size, the wheel is withdrawn from the hole and automatically dressed before the next hole is ground.

Internal grinding may also be performed on the universal cylindrical grinder, the cutter and tool grinder, and the lathe. Since these machines are not designed primarily for internal grinding, they are not as efficient as the standard internal grinder. For most internal grinding operations, the work is rotated in a chuck mounted on the workhead spindle.

Internal grinding may also be performed on the universal cylindrical grinder, the cutter and tool grinder, and the lathe. Since these machines are not designed primarily for internal grinding, they are not as efficient as the standard internal grinder. For most internal grinding operations, the work is rotated in a chuck mounted on the workhead spindle. Work may also be mounted on a faceplate, collet chucks, or special fixtures. When work is too large to be rotated, internal diameters may be ground by using a planetary grinder. In this operation, the grinding wheel is guided in a circular motion about the axis of the hole and is fed out to the required diameter. The work or the grinding head is fed parallel to the wheel spindle to provide a smooth, uniform surface.

INTERNAL GRINDING ON A UNIVERSAL CYLINDRICAL GRINDER

Although the universal cylindrical grinder is not designed primarily as an internal grinder, it is used extensively in toolrooms for this purpose. On most universal cylindrical grinders, the internal grinding attachment is mounted on the wheelhead column and is easily swung into place when required. One advantage of this machine is that the outside and inside diameters of a workpiece may often be finished in one setup. Although the grade of the wheel used for internal grinding will depend on the type of work and the rigidity of the machine, the wheels used for internal grinding are generally softer than those for external grinding, for the following reasons.

1. There is a larger area of contact between the wheel and the workpiece during the internal grinding operation.

2. A soft wheel requires less pressure to cut than a hard wheel; thus the spindle pressure and spring is reduced.

To Grind a Parallel Internal Diameter on a Universal Cylindrical Grinder

1. Mount the workpiece in a universal chuck, a collet chuck, or on a faceplate. Care must be taken not to distort thin workpieces.

2. Swing the internal grinding attachment into place and mount the proper spindle in the quill. For maximum rigidity, the spindle should be as large as possible, with the shortest overhang.

3. Mount the proper grinding wheel, as large as possible, for the job.

4. Adjust the spindle height until its center is in line with the center axis of the hole in the workpiece.

5. True and dress the grinding wheel.

6. Set the wheelspeed to 5000 to 6500 sf/min (1524 to 1981 m/min).

7. Set the workspeed to 150 to 200 sf/min (46 to 61 m/min).

8. Adjust the table dogs so that *only* one-third of the wheel width overlaps the ends of the work at each end of the stroke. On blind holes, the dog should be set to reverse the table just as the wheel clears the undercut at the bottom of the hole.

NOTE: In order to prevent bell-mouthing, the wheel must *never* overlap the end of the work by more than one-half the width of the wheel.

9. Start the work and the grinding wheel.

TABLE 81-1   Cylindrical Grinding Faults, Causes, and Remedies

FAULT	CAUSE	POSSIBLE REMEDY
"Barber Pole" finish	Work loose on centers. Centers are a poor fit in the spindle.	Adjust center tension. Use properly fitting centers.
Burnt work, cracked surfaces	Grinding wheel not trued. Grinding wheel too hard. Grinding wheel structure too dense. Incorrect bond. Grinding wheel too fast. Work revolving too slowly. Too heavy cut. Insufficient coolant. Wrong type of coolant. Dull diamond dresser.	True and dress the wheel. Use a softer wheel. Use a more open wheel. Consult manufacturer's handbook. Adjust the speed. Increase the speed. Reduce the depth of cut. Increase coolant supply. Try other type. Turn diamond in holder.
Chatter marks	Too heavy a cut. Grinding wheel too hard. Work too slender. Vibrations in machine. External vibrations transferred to machine.	Try a lighter cut. Use a softer wheel. Increase work speed. Reduce wheel speed. Use a steadyrest. Locate the vibrations and correct. Isolate the machine to prevent vibrations.
Diamond truing lines	Fast dressing feed. Diamond too sharp.	Slow dressing feed. Use a cluster-type nib.
Feed lines	Wrong wheel structure. Improper traverse speed when finish grinding. Coolant not directed properly. Improperly adjusted steadyrest.	Change to suit. Change work speed and wheel speed. Adjust nozzle. Check and adjust.
Intermittent cutting action	Work too tight on centers. Work too hot. Work out of balance.	Adjust center tension. Use coolant. Counterbalance as required.
Out-of-round work	Work center holes damaged or dirty. Work loose on centers. Loose machine centers. Machine centers worn. Loose gibs on the table.	Hone and lap centers. Adjust the center tension. Clean and reset. Regrind centers. Adjust the gibs.
Rough finish	Wheel too coarse. Wheel too hard. Diamond too sharp. Wheel rough-dressed. Table traverse too fast. Work speed too fast. Work springing.	Use a finer wheel. Use a softer wheel. Use a cluster-type nib. Finish-dress wheel. Slow to suitable feed. Slow to suitable speed. Use a steadyrest.
Wavy marks (long)	Wheel out of balance. Coolant has been directed against a stationary wheel.	Redress and balance wheel. Always shut off coolant well before stopping the wheel.
Wavy marks (short)	Vibrations caused by unmatched belts.	Replace belts with a matched set. Check motor and pulleys for balance.

Fig. 81-3 A centerless grinder. (Courtesy Cincinnati Milacron Incorporated.)

Fig. 81-4 Principle of the centerless grinder.

10. Touch the grinding wheel to the diameter of the hole.
11. Turn on the coolant.
12. Grind until the hole just cleans up, feeding the wheel in no more than 0.002 in. (or 0.05 mm) per table reversal.
13. Check the hole size and set the automatic feed (if the machine is so equipped) to disengage when the work is roughed to within 0.001 in. (0.02 mm) of size.
14. Reset the automatic infeed to 0.0002 in. (0.005 mm) per table reversal.
15. Finish-grind the work and let the wheel spark out.
16. Move the table longitudinally and withdraw the wheel and spindle from the workpiece.
17. Check the hole diameter and finish grind if necessary.

TO GRIND A TAPERED HOLE

When grinding tapered holes, the same procedures and precautions should be observed as for grinding parallel holes. However, the workhead must be set to one-half the included angle of the taper. It is *very* important that the centerlines of the grinding wheel and the hole be set at the same height in order to produce the correct taper.

CENTERLESS GRINDERS

The production of cylindrical, tapered, and multidiameter workpieces may be achieved on a centerless grinder (Fig. 81-3). As the name suggests, the work is not supported on centers but rather by a work rest blade, a regulating wheel, and a grinding wheel (Fig. 81-4).

On a centerless grinder, the work is supported on the work rest blade, which is equipped with suitable guides for the type of workpiece. The rotation of the grinding wheel forces the workpiece onto the work rest blade and against the regulating wheel, while the regulating wheel controls the speed of the work and the longitudinal feed movement. To provide longitudinal feed to the work, the regulating wheel is set at a slight angle. The rate of feed may be varied by changing the angle and the speed of the regulating

wheel. The regulating and grinding wheels rotate in the same direction, and the center heights of these wheels are fixed. Because the centers are fixed, the diameter of the workpiece is controlled by the distance between the wheels and the height of the work rest blade.

The higher the workpiece is placed above the centerlines of the wheels, the faster it will be ground cylindrical. However, there is a limit to the height at which it may be placed, since the work will eventually be lifted periodically from the work rest blade. There is one exception to placing the work above center: when removing slight bends in long, small-diameter work, the center of the piece is placed below the centerline of the wheels and the rate of traverse is high. This operation eliminates whipping and chattering that might result from bent work and is used primarily for straightening the workpiece. After the work has been straightened, it is ground in the normal manner above centers.

METHODS OF CENTERLESS GRINDING

There are three methods of centerless grinding: thru-feed, infeed, and endfeed.

THRU-FEED CENTERLESS GRINDING

Thru-feed centerless grinding (Fig. 81-5) consists of feeding the work between the grinding and regulating wheels. The cylindrical surface is ground as the work is fed by the regulating wheel past the grinding wheel. The speed at which the work is fed across the grinding wheel is controlled by the speed and angle of the regulating wheel.

INFEED CENTERLESS GRINDING

Infeed centerless grinding, a form of plunge grinding, is used when the work being ground has a shoulder or head (Fig. 81-6). Several diameters of a workpiece may be finished simultaneously by infeed grinding. Tapered, spherical, and other irregular profiles are ground efficiently by this method.

With infeed grinding, the work rest blade and the regulating wheel are clamped in a fixed relation to each other. The work is placed on the rest, against the regulating wheel, and is fed into the grinding wheel by moving the infeed

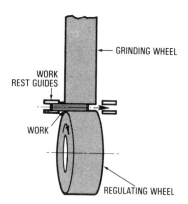

Fig. 81-5 Principle of thru-feed centerless grinding. (Courtesy Cincinnati Milacron Incorporated.)

Fig. 81-6 Principle of infeed centerless grinding. (Courtesy Cincinnati Milacron Incorporated.)

Fig. 81-7 Principle of endfeed centerless grinding. (Courtesy Cincinnati Milacron Incorporated.)

lever through a 90° arc. When the lever is at the full end of the travel, the predetermined size has been reached and the part is the desired size. When the lever is reversed, the regulating wheel and work rest move back and the part is ejected either manually or automatically.

If the part to be ground is longer than the wheels, one end is supported on the work rest and the other on rollers mounted on the machine.

ENDFEED CENTERLESS GRINDING

The *endfeed method* (Fig. 81-7) is used mainly for grinding tapered work. The grinding wheel, the regulating wheel, and the work rest all remain in a fixed position. The work is then fed in from the front, manually or mechanically, up to a fixed stop. When the machine is prepared for endfeed grinding, the grinding wheel and the regulating wheel are often dressed to the required taper. In some cases where only a few parts are required, only the regulating wheel may be dressed.

ADVANTAGES OF CENTERLESS GRINDING

1. There is no limit to the length of work being ground.
2. There is no axial thrust on the workpiece, permitting the grinding of long workpieces which would be distorted by other methods.
3. For truing purposes, less stock is required on the workpiece than if the work is held between centers. This is due to the fact that the work "floats" in the centerless grinder. Work held between centers may run eccentrically and require more stock for truing up.
4. Because there is less stock to be removed, there is less wheel wear and less grinding time is required.

TABLE 81-2 Internal Grinding Problems, Causes, and Remedies		
PROBLEM	CAUSE	REMEDY
Bell-mouthed hole	Stroke is too long and wheel overlaps hole too much. Centerlines of workpiece and wheel spindle are at different heights.	Reduce overlap of wheel at each end of hole. Align the centers before setting up work.
Burning or discoloration of work	Wheel too hard. Insufficient coolant.	Use a softer wheel. Increase work speed. Decrease diameter of wheel. Use a narrower wheel. Coarse dress the wheel. Increase coolant supply and direct it at the point of grinding contact.
Chatter marks	Worn spindle bearings. Belt slipping. Defective belts. Wheel out of balance. Wheel not true. Wheel too hard. Incorrect work speed.	Adjust if possible or replace. Adjust tension. Replace the complete set of belts. Balance the wheel. True and dress. Use a softer wheel. Adjust.

TABLE 81-2 (Continued)

PROBLEM	CAUSE	REMEDY
Feed lines or spirals	Improper dressing. Wheel too hard. Edges of wheel too sharp. Feed too coarse. Wheelhead is tipped or swung.	Dress the wheel carefully using a sharp diamond. Use a softer wheel. Round edges slightly with an abrasive stick. Reduce feed on final passes. Align wheelhead and spindle.
Out-of-round hole	Work is distorted when mounting it in the chuck or holding device. Work overheated when rough grinding.	Use extreme care when mounting work. Reduce depth of cut and feed. If work is mounted on a faceplate, loosen each clamp slightly and re-tighten evenly.
Scratches on ground surface	Wheel too soft and abrasive grains are caught between work surface and wheel. Improperly dressed wheel. Dirty coolant deposits particles between the wheel and the work. Wheel too coarse.	Use a harder wheel. Carefully dress the wheel. Clean coolant tank and replace coolant. Use a finer wheel.
Tapered hole	Workhead set at a slight angle. Wheel too soft to hold size. Feed too fast.	Align workhead. Use a harder wheel. Reduce feed.
Wheel glazing	Wheel too hard. Wheel too dense. Improper dressing.	Use a softer wheel. Increase work speed. Use a more open wheel. Use a sharp diamond and rough-dress.
Wheel loading	Wheel too hard. Wheel too fine. Dirty coolant. Truing diamond is dull.	Use a softer wheel. Increase work speed. Increase traverse feed. Use a coarser grain. Clean coolant system and replace coolant. Use a sharp diamond and coarse-dress the wheel.

REVIEW QUESTIONS

CYLINDRICAL GRINDERS

1. Name two types of cylindrical grinders.
2. List the main parts of a cylindrical grinder and state the purpose of each.
3. What precautions should be observed when truing and dressing the grinding wheel of a cylindrical grinder?

TO PARALLEL-GRIND THE OUTSIDE DIAMETER

4. List six precautions which must be taken to ensure the utmost accuracy when parallel grinding the outside diameter of a workpiece.

TO GRIND A TAPERED WORKPIECE

5. List three ways in which tapers may be ground on a universal cylindrical grinder.

CYLINDRICAL GRINDING FAULTS, CAUSES, AND REMEDIES

6. List three causes and three remedies for the following cylindrical grinding faults.
   (a) burnt work
   (b) chatter marks
   (c) feed lines
   (d) work out-of-round
   (e) rough finish
   (f) wavy marks

INTERNAL GRINDING

7. Describe internal grinding and name four machines on which it may be performed.

INTERNAL GRINDING ON A UNIVERSAL CYLINDRICAL GRINDER

8. Why are grinding wheels used for internal grinding generally softer than those used for external grinding?

9. What precautions must be taken when setting up and grinding a parallel hole in a workpiece?

TO GRIND A TAPERED HOLE

10. At what height should the grinding wheel be set when grinding a taper? Why is this necessary?

INTERNAL GRINDING PROBLEMS, CAUSES, AND REMEDIES

11. List four important problems encountered when internal grinding. State at least two causes and two remedies for each of the problems listed.

CENTERLESS GRINDING

12. Describe the principle of centerless grinding. Illustrate by means of a suitable sketch.

13. How is the work held when centerless grinding to make it cylindrical as quickly as possible?

METHODS OF CENTERLESS GRINDING

14. Name three methods of centerless grinding and illustrate each of these methods by a suitable labeled sketch.

15. List four advantages of centerless grinding.

UNIVERSAL CUTTER AND TOOL GRINDER

UNIT 82

The universal cutter and tool grinder is designed primarily for the grinding of cutting tools such as milling cutters, reamers, and taps. Its universal feature and various attachments permit a variety of other grinding operations to be performed. Other operations which may be performed are internal, cylindrical, taper, and surface grinding, single-point tool grinding, and cutting-off operations. Most of the latter operations require additional attachments or accessories.

OBJECTIVES

After completing this unit you will be able to:
1. Identify and state the purpose of the main parts of a cutter and tool grinder
2. Grind clearance angles on helical and staggered tooth cutters
3. Grind a form-relieved cutter
4. Set up the grinder for cylindrical and internal grinding

PARTS OF THE UNIVERSAL CUTTER AND TOOL GRINDER (Fig. 82-1)

The *base* is of heavy, cast-iron, boxlike construction that provides rigidity. The top of the base is machined to provide the *ways* (which are generally hardened) for the saddle.

The *wheelhead* is mounted on a column at the back of the base. It may be raised or lowered by the wheelhead handwheels located on either side of the base. The wheelhead may be swiveled through 360°. The wheelhead spindle is mounted in antifriction bearings and is tapered and threaded at both ends to receive grinding wheel collets. The spindle speed may be varied by stepped pulleys to suit the size of the wheel being used.

The *saddle* is mounted on the ways of the base and is moved in and out by the crossfeed handwheels located at the front and back of the machine. The upper part of the saddle has machined and hardened ways at right angles to the ways on top of the base.

The *table* is composed of two units, the *upper* and *lower* table. The lower table, mounted on the upper ways of the saddle, rests and moves on antifriction bearings. The upper table is fastened to the lower table and may be swiveled for grinding tapers. The table unit (upper and lower tables) may be moved longitudinally by three *table traverse knobs*, one located at the front of the machine and two at the back. The table may also be traversed slowly by means of the *slow table traverse crank*. The table may be locked in place laterally and longitudinally with locking screws.

Stop dogs, mounted in a T-slot on the front of the table, control the length of the table traverse. Each dog has a positive stop pin on one side and a spring-loaded plunger on the other. They are reversible to provide for a positive or a cushioned stop for the table, as desired.

ACCESSORIES AND ATTACHMENTS

The right- and left-hand tailstocks are mounted in the T-slot of the upper table and support the work for certain grinding operations. They may be placed at any point along the table.

The universal workhead (Fig. 82-1) or headstock is mounted on the left side of the table and used for supporting end mills and face mills for grinding. It may also be equipped with a pulley and motor (motorized headstock, Fig. 82-19) and used for cylindrical grinding. A chuck may be mounted in the workhead to hold work for internal and cylindrical grinding, as well as cutting-off operations (Fig. 82-20).

The centering gage (Fig. 82-13) is used to align quickly the tailstock center with the center of the wheelhead spindle. It is also used to align the cutter tooth on center in some grinding setups.

The adjustable tooth rest supports the cutter tooth and may be fastened to the wheelhead or table, depending on the type of cutter being ground. Another form of tooth rest is the *universal micrometer flicker type*, which has a micrometer adjustment for small vertical movements of the tooth rest.

Plain tooth rest blades (Fig. 82-2A) are used for grinding straight-tooth milling cutters.

Rounded tooth rest blades (Fig. 82-2B) are used for sharpening shell end mills, small end mills, taps, and reamers.

Offset tooth rest blades (Fig. 82-2C) are a universal type

Fig. 82-1 A universal cutter and tool grinder. (Courtesy Cincinnati Milacron Incorporated.)

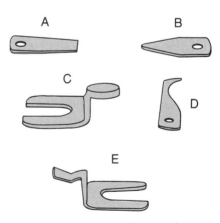

Fig. 82-2 Various shaped tooth rest blades.

suitable for most applications, such as coarse-pitch helical milling cutters and large face mills with inserted blades.

Hook or L-shaped tooth rest blades (Fig. 82-2D) are used for sharpening slitting saws, straight-tooth plain milling cutters with closely spaced teeth, and end mills.

Inverted V-tooth rest blades (Fig. 82-2E) are used for grinding the periphery of staggered-tooth cutters.

Cutter grinding mandrels and arbors (Fig. 82-3A, B). It is most important when grinding a milling cutter to hold it in the same manner as it is held for milling. For example, shell end mills should be sharpened on the same arbor as that used for milling.

Plain milling and side facing cutters, which are held on the standard milling machine arbor, should be held on a grinding mandrel (Fig. 82-3A) or a cutter grinding arbor (Fig. 82-3B).

A grinding mandrel rather than a lathe mandrel should be used to hold the cutter. This is necessary since a lathe mandrel will hold the cutter only at one end. The straight length of a grinding mandrel is a sliding fit into the cutter, and the slightly tapered end will hold the cutter securely for grinding.

Where considerable cutter grinding is done, a cutter grinder arbor will be found useful.

MILLING CUTTER NOMENCLATURE

To grind cutters correctly, the cutter parts and their function should be understood. Milling cutter parts are shown in Fig. 82-4. A brief description of the various parts follows.

Primary clearance is the clearance ground on the land adjacent to the tooth face. It is the angle formed between the slope of the land and a line tangent to the periphery. Primary clearance prevents the land behind the cutting edge from rubbing on the work. The amount of primary clearance on a cutter will vary with the type of material being cut.

Secondary clearance is ground behind the primary clearance and gives additional clearance to the cutter behind the tooth face. When grinding the clearance on a milling cutter, the primary clearance is always ground first. The secondary clearance is then used to control the width of the land.

The cutting edge is formed by the intersection of the face of the tooth with the land. This angle formed by the face of the tooth and the primary clearance is called the *angle of keenness.*

On side milling cutters, the cutting edges may be on one or both sides as well as on the periphery. When the teeth are straight, the cutting edge engages along the full width of the tooth at the same moment. This creates a gradual buildup of pressure as the tooth cuts into the work and a sudden release of this pressure as the tooth breaks through, causing a vibration or chatter. This type of cutter produces a poor finish and does not retain its sharp cutting edge as long as a helical cutter does.

When the teeth are helical, the length of the cutting edge contacting the work varies with the helix angle. The number

Fig. 82-3 (A) A cutter grinding mandrel; (B) a cutter grinding arbor.

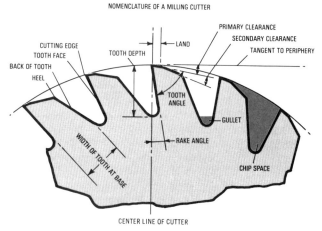

Fig. 82-4 Nomenclature of a milling cutter.

of teeth in contact with the work will vary with the size of the surface being machined, the number of teeth in the cutter, the cutter diameter, and the helix angle. Helical cutters produce a shearing action on the material being cut.

The helix angle, sometimes called the *shear angle,* is the angle formed by the angle of the teeth and the centerline of the cutter. It may be measured with a protractor or by bluing the edge of the cutter teeth and rolling the cutter against a straightedge over a sheet of paper (Fig. 82-5). The marks left by the teeth can easily be measured in relation to the axis of the cutter to determine the helix angle.

The land is the narrow surface behind the cutting edge on the primary clearance produced when the secondary clearance is ground on the cutter. The width of the land varies from about 1/64 in. (0.39 mm) on small cutters to about 1/16 in. (1.58 mm) on large cutters. On face mills, the land is more correctly called the *face edge.*

The tooth angle is the included angle between the face of the tooth and the land caused by grinding the primary clearance. This angle should be as large as possible to provide maximum strength at the cutting edge and better dissipation of heat generated during the cutting process.

The tooth face is the surface on which the metal being cut forms a chip. This face may be flat, as in straight-tooth plain milling cutters and inserted face-tooth mills, or curved, as in helical milling cutters.

CUTTER CLEARANCE ANGLES

To perform efficiently, a milling cutter must be ground to the correct clearance angle. The proper clearance angle on a milling cutter may only be determined by the "cut-and-try" method. The clearance angle will be influenced by such factors as finish, the number of pieces per sharpening,

Fig. 82-5 One method of measuring the helix angle of a milling cutter. (Courtesy Kostel Enterprises Limited.)

the type of material, and the condition of the machine. Excessive primary clearance produces chatter, causing the cutter to dull quickly.

A general rule followed by a large machine tool manufacturer for grinding cutter clearance angles is on high-speed steel cutters, 5° primary clearance plus an additional 5° for the secondary clearance. Thus the primary clearance is 5° and the secondary clearance is 10° when cutting machine steel. Carbide cutters used on machine steel are ground to 4° primary clearance plus an *additional* 4° (8°) for the secondary clearance.

Table 82-1 provides a rule of thumb for grinding cutter clearance angles on high-speed steel milling cutters. Table 82-2 gives the angles for carbide cutters. It should be remembered that this is a guide only. If the cutter does not perform satisfactorily with these angles, adjustments will have to be made to suit the job.

METHODS OF GRINDING CLEARANCE ON CUTTERS

Clearance may be ground on cutters by the following three methods. The type of cutter being ground will determine the method used.

CLEARANCE GRINDING

Clearance grinding (Fig. 82-6) produces a flat surface on the land. A 4-in. (101.6-mm) flared-cup wheel is used for this method and is offset slightly to permit long cutters to clear the opposite side of the wheel. When clearance grinding, the tooth rest may be set between the center and the top of the wheel, but never below center. The higher the tooth rest is placed, the less will be the clearance between

TABLE 82-1 Clearance Angles for High-Speed Steel Cutters		
MATERIAL TO BE MACHINED	PRIMARY CLEARANCE ANGLE	SECONDARY CLEARANCE ANGLE
High-carbon and alloy steels	3–5°	6–10°
Machine steel	3–5°	6–10°
Cast iron	4–7°	7–12°
Medium and hard bronze	4–7°	7–12°
Brass and soft bronze	10–12°	13–17°
Aluminum, magnesium, and plastics	10–12°	13–17°

TABLE 82-2 Primary Clearance Angles for Cemented-Carbide Cutters									
	PERIPHERY			CHAMFER			FACE		
TYPE OF CUTTER	STEEL	CAST IRON	ALUMI-NUM	STEEL	CAST IRON	ALUMI-NUM	STEEL	CAST IRON	ALUMI-NUM
Face or side	4–5°	7°	10°	4–5°	7°	10°	3–4°	5°	10°
Slotting	5–6°	7°	10°	5–6°	7°	10°	3°	5°	10°
Sawing	5–6°	7°	10°	5–6°	7°	10°	3°	5°	10°

Fig. 82-6   Setup for clearance grinding using a flaring-cup wheel.

Fig. 82-7   The periphery of the wheel is used for hollow grinding. (Courtesy Cincinnati Milacron Incorporated.)

Fig. 82-8   The work is revolved against the grinding wheel for circle grinding.

the cutter and the opposite edge of the wheel. When clearance grinding, the tooth rest may be attached to the table or the wheelhead, depending on the type of cutter being ground. For straight-tooth cutters, it may be mounted on the table, while for helical teeth, it must be mounted on the wheelhead.

HOLLOW GRINDING

The land produced by hollow grinding (Fig. 82-7) is concave. A 6-in.-diameter (152.4-mm) dish wheel or a 6-in.-diameter cutoff wheel is desirable. The cutoff wheel generally produces a better finish and breaks down more slowly because of the resinoid bond. Since all grinding wheels break down in use, it is better to grind diagonally opposite teeth in rotation and to take light cuts. In hollow grinding, the wheel and cutter centers must be aligned. Then, clearance is obtained by raising or lowering the wheel, depending on the method used to set up the cutter.

CIRCLE GRINDING

Circle grinding (Fig. 82-8) provides only a minute amount of clearance and is used mainly for reamers. The reamer is mounted between centers and is rotated *backward* so that the heel of the tooth contacts the wheel first. As the tooth rotates against the grinding wheel, the pressure of the wheel causes the cutter to spring back slightly as each cut progresses. Thus a very small amount of clearance is produced between the cutting edge and the heel of the tooth. The grinding wheel should be set on center when circle grinding. Secondary clearance must be obtained by clearance or hollow grinding. Circle grinding is also used to obtain

concentricity of milling cutters prior to clearance or hollow grinding.

METHODS OF CHECKING CUTTER CLEARANCE ANGLES

There are three methods of determining tooth clearance on a milling cutter. They are as follows:

1. Dial indicator
2. Brown and Sharpe cutter clearance gage
3. Starrett cutter clearance gage

TO CHECK CUTTER CLEARANCE WITH A DIAL INDICATOR

When a dial indicator is being used, clearance is determined by the movement of the indicator needle from the front to the back of the cutter land (Fig. 82-9). The basic rule used to determine the clearance by this method is as follows.

For a land of $\frac{1}{16}$ in. (or 1.58 mm) width, 1° of clearance is equivalent to 0.001 in. (or 0.02 mm) on the dial indicator. Thus 4° of clearance on a $\frac{1}{16}$-in. (or 1.58 mm) land would register 0.004 in. (or 0.10 mm) on the dial indicator. The cutter diameter does not affect the measurement.

Fig. 82-9   Measuring the clearance angle with a dial indicator.

Fig. 82-10   Checking the clearance angle with a Brown and Sharpe cutter clearance gage.

Fig. 82-11   Checking the clearance angle with a Starrett cutter clearance gage.

TO CHECK CUTTER CLEARANCE WITH A BROWN AND SHARPE CLEARANCE GAGE

When the Brown and Sharpe clearance gage is being used (Fig. 82-10), the inside surfaces of the hardened arms (which are at 90°) are placed on top of two teeth of the cutter. The cutter is revolved sufficiently to bring the face of the tooth in contact with the angle ground on the end of the hardened center blade. The clearance angle of the tooth should correspond with the angle marked on the end of the blade. Two gage blades are furnished with each gage and are stamped at each end with the diameters of the cutters for which they are intended. This cutter clearance gage measures all cutters from ½ to 8 in. (12.7 to 203.2 mm) in diameter, except those with less than eight teeth.

CHECKING THE CUTTER CLEARANCE WITH A STARRETT CUTTER CLEARANCE GAGE

The Starrett gage (Fig. 82-11) may be used to check the clearance on all types of inch cutters from 2 to 30 in. (50.8 to 762 mm) in diameter, and on small cutters and end mills from ½ to 2 in. (12.7 to 50.8 mm), providing the teeth are evenly spaced. This gage consists of a frame graduated from 0 to 30°, a fixed foot, and a beam. An adjustable foot slides along the beam extension. A blade, which may be adjusted angularly and vertically, is used to check the angle of the land on a tooth. When in use, the feet are positioned on two alternate teeth of the cutter with the gage at right angles to the tooth face. The adjustable blade is then lowered onto the top of the middle tooth and adjusted until the angle corresponds to the angle of the land being checked. The land angle is indicated on the protractor on the top of the frame.

CUTTER GRINDING OPERATIONS AND SETUPS

It is most important that milling cutters be ground properly and to the correct clearance angles. Otherwise, the cutter will not cut efficiently and its life will be shortened considerably.

TO GRIND A PLAIN HELICAL MILLING CUTTER

Primary Clearance

1. Mount a parallel-ground test bar between the tailstock centers and check the alignment with an indicator. This will ensure that the table travel is parallel to the grinding edge of the wheel.
2. Remove the test bar.
3. Mount a 4-in. (101.6 mm) flaring-cup wheel (A 60-L 5-V BE) on the grinding head spindle so that the wheel rotates in a counterclockwise direction.
4. Adjust the machine so that the wheel revolves at the proper speed.
5. True the face of the wheel and dress the cutting edge so that it is no more than ¹/₁₆ in. (1.58 mm) wide.
6. Swivel the wheelhead to 89° so that the wheel will touch the cutter on the left side of the wheel only.
7. Using a centering gage, adjust the wheelhead spindle to the height of the tailstock centers (Fig. 82-12). Lock the wheelhead spindle.
8. Mount the cutter on a mandrel and place it temporarily between the footstock centers on the machine table.
9. Set up the tooth rest, on which an offset tooth rest blade (Fig. 82-2C) has been mounted, on the wheelhead

Fig. 82-12   Setting the wheelhead spindle to center height with a centering gage. (Courtesy Cincinnati Milacron Incorporated.)

Fig. 82-13   Centering the bearing point on the tooth rest.

Fig. 82-14   The cutter clearance dial indicates the amount of clearance being ground on the cutter.   (Courtesy Kostel Enterprises Limited.)

housing. Adjust the top of the tooth rest to approximately center height.

10. Move the table until the cutter is near the tooth rest.

11. Adjust the tooth rest between two teeth at the approximate helix angle of the cutter teeth.

12. Chalk or blue the top of the tooth rest blade.

13. Move the cutter over the tooth rest blade and rotate it until the tooth rests on top of the blade.

14. While holding the tooth face against the rest, traverse the table back and forth to mark the point where the tooth bears on the tooth rest blade.

15. Remove the mandrel and cutter from between the centers.

16. Using the centering gage, adjust the tooth rest so that the *center* of the marked bearing point on the tooth rest is at center height and also in the center of the grinding surface of the wheel (Fig. 82-13). The tooth rest blade must be as close to the wheel as possible without touching it.

NOTE: At this point, the center of the grinding spindle, the footstock centers, and the bearing point on the tooth rest blade are in line.

17. Place a dog on the end of the grinding mandrel and mount the work between the tailstock centers.

18. Adjust a cutter tooth onto the top of the tooth rest blade.

19. Set the cutter clearance setting dial (Fig. 82-14) to zero and lock. Adjust the dog into the pin of the cutter clearance gage.

20. Set the wheelhead graduated collar to zero.

21. Loosen the wheelhead lock and the cutter clearance setting dial lock.

22. Holding the cutter tooth on the tooth rest blade, carefully *lower* the wheelhead until the required clearance is shown on the cutter clearance dial. See Table 82-3 for the wheelhead adjustment for the required angle.

When using a flaring-cup wheel, the distance to lower the wheelhead may also be calculated by either of the following methods:

a. Distance = 0.0087 × clearance angle × diameter of cutter.

b. Distance = sine of clearance angle × diameter of cutter ÷ 2

If the cutter is being hollow ground, the distance to lower the wheelhead is

Distance = 0.0087 × clearance angle × diameter of wheel

23. Remove the dog from the end of the mandrel and unlock the table.

24. Adjust the table stops so that the wheel clears the cutter sufficiently at each end to permit indexing for the next tooth.

25. Start the grinding wheel.

26. Carefully feed the cutter in until it just touches the wheel.

27. Standing at the rear of the machine, turn the table traverse knob with the left hand. At the same time, with the right hand, hold the arbor firmly enough to keep the cutter tooth on the tooth rest (Fig. 82-6).

28. Grind one tooth for the full length and return to the starting position, being careful *at all times* to keep the tooth tight against the tooth rest.

29. Traverse the table until the cutter is clear of the tooth rest and rotate the cutter until the diagonally opposite tooth comes in line with the tooth rest blade.

30. Grind this tooth without changing the infeed setting.

31. Check for taper by measuring both ends of the cutter with a micrometer.

32. Remove any taper, if necessary, by loosening the holding nuts on the upper table and adjusting the table.

33. Grind the remaining teeth.

34. Finish-grind all teeth by using a 0.0005-in. (0.01-mm) depth of cut.

35. If the land is over 1/16 in. (1.58 mm) for larger cutters grind the secondary clearance.

To Grind the Secondary Clearance of a Plain Helical Milling Cutter

1. Reset the dog on the mandrel as in step 17 for grinding the primary clearance.

2. Loosen the clearance dial setscrew.

TABLE 82-3 Vertical Wheelhead Adjustment for Cutter Clearance Angles

CUTTER DIAMETER, in in.	CLEARANCE ANGLE AND DISTANCE							
	4°		5°		6°		7°	
	in.	mm	in.	mm	in.	mm	in.	mm
½	0.017	0.45	0.022	0.55	0.026	0.65	0.031	0.80
¾	0.026	0.65	0.033	0.85	0.040	1.00	0.046	1.15
1	0.035	0.90	0.044	1.10	0.053	1.35	0.061	1.55
1¼	0.044	1.10	0.055	1.40	0.066	1.65	0.077	1.95
1½	0.053	1.35	0.066	1.65	0.079	2.00	0.092	2.35
1¾	0.061	1.55	0.076	1.95	0.092	2.35	0.108	2.75
2	0.070	1.75	0.087	2.20	0.105	2.65	0.123	3.10
2½	0.087	2.20	0.109	2.75	0.131	3.30	0.153	3.90
2¾	0.097	2.45	0.120	3.05	0.144	3.65	0.168	4.25
3	0.105	2.65	0.131	3.30	0.158	4.00	0.184	4.65
3½	0.122	3.10	0.153	3.90	0.184	4.65	0.215	5.45
4	0.140	3.55	0.195	4.95	0.210	5.35	0.245	6.20
4½	0.157	4.00	0.197	5.00	0.237	6.00	0.276	7.00
5	0.175	4.45	0.219	5.55	0.263	6.70	0.307	7.80
5½	0.192	4.85	0.241	6.10	0.289	7.35	0.338	8.60
6	0.210	5.35	0.262	6.65	0.315	8.00	0.368	9.35
6½	0.226	5.75	0.283	7.20	0.339	8.60	0.396	10.05
7	0.244	6.20	0.305	7.45	0.365	9.25	0.426	10.80
7½	0.261	6.60	0.326	8.30	0.392	9.95	0.457	11.60
8	0.278	7.05	0.348	8.85	0.418	10.60	0.487	12.35
8½	0.296	7.50	0.370	9.40	0.444	11.25	0.519	13.20
9	0.313	7.95	0.392	9.95	0.470	11.95	0.548	13.90
9½	0.331	8.40	0.413	10.50	0.496	12.60	0.579	14.70
10	0.348	8.85	0.435	11.05	0.522	13.25	0.609	15.45
11	0.383	9.70	0.479	12.15	0.574	14.55	0.670	17.00
12	0.418	10.60	0.522	13.25	0.626	15.90	0.731	18.55
13	0.452	11.50	0.566	14.35	0.679	17.25	0.792	20.10
14	0.487	12.35	0.609	15.45	0.731	18.55	0.853	21.65
15	0.522	13.25	0.653	16.60	0.783	19.90	0.914	23.20
16	0.557	14.15	0.696	17.65	0.835	21.20	0.974	24.75

Note: Inch and metric wheelhead adjustments are approximate equivalents.

3. Hold the cutter tooth against the tooth rest and lower the wheelhead until the required secondary clearance is shown on the clearance setting dial.

4. Lock the dial, remove the dog, and proceed to grind the secondary clearance in the same manner as for primary clearance.

5. Grind the secondary clearance until the land is the required width.

TO GRIND A STAGGERED TOOTH CUTTER

To grind the primary clearance on the periphery of a staggered-tooth cutter (Fig. 82-15) proceed as follows.

1. Carry out steps 1 to 7 as for grinding primary clearance on a plain helical milling center.

2. Mount a staggered-tooth cutter tooth rest blade (Fig. 82-2E) in the holder and mount the unit on the wheelhead.

3. Place the high point of the inverted V *exactly* in the

Fig. 82-15 Setup for clearance grinding a staggered-tooth cutter. (Courtesy Kostel Enterprises Limited.)

center of the width of the grinding wheel cutting face and at center height.

4. Place the centering gage on the table and adjust the wheelhead height until the highest point of the tooth rest blade is at center height.

5. Mount the cutter between centers with the dog loosely on the mandrel and adjust the table until one cutter tooth rests on the blade. Lock the table in position.

6. Set the cutter clearance dial (Fig. 82-14) to zero and tighten the dog on the mandrel.

7. Loosen the cutter clearance dial lock and the wheelhead lock.

8. Lightly holding the cutter tooth onto the tooth rest blade, lower the wheelhead until the required clearance shows on the clearance setting dial.

9. Remove the clearance setting dog and unlock the table.

10. Set the stop dogs so that the wheel clears both sides of the cutter enough to allow indexing for the next tooth.

11. Start the grinding wheel.

12. Adjust the saddle until the cutter just touches the grinding wheel.

13. Grind one tooth and move the cutter clear of the tooth rest.

14. Rotate the next tooth, which is offset in the opposite direction, onto the tooth rest and grind it on the return stroke.

15. After grinding two teeth, check them with a dial indicator to see that they are the same height. If not, adjust the blade slightly toward the high side and grind the next two teeth. Repeat the process until the teeth are within 0.0003 in. (0.007 mm).

Secondary Clearance

Because it is necessary to provide adequate chip clearance when milling deep slots, a secondary clearance of 20 to 25° on staggered-tooth cutters is recommended. It is also suggested that enough secondary clearance be ground to reduce the width of the land to approximately ⅟₃₂ in. (0.79 mm). This will permit regrinding of the primary clearance at least once without the need for grinding the secondary clearance.

To Grind the Secondary Clearance on a Staggered-Tooth Cutter

1. Remove the tooth rest from the wheelhead and mount it on the table between the tailstocks. A universal micrometer flicker-type tooth rest and a straight blade (Fig. 82-16) should be used to permit the cutter to be rotated.

2. Place the centering gage on the table and bring the *center* of one tooth to center height. Mark this tooth with layout dye or chalk.

3. Locate the dog on the clearance setting dial pin and tighten it on the mandrel.

4. Rotate the cutter to the desired amount of clearance using the clearance setting dial.

5. Adjust the tooth rest under, or on the side of, the marked tooth.

6. Swivel the table sufficiently to the right or the left

Fig. 82-16 A flicker-type tooth rest is used when grinding the clearance on the sides of the teeth. (Courtesy Kostel Enterprises Limited.)

(depending on the helix angle of the tooth being ground) to grind a straight land.

7. Grind the secondary clearance on this tooth until the land is ⅟₃₂ in. (0.79 mm) wide.

8. Grind all remaining teeth having the same slope or helix.

9. Swivel the table in the opposite direction and follow steps 6, 7, and 8 to set up and grind the remaining teeth.

Side Clearance

The side of the teeth of any milling cutter should not be ground unless absolutely necessary, since this reduces the width of the cutter. If the teeth must be ground, proceed as follows.

1. Mount the cutter on a stub arbor in the workhead (Fig. 82-16).

2. Mount a flaring-cup wheel.

3. Tilt the workhead to the desired primary clearance angle. This is generally 2 to 4°. The secondary clearance is about 12°.

4. Place the centering gage on the wheelhead and adjust one tooth of the cutter until it is on center and level. Clamp the workhead spindle.

5. Mount the tooth rest on the workhead using a flicker-type rest and a plain blade.

6. Raise or lower the wheelhead so that the grinding wheel only contacts the tooth resting on the blade.

7. Grind the primary clearance on all teeth.

8. Tilt the workhead to the required angle for the secondary clearance and grind all teeth.

TO GRIND A FORM-RELIEVED CUTTER

Unlike other types of milling cutters, form-relieved cutters are ground on the face of the teeth rather than on the periphery; otherwise the form of the cutter will be changed when it is sharpened.

When grinding formed cutters for the first time, grind the backs of the teeth before grinding the cutting face, to

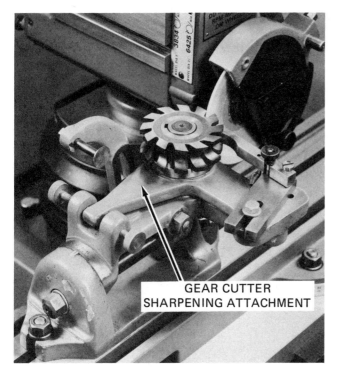

Fig. 82-17 Grinder setup to sharpen a form-relieved cutter

Fig. 82-18 The back of the tooth is set parallel to the wheel and held in place by the pawl. (Courtesy Cincinnati Milacron Incorporated.)

ensure that all teeth are the same thickness. This is necessary because the locating pawl on the grinding fixture bears against *the back of the tooth* when the cutter is being ground.

PROCEDURE

1. Swing the wheelhead so that the spindle is at 90° to the table travel.
2. Mount a dish wheel and the proper wheel guard.
3. Mount a gear cutter sharpening attachment on the table to the left of the grinding wheel (Fig. 82-17).
4. Place the gear cutter on the stud of the attachment so that the back of each tooth may be ground.

NOTE: This operation is only necessary when the cutter is being sharpened for the first time.

5. Place the centering gage on the wheelhead and adjust the wheelhead until the center of the tooth face is on center.
6. Move the table in until the back edge of a tooth is near the grinding wheel. At the same time, rotate the cutter until the back of the tooth is parallel with the face of the wheel (Fig. 82-18).
7. Engage the edge of the pawl on the face of the tooth and clamp the pawl in place (Fig. 82-18).
8. Grind the back of this tooth.
9. Move the table to the left so that the cutter is clear of the grinding wheel.
10. Index the cutter so that the pawl will bear against the next tooth. Hold the tooth face against the pawl when grinding.
11. Grind the backs of all teeth.

12. Reverse the cutter on the stud and adjust the pawl against the back of the tooth, after the face of the tooth has been brought to bear against the centering gage fastened to the attachment. Swing the centering gage out of the way.
13. Adjust the saddle to bring the face of one tooth in line with the grinding wheel. After this time, adjust the saddle only to compensate for wheel wear.
14. Loosen one setscrew and tighten the other to rotate the cutter against the grinding wheel.
15. Grind one tooth, traverse the table and index for the next tooth.
16. Grind all tooth faces.

CYLINDRICAL GRINDING

With the aid of a motorized workhead, the cutter and tool grinder may be used for cylindrical and plunge grinding. Work may be ground between centers or held in a chuck, depending on the type of work.

To Grind Work Parallel Between Centers

1. Mount the motorized workhead on the left end of the table (Fig. 82-19).
2. Examine the centers of the machine and the work to see that they are in good condition.
3. Using the centering gage on the wheelhead, adjust the wheelhead to the tailstock center height.
4. Mount a 6-in. (152.4 mm) straight grinding wheel on the wheelhead spindle so that the wheel rotates downward at the front of the wheel. This will deflect the sparks downward.
5. Mount a parallel, hardened and ground test bar between centers.
6. Using a dial indicator, align the centers for height and

Fig. 82-19 Setup for cylindrical grinding on a cutter and tool grinder.

Fig. 82-20 Setup for internal grinding on a cutter and tool grinder. (Courtesy Cincinnati Milacron Incorporated.)

then align the side of the bar parallel with the table travel. Remove the bar and indicator.

7. Mount the work between centers.

8. Set the stop dogs so that the wheel overlaps the work by one-third the width of the wheelface at each end.

9. Start the grinding wheel and workhead. The workpiece should revolve in the opposite direction to that of the grinding wheel.

10. Bring the revolving work up until it just touches the grinding wheel.

11. Traverse the table slowly and clean up the workpiece. The traverse speed should be such that the work travels approximately one-quarter the width of the wheel for each revolution of the work.

12. Measure each end of the workpiece for size and taper. If a taper exists, adjust as required.

13. After the work is parallel, set the crossfeed graduated collar to zero.

14. Feed the work into the grinding wheel approximately 0.001 in. (0.02 mm) per pass until the work is within 0.001 in. (0.02 mm) of finished size. Use 0.0002-in. (0.005-mm) cuts for finishing.

15. Feed in the work until the graduated collar indicates it is to the proper size.

NOTE: Since work expands during grinding, it should never be measured for accurate size when warm.

16. Traverse the table several times to permit the wheel to spark out.

The same procedure is followed for taper grinding, except that the table must be swung to half the angle of the taper. After the work is cleaned up, the taper should be carefully checked for size and accuracy, and adjustments made as required. When taper grinding, it is most important that the center height of the wheel and the workpiece be in line.

INTERNAL GRINDING

Light internal grinding may be performed on the tool and cutter grinder by mounting the internal-grinding attachment on the wheelhead. The workpiece is held in a chuck mounted on the motorized workhead (Fig. 82-20).

1. Mount a test bar in the workhead spindle and align it both vertically and horizontally. When grinding a tapered hole, the workhead spindle must be aligned vertically and then swung to half the included angle of the taper.

2. Mount the internal-grinding attachment on the workhead.

3. Center the grinding wheel spindle using the centering gage.

4. Mount the proper grinding wheel on the spindle.

5. Mount a chuck on the motorized workhead.

6. Mount the work in the chuck. Care must be taken not to distort the workpiece by gripping it too tightly.

7. Set the rotation of the workhead in an opposite direction to that of the grinding spindle.

8. Start the grinding wheel and the workpiece.

9. Carefully bring the wheel into the hole of the workpiece.

10. Set the table travel so that only one-third of the wheel overlaps the hole at each end.

11. Clean up the inside of the hole and check for size, parallelism, and bell-mouthing. Correct as required.

12. Set the crossfeed graduated collar to zero and determine the amount of material to be removed.

13. Feed the grinding wheel in about 0.0005 in. (0.01 mm) per pass.

14. When work is close to the finished size, let the wheel spark out to improve the finish and remove the spring from the spindle.

15. Finish-grind the hole to size.

REVIEW QUESTIONS

PARTS OF THE UNIVERSAL CUTTER AND TOOL GRINDER

1. Name five main parts of the universal cutter and tool grinder.

2. How many controls are there for each of the following parts of the universal cutter and tool grinder?

(a) wheelhead
(b) saddle
(c) table

3. Name five accessories used with the universal cutter and tool grinder and state the purpose of each.

4. Sketch and name five types of tooth rest blades and indicate the purpose for which each is used.

CUTTER GRINDING MANDRELS AND ARBORS

5. Make a suitable sketch of a cutter grinding mandrel and a cutter grinding arbor.

6. How does a cutter grinding mandrel differ from a lathe mandrel?

MILLING CUTTER NOMENCLATURE

7. Make a suitable sketch of at least two teeth on a milling cutter and indicate the following parts:
(a) primary clearance angle
(b) secondary clearance angle
(c) cutting angle
(d) land
(e) tooth angle
(f) tooth face

8. Why are cutters with widely spaced teeth more efficient than those with closely spaced teeth?

9. What factors influence the cutter clearance angle?

METHODS OF GRINDING CLEARANCE ON MILLING CUTTERS

10. Name and describe briefly three methods of grinding clearance on cutting tools.

CHECKING CUTTER CLEARANCE WITH A DIAL INDICATOR

11. Describe the principle by which cutter clearance is checked using a dial indicator.

12. Name and describe two other methods of checking cutter clearance angles and state the limitations of each method.

CUTTER GRINDING OPERATIONS AND SETUPS

13. Describe how the tooth edge of a milling cutter is placed on center.

14. Describe how the correct cutter clearance is set on the machine using the clearance setting dial.

15. Explain two methods to determine how far to lower the wheelhead, for the proper cutter clearance angle.

16. How far would the wheelhead be lowered to grind the proper clearance on the following cutters?
(a) 5° clearance angle on a 3-in.-diameter cutter using a 6-in. straight wheel.
(b) 5° clearance angle on a 75-mm-diameter cutter using a flaring-cup wheel

17. List four precautions to be taken when grinding a helical milling cutter.

18. Where is the tooth rest mounted when grinding cutters having:
(a) helical teeth?
(b) straight teeth?

TO GRIND A STAGGERED-TOOTH CUTTER

19. How does the grinding of a staggered-tooth milling cutter differ from the grinding of a plain helical milling cutter?

TO GRIND A FORM-RELIEVED CUTTER

20. On what surface are form-relieved cutters ground when being sharpened? Explain why.

21. Why is it advisable to grind the backs of the teeth on a new gear cutter?

22. What type of grinding wheel is used for sharpening gear cutters?

CYLINDRICAL GRINDING

23. List the steps required to *set up* the cutter and tool grinder for cylindrical grinding.

24. Why is it advisable to permit the grinding wheel to spark out when cylindrical grinding?

25. How should the stop dogs be set when cylindrical grinding?

26. How fast should the table be traversed when cylindrical grinding on a cutter and tool grinder?

27. What precautions must be taken when setting up the machine for taper grinding?

INTERNAL GRINDING

28. What precautions should be taken when setting up the work for internal grinding?

29. List briefly the steps required to grind the inside diameter of a bushing using a cutter and tool grinder.

CUTTING-TOOL MATERIALS

SECTION 17

Single-point cutting tools have always been widely used for metal removal. There has been a constant search to find new cutting-tool materials to increase production. Over the past 100 years, high-carbon steel, high-speed steel, cemented carbide, ceramic, and diamond cutting tools have been developed. During this time, the rate of cutting metal has increased from 10 to 15 surface feet per minute (sf/min) (3.04 to 4.57 m/min) to as high as 1000 sf/min (3048 m/min) in some cases. Cutting speeds of 300 to 1000 sf/min (91 to 305 m/min) are quite common in industry now. Recently a ceramic cutting tool was introduced which can run at 4000 sf/min (1219 m/min), and can take a depth of cut of 0.400 in. (10.16 mm) at a feed rate up to 0.025 in. (0.64 mm). Better designs in cutting tools and machines will lead to definitely higher production in industry.

CARBIDE AND BORAZON
CUTTING TOOLS
UNIT 83

Carbide was first used for cutting tools in Germany during World War I as a substitute for badly needed diamonds. During the 1930s, various additives were discovered which generally improved the quality and performance of carbide tools. Since that time, various types of cemented (sintered) carbides have been developed to suit different materials and machining operations. Cemented carbides are similar to steel in appearance, but they are so hard that the diamond is almost the only material which will scratch them. They have been accepted by industry because they have good wear resistance and can operate efficiently at cutting speeds ranging from 150 to 1200 sf/min (46 to 366 m/min). Carbide tools can machine metals at speeds that cause the cutting edge to become red hot without losing its hardness or sharpness.

Borazon (cubic boron nitride), a hard abrasive material used as a coating on a cemented-carbide base, has led to increased cutting speeds and more workpieces per cutting edge. This, along with the fact that Borazon is capable of withstanding very high machining temperatures, makes it a welcome addition to the cutting tool family.

OBJECTIVES

After completing this unit you will be able to:
1. Identify and state the purpose of the two main types of carbide grades
2. Select the proper grade of carbide for various workpiece materials
3. Select the proper speeds and feeds for carbide tools

MANUFACTURE OF CEMENTED CARBIDES

Cemented carbides are products of the powder-metallurgy process; they consist primarily of minute particles of tungsten and carbon powders cemented together under heat by a metal of lower melting point, usually cobalt. Powdered metals such as tantalum, titanium, and niobium are also used in the manufacture of cemented carbides to provide cutting tools with various characteristics. The entire operation of producing cemented-carbide products is illustrated in Fig. 83-1.

BLENDING

Five types of powders are used in the manufacture of cemented-carbide tools: tungsten carbide, titanium carbide, tantalum carbide, niobium carbide, and cobalt. One or any combination of these carbide powders and cobalt (the binder) are blended together in different proportions, de-

pending on the grade of carbide desired. This powder is mixed in alcohol; the mixing process takes anywhere from 24 to 190 h. After the powder and alcohol have been thoroughly mixed, the alcohol is drained, and paraffin is added to simplify the pressing operation.

COMPACTION

After the powders have been thoroughly mixed, they must be molded to shape and size. Five different methods may be used to compact the powder to shape (Fig. 83-2): the extrusion process, the hot press, the isostatic press, the ingot press, or the pill press. The green (pressed) compacts are soft and must be presintered to dissolve the paraffin and lightly bond the particles together so that they may be handled easily.

PRESINTERING

The green compacts are heated to about 1500°F (815°C) in a furnace under a protective atmosphere of hydrogen. After

Fig. 83-1 The process of manufacturing cemented-carbide products. (Courtesy General Electric Co. Ltd.)

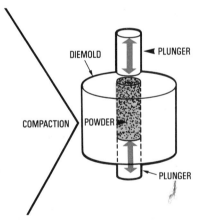

Fig. 83-2 Compacting carbide powders to shape and size. (Courtesy General Electric Co. Ltd.)

this operation, carbide blanks have the consistency of chalk and may be machined to the required shape and approximately 40 percent oversize to allow for the shrinkage which occurs during final sintering.

SINTERING

Sintering, the last step in the process, converts the presintered machined blanks into cemented carbide. Sintering is carried out in either a hydrogen atmosphere or a vacuum, depending on the grade of carbide manufactured, at temperatures between 2550 and 2730°F (1400 and 1500°C). During the sintering operation, the binder (cobalt) unites and cements the carbide powders into a dense structure of extremely hard carbide crystals.

CEMENTED-CARBIDE APPLICATIONS

Because of the extreme hardness and good wear-resistance properties of cemented carbide, it has been used extensively in the manufacture of metal-cutting tools. Drills, reamers, milling cutters, and shaper and lathe cutting tools are only a few examples of uses for cemented carbides. These tools may have cemented-carbide tips, either brazed or held mechanically, on the cutting edge.

Cemented carbides were first used successfully in machining operations as lathe cutting tools. The majority of cemented-carbide tools in use are single-point cutting tools used on machines such as lathes, shapers, and milling machines.

TYPES OF CARBIDE LATHE CUTTING TOOLS

Cemented-carbide lathe cutting tools are available in two types: the brazed-tip type and the throwaway insert type.

Brazed-Tip Carbide Tools Cemented-carbide tips can be brazed to steel shanks and are available in a wide variety

Fig. 83-3   A variety of brazed carbide tools.   (Courtesy General Electric Co. Ltd.)

Fig. 83-4   A variety of indexible throw away cemented-carbide inserts.   (Courtesy General Electric Co. Ltd.)

Fig. 83-5   A toolholder for indexible throw-away cemented-carbide inserts.   (Courtesy General Electric Co. Ltd.)

of styles and sizes (Fig. 83-3). Brazed-tip carbide tools are rigid and are generally used for production turning purposes.

Throwaway Inserts   Cemented-carbide indexible throwaway inserts (Fig. 83-4) are made in a wide variety of shapes, such as triangular, square, diamond, round, etc. These inserts are held mechanically in a special holder (Fig. 83-5), and when one cutting edge becomes dull, it may be quickly indexed or turned in the holder and a new cutting edge will be presented. A triangular insert has three cutting edges on the top surface and three on the bottom, for a total of six cutting edges. When all six cutting edges are dull, the carbide insert is discarded and replaced with a new insert.

Cemented-carbide throwaway inserts are becoming more popular than the brazed-tip carbide tools because:

1. Less time is required to change to a new cutting edge.
2. The amount of machine downtime is reduced considerably and production is thus increased.
3. The time normally spent in regrinding a tool is eliminated.

4. Faster speeds and feeds can be used with throwaway inserts.
5. The cost of diamond wheels, required for grinding carbide tools, is eliminated.
6. Throwaway inserts are cheaper than brazed-tip tools.

CEMENTED-CARBIDE INSERT IDENTIFICATION

The American Standards Association has developed a system by which throwaway inserts can be identified quickly and accurately. This system has been generally adopted by manufacturers of cemented-carbide inserts (Table 83-1).

GRADES OF CEMENTED CARBIDES

There are two main groups of carbides from which various grades can be selected: the straight tungsten carbide grades and the crater-resistant grades containing titanium and/or tantalum carbide.

The straight tungsten carbide grades, containing only tungsten carbide and cobalt, are the strongest and most wear resistant. Generally they are used for machining cast iron and nonmetals. Tungsten carbide grades are not very satisfactory for steel because of their tendency to crater and rapid tool failure.

The size of the tungsten carbide particles and the percentage of cobalt used determine the qualities of tungsten carbide tools.

1. the finer the grain particles, the lower the tool toughness
2. the finer the grain particles, the higher the tool hardness
3. the higher the hardness, the greater the wear resistance
4. the lower the cobalt content, the lower the tool toughness

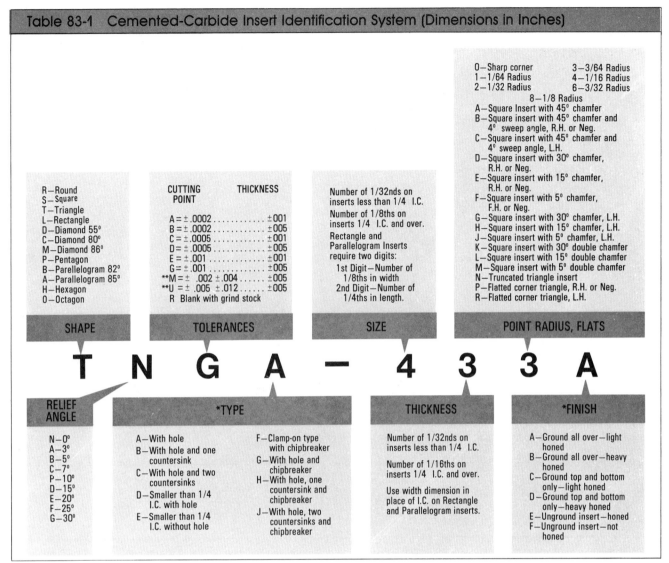

SHAPE	TOLERANCES	SIZE	POINT RADIUS, FLATS

	CUTTING POINT	THICKNESS			

SHAPE

R—Round
S—Square
T—Triangle
L—Rectangle
D—Diamond 55°
C—Diamond 80°
M—Diamond 86°
P—Pentagon
B—Parellelogram 82°
A—Parallelogram 85°
H—Hexagon
O—Octagon

TOLERANCES

CUTTING POINT .... THICKNESS

A = ±.0002 .... ±001
B = ±.0002 .... ±005
C = ±.0005 .... ±001
D = ±.0005 .... ±005
E = ±.001 .... ±001
G = ±.001 .... ±005
**M = ± .002 ±.004 ...... ±005
**U = ± .005 ±.012 ...... ±005
R Blank with grind stock

SIZE

Number of 1/32nds on inserts less than 1/4 I.C.
Number of 1/8ths on inserts 1/4 I.C. and over.
Rectangle and Parallelogram Inserts require two digits:
1st Digit—Number of 1/8ths in width
2nd Digit—Number of 1/4ths in length.

POINT RADIUS, FLATS

0—Sharp corner
1—1/64 Radius
2—1/32 Radius
3—3/64 Radius
4—1/16 Radius
6—3/32 Radius
8—1/8 Radius
A—Square Insert with 45° chamfer
B—Square insert with 45° chamfer and 4° sweep angle, R.H. or Neg.
C—Square insert with 45° chamfer and 4° sweep angle, L.H.
D—Square insert with 30° chamfer, R.H. or Neg.
E—Square insert with 15° chamfer, R.H. or Neg.
F—Square insert with 5° chamfer, F.H. or Neg.
G—Square insert with 30° chamfer, L.H.
H—Square insert with 15° chamfer, L.H.
J—Square insert with 5° chamfer, L.H.
K—Square insert with 30° double chamfer
L—Square insert with 15° double chamfer
M—Square insert with 5° double chamfer
N—Truncated triangle insert
P—Flatted corner triangle, R.H. or Neg.
R—Flatted corner triangle, L.H.

T N G A — 4 3 3 A

RELIEF ANGLE	*TYPE	THICKNESS	*FINISH

RELIEF ANGLE

N—0°
A—3°
B—5°
C—7°
P—10°
D—15°
E—20°
F—25°
G—30°

*TYPE

A—With hole
B—With hole and one countersink
C—With hole and two countersinks
D—Smaller than 1/4 I.C. with hole
E—Smaller than 1/4 I.C. without hole
F—Clamp-on type with chipbreaker
G—With hole and chipbreaker
H—With hole, one countersink and chipbreaker
J—With hole, two countersinks and chipbreaker

THICKNESS

Number of 1/32nds on inserts less than 1/4 I.C.
Number of 1/16ths on inserts 1/4 I.C. and over.
Use width dimension in place of I.C. on Rectangle and Parallelogram inserts.

*FINISH

A—Ground all over—light honed
B—Ground all over—heavy honed
C—Ground top and bottom only—light honed
D—Ground top and bottom only—heavy honed
E—Unground insert—honed
F—Unground insert—not honed

* SHALL BE USED ONLY WHEN REQUIRED.
** EXACT TOLERANCE IS DETERMINED BY SIZE OF INSERT.

Source: General Electric Co. Ltd.

---

5. the lower the cobalt content, the higher the hardness

For maximum tool life, always select a tungsten carbide grade with the lowest cobalt content and the finest grain size possible which gives satisfactory performance without breakage.

Crater-resistant grades contain titanium carbide and tantalum carbide in addition to the basic components of tungsten carbide and cobalt. These grades are used for machining most steels.

The additions of tantalum carbide and/or titanium carbide provide tools with various characteristics.

1. The addition of titanium carbide provides resistance to tool cratering. The higher the titanium content, the greater the resistance to cratering.
2. As the titanium carbide content is increased, the toughness of the tool is decreased.
3. As the titanium carbide content is increased, the abrasive wear resistance at the cutting edge is lowered.

4. Tantalum carbide additions have effects similar to tungsten carbide on the resistance to cratering and strength.
5. Tantalum carbide gives good crater resistance without affecting the abrasive wear resistance.
6. The addition of tantalum carbide increases the tool's resistance to deformation.

Because of the wide variety of cemented-carbide compositions available and the rapid development of new, improved types, no standard grade classification system has been developed. It is wise to follow the manufacturer's recommendations as to the type and grade of carbide for each specific application. A few general rules will assist in the selection of the proper cemented-carbide grade.

1. Always use a grade with the lowest cobalt content and the finest grain size (strong enough to eliminate breakage).
2. To combat abrasive wear only, use straight tungsten carbide grades.
3. To combat cratering, seizing, welding, and galling, use

titanium carbide grades.

4. For crater and abrasive wear resistance, use tantalum carbide grades.

5. For heavy cuts in steel, when heat and pressure might deform the cutting edge, use tantalum carbide grades.

COATED CARBIDE INSERTS

A relatively recent development in the search for better tooling has been the coating of carbide cutting tools with titanium nitride. Coated carbide inserts have proved to give longer tool life, greater productivity, and freer-flowing chips. The coating acts as a permanent lubricant greatly reducing the cutting forces, heat generation, and wear. This permits higher speeds to be used during the machining process, particularly when a good surface finish is required. The lubricity and antiweld characteristics of the coating greatly reduce the amount of heat and stress generated when taking the cut. The coating also greatly improves the tool life.

TOOL GEOMETRY

The geometry of cutting tools refers to the various angles and clearances machined or ground on the tool faces. Although the terms and definitions relating to single-point cutting tools vary greatly, the ones adopted by the American Society of Mechanical Engineers (ASME) and currently in general use are illustrated in Fig. 83-6.

CUTTING-TOOL TERMS

Front, or End, Relief (Clearance) (Fig. 83-6) The front relief, or clearance, allows the end of the cutting tool to enter the work. This clearance should be just enough to prevent the tool from rubbing. Too much front clearance will reduce the support under the point and cause rapid tool failure.

Side Relief (Clearance) (Fig. 83-6) This angle permits the side of the tool to advance into the work. Too little side relief, or clearance, will prevent the tool from cutting, and excessive heat will be generated by the rubbing action. Too much side clearance will weaken the cutting edge and cause it to chip.

Side Cutting Edge Angle The angle of the side of the cutting edge that meets the work may be either positive or negative. A negative side cutting angle (Fig. 83-6) is preferred because it protects the point of the tool at both the start and the end of a cut; this is especially useful on work that has a hard abrasive scale.

Nose Radius The nose radius strengthens the finishing point of the tool and improves the surface finish on the work. The nose radius of most cutting tools should be approximately twice the amount of feed per revolution. Too large a nose radius may cause chatter, while too small a radius weakens the point of the cutting tool.

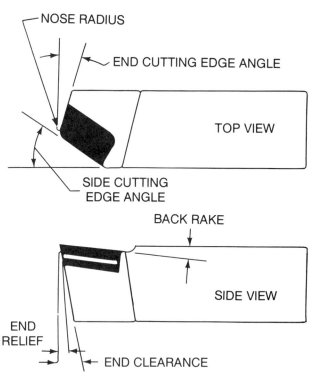

Fig. 83-6 Single-point tool nomenclature.

To obtain maximum efficiency with carbide tools, keep the nose radius as small as possible. Use Table 83-2 to find the proper nose radius for the depth of cut and feed being used.

Side Rake The side rake angle should be as large as possible, without weakening the cutting edge, to allow the chips to escape readily (Fig. 83-7A). The amount of side rake will be determined by the type and grade of the cutting tool, the type of material being cut, and the feed per revolution. The included angle formed by the side rake and side clearance is called the angle of keenness. This angle will vary, depending on the material being cut. For difficult-to-machine metals, it may be advisable to use a small side rake angle or at times even negative side rake (Fig. 83-7B). Figure 83-8 shows suggested side rakes for a variety of materials.

Back Rake The back rake angle is the angle formed between the top face of the tool and the top of the tool

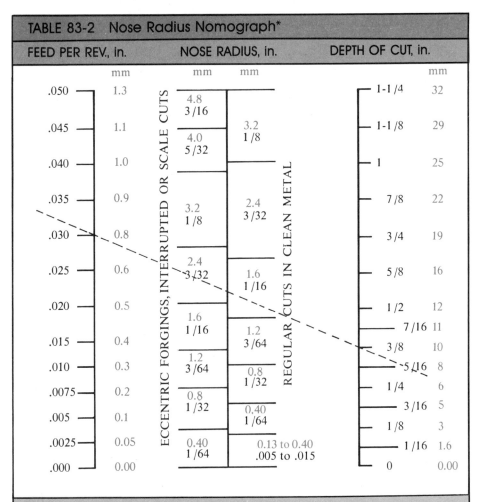

TABLE 83-2   Nose Radius Nomograph*

FEED PER REV., in.		NOSE RADIUS, in.		DEPTH OF CUT, in.	
	mm	mm	mm		mm
.050	1.3			1-1/4	32
		4.8 3/16			
.045	1.1		3.2 1/8	1-1/8	29
		4.0 5/32			
.040	1.0			1	25
.035	0.9	3.2 1/8	2.4 3/32	7/8	22
.030	0.8			3/4	19
.025	0.6	2.4 3/32	1.6 1/16	5/8	16
.020	0.5			1/2	12
		1.6 1/16	1.2 3/64	7/16	11
.015	0.4			3/8	10
.010	0.3	1.2 3/64	0.8 1/32	5/16	8
.0075	0.2	0.8 1/32		1/4	6
.005	0.1		0.40 1/64	3/16	5
				1/8	3
.0025	0.05	0.40 1/64	0.13 to 0.40 .005 to .015	1/16	1.6
.000	0.00			0	0.00

(Vertical labels between columns: "ECCENTRIC FORGINGS, INTERRUPTED OR SCALE CUTS" and "REGULAR CUTS IN CLEAN METAL")

* To obtain maximum efficiency with carbide tools, the nose radius should be kept small. Use chart as a guide. Use straightedge to join feed with depth of cut. Use nose radius where line crosses.
Source: General Electric Co. Ltd.

Fig. 83-7   (A) Positive rake side angle on a cutting tool; (B) negative rake side angle on a cutting tool.   (Courtesy A.C. Wickman Limited.)

shank. It may be positive, negative, or neutral. When a tool has *negative back rake,* the top face of the tool slopes upward away from the point (Fig. 83-9A). Negative back rake protects the tool point from the cutting pressure and also from the abrasive action of hard materials and scale. When a tool has *positive back rake* (Fig. 83-9B), the top face of the tool slopes downward away from the point. This allows the chips to flow away freely from the cutting edge.

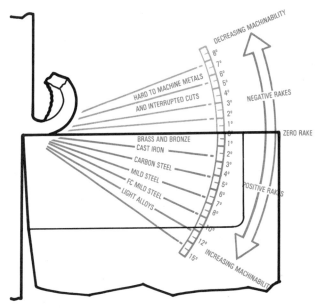

Fig. 83-8   Recommended side rake angles for various workpiece materials.   (Courtesy A.C. Wickman Limited.)

Fig. 83-9 (A) Negative back rake; (B) positive back rake.

CEMENTED-CARBIDE CUTTING-TOOL ANGLES AND CLEARANCES

The angles and clearances of single-point carbide tools vary greatly and generally depend on three factors.

1. The hardness of the cutting tool
2. The workpiece material
3. The type of cutting operation

Table 83-3 lists recommended cutting-tool angles and clearances for a variety of materials. These may have to be altered slightly to suit various conditions encountered while machining.

CUTTING SPEEDS AND FEEDS

Many variables influence the speeds, feeds, and depth of cut which should be used with cemented-carbide tools. Some of the most important factors are:

1. The type and hardness of the work material
2. The grade and shape of the cutting tool
3. The rigidity of the cutting tool
4. The rigidity of the work and machine
5. The power rating of the machine

Table 83-4 gives the recommended cutting speeds and feeds for single-point carbide tools. These should be used as a guide and may have to be altered slightly to suit the machining operation.

Table 83-5 illustrates the nomograph method of determining the cutting speed in feet or meters per minute when the Brinell hardness of the steel is known.

MACHINING WITH CARBIDE TOOLS

To obtain maximum efficiency with cemented-carbide cutting tools, certain precautions in machine setup and the cutting operation should be observed. The machine being used should be rigid, equipped with heat-treated gears, and have sufficient power to maintain a constant cutting speed. The work and cutting tool should be held as rigidly as possible to avoid chatter or keep it to a minimum.

Single-point carbide cutting tools are more commonly used on a lathe, and therefore the setups and precautions for this machine will be outlined. The same basic precautions and setups should be applied when carbide tools are used on other machines.

SUGGESTIONS FOR USING CEMENTED-CARBIDE CUTTING TOOLS

WORK SETUP

1. Work mounted in a chuck or other work-holding device must be held firmly enough to prevent it from slipping or chattering.
2. A revolving center should be used in the tailstock for turning work between centers.
3. The tailstock spindle should be extended only a minimum distance and locked securely to ensure rigidity.
4. The tailstock should be clamped firmly to the lathe bed to prevent loosening.

TOOL GEOMETRY

1. Use a cutting tool with the proper rake and clearances for the material being cut.
2. Hone the cutting edge for good performance and longer tool life.
3. If the work permits, use a side cutting edge angle (Fig. 83-10A) large enough that the tool can be eased into the work. This helps to protect the nose of the cutting tool (weakest part) from shock and wear as it enters or leaves the work.
4. Use the largest nose radius (Fig. 83-10B) which oper-

TABLE 83-3 Recommended Angles for Single-Point Carbide Tools				
MATERIAL	END RELIEF (FRONT CLEARANCE)	SIDE RELIEF (SIDE CLEARANCE)	SIDE RAKE	BACK RAKE
Aluminum	6 to 10°	6 to 10°	10 to 20°	0 to 10°
Brass, bronze	6 to 8°	6 to 8°	+8 to −5°	0 to −5°
Cast iron	5 to 8°	5 to 8°	+6 to −7°	0 to −7°
Machine steel	5 to 10°	5 to 10°	+6 to −7°	0 to −7°
Tool steel	5 to 8°	5 to 8°	+6 to −7°	0 to −7°
Stainless steel	5 to 8°	5 to 8°	+6 to −7°	0 to −7°
Titanium alloys	5 to 8°	5 to 8°	+6 to −5°	0 to −5°

Note: Use the lower range of these figures for hard-to-machine metals and interrupted cuts.

SIDE CUTTING EDGE ANGLE

A

WEAK   STRONG

B

Fig. 83-10 (A) A side cutting edge angle protects the point of the cutting tool. (B) A large nose radius strengthens the tool point and produces a better surface finish. (Courtesy General Electric Co. Ltd.)

TABLE 83-4	Recommended Cutting Speeds and Feeds for Single-Point Carbide Tools					
	DEPTH OF CUT		FEED PER REVOLUTION		CUTTING SPEED	
MATERIAL	in.	mm	in.	mm	ft/min	m/min
Aluminum	0.005–0.015	0.15–0.40	0.002–0.005	0.05–0.15	700–1000	215–305
	0.020–0.090	0.50–2.30	0.005–0.015	0.15–0.40	450–700	135–215
	0.100–0.200	2.55–5.10	0.015–0.030	0.40–0.75	300–450	90–135
	0.300–0.700	7.60–17.80	0.030–0.090	0.75–2.30	100–200	30–60
Brass, bronze	0.005–0.015	0.15–0.40	0.002–0.005	0.05–0.15	700–800	215–245
	0.020–0.090	0.50–2.30	0.005–0.015	0.15–0.40	600–700	185–215
	0.100–0.200	2.55–5.10	0.015–0.030	0.40–0.75	500–600	150–185
	0.300–0.700	7.60–17.80	0.030–0.090	0.75–2.30	200–400	60–120
Cast iron (medium)	0.005–0.015	0.15–0.40	0.002–0.005	0.05–0.15	350–450	105–135
	0.020–0.090	0.50–2.30	0.005–0.015	0.15–0.40	250–350	75–105
	0.100–0.200	2.55–5.10	0.015–0.030	0.40–0.75	200–250	60–75
	0.300–0.700	7.60–17.80	0.030–0.090	0.75–2.30	75–150	25–45
Machine steel	0.005–0.015	0.15–0.40	0.002–0.005	0.05–0.15	700–1000	215–305
	0.020–0.090	0.50–2.30	0.005–0.015	0.15–0.40	550–700	170–215
	0.100–0.200	2.55–5.10	0.015–0.030	0.40–0.75	400–550	120–170
	0.300–0.700	7.60–17.80	0.030–0.090	0.75–2.30	150–300	45–90
Tool steel	0.005–0.015	0.15–0.40	0.002–0.005	0.05–0.15	500–750	150–230
	0.020–0.090	0.50–2.30	0.005–0.015	0.15–0.40	400–500	120–150
	0.100–0.200	2.55–5.10	0.015–0.030	0.40–0.75	300–400	90–120
	0.300–0.700	7.60–17.80	0.030–0.090	0.75–2.30	100–300	30–90
Stainless steel	0.005–0.015	0.15–0.40	0.002–0.005	0.05–0.15	375–500	115–150
	0.020–0.090	0.50–†2.30	0.005–0.015	0.15–0.40	300–375	90–115
	0.100–0.200	2.55–5.10	0.015–0.030	0.40–0.75	250–300	75–90
	0.300–0.700	7.60–17.80	0.030–0.090	0.75–2.30	75–175	25–55
Titanium alloys	0.005–0.015	0.15–0.40	0.002–0.005	0.05–0.15	300–400	90–120
	0.020–0.090	0.50–2.30	0.005–0.015	0.15–0.40	200–300	60–90
	0.100–0.200	2.55–5.10	0.015–0.030	0.40–0.75	175–200	55–60
	0.300–0.700	7.60–17.80	0.030–0.090	0.75–2.30	50–125	15–40

Note: Millimeter and inch speeds are approximate equivalents.
Source: General Electric Co. Ltd.

ating conditions will permit. Too large a nose radius causes chattering, while too small a nose radius causes the point to break down quickly. Use Table 83-2 to determine the correct nose radius to use for the amount of feed and depth of cut being used.

TOOL SETUP

1. Carbide tools should preferably be held in a sturdy, turret-type holder (Fig. 83-11). The amount of tool overhang should be just enough for chip clearance.
2. The cutting tool should be set exactly on center; when it is above or below center, the tool angles and clearances in relation to the job change and result in poor cutting action.
3. Carbide tools are designed to operate while the bottom of the tool shank is in a *horizontal position* (Fig. 83-12).

4. If a rocker-type toolpost is being used (Fig. 83-13):
   a. Remove the rocker.
   b. Invert the rocker base.
   c. Shim the tool to the correct height.
   d. Use a special carbide toolholder (having no rake) when machining with carbide toolbits (Fig. 83-14).
5. When setting up a carbide tool, always keep it from touching the work and machine parts to avoid damaging the tool point.

MACHINE SETUP

1. Always make sure that the machine has an adequate power rating for the machining operation and that there is no slippage in the clutch and belts.
2. Set the correct speed for the material being cut and the operation being performed.

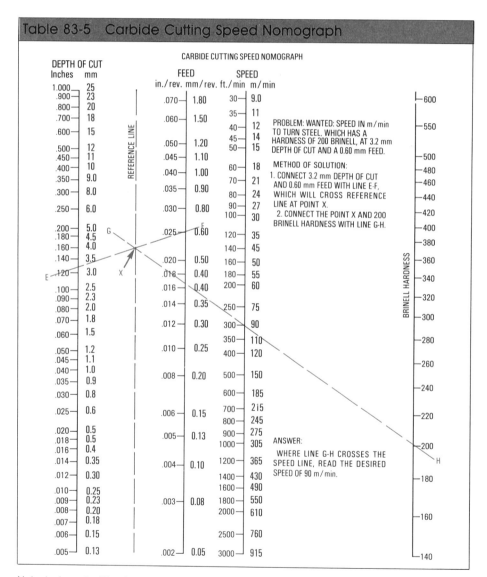

Table 83-5 Carbide Cutting Speed Nomograph

CARBIDE CUTTING SPEED NOMOGRAPH

PROBLEM: WANTED: SPEED IN m/min TO TURN STEEL, WHICH HAS A HARDNESS OF 200 BRINELL, AT 3.2 mm DEPTH OF CUT AND A 0.60 mm FEED.

METHOD OF SOLUTION:
1. CONNECT 3.2 mm DEPTH OF CUT AND 0.60 mm FEED WITH LINE E-F, WHICH WILL CROSS REFERENCE LINE AT POINT X.
2. CONNECT THE POINT X AND 200 BRINELL HARDNESS WITH LINE G-H.

ANSWER:
WHERE LINE G-H CROSSES THE SPEED LINE, READ THE DESIRED SPEED OF 90 m/min.

Note: Inch and millimeter speeds are approximate equivalents.
Source: General Electric Co. Ltd.

Fig. 83-11   Turret-type toolholders hold carbide tools securely.

Fig. 83-12   The heavy-duty toolpost permits rigid setting of the cutting tool for heavy cuts.

Fig. 83-13   To hold carbide cutting tools in a rocker-type toolpost, remove the rocker, invert the rocker base, and shim the tool to the correct height.

FLAT SIDE OF COLLAR TURNED UP

NO ROCKER BASE

Fig. 83-14   The hole in a carbide toolholder is straight and not slanted as in standard toolholders.   (Courtesy J.H. Williams & Co.)

    a. Too high a speed will cause rapid tool failure, while too low a speed will result in inefficient cutting action and poor production rates.
    b. Use Table 83-4 or 83-5 to calculate the correct speed for the type of material being machined.
3. Set the machine at a feed which will give good metal removal rates and still provide the surface finish desired.
    a. Too light a feed causes rubbing which may result in hardening of the material being cut.
    b. Too coarse a feed slows down the machine, creates excessive heat, and results in premature tool failure.

CUTTING OPERATION

1. Never bring the tool point against work that is stationary; to do this will damage the cutting edge.
2. Always use the heaviest depth of cut possible for the machine and size of cutting tool.
3. Never stop a machine while the feed is engaged; this will break the cutting edge. Always stop the feed and allow the tool to clear itself before stopping the machine.
4. Never continue to use a dull cutting tool.
5. A dull cutting tool may be recognized by:
    a. Work being produced oversize and with a glazed finish
    b. A rough and ragged finish
    c. A change in the shape or color of the chips
6. Apply cutting fluid only if:
    a. It can be applied under pressure
    b. It can be directed at the point of cutting and kept there at all times

GRINDING CEMENTED-CARBIDE TOOLS

The efficiency of a cutting tool determines to a large extent the efficiency of the machine tool on which it is being used. A tool that has been improperly ground cannot perform well, and the cutting edge will soon break down. In order to grind cemented-carbide cutting tools successfully, the type of grinding wheel used and the grinding procedure followed are important.

Fig. 83-15 A crown dressed on a silicon carbide wheel reduces heat when grinding carbide tools.

GRINDING WHEELS

1. An 80-grit silicon carbide wheel should be used for rough grinding carbides.
2. A 100-grit silicon carbide wheel should be used for finish grinding carbides. The silicon carbide wheels should be dressed with a 1/16 in. (1.58 mm) crown (Fig. 83-15) to minimize the amount of heat generated during grinding. These wheels should be used to grind only the carbide, not the steel tool shank of a brazed carbide toolbit.
3. Aluminum oxide grinding wheels should be used if it is necessary to grind the steel shank of a brazed carbide tool.
4. Diamond grinding wheels (100 grit) are excellent for finish grinding of carbides for general work. Where high finishes on the tool and work are desired, a 220-grit diamond wheel is recommended.

TYPE OF GRINDER

1. A heavy-duty grinder should be used for grinding carbides because the cutting pressures required to remove carbide are from 5 to 10 times as great as when high-speed steel tools are ground.
2. The grinder should be equipped with an adjustable table and a protractor (Fig. 83-16) so that the necessary tool angles and clearances may be ground accurately.

TOOL GRINDING

1. Regrind the cutting tool to the same angles and clearances recommended by the manufacturer.
2. Use silicon carbide wheels for rough grinding. Diamond wheels should be used where high surface finishes are required.
3. When grinding, move the carbide tool back and forth over the grinding wheel face to keep the amount of heat generated to a minimum.
4. *Never quench carbide tools* which become hot during grinding; allow them to cool gradually. The shock of quenching carbide tools creates heat checks and results in rapid tool failure.

HONING

After carbide tools have been ground, it is important that the cutting edge be honed. The purpose of honing is to remove the fine, ragged edge left by the grinding wheel. This fine, nicked edge is fragile and will break down quickly under machining conditions. Honing often means the difference between tool success or failure.

Fig. 83-16 A carbide tool grinder equipped with an adjustable table and protractor.

Fig. 83-17 Honing a slight chamfer on the edge of a carbide tool will produce a stronger cutting edge.

1. A 320-grit silicon carbide or diamond hone is recommended for carbide tools.
2. On carbides used for cutting steel, a 45° chamfer (Fig. 83-17) 0.002 to 0.004 in. (0.05 to 0.10 mm) wide should be honed on the cutting edge.
3. Carbide tools used for aluminum, magnesium, and plastics should not be chamfered with a hone. The fine, nicked edge should be honed to produce a sharp, keen cutting edge.

CEMENTED CARBIDE TOOL PROBLEMS

When problems occur while machining with carbide cutting tools, consult Table 83-6 for their possible causes and remedies. It is wise to change only one thing at a time until the problem is corrected. In this way, the real cause of the

TABLE 83-6 Carbide Tool Problems, Causes, and Remedies

PROBLEM	CAUSE	REMEDY
Brazing failure and cracks	Improper braze material Too much heat Improper cooling Tip not wiped on Contact surfaces dirty	Use the sandwich braze. Silver solder at 1400°F (760°C). Cool slowly. Slide tip back and forth and tap gently. Thoroughly clean both surfaces.
Built-up edge (on the tool)	Cutting speed too slow Insufficient rake angle Wrong grade of carbide	Increase the speed. Increase rake angle. Change to titanium and/or tantalum grade.
Chipped and broken cutting edge	Incorrect tool geometry Improper carbide grade Cutting edge not honed Improper speed, feed or depth of cut Too much tool overhang Lack of rigidity in setup or machine Machine stopped during cut	Increase side cutting edge angle. Decrease end cutting edge angle. Use a negative rake. Increase the nose radius. Decrease the relief angles. Use a tougher grade. Hone cutting edge at 45°. Change one or all as required. Reduce overhang or use a larger shank. Locate and correct. Disengage feed before stopping the machine.
Cratering	Improper grade Speed too high Wrong tool geometry	Select harder or crater resistant grade. Reduce the speed. Increase side rake.
Grinding cracks	Improper wheel Wheel glazed Tool quenched while hot Grinding carbide and steel shank with the same wheel	Select proper wheel. Dress often to a 1/16-in. crown. Do not quench; allow tool to cool slowly. Relieve the steel shank with an aluminum oxide wheel first.
Tool wear (excessive)	Improper carbide grade Cutting speed too high Feed too light Too little relief angle	Select a wear-resistant grade. Reduce the speed. Increase the feed rate. Increase the relief angle.

problem can be determined and steps taken to guard against its recurrence.

CUBIC BORON NITRIDE (BORAZON) TOOLS

High-speed turning of hardened alloy steels and tough superalloys, which up until now has been impossible because of rapid cutting edge wear, is now possible with Borazon [cubic boron nitride (CBN)] indexable inserts.

Borazon CBN inserts are made by bonding a layer of polycrystalline CBN to a cemented-carbide base. The CBN layer is about 0.035 in. (0.88 mm) thick and the total thickness of the insert is about 1/4 in. (6.35 mm). The CBN layer does the cutting, and the carbide provides a strong, tough, shock-resistant base for the cutting edge which can withstand temperatures up to 1832°F (1000°C) before breaking down.

The high-temperature resistance to oxidation makes it possible to machine super-alloys which lose some strength at high temperatures, thus making them easier to machine.

In many cases, turning workpieces with CBN inserts has

replaced some grinding operations, and in most cases it has proved much more satisfactory than carbides for those difficult-to-machine alloys. In certain test applications, carbide toolbits produced three pieces per cutting edge, whereas CBN inserts machined 1000 pieces per cutting edge.

Borazon CBN inserts produce good surface finishes—24.4 to 62.9 microinches (0.62 to 1.6 μm). Out-of-roundness on tough alloys has been reduced from 0.0012 to 0.002 in. (0.03 to 0.05 mm), resulting in better accuracy. As a result of the good surface finishes and the geometric accuracy produced by Borazon CBN inserts, the grinding required on workpieces has been reduced.

Although the application of CBN for indexible inserts is relatively new, CBN is becoming more popular and is being used for many machining applications.

REVIEW QUESTIONS

CARBIDE CUTTING TOOLS

1. State four reasons why cemented-carbide tools have been widely accepted by industry.

MANUFACTURE OF CEMENTED CARBIDES

2. Name five powders used in the manufacture of cemented carbides.
3. What purpose does cobalt serve?
4. Describe the process of compaction.
5. What is the purpose of presintering?
6. Describe the sintering process.

CEMENTED-CARBIDE APPLICATIONS

7. Why is cemented carbide used extensively in the manufacture of cutting tools?
8. Name two types of cemented-carbide lathe cutting tools, and state the advantages of each.
9. Identify the following throwaway insert: SNG—321—A.

GRADES OF CEMENTED CARBIDES

10. Describe straight tungsten carbide tools, and state the purpose for which they are used.
11. Explain how the size of the tungsten carbide particles and the percentage of cobalt affect the grade of carbide.
12. How does the addition of titanium carbide affect the cutting tool?
13. What properties does the addition of tantalum carbide provide?
14. List four important rules to observe for selecting a cemented-carbide grade.

TOOL GEOMETRY

15. Define and state the purpose of:
    (a) front or end relief
    (b) side relief
    (c) side rake

16. (a) What is the purpose of the side cutting edge angle?
    (b) How is the angle of keenness formed?
17. State the purpose of the nose radius on a cutting tool. How large should it be?
18. Name two types of back rake and state the purpose of each.
19. What three factors determine the angles and clearances of single-point carbide tools?

CUTTING SPEEDS AND FEEDS

20. Name four important factors that influence the speed, feed, and depth of cut for carbide tools.
21. Use Table 83-5 to determine what cutting speed should be used for:
    (a) taking a $\frac{3}{32}$-in. depth of cut on 260 Brinell steel with a 0.015 in. feed.
    (b) a cut 1.50 mm deep on 300 Brinell steel using a 0.20-mm feed.

MACHINING WITH CARBIDE TOOLS

22. List two important precautions that should be observed when work is being set up on a lathe.
23. Why should the cutting edge of a carbide tool be honed?
24. What occurs if the nose radius on the cutting tool is
    (a) too large?   (b) too small?
25. Using Table 83-2, determine the nose radius for a carbide tool taking:
    (a) a $\frac{1}{4}$-in.-deep cut at .020-in. feed.
    (b) a 3.2-mm depth of cut using a 0.40-mm feed.
26. How should carbide tools be set up for machining?
27. Explain how carbide tools should be set up in a rocker-type toolpost.
28. What precautions should be taken when setting up a machine for cutting with carbide tools?
29. Discuss the effects of:
    (a) too light feed   (b) too coarse feed
30. Why should a machine never be stopped while the tool is engaged in a cut?
31. Explain how a dull cutting tool may be recognized.

GRINDING CEMENTED-CARBIDE TOOLS

32. What type of grinding wheels are used to grind carbide tools?
33. How should silicon carbide wheels be dressed? Explain why.
34. List three important points that should be observed in the grinding of carbide tools.
35. Why is it important not to quench carbide tools?
36. What is the purpose of honing carbide tools?
37. How should carbide tools used for steel be honed?

CEMENTED-CARBIDE TOOL PROBLEMS

38. List the factors that would cause the following problems:
    (a) cratering
    (b) grinding cracks
    (c) chipped or broken cutting edge

CERAMIC AND DIAMOND CUTTING TOOLS

UNIT 84

The first ceramic (cemented-oxide) cutting-tool inserts were put on the market in 1956; they were the result of many years of research. At first, these ceramic inserts were weak; inconsistent and unsatisfactory results were obtained because of lack of knowledge and improper use. Since then, the strength of ceramic cutting tools has nearly doubled and the uniformity and quality have greatly improved; they are now widely accepted by industry. Ceramic cutting tools are being used successfully in the machining of hard ferrous materials and cast iron; lower costs, increased productivity, and better results are being gained. In some operations, ceramic toolbits can be operated at three to four times the speed of carbides.

Single-point diamond-tipped cutting tools are used to a limited degree to machine nonferrous metals and nonmetallic materials where extremely high finishes and close tolerances are required.

OBJECTIVES

After completing this unit you will be able to:
1. Identify and know the purpose of two types of ceramic cutting tools
2. Know the purpose and applications of diamond cutting tools

(Courtesy The Carborundum Company.)

MANUFACTURE OF CERAMIC TOOLS

Most ceramic or cemented-oxide cutting tools are manufactured primarily from aluminum oxide. Bauxite (a hydrated alumina form of aluminum oxide) is chemically processed and converted into a denser, crystalline form called *alpha alumina*. Fine grains (micron size) are obtained from the precipitation of the alumina or from the precipitation of the decomposed alumina compound.

Ceramic tool inserts are produced by either *cold or hot pressing*. In cold pressing, the fine alumina powder is compressed into the required form and then sintered in a furnace at 2912 to 3092°F (1600° to 1700°C). Hot pressing combines forming and sintering, with pressure and heat being applied simultaneously. Certain amounts of titanium oxide or magnesium oxide are added for certain types of ceramics to aid in the sintering process and to retard growth. After the inserts have been formed, they are finished with diamond-impregnated grinding wheels.

Further research has led to the development of stronger ceramic cutting tools. Aluminum oxide (Al_2O_3) and zirconium oxide (ZrO_2) are mixed in powder form, cold-pressed into the required shape, and sintered. This white ceramic insert has proved successful in machining high-tensile-strength materials (up to 42 Rockwell C hardness) at speeds up to 2000 sf/min. However, the most common applications of ceramic inserts are in the general machining of steel where there are no heavy interrupted cuts and where negative rakes can be used. This type of toolbit has the highest hot-hardness strength of any cutting-tool material and gives excellent surface finish. No coolant is required with ceramic tools since most of the heat goes into the chip and not into the workpiece. Ceramic inserts should be used with toolholders which have a fixed or adjustable chipbreaker.

The most common ceramic cutting tool is the *throwaway insert* (Fig. 84-1), which is fastened in a mechanical holder. Throwaway inserts are available in many styles, such as triangular, square, rectangular, and round. These inserts are indexible; when a cutting edge becomes dull, a sharp edge can be obtained by indexing (turning) the insert in the holder.

Cemented ceramic tools (Fig. 84-2) are the most economical, especially if the tool shape must be altered from the standard shape. The ceramic insert is bonded to a steel shank with an epoxy glue. This method of holding the ceramic inserts almost eliminates the strains caused by clamping in mechanical holders.

Much research and experimentation has been done to reduce the brittleness of ceramics and to improve their qualities. *Ceramic-metal combinations (cermets)* are the most promising. Iron, chromium, titanium, and other metals have been combined with aluminum oxide and boron carbide to form ceramic-metal cutting tools.

Many combinations show definite advantages, and one such combination is that of aluminum oxide, titanium carbide, and zirconium oxide (Al_2O_3 + TiC + ZrO_2). Fine powders of the materials are mixed, compacted, and pressed simultaneously under intense heat and pressure. The resultant black material is ceramic particles bonded with metal. Inserts made of this cermet material are capable

Fig. 84-1 Throwaway ceramic inserts are available in a variety of shapes.

Fig. 84-2 Ceramic tips may be cemented to the toolshank.

of speeds up to 2000 sf/min, depending on the material being machined. Cermets may be used to machine stainless steel, high-temperature alloys, heat-treated materials, high-tensile steel (up to 66 Rc) and chilled cast iron. Coolant is generally used with these inserts only on finish cuts to provide a better surface finish.

Another form of cermet is the combination of titanium nitride (TiN) and titanium carbide (TiC). The powders of these materials are mixed, pressed into shape, and sintered without pressure producing a cermet which is gray in color. Cermets of this type may be used to machine harder steels and stainless steel up to 45 Rc.

CERAMIC TOOL APPLICATIONS

Ceramic tools were intended to supplement rather than replace carbide tools. They are extremely valuable for specific applications and must be carefully selected and used. Ceramic tools can be used to replace carbide tools which wear rapidly in use but should never be used to replace carbide tools which are breaking.

Ceramics are being used successfully for:

1. High-speed, single-point turning, boring, and facing operations, with continuous cutting action
2. Finishing operations on ferrous and nonferrous materials
3. Light, interrupted finishing cuts on steel or cast iron; heavy, interrupted cuts on cast iron only if there is adequate rigidity in the machine and tool
4. Machining castings when other tools break down, because of the abrasive action of sand, inclusions, or hard scale
5. Cutting hard steels up to a hardness of Rockwell C 66, which previously could be machined only by grinding
6. Any operation in which size and finish of the part must be controlled and in which previous tools have not proved satisfactory

FACTORS AFFECTING CERAMIC TOOL PERFORMANCE

As a result of an extensive research and testing program, several factors have been found which significantly affect the performance of ceramic tools. The following factors must be considered for optimum results to be derived from ceramic cutting tools.

1. Accurate and rigid machine tools are essential when ceramic tools are used. Machines with loose bearings, inaccurate spindles, slipping clutches, or any imbalance will result in the ceramic tools chipping and in premature failure.
2. The machine tool must be equipped with ample power and be capable of maintaining the high speeds necessary for ceramics.
3. The tool mounting and the toolholder rigidity are as important as the machine rigidity. An intermediate plate between the toolholder clamp and ceramic insert is recommended to distribute the clamping pressure.
4. The overhang of the toolholder should be kept to a minimum; no more than 1½ times the shank thickness.
5. Negative rake inserts give the best results because less force is applied directly to the ceramic tip.
6. A large nose radius and a large side cutting edge angle on the ceramic insert reduces the tendency to chip.
7. Cutting fluids are generally not required because ceramics remain cool during the machining operation. If cutting fluids are required, a continuous and copious flow should be used to prevent thermal shock to the tool.
8. As the cutting speed or the hardness of the workpiece increases, the ratio of feed to the depth of cut must be checked. Always take a deeper cut with a light feed rather than a light cut with a heavy feed. Most ceramic tools are capable of cuts as deep as one-half the width of the cutting surface on the insert.
9. Toolholders with fixed or adjustable chipbreakers are best for use with ceramic inserts. Adjust the chipbreakers to produce a figure 6 or 9 curled chip.

ADVANTAGES AND DISADVANTAGES OF CERAMIC TOOLS

ADVANTAGES

Ceramic cutting tools, when properly mounted in suitable holders and used on accurate rigid machines, offer many advantages.

1. Machining time is reduced because of the higher cutting speeds possible. Speeds ranging from 50 to 200 percent higher than those used for carbides are quite common.
2. High stock removal rates and increased productivity result because heavy depths of cut can be made at high surface speeds.
3. A ceramic tool used under proper conditions lasts from 3 to 10 times longer than a plain carbide tool and will exceed the life of coated carbide tools.
4. Ceramic tools retain their strength and hardness at high machining temperatures [in excess of 2000°F (1093°C)].
5. More accurate size control of the workpiece is possible because of the greater wear resistance of ceramic tools.
6. Ceramic cutting tools withstand the abrasion of sand and of inclusions found in castings.
7. A better surface finish is produced than is possible with other types of cutting tools.
8. Heat-treated materials as hard as Rockwell C 66 can be readily machined.

DISADVANTAGES

Although ceramic cutting tools have many advantages, they have the following limitations.

1. Ceramic tools are brittle and therefore tend to chip easily.
2. They are satisfactory for interrupted cuts only under ideal conditions.
3. The initial cost of ceramics is higher than that of carbides.
4. A more rigid machine is required than is necessary for other cutting tools.
5. Considerably more power is required for ceramics to cut efficiently.

CERAMIC TOOL GEOMETRY

The geometry of ceramic tools depends upon five main factors.

1. The material to be machined
2. The operation being performed
3. The condition of the machine
4. The rigidity of the work setup
5. The rigidity of the toolholding device

Although the final geometry of a ceramic cutting tool depends upon these five main factors, some general considerations can also be mentioned.

TABLE 84-1 Suggested Rake and Relief Angles for Ceramic Tools*

WORKPIECE MATERIAL	RAKE ANGLES (degrees)	RELIEF ANGLES (degrees)
Carbon and alloy steels: Annealed and heat-treated	Neg. 2 to 7	2 to 7
Cast iron: Hard or chilled Gray or ductile	0 to Neg. 7	2 to 7
Nonferrous: Hard or soft	0 to Neg. 7	2 to 7
Nonmetallics: Wood, paper, green ceramics, fiber, asbestos, rubbber, carbon, graphite	0 to Pos. 10	6 to 18

* For uninterrupted turning with tool point on centerline of workpiece.
Source: Courtesy The Carborundum Company

Rake Angles Because of their brittleness, negative rake angles are generally preferred for ceramic tools. With the use of a negative rake angle, the shock of the cutting force is absorbed behind the tip and thus the cutting edge is protected. Negative rake angles from 2 to 30° are used for machining ferrous and nonferrous metals. Positive rake angles are used for machining nonmetals such as rubber, graphite, and carbon (Table 84-1).

Side Clearance A side clearance angle is desirable for ceramic cutting tools whenever possible. The side clearance angle must not be too great, otherwise the cutting edge will be weakened and tend to chip.

Front Clearance The front clearance angle should be only large enough to prevent the tool from rubbing on the workpiece. If this angle is too great, the ceramic tool becomes susceptible to chipping.

End Cutting Edge Angle This angle governs the strength of the tool and the area of contact between the work and the end of the cutting tool. If properly designed, it will remove the crests resulting from feed lines (Fig. 84-3) and improve the surface finish.

Nose Radius The nose radius (Fig. 84-3) has two important functions: to strengthen the weakest part of the tool, and to improve the surface finish of the workpiece. It should be as large as is possible without chatter or vibration resulting.

Cutting Edge Chamfer A small chamfer, or radius, on the cutting edge is recommended for ceramic tools, especially on heavy cuts and hard materials. This strengthens and protects the cutting edge. A 0.002- to 0.008-in. (0.05- to 0.20-mm) radius, or chamfer, is recommended for machining steel. For heavy roughing cuts and hard materials,

Fig. 84-3 The relationship of the nose radius and end cutting edge angle to the surface finish produced.

a 0.030- to 0.060-in. (0.76- to 1.52-mm) chamfer gives satisfactory results.

CUTTING SPEEDS

When ceramics are used in machining, the highest cutting speed possible, considering the machine tool limitations, which gives reasonable tool life should be used. There is less heat generated when machining is done with ceramics than with any other cutting tool, because there is a lower coefficient of friction between the chip, work, and tool surface. Since most of the heat generated escapes with the chip, the cutting speed can be from 2 to 10 times as high as with other cutting tools. Table 84-2 lists the recommended speeds for various materials under ideal conditions. Should lower speeds be necessary to suit various machine or setup conditions, ceramic cutting tools will still perform well.

CERAMIC TOOL PROBLEMS

Ceramic tools remove metal faster than any other type of cutting tool and therefore should be selected carefully for the type of material and the operation. Some points to consider when selecting ceramic tools are:

CUTTING-TOOL MATERIALS

TABLE 84-2 Recommended Cutting Speeds for Ceramic Cutting Tools

WORKPIECE MATERIAL	MATERIAL CONDITION OR TYPE	ROUGHING CUT		FINISHING CUT		RECOMMENDED TOOL GEOMETRY (TYPE OF RAKE) ANGLE)	RECOMMENDED COOLANT
		DEPTH > 0.062 in. FEED 0.015–0.030 in.	DEPTH > 1.60 mm FEED 0.40–0.75 mm	DEPTH < 0.062 in. FEED 0.010 in.	DEPTH < 1.60 mm FEED 0.25 mm		
Carbon and tool steels	Annealed Heat-treated Scale	300–1500 300–1000 300–800	90–455 90–305 90–245	600–2000 500–1200	185–610 150–365	Neg. Neg. Neg. honed edge	None
Alloy steels	Annealed Heat-treated Scale	300–800 300–800 300–600	90–245 90–245 90–185	400–1400 300–1000	120–425 90–305	Neg. Neg. honed edge Neg. honed edge	None
High-speed steel	Annealed Heat-treated Scale	100–800 100–600 100–600	30–245 30–185 30–185	100–1000 100–600	30–305 30–385	Neg. Neg. honed edge Neg. honed edge	None
Stainless steel	300 series 400 series	300–1000 300–1000	90–305 90–305	400–1200 400–1200	120–365 120–365	Pos. and neg. Neg.	Sulfur base oil
Cast iron	Gray iron Pearlitic Ductile Chilled	200–800 200–800 200–600 100–600	60–245 60–245 60–185 30–185	200–2000 200–2000 200–1400 200–1400	60–610 60–610 60–427 60–427	Pos. and neg. Neg. Neg. Neg. honed edge	None
Copper and alloys	Pure Brass Bronze	400–800 400–800 150–800	120–245 120–245 45–245	600–1400 600–1200 150–1000	185–425 185–365 45–305	Pos. and neg. Pos. and neg. Pos. and neg.	Mist coolant Mist coolant Mist coolant
Aluminum alloys*		400–2000	120–610	600–3000	185–915	Pos.	None
Magnesium alloys		800–10,000	245–3050	800–10,000	245–3050	Pos.	None
Non-metallics	Green ceramics Rubber Carbon	300–600 300–1000 400–1000	90–185 90–305 120–305	500–1000 400–1200 600–2000	150–305 120–365 185–610	Pos. Pos. Pos.	None None None
Plastics		300–1000	90–305	400–1200	120–365	Pos.	None

*Alumina-based cutting rods have a tendency to develop a built-up cutting edge on certain aluminum alloys.
Source: Courtesy A. C. Wickman Limited.

1. The tool should be large enough for the job. It cannot be too large, but can easily be too small.
2. The style (tool geometry) should be right for the type of operation and material. A tool designed for one job is not necessarily right for another.

Some of the more common problems which occur with ceramic cutting tools and their possible causes are listed in Table 84-3.

GRINDING CERAMIC TOOLS

It is not recommended that ceramic tools be ground; however, with the proper care, these tools may be resharpened successfully. Resinoid-bonded, diamond-impregnated wheels are recommended for grinding ceramic tools. A coarse-grit wheel should be used for rough grinding, while at least a 220-grit wheel should be used for finish grinding. Because ceramic tools are extremely notch-sensitive, all

surfaces should be ground as smoothly as possible to avoid notches or grinding lines at the cutting edge. The cutting edge should be honed or lapped after grinding to remove any notches left by grinding and to avoid the wedging action which would occur when the material entered these notches.

FUTURE DEVELOPMENTS

Because of the high melting point, the excellent hardness, especially at elevated cutting temperatures, and the resistance to abrasion wear and cratering, ceramics will continue to be used and developed through continued research. Many new types of ceramic and cermet cutting tools are expected to be developed as the result of experimentation in the use of titanium, tantalum, zirconium carbides and nitrides, along with aluminum oxides.

It is possible in the near future that to accommodate the more sophisticated machines, ceramic and cermet cutting tools will be harder and tougher with sufficient shock re-

TABLE 84-3 Ceramic tool problems and Possible Causes

PROBLEM	POSSIBLE CAUSES
Chatter	1. Tool not on center 2. Insufficient end relief and/or clearance 3. Too much rake angle 4. Too much overhang or tool too small 5. Nose radius too large 6. Feed too heavy 7. Lack of rigidity in the tool or machine 8. Insufficient power or slipping clutch
Chipping	1. Lack of rigidity 2. Saw-toothed or too keen a cutting edge 3. Chipbreaker too narrow or too deep 4. Chatter 5. Scale or inclusions in the workpiece 6. Improper grinding 7. Too much end relief 8. Defective toolholder
Cracking or Breaking	1. Insert surfaces not flat 2. Insert not seated solidly 3. Stopping workpiece while tool is engaged 4. Worn or chipped cutting edges 5. Feed too heavy 6. Improperly applied coolant 7. Too much rake angle or end relief 8. Too much overhang or tool too small 9. Lack of rigidity in setup 10. Speed too slow 11. Too much variation in cut for size of tool 12. Chatter 13. Grinding cracks
Cratering	1. Chipbreaker set too close to the edge 2. Nose radius too large 3. Side cutting edge angle too great
Torn finish	1. Lack of rigidity 2. Dull tool 3. Speed too slow 4. Chipbreaker too narrow or too deep 5. Improper grinding
Wear	1. Speed too high or feed too light 2. Nose radius too large 3. Improper grinding

sistance to withstand interrupted cuts at higher speeds and feeds and high tip temperatures.

DIAMOND CUTTING TOOLS

Diamond-tipped cutting tools (Fig. 84-4) are used to machine nonferrous and nonmetallic materials which require a high surface finish and extremely close tolerances. They are used primarily as finishing tools because they are brittle and do not resist shock or cutting pressure as well as carbide cutting tools. Single-point, diamond-tipped cutting tools are available in various shapes for turning, boring, grooving, and special forming.

The main advantages of diamond-tipped cutting tools are:

1. They can be operated at high cutting speeds and pro-

Fig. 84-4 Diamond cutting tools are used efficiently for shallow cuts at very high speeds. (Courtesy Precision Diamond Tool Co.)

duction can be increased to 10 to 15 times that of other cutting tools.

2. Surface finishes of 5 microinches (0.127 μm) or less can be obtained easily. In many cases, the necessity of finishing operations on the workpiece is eliminated.

3. They are very hard and resist abrasion. Much longer runs are therefore possible on abrasive materials.

4. Closer-tolerance work can be produced with diamond-tipped cutting tools.

5. Minute cuts, as low as 0.0005 in. (0.012 mm) deep, can be taken from the inside or outside diameter.

6. Metallic particles do not build up (weld) on the cutting edge.

USE OF DIAMOND CUTTING TOOLS

Some of the most successful applications of diamond cutting tools have been in the turning of metallic (nonferrous) and nonmetallic materials. The commonest of these materials are listed.

Metallic Materials

1. *Light metals,* such as aluminum, duraluminum, magnesium alloys, etc.
2. *Soft metals,* such as copper, brass, and zinc alloys
3. *Bearing metals,* such as bronze and babbitt
4. *Precious metals,* such as silver, gold, and platinum

For finishing cuts in these metallic materials, diamond tools can be expected to increase production from 10 to 15 times that possible with any other cutting tool.

Nonmetallic Materials

Some of the most common materials machined with diamond cutting tools are hard and soft rubber, all types of plastics, carbon, graphite, and ceramics. In some cases, diamond tools will increase production 20 to 50 times that of carbide tools.

CUTTING SPEEDS AND FEEDS

In general, diamond-tipped cutting tools operate most efficiently with shallow cuts at high cutting speeds and fine feeds. They are not recommended for materials in which the temperature of the chip or the heat generated at the chip-tool interface exceeds 1400°F (760°C). High cutting speeds, fine feeds, and shallow cuts are used. The heat generated is entirely dissipated in the chip as it leaves the cutting edge. Low cutting speeds and heavy cuts create more heat and damage the diamond tip.

There is an ideal cutting speed for each type of material-machine combination. The minimum cutting speed for diamond tools should be 250 to 300 sf/min (76 to 91 m/min).

TABLE 84-4 Diamond Cutting Tool Data						
MATERIAL	CUTTING SPEED		FEED (per rev.)		DEPTH OF CUT	
	ft/min	m/min	in.	mm	in.	mm
Metallic (nonferrous)	250–10,000	75–3050	0.0008–0.004	0.002–0.10	0.0005–0.024	0.001–0.60
Nonmetallic	250–3300	75–1005	0.0008–0.024	0.002–0.60	0.0008–0.060	0.002–1.50

Fig. 84-5 Diamond-tipped cutting tool angles and clearances.

The conditions of the machine will determine the maximum cutting speed for each job. Cutting speeds as high as 10,000 sf/min (3048 m/min) have been used for some applications. Table 84-4 lists cutting speed, feed, and depth of cut ranges for various material groups.

HINTS ON THE USE OF DIAMOND TOOLS

Diamond cutting tools will perform more efficiently and have longer life if the following hints and precautions are observed.

1. Diamond-tipped points should be designed with maximum included point angle and radius for added strength.
2. Diamond tools should always be handled with care, especially when setups are being made. The cutting edges should *never* be checked with a micrometer or bumped with a height gage.
3. Diamond tools should always be stored in separate containers, with rubber protectors over the tips, so that they will not be fractured by coming into contact with other tools.
4. The machine tool should be as free of vibration as possible. Any vibration can result in tool failure.
5. A very rigid setup, with the diamond tip set exactly on center, should be used.
6. The work should be roughed out with a carbide tool. This step will establish an even work surface and bring the work close to size.

7. Diamond tools should always be fed into the work while the work is revolving. Never stop a machine during a cut.
8. Interrupted cuts, especially on hard metals, will shorten tool life.

FUTURE DEVELOPMENTS

Much research and testing remains to be done in order to establish the exact role the diamond-tipped cutting tool will play in the future. It is reasonable to assume that as new and exotic materials are developed and as close tolerances and high surface finishes are required, the use of diamond cutting tools will increase. When, in the future, machine tools are designed for high-speed machining, it is felt that diamond cutting tools will be used to precision-finish steel parts.

REVIEW QUESTIONS

MANUFACTURE OF CERAMIC TOOLS

1. Briefly explain how cemented oxide cutting tools are manufactured.
2. What is the difference between ceramic and cermet cutting tools?

CERAMIC TOOL APPLICATIONS

3. Name four important applications for ceramic tools.
4. List four main factors which affect the performance of ceramic tools.

ADVANTAGES AND DISADVANTAGES OF CERAMIC TOOLS

5. Name five advantages of ceramic tools.
6. What are the disadvantages of using ceramic cutting tools?

CERAMIC TOOL GEOMETRY

7. List five factors that determine the geometry of ceramic tools.

8. Fully explain the rake angle of ceramic tools.
9. Define ''cutting edge chamfer'' and state its purpose.

CUTTING SPEEDS

10. Explain why high cutting speeds can be used with ceramic tools.

CERAMIC TOOL PROBLEMS

11. What two points should be kept in mind when ceramic tools are being selected?
12. List the factors that may cause the following problems.
    (a) chipping
    (b) chatter
    (c) wear

GRINDING CERAMIC TOOLS

13. Explain how ceramic tools should be ground.
14. What is the purpose of honing?

DIAMOND CUTTING TOOLS

15. List four main advantages of diamond cutting tools.
16. Explain where diamond cutting tools may be used successfully.

CUTTING SPEEDS AND FEEDS

17. What speeds, feeds, and depth of cut should be used for diamond tools? Explain why.
18. Why should the diamond tool be as rigid as possible and the machine free of vibration?

CUTTING FLUIDS
SECTION 18

Cutting fluids are essential in most metal-cutting operations. During a machining process, considerable heat and friction are created by the plastic deformation of metal occurring in the shear zone when the chip slides along the chip-tool interface. This heat and friction cause metal to adhere to the cutting edge of the tool and the tool may break down; the result is a poor finish and inaccurate work.

The use of cutting fluids is not new, some types having been used for hundreds of years. Centuries ago it was discovered that water dripped onto a grindstone kept the stone from glazing and a better surface finish was produced on the part being ground. The major problem was that the water caused the part ground to rust. About 100 years ago, machinists found that wiping tallow on parts before machining helped to produce smoother and more accurate parts. Tallow, the forerunner of cutting fluids, lubricated but did not cool. Lard oils, developed later, lubricated well and had some cooling properties but became rancid quickly. In the early 20th century, soaps were added to water to improve cutting, prevent rust, and provide some form of lubrication.

In 1936, the development of soluble oils was a great improvement over the previously used cutting oils. These milky-white emulsions combined the high cooling ability of water with the lubricity of petroleum oil. Although economical, they had relatively poor rust control and tended to become rancid.

Chemical cutting fluids were introduced about 1944. They contained relatively little oil, being dependent upon chemical agents for lubrication and friction reduction. The chemical emulsions mix easily with water and reduce as well as remove the heat created during machining. Chemical cutting fluids are rapidly increasing in popularity because they provide good rust resistance, do not become rancid quickly, and have good cooling and lubricating qualities.

REAMING

GRINDING

MILLING

BROACHING

TAPPING

TURNING

DRILLING

POWER SAWING

CUTTING FLUIDS—TYPES AND APPLICATIONS

UNIT 85

During a machining process, considerable heat and friction are created. The correct selection and application of cutting fluids can prevent these results by effectively cooling the work and reducing the friction. Cutting fluids cool and lubricate the tool and workpiece. Their use can result in the following economic advantages.

1. *Reduction of tool costs.* Cutting oils reduce tool wear; they last longer and less time is spent in resharpening and resetting.
2. *Increased speed of production.* Because cutting oils help reduce heat and friction, higher cutting speeds can be used for machining operations.
3. *Reduction of labor costs.* As cutting tools last longer and require less regrinding when cutting oils are used, there is less stoppage of production, reducing the labor cost per part.
4. *Reduction of power costs.* Since friction is reduced by a cutting fluid, less power is required for machining operations and a corresponding saving in power costs is possible.

OBJECTIVES

After completing this unit you will be able to:
1. State the importance and function of cutting fluids
2. Identify three types of cutting fluids and state the purpose of each
3. Apply cutting fluids efficiently for a variety of machining operations

CHARACTERISTICS OF A GOOD CUTTING FLUID

For a cutting fluid to function effectively, it should possess the following desirable characteristics.

1. *Good cooling capacity,* to reduce the cutting temperature, increase tool life and production, and improve dimensional accuracy.
2. *Good lubricating qualities,* to prevent metal from adhering to the cutting edge, forming a built-up edge, resulting in a poor surface finish.
3. *Rust resistance.* It should not cause stain, rust, or corrosion to the workpiece or machine.
4. *Stability (long life),* both in storage and in use.
5. *Resistance to rancidity.* It should not become rancid easily.
6. *Nontoxic.* It should not cause skin irritation to the operator.
7. *Transparent,* so that the operator can clearly see the work during machining.
8. *Relatively low viscosity,* to permit the chips and dirt to settle quickly.

9. *Nonflammable.* It should not burn easily, and should preferably be noncombustible. In addition, it should not smoke excessively, form gummy deposits which may cause machine slides to become sticky, or clog the circulating system.

TYPES OF CUTTING FLUIDS

The need for a cutting fluid which possesses as many of the desirable characteristics of a good cutting fluid as possible has resulted in the development of many different types. The most commonly used cutting fluids are either aqueous- (water-) based solutions or cutting oils. These fluids fall into three categories: cutting oils, emulsifiable oils, and chemical (synthetic) cutting fluids.

CUTTING OILS

Cutting oils are classified under two types: active or inactive. These terms relate to the oil's chemical activity or ability to react with the metal surface at elevated temperatures to protect it and improve the cutting action.

Active cutting oils may be defined as those that will darken a copper strip immersed in it for 3 h at a temperature of 212°F (100°C). These oils are generally used when steel is being machined and may be either dark or transparent. Dark oils usually contain more sulfur than the transparent types and are considered better for heavy-duty jobs.

Active cutting oils fall into three general categories.

1. *Sulfurized mineral oils* contain from 0.5 to 0.8 percent sulfur. They are generally light-colored, transparent, and have good cooling, lubricating, and antiweld properties. They are useful for the cutting of low-carbon steels and tough, ductile metals. Sulfurized mineral oils stain copper and its alloys, and, therefore, are not recommended for these metals.

2. *Sulfochlorinated mineral oils* contain up to 3 percent sulfur and 1 percent chlorine. These oils prevent excessive built-up edges from forming and prolong the life of the cutting tool. Sulfochlorinated mineral oils are more effective than sulfurized mineral oils in cutting tough low-carbon and chrome-nickel alloy steels. They are extremely valuable in cutting threads in soft, draggy steel.

3. *Sulfochlorinated flatty oil blends* contain more sulfur than the other types and are effective cutting fluids for heavy-duty machining.

Inactive cutting oils may be defined as those which will not darken a copper strip immersed in it for 3 h at 212°F (100°C). The sulfur contained in an inactive oil is the natural sulfur of the oil and has no chemical value in the cutting fluid's function during machining. These fluids are termed inactive because the sulfur is so firmly attached to the oil that very little is released to react with the work surface during the cutting action.

Inactive cutting oils fall into four general categories.

1. *Straight mineral oils,* because of their low viscosity, have faster wetting and penetrating factors. They are used for the machining of nonferrous metals, such as aluminum, brass, and magnesium, where lubricating and cooling properties are not essential. Straight mineral oils are also recommended for use in cutting leaded (free-machining) metals and the tapping and threading of white metal.

2. *Fatty oils,* such as lard and sperm oil, once widely used, find limited applications as cutting fluids today. They are generally used for severe cutting operations on tough nonferrous metals where a sulfurized oil might cause discoloration.

3. *Fatty and mineral oil blends* are combinations of fatty and mineral oil, resulting in better wetting and penetrating qualities than straight mineral oils. These factors result in better surface finishes on both ferrous and nonferrous metals.

4. *Sulfurized fatty-mineral oil blends* are made by sulfur being combined with fatty oils and then mixed with certain mineral oils. Oils of this type provide excellent antiweld properties and lubricity when cutting pressures may be high and tool vibration excessive. Most sulfurized fatty-mineral oil blends can be used when nonferrous metals are cut to produce high surface finishes. They may also be used on machines when ferrous and nonferrous metals are machined at the same time.

EMULSIFIABLE (SOLUBLE) OILS

An effective cutting fluid should possess high heat conductivity; neither mineral nor fatty oils are very effective as coolants. Water is the best cooling medium known; however, used as a cutting fluid, water alone would cause rust and have little lubricating value. By adding a certain percentage of soluble oil to water, it is possible to add rust resistance and lubrication qualities to the excellent cooling capabilities of water.

Emulsifiable, or soluble oils are mineral oils containing a soaplike material (emulsifier) which makes them soluble in water. These emulsifiers break the oil into minute particles and keep them separated in the water for a long period of time. Emulsifiable, or soluble, oils are supplied in concentrate form. From *1 to 5 parts* of this concentrate are added to *100 parts* of water. Lean mixtures are used for light machining operations and when cooling is essential. Denser mixtures are used when lubrication and rust prevention are essential.

Soluble oils, because of their good cooling and lubricating qualities, are used when machining is done at high cutting speeds, at low cutting pressure, and when considerable heat is generated.

Emulsifiable, or soluble, oils are manufactured in three types for various machining conditions.

1. *Emulsifiable mineral oils* are mineral oils to which various compounds have been added to make the oil soluble in water. These oils are low in cost, provide good cooling and lubrication qualities, and are widely used for general cutting applications.

2. *Superfatted emulsifiable oils* are emulsifiable mineral oils to which some fatty oil has been added. These mixtures provide better lubrication qualities and are, therefore, used for tougher machining operations. Often these soluble oils are used when aluminum is machined.

3. *Extreme-pressure emulsifiable oils* contain sulfur, chlorine, and phosphorus, as well as fatty oils, to provide the added lubrication qualities required for tough machining operations. Extreme-pressure oils are generally mixed with water at the rate of *1 part* oil to *20 parts* water.

CHEMICAL CUTTING FLUIDS

Chemical cutting fluids, sometimes called *synthetic fluids,* have been widely accepted since they were first introduced about 1945. They are stable, pre-formed emulsions which contain very little oil and mix easily with water. Chemical cutting fluids depend upon chemical agents for lubrication and friction reduction. Some types of chemical cutting fluids contain *extreme-pressure* (EP) lubricants which react with freshly machined metal under the heat and pressure of a cut to form a solid lubricant. Fluids containing EP lubricants reduce both the *heat of friction* between the chip and tool face and the *heat caused by plastic deformation* of the metal.

The chemical agents found in most synthetic fluids include the following.

1. *Amines* and *nitrites* for rust prevention
2. *Nitrates* for nitrite stabilization

559

3. *Phosphates* and *borates* for water softening
4. *Soaps* and *wetting agents* for lubrication
5. *Phosphorus, chlorine,* and *sulfur compounds* for chemical lubrication
6. *Glycols* to act as blending agents
7. *Germicides* to control bacteria growth

As a result of the chemical agents which are added to the cooling qualities of water, synthetic fluids provide the following advantages.

1. Good rust control
2. Resistance for long periods of time to becoming rancid
3. Reduction of the amount of heat generated during a cutting action
4. Excellent cooling qualities
5. Longer durability than cutting or soluble oils
6. Nonflammable, nonsmoking
7. Nontoxic
8. Easy separation from the work and chips, which makes them clean to work with
9. Quick settling of grit and fine chips so they are not recirculated in the cooling system
10. No clogging of the machine cooling system due to detergent action of the fluid

Chemical cutting fluids are manufactured in three types: *true solution fluids, wetting-agent types,* and *wetting-agent types with EP lubricants.*

1. *True solution fluids* contain mostly rust inhibitors and are used primarily to prevent rust and provide rapid heat removal in grinding operations. These are generally clear solutions (sometimes a dye is added to color the water) and are mixed *1 part* solution to *50 to 250 parts* water, depending on the application.

Some of the true solution types have a tendency to form hard, crystalline deposits when the water evaporates. These deposits may interfere with the operation of chucks, slides, and moving parts.

2. *Wetting-agent types* contain agents which improve the wetting action of water, providing more uniform heat dissipation and antirust action. They also contain mild lubricants, water softeners, and antifoaming agents.

Wetting-type chemical cutting fluids are versatile; they have excellent lubricating qualities and provide rapid heat dissipation. They can be used when machining is done with either high-speed steel or carbide cutting tools.

3. *Wetting-agent types with EP lubricants* are similar to wetting-agent types, but have chlorine, sulfur, or phosphorus additives to provide EP or boundary lubrication effects. They are used for tough machining jobs with either high-speed steel or carbide cutting tools.

Caution: Although chemical cutting fluids have been widely accepted and used for many types of metal-cutting operations, there are certain precautions which should be observed regarding their use. Chemical cutting fluids are generally used on ferrous metals; however, many aluminum alloys can be machined successfully with them. Most chemical cutting fluids are not recommended for use on alloys of magnesium, zinc, cadmium, or lead. Certain types of paint (generally poor quality) may be affected by some chemical cutting fluids which may mar the machine's appearance and allow paint to get into the coolant. Before changing to any type of cutting fluid, it is wise to contact suppliers for the right cutting fluid for the machining operation and the metal being cut.

FUNCTIONS OF A CUTTING FLUID

The prime functions of a cutting fluid are to provide both cooling and lubrication. In addition, good cutting fluids prolong the cutting-tool life, possess good resistance to becoming rancid, and provide rust control.

COOLING

Laboratory tests have proved that the heat produced during machining has a definite bearing on cutting-tool wear. Temperature reduction of the cutting tool is important to tool life. Even a small reduction in temperature will result in a large extension in the life of the cutting tool. For example, if the tool temperature were reduced only 50°F (27.7°C) from 950 to 900°F (510 to 482.3°C), the cutting-tool life would be increased five times, from 19.5 to 99 min.

There are two sources of heat generated during a cutting action.

1. The plastic deformation of the metal which occurs immediately ahead of the cutting tool accounts for approximately two-thirds to three-quarters of the heat generated.
2. The friction resulting from the chip sliding along the cutting-tool face also produces heat.

To date, water has been found to be the most effective agent for reducing the heat generated during machining. Since water alone causes rusting, soluble oils or chemicals which prevent rust and provide other essential qualities are added to make it a good cutting fluid.

An abundant supply of cutting fluid should be applied to the machining area at a very low pressure. This will assure that the machining area will be well covered and that little splashing will occur. The flow of the cutting fluid will help to wash away the chips from the machining area.

LUBRICATING

The lubricating function of a cutting fluid is as important as its cooling function. Heat is generated by the plastic deformation of the metal and by the friction between the chip and tool face. The plastic deformation of metal occurs along the shear plane (Fig. 85-1). Any way of shortening the length of the shear plane would result in a reduction in the amount of heat that is generated.

The only known method of shortening the length of the shear plane for any given shape of a cutting tool and work material is to reduce the friction between the chip and tool face. Fig 85-2 illustrates a chip sliding along the face of the cutting tool. The enlarged illustration shows the irregularities on the tool face which create areas of friction and tend to cause a built-up edge to form. Note also that because of this friction there is a long shear plane and a small shear

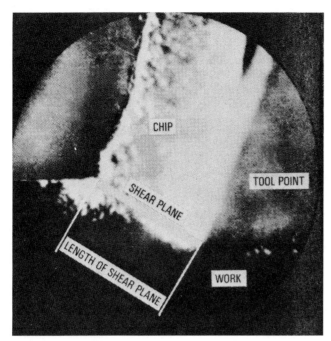

Fig. 85-1 During the cutting process the metal is deformed along the shear plane, producing heat. (Courtesy Cincinnati Milacron Incorporated.)

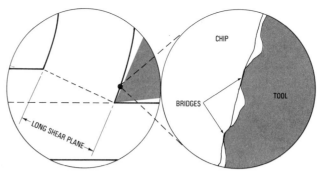

Fig. 85-2 A long shear plane results in great heat at the shear zone. (Courtesy Cincinnati Milacron Incorporated.)

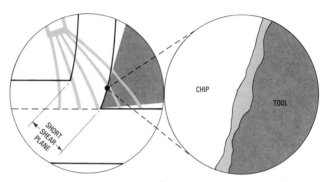

Fig. 85-3 Cutting fluid reduces friction and produces a shorter shear plane. (Courtesy Cincinnati Milacron Incorporated.)

angle. Most heat is created at the cutting edge when there is a small shear angle and a long shear plane.

Figure 85-3 illustrates the same depth of cut as Fig. 85-2, but shows cutting fluid being used to reduce the friction at the chip-tool interface. As soon as the friction is reduced, the shear plane becomes shorter, and the area where plastic deformation occurs is correspondingly smaller. Therefore, by reduction of the friction at the chip-tool interface, both sources of heat (plastic deformation and friction at the chip-tool interface) can be reduced.

The effective life of a cutting tool can be greatly lengthened if the friction and the resultant heat generated are reduced. When steel is machined, the temperature and pressure at the chip-tool interface may reach 1000°F (537.7°C) and 200,000 psi (1,379,000 kPa), respectively. Under such conditions, some oils and other liquids tend to vaporize or be squeezed out from between the chip and tool. Extreme-pressure lubricants reduce the amount of heat-producing friction. The EP chemicals of synthetic fluids combine chemically with the sheared metal of the chip to form solid compounds. These solid compounds or lubricants can withstand high pressure and temperature and allow the chip to slide up the tool face easily even under these conditions.

CUTTING-TOOL LIFE

Heat and friction are the prime causes of cutting-tool breakdown. Decreasing the amount of heat and friction created during a machining operation can greatly increase the life of a cutting tool. Laboratory tests have proved that if the temperature at the chip-tool interface is reduced by as little as 50°F (27.7°C), the life of the cutting tool increases fivefold. As a result, when cutting fluids are used, faster speeds and feeds can be used; increased production and a reduction in the cost per piece result.

During a cut, pieces of metal tend to weld themselves to the tool face, causing a built-up edge to form (Fig. 85-4). If the built-up edge becomes large and flat along the tool face, the effective rake angle of the cutting tool is decreased and more power is required to cut the metal. The built-up edge keeps breaking off and reforming; the result is a poor surface finish, excessive flank wear, and cratering of the tool face. Almost all the roughness of a machined surface is caused by tiny fragments of metal which have been left behind by the built-up edge.

The use of an effective cutting fluid affects the action of a cutting tool in the following ways.

1. It lowers the heat created by the plastic deformation of the metal; thereby it increases cutting-tool life.
2. The friction at the chip-tool interface is decreased; the resultant heat is reduced.
3. Less power is required for machining because of the reduced friction.
4. It prevents a built-up edge from forming; longer tool life is the result.
5. The surface finish of the work is greatly improved.

RUST CONTROL

Cutting fluids used on machine tools should prevent rust from forming; otherwise the parts of machines and the workpieces will be damaged. Cutting oil prevents rust from forming but does not cool as effectively as water. Water is the best and most economical coolant but causes parts to rust unless rust inhibitors are added.

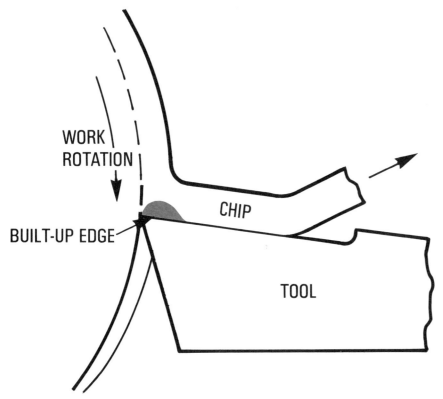

Fig. 85-4 A built-up edge is caused by chip fragments pressure welding to the cutting-tool face.

RELATIVE + CHARGE

POLAR FILM   PASSIVATING FILM

Fig. 85-5 Polar and passivating films on metal surfaces prevent rust. (Courtesy Cincinnati Milacron Incorporated.)

Rust is oxidized iron, or iron that has reacted chemically with oxygen, water, and minerals in the water. Water alone on a piece of steel or iron acts as a medium for the electro-chemical process to start causing corrosion or rust.

Chemical cutting fluids contain rust inhibitors which prevent the electro-chemical process of rusting. Some types of cutting fluids form a *polar film* on all metals they contact which prevents rusting. This polar film (Fig. 85-5) consists of negatively charged, long, thin molecules which are attracted and firmly bond themselves to the metal. This invisible film, only molecules thick, is sufficient to prevent the electro-chemical action of rusting. Other types of cutting fluids contain rust inhibitors which form an insulating blanket known as a *passivating film* on the metal surface.

These inhibitors combine chemically with the metal and form a nonporous, protective coating which prevents rust.

RANCIDITY CONTROL

When lard oil was the only cutting fluid used, after a few days it would start to spoil and give off an offensive odor. This rancidity is caused by bacteria and microscopic organisms growing and multiplying almost anywhere and eventually causing bad odors to form. Today, any cutting fluid that has an offensive smell is termed *rancid*.

Most cutting fluids contain bactericides which control the growth of bacteria and make the fluids more resistant to rancidity. The bactericide, which is added to the fluid by the manufacturers, must be strong enough to control the growth of bacteria, but weak enough not to be harmful to the skin of the operator.

APPLICATION OF CUTTING FLUIDS

Cutting-tool life and the machining operation are greatly influenced by the way that the cutting fluid is applied. It should be supplied in a copious stream under low pressure so that the work and cutting tool are well covered. The rule of thumb is that the inside diameter of the supply nozzle should be about three-quarters the width of the cutting tool. The fluid should be directed to the area where the chip is being formed to reduce and control the heat created during the cutting action and to prolong tool life.

Fig. 85-7 Bottom and top nozzles are used to supply cutting fluid for heavy-duty turning operations.

"A" INSIDE DIAMETER OF PIPE OR TUBE

Fig. 85-6 Cutting fluid supplied by one nozzle for general turning and facing operations.

LATHE-TYPE OPERATIONS

On horizontal-type turning and boring machines, cutting fluid should be applied to that portion of the cutting tool which is producing the chip. For general turning and facing operations, cutting fluid should be supplied directly over the cutting tool, close to the zone of chip formation (Fig. 85-6). In heavy-duty turning and facing operations, it is recommended that cutting fluid be supplied by two nozzles, one directly above and the other directly below the cutting tool (Fig. 85-7).

DRILLING AND REAMING

The most effective method of applying cutting fluids for these operations is to use "oil-feed" drills and hollow-shank reamers. Tools of this type transmit the cutting fluid directly to the cutting edges and at the same time flush the chips out of the hole (Fig. 85-8). When conventional drills and reamers are used, an abundant supply of fluid should be applied to the cutting edges.

MILLING

In *slab milling*, cutting fluid should be directed to both sides of the cutter by fan-shaped nozzles approximately three-quarters the width of the cutter (Fig. 85-9).

For *face milling*, a ring-type distributor (Fig. 85-10) is recommended in order to flood the cutter completely. Keeping each tooth of the cutter immersed in cutting fluid at all times can increase cutter life almost 100 percent.

GRINDING

Cutting fluid is very important in a grinding operation; it cools the work and keeps the grinding wheel clean. Cutting fluid should be applied in large quantities and under very little pressure.

1. *Surface Grinding*. Three methods may be used to apply the cutting fluid for surface grinding operations.

Fig. 85-8 Cutting fluid being supplied through a hole in the center of a reamer. (Courtesy Cincinnati Milacron Incorporated.)

Fig. 85-9 Cutting fluid being supplied to both sides of the cutter in slab milling. (Courtesy Cincinnati Milacron Incorporated.)

a. The *flood method* is most commonly used and consists of a steady flow of cutting fluid being applied through a nozzle. Because of the reciprocating table action, most surface-grinding operations can be greatly improved if the fluid is supplied through two nozzles as in Fig. 85-9.

563

TABLE 85-1 Recommended Cutting Fluids for Various Materials

MATERIAL	DRILLING	REAMING	THREADING	TURNING	MILLING
Aluminum	Soluble oil Kerosene Kerosene and lard oil	Soluble oil Kerosene Mineral oil	Soluble oil Kerosene and lard oil	Soluble oil	Soluble oil Lard oil Mineral oil Dry
Brass	Dry Soluble oil Kerosene and lard oil	Dry Soluble oil	Soluble oil Lard oil	Soluble oil	Dry Soluble oil
Bronze	Dry Soluble oil Lard oil	Dry Soluble oil Lard oil	Soluble oil Lard oil	Soluble oil	Dry Soluble oil Lard oil
Cast iron	Dry Air jet Soluble oil	Dry Soluble oil Mineral lard oil	Dry Sulfurized oil Mineral lard oil	Dry Soluble oil	Dry Soluble oil
Copper	Dry Soluble oil Mineral lard oil Kerosene	Soluble oil Lard oil	Soluble oil Lard oil	Soluble oil	Dry Soluble oil
Malleable iron	Dry Soda water	Dry Soda water	Lard oil Soda water	Soluble oil	Dry Soda water
Monel metal	Soluble oil Lard oil	Soluble oil Lard oil	Lard oil	Soluble oil	Soluble oil
Steel alloys	Soluble oil Sulfurized oil Mineral lard oil	Soluble oil Sulfurized oil Mineral lard oil	Sulfurized oil Lard oil	Soluble oil	Soluble oil Mineral lard oil
Steel, machine	Soluble oil Sulfurized oil Mineral lard oil	Soluble oil Mineral lard oil	Soluble oil Mineral lard oil	Soluble oil	Soluble oil Mineral lard oil
Steel, tool	Soluble oil Sulfurized oil Mineral lard oil	Soluble oil Sulfurized oil Lard oil	Sulfurized oil Lard oil	Soluble oil	Soluble oil Lard oil

Note: Chemical cutting fluids can be used successfully for most of the above cutting operations. These concentrates are diluted with water in proportions ranging from 1 part cutting fluid to 15 and as high as 100 parts of water, depending upon the metal being cut and the type of machining operation. When using chemical cutting fluids, it is wise to follow the manufacturer's recommendations for use and mixture.
Source: Courtesy Cincinnati Milacron Incorporated.

b. *Through-the-wheel* is a method in which the coolant is fed to a special wheel flange and is forced to the periphery of the wheel and to the area of contact by centrifugal force.

c. *Mist spray,* where the coolant is siphoned from a reservoir by a stream of air and a fine spray is directed through a nozzle to the area of contact.

2. *Cylindrical Grinding.* For cylindrical grinding operations, it is important that the entire contact area between the wheel and work be flooded with a steady stream of clean, cool cutting fluid. A fan-shaped nozzle (Fig. 85-11) somewhat wider than the wheel should be used to direct the cutting fluid.

3. *Internal Grinding.* During internal grinding, the cutting fluid must flush the chips and the abrasive wheel particles out of the hole being ground. Because internal grinding practices call for using as large a wheel as possible, it is sometimes difficult to get enough cutting fluid into the hole. A compromise must be made between the grinding wheel size and the amount of fluid entering the hole. As much fluid as possible should be applied during internal grinding operations.

Fig. 85-10 Cutting fluid being applied by a ring-type distributor for a face-milling operation.

Fig. 85-11 Cutting fluid being applied to the contact zone during cylindrical grinding. (Courtesy Cincinnati Milacron Incorporated.)

REVIEW QUESTIONS

1. What is the function of a modern cutting fluid?
2. Briefly trace the development of cutting fluids.

PURPOSE AND ADVANTAGES

3. Explain the causes of heat and friction during a machining process.
4. Name four economical advantages which result from applying correct cutting fluids.

CHARACTERISTICS OF A GOOD CUTTING FLUID

5. List six *important* characteristics that a good cutting fluid should possess.

TYPES OF CUTTING FLUIDS

6. Name three categories into which cutting fluids fall.
7. Describe active and inactive cutting oils.
8. What type of cutting oil should be used for the following:
   (a) tough, ductile metals?
   (b) heavy-duty machining?
   (c) nonferrous metals?
   (d) threading white metal?
9. Describe the composition of an emulsifiable oil and state its advantages.
10. State the purpose of:
    (a) emulsifiable mineral oil
    (b) EP emulsifiable oil
11. Discuss and state six important advantages of chemical cutting fluids.
12. State the purpose of:
    (a) true solution fluids
    (b) wetting-agent types with EP lubricants

FUNCTIONS OF A CUTTING FLUID

13. Name five functions of a cutting fluid.
14. Discuss the importance of lubricating and cooling as applied to cutting fluids.
15. Explain how a cutting fluid can change the length of the shear plane.
16. For what purpose are EP lubricants used?
17. What occurs at the cutting-tool face during a cut?
18. What is the main cause of surface roughness?
19. How does the application of cutting fluid affect the cutting tool?
20. Why is the control of rust important?
21. Describe a *polar film*, a *passivating film*.
22. Define *rancidity* and state the purpose of bactericides.

APPLICATION OF CUTTING FLUIDS

23. What is the general recommendation for applying cutting fluids?
24. How should cutting fluid be applied for lathe operations?
25. State two methods of applying cutting fluid when drilling or reaming is done.
26. Explain how cutting fluid should be applied for:
    (a) slab milling
    (b) face milling
    (c) cylindrical grinding
27. Describe three methods of applying cutting fluid for surface-grinding operations.
28. Why is it sometimes difficult to apply cutting fluid during internal grinding operations?

METALLURGY
SECTION 19

Understanding the properties and heat treatment of metals has become increasingly important to machinists during the past two decades. Study of metal properties and development of new alloys have facilitated reduction in mass and increase in strength of machines, automobiles, aircraft, and many present-day commodities.

The most commonly used metals today are ferrous metals, or those which contain iron. The composition and properties of ferrous materials may be changed by the addition of various alloying elements during manufacture, to impart the desired qualities to the material. Cast iron, machine steel, carbon steel, alloy steel, and high-speed steel are all ferrous metals, each having different properties.

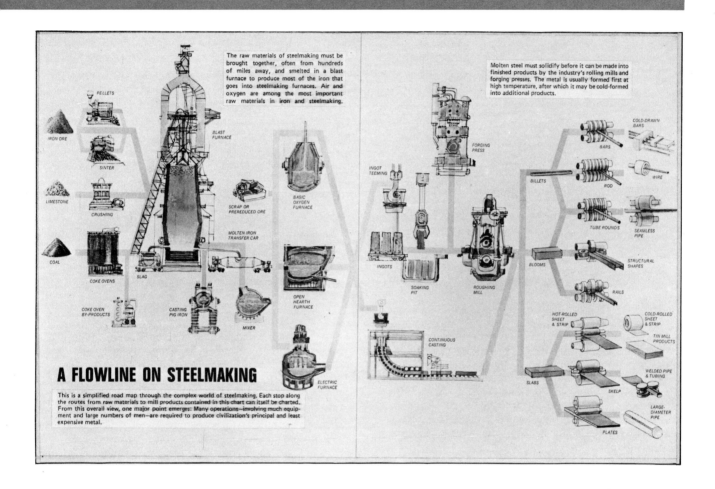

The raw materials of steelmaking must be brought together, often from hundreds of miles away, and smelted in a blast furnace to produce most of the iron that goes into steelmaking furnaces. Air and oxygen are among the most important raw materials in iron and steelmaking.

Molten steel must solidify before it can be made into finished products by the industry's rolling mills and forging presses. The metal is usually formed first at high temperature, after which it may be cold-formed into additional products.

PELLETS

IRON ORE

SINTER

LIMESTONE

CRUSHING

COAL

COKE OVENS

COKE OVEN BY-PRODUCTS

SLAG

CASTING PIG IRON

MIXER

BLAST FURNACE

SCRAP OR PREREDUCED ORE

MOLTEN IRON TRANSFER CAR

BASIC OXYGEN FURNACE

OPEN HEARTH FURNACE

ELECTRIC FURNACE

INGOT TEEMING

INGOTS

SOAKING PIT

CONTINUOUS CASTING

FORGING PRESS

ROUGHING MILL

BILLETS

BLOOMS

SLABS

SKELP

PLATES

BARS

ROD

TUBE ROUNDS

RAILS

HOT-ROLLED SHEET & STRIP

COLD-DRAWN BARS

WIRE

SEAMLESS PIPE

STRUCTURAL SHAPES

COLD-ROLLED SHEET & STRIP

TIN MILL PRODUCTS

WELDED PIPE & TUBING

LARGE-DIAMETER PIPE

A FLOWLINE ON STEELMAKING

This is a simplified road map through the complex world of steelmaking. Each stop along the routes from raw materials to mill products contained in this chart can itself be charted. From this overall view, one major point emerges: Many operations—involving much equipment and large numbers of men—are required to produce civilization's principal and least expensive metal.

Metals such as iron, aluminum, and copper are among the most common elements found in nature. Iron, which is found in most parts of the world, was considered a rare and precious metal in ancient times. Throughout the ages, iron was transformed into steel, and today it is one of our more versatile metals. Almost every product made today contains some steel or was manufactured by tools made of steel. Steel can be made hard enough to cut glass, pliable as the steel in a paper clip, flexible as steel in springs, or strong enough to withstand enormous stress. Metals may be shaped in six ways: casting, forging, rolling and bending, drawing and forming, cutting, and joining. To better understand this versatile metal, a knowledge of its properties and manufacture is desirable.

OBJECTIVES

After completing this unit you will be able to:
1. Identify six properties found in metals
2. Understand the processes by which iron and steel are made
3. Understand the effect of alloying elements on steel

Physical metallurgy is the science concerned with the physical and mechanical properties of metals. The properties of metals and alloys are affected by three variables:

1. *Chemical Properties*—those that a metal attains through the addition of various chemical elements
2. *Physical Properties*—those that are not affected by outside forces such as color, density, conductivity, or melting temperature
3. *Mechanical Properties*—those that are affected by outside forces such as rolling, forming, drawing, or machining

PHYSICAL PROPERTIES OF METALS

To better understand the use of the various metals, one should be familiar with the following terms:

Brittleness (Fig. 86-1A) is that property of a metal which permits no permanent distortion before breaking. Cast iron is a brittle metal; it will break rather than bend under shock or impact.
Ductility (Fig. 86-1B) is the ability of the metal to be permanently deformed without breaking. Metals such as copper and machine steel, which may be drawn into wire, are ductile materials.
Elasticity (Fig. 86-1C) is the ability of a metal to return to its original shape after any force acting upon it has been removed. Properly heat treated springs are good examples of elastic materials.
Hardness (Fig. 86-1D) may be defined as the resistance to forceable penetration or plastic deformation.
Malleability (Fig. 86-1E) is that property of a metal which permits it to be hammered or rolled into other sizes and shapes.
Tensile strength (Fig. 86-1F) is the maximum amount of pull that a material will withstand before breaking. It is expressed as the number of pounds per square inch (on inch testers) or in kilograms per square centimeter (on metric testers) of pull required to break a bar having a one-square-inch or one-square-centimeter cross section.
Toughness is the property of a metal to withstand shock or impact. Toughness is the opposite condition to brittleness.

MANUFACTURE OF FERROUS METALS

PIG IRON

Production of pig iron in the blast furnace (Fig. 86-3) is the first step in the manufacture of cast iron or steel.

Fig. 86-1 (A) Brittle metals will not bend but break easily. (B) Ductile metals are easily deformed. (C) Elastic metals return to their original shape after the load is removed. (D) Hard metals resist penetration. (E) Malleable metals may be easily formed or shaped. (F) Tensile strength is the amount that a metal will resist a direct pull.

Raw Materials

Iron ore is the chief raw material used to make iron and steel. The main sources of iron ore in North America are at Steep Rock, the Ungava district near the Quebec-Labrador border, and the Mesabi range, situated at the western end of Lake Superior. The most important iron ores are:

Hematite which contains about 70 percent iron and varies in color from black to brick red.
Limonite is a brownish ore similar to hematite, but contains water. When the water has been removed by roasting, the ore resembles hematite.
Magnetite, a rich, black ore, contains a higher percentage of iron than any other ore but is not found in large quantities.
Taconite, a low-grade ore containing about 25 to 30 percent iron, must be specially treated before it is suitable for reduction into iron.

Pelletizing Process

Low-grade iron ores are uneconomical to use in the blast furnace and, as a result, go through a pelletizing process, where most of the rock is removed and the ore is brought to a higher iron concentration. Some steelmaking firms are now pelletizing most ores to reduce transportation costs and the problems of pollution and slag disposal at the steel mills.

The crude ore is crushed and ground into a powder and passed through magnetic separators where the iron content is increased to about 65 percent (Fig. 86-2A). This high-grade material is mixed with clay and formed into pellets about ½ to ¾ in. (12.7 to 19.05 mm) in diameter in a pelletizer. The pellets at this stage are covered with coal dust and sintered (baked) at 2354°F (1290°C) (Fig. 86-2B). The resultant hard, highly concentrated pellets will remain intact during transportation and loading into the blast furnace.

Coal, after being converted to coke, is used to supply the heat to reduce the iron ore. The burning coke produces carbon monoxide which removes the oxygen from the iron ore and reduces it to a spongy mass of iron.

Limestone is used as a flux in the production of pig iron to remove the impurities from the iron ore.

MANUFACTURE OF PIG IRON

In the blast furnace (Fig. 86-3) iron ore, coke, and limestone are fed into the top of the furnace by means of a skip car. Hot air at 1000°F (537.7°C) is fed into the bottom of the furnace through the bustle pipe and tuyeres. After the coke is ignited, the hot air makes it burn vigorously. Carbon monoxide, produced by the burning coke, combines with the oxygen in the iron ore, reducing it to a spongy mass of iron. The iron gradually seeps down through the charge and collects in the bottom of the furnace. During this

Crusher Magnetic Separator Balling Drum Pellet Hardening Furnace

3 TONS
25% IRON
A

1 TON
WASTE
2 TONS
40% IRON

1 TON
65.5% IRON
B

1.1 TON
63.25% IRON

Fig. 86-2 (A) The first step in pelletizing separates the iron ore from the rock. (B) Iron ore pellets are hardened in the furnace.

SKIP CAR DUMPING

RECEIVING HOPPER

SMALL BELL

LARGE BELL

HOT GAS

HOT GAS

DOWNCOMER

GAS

205°C

DUST CATCHER

CHECKERS

TYPICAL REFRACTORY BRICK

COMBUSTION CHAMBER

SKIP HOIST

STOVE

STOVE

DUST

STOVE
HEATING

STOVE
BLOWING

BLAST FURNACE

TO STACK

COLD AIR BLAST

650°C

HOT AIR BLAST

GAS MAIN

BUSTLE PIPE

1650°C

TUYERES

CINDER NOTCH
TAPPING HOLE

SLAG
IRON

SLAG RUNNER

HOT METAL RUNNER

HOT METAL CAR

CAST HOUSE

SLAG LADLE

Fig. 86-3 A blast furnace produces pig iron. (Courtesy American Iron and Steel Institute.)

process, the decomposed limestone acts as a flux and unites with the impurities (silica and sulfur) in the iron ore to form a slag, which also seeps to the bottom of the furnace. Since the slag is lighter, it floats on the top of the molten iron. Every 6 h, the furnace is tapped. The slag is drawn off first and then the molten iron is poured into ladles. The iron may be further processed into steel or cast into *pigs,* which are used by foundries in the manufacture of castings.

The manufacture of pig iron is a continuous process, and the blast furnaces are shut down only for repair or rebricking.

MANUFACTURE OF CAST IRON

Most of the pig iron manufactured in a blast furnace is used to make steel. However, a considerable amount is used to manufacture cast-iron products. Cast iron is manufactured in a cupola furnace which resembles a huge stovepipe (Fig. 86-4).

Layers of coke, solid pig iron, and scrap iron are charged into the top of the furnace. After the furnace is charged, the fuel is ignited and air is forced in near the bottom to aid combustion. When the iron is melted, it settles to the bottom of the furnace and is then tapped into ladles. The molten iron is poured into sand molds of the required shape and the metal assumes the shape of the mold. After the metal has cooled, the castings are removed from the molds.

The principal types of cast-iron castings are:

Gray-iron castings, made from a mixture of pig iron and steel scrap, are the most widely used. They are made into a wide variety of products including bathtubs, sinks, and parts for automobiles, locomotives, and machinery.

Chilled-iron castings are made by pouring molten metal into metal molds so that the surface cools rapidly. The surface of such castings becomes very hard, and the castings are used for crusher rolls or other products requiring a hard, wear-resistant surface.

Alloyed castings contain certain amounts of alloys such as chromium, molybdenum, nickel, etc. Castings of this type are used extensively by the automobile industry.

Malleable castings are made from a special grade of pig iron and foundry scrap. After these castings have solidified, they are annealed in special furnaces. This makes the iron malleable and resistant to shock.

MANUFACTURE OF STEEL

Before molten pig iron from the blast furnace can be converted into steel, most of the impurities must be burned out. This may be done in one of four types of furnaces: the *open-hearth furnace,* the *Bessemer converter,* the *basic oxygen process furnace,* and the *electric furnace.*

Open-Hearth Process

For several decades, the open-hearth furnace was the ''work horse'' of the steel industry. At one time, almost 90 percent of the steel produced in North America was made in the open-hearth furnace. Although many steel companies still use the modified open-hearth furnace, this process is grad-

Fig. 86-4 A cupola furnace is used to make cast iron.

ually giving way to more efficient methods of producing steel. The furnace is a large, rectangular, brick structure, 40 to 50 ft (12.19 to 15.24 m) long and 15 to 17 ft (4.57 to 5.18 m) wide (Fig. 86-5). The *hearth* is a large dish-shaped structure about 25 to 35 in. (63.5 to 88.9 cm) deep. Above each end of the hearth is a *burner* and an inlet for hot air from the *checkers,* which are large chambers filled with firebrick placed in a checkerboard fashion.

When the furnace is charged, limestone is put in and steel scrap is then added. After the scrap has been melted, molten pig iron is poured into the hearth. Large tongues of flame, produced by the union of fuel and the hot gas from the checkers, sweep over the molten metal from the right side (Fig. 86-5) and burn the impurities out of it. The burned gases from this operation are drawn off through the checkers on the left side and heat these checkers. At intervals of about 10 to 15 min, the direction of the air and the flame are reversed and pass from left to right over the molten metal. During this operation, the limestone unites with the impurities and rises to the top as slag.

After 7 to 12 h, the furnace is tapped and the molten steel is run into a ladle which is just large enough to hold the batch (heat) of steel; the slag overflows through a spout in the ladle. Alloying materials, such as silicon, manganese, or other elements, are then added to the molten steel to give it the desired properties. The steel is poured, or *teemed,* into ingot molds and after solidification the ingots are removed from the molds and transferred to soaking pits. Here they are brought to a uniform temperature prior to rolling into the desired shapes and sizes. Open-hearth furnaces have a capacity of up to 550 tons (498.85 t) of steel.

Fig. 86-5 A schematic cross section of an open-hearth furnace.   (Courtesy American and Steel Institute.)

Since the introduction of the basic oxygen furnace, steel-makers have found that the addition of oxygen to any steelmaking process will increase production considerably. Consequently, many open-hearth furnaces have been updated with the addition of an oxygen lance (Fig. 86-5), which directs almost pure oxygen at a high velocity onto the top of the molten steel in the hearth. The oxygen burns the impurities out of the steel much quicker than was possible by the straight open-hearth process. As a result of this modification, the production of these furnaces has been greatly increased.

Bessemer Converter

At one time the Bessemer converter was used to produce almost all the steel made. Now, however, the open-hearth furnace introduced in 1908 produces most of the steel, and the Bessemer converter makes only 3 percent of it.

Molten iron from the blast furnace is poured into the converter which is tipped on its side to receive it. The converter is then brought to a vertical position and a blast of air is blown through the holes in the bottom of it. Oxygen in the air burns out the impurities in the hot iron. After 10 to 15 min, the converter is tilted forward and the steel flows into a ladle. Alloying materials are added to the steel in the ladle to bring it to the correct chemical composition required. From the ladle the steel is teemed into ingots.

Basic Oxygen Process

One of the newest developments in steel-making is the basic oxygen process.

The basic oxygen furnace (Fig. 86-6) resembles a Bessemer converter but does not have the air chamber and tuyeres at the bottom to admit air through the charge. Instead of air being forced through the molten metal, as in the Bessemer process, a high-pressure stream of pure oxygen is directed onto the top of the molten metal.

The furnace is tilted and is first charged with scrap metal (about 30 percent of the total charge). Molten pig iron is poured into the furnace, after which the required fluxes are added. An oxygen lance with a water-cooled hood is then lowered into the furnace until the tip is within 60 to 100 in. (152 to 254 cm) above the surface of the molten metal, depending on the blowing qualities of the iron and the type of scrap used. The oxygen is turned on and flows at the rate of 5000 to 6000 ft^3/min (141.5 to 169.9 m^3/min) at a pressure of 140 to 160 psi (965.3 to 1103.2 kPa).

The introduction of the oxygen causes the temperature of molten steel (batch) to rise, at which time lime and fluospar may be added to help carry off the impurities in the form of slag. A recent development in this process is the Lance Bubbling Equilibrium (LBE). With this process, inert gases such as argon and nitrogen are introduced by lances through the bottom of the furnace. These gases bubble up through the molten metal increasing the contact between the metal and the slag and speed mixing. This results in an increase in yield and a reduction in alloying elements such as aluminum or silicon which are used to reduce the oxygen and carbon content in the steel.

The force of the oxygen starts a high-temperature churning action and burns out the impurities. After all the impurities have been burned out, there will be a noticeable

CHARGING SCRAP INTO FURNACE [A]

CHARGING MOLTEN IRON [B]

ADDITION OF BURNT LIME [C]

[D]

BLOWING WITH OXYGEN

TAPPING THE FURNACE [E]

POURING THE SLAG [F]

Fig. 86-6   The basic oxygen process.   (Courtesy Inland Steel Corporation.)

ELECTRODES

ELECTRIC ARC

SLAG

MOLTEN STEEL

POURING POSITION

STEEL LADLE

Fig. 86-7   Diagram of the electric furnace.

drop in the flame and a definite change in sound. The oxygen is then shut off and the lance removed.

The furnace is now tilted (Fig. 86-6) and the molten steel flows into a ladle or taken directly to the strand casting machine. The required alloys are added, after which the molten metal is teemed into ingots. The refining process takes only about 50 min and about 300 tons 272.1 t) of steel can be made per hour.

Electric Furnace

The electric furnace (Fig. 86-7) is used primarily to make fine alloy and tool steels. The heat, the amount of oxygen, and the atmospheric condition can be regulated at will in the electric furnace; this furnace is therefore used to make steels that cannot be readily produced in any other way.

Carefully selected steel scrap, containing smaller amounts of the alloying elements than are required in the finished steel, is loaded into the furnace. The three carbon electrodes are lowered until an arc jumps from them to the scrap. The heat generated by the electric arcs gradually melts all the steel scrap. Alloying materials such as chromium, nickel, tungsten, etc., are then added to make the type of alloy steel required. Depending upon the size of the furnace, it takes from 4 to 12 h to make a heat of steel. When the metal is ready to be tapped, the furnace is tilted forward and the steel flows into a large ladle. From the ladle, the steel is teemed into ingots.

Steel Processing

After the steel has been properly refined in any of the furnaces, it is tapped into ladles where alloying elements and deoxidizers may be added. The molten steel may then be teemed into *ingots* weighing as much as 20 tons

(18.14 t), or it may be formed directly into slabs by the *continuous-casting process* (Fig 86-8).

The steel is teemed into ingot molds and allowed to solidify. The ingot molds are then removed or *stripped* and the hot ingots are placed in soaking pits at 2200°F (1204°C) for up to 1.5 h to make them a uniform temperature throughout. The ingots are then sent to the rolling mills where they are rolled and reduced in cross section to form blooms, billets, or slabs.

Blooms are generally rectangular or square and are larger than 36 in.³ in cross section. They are used to manufacture structural steel and rails.

Billets may be rectangular or square, but are less than 36 in.² in cross-sectional area. They are used to manufacture steel rods, bars and pipes.

Slabs are usually thinner and wider than billets. They are used to manufacture plate, sheet, and strip steel.

Strand or Continuous Casting

Strand or continuous casting, Fig. 86-8, is the most modern and efficient method of converting molten steel into semi-finished shapes such as slabs, blooms, and billets. This process eliminates ingot teeming, stripping, soaking, and rolling. It also produces higher quality steel, reduces energy consumption, and has increased overall productivity by over 13 percent.

Continuous casting is fast becoming the major method of producing steel slabs and billets. This, combined with the much faster production of molten steel by the basic oxygen process, has greatly improved the efficiency of steel production. In 1982, 31 percent of the steel produced in the United States was by the strand or continuous casting method. In the same year, 53 percent of the steel produced in Europe and 85 percent of Japanese steel was produced by this method.

Molten steel from the furnace is taken in a ladle to the top of the strand or continuous caster and poured into the tundish. The tundish provides an even pool of molten metal to be fed into the casting machine. It also acts as a reservoir permitting the empty ladle to be removed and the full ladle to be poured without interrupting the flow of molten metal to the caster. The steel is stirred continuously by a nitrogen lance or by electromagnetic divices.

The molten steel drops in a controlled flow from the tundish into the mold section. Cooling water quickly chills the outside of the metal to form a solid skin which becomes thicker as the steel strand descends through the cooling system. As the strand reaches the bottom of the machine, it becomes solid throughout. The solidified steel is moved in a gentle curve by bending rolls until it reaches a horizontal position. The strand is then cut by a traveling cutting torch into required lengths. In some strand casting machines, the solidified steel is cut when it is in the vertical position. The slab or billet then topples to the horizontal position and is taken away.

Continuous casters are capable or producing strands at up to 15 ft per min (4.57 m per min). This process is designed for high tonnage operations or small batches, as required. When small batches are required, the molten steel is generally produced by methods other than the basic oxygen or open hearth furnaces.

After the metal has been made into blooms or slabs by either of the forementioned processes, it is further rolled into billets and then, while still hot, into the desired shape, such as round, flat, square, hexagonal, etc. (Fig. 86-9). These rolled products are known as *hot-rolled steel* and are easily identified by the bluish-black scale on the outside.

Hot-rolled steel may be further processed into *cold-rolled* or *cold-drawn steel* by removing the scale in an acid bath and passing the metal through rolls or dies of the desired shape and size.

One of the newest processes in the production of sheet steel is the Continuous Annealing Line (CAL) which produces high-strength and ultra-high-strength cold roll steel without increased weight, mainly for the automotive industry. This process is capable of altering the steel strength without changing its chemistry. The sheet steel produced by this method is virtually free of defects such as edge damage or sticker breaks. Continuous annealing reduces the time from 5 to 6 days required for batch annealing to a mere 10 min. It also permits superior steel to be made with lesser amounts of costly alloying elements. Advanced computer controls permit the heating-cooling sequence to ''lock in'' the grain structure at the desired strength level. It can produce sheet steel with yield strength of up to 220,000 psi.

Vacuum Processing of Molten Steel Steel used in space and nuclear projects is often processed and solidified in a vacuum to remove oxygen, nitrogen, and hydrogen, and thus produce a high-quality steel.

CHEMICAL COMPOSITION OF STEEL

Although iron and carbon are the main elements in steel, certain other elements may be present in varying quantities. Some are present because they are difficult to remove, and others are added to impart certain qualities to the steel. The elements found in plain carbon steel are carbon, manganese, phosphorous, silicon, and sulfur.

Carbon is the element which has the greatest influence on the property of the steel, since it is the hardening agent. The hardness, hardenability, tensile strength, and wear resistance will be increased as the percentage of carbon is increased up to about 0.83 percent. After this point has been reached, additional carbon does not noticeably affect the hardness of the steel but increases wear resistance and hardenability.

Manganese, when added in small quantities (0.30 to 0.60 percent) during the manufacture of steel, acts as a deoxidizer or purifier. Manganese helps to remove the oxygen which, if it remained, would make the steel weak and brittle. Manganese also combines with sulfur, which in most cases is considered an undesirable element in steel. The addition of manganese increases the strength, toughness, hardenability, and shock resistance of steel. It will also slightly lower the critical temperature and increase ductility.

When manganese is added in quantities above 0.60 percent, it is considered an alloying element and will impart certain properties to the steel. When 1.5 to 2 percent man-

CONTINUOUS CASTING

1. Molten steel pours from a ladle into a reservoir called a tundish.

2. The metal flows out the bottom of the tundish at a carefully regulated rate into the mold, which is moving up and down to prevent the hot metal from sticking. The interior of the mold is hollow—just the size, in width and thickness, of the slab to be formed. Lining the walls are pipes through which water flows, chilling the metal. A thin shell of steel begins to solidify around the molten metal.

3. The gradually solidifying slab moves down through the secondary cooling zone. A series of rollers support the slab and gradually turn it into a horizontal position. Sprays of water under high pressure cool and harden the metal still further.

4. The ribbon of steel moves on to a level table.

5. A flame-cutting torch slices down through the metal. When the slab is cut off, it is carried on rollers to a cooling bed. The entire trip from the ladle has taken less than one-half hour.

Fig. 86-8   The continuous-casting process for making steel blooms or slabs.

Fig. 86-9   Various shapes of steel are produced by rolling.

TABLE 86-1 The Effect of Alloying Elements on Steel

EFFECT	ELEMENT											
	CARBON	CHROMIUM	COBALT	LEAD	MANGANESE	MOLYBDENUM	NICKEL	PHOSPHORUS	SILICON	SULFUR	TUNGSTEN	VANADIUM
Increases tensile strength	X	X			X	X	X					
Increases hardness	X	X										
Increases wear resistance	X	X			X		X				X	
Increases hardenability	X	X			X	X	X					X
Increases ductility					X							
Increases elastic limit		X				X						
Increases rust resistance		X					X					
Increases abrasion resistance		X			X							
Increases toughness		X				X	X					X
Increases shock resistance		X					X					X
Increases fatigue resistance												X
Decreases ductility	X	X										
Decreases toughness			X									
Raises critical temperature		X	X								X	
Lowers critical temperature					X		X					
Causes hot shortness										X		
Causes cold shortness								X				
Imparts red hardness			X			X					X	
Imparts fine grain structure					X							X
Reduces deformation					X	X						
Acts as deoxidizer					X				X			
Acts as desulphurizer					X							
Imparts oil hardening properties		X			X	X	X					
Imparts air hardening properties					X	X						
Eliminates blow holes								X				
Creates soundness in casting									X			
Facilitates rolling and forging					X					X		
Improves machinability				X						X		

ganese is added to high-carbon steel it will produce deep-hardening, nondeforming steel, which must be quenched in oil. Hard, wear-resistant steels, suitable for use in power shovel scoops, rock crushers, and grinding mills, are produced when up to 15 percent manganese is added to high-carbon steel.

Phosphorus is generally considered an undesirable element in carbon steel when present in amounts over 0.6 percent, since it will cause the steel to fail under vibration or shock. This condition is termed *cold-shortness.* Small amounts of phosphorus (about 0.3 percent) tend to eliminate blow holes and decrease shrinkage in the steel. Phosphorus and sulfur may be added to low-carbon steel (machine steel) to improve the machinability.

Silicon, present in most steels in amounts from 0.10 to 0.30 percent, acts as a deoxidizer and makes steel sound when it is cast or hot-worked. Silicon, when added in larger amounts (0.60 to 2 percent) is considered an alloying element. It is never used alone or simply with carbon; some deep-hardening element, such as manganese, molybdenum, or chromium, is usually added with silicon. When added as an alloying element, silicon increases the tensile strength, toughness, and hardness penetration of steel.

Sulfur, generally considered an impurity in steel, causes the steel to crack during working (rolling or forging) at high temperatures. This condition is known as *hot-shortness.* Sulfur may be added purposely to low-carbon steel in quantities ranging from 0.07 to 0.30 percent, to increase its machinability. Sulfurized, free-cutting steel is known as *screw stock* and is used in automatic screw machines.

CLASSIFICATION OF STEEL

Steel may be classified into two groups, *carbon steels* and *alloy steels.*

PLAIN CARBON STEELS

Plain carbon steels may be classified as those which contain only carbon and no other major alloying element. They are divided into three categories: low-carbon steel, medium-carbon steel, and high-carbon steel.

Low-carbon steel contains from 0.02 to 0.30 percent carbon by mass. Because of the low carbon content, this type of steel cannot be hardened, but can be case-hardened. *Machine steel* and *cold-rolled steel,* which contain from 0.08 to 0.30 percent carbon, are the most common low-carbon steels. These steels are commonly used in machine shops for the manufacture of parts that do not have to be hardened. Such items as bolts, nuts, washers, sheet steel, and shafts are made of low-carbon steel.

Medium-carbon steel contain from 0.30 to 0.60 percent carbon and is used where greater tensile strength is required. Because of the higher carbon content, this steel can be hardened, which makes it ideal for steel forgings. Tools such as wrenches, hammers, and screwdrivers are drop-forged from medium-carbon steel and later heat-treated.

High-carbon steel, also known as tool steel, contains over 0.60 percent carbon and may range as high as 1.7

percent. This type of steel is used for cutting tools, punches, taps, dies, drills and reamers. It is available in hot-rolled stock or in finish-ground flat stock and drill rod.

ALLOY STEELS

Often certain steels are needed which have special characteristics that a plain carbon steel would not possess. It is then necessary to choose an *alloy steel.*

Alloy steel may be defined as steel containing other elements, in addition to carbon, which produce the desired qualities in the steel.

The addition of alloying elements may impart one or more of the following properties to the steel.

1. Increase in tensile strength
2. Increase in hardness
3. Increase in toughness
4. Alteration of the critical temperature of the steel
5. Increase in wear abrasion
6. Red hardness
7. Corrosion resistance

High-Strength, Low-Alloy Steels

A recent development in the steelmaking industry is that of the high-strength, low-alloy (HSLA) steels. These steels, containing a maximum carbon content of 0.28 percent and small amounts of vanadium, columbium, copper, and other alloying elements, offer many advantages over the regular low-carbon construction steels. These advantages are:

1. Higher strength than medium-carbon steels.
2. Less expensive than other alloy steels.
3. Strength properties are "built into" the steel and no further heat treating is needed.
4. Bars of smaller cross sections can do the work of larger regular carbon steel bars.
5. Higher hardness, toughness (impact strength), and fatigue failure limits than carbon steel bars.
6. May be used unpainted because it develops a protective oxide coating on exposure to the atmosphere.

Effects of the Alloying Elements

The following elements such as chromium, cobalt, manganese, molybdenum, nickel, phosphorus, silicon, sulphur, tungsten, and vanadium may be added to steel to give it the desired properties.

The properties imparted to the steel by these elements are given in Table 86-1.

REVIEW QUESTIONS

1. What effect has the study of metals had on modern living?

2. Compare hardness, brittleness, and toughness.

MANUFACTURING OF PIG IRON

3. Briefly describe the manufacture of pig iron.

MANUFACTURING OF CAST IRON

4. Explain how cast iron is manufactured.

5. Name four types of cast iron and state one purpose for each.

THE OPEN-HEARTH PROCESS

6. What is the purpose of the checkers in an open-hearth furnace?

7. How are the impurities removed from the molten metal in an open-hearth furnace?

8. What modification has been made to most open-hearth furnaces and what has been its result?

THE BESSEMER PROCESS AND THE BASIC OXYGEN PROCESS

9. What are the chief differences between the Bessemer and the basic oxygen process?

10. Briefly describe the basic oxygen process.

THE CHEMICAL COMPOSITION OF STEEL

11. What effect does carbon in excess of 0.83 percent have on the steel?

12. What is the purpose of adding small quantities of manganese to steel?

13. What effect will the addition of larger quantities of manganese (1.5 to 2.0 percent) have on steel?

14. Why is phosphorus considered an undesirable element in steel?

15. How will steel be affected by the addition of silicon
  (a) in small amounts?   **(b)** in large amounts?

16. Why is sulfur considered an undesirable element in steel?

17. Why are sulfur and phosphorus sometimes added to steel?

CLASSIFICATION OF STEEL

18. State the carbon content of and two uses for:
  (a) low-carbon steel
  (b) high-carbon steel

19. List six properties that alloying elements may impart to steel.

In order for a steel component to function properly, it is often necessary to heat-treat it. Heat treating is the process of heating and cooling a metal in its solid state in order to obtain the desired changes in its physical properties. One of the most important mechanical properties of steel is its ability to be hardened to resist wear and abrasion or be softened to improve ductility and machinability. Steel may also be heat-treated to remove internal stresses, reduce grain size, or increase its toughness. During manufacture, certain elements are added to steel to produce special results when the metal is properly heat treated.

OBJECTIVES

After completing this unit you will be able to:
1. Select the proper grade of tool steel for a workpiece
2. Harden and temper a carbon steel workpiece
3. Case-harden a piece of machine steel

HEAT-TREATING EQUIPMENT

The heat treating of metal is carried out in specially controlled furnaces which may employ gas, oil, or electricity to provide heat. These furnaces must also be equipped with certain safety devices, as well as control and indicating devices to maintain the temperature required for the job. All furnace installations should be equipped with a fume hood and exhaust fan to take away any fumes resulting from the heat-treating operation, or in case of gas installation, to exhaust the gas fumes.

In most gas installations, the exhaust fan, when running, will actuate the air switch in the exhaust duct. The air switch in turn operates a solenoid valve which permits the main gas valve to be opened. Should the exhaust fan fail for any reason, the air switch will also fail and the main gas supply will close down.

The furnace temperature is controlled by a thermocouple and an indicating pyrometer (Fig. 87-1). After the furnace has been started, the desired temperature is set on the indicating pyrometer. This pyrometer is connected on one side to a thermocouple, and on the other side to a solenoid valve which controls the flow of gas to the furnace.

The thermocouple is made. up of two dissimilar metal wires twisted together and welded at the end. The thermocouple is generally mounted in the back of the furnace in a refectory tube to prevent damage and oxidation to the thermocouple wires.

As the temperature in the furnace rises, the thermocouple becomes hot, and due to the dissimilarity of the wires, a small electrical current is produced. This current is con-ducted to the pyrometer on the wall and causes the pyrometer needle to indicate the temperature of the furnace. When the temperature in the furnace reaches the amount set on the pyrometer, a solenoid valve connected to the gas supply is actuated and the flow of gas to the furnace is restricted. When the furnace temperature drops below the temperature indicated on the pyrometer, the solenoid valve opens, permitting a full flow of gas.

TYPES OF FURNACES

For most heat-treating operations, it is advisable to have a *low-temperature furnace* capable of temperatures up to 1300°F (704.4°C), a *high-temperature furnace* capable of temperatures up to 2500°F (1371.1°C), and a *pot-type furnace* (Fig. 87-2). The pot-type furnace may be used for hardening and tempering by immersing the part to be heat-treated in the molten heat-treating medium, which may be

Fig. 87-1 A thermocouple and pyrometer are used to indicate and control the temperature of a heat-treating furnace.

Fig. 87-2 Types of heat-treating furnaces. (Courtesy Charles A. Hones Inc.)

salt or lead. Cyanide-type mixtures are used for case-hardening operations. One advantage of this type of furnace is that the parts, when being heated, do not come in contact with the air. This eliminates the possibility of oxidation or scaling. The temperature of the pot furnace is controlled by the same method as the high-temperature and low-temperature furnaces.

HEAT TREATMENT TERMS

Plain carbon steel is composed of alternate layers of iron and iron carbide. In this unhardened state, it is known as pearlite (Fig. 87-3A). Before we get involved in the theory of heat treating, it is well to study and understand a few terms connected with this topic.

Heat treatment—the heating and subsequent cooling of metals to produce the desired mechanical properties.

Decalescence point—that temperature at which carbon steel, when being heated, transforms from pearlite to austenite; this is generally at about 1330°F (721°C) for 0.83 percent carbon steel.

Recalescence point—that temperature at which carbon steel, when being slowly cooled, transforms from austenite to pearlite.

Lower critical temperature point—the lowest temperature at which steel may be quenched in order to harden it. This temperature coincides with the decalescence point.

Upper critical temperature point—the highest temperature at which steel may be quenched in order to attain maximum hardness and the finest grain structure.

Critical range—the temperature range bounded by the upper and the lower critical temperatures.

Hardening—the heating of steel above its lower critical temperature and quenching in the proper medium (water, oil, or air) to produce martensite (Fig. 87-3B).

Tempering (Drawing)—reheating hardened steel to a desired temperature below its lower critical temperature, followed by any desired rate of cooling. Tempering removes the brittleness and toughens the steel. Steel in

this condition is called tempered martensite (Fig. 87-3C).

Annealing (Full)—heating metal to just above its upper critical point for the required period of time, followed by slow cooling in the furnace, lime, or sand. Annealing will soften the metal, relieve the internal stresses and strains, and improve its machinability.

Process annealing—heating the steel to just below the lower critical temperature, followed by any suitable cooling method. This process is often used on metals which have been work-hardened. Process annealing will soften it sufficiently for further cold working.

Normalizing—heating the steel to just above its upper critical temperature and cooling it in still air. Normalizing is done to improve the grain structure and remove the stresses and strains. In general, it brings the metal back to its normal state.

Spheroidizing—the heating of steel to just below the lower critical temperature for a prolonged period of time followed by cooling in still air. This process produces a grain structure with globular-shaped particles (spheroids) of cementite rather than the normal needlelike structure, which improves the machinability of the metal.

Alpha iron—the state in which iron exists below the lower critical temperature. In this state, the atoms form a body-centered cube.

Gamma iron—the state in which iron exists in the critical range. In this state the molecules form face-centered cubes. Gamma iron is nonmagnetic.

Pearlite—a laminated structure of ferrite (iron, and cementite (iron carbide); usually the condition of steel before heat treatment (Fig. 87-3A).

Cementite—a carbide of iron (Fe_3C), which is the hardener in steel.

Austenite—a solid solution of carbon in iron which exists between the lower and upper critical temperatures.

Martensite—the structure of fully hardened steel obtained when austenite is quenched. Martensite is characterized by its needlelike pattern (Fig. 87-3B).

Tempered martensite—the structure obtained after martensite has been tempered (Fig. 87-3C). Tempered martensite was formerly known as *troosite* and *sorbite*.

Eutectoid steel—steel containing just enough carbon to dissolve completely in the iron when the steel is heated to its critical range. Eutectoid steel contains fom 0.80 to 0.85 percent carbon. This may be likened to a saturated solution of salt in water.

Hypereutectoid steel—steel containing more carbon than will completely dissolve in the iron when the steel is heated to the critical range. This is similar to a supersaturated solution.

Hypoeutectoid steel—containing less carbon than can be dissolved by the iron when the steel is heated to the critical range. Here there is an excess of iron. This is similar to an unsaturated solution.

SELECTION OF TOOL STEEL

The proper selection and proper heat treatment of a tool steel are both essential if the part being made is to perform

A   B   C

Fig. 87-3   (A) Pearlite is usually the condition of carbon steel before heat treating. (B) Martensite is the condition of the hardened carbon steel. (C) Martensite structure of steel can be altered by tempering.

efficiently. Many problems may arise in the selection and heat treatment of the tool steel.

1. The steel may not be tough or strong enough for the job requirements.
2. It may not offer sufficient abrasion resistance.
3. It may not have sufficient hardening penetration.
4. It may warp during heat treatment.

Because of these problems, the steel producers have been forced to manufacture many types of alloy steels to cover the range of most jobs.

To select the correct steel for a job, consult the handbook provided by the steel company for the specifications and application of the various steels it produces. This handbook will also outline the proper heat treatment of the particular material chosen. These instructions should be carefully followed.

Tool steels are generally classified as *water-hardening, oil-hardening, air-hardening,* or *high-speed steels.* They are usually identified by each manufacturer by a trade name such as Alpha 8, Keewatin, Nutherm, or Nipigon, which are some of the trade names of the Atlas Steel Company products.

WATER-HARDENING TOOL STEEL

Water-hardening tool steels generally contain from 0.50 to 1.3 percent carbon, along with small amounts (about 0.20 percent) of silicon and manganese. The addition of silicon facilitates the forging and rolling of the material, while manganese helps to make the steel more sound when it is first cast into the ingot. Further addition of silicon (above 0.20 percent) will reduce the grain size and increase the toughness of water-hardening steel.

Most water-hardening steels achieve the maximum hardness for a depth of about 1/8 in. (3.17 mm); the inner core remains softer but still tough. Chromium or molybdenum is sometimes added to increase the hardenability (hardness penetration), toughness, and wear resistance of water-hardening steels. Water-hardening steels are heated to around 1450 to 1500°F (787.7 to 815.5°C) during the hardening process. These steels are used where a dense, fine-grained

outer casing with a tough inner core is required. Typical applications are drills, taps, reamers, punches, jig bushings, and dowel pins.

The problems connected with water-hardening steels are those of distortion and cracking when the material is quenched. Should these problems occur, it would then be wise to select an oil-hardening steel.

OIL-HARDENING STEELS

A typical oil-hardening steel contains about 0.90 percent carbon, 1.6 percent manganese, and 0.25 percent silicon.

The addition of manganese in quantities of 1.5 percent or more increases the hardenability (hardness peneration) of the steel up to about 1 in. (25.4 mm) from each surface. During the quenching of steels with a higher manganese content, the hardening is so rapid that a less severe quenching medium (oil) must be used. The use of oil as a quenching medium retards the cooling rate and reduces the stresses and strains in the steel which cause warping and cracking. Chromium and nickel, in varying quantities, may be added to oil-hardening steel to increase its hardness and wear resistance. Higher hardening temperatures, fom 1500 to 1550°F (815.5 to 843.3°C), are required for these latter alloy steels.

Often, due to the intricate shape of the part, it may not be possible to eliminate warping or cracking during the quench, and it will be necessary to select air-hardening steel for the particular part. Typical applications of oil-hardening steels are blanking, forming, punching dies, precision tools, broaches, and gages.

AIR-HARDENING STEELS

Due to the slower cooling rate of air-hardening steels, the stresses and strains which cause cracking and distortion are kept to a minimum. Air-hardening steels are also used on parts having large cross sections, where full hardness throughout could not be obtained by using water- or oil-hardening steels.

A typical air-hardening steel will contain about 1.00 percent carbon, 0.70 percent manganese, 0.20 percent silicon, 5.00 percent chromium, 1.00 percent molybdenum, and 0.20 percent vanadium. Air-hardening steels require higher hardening temperatures, from 1600 to 1775°F (871.1 to 968.3°C).

Typical applications of this steel are large blanking, forming, trimming and coining dies, rolls, long punches, precision tools and gages.

HIGH-SPEED STEELS

High-speed steels are used in the manufacture of cutting tools such as drills, reamers, taps, milling cutters, and lathe cutting tools. The analysis of a typical high-speed steel could be as follows: 0.72 percent carbon, 0.25 percent manganese, 0.20 percent silicon, 4 percent chromium, 18 percent tungsten, and 1 percent vanadium. Tools of this type will retain their hardness and cutting edges even when operating at red heat.

During heat treatment, high-speed steels must be preheated slowly to 1500 to 1600°F (815.5 to 871.1°C) in a neutral atmosphere and then transferred to another furnace and quickly brought up to 2300 to 2400°F (1260 to 1315.5°C). They are generally quenched in oil, but small, intricate sections may be air cooled.

CLASSIFICATION OF STEEL

In order to ensure that the composition of various types of steel remains constant and that a certain type of steel will meet the required specifications, the Society of Automotive Engineers (SAE) and the American Iron and Steel Institute (AISI) have devised similar methods of identifying different types of steel, and both are widely used.

THE SAE-AISI CLASSIFICATION SYSTEMS

The systems designed by the SAE and the AISI are similar in most respects. They both use a series of four or five numbers to designate the type of steel.

The first digit in these series indicates the predominant alloying element. The last two digits (or sometimes three in certain corrosion- or heat-resisting alloys) indicate the average carbon content in points (hundredths of 1 percent) (Fig. 87-4).

The main difference in the two systems is that the AISI system indicates the steel-making process used by the following prefixes:

A—basic open-hearth alloy steel
B—acid-Bessemer carbon steel
C—basic open-hearth carbon steel
D—acid-open-hearth carbon steel
E—electric furnace steel

In the classification charts, the various types of steels are indicated by the first number in the series as follows:

1. Carbon
2. Nickel
3. Nickel-chrome
4. Molybdenum
5. Chromium
6. Chromium-vanadium
8. Triple alloy
9. Manganese-silicon

Table 87-1 indicates the SAE classification of the various steels and alloys. The number 7 does not appear on the chart. It formerly represented tungsten steel which is no longer listed in this chart since it is now considered a special steel.

Examples of Steel Identification

Determine the types of steel indicated by the following numbers: 1015, A2340, 4170.

1015—1 indicates plain carbon steel
—0 indicates there are no major alloying elements
—15 indicates that there is between 0.01 and 0.20 percent carbon content

NOTE: This steel would naturally contain small quantities of manganese, phosphorus, and sulfur.

A2360—A indicates an alloy steel made by the basic open-hearth process.
—23 indicates the steel contains 3.5 percent nickel (see Table 87-1).
—60 indicates 0.60 percent carbon content.

4170—41 indicates a chromium-molybdenum steel.
—70 indicates 0.70 percent content.

HEAT TREATMENT OF CARBON STEEL

The proper performance of a steel part depends not only on the correct selection of steel but also upon the correct heat-treating procedure and an understanding of the theory behind it. When steel is heated from room temperature to the upper critical temperature and then quenched, several changes take place in the steel. These may be more easily understood if the changes which take place in water, from the frozen state until it is transformed into steam, are considered.

By referring to Fig. 87-5, it is noted that water exists as a solid at or below 32°F (0°C). If the ice is heated, the temperature will remain 32°F (0°C) until the ice completely melts. If the water is heated further, it will turn into steam at 212°F (100°C). Again the water remains at this temperature for a short time before turning into steam. It should also be noticed that if the process were reversed and the steam cooled, it would form water at 212°F (100°C) and ice at 32°F (0°C). The points where water transforms to another state are known as the critical points of water.

Steel, like water, has critical points which, when determined, will lead to successful heat treatment of the metal.

TO DETERMINE THE CRITICAL POINTS OF 0.83 PERCENT CARBON STEEL

A simple experiment may be performed to illustrate the critical points and the changes that take place in a piece of carbon steel when heated and slowly cooled.

A, B, C, D, E

PREFIX

10XXX

% OF CARBON CONTENT

% OF MAJOR ALLOY

TYPE OF STEEL

1—CARBON
2—NICKEL
3—NICKEL-CHROME
4—MOLYBDENUM

5—CHROMIUM
6—CHROME-VANADIUM
8—TRIPLE ALLOY
9—SILICON-MANGANESE

AISI PREFIXES

A—BASIC OPEN HEARTH STEEL
B—ACID BESSEMER CARBON STEEL
C—BASIC OPEN HEARTH CARBON STEEL
D—ACID OPEN HEARTH CARBON STEEL
E—ELECTRIC FURNACE STEEL

Fig. 87-4 The SAE and AISI numbering system.

TABLE 87-1 SAE Classification of Steels	
CARBON STEELS	1xxx
Plain carbon	10xx
Free-cutting (resulfurized screw stock)	11xx
Free-cutting, manganese	X13xx
HIGH-MANGANESE	T13xx
NICKEL STEELS	2xxx
0.50% nickel	20xx
1.50% nickel	21xx
3.50% nickel	23xx
5.00% nickel	25xx
NICKEL-CHROMIUM STEELS	3xxx
1.25% nickel, 0.60% chromium	31xx
1.75% nickel, 1.00% chromium	32xx
3.50% nickel, 1.50% chromium	33xx
3.00% nickel, 0.80% chromium	34xx
Corrosion- and heat-resisting steels	30xxx
MOLYBDENUM STEELS	4xxx
Chromium-molybdenum	41xx
Chromium-nickel-molybdenum	43xx
Nickel-molybdenum	46xx and 48xx
CHROMIUM STEELS	5xxx
Low-chromium	51xx
Medium-chromium	52xxx
CHROMIUM-VANADIUM STEELS	6xxx
TRIPLE-ALLOY STEELS (NICKEL, CHROMIUM, MOLYBDENUM)	8xxx
MANGANESE-SILICON STEELS	9xxx

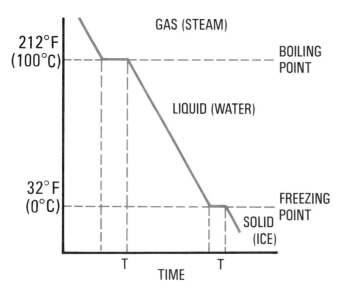

Fig. 87-5   The critical points of water.

ELECTRIC FURNACE

Fig. 87-6   The setup to determine the critical points of steel.

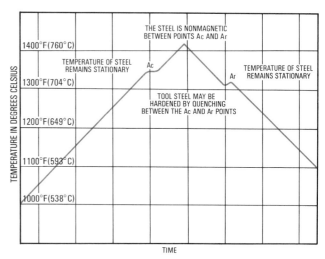

Fig. 87-7   A graph to illustrate the critical points of steel.

PROCEDURE

1. Select a piece of 0.83 percent (eutectoid) carbon steel about 1½ in. × 1½ in. × 2 in. (38.1 mm × 38.1 mm × 50.8 mm) long and drill a small hole in the end for most of the length.

2. Insert a thermocouple in the hole as shown and seal the end of the hole with fireclay.

3. Place the block in a furnace and run the thermocouple wire to a voltmeter (Fig. 87-6).

4. Light the furnace and set the temperature for about 1425°F (773.8°C) on the pyrometer.

5. Plot the readings of the voltmeter needle at regular time intervals (Fig. 87-7).

6. When the furnace reaches 1425°F (773.8°C), shut it down and let it cool.

7. Continue to plot the readings until the temperature in the furnace drops to approximately 1000°F (537.7°C).

OBSERVATIONS AND CONCLUSIONS

Steel, at room temperature, consists of laminated layers of ferrite (iron) and cementite (iron carbide). This structure is called pearlite (Fig. 87-8). As the steel is heated from room temperature, the curve climbs uniformly until a temperature of about 1333°F (722.7°C) is reached. At this point, Ac (Fig. 87-7), the temperature of the steel drops slightly although the temperature of the furnace is rising.

The point Ac indicates the *decalescence point,* and it is here that several changes take place in the steel.

1. If the steel were observed in the furnace at this time, it would be noticed that the dark shadows in the steel disappear.

2. The steel would be nonmagnetic when tested with a magnet.

3. These changes were caused by a change in the atomic structure of the steel. The atoms rearrange themselves from body-centered cubes (Fig. 87-9A) to face-centered cubes (Fig. 87-9B). When the atoms are rearranged, the energy (heat) required for this change is drawn from the metal; thus a slight drop in the temperature of the workpiece is recorded at the decalescence point. The layers of iron carbide completely dissolve in the iron to form a solid solution known as *austenite.* Thus the decalescence point marked the transformation point from *pearlite* to *austenite,* or from body-centered cubes to face-centered cubes.

4. It is at this point that the steel, if quenched in water, would also show the first signs of hardening.

5. If the steel could be examined under a microscope, it would be noticed that the grain structure starts to get smaller. As the curve progresses upward past Ac, the grain size would become progressively smaller until the upper critical temperature [1425°F (773.8°C)] were reached.

As the steel cools, the curve would continue uniformly down and the grain size would gradually get larger until the point Ar, at about 1300°F (704.4°C), is reached. This is the *recalescence point* and here the needle on the voltmeter would show a slight rise in temperature, although the furnace is cooling. This process is obviously the reverse of the phenomenon which occurs at the decalescence point;

A

B

Fig. 87-8 (A) A pearlite grain is composed of alternate layers of iron (ferrite) and iron carbide (cementite). (B) A photomicrograph of high carbon steel. The pearlite grains are surrounded by iron carbide shown as white lines between the grains. (Courtesy Linde Division, Union Carbide Corporation.)

A

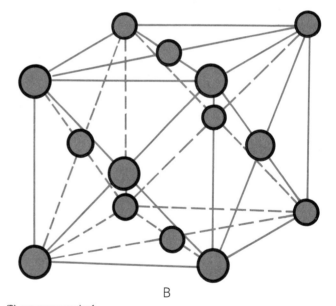

B

Fig. 87-9 (A) Arrangement of the atoms in a body-centered cube; (B) arrangement of the atoms in a face-centered cube.

the austenite reverts to pearlite, the atoms rearrange themselves into body-centered cubes, and the steel again becomes magnetic.

Another experiment, which demonstrates the decalescence and recalescence points, is as follows.

Decalescence Point

1. Place a magnet on a firebrick.
2. Select a ½ to ⅝ in. (12.7 to 15.87 mm) round piece of 0.90 to 1.00 carbon steel and place it on the magnet (Fig. 87-10).
3. Place a can of cold water under the magnet ends.
4. Heat the piece held to the magnet using a small flame.

NOTE: Do not allow the flame to come in contact with the magnet.

5. When the temperature reaches its critical point, the steel will drop into the water and become hardened.

Recalescence Point

1. Remove the can of water from under the magnet.
2. Place a flat plate under the work held on the magnet.
3. Heat the steel until it drops from the magnet onto the plate.
4. When the steel cools, it will become attracted by the magnet.

Fig. 87-10   Decalescence point experiment.

Summary

When the steel loses its magnetic value (decalescence point), it drops into the water and the change in the steel is trapped or stopped. The steel then hardens because it does not have time to revert to another state.

Conversely, when the steel is not quenched but is allowed to cool gradually from the decalescence point, it regains its magnetic value when it has cooled slightly (recalescence point). The steel does not change from magnetic to non-magnetic; it merely acquires temporary characteristics of being attracted or *not* attracted to the magnet.

HARDENING OF 0.83 PERCENT CARBON STEEL

Once the critical temperature of a steel has been determined, the proper quenching temperature, which is about 50°F (27.7°C) over the upper critical temperature, can be determined. Not all steels have the same critical temperature; Fig. 87-11 indicates that the critical temperature of a steel drops as the carbon content increases up to 0.83 percent, after which it does not change. As a result, steels containing over 0.83 percent carbon (hypereutectoid) only need be heated to just above the lower critical temperature Ac$_1$ (Fig. 87-11) to obtain maximum hardness. This makes it possible to use a lower hardening temperature for hypereutectoid steels, thus decreasing the possibility of warping.

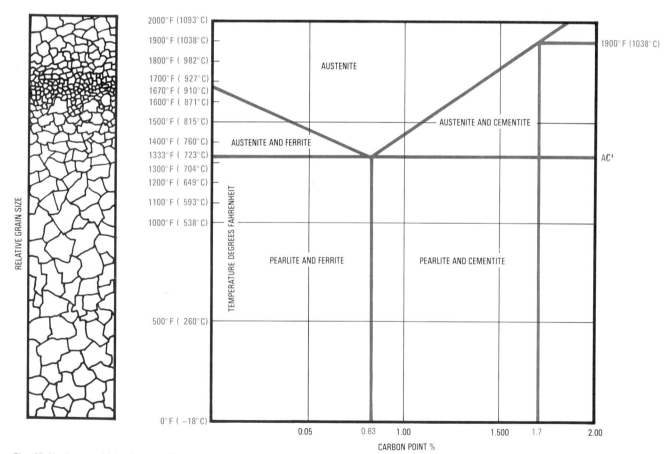

Fig. 87-11   Iron carbide diagram to illustrate the relationship between the carbon content of steel and the critical temperature.

The increase in the carbon content beyond 0.83 percent will not increase the hardness of the steel; however, it does increase the wear resistance of the steel considerably.

In order for the steel to be hardened properly, it must be heated uniformly to about 50°F (27.7°C) above the upper critical temperature and held at this temperature long enough to allow sufficient carbon to dissolve and form a solid solution which permits maximum hardness. At this point the steel will have the smallest grain size and, when quenched, will produce the maximum hardness.

The critical temperature of a steel is also affected by alloying elements such as manganese, nickel, chromium, cobalt, and tungsten.

QUENCHING

When steel has been properly heated throughout, it is quenched in brine, water, or oil (depending on the type of steel), to cool it rapidly. During this operation, the austenite is transferred into *martensite,* a hard brittle metal. Because the steel is cooled rapidly, the austenite is prevented from passing through the recalescence point (Ar), as in the case of slow cooling, and the small grain size of the austenite is retained in the martensite (Fig. 87-12).

The rate of cooling affects the hardness of steel. If a water-hardening steel is quenched in oil, it cools more slowly and does not attain the maximum hardness. On the other hand, if an oil-hardening steel is cooled too quickly by quenching it in water, it may crack. Cracking may also occur when the quenching medium is too cold.

The method of quenching greatly affects the stresses and strains set up in the metal, which cause warping and cracking. For this reason, long, flat pieces should be held vertically above the quenching medium and plunged straight

into the liquid. After immersion, the part should be moved around in a figure 8 motion. This keeps the liquid at a uniform temperature and prevents air pockets forming on the steel, which would affect the uniformity of hardness.

METCALF'S EXPERIMENT

This simple experiment demonstrates the effect that various degrees of heat have on the grain structure, hardness, and strength of tool steel.

1. Select a piece of SAE 1090 (tool steel) about ½ in. (12.7 mm) in diameter and about 4 in. (101.6 mm) long.
2. With a sharp, pointed tool, cut shallow grooves approximately ½ in. (12.7 mm) apart.
3. Number each section (Fig. 87-13).
4. Heat the bar with an oxyacetylene torch, bringing section 1 to a white heat.
5. Keep section 1 at white heat, and heat sections 4 and 5 to a cherry red. *DO NOT* apply heat to sections 6 to 8.
6. Quench in cold water or brine.
7. Test each section with the edge of a file for hardness.
8. Break off the sections and examine the grain structure under a microscope.

Results

Sections 1 and 2 have been overheated. They break easily and the grain structure is very coarse (Fig. 87-14A).

Section 3 requires more force to break and the grain structure is somewhat finer. Sections 4 and 5 have the greatest strength and resistance to shock. These sections have the finest grain structure (Fig. 87-14B).

Sections 6 to 8, where the metal was underheated, require greatest force to break, and bending occurs. Note that the grain structure becomes coarser toward section 8 (Fig. 87-14C). This section is the original structure of unheated steel (pearlite).

TEMPERING

Tempering is the process of heating a hardened carbon or alloy steel below its lower critical temperature and cooling it by quenching in a liquid or in air. This operation removes many of the stresses and strains set up when the metal was hardened. Tempering imparts toughness to the metal but decreases the hardness and tensile strength. The tempering process modifies the structure of the martensite, changing it to *tempered martensite* which is somewhat softer and tougher than martensite.

The tempering and drawing temperature is not the same for each type of steel and is affected by several factors:

1. The toughness required for the part
2. The hardness required for the part
3. The carbon content of the steel
4. The alloying elements present in the steel

The hardness obtained after tempering depends on the temperature used and the length of time the workpiece is held at this temperature. Generally, hardness decreases and the toughness increases as the temperature is increased (Fig. 87-15).

Fig. 87-12 Carbon steel, when quenched at the upper critical temperature, retains the smallest grain size.

Fig. 87-13 A test piece for Metcalf's experiment.

Fig. 87-14 (A) Toolsteel overheated; (B) toolsteel heated to the proper temperature; (C) toolsteel underheated. (Courtesy Kostel Enterprises Limited.)

Fig. 87-15 As the tempering heat increases, the hardness decreases.

As the length of the tempering time is increased for a specific part, the hardness of the metal decreases. On the other hand, if the tempering time is too short, the stresses and strains set up by hardening are not totally removed and the metal will be brittle. The cross-sectional size of the workpiece affects its tempering time. The tempering time and temperatures for various steels are always supplied in the steel manufacturer's handbook; these recommended times and temperatures should be followed to obtain the best results.

Tempering Colors

When a piece of steel is heated from room temperature to a red heat, it passes through several color changes, caused by the oxidation of the metal. These color changes indicate the approximate temperature of the metal and are often used as a guide when tempering (Table 87-2).

ANNEALING

Annealing is a heat-treating operation used to soften metal and to improve its machinability. Annealing also relieves the internal stresses and strains caused by previous operations, such as forging or rolling.

PROCEDURE

1. Set the pyrometer approximately 30°F (16.6°C) above the upper critical temperature and start the furnace (Fig. 87-16).
2. Place the part in the furnace. After the required temperature has been reached, allow it to soak for 1 h per inch (25.4 mm) of workpiece thickness.
3. Shut off the furnace and allow the part to cool slowly in the furnace; or remove the part from the furnace and pack it immediately in lime or ashes and leave it covered for several hours, depending on the size, until it is cool.

NORMALIZING

Normalizing is performed on metal to remove internal stresses and strains and to improve its machinability.

PROCEDURE

1. Set the pyrometer approximately 30°F (16.6°C) above the upper critical temperature of the metal and start the furnace (Fig. 87-16).
2. Place the part in the furnace. After the required temperature has been reached, allow the workpiece to soak for 1 h per inch (25.4 mm) of thickness.
3. Remove the part from the furnace and allow it to cool slowly in still air. Thin workpieces may cool too rapidly and may harden if normalized in air. It may be necessary to pack them in lime to retard the cooling rate.

TABLE 87-2 Tempering Colors and Approximate Temperatures for Carbon Steel

COLOR	TEMPERATURE		USE
	°F	°C	
Pale yellow	430	220	Lathe tools, shaper tools
Light straw	445	230	Milling cutters, drills, reamers
Dark straw	475	245	Taps and dies
Brown	490	255	Scissors, shear blades
Brownish-purple	510	265	Axes and wood chisels
Purple	525	275	Cold chisels, center punches
Bright blue	565	295	Screwdrivers, wrenches
Dark blue	600	315	Woodsaws

SPHEROIDIZING

Spheroidizing is a process of heating metal for an extended period to just below the lower critical temperature. This process produces a special kind of grain structure whereby the cementite particles become spherical in shape. Spheroidizing is generally done on high-carbon steel to improve the machinability.

PROCEDURE

1. Set the pyrometer approximately 30°F (16.6°C) below the lower critical temperature of the metal and start the furnace (Fig. 87-16).
2. Place the part in the furnace and allow it to soak for several hours at this temperature.
3. Shut down the furnace and let the part cool slowly to about 1000°F (537.7°C).
4. Remove the part from the furnace and cool it in still air.

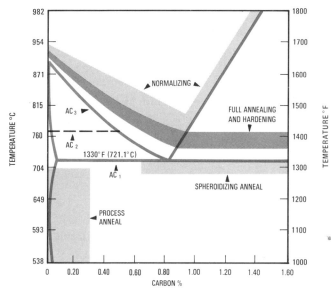

Fig. 87-16 Temperature ranges for various heat-treating operations.

CASE-HARDENING METHODS

When hardened parts are required, they may be made from carbon steel and heat-treated to the specifications. Often these parts can be made more cheaply from machine steel and then case-hardened. This process produces a hard outer case with a soft inner core, which is often preferable to through-hard parts made from carbon steel. Case hardening may be performed by several methods, such as carburizing, carbonitriding, and nitriding processes.

CARBURIZING

Carburizing is a process whereby low-carbon steel, when heated with some carbonaceous material, absorbs carbon into its outer surface. The depth of penetration depends on the time, the temperature, and the carburizing material used. Carburizing may be performed by three methods: pack carburizing, liquid carburizing, and gas carburizing.

Pack carburizing is generally used when a hardness depth of penetration of 0.060 in. (1.52 mm) or more is required. The parts to be carburized are packed with a carbonaceous material such as activated charcoal in a sealed steel box.

PROCEDURE

1. Place a 1- to 1½-in. (25.4- to 38.1-mm) layer of carbonaceous material in the bottom of a steel box which will fit into the furnace.
2. Place the parts to be carburized in the box leaving about 1½ in. (38.1 mm) between parts.
3. Pack the carburizer around the parts and cover the parts with about 1½ in. (38.1 mm) of material.
4. Tap the sides of the box to settle the material and to pack it around the workpieces. This will exclude most of the air.
5. Place a metal cover over the box and seal around the joint with fireclay.
6. Place the box in the furnace and bring the temperature up to about 1700°F (926.6°C).
7. Leave the box in the furnace long enough to give the required penetration. The rate of penetration is generally about 0.007 to 0.008 in./h (0.17 to 0.20 mm/h), however this decreases as the depth of penetration increases. The

Fig. 87-17 Relationship between temperature, time, and depth of case.

Fig. 87-18 A broken cross section of a piece of case-hardened machine steel showing the hardened outer case.

proper time (and temperature) for any depth of penetration is usually given in the literature supplied by the manufacturer of the carburizing material (Fig. 87-17).

8. Shut down the furnace and leave the box in the furnace until it cools. This may take 12 to 16 h.

9. Remove the box from the furnace and take out the parts and clean them.

NOTE: The surfaces of the workpieces have now been transformed into a thin layer of carbon steel which is soft, due to the slow cooling of the parts in the carburizing material.

10. Heat the parts to the proper critical temperature in a furnace and quench in oil or water, depending on the shape of the parts and the hardness required. The parts are now surrounded with a hard layer of carbon steel and have a soft inner core (Fig. 87-18).

Liquid carburizing is generally used to produce a thin layer of carbon steel on the outside of low-carbon steel parts. The parts are not usually machined after liquid carburization.

PROCEDURE

1. Place the carburizing material into a pot furnace. Heat until it is molten and reaches the proper temperature.

2. Preheat the parts to be carburized to approximately 800°F (426.6°C) in the low-temperature drawing furnace. This eliminates the possibility of an explosion due to water or oil on the work when the parts are immersed in the molten liquid.

3. Suspend the parts in the liquid carburizer and leave them for the time required to give the desired penetration. The depth of penetration (depending on the temperature of the liquid) may be from 0.015 to 0.020 in. (0.38 to 0.50 mm)

for the first hour and about 0.010 in. (0.25 mm) for each succeeding hour.

4. Use dry tongs to remove the parts and quench immediately in water.

CAUTION: Since most liquid carburizers contain cyanide, extreme care must be taken when using these materials.

1. Avoid letting *any moisture* come into contact with the liquid carburizer. This will cause an explosion.

2. Heat the jaws of the tongs before using to remove any moisture or oil.

3. Avoid inhaling the fumes; they are toxic.

4. Wear protective clothing (gloves, face and arm shields) when removing and quenching parts.

Gas carburizing, like pack carburizing, is used on parts where over 0.060 in. (1.52 mm) depth of case is required, and where it is necessary to grind the parts after carburizing. This method is generally done by specialized heat-treating firms since it requires special types of furnaces.

The parts are placed in a sealed drum into which natural gas or propane is introduced and circulated. The workpieces are heated to a carburizing temperature in the gas atmosphere. The gas exhausts at one end of the drum and is burned to prevent air from entering the chamber. The exterior of the drum is heated by a source such as gas or oil. In this process the carbon from the gas is absorbed by the workpiece.

The parts remain in the drum for the time required to give the desired penetration. Depths of 0.020 to 0.030 in. (0.50 to 0.76 mm) are obtained in about 4 h at a temperature of 1700°F (926.6°C). The parts may then be removed and quenched or allowed to cool, after which they are reheated to the critical temperature and quenched.

CARBONITRIDING PROCESSES

Carbonitriding is a process whereby both carbon and nitrogen are absorbed by the surface of a steel workpiece when it is heated to the critical temperature to produce a hard, shallow outer case. Carbonitriding may be done by liquid or gas methods.

Cyaniding (liquid carbonitriding) is a process that uses a salt bath composed of cyanide-carbonate-chloride salts with varying amounts of cyanide, depending on the application. Liquid cyaniding is generally carried out in a pot-type furnace, and since cyanide fumes are poisonous, extreme care must be taken when using this method.

The parts are suspended in a liquid cyanide bath which must be at a temperature above the lower critical point of the steel being used. The depth of penetration when using this method is about 0.005 to 0.010 in. (0.12 to 0.25 mm) in 1 h at 1550°F (843.3°C). A depth of about 0.015 in. (0.38 mm) may be obtained in 2 h at the same temperature. The parts may then be quenched in water or oil, depending on the steel being used. After the parts have been hardened, they should be thoroughly washed to remove all traces of the cyanide salt.

Carbonitriding (gas cyaniding) is carried out in a special furnace similar to the gas carburizing furnace. The workpieces are put into the inner drum of the furnace. A mixture of ammonia and a carburizing gas is introduced into and circulated through this chamber, which is heated externally to a temperature of 1350 to 1700°F (732.2 to 926.6°C). During this process, the workpiece absorbs carbon from the carburizing gas and nitrogen from the ammonia.

The parts are removed from the furnace and quenched in oil, which gives the part maximum hardness and minimum distortion. The depth of case produced by this method is relatively shallow. Depths of about 0.030 in. (0.76 mm) are obtained in 4 to 5 h at a temperature of 1700°F (926.6°C).

NITRIDING PROCESSES

Nitriding is used on certain alloy steels to provide maximum hardness. Most carbon alloy steels can only be hardened to about 62 Rockwell C by conventional means, whereas readings of 70 Rockwell C may be obtained on certain vanadium and chromium alloy steels using a nitriding process. Nitriding may be done in a protected atmosphere furnace or in a salt bath.

Gas nitriding. The parts to be nitrided are placed in an airtight drum, which is heated externally to a temperature of 900 to 1150°F (482.2 to 621.1°C). Ammonia gas is circulated through the chamber. The ammonia, at this temperature, decomposes into nitrogen and hydrogen. The nitrogen penetrates the outer surface of the workpiece and combines with the alloying elements to form hard nitrides. Gas nitriding is a slow process requiring approximately 48 h to obtain a case depth of 0.020 in. (0.50 mm). Because of the low operating temperatures used in this process, and since no quenching of the part is required, there is little or no distortion. This method of increasing hardness is used on parts that have been

hardened and ground. No further finishing is required on such parts.

Salt bath nitriding. Nitriding may also be carried out in a salt bath containing nitriding salts. The hardened part is suspended in the molten nitriding salt, which is held at a temperature from 900 to 1100°F (482.2 to 593.3°C) depending on the application. Parts such as high-speed taps, drills, and reamers are nitrided to increase surface hardness, which improves durability.

SURFACE HARDENING OF MEDIUM-CARBON STEELS

When selected areas of a part are to be surface-hardened, to increase wear resistance and retain a soft inner core, the part must have a medium carbon content. It may be surface hardened by flame or induction hardening, depending on the size of the part and its application. In both processes, the steel must contain carbon, since no external carbon is added as in other case-hardening methods.

INDUCTION HARDENING

In this process, the part is surrounded by a coil through which a high-frequency electrical current is passed. The current heats the surface of the steel to above the critical temperature in a few seconds. An automatic spray of water, oil, or compressed air is used to quench and harden the part, which is held in the same position as for heating. Since only the surface of the metal is heated, the hardness is localized at the surface. The depth of hardness is governed by the current frequency and heating cycle duration.

The current frequencies vary from 1 kHz to 2 MHz. Higher frequencies produce shallow hardening depths. Lower frequencies produce hardening depths up to $\frac{1}{4}$ in. (6.35 mm).

Induction hardening may be used for the selective hardening of gear teeth, splines, crank shafts, camshafts, and connecting rods.

FLAME HARDENING

Flame hardening is used extensively to harden ways on lathes and other machine tools, as well as gear teeth, splines, crank shafts, etc.

The surface of the metal is heated very rapidly to above the critical temperature and is hardened quickly by a quenching spray. Large surfaces such as lathe ways are heated by a special oxyacetylene torch which is moved automatically along the surface, followed by a quenching spray. Smaller parts are placed under the flame, and spray-quenched automatically.

Flame-hardened parts should be tempered immediately to remove strains created by the hardening process. Large surfaces are tempered by a special low-tempering torch which follows the quenching nozzle as it moves along the work. The depth of flame hardening varies from $\frac{1}{16}$ to $\frac{1}{4}$ in. (1.58 to 6.35 mm), depending on the speed at which the surface is brought up to the critical temperature.

REVIEW QUESTIONS

HEAT-TREATING EQUIPMENT

1. Describe a thermocouple and explain how it functions.
2. Sketch a furnace installation showing the furnace, thermocouple, and pyrometer.
3. What is the purpose of:
   (a) the pyrometer?
   (b) the solenoid valve?
   (c) the air switch?

HEAT TREATMENT TERMS

4. What is the difference between the decalescence point and the recalescence point of a piece of steel?
5. At what point in relation to the upper and lower critical temperatures are the following heat-treating operations performed?
   (a) hardening
   (b) tempering
   (c) annealing
   (d) normalizing
   (e) spheroidizing
6. What is the difference between hypereutectoid and hypoeutectoid steel?

WATER-HARDENING TOOL STEEL

7. State two problems which often occur with water-hardening steel.

OIL-HARDENING TOOL STEEL

8. Why is oil used to advantage as a quenching medium?

AIR-HARDENING STEELS

9. What are the advantages of air-hardening steels?
10. What elements will give red hardness quality to a high-speed steel toolbit?
11. Explain the difference in the hardening procedures between plain carbon steel and high-speed steel.

THE SAE AND AISI AND CLASSIFICATION SYSTEMS

12. How does the AISI system differ from the SAE system of identifying steel?
13. What is the composition of the following steels?
   (a) 2340
   (b) 1020
   (c) E4340

HEAT TREATMENT OF CARBON STEEL

14. Explain how the critical points of water are determined.
15. List five changes that take place in steel when it is heated to (and above) the decalescence point from room temperature.
16. Describe the changes that take place in a piece of steel when it is cooled to the recalescence point from the upper critical temperature.

HARDENING OF 0.83 PERCENT CARBON STEEL

17. What is the advantage of using hypereutectoid steel?
18. To what temperature should steel be heated prior to quenching, in order to produce the best results?

QUENCHING

19. Name two quenching media and state the purpose of each.
20. Describe the proper method of quenching long, thin workpieces.

TEMPERING

21. What is the purpose of tempering a piece of steel?
22. What factors will affect the temperature at which a piece of steel is tempered?
23. What will happen to the steel if the tempering time is
   (a) too long?
   (b) too short?

ANNEALING, NORMALIZING, SPHEROIDIZING

24. What is the difference between annealing, normalizing, and spheroidizing?

CARBURIZING

25. Describe briefly how pack carburizing is performed.
26. What precautions should be observed when using the liquid carburizing process?
27. Describe the gas carburizing process.

CARBONITRIDING PROCESSES

28. Explain the difference between the carburizing and the carbonitriding processes.

SURFACE HARDENING OF CARBON STEELS

29. What type of steel must be used for the flame-hardening and induction-hardening processes?

INDUCTION HARDENING

30. Describe briefly the induction-hardening principle.

FLAME HARDENING

31. Why is it desirable to temper parts immediately after flame hardening?
32. How may this be done on large surfaces?

TESTING OF METALS AND NONFERROUS METALS

UNIT 88

After heat treating, certain tests may be performed to determine the properties of the metal. These tests fall into two categories:

1. *Nondestructive testing,* whereby the test may be performed without damaging the sample

2. *Destructive testing,* whereby a sample of the material is broken to determine the qualities of the metal

Nonferrous metals find a variety of uses in the machine trade. Since nonferrous metals contain little of no iron, they are used as bearings to prevent two like metal parts from being in contact with each other, where rust and corrosion is a factor, and where the weight of the product is important.

OBJECTIVES

After completing this unit you will:
1. Understand three methods of hardness testing
2. Be able to perform a Rockwell C hardness test on a workpiece
3. Be able to perform tensile strength and impact tests on a workpiece
4. Be familiar with several nonferrous metals used in industry

HARDNESS TESTING

This is the most common form of nondestructive testing and is used to determine the hardness of a metal. Hardness in steel may be defined as its capacity to resist wear and deformation. The term hardness applied to metal is relative and indicates some of its properties. For example, if a piece of steel is hardened, the tensile strength increases, but the ductility of the steel is reduced. If the hardness of a metal is known, the properties and performance of the metal can be predicted accurately.

Two types of testing machines are used to measure the hardness of a metal.

1. Those which measure the depth of penetration made by a penetrator under a known load. Rockwell, Brinell, and Vickers hardness testers are examples of this type.
2. Those which measure the height of rebound of a small mass dropped from a known height. The scleroscope employs this principle.

ROCKWELL HARDNESS TESTER

The Rockwell hardness tester (Fig. 88-1) indicates the hardness value by the depth that a penetrator advances into the metal under a known pressure. When testing hard materials, a 120° conical diamond penetrator (brale) is used. A $\frac{1}{16}$- or $\frac{1}{8}$-in.-diameter (1.58- or 3.16-mm) ball is used as a penetrator for soft materials (Fig. 88-2).

Rockwell hardness is designated by various letters and numbers. The scales are indicated by the letters A, B, C, and D. The C scale, which is the outside scale on the dial, is used in conjunction with the 120° diamond penetrator and a 330-lb (150-kg) major load for testing hardened metals. The B scale, or the red inner scale, is read when the $\frac{1}{16}$-in. (1.58-mm) ball penetrator is used along with the 220-lb (100-kg) load for testing soft metals. The other letters A and D are special scales and are not used as often as the B and C scales. Rockwell superficial hardness scales are used when testing the hardness of thin materials and case-hardened parts.

To Perform a Rockwell C Hardness Test

Although the various Rockwell-type testers may differ slightly in construction, they all operate on the same principle.

1. Select the proper penetrator for the material to be tested. Use a 120° diamond for hardened materials. Use a $\frac{1}{16}$-in.

Fig. 88-1   Rockwell hardness tester showing various anvils.   (Courtesy Wilson Instrument Division, American Chain and Cable Company.)

(1.58-mm) ball for soft steel, cast iron, and nonferrous metals.

2. Mount the proper anvil for the shape of the part being tested.

3. Remove the scale or oxidation from the surface on which the test is to be made. Usually an area of about ½ in. (12.7 mm) in diameter is sufficient.

4. Place the workpiece on the anvil and apply the minor load (10 kg) by turning the handwheel until the small needle is in line with the red dot on the dial.

5. Adjust the bezel (outer dial) to zero.

6. Apply the major load (150 kg).

7. After the large hand stops, remove the major load.

8. When the hand ceases to move backward, note the hardness reading on the C scale (black). This reading indicates the difference in penentration of the brale between the minor and major loads (Fig. 88-3) and indicates the Rockwell C (Rc) hardness of the material.

9. Release the minor load and remove the specimen.

NOTE: For accurate results, two or three readings should be taken and averaged. If either the brale or the anvil has been removed and replaced, a "dummy run" should be made to properly seat these parts before performing a test. A test piece of known hardness should be tested occasionally to check the accuracy of the instrument.

BRINELL HARDNESS TESTER

The Brinell hardness tester is operated by pressing a 10-mm hardened steel ball under a load of 3000 kg into the surface of the specimen and measuring the diameter of the impression with a microscope. When the diameter of the impression and the applied load is known, the Brinell hardness number (BHN) can be obtained from Brinell hardness tables. The Brinell hardness value is determined by dividing

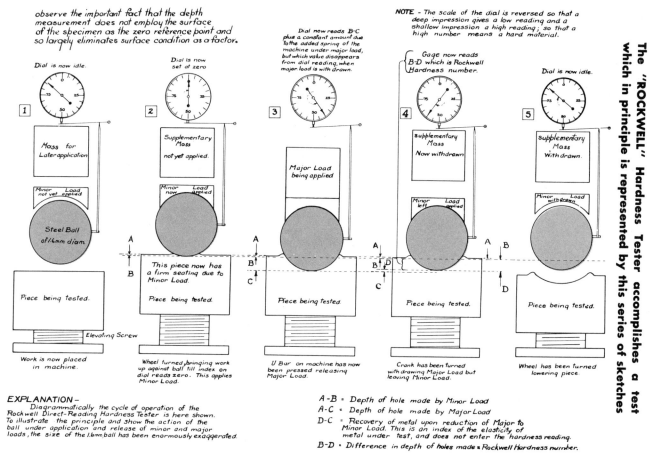

observe the important fact that the depth measurement does not employ the surface of the specimen as the zero reference point and so largely eliminates surface condition as a factor.

1 Dial is now idle.

Mass for Later application

Minor Load not yet applied

Steel Ball of 1.6mm diam.

Piece being tested.

Elevating Screw

Work is now placed in machine.

2 Dial is now set at zero.

Supplementary Mass not yet applied.

Minor Load now applied

A
B

This piece now has a firm seating due to Minor Load.

Piece being tested.

Wheel turned bringing work up against ball till index on dial reads zero. This applies Minor Load.

3 Dial now reads B-C plus a constant amount due to the added spring of the machine under major load, but which value disappears from dial reading, when major load is withdrawn.

Major Load being applied

A
B
C

Piece being tested.

U Bar on machine has now been pressed releasing Major Load.

NOTE - The scale of the dial is reversed so that a deep impression gives a low reading and a shallow impression a high reading; so that a high number means a hard material.

4 Gage now reads B-D which is Rockwell Hardness number.

supplementary Mass Now withdrawn

Minor Load left applied

A
B D
C

Piece being tested.

Crank has been turned withdrawing Major Load but leaving Minor Load.

5 Dial is now idle.

supplementary Mass Withdrawn

Minor Load withdrawn

A
B
D

Piece being tested.

Wheel has been turned lowering piece.

The "ROCKWELL" Hardness Tester accomplishes a test which in principle is represented by this series of sketches

EXPLANATION –
   Diagrammatically the cycle of operation of the Rockwell Direct-Reading Hardness Tester is here shown. To illustrate the principle and show the action of the ball under application and release of minor and major loads, the size of the 1.6mm ball has been enormously exaggerated.

A-B = Depth of hole made by Minor Load
A-C = Depth of hole made by Major Load
D-C = Recovery of metal upon reduction of Major to Minor Load. This is an index of the elasticity of metal under test, and does not enter the hardness reading.
B-D = Difference in depth of holes made = Rockwell Hardness number.

Fig. 88-2   Operating principle of a Rockwell hardness tester—steel-ball type.

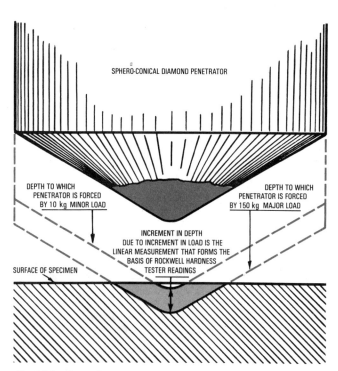

SPHERO-CONICAL DIAMOND PENETRATOR

DEPTH TO WHICH PENETRATOR IS FORCED BY 10 kg MINOR LOAD

DEPTH TO WHICH PENETRATOR IS FORCED BY 150 kg MAJOR LOAD

INCREMENT IN DEPTH DUE TO INCREMENT IN LOAD IS THE LINEAR MEASUREMENT THAT FORMS THE BASIS OF ROCKWELL HARDNESS TESTER READINGS

SURFACE OF SPECIMEN

Fig. 88-3   Operating principle of a Rockwell hardness tester— diamond-cone type.

the area of the impression (in square millimeters) by the load in kilograms which was applied to the penetrator.

A standard load of 500 kg is used for testing nonferrous metals. The impression made in softer metals is larger and the Brinell hardness number is lower.

SCLEROSCOPE HARDNESS TESTER

The scleroscope hardness tester is operated on the principle that a small diamond-tipped hammer, when dropped from a fixed height, will rebound higher from a hard surface than from a softer one. The height of the rebound is converted to a hardness reading.

Scleroscopes are available in several models. On some models, the hardness reading is taken directly from the vertical barrel or tube (Fig. 88-4). On others, the hardness numbers are marked on the dial (Fig. 88-5). These models may also show the corresponding Brinell and Rockwell numbers. See Table 14 in appendix for hardness conversion chart.

DESTRUCTIVE TESTING

There is a close relationship between the various properties of a metal; for example, the tensile strength of a metal

595

Fig. 88-4   A vertical scale scleroscope.   (Courtesy The Shore Instrument and Manufacturing Company.)

Fig. 88-5   A dial recording scleroscope.

increases as the hardness increases, and the ductility decreases as the hardness increases. Thus the tensile strength of a metal may be determined with reasonable accuracy if the hardness and the composition of the metal are known. A more accurate method of determining the tensile strength of a material is that of tensile testing. The tensile strength, or the maximum amount of pull (force) that a material can withstand before breaking, is determined on a tensile testing machine (Fig. 88-6).

A sample of the metal is pulled or stretched in a machine until the metal fractures. This test indicates not only the tensile strength but also the elastic limit, the yield point, the percentage of area reduction, and the percentage of elongation of the material.

TENSILE TESTING (INCH EQUIPMENT)

The tensile strength of a material is expressed in terms of pounds per square inch and is calculated as follows:

$$\text{Tensile strength} = \frac{\text{load, lb}}{\text{area, in.}^2}$$

Machines capable of extremely high tension loads are required to test a steel sample having a cross-sectional area of 1 in.^{2} For this reason, most samples are reduced to a definite cross-sectional area which can be used conveniently in calculating the tensile strength. For example, most samples are machined to 0.505 in. in diameter (Fig. 88-7), which is 0.2 in.^{2} Smaller tensile testing machines use a sample that has been machined to 0.251 in. in diameter (Fig. 88-8), which is an area of 1/20 in.^{2}

The tensile strength of these samples is as follows:

Load of 10,000 lb using 0.505-in.-diameter sample (0.2 in.^{2})

$$\text{Tensile strength} = \frac{10,000}{0.2}$$
$$= 50,000 \text{ psi}$$

Load of 3000 lb using 0.251-in.-diameter sample (1/20 in.^{2})

$$\text{Tensile strength} = \frac{3000}{1/20}$$
$$= 60,000 \text{ psi}$$

TO DETERMINE THE TENSILE STRENGTH OF STEEL

1. Turn a sample of the steel to be tested to the dimensions shown in Fig. 88-8, and place on it two center-punch marks exactly 2 in. apart.

2. Mount the specimen in the machine (Fig. 88-9A), and make sure that the jaws grip the sample properly by jogging the motor switch until the black needle just starts to move. If necessary, remove any tension by reversing the motor.

3. Turn the red hand back until it bears against the black hand on the dial.

4. Set a pair of dividers to the center-punch marks on the sample (2 in.).

5. Start the machine and apply the load to the specimen.

6. Observe and record the readings at which there are any changes in the uniform movement of the needle.

Fig. 88-6　A tensile testing machine.

2-in. GAGE LENGTH

Fig. 88-7　A test sample 0.505 in. in diameter.

2-in. GAGE LENGTH

Fig. 88-8　A test sample 0.251 in. in diameter.

NOTE:　At this time it is possible to determine the elastic limit of the metal, or the maximum stress which can be developed without causing permanent deformation. This is done by checking the distance between the two center-punch marks with the preset dividers. Increased loads must be applied to the specimen and removed several times. After each load has been removed check that the distance between the center-punch marks remains at 2 in. When this distance increases even slightly, the elastic limit of the metal has been reached. An *extensometer* may also be used to indicate the elastic limit.

7. Continue to exert the pull on the sample until it "necks down" (Fig. 88-9B) and finally breaks (Fig. 88-9C). From this procedure, several properties of the metal may be determined (Fig. 88-10).

8. Remove the sample pieces, place the broken ends together, and clamp them in this position.

9. Measure the distance between the center-punch marks to determine the amount of elongation.

10. Measure the diameter of the specimen at the break to determine the reduction in diameter.

Observations (refer to Fig. 88-10)

1. The needle continued to move uniformly until about 3600 lb showed on the scale, after which it slowed down slightly. The point at which it began to slow down indicated the *proportional limit*. It is here that the metal reached its *elastic limit* and no longer returned to its original size or shape. At this point, the stress and strain were no longer proportional and the curve changed.

The *yield point* was reached just beyond the proportional limit and here the metal started to stretch or yield; the strain increased without a corresponding stress increase.

2. The needle continued to move slowly up to about 6300 lb and then it remained stationary.

3. After a short time, the metal began to show a reduction in diameter (necking down) at which time the needle began to show a rapid drop. The highest travel of the needle indicated the *ultimate strength* or the *tensile strength* of the metal. This was the maximum pull to which a metal may be subjected before breaking.

A   B   C

Fig. 88-9 (A) A 0.251-in.-diameter specimen mounted ready for testing; (B) specimen "necking down"; (C) fractured specimen.

Fig. 88-10 Stress-strain graph.

4. The needle continued to move backward (leaving the red hand stationary) and suddenly the metal broke at the point where it had necked down. The point at which it broke is known as the *breaking stress*.

5. The position of the red hand was at 6300 lb. This indicated the load required to break a cross-sectional area of 1/20 in.² The ultimate strength of the metal or the tensile strength was 6300 × 20 = 126,000 lb.

6. When the pieces were placed back together, clamped, and measured, the distance between the center-punch marks was about 2.185 in. This was an elongation of about 9 percent.

7. When the diameter of the metal at the fracture was measured, it was found to be 0.170 in. This indicated a reduction in area of 0.080 in. or about 32 percent.

TENSILE TESTING (METRIC EQUIPMENT)

Metric tensile testers are graduated in kilograms per square centimeter and the cross-sectional area of the sample is calculated in square centimeters or square millimeters. Metric extensometers are graduated in millimeters.

The calculations involved in determining the tensile strength, percentage elongation, and area of reduction are the same as for inch calculations.

Although not the accepted SI unit of pressure, most metric tensile testing machines available and in use at the time of publication were graduated in kilograms per square centimeter (kg/cm²). For conversion to pascals, use the formula 1 kg/cm² = 980.6 Pa.

EXAMPLE

A sample piece is 1.3 cm in diameter. The ultimate pull exerted on it during a tensile test was 4650 kg. What was the tensile strength of this metal?

$$\text{Tensile strength, kg/cm}^2 = \frac{\text{load, kg}}{\text{area, cm}^2}$$
$$= \frac{4650}{1.327}$$
$$= 3504 \text{ kg/cm}^2$$

Percentage of Elongation

If the punch marks were 50 mm apart at the start of the pull and 54 mm after the sample was broken, what was the percentage elongation?

$$\% \text{ elongation} = \frac{\text{amount of elongation}}{\text{original length}} \times 100$$
$$= \frac{4}{50} \times 100$$
$$= 8\%$$

Percentage Reduction of Area

The original diameter was 1.3 cm and after breaking, the diameter was 0.95 cm. What was the percentage of area of reduction?

Fig. 88-11   Principle of the Charpy impact test.

$$\text{Amount of reduction} = 1.3 - 0.95$$
$$= 0.35 \text{ cm}$$
$$\% \text{ of reduction} = \frac{0.35}{1.3} \times 100$$
$$= 27\%$$

IMPACT TESTS

The toughness of metal, or its ability to withstand a sudden shock or impact may be measured by the *Charpy impact test* or the *Izod test*.

In both tests a 10-mm-square specimen is used which may be notched or grooved, depending on the test used.

Both tests use a swinging pendulum of a fixed mass, which is raised to a standard height, depending on the type of specimen. The pendulum is released and as it swings through an arc, it strikes the specimen in the pendulum's path.

In the Charpy test (Fig. 88-11), the specimen is mounted in a fixture and supported at both ends. The V or notch is placed on the side opposite to the direction of the pendulum's swing. When the pendulum is released, the knife edge strikes the sample in the center, reducing the travel of the pendulum. The difference in height of the pendulum at the beginning and end of the stroke is shown on the gage and this indicates the amount of energy used to fracture the specimen.

The Izod test (Fig. 88-12) is similar in principle to the Charpy test. One end of the work is gripped in a clamp with the notched side toward the direction of the pendulum's swing. The amount of energy required to break the specimen is recorded on the scale (Fig. 88-13).

NONFERROUS METALS AND ALLOYS

Nonferrous metals, as the name implies, contain little or no iron and are generally nonmagnetic. Since all pure nonferrous metals do not offer the qualities required for industrial applications, they are often combined to produce alloys having the desired qualities for a particular job. The most widely used nonferrous metals for industrial use are aluminum, copper, lead, magnesium, nickel, tin, and zinc.

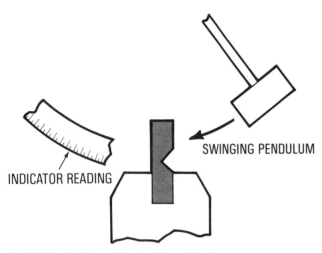

Fig. 88-12 Principle of the Izod impact test.

ALUMINUM

Aluminum is a light, soft, white metal produced from bauxite ore. It is resistant to atmospheric corrosion and is a good conductor of electricity and heat. It is malleable and ductile and can easily be machined, forged, rolled, and extruded. It has a low melting point of 1220°F (660°C) and can be cast easily. It is used extensively in transportation vehicles of all types, the construction industry, transmission line, cooking utensils, and hardware.

Aluminum is not generally used in a pure state, since it is too soft and weak. It is alloyed with other metals to form strong alloys used extensively in industry.

Aluminum-base alloys

Duralumin, an alloy of 95 percent aluminum and 4 percent copper, 0.05 percent manganese, and 0.05 percent magnesium, is widely used in the aircraft and transportation industries. This is a natural-aging alloy, that is, it hardens as it ages. Due to this peculiarity, duralumin must be kept at a subzero temperature until ready for use. When it is brought to room temperature, the hardening process begins.

Other alloys contain varying amounts of copper, magnesium, and manganese.

Aluminum-manganese alloys have good formability, good resistance to corrosion, and good weldability. These alloys are used for utensils, gasoline and oil tanks, pressure vessels, and piping.

Aluminum-silicon alloys are easily forged and cast. They are used for forged automotive pistons, intricate castings, and marine fittings.

Aluminum-magnesium alloys have good corrosion resistance and moderate strength. They are used for architectural extrusions and automotive gas and oil lines.

Aluminum-silicon-magnesium alloys have excellent corrosion resistance, are heat-treatable, and are easily worked and cast. They are used for small boats, furniture, bridge railings, and architectural applications.

Auminum-zinc alloys contain zinc, magnesium, and copper with smaller amounts of manganese and chromium.

Fig. 88-13 Impact testing machine.   (Courtesy Ametek Testing Equipment.)

They have high tensile strength, good corrosion resistance, and may be heat-treated. They are used for aircraft structural parts when great strength is required.

COPPER

Copper is a heavy, soft, reddish-colored metal refined from copper ore (copper sulfide). It has high electrical and thermal conductivity, good corrosion resistance, strength, and is easily welded, brazed, or soldered. It is very ductile and is easily drawn into wire and tubing. Since copper work hardens readily, it must be heated at about 1200°F (648.8°C) and quenched in water to anneal.

Because of its softness, copper does not machine well. The long chips produced in drilling and tapping tend to clog the flutes of the cutting tool and must be cleared frequently. Sawing and milling operations require cutters with good chip clearance. Coolant should be used to minimize heat and aid the cutting action.

Copper-base alloys

Brass, an alloy of copper and zinc, has good corrosion resistance and is easily formed, machined, and cast. There are several forms of brass. *Alpha* brasses containing up to 36 percent zinc are suitable for cold working. *Alpha + beta* brasses containing 54 to 62 percent copper are used in hot working of this alloy. Small amounts of tin or antimony are added to alpha brasses to minimize the pitting effect of salt water on this alloy.

Brass alloys are used for water and gasoline line fittings, tubing, tanks, radiator cores, and rivets.

Bronze, the term which originally referred to an alloy of copper and tin, has now been extended to include all alloys except copper-zinc alloys, which contain up to 12 percent of the principal alloying element.

Phospor-bronze contains about 90 percent copper, 10 percent tin, and a very small amount of phosphorus, which acts as a hardener. This metal has high strength, toughness, and corrosion resistance and is used for lock washers, cotter pins, springs, and clutch discs.

Silicon-bronze (a copper-silicon alloy) contains less than 5 percent silicon and is the strongest of the work-hardenable copper alloys. It has the mechanical properties of machine steel and the corrosion resistance of copper. It is used for tanks, pressure vessels, and hydraulic pressure lines.

Aluminum-bronze (a copper-aluminum alloy) contains between 4 and 11 percent aluminum. Other elements such as iron, nickel, manganese, and silicon are added to aluminum bronzes. Iron (up to 5 percent) increases the strength and refines the grain. Nickel, when added (up to 5 percent), has similar effects to iron. Silicon (up to 2 percent) improves machinability. Manganese promotes soundness in casting.

Aluminum-bronzes have good corrosion resistance and strength and are used for condenser tubes, pressure vessels, nuts, and bolts.

Beryllium-bronze (copper and beryllium), containing up to about 2 percent beryllium, is easily formed in the annealed condition. It has high tensile strength and fatigue strength in the hardened condition. Beryllium-bronze is used for surgical instruments, bolts, nuts, and screws.

LEAD

Lead is a soft, heavy metal that has a bright, silvery color when freshly cut but turns gray quickly when exposed to air. It has a low melting point, low strength, low electrical conductivity, and high corrosion resistance. It is used extensively in the chemical and plumbing industries. Lead is also added to bronzes, brasses, and machine steel to improve their machinability.

LEAD ALLOYS

Antimony and *tin* are the most common alloying elements of lead. *Antimony,* when added to lead (up to 14 percent), increases its strength and hardness. This alloy is used for battery plates and cable-sheathing.

The most common lead-tin alloy is solder, which may be composed of 40 percent tin and 60 percent lead, or 50 percent of each. Antimony is sometimes added as a hardener.

Lead-tin-antimony alloys are used as type metals in the printing industry.

MAGNESIUM

Magnesium is a light-weight element which, when alloyed, produces a light, strong metal used extensively in the air-craft and missile industries. Magnesium plates are used to prevent corrosion by salt water in underwater fittings on ship hulls. Magnesium rods, when inserted in galvanized domestic water tanks, will prolong the life of the tank. Other uses for this metal are in photographic flash bulbs and thermite bombs.

NICKEL

Nickel, a whitish metal, is noted for its resistance to corrosion and oxidation. It is used extensively for electroplating, but its most important application is in the manufacture of stainless and alloy steels.

Nickel Alloys

Nickel-chromium-Iron-base alloys (containing about 60 percent nickel, 16 percent chromium, and 24 percent iron) are widely used for electric heating elements in toasters, percolators, and water heaters.

Monel metal, containing about 60 percent nickel, 38 percent copper, and small amounts of manganese or aluminum, is a tough, ductile metal with good machining qualities. It is corrosion-resistant, nonmagnetic, and is used in valve seats, chemical marine pumps, and in nonmagnetic aircraft parts.

Hasteloy, containing about 87 percent nickel, 10 percent silicon, and 3 percent copper, is widely used in the chemical industry because of its noncorrosive qualities.

Inconel, a strong, tough alloy containing about 76 percent nickel, 16 percent chromium, and 8 percent iron is used in food-processing equipment, milk pasteurizers, exhaust manifolds for aircraft, and in heat-treating furnaces and equipment.

TIN

Pure tin has a silvery-white appearance, good corrosion resistance, and melts at about 450°F (232.2°C). It is used as a coating on other metals, such as iron, to form tin plate.

Tin Alloys

As already mentioned, tin when alloyed with lead forms solder.

Babbitt is an alloy containing tin, lead, and copper.

Pewter, another tin-base alloy, is composed of 92 percent tin and 8 percent antimony and copper.

ZINC

Zinc is a coarse, crystalline, brittle metal used mainly for die-casting alloys and as a coating for sheet steel, chain, wire, screws, and piping. Zinc alloys, containing approximately 90 to 95 percent zinc, 5 percent aluminum, and small amounts of copper or magnesium, are widely used in the die-casting field to produce automotive parts, building hardware, padlocks, and toys.

BEARING METALS

Bearing metals may be divided into two groups: leaded bronzes and babbitt.

Leaded Bronzes

The composition of bronze bearings varies according to their use. Bearings used to support heavy loads contain about 80 percent copper, 10 percent tin, and 10 percent lead. For lighter loads and faster speeds, the lead content is increased. A typical bearing of this type might contain 70 percent copper, 5 percent tin, and 25 percent lead.

Babbitt

Babbitt bearing materials may be divided into two groups: lead base and tin base.

Lead-base babbitt may contain 75 percent lead, 10 percent tin, and 15 percent antimony, depending upon the application. A small amount of arsenic is often added to permit the bearing to carry heavier loads. Applications of lead-base babbitt bearings are in automotive connecting rods, main and crankshaft bearings, and diesel engine bearings.

Tin-base babbitt may contain up to 90 percent tin with copper and antimony added, to 65 percent tin, 15 percent antimony, 2 percent copper, and 18 percent lead. Since tin has become less plentiful, tin-base babbitts are used in high-grade bearing applications such as steam turbines.

REVIEW QUESTIONS

TESTING OF METALS

1. Explain the difference between nondestructive and destructive testing.
2. What information can be determined by hardness testing?

ROCKWELL HARDNESS TESTER

3. Name two scales which are found on a Rockwell-type hardness tester and state the penetrator used in each case.
4. Briefly describe how a Rockwell C hardness test is performed.

BRINELL HARDNESS TESTER

5. Compare the principles of the Brinell hardness tester and the Rockwell hardness tester.

SCLEROSCOPE

6. Describe the principle of the scleroscope.

DESTRUCTIVE TESTING

7. What effect does hardening have on the tensile strength and ductility of a metal?
8. Explain the principle of tensile testing.
9. What other properties may be determined by a tensile test?
10. How is the tensile strength of a metal calculated?
11. Explain the procedure for performing a tensile test.
12. How may the elastic limit of a metal be determined?
13. Define: proportional limit, yield point, ultimate strength, breaking stress.
14. Describe the principle of impact testing.

NONFERRROUS METALS AND ALLOYS

15. Define a nonferrous metal.
16. Name four nonferrous metals commonly used as base metals in alloys.
17. Name four aluminum-base alloys and state one application of each.
18. Name two types of brasses and state the composition of each.
19. What is the difference between brass and bronze alloys?
20. Why is lead added to steel?
21. Where are magnesium alloys used extensively?
22. Name three nickel-base alloys and state two applications of each.
23. Name two tin-base alloys.
24. For what purpose are zinc alloys used extensively?
25. What are the three metals used for bearing alloys?
26. What is the basic difference between bronze and babbitt?
27. Explain the use of the following:
    (a) lead-base babbitt
    (b) tin-base babbitt

HYDRAULICS
SECTION 20

A hydraulic system is a method of transmitting force or motion by applying pressure on a confined liquid. It can be used to substitute liquid linkage between moving parts for mechanical linkages such as shafts, gears, connecting rods, etc. The science of hydraulics dates back several thousand years when dams, sluice gates, and waterwheels were used to control water flow for domestic use and irrigation purposes.

Hydraulic power is used in nearly every phase of industry. Machine tools, cars, airplanes, jacks, missiles, and boats are only a few of the areas where hydraulics is applied. Hydraulic systems, because of their extreme versatility, are adaptable to almost every type of machine tool to provide feeds and drives.

No other medium combines the same degree of accuracy, positiveness, and flexibility of control with the ability to transmit a maximum of motion in a minimum of bulk and mass. Oil is the fluid most frequently used in hydraulic systems because it has all these characteristics. Oil also lubricates and gives anticorrosion properties to the mechanical parts of the system.

HYDRAULIC CIRCUITS AND COMPONENTS

UNIT 89

The wide acceptance of hydraulic applications in the machine tool industry is due to the following characteristics of fluids.
1. A fluid is one of the most versatile means of transmitting power and modifying motion.
2. A fluid is infinitely flexible, yet as unyielding as steel, due to its incompressibility.
3. It adjusts readily to changes in diameter and shape.
4. It can be divided to perform work in different areas simultaneously. The hydraulic brake system in an automobile is an example of this.

OBJECTIVES

After completing this unit you will:
1. Be familiar with Pascal's law and the hydraulic principle
2. Be able to identify and state the purpose of the main parts in a hydraulic system
3. Understand the application of constant- and variable-volume systems as applied to machines

THE HYDRAULIC PRINCIPLE

The extensive use of hydraulics is due to the fact that liquids possess two properties which make them especially useful in power transmission. Liquids cannot be compressed and they have the ability to multiply force. In the 17th century, the French scientist Pascal studied the physical properties of liquids and formulated the basic law of hydraulics. Pascal's law states that: *pressure at any one point in a static liquid is the same in every direction, and pressure exerted on an enclosed liquid is transmitted undiminished in every direction, and acts with equal force on equal areas* (Fig. 89-1).

The two types of hydraulic systems are the *hydrostatic* and the *hydrodynamic*. A basic comparison of these two principles is shown in Fig. 89-2.

HYDROSTATIC PRINCIPLE

Most industrial systems work on the hydrostatic principle, which can be easily understood by referring to Fig. 89-1.

Fig. 89-2   (A) The static principle: A heavy mass applies constant pressure to split a log.   (B) The dynamic principle: The force of a sledge hammer is used on the wedge to split the log.

1-lb PUSH

FLUID PRESSURE THE SAME AT ALL POINTS

PISTON WITH CROSS-SECTIONAL AREA OF 1 in.²

Fig. 89-1   An illustration of Pascal's law.

Fig. 89-3 The hydrostatic principle is used to increase force or pressure.

Fig. 89-4 Hydrodynamic force is used to turn a paddle wheel.

Suppose a piston with a cross-sectional area of 1 square inch (in.^{2}) is tightly fitted in a liquid-filled cylinder. If a 1-lb force is applied to the piston, a pressure of 1 psi is created and this pressure is transmitted to all points in the cylinder. If there were no variations of pressure in the cylinder due to gravity, frictional losses, etc., a gage installed anywhere in the cylinder would record a pressure of 1 psi (Fig. 89-1).

The hydrostatic principle is often used to increase force or pressure. This can best be understood by connecting the piston-cylinder combination in Fig. 89-1 to a piston-cylinder combination which has a cross-sectional area of 10 in.^{2} (Fig. 89-3). By applying a 1-lb force on the small piston, this 1-lb force is transmitted to the 10 in.^{2} piston area, creating a total force of 10 lb (Fig. 89-3). This increase in force is gained at the expense of the distance moved. For example, it would be necessary to move the small piston 10 in. in order to move the large piston 1 in. By applying pressure on the large piston, it is possible to reduce the total force and increase the distance the small piston moves.

Nearly all industrial hydraulic systems using the hydrostatic principle have pumps which apply constant pressure to the fluid. By using various devices to control the pressure, rate, and direction of flow, extremely flexible and vesatile machines can be designed.

HYDRODYNAMIC PRINCIPLE

Dynamic force uses *kinetic energy* stored in a moving body to perform work. A log can be split *dynamically* by hitting the wedge with a sledge hammer (Fig. 89-2B). The amount of work performed depends on the speed and mass of the sledge hammer. Increasing either, or both, increases the amount of force exerted.

Hydraulic fluid can be used dynamically by directing a stream of fluid against a paddle wheel (Fig. 89-4). Through impact, much of the kinetic energy of the moving fluid is conveyed to the paddle wheel. Hydraulic couplings and torque converters work on the hydrodynamic principle.

FUNDAMENTAL HYDRAULIC CIRCUIT

All hydraulic systems are basically simple, with six essential elements (Fig. 89-5).

1. A reservoir for the oil supply
2. A pump to move the oil under pressure

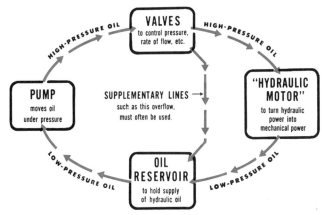

Fig. 89-5 A fundamental hydraulic circuit.

3. One or more control valves to regulate the flow
4. A piston and cylinder (hydraulic motor) to convert hydraulic power into mechanical power
5. Pipe or tubing to connect the various parts
6. Hydraulic fluid, which is the ''life blood'' of the system

There is no such thing as an all-hydraulic system. At certain points in any hydraulic circuit, mechanical devices must be used to control the fluid. A hydraulic system is a collection of mechanical elements joined together by *fluid connecting rods* and in many cases by *fluid levers*.

HYDRAULIC SYSTEM COMPONENTS

Certain basic parts must be used in every hydraulic system; other parts are sometimes used to refine the system or to cope with special conditions. Each basic part will be discussed in detail, to show how a hydraulic system operates.

OIL RESERVOIR

Every hydraulic system requires a reservoir for storing the oil used in the system. A separate tank (Fig. 89-6) is generally used for large hydraulic systems. On smaller machines, the oil reservoir may be incorporated in the base of the machine. The built-in reservoirs are generally compact but are hard to get at and difficult to clean.

PUMPS

The pump is an important part of a hydraulic system; its purpose is to transmit force by causing oil to flow. A simple

Fig. 89-6   The construction of a hydraulic oil reservoir.

Fig. 89-7   The principle of a reciprocating hydraulic pump.

Fig. 89-8   The principle of a gear pump.

Fig. 89-9   The principle of an internal pump.

Fig. 89-10   The principle of a screw pump.

example of a hydraulic pump may be found in the hydraulic brake system of a car. By depressing the brake pedal, the piston in the master brake cylinder moves toward the discharge end, causing the fluid under pressure to flow to the four wheel cylinders. By applying force at one end, this force is transmitted from one point (the brake pedal) to four points (the wheel cylinders). As soon as the brake shoes contact the brake drums, oil stops flowing. The primary purpose of the master cylinder, which is a very simple pump, is to transmit force hydraulically. In more complex hydraulic systems, the pump must provide a steady flow of liquid under pressure. Therefore the pump must provide a means of transmitting both force and motion.

Some of the more common types of pumps used in hydraulic systems will be described.

Reciprocating pumps (Fig. 89-7) have limited use because the oil pulsates too much to provide smooth operation. However, by using several multipiston types, a fairly steady flow can be provided. Reciprocating pumps are quite rugged and work well under adverse conditions such as low temperatures.

Gear pumps (Fig. 89-8) are simple in design. Because of their almost nonpulsating flow, they are well suited to applications requiring smooth operation. Simple gear pumps have spur gears; however, for quieter and smoother operation, some types contain helical or herringbone gears.

Internal gear pumps (Fig. 89-9) are basically the same as gear pumps except that they have an internal gear. They are generally not as efficient as gear pumps and

lose their efficiency rapidly as soon as mating parts become worn.

Screw pumps (Fig. 89-10) are usually used to transfer oil from one point to another. They are seldom used to provide hydraulic power. Screw pumps are simple, rugged, and capable of operating at high speeds.

Vane pumps (Fig. 89-11) are used extensively to provide hydraulic power because they produce a reasonably

1. As this rotor turns counter-clockwise . . .

2. these vanes are thrown out by centrifugal force and oil pressure . . .

3. against this elliptical housing.

5. Conversely, over areas covered by these discharge ports, the capacity of spaces between adjacent vanes decreases as rotor turns, and thus oil is forced out of pump.

4. Over areas covered by these intake ports, the capacity of spaces between adjacent vanes increases as rotor turns, and thus oil is drawn into pump.

Fig. 89-11   The principle of a vane pump.

3. in this circular housing.

2. this rotating cylinder block mounted eccentrically . . .

1. Pump pistons reciprocate as they are carried around by . . .

4. Assuming clockwise rotation, oil is drawn in through these ports . . .

These parts do not rotate.

5. and discharged through these ports.

Fig. 89-12   The principle of a radial-piston pump.

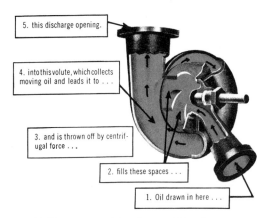

5. this discharge opening.

4. into this volute, which collects moving oil and leads it to . . .

3. and is thrown off by centrifugal force . . .

2. fills these spaces . . .

1. Oil drawn in here . . .

Fig. 89-13   The principle of a centrifugal pump.

Flow in this direction only

Fig. 89-14   A ball valve permits flow in one direction only.

steady flow. Wear of parts does not affect the efficiency of these pumps greatly, since the vanes can always maintain close contact with the ring in which they rotate.

Radial-piston pumps (Fig. 89-12) are compact, rugged, and used in applications requiring a flexible source of hydraulic power. Through the use of ring-shifting mechanisms, the output of radial-piston pumps can be varied between zero and full volume.

Centrifugal pumps (Fig. 89-13) provide a nonpulsating flow and are used in moving large quantities of oil. This pump cannot be overloaded because the oil in the pump begins to rotate with the rotor as soon as the output pressure becomes too great. The output pressure is difficult to control; therefore centrifugal pumps are not generally used to provide hydraulic power.

VALVES

Once the pump begins to move the oil under pressure, valves are usually required to *control pressure* and *to direct and control oil flow*. Valves were the only method of controlling flow and pressure until recently. Successful pumps have been developed during the past few years which are proving popular because they have the means of varying both the flow and pressure. However, valves are still the most important devices for providing flexibility in a hydraulic system, and a few of the more common types will be discussed.

Ball valves (Fig. 89-14) are used in small reciprocating pumps and in small oil lines to permit oil flow in one direction only. Ball valves are usually spring loaded and will open when the pressure in the system becomes greater than pressure exerted by the spring. Spring-loaded ball valves are often used as pressure-relief valves. They are placed in a line connecting the hydraulic pump with the oil reservoir to prevent the buildup of excessive pressure in the system.

Rotary valves (Fig. 89-15) are used to control the direction of oil flow. They are generally used as pilot valves to control the movement of spool valves. The extra ports and passages in the rotating part of the valve allow the oil flow in several lines to be controlled by one valve. In a hydraulic surface grinder, the reciprocating and crossfeed table movements are controlled by one rotary valve.

To Reservoir   To Valve   To Reservoir   To Valve   To Reservoir   From Valve

From Pump   From Valve   From Pump   From Valve   From Pump   To Valve

Fig. 89-15   The rotary valve may control the flow of oil to the spool valve or to the cylinder.

Fig. 89-16 The spool valve controls the flow of oil to and from the cylinder.

Flow to one end of hydraulic cylinder

From pump

From cylinder

From pump

From cylinder

Flow to other end of hydraulic cylinder

To reservoir

Valved oil lines connected here provide hydraulic control of valves

From pump

From cylinder

To reservoir

Pump flow shut off

To reservoir

Fig. 89-17 A cock valve.

Spool valves (Fig. 89-16) are used extensively to control the direction of oil flow because of their quick positive action. Spool valves are versatile and can control flow through several parts of a hydraulic system. The mating surfaces of this type of valve must be accurately machined and fitted to prevent leakage and to ensure efficient operation. Oil grooves are often machined around the pistons to keep the spool centered and lubricated. Spool valves can be operated mechanically, electrically, or hydraulically.

Cock valves (Fig. 89-17) are simple valves used to bleed cylinders, eliminate air pockets, and control the operation of pressure gauges. They are generally built in small sizes and are used in systems using moderate pressure.

Needle valves (Fig. 89-18) provide a means of regulating oil flow precisely. They are operated manually and can be used in any position from fully opened to fully closed. Their design prevents an abrupt change in oil flow; they are therefore often used in lines connecting sensitive devices which could be damaged by a sudden surge of pressure.

HYDRAULIC MOTOR

A hydraulic motor, generally a simple piston-and-cylinder combination, converts the oil flow into mechanical motion. Hydraulic motors alone can produce only two types of motion: straight-line (piston and cylinder type) and rotary (torque converter). When more complex motions are required, mechanical elements must be used in conjunction with the hydraulic motor.

The *piston motor* (Fig. 89-19) is the most commonly used hydraulic motor. Its design is simple and it can be made to withstand any pressure. When oil flows into the

Fig. 89-18 A needle valve.

cylinder from the right side, the piston is moved to the left a certain distance in a given time. When the oil flow is reversed, the piston moves to the right at a slow rate because, with no piston rod to take up some of the cylinder volume, more oil must flow into the cylinder to move the piston the same distance. This type of piston motor, commonly called a *differential piston motor* (Fig. 89-19), is often used in machines which require a powerful working stroke and a fast, but less powerful return stroke.

THE HYDRAULIC SYSTEM

A hydraulic system is produced when the various components such as reservoir, pump, valve, and motor are combined and joined together with the necessary pipe, tubing, and fittings. The hydraulic system illustrated in Fig. 89-20 is quite simple; however, when it becomes necessary to vary speeds, motions, etc., in a fixed or variable sequence, the complexity of the system increases. The most complex system still contains the basic components found in a simple system.

With flow through these lines as shown, piston moves to the left. When flow is reversed by control valve, piston moves to the right at a lower speed but with greater force.

Fig. 89-19   A simple hydraulic piston motor.

Two types of hydraulic systems, the *constant volume* and the *variable volume,* are generally used.

Constant-volume systems contain pumps which have a fixed output. The pump operates whenever the machine is in operation. If the full flow of oil delivered by the pump is not required in the system, it is bypassed through valves back to the oil reservoir.

Variable-volume systems contain pumps whose output can be varied to suit different requirements. Combinations of mechanical, electrical, and pneumatic controls may be used to regulate the flow of oil. Variable-volume systems generally do not require certain bypass and flow-control valves which are needed in a constant-volume system. Variable-volume systems are widely used when the output power and speed vary over a wide range.

Many machine tools use both the constant-volume and the variable-volume systems. Each system may be used individually to operate different parts of a machine, or they may be combined to alternately control certain portions of the operating cycle. Examples of a few hydraulic systems found on machine tools will be explained in the following section.

CONSTANT-VOLUME SYSTEM

The shaper (Fig. 89-21) is an example of a constant-volume system applied to a machine tool. A brief explanation of how this system operates follows.

1. The gear pump develops moderate oil pressure which it transmits through piping and a reversing valve to the operating cylinder.
2. The oil acts against the piston, moving it and the shaper ram forward on its cutting stroke.

4. the center opening of this control valve, which can be operated manually or automatically to direct oil flow to either right or left end of . . .

3. and delivered to this valve, which can be used to shut off flow and to control the rate of flow. From here, oil flows to . . .

2. is picked up by this pump . . .

1. Oil stored in this reservoir . . .

5. this hydraulic motor. In position shown, high-pressure oil is flowing to right end, and oil under lower pressure is being forced back through control valve to reservoir.

6. When pressure on output side of pump rises above a predetermined level, this relief valve opens and oil flow is shunted back to reservoir.

Fig. 89-20   A simple hydraulic system.

Fig. 89-21 The constant-volume hydraulic system of a shaper. (Courtesy Mobil Oil Corporation.)

3. As the piston moves forward, the oil ahead of the piston is expelled and returns to the reservoir.

4. At the end of the cutting stroke, a stop engages the reversing lever, which actuates a reversing valve, causing the oil to flow in the ópposite direction.

5. This reversal of oil flow moves the piston and the shaper ram back to its starting position.

6. At the end of the return stroke, a second stop engages the reversing lever, actuating the reversing valve and again reversing the oil flow.

7. The length of the shaper stroke can be controlled by adjusting the position of the stops.

8. The speed of the forward (cutting) stroke can be controlled by a throttling valve which regulates the flow of oil back to the reservoir.

The operation of the constant-volume hydraulic system of a surface grinder (Fig. 89-22) is as follows.

1. A constant-volume or a constant-discharge gear pump delivers moderate oil pressure through various valves to the main operating cylinder.

2. The oil acts on the piston to reciprocate the worktable.

3. Excess oil pressure is released through a spring-loaded relief valve and returns to the reservoir.

4. At each reversal of the worktable, cams actuate the reversing valve, directing the oil through a wheel-infeed cylinder, then alternately to opposite ends of the main operating cylinder.

5. The wheel-infeed cylinder also actuates the ratchet mechanism of the wheel feed.

6. The rate of table travel is controlled by adjusting the throttling valve.

VARIABLE-VOLUME SYSTEM

The milling machine requires a slow feed during the cutting stroke, followed by a rapid return stroke which requires

Fig. 89-22 The constant-volume hydraulic system of a surface grinder.

very little oil pressure. The lock-feed hydraulic system of a milling machine using three pumps is illustrated in Fig. 89-23. A brief explanation of the operation of this system follows.

1. The simple gear pump draws oil from the reservoir and supplies it at low pressure to the booster pump.

2. Oil from the booster pump, now under high pressure, travels through a reversing valve to one end of the working cylinder.

3. As the piston moves forward, the machine table, attached to the piston, also moves at a slow, steady rate.

4. A variable-discharge metering pump, connected to both ends of the cylinder, draws a specific amount of oil from the cylinder *ahead* of the motion of the piston.

5. The metering pump holds back and regulates the advance of the piston, which determines the rate of table travel.

6. When the reversing valve is actuated, the discharge and suction ports of the metering pump are interconnected.

7. The booster pump then discharges oil into the reservoir and the gear pump supplies low-pressure oil to the cylinder for the rapid return stroke.

Fig. 89-23 The variable-volume system of a manufacturing-type milling machine.

HYDRAULIC COUPLINGS AND TORQUE CONVERTERS

In the past few years, great strides have been made in perfecting two *hydrodynamic* devices: the hydraulic coupling and the hydraulic torque converter. Hydraulic couplings and hydraulic torque converters are not new but have only recently come into widespread use because of refinements in both.

A hydraulic coupling, sometimes called a fluid clutch (Fig. 89-24), is used to join two mechanisms hydraulically. An engine or motor at the power-input end may be joined hydraulically to another mechanism at the output end, without the use of mechanical connections. A hydraulic coupling transfers the torque (twist) delivered to it very smoothly, and at the same time protects the mechanisms against damage from vibration or shock loads.

A hydraulic torque converter, sometimes called a hydraulic gearbox (Fig. 89-25), is both a clutch and transmission source. It provides a means of coupling two mechanical elements and also of varying the ratio of the input torque to the output torque. Hydraulic torque converters provide a smooth flow of power, along with high torque at low speed, low torque at high speed, or any combination of the two within the capacity of the equipment.

HYDRAULIC OIL REQUIREMENTS

A hydraulic system contains an assembly of valves, cams, operating cylinders, etc., which all have close fitting tolerances. To obtain maximum efficiency from the system and protect the intricate moving parts, it is important that

4. this driven member. Here the oil's direction of flow is again changed and it flows back to the driving member.

5. In effect, oil swirl thus formed is an "endless corkscrew of oil" that moves simultaneously in several directions: it spins as shown by the arrows and also rotates about the axis of the coupling.

3. but this curvature causes oil to make a turn and flow toward . . .

2. centrifugal force causes oil to flow toward the circumference at right angles to the axis of rotation . . .

1. As this driving member rotates . . .

Fig. 89-24 The principle of a hydraulic coupling.

Fig. 89-25   The principle of a hydraulic torque converter.

the proper oil be selected for use in hydraulic systems. A hydraulic oil must perform two important functions:

1. It must transmit power efficiently.
2. It must lubricate adequately.

For a hydraulic oil to perform these two important functions satisfactorily, it must have the following characteristics.

1. *Proper viscosity,* to assure a ready flow of oil at all times and to minimize leakage.
2. *Film strength and lubricity,* to provide adequate lubrication between closely fitted sliding parts.
3. *Resistance to oxidation,* to prevent intricate moving parts from damage due to oxidation of the oil (sludge formation).
4. *Resistance to emulsification,* to separate quickly from any water which may enter the system.
5. *Resistance to corrosion and rusting,* to prevent damage to the closely fitting parts.
6. *Resistance to foaming,* to free itself readily of air drawn into the hydraulic system.

VISCOSITY

Viscosity is one of the most important qualities that a hydraulic oil must possess. Viscosity should be thought of as a measure of the resistance of oil to flow. Low-viscosity oils flow freely; high-viscosity oils flow sluggishly. The viscosity of oil changes with the temperature; low temperatures cause the oil to flow sluggishly, while heat causes the oil to run freely.

The viscosity of oil has a direct bearing on the efficient transmission of power and the adequate lubrication of the parts in a hydraulic system. If hydraulic oil does not have the right viscosity, it cannot perform satisfactorily. Viscosity of an oil affects the operation of a hydraulic system in the following ways.

1. If an oil is too light (low viscosity), it usually results in:
   a. excessive leakage
   b. lower efficiency at the pump
   c. increased wear on parts
   d. loss of oil pressure

e. lack of positive hydraulic control
f. lower overall efficiency

2. If an oil is too heavy (high viscosity), it usually results in:
   a. increased pressure drop
   b. higher oil temperatures
   c. sluggish operation
   d. lower mechanical efficiency
   e. higher power consumption

FILM STRENGTH AND LUBRICITY

When parts of the hydraulic pump, cylinder, and control valves become worn, loss of pressure, leakage, and less accurate control may result. An important quality of a hydraulic oil should be its ability to minimize wear under the *thin-film* or *boundary lubrication* conditions that exist between the closely fitted sliding parts. A suitable oil maintains strong films that resist being squeezed out or pushed aside.

Both light (low-viscosity) and heavy (high-viscosity) oils have antiwear or film-strength qualities. High-viscosity oils have these qualities to a greater extent than low-viscosity oils. Heavy oils, because of their greater resistance to leakage, tend to resist displacement from lubricated surfaces better than light oils. Therefore, an oil with as high a viscosity as possible, which fulfills all other requirements, should be used to minimize wear.

RESISTANCE TO OXIDATION

Hydraulic oil comes into contact with warm air in the reservoir. Because of this contact, the oil tends to oxidize, that is, combine chemically with the oxygen of the air. The tendency to oxidize is greatly increased at high operating temperatures and pressures, and by excessive agitation or splashing of oil.

A slight oxidation is not harmful, but if an oil has poor resistance to this chemical change, oxidation may become too great. If this occurs, substantial amounts of both soluble and insoluble oxidation products are created and the oil gradually increases in viscosity.

Insoluble oxidation products may be deposited in the form of gum and sludge in the oil passages, pump parts, and valve parts, where it will restrict the flow of oil. This may slow down the movement of pump vanes and spool- and piston-type valves and cause seated valves to leak.

RESISTANCE TO EMULSIFICATION

Regardless of the precautions taken, some water gets into hydraulic systems through leaks and the condensation of atmospheric moisture. Most of the condensation occurs above the oil in the reservoir as the machine cools during idle periods. Any appreciable amount of water in a hydraulic system will promote rusting and decrease the lubricating value of the oil. This will eventually cause leakage, erratic pump action, and a breakdown in the system.

Where the percentage of water in a system is high, it is very important to use a hydraulic oil that has the ability to separate quickly and completely from the water. Oils that have exceptionally high oxidation resistance retain good water-separating abilities for a long time.

RESISTANCE TO CORROSION AND RUSTING

Water and oxygen can cause rusting of ferrous metal parts in a hydraulic system. Air is always present, oxygen is available, and some water is usually present; these conditions promote rusting. The danger of rusting is greatest when the machine is idle and surfaces usually covered with oil are left unprotected.

Rusting causes surface deterioration of metal parts, and rust particles may eventually be carried into the hydraulic system. This contributes to the formation of sludge-like deposits which can cause serious abrasions and interfere with the operation of cylinders, valves, and pumps.

To guard against rusting, it is important that a hydraulic system be kept as free from water as possible. High-quality hydraulic fluids contain a rust inhibitor which forms a film that resists displacement by water, thereby protecting the surfaces from contact with the water.

RESISTANCE TO FOAMING

Oil in a hydraulic system usually dissolves a certain amount of air, which causes a foaming (bubbling) action. Air can enter a system by oil falling from a considerable height into the reservoir, through leaks, or by suction of the pump. A hydraulic system can absorb the amount of air bubbles normally present; however, if an excessive amount of air is present and the bubbles persist at operating temperatures, pumping will be noisy and irregular, and the action of motors and controls will be erratic.

Foaming can be controlled by taking the following precautions.

1. Use of a premium-quality hydraulic oil which has good resistance to foaming.
2. The oil returning from the system should not fall from a great height into the reservoir.
3. Vent valves should be installed at high points in the hydraulic system.
4. The reservoir should be properly designed to allow the air to free itself readily at the oil surface.

ADVANTAGES OF HYDRAULIC SYSTEMS

The development of suitable hydraulic oils and the precision tolerances maintained by manufacturers have led to the wide use of hydraulics in machine tool manufacture. The reason for this widespread use is that hydraulic power transmission offers many advantages.

1. A complex combination of cams, gears, levers, etc., can often be replaced by a simpler combination of pumps, valves, lines, and hydraulic motors.
2. Speeds and feeds can be varied infinitely without stopping the machine.
3. Smooth, steady cutting action, with a minimum amount of chatter or vibration, is easily obtained.
4. Even, positive motion is provided at all loads.
5. The rate of oil pressure or flow can easily be controlled by simple valves.
6. Relief valves prevent overloads which may cause extensive damage.
7. Power costs and friction losses can be held to a minimum.
8. All moving parts in the system are constantly lubricated by the hydraulic oil.
9. The operation of a hydraulic system is comparatively quiet.
10. Rapid reversals at the end of each reciprocating stroke, such as on a shaper or surface grinder, are cushioned (smooth and shockless).
11. Component parts can be located close to the moving members because the power is transmitted through pipes which can run in any direction.
12. Intricate operations can be made almost completely automatic.

REVIEW QUESTIONS

1. Define hydraulics.
2. List four reasons why hydraulic applications have found wide acceptance in the machine tool industry.

THE HYDRAULIC PRINCIPLE

3. Name the two important properties that liquids possess.
4. State and illustrate Pascal's law.
5. Explain the hydrostatic principle and state the purpose for which it is used.
6. Compare the hydrostatic and the hydrodynamic principles. Illustrations should be used to supplement the answer.
7. What hydraulic components operate on the hydrodynamic principle?

THE FUNDAMENTAL HYDRAULIC CIRCUIT

8. Name the six essential elements of a hydraulic system.

HYDRAULIC SYSTEM COMPONENTS

9. Give an example of a simple hydraulic pump and explain how it operates.

10. What purpose does a pump serve in a hydraulic system?
11. Describe briefly the operation of the following.
   (a) gear pumps   (b) vane pumps
   (c) radial-piston pumps
12. What purpose do valves serve in a hydraulic system?
13. For what purpose are the following used?
   (a) ball valves   (b) rotary valves
   (c) spool valves   (d) needle valves
14. Explain how a simple hydraulic motor operates.

THE HYDRAULIC SYSTEM

15. Compare a constant-volume and a variable-volume hydraulic system.
16. Explain briefly the operation of the constant-volume hydraulic system of a surface grinder.
17. Explain briefly the operation of the variable-volume hydraulic system of a milling machine.

HYDRAULIC COUPLINGS AND TORQUE CONVERTERS

18. State the purpose of a hydraulic coupling and briefly explain its operation.

19. State the purpose of a hydraulic torque converter and explain how it operates.

HYDRAULIC OIL REQUIREMENTS

20. What two important functions must a hydraulic oil perform?
21. List six characteristics that a hydraulic oil should possess in order to perform satisfactorily.
22. Discuss the importance of viscosity.
23. What would happen if either too light or too heavy an oil were used in a hydraulic system?
24. What effect does oxidation have on a hydraulic system?
25. Explain how water may enter a hydraulic system.
26. What effect does water have on a hydraulic system?
27. Discuss the effects of rusting and explain how it can be controlled.
28. What effect does foaming have on a hydraulic system?
29. How can foaming be minimized?

ADVANTAGES OF HYDRAULIC SYSTEMS

30. List six important advantages of hydraulic power transmission.

SPECIAL PROCESSES
SECTION 21

With the introduction of unique mechanisms and exotic materials in recent years, it has been found necessary to develop new methods of efficiently machining metals. Parts made out of cemented carbide or difficult-to-machine metals were previously shaped by the costly process of diamond wheel grinding. *Electro-chemical machining, electrical discharge machining,* and *electrolytic grinding* are three methods which have been developed recently. All remove metal by some form of electrical discharge.

Production of other hard-to-machine parts has been made possible by high-energy forming using either the explosive or electromagnetic process. Another widely used nonmachining process is that of powder metallurgy whereby parts are made from compressed, heat-treated powder metals. Powder-metallurgy parts have compared mechanically with machined parts and are usually cheaper to produce.

In recent years, the laser has proved invaluable for many industrial applications. It is widely used to cut holes in any material as well as for special welding applications not possible by other methods. Extremely fine measurements as well as accurate machine part alignment is possible with the laser beam.

ELECTRO-CHEMICAL MACHINING AND ELECTROLYTIC GRINDING

UNIT 90

Electro-chemical machining, more commonly called ECM, differs from conventional metal-cutting techniques in that electrical and chemical energy are used as the cutting tools. This process machines metal easily, regardless of the work hardness, and is characterized by its "chipless" operation. A nonrotating tool the shape of the cavity required is the *cutting tool;* therefore square or difficult-to-machine shapes can easily be cut in a workpiece. The wear on the cutting tool is hardly noticeable since the tool *never* touches the work. Electro-chemical machining is particularly suitable for producing round through holes, square through holes, round or square blind holes, simple cavities which have straight, parallel sides, and planing operations. Electro-chemical machining is especially valuable when metals that exceed a hardness of 42 Rockwell C (400 BHN) are machined. Sharp corners, flat bottom sections, or true radii are difficult to maintain because of the slight overcut which occurs during this process. A significant advantage of ECM is that the surfaces and edges of workpieces are not deformed and are left burr-free.

Although *electrolytic grinding* uses the principle of ECM, its application is different in that a grinding wheel is used instead of an electrode.

OBJECTIVES

After completing this unit you will understand:
1. The principle and purpose of ECM and electrolytic grinding
2. The various components and their function in each of these systems

THE PROCESS

For years, metal in a solution has been transferred from one metal to another by means of electro-plating baths. Since ECM evolved from this process, it may be wise to examine the electro-plating principle (Fig. 90-1).

1. Two bars of unlike metal are immersed in an electrolyte solution.
2. One bar is fastened to a negative lead on a battery while the other is fastened to the positive lead.
3. When the circuit is closed, direct current passes through the electrolyte between the two bars of metal.
4. Chemical reaction transfers metal from one bar to the other.

Electro-chemical machining differs from the plating process in that an electro-chemical reaction dissolves metal *from a workpiece* into an electrolyte solution. A direct current is passed through an electrolyte solution between the electrode tool (the shape of the cavity desired), which is negative, and the workpiece, which is positive. This causes metal to be removed ahead of the electrode tool as the tool is fed toward the work. Chemical reaction caused by the direct current in the electrolyte dissolves the metal from the workpiece (Fig. 90-2).

The *electrode* for ECM is not a simple bar of metal, but a precision insulated tool which has been made to a specific shape and exact size. The electrode (tool) and workpiece, although located within 0.002 to 0.003 in. (0.05 to 0.07 mm), never contact each other. The *electrolyte solution* is a controlled, swiftly flowing stream which carries the current. The *direct current* used may at times be as high as 10,000 A/in.^{2} (1550) A/cm^2) of work material. All these factors affect the operation of the electro-chemical machining process and will be discussed in greater detail.

Fig. 90-1 Principle of the electroplating process.

Fig. 90-2 Schematic diagram of a typical electro-chemical machining system.

THE ELECTROLYTE

The electrolyte is a solution of water to which salt, mineral acid, caustic-potash, or caustic soda has been added to increase the electrical conductivity. A weak supply of the electrolyte solution will result in two disadvantages:

1. Metal removal rates will be low.
2. Excessive heat will destroy the effectiveness of the solution.

Electrical energy which starts a chemical reaction in the electrolyte solution results in the formation of gas between the tool and workpiece and dissolves metal from the workpiece. The gas escapes into the atmosphere while the dissolved metal is carried away in the solution. As there is some resistance to current flow and chemical reactions occur, heat is generated in the machining area. The electrolyte is introduced into the machining area in great quantities to dissipate the heat and wash away the dissolved metal. Filters placed in an ECM system remove the dissolved metal from the electrolyte and ensure a fresh flow of solution to the machining area.

The electrolyte is introduced to the machining area *through* the electrode tool; therefore, the amount of flow is affected by the electrode length, diameter, and shape. As much flow as possible is desirable, and some applications have been known to use as much as 200 gal/min (747 l/min) at pressures up to 300 psi (2068.5 kPa).

THE ELECTRODE

The *electrode tool,* which is always the *negative terminal* of the electrical circuit, is an insulated tool made to the size and shape of the cavity desired. The electrolyte solution is fed to the machining area by a hole through the center of the electrode. Because it is necessary that the electrolyte flow completely around the electrode and allowance be made for the overcut which occurs during the ECM process, the tool is made about 0.005 in (0.12 mm) per side smaller than the hole it produces. The periphery of the electrode is insulated (Fig. 90-3) to prevent the sides from cutting as the tool extends deeper into the hole.

One of the chief purposes of the electrode tool is to impart its shape to the workpiece. For example, a square electrode will produce a square hole, while a round hole is

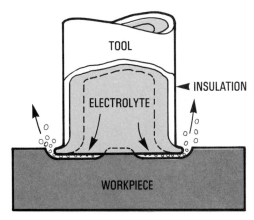

Fig. 90-3 Electrolyte flowing through the electrode tool.

produced by a round electrode. The shape of the cavity which can be produced through ECM is limited only by the shape which can be cut on the electrode. The material used to make the electrode tool should possess the following characteristics.

1. It must be machinable.
2. It must be rigid.
3. It must be a good conductor of electricity.
4. It must be able to resist corrosion.

Copper, brass, and stainless steel have generally been found to be good electrode material for ECM.

Copper, an excellent conductor of electricity, is not recommended for manufacturing electrodes for thin-walled sections or deep holes. Because of its softness and tendency to bend, it is difficult to machine long or thin sections.

Brass, while not as good a conductor of electricity as copper, can be used for electrode material with good results. The greater strength of brass, its ease in machining, and its lower cost make it an ideal material for most electrodes.

Stainless steel is used when large volume flow and high pressures of electrolyte are necessary. It is stronger than the other two materials, but its high initial cost and difficulties encountered in machining limit its use as an electrode material.

METAL REMOVAL

In the electro-chemical process, the distance between the electrode and the work (the machining gap) (Fig. 90-3) is important. In order to encourage efficient electrical transmission, the tool and the work must be as close as possible to each other yet never come into contact. For most conditions, this gap will vary from 0.001 to 0.003 in. (0.02 to 0.07 mm). Because of the high current levels used, serious damage to both the electrode tool and the work will occur if there is any physical contact between them.

The rate of metal removal is directly proportional to the current passing between the tool and the workpiece. Current densities ranging from a maximum of 1000 A/in.^{2} (155 A/cm^2) to a maximum of 10,000 A/in.^{2} (1550 A/cm^2) have been used in ECM applications. The use of high current will result in a high rate of metal removal, while low current will result in a low rate of metal removal.

The amount of overcut (the difference between the tool size and the hole produced) depends upon cutting conditions and may vary from 0.008 to 0.012 in. (0.20 to 0.30 mm). Once the amount of overcut is known for a given tool, hole sizes will be repeated to within at least 0.0015 to 0.005 in. (0.035 to 0.01 mm) of roundness.

The rate of penetration varies with the type of operation being performed, the type of work material, the cross section of the electrode, and the current density used. ECM rates of penetration may vary from 0.250 to 0.430 in. (6.35 to 10.92 mm).

ADVANTAGES OF ECM

Electro-chemical machining is one of the metal-cutting processes which has contributed to the machining of space-age metals. Some of its characteristics and advantages are:

1. Metal of any hardness can be machined.
2. It competes with drilling and some through-hole milling operations, especially if the work exceeds 42 Rockwell C hardness.
3. No heat is created during the machining process; therefore, there is no work distortion.
4. It machines metal without tool rotation.
5. Tool wear is insignificant as the tool never touches the work.
6. Because the tool never touches the work, thin fragile sections can be machined without distortion.
7. The workpiece is left burr-face.
8. Intricate forms, difficult to machine by other processes, can be produced easily (Fig. 90-4).
9. It is suitable for production-type work where multiple holes or cavities may be machined at the same time.
10. Surface finishes of 25 μin. (0.63 μm) or less may be obtained.

ELECTROLYTIC GRINDING

Electrolytic grinding has been a boon to the machining of thin, fragile metal products and tough, difficult-to-machine space-age alloys. In electrolytic grinding, the metal is removed from the work surface by a combination of electro-

Fig. 90-4 Electro-chemical machining applications.

Fig. 90-5 Electrolytic grinding principle. (Courtesy Cincinnati Milacron Incorporated.)

chemical action and the action of a metal-bonded abrasive grinding wheel (Fig. 90-5). Approximately 90 percent of the metal removed from the work surface is the result of this electro-chemical deplating action, while the remaining 10 percent is "wiped away" by the grinding wheel. The electrolytic grinding process is very similar to the process used for ECM.

THE ELECTROLYTIC GRINDING PROCESS

The metal-bonded grinding wheel and the electrically conductive workpiece are both connected to a *direct current* power supply and are separated by the protruding abrasive particles of the wheel. An electrolytic grinding solution (electrolyte) is injected into the gap between the wheel and the work, completing the electrical circuit and producing the necessary deplating action which decomposes the work material. This decomposed material is removed by the action of the revolving grinding wheel and is washed away in the solution. In electrolytic grinding, the wheel never comes into actual contact with the work material.

THE GRINDING WHEEL

Metal-bonded, electrically conductive, abrasive grinding wheels are used for the electrolytic grinding process. Brass-, bronze-, and copper-bonded wheels which can be dressed to various forms are common in electrolytic grinding. The grinding wheel is the cathode ($-$) of the electrical circuit. The wheel spindle is connected to a *direct current* power supply through a set of contact brushes and is isolated from the rest of the machine by an insulating sleeve (Fig. 90-5). Metal-bonded diamond wheels are recommended for the grinding of tungsten carbides. Metal-bonded aluminum oxide wheels are used for grinding all other electrically conductive materials.

The wheel abrasive performs an important function in the electrolytic grinding process. The abrasive particles, protruding about 0.0005 to 0.001 in (0.01 to 0.02 mm) beyond the metal-bonded wheel, act as nonconductive spacers, maintaining the necessary gap between the wheel and the workpiece. They also provide thousands of little pockets filled with electrolyte solution which completes the electrical circuit. The accuracy of the working gap (the distance between the metal bond of the wheel and the workpiece) is determined by the amount these abrasive particles protrude from the wheel (cathode).

TRUING

It is important that the grinding wheel run true within 0.0005 in. (0.01 mm) to maintain grinding accuracy and also to provide maximum passage for the electrolyte solution to achieve maximum stock removal. Metal-bonded aluminum oxide wheels can be dressed or trued with a single-point diamond or a commercial brake-truing attachment. Diamond wheels should be trued by the use of a dial indicator. After each dressing or truing operation, the electrolytic machining process is reversed by switching the electrical leads on the power supply, causing a deplating action of the metal wheel bond. This removes a slight amount of the metal wheel bond and causes the abrasive particles to protrude.

THE WORKPIECE

The work, which must be electrically conductive, is the anode ($+$) of the circuit. It is electrically connected to the *direct current* power supply through the table of the machine. Any material, even tungsten carbide and the difficult-to-machine space-age alloys, can be readily ground by this process, providing that the material is an electrical conductor.

THE CURRENT

A direct current of relatively low voltage (approximately 4 to 16 V) and high amperage (300 to 1000 A or higher) is used in the electrolytic grinding process. The amount of current flow depends upon the size of the area on which

the cutting action is occurring. A rule of thumb for stock removal rates is 0.010 in.^{3}/min (0.16 cm^3/min) for every 100 A of electrolytic grinding current. Each work material has a saturation point which limits the current it can accept. Machine tool manufacturers of electrolytic grinders generally provide charts with their machines showing the stock removal rates for various materials.

THE ELECTROLYTE

The electrolyte, generally a saline solution, serves two important functions:

1. It acts as a current conductor between the work and the wheel.
2. It combines chemically with the decomposed work material.

As the current flows from the workpiece (positive pole) through the electrolytic solution to the grinding wheel (negative pole), pockets of the solution act as electro-chemical cells decomposing the work surface. The current combines with the electrolyte to form a soft oxide film on the work surface which will not allow the current to flow. The wiping action of the revolving grinding wheel removes these oxides, allowing the current to flow again. The rate of this decomposition is in direct proportion to the amount of current flowing from the work to the wheel.

As the solution becomes contaminated with metal particles, it is flushed away into a storage tank, then passed through filters to remove these particles. A freshly filtered solution is necessary for efficient electrolytic grinding. An uneven supply of the electrolyte will result in excessive wheel wear, while a weak supply retards the metal removal rates.

The machine surfaces coming into contact with the electrolyte must be chrome-plated to overcome the corrosive action of the electrolytic solution.

SURFACE FINISH

The surface finish obtained by electrolytic grinding ranges from 8 to 20 μin. (0.2 to 0.5 μm) when steels and various alloys are ground. As a rule, the higher the alloy, the better the surface finish obtainable. When surface or traverse grinding tungsten carbide, a surface finish of 10 to 12 μin. (0.25 to 0.3 μm) may be expected. A surface finish 8 to 10 μin. (0.2 to 0.25 μm) is generally the rule when plunge grinding carbides. These surface finishes can be obtained at maximum stock removal rates with no finish cut required.

ELECTROLYTIC GRINDING METHODS

Grinding methods, such as cylindrical, form, plunge, surface, traverse, etc., all lend themselves to electrolytic grinding (Fig. 90-6). With all types of grinding methods, one important fact must be kept in mind, that is, the area of work-wheel contact along the cutting path of the wheel should never exceed 3/4 in. (19.05 mm) under normal conditions. If it is necessary to exceed this 3/4-in. (19.05-mm) dimension, a much slower feed rate must be used or the electrolyte must be supplied to the grinding area by auxiliary methods.

Fig. 90-6 Examples of types of electrolytic grinding.   (Courtesy Avco Bay State Abrasive Company.)

Cylindrical Grinding

All the advantages of electrolytic grinding apply to this form of grinding, except high stock removal rates. The reason for the lower stock removal is that the area of contact between the work and wheel is small, allowing little current to flow. It is recommended that plunge grinding be used to rough out work to within a few thousandths of size and then that work be finished by traverse grinding.

Form Grinding

Metal-bonded aluminum oxide wheels are usually employed for form-grinding operations. They are easily formed to the desired shape by a conventional diamond truing attachment. Formed diamond wheels for grinding carbides are expensive to change from form to form and difficult to true. Therefore, metal-bonded abrasive wheels are generally used for the production grinding of carbides.

Plunge Grinding

When an area is plunge-ground, there is no traverse motion of the wheel, and the wheel is fed straight into the surface. Either the face or the side of the wheel may be used for plunge grinding. Single-point tools, face-milling cutters, straight side milling cutters, and any other single-plane surfaces within the range of the wheel sizes may be effectively ground by this method. Fine manual feeds should be used for grinding individual pieces, while automatic feeds are recommended for production grinding.

Surface Grinding

Since the area of contact in surface grinding varies with the depth of cut for each wheel diameter, the full depth of cut per pass is recommended for maximum results. Always use the largest wheel possible, depending on the capacity of the machine. The rate of traverse is not dependent upon the width of the wheel as long as there is sufficient current available. Whenever possible, automatic power feeds are advisable since too slow a feed rate results in excessive overcut and too fast a feed results in excessive wheel wear. At no time should the work and wheel contact exceed 3/4 in. (19.05 mm). Machine tool manufacturers generally supply feed rate charts with their machines. The charts consider grinding current, wheel size, depth of cut, and area of contact.

Traverse Grinding

Most cutter-sharpening operations are performed by traverse grinding with a cup- or flare-type wheel. Helical side milling cutters and others that are impractical to plunge-grind are sharpened by this method.

Edge grinding with cup wheels is impractical as the work-wheel contact is small and allows little current to flow. For efficiency in traverse grinding, the wheel is swiveled slightly or an angle of 1 to 2° is dressed on the wheel surface to increase the area of work-wheel contact. Traverse feed rates depend upon the swivel angle, width of the wheel face, and the amount of current available.

Advantages of Electrolytic Grinding

Electrolytic grinding offers many advantages over conventional machining methods for metal removal.

1. It saves wheel costs, especially in metal-bonded diamond wheels, since only 10 percent of the metal is removed by abrasive action.
2. There is a high ratio of stock removal in relation to the wheel wear.
3. Higher production rates are possible because of the longer runs between wheel truing.
4. No heat is generated during the grinding operation: therefore, there is no burning or heat distortion of the work.
5. The process is burr-free, thereby eliminating deburring operations.
6. Thin, fragile workpieces can be ground without distortion since the wheel never touches the work.
7. Tungsten carbide and super-tough alloys can be ground quickly and with ease.
8. Exotic materials, such as zirconium, beryllium, etc., can be cut regardless of their hardness, fragility, or thermal sensitivity.
9. Dissimilar metals can be ground, providing they are electrically conductive.
10. Cutters may be reground in one pass, eliminating the necessity of finishing cuts.
11. No stresses are created in the work material.
12. No work hardening occurs during this process.

Disadvantages or Limitations

While there are many advantages of electrolytic grinding, there are also some disadvantages or limitations.

1. Only electrically conductive work can be ground.
2. Grinding wheels, especially the metal-bonded diamond wheels, are more expensive.
3. Inside corners cannot be ground sharper than a 0.010 to 0.015-in. (0.25 to 0.40-mm) radius because of the overcut occurring during the electrochemical action.
4. Accuracy is possible to within 0.0005 in. (0.01 mm) only.
5. The electrolytic solution is corrosive; therefore, machine parts coming into contact with the electrolyte must be chrome-plated.
6. The wheel contact should not exceed ¾ in. (19.05 mm).

Conclusion

At present, this method of electrical machining simplifies production methods and provides savings in both machining time and grinding wheel costs. As new types of metal-bonded wheels, power supplies, and electrolytes are developed, this grinding process should find many more applications.

REVIEW QUESTIONS

ELECTRO-CHEMICAL MACHINING

1. How does ECM differ from conventional machining techniques?
2. Describe briefly the process of ECM.
3. Name four suitable electrolytes.
4. Explain fully the purpose of the electrolyte solution.
5. Explain what part the electrolyte plays in the ECM process.
6. What are the characteristics of a good electrode tool?
7. Name three electrode materials suitable for ECM and state the advantage of each.
8. Define:
   (a) machining gap
   (b) overcut
   (c) rate of penetration
9. State seven main advantages of ECM.

ELECTROLYTIC GRINDING

10. Describe briefly the electrolytic grinding process.
11. Describe the type and function of the electrolytic grinding wheel.
12. What purpose does the wheel abrasive serve in electrolytic grinding?
13. Discuss the importance and operation of truing a grinding wheel.
14. What function does the workpiece serve in the electrolytic grinding circuit?
15. What type of current is used for this process?
16. Name two functions of the electrolyte solution.
17. Explain what occurs to the electrolyte when the current flows.
18. What surface finishes are possible to obtain by electrolytic grinding?
19. Discuss plunge and surface grinding.
20. Name eight main advantages of the electrolytic grinding process.
21. List four disadvantages of electrolytic grinding.

ELECTRICAL DISCHARGE MACHINING
UNIT 91

Electrical discharge machining, commonly known as EDM, is a process that is used to remove metal through the action of an electrical discharge of short duration and high current density between the tool and the workpiece (Fig. 91-1A). This principle of removing metal by an electric spark has been known for quite some time. In 1889, Paschen explained the phenomenon and devised a formula which would predict its arcing ability in various materials. The EDM process can be compared to a miniature version of a lightning bolt striking a surface, creating a localized intense heat, and melting away the work surface.

OBJECTIVES

After completing this unit you should be able to:
1. Understand the principle of electrical discharge machining
2. Name and state the purpose of the main components of an electrical discharge machine
3. State the advantages and applications of the electrical discharge process

Electrical discharge machining has proved especially valuable in the machining of super-tough, electrically conductive materials such as the new space-age alloys. These metals would have been difficult to machine by conventional methods, but EDM has made it relatively simple to machine intricate shapes that would be impossible to produce with conventional cutting tools. This machining process is continually finding further applications in the metal-cutting industry. It is being used extensively in the plastic industry to produce cavities of almost any shape in the steel molds.

PRINCIPLE OF EDM

Electrical discharge machining is a controlled metal removal technique whereby an electric spark is used to cut (erode) the workpiece, which takes a shape opposite to that of the cutting tool or electrode (Fig. 91-1B). The *cutting tool (electrode)* is made from electrically conductive material, usually carbon. The electrode, made to the shape of the cavity required, and the workpiece are both submerged in a *dielectric fluid,* which is generally a light lubricating oil. This dielectric fluid should be a nonconductor (or poor conductor) of electricity. A *servo mechanism* maintains a gap of about 0.0005 to 0.001 in. (0.01 to 0.02 mm) between

the electrode and the work, preventing them from coming into contact with each other. A *direct current* of low voltage and high amperage is delivered to the electrode at the rate of approximately 20,000 hertz (Hz). These electrical energy impulses become sparks which jump the gap (Fig. 91-2) between the electrode and the workpiece through the dielectric fluid. Intense heat is created in the localized area of the spark impact; the metal melts and a small particle of molten metal is expelled from the surface of the workpiece. The dielectric fluid, which is constantly being circulated, carries away the eroded particles of metal and also assists in dissipating the heat caused by the spark.

TYPES OF ELECTRICAL DISCHARGE CIRCUITS

Several types of electrical discharge power supply have been used for EDM. Although there are many differences between them, each type is used for the same basic purpose, that is, the precise, economical removal of metal by electrical spark erosion.

The following types of electrical power supplies have been used.

1. Resistance-capacitance power supply

Fig. 91-1 (A) A controlled spark removes metal during electrical discharge machining.
(B) Basic elements of an electro-discharge system.

Fig. 91-2 Stages of single spark.

Fig. 91-3 Electro-discharge power supply circuits.

2. Pulse-type power supply
3. Rotary impulse generator power supply
4. Static impulse generator power supply

The resistance-capacitance and the pulse-type direct current power supplies are most commonly used; therefore, only these two will be explained in greater detail.

Resistance-capacitance power supply, also known as the *relaxation-type power supply,* was widely used on the first EDM machines. It is still the power supply used on many of the machines of foreign manufacture.

As illustrated in Fig. 91-3A, the capacitor is charged through a resistance from a direct current voltage source that is generally fixed. As soon as the voltage across the capacitor reaches the breakdown value of the dielectric fluid in the gap, a spark occurs. A relatively high voltage (125 V), high capacitance of over 100 μF (microfarads) for roughing cuts, low spark frequency, and high amperage are characteristics of the resistance-capacitance power supply.

In resistance-capacitance circuits, an increase in metal removal rates depends more on larger amperage and capacitance than on increasing the number of discharges per second. The combination of low frequency, high voltage, high capacitance, and high amperage results in:

1. A rather coarse surface finish
2. Large overcut around the electrode (tool)
3. Larger metal particles being removed and more space being required to flush out particles

The advantages of the resistance-capacitance power supply are:

1. The circuit is simple and reliable.
2. It works well at low amperages, especially with milliampere currents required for holes under 0.005 in. (0.12 mm) in diameter.

Pulse-type power supply is used almost exclusively by American manufacturers. It is similar to the resistance-capacitance type; however, vacuum tubes or solid-state devices are used to achieve an extremely fast pulsing switch effect (Fig. 91-3B). The pulse width and intervals may also be accurately controlled by switching devices. The switching is extremely fast and the discharges per second are 10 or more times greater than with the resistance-capacitance power supply at low frequencies. The results of more discharges per second are illustrated in Fig. 91-4. With more discharges per second, and using the same current (10 A), it is clear that smaller craters are created, producing a finer surface finish while still maintaining the same metal removal rate.

Pulse-type power supply circuits are usually operated on low voltages (70 to 80 V), high frequency (sparks at the rate of 260,000 Hz), low capacitance (50 μF or less), and low-energy spark levels.

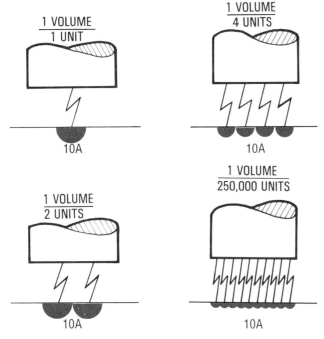

Fig. 91-4 Effects on cratering and surface finish using various frequencies of spark discharge.

The main advantages of the pulse-type circuit are:

1. It is extremely versatile and can be accurately controlled for roughing and finishing cuts.
2. Better surface finish is produced as less metal is removed per spark since there are many sparks per unit of time.
3. There is less overcut around the electrode (tool).

THE ELECTRODE

The *electrode* in EDM is formed to the shape of the cavity desired. As in conventional machining, some materials have better cutting and wearing qualities than others. Electrode materials must, therefore, have the following characteristics.

1. Be good conductors of electricity and heat
2. Be easily machined to shape at a reasonable cost
3. Produce efficient metal removal from the workpiece
4. Resist deformation during the erosion process
5. Exhibit low electrode (tool) wear rates

Much experimentation has been carried out to find a good, economical material for the manufacture of EDM electrodes. Tungsten, carbide, copper tungsten, silver tungsten, yellow brass, copper, chrome-plated materials, graphite, and zinc alloys are some of the materials which have found certain applications as electrode materials. None of these electrode materials has general-purpose application; each machining operation dictates the selection of the electrode material. *Yellow brass* has been used primarily as electrode material for pulse-type circuits because of its good machinability, electrical conductivity, and relatively low cost. *Copper* produces better results in the resistance-capacitance circuits where higher voltages are employed.

High-density and high-purity carbon, or *graphite,* is a relatively new electrode material which is gaining wide acceptance. It is commercially available in various shapes and sizes, is relatively inexpensive, can be machined easily, and makes an excellent electrode. Its tool wear rate is much less and its high metal removal rate is almost double that of any other metal. Development of superior graphite electrodes has increased the use of graphite electrodes because of their low cost and ease of fabrication.

THE EDM PROCESS

The use of EDM is increasing as more and more applications are found for this process. As new, important technological advances in equipment and application techniques become available, more industries are adopting the process. It is necessary, therefore, to discuss the various aspects of EDM in more detail.

THE SERVO MECHANISM

It is important that there is no physical contact between the electrode (tool) and the workpiece; otherwise arcing will occur, causing damage to both the electrode and the workpiece. Electrical discharge machines are equipped with a servo control mechanism that automatically maintains a constant gap of approximately 0.0005 to 0.001 in. (0.01 to 0.02 mm.) between the electrode and the workpiece. The mechanism also advances the tool into the workpiece as the operation progresses and senses and corrects any shorted condition by rapidly retracting and returning the tool. Precise control of the gap is essential to a successful machining operation. If the gap is too large, ionization of the dielectric fluid does not occur and machining cannot take place. If the gap is too small, the tool and workpiece may weld together.

Precise gap control is accomplished by a circuit in the power supply comparing the average gap voltage to a preselected reference voltage. The difference between the two voltages is the input signal which tells the servo mechanism how far and how fast to feed the tool and when to retract it from the workpiece.

When chips in the spark gap reduce the voltage below a critical level, the servo mechanism causes the tool to withdraw until the chips are flushed out by the dielectric fluid. The servo system should not be too sensitive to "short-lived" voltages caused by chips being flushed out; otherwise the tool would be constantly retracting, thereby seriously affecting machining rates.

Servo feed control mechanisms can be used to control the vertical movement of the electrode (tool) for sinking cavities. It can also be applied to the table of the machine for work requiring horizontal movement of the electrode (tool).

CUTTING CURRENT (AMPERAGE)

The EDM power supply provides the direct current electrical energy for the electrical discharges which occur between the tool and the workpiece. As the pulse-type power supply is the more commonly used in North America, only the characteristics of this type will be discussed.

Characteristics of Pulse-Type Circuits

1. Low voltages (normally about 70 V which drop to about 20 V after the spark is initiated)
2. Low capacitance (about 50 μF or less)
3. High frequencies (usually 20,000 to 30,000 Hz but may be as high as 260,000 Hz)
4. Low-energy spark levels

THE DISCHARGE PROCESS

Upon application of sufficient electrical energy between the electrode (cathode) and the workpiece (anode), the dielectric fluid changes into a gas, allowing a heavy discharge of current to flow through the ionized path and strike the workpiece. The energy of this discharge vaporizes and decomposes the dielectric fluid surrounding the column of electric conduction. As the conduction continues, the diameter of the discharge column expands and the current increases. The heat between the electrode and the work surface causes a small pool of molten metal to be formed on the work surface. When the current is stopped, usually only for microseconds, the molten metal particles solidify and are washed away by the dielectric fluid.

These electrical discharges occur at the rate of 20,000 to 30,000 Hz between the electrode and the workpiece. Each discharge removes a minute amount of metal. Since the voltage during discharge is constant, the amount of metal removed will be proportional to the amount of charge between the electrode and the work. For fast metal removal, high amounts of current should be delivered as quickly as possible to melt the maximum amount of metal. This, however, produces large craters in the workpiece, resulting in rough surface finish. To obtain smaller craters and therefore finer surface finishes, smaller charges of energy can be used. This results in slower stock removal rates. If the current is maintained but the number of hertz is increased, this also results in smaller craters and better surface finish. The surface finish is proportional to the number of electrical discharges (cycles) per second (Fig. 91-4).

THE DIELECTRIC FLUID

The dielectric fluid used in the electrical discharge process serves several main functions.

1. Helps to initiate the spark between the electrode and the workpiece
2. Confines the spark path to a narrow channel
3. Serves as an insulator between the tool and the workpiece
4. Flushes away the metal particles to prevent shorting
5. Acts as a coolant for both the electrode and the workpiece

Types of Dielectrics

Many types of fluids have been used as dielectrics, resulting in various rates of metal removal. They must be able to ionize (vaporize) and deionize rapidly and have a low viscosity that will allow them to be pumped through the narrow machining gap. The most commonly used EDM fluids, proven to be satisfactory dielectrics, have been various petroleum products, such as light lubricating oils, transformer oils, silicon-base oils, and kerosene. These all perform reasonably well, especially with graphite electrodes, and are reasonable in cost. In certain cases, dielectrics such as carbon tetrachloride and certain compressed gases have been used.

The selection of the dielectric is important to the EDM process since it affects the metal removal rate and electrode wear. Industry is continually searching for new and better dielectric fluids for this process. A fluid consisting of triethylene glycol, water, and monoethyl ether of ethylene glycol has been used in research with superior results, especially with metallic electrodes. It is quite likely that many new dielectric fluids will be developed to improve the EDM process.

Methods of Circulating Dielectrics

The dielectric fluid must be circulated under constant pressure if it is to flush away efficiently the metal particles and assist in the machining process. The pressure used generally begins with 5 psi (34.4 kPa) and is increased until optimum cutting is attained. Too much dielectric fluid will remove the chips before they can assist in the cutting action and thereby cause slower machining rates. Too little pressure will not remove the chips quickly enough and thereby cause short circuits.

Four methods are generally used to circulate the dielectric fluid. All methods must use fine filters in the system to remove the metal particles so that they are not recirculated.

Down through the Electrode (Fig. 91-5A): A hole or holes are drilled through the electrode, and the dielectric fluid is forced through the electrode and between it and the workpiece. This rapidly washes away the metal particles from the machining area. On cavities, a small standing slug or core remains which must be ground away after the machining operation has been completed.

Up through the Workpiece (Fig. 91-5B): Another common method is to cause the fluid to be circulated up through the workpiece. This type of flushing is limited to through-hole cutting applications and to cavities having holes for core or ejector pins.

Vacuum Flow (Fig. 91-5C): A negative pressure (vacuum) is created in the gap which causes the dielectric to flow through the normal 0.001-in. (0.02-mm) clearance between the electrode and the workpiece. The flow can be either up through a hole in the electrode or down through a hole in the workpiece. Vacuum flow has several advantages over other methods in that it improves machining efficiency, reduces smoke and fumes, and helps to reduce or eliminate taper in the workpiece.

Vibration (Fig. 91-5D): A pumping and sucking action is used to cause the dielectric to disperse the chips from

Fig. 91-5 (A) Circulating dielectric fluid down through the electrode; (B) circulating dielectric fluid up through the workpiece; (C) circulating the dielectric fluid down through the workpiece by suction; (D) circulating the dielectric fluid by vibration.

the spark gap. The vibration method is especially valuable for very small holes, deep holes, or blind cavities where it would be impractical to use other methods.

METAL REMOVAL RATES

Metal removal rates for EDM are somewhat slower than with conventional machining methods. The rate of metal removal is dependent upon the following factors.

1. Amount of current in each discharge
2. Frequency of the discharge
3. Electrode material
4. Workpiece material
5. Dielectric flushing conditions

The normal metal removal rate is approximately 1 in.^{3} (16.38 cm^3) of work material per hour for every 20 A of machining current. However, metal removal rates of up to 15 in.^{3}/h (245.8 cm^3/h) are possible for roughing cuts with special power supplies.

ELECTRODE (TOOL) WEAR

During the discharge process, the electrode (tool) as well as the workpiece is subject to wear or erosion. As a result, it is difficult to hold close tolerances as the tool gradually loses its shape during the machining operation. At times it is necessary to use as many as five electrodes to produce a cavity of the required shape and tolerance. For through-hole operations, stepped electrodes are used to produce roughing and finishing cuts in one pass.

The rate at which the tool wears is fortunately considerably less than that of the workpiece. An *average wear ratio* of the workpiece to the electrode is 3:1 for metallic tools, such as copper, brass, zinc alloys, etc. With graphite electrodes, this wear ratio can be greatly improved to 10:1.

Much development and research remains to be done to reduce the wear ratio of the electrode. *Reverse-polarity machining,* a relatively new development, promises to be a major breakthrough in reducing electrode wear. With this method, molten metal from the workpiece is deposited on a graphite electrode about as fast as the electrode is worn

Fig. 91-6 Example of overcut produced by different voltages in EDM.

away. Thus minute electrode wear is continually being replaced by a deposit of the work material. Reverse-polarity machining operates best on low spark-discharge frequencies and high amperage. It improves the metal removal rates and greatly reduces electrode wear.

OVERCUT

Overcut is the amount the cavity in the workpiece is cut larger than the size of the electrode used in the machining process. The distance between the surface of the work and the surface of the electrode (overcut) is equal to the length of the sparks discharged, which is constant over all areas of the electrode.

The amount of overcut in EDM ranges from 0.0002 to 0.007 in. (0.005 to 0.1 mm) and is dependent upon the amount of gap voltage. As illustrated in Fig. 91-6, the overcut distance increases with the increased gap voltage. Therefore, the amount of overcut can be controlled fairly accurately. The amount of overcut is generally varied to suit the metal removal rate and the surface finish required, which in turn determines the size of the chip removal.

Most manufacturers of EDM machines provide overcut charts to show the amount of clearance produced with various currents. The charts make it possible to determine accurately the electrode size required to machine an opening to within 0.0001 in. (0.002 mm).

The size of the chips removed is an important factor in setting the amount of overcut because:

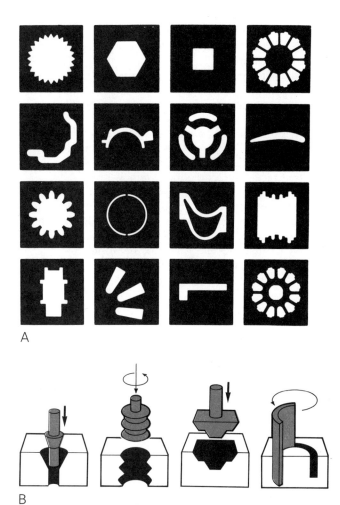

A

B

Fig. 91-7   (A) Examples of work produced by EDM;   (B) tool movement required to produce various-shaped cavities.

1. Chips in the space between the electrode and the work serve as conductors for the electrical discharges.
2. Large chips produced with higher amperages require a larger gap to enable them to be flushed out effectively.

Therefore, overcut depends upon the gap voltage and the chip size, which vary with the amperage used.

SURFACE FINISH

In the last few years, major advances have been made with regard to the surface finishes that can be produced. With the low metal removal rates, surface finishes of 2 to 4 μin. (0.05 to 0.10 μm) are possible. With high metal removal rates [as much as 15 in.^{3}/h (245.8 cm^3/h)] finishes of 1000 μin. (25 μm) are produced.

The type of finish required determines the number of amperes which can be used. the capacitance, frequency, and the voltage setting. For fast metal removal (roughing cuts), high amperage, low frequency, high capacitance, and minimum gap voltage are required. For slow metal removal (finish cut) and good surface finish, low amperage, high frequency, low capacitance, and the highest gap voltage are required.

ADVANTAGES OF THE EDM PROCESS

Electrical discharge machining has many advantages over conventional machining processes.

1.   Any material that is electrically conductive can be cut, regardless of its hardness. It is especially valuable for cemented carbides and the new supertough space-age alloys that are extremely difficult to cut by conventional means.
2.   Work can be machined in a hardened state, thereby overcoming the deformation caused by the hardening process.
3.   Broken taps or drills can readily be removed from workpieces.
4.   It does not create stresses in the work material since the tool (electrode) never comes in contact with the work.
5.   The process is burr-free.
6.   Thin, fragile sections can be easily machined without deforming.
7.   Secondary finishing operations are generally eliminated for many types of work.
8.   The process is automatic in that the servo mechanism advances the electrode into the work as the metal is removed.
9.   One person can operate several EDM machines at one time.
10.   Intricate shapes, impossible to produce by conventional means, are cut out of a solid with relative ease (Fig. 91-7A and B).
11.   Better dies and molds can be produced at lower costs.
12.   A die punch can be used as the electrode to reproduce its shape in the matching die plate, complete with the necessary clearance.

LIMITATIONS

Electrical discharge machining has found many applications in the machine trade; however, it does have some limitations.

1.   The metal removal rates are low.
2.   The material to be machined must be electrically conductive.
3.   Cavities produced are slightly tapered, but can be controlled for most applications to as little as 0.0001 in. (0.002 mm) in every ¼ in. (6.35 mm).
4.   Rapid electrode wear can become costly in some types of EDM equipment.
5.   Electrodes smaller than 0.003 in. (0.07 mm) in diameter are impractical.
6.   The work surface is damaged to a depth of 0.0002 in. (0.005 mm), but it is easily removed.
7.   A slight case hardening occurs. This, however, may be classed as an advantage in some cases.

CONCLUSION

Since EDM was first put to practical application, its use has been expanding by virtue of its ability to perform economically jobs which are extremely difficult or impossible to perform by conventional machining methods. Research and development are continuing to improve tool wear ratios, to increase metal removal rates without the surface finish suffering, and to improve power supplies and machine tool components. Wider recognition of this process and its possibilities will lead to increasing the importance of EDM in the future.

REVIEW QUESTIONS

ELECTRICAL DISCHARGE MACHINING

1. Describe briefly the principle of EDM.
2. Explain the operation and advantages of the pulse-type power supply.

3. List the characteristics of a good electrode material.
4. Explain why graphite is gaining wide acceptance as an electrode material.
5. What is the purpose of the servo mechanism, and how does it operate?
6. Briefly explain what occurs during the discharge process?
7. What purpose does a dielectric serve?
8. Discuss the four methods of circulating dielectric fluids, and state the advantages of each method.
9. What factors affect the metal removal rate?
10. Explain the principle of reverse-polarity machining.
11. Define *overcut,* and explain how it can be controlled.
12. What surface finishes are possible by EDM, and how are these achieved?
13. List six main advantages of the EDM process.
14. Name four limitations of EDM.

FORMING PROCESSES
UNIT 92

Forming large metal parts has become important as a result of the missile and aerospace program. Small parts may be easily formed to shape in conventional presses; however, these presses do not have the capacity or the force necessary to form materials with higher elastic strength or large parts made of common material. Formed parts as large as 10 ft (3.04 m) in diameter were required by the aerospace program; to produce them in conventional presses would have been economically impractical due to the cost and size of the press that would be required.

High-energy-rate (explosive) forming has received considerable attention in recent years because of its ability to apply great amounts of pressure [as much as 100,000 psi (689.5 MPa) or higher] by controlled explosions to form metal to shape. In this type of metal forming, only the cavity part of the die is required, and the force created by the controlled explosion replaces the die punch. Although there are many variations of explosive forming, only chemical explosive, electric spark discharge, and electromagnetic will be briefly discussed to give the reader an insight into the principles involved.

Powder metallurgy is the process of producing metal parts by:
1. Blending powdered metals and alloys
2. Compressing the powders in a die which is the shape of the part to be produced
3. Subjecting the shaped form to elevated temperatures (sintering) which causes the metal particles to "weld" together and form a solid part

The process of fabricating useful objects by forming metal powders is not new. The ancient Egyptians practiced a form of powder metallurgy, and since that time experimentation has continued to perfect the process. Since World War I, research and development of this process have made considerable progress. Today powder metallurgy is used to produce cams, gears, self-oiling bearings, light filaments, levers, cemented-carbide tools, automotive filters, etc., which were previously produced by conventional machining methods (Fig. 92-5).

OBJECTIVES

After you have completed this unit you will:
1. Be familiar with processes of explosive and electromagnetic forming
2. Understand the powder metallurgy process of producing parts without machining

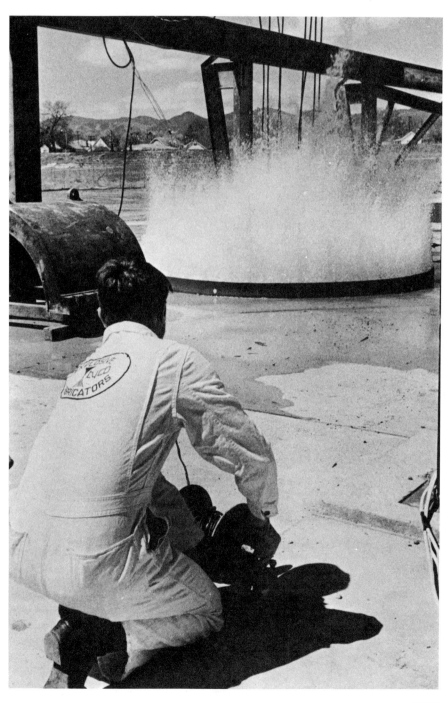

EXPLOSIVE FORMING

Chemical explosives have proved to be a compact and inexpensive source of high energy. Only a small amount of explosive is required to create a large force to form metal. For example, 2 oz (57 g) of explosives in a 5-ft diameter (1.52-m) water tank can produce the same force as a hydraulic press rated at 1000 tons (907 t). If 5 lb (2268 g) of explosives were used, this would far exceed the capacity of any existing press.

THE PROCESS

1. The workpiece (blank) to be formed is clamped to the top of the die cavity by a clamping ring (Fig. 92-1).
2. The explosive is mounted over the blank at a predetermined distance away from its surface.
3. The entire assembly is lowered into a water tank.
4. A vacuum is created in the die (the space between the workpiece and the die cavity) (Fig. 92-1).
5. The explosive charge is set off.
6. The shock wave which is created causes the metal to form to the shape of the die cavity.

The detonation of the explosive under water creates a shock wave that strikes the workpiece, forcing the metal into contact with the die. The pressures exerted on the work are high; however, the intensity and duration of the pressure can be controlled so that only slightly more energy than required is exerted. If too much pressure is exerted, it may cause the workpiece to rupture.

ADVANTAGES

Many advantages are offered by the use of explosives for forming large metal sections.

1. Since only one-half the die is required, the cost of die manufacture is reduced.
2. The cost of the equipment required is relatively low.
3. Parts difficult or impossible to form by mechanical means can be formed.
4. A better surface finish on the work is created than is possible by conventional die sets.
5. Better forming accuracy is possible because there is little or no springback in the workpiece.
6. Annealing operations required for deep forming by conventional means are eliminated.

LIMITATIONS AND DISADVANTAGES

Certain limitations or disadvantages must be considered when using explosives to form metals.

1. Employees must be trained in the safe use of explosives.
2. Insurance rates are generally higher because of the use of explosives.
3. The forming must be done in a remote area, which increases transportation and handling costs.
4. Elevated temperatures which could assist in forming the workpiece cannot be used because of the cooling effects of water.

Fig. 92-1 Explosive forming setup.

CONCLUSION

The use of explosives has provided industry with a relatively inexpensive method of forming large metal parts. This process should be considered for parts difficult or impossible to form by conventional means or if costs are lower.

ELECTRIC SPARK DISCHARGE FORMING

Shock waves, comparable to those produced by explosives, created by an underwater electrical discharge of high voltage have been used successfully for metal forming. The amount of electrical energy required for forming workpieces depends upon the following factors:

1. Diameter and depth of the die cavity
2. Distance from the spark to the surface of the water
3. Width of the spark gap
4. Thickness of the workpiece

Electrospark forming cannot and was not intended to compete with conventional presses in the forming of simple parts. It is generally used for forming parts which would be difficult or impossible to manufacture by conventional methods. Operations such as forming, blanking, piercing, embossing, flaring, trimming, and forming tubular parts either by expansion or contraction are possible with electrospark forming.

THE PROCESS

1. The piece to be formed is fastened securely to the top of the die by means of a clamp ring (Fig. 92-2).
2. The die is then submerged in water.
3. A vacuum is created in the die cavity.
4. The electrodes are positioned at a predetermined distance above the workpiece (Fig. 92-2).
5. The stored energy from the high-voltage capacitor bank is released between the submerged electrodes.

Fig. 92-2 Electrospark forming principle.

Fig. 92-3 Examples of electromagnetic forming.

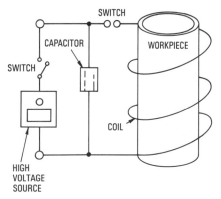

Fig. 92-4 Electromagnetic forming principle.

6. The spark discharge between the electrodes creates a shock wave, causing the workpiece to form to the shape of the die cavity.

The release of high energy between the submerged electrodes heats and vaporizes a thin channel of water, rapidly expanding it and creating a shock wave. When this shock wave reaches the workpiece, there is an imbalance of forces on the workpiece caused by low pressure in the die cavity as a result of the vacuum and high pressure of the shock wave. Because of this imbalance of forces, the work is forced in the direction of the low pressure and assumes the shape of the die cavity.

Often the path of the spark is controlled by connecting the space between the two electrodes with a thin wire. During discharge, the spark is confined to a specific path; however, the wire is vaporized and must be replaced for each part formed.

CONCLUSION

Most advantages and disadvantages listed for explosive forming apply also to electrospark forming. The differences between the two processes are:

1. The cost of the equipment required to initiate the discharge is considerably higher than with explosive forming.
2. Electrospark forming is a much safer process and does not have to be performed in a remote area.
3. Higher production rates are possible.
4. The path of the discharge can be more accurately controlled.

5. The amount of discharge is limited to the capacity of the electrical power bank.

ELECTROMAGNETIC FORMING

Electromagnetic forming is one of the newest methods of high-energy metal forming. The magnetic field of force, created by passing a high current through a coil around either the outside or inside of a workpiece, is used to form parts to the desired shape. The best results with this method of high-energy metal forming have been obtained when forming parts made of copper and aluminum which are good electrical conductors. Electromagnetic forming is primarily used for swaging or expanding tubular shapes. It may also be used for embossing, punching, forming, and shrinking operations (Fig. 92-3).

THE PROCESS

In electromagnetic forming, the electrical energy from the capacitor bank (Fig. 92-4) is passed through a coil rather than between two electrodes as is the case in electrospark forming. A large magnetic field builds up around the coil, inducing a voltage in the workpiece. The resultant high current builds up its own magnetic field. These two magnetic fields of force are opposite in direction and repel each other, causing the deformation of the workpiece. If the coil is placed on the inside of a tubular workpiece, the magnetic force will cause the workpiece to bulge and assume the shape of the die cavity. Workpieces may be shrunk onto formed mandrels by placing the coil around the outside of the piece to be formed (Fig. 92-4). Field-shaping coils may be used to concentrate the magnetic field of force when irregular shapes must be formed.

Experiments have been conducted using variations of electrical energy. The best results for electromagnetic forming have been obtained with a capacitor bank of low voltage and high capacitance. The amount of work deformation increases in relation to the amount of stored electrical energy released. Deformation of the workpiece decreases when the distance between the coil and the work is increased. Parts made of poor or nonconductive materials can be formed by placing a conductive foil around or inside the parts to be formed. The foil acts as would a good electrically conductive material and causes the workpiece to assume the shape of the die cavity or formed mandrel.

ELECTROMAGNETIC FORMING FACTORS

Although electromagnetic forming is a relatively new method of metal forming, it has proved invaluable for many operations which were difficult or impossible to perform by other methods. The following factors must be carefully considered since they affect the efficiency of the electromagnetic forming process.

1. The amount of electrical energy employed must be sufficient to form the part completely.
2. The coil should be designed so that it is stronger than the part to be formed; otherwise, the electromagnetic force may deform the coil rather than the workpiece.
3. The size of the wire and the number of turns in the coil are very important because they determine the strength of the electromagnetic field which can be created. The coil must be placed at a specific distance from the part to be formed.
4. The electrical conductivity of the work material is an important factor in this process.
5. The thickness of the work material determines the location of the coil and the amount of electrical energy required.

ADVANTAGES

In electromagnetic forming, the size of the work that can be formed is controlled by the amount of electrical power that can be directed to the forming coil. However, it has certain advantages over other methods of high-energy forming.

1. The amount of electrical energy can be accurately controlled.
2. An equal amount of force is applied to all areas of the part.
3. No forces are set up unless a part is in the magnetic field.
4. The work may be preheated because no water or liquid is required for the forming process.
5. There are no moving parts in the forming equipment.
6. The operation can be automated.

7. Forming can be performed in a vacuum or in an inert atmosphere.
8. It provides a low-cost method of assembling where bands of metal must be formed around other parts.

CONCLUSION

Electromagnetic forming, although a relatively new process, has been used for a great variety of industrial applications ranging from the aerospace program to the assembly of electronic components. As electromagnetic pulse equipment of greater capacities becomes available, this method of metal forming should assume more importance in manufacturing processes.

POWDER METALLURGY

COMPARISON OF THE POWDER METALLURGY PROCESS

Powder metallurgy is a radical departure from the conventional machining process (Fig. 92-5). With conventional machining methods, a piece of steel or bar stock larger than the finished part is selected and the required form is machined. The material cut away in the machining process is in the form of steel chips which are considered *scrap loss* (Fig. 92-6). In powder metallurgy, the correct types of powders are blended together and formed into the required shape in a die. Since machining is not required, scrap loss is minute as only the required amount of powder is used for each part.

The unique features of the powder metallurgy process are:

1. Handling of molten metals is not involved.
2. Machining or finishing operations are rarely required.
3. It permits rapid mass production of steel and other high melting metal shapes.
4. Beryllium, molybdenum, and tungsten parts can be produced easily, when this would be impractical or uneconomical by other methods.
5. It permits combinations of metal and nonmetals as well as alloys which are not possible by any other method.
6. It permits accurate control over the density or the porosity of the finished part.

THE POWDER METALLURGY PROCESS

A part made by the powder metallurgy process begins with metal powders and goes through four main stages before it becomes a finished product (Fig. 92-7). These include securing raw powders, blending, compacting, and sintering.

Raw Powders

Almost any type of metal can be produced in powder form; however, only a few have the desired characteristics and properties necessary for economical production. Iron-and copper-base powders are the two main types which lend themselves well to the powder metallurgy process. Alu-

Fig. 92-5   Examples of parts produced by powder metallurgy.   (Courtesy The Powder Metallurgy Parts Mfg. Association.)

MACHINING
SCRAP LOSS

BAR
STOCK

MACHINED
PART

IRON POWDER

POWDER METALLURGY PART

Fig. 92-6   Comparison of conventional machining and powder metallurgy proceses.

minum, nickel, silver, and tungsten powders are not widely used; however, they have some important applications.

Some of the more common methods of manufacturing powders are:

1. *Atomization* or metal spraying is a means of producing powders from low-melting temperature metals, such as aluminum, lead, tin, and zinc. This process produces powder particles, irregular in shape.

2. *Electrolytic deposition* is the common method used to produce copper, iron, silver, and tantalum powders.

3. *Granulation* is used to convert a few metals into powder. The metal is stirred rapidly while it is being cooled. This process depends upon the formation of oxides on the metal particles during stirring.

4. *Machining* produces coarse powders and is primarily used for producing magnesium powders.

5. *Milling* involves using various types of crushers, rollers, or presses to break the metal into powders.

6. *Reduction* is used to transform metal oxides to powder form by contact with gas at temperatures below their melting point. Cobalt, iron, molybdenum, nickel, and tungsten powders are produced by the reduction process.

7. *Shotting* is the operation of passing molten metal through an orifice or sieve and having the particles drop into water. Most metals can be powdered by this method; however, the particles are generally large.

Blending

The powder for a specific product must be carefully selected to ensure economical production and so that the finished part will have the required characteristics. The following information regarding the powder should be considered during its selection.

1. Shape of the powder particle
2. Particle size
3. Flowability (its ability to flow readily)
4. Compressibility (its ability to maintain a form)
5. Sintering ability (its ability to fuse or bond together)

Once the correct powders have been selected, the mass of each is carefully determined in the proportion required for the finished component, and a die lubricant, such as powdered graphite, zinc stearate, or stearic acid, is added. The purpose of this lubricant is to assist the flow of powder in the die, prevent scoring of the die walls, and permit easy ejection of the compressed part. This composition is then carefully mixed or blended (Fig. 92-8) to ensure homogeneous grain distribution in the finished product.

Compact (Briquetting)

The powder blend is then fed into a precision die which is the shape and size of the finished product. The die generally consists of a die shell, and upper punch, and a lower punch (Fig. 92-8). These dies are usually mounted on hydraulic or mechanical presses where pressures from about 3000 psi (20.6 MPa) to as much as 200,000 psi (1379 MPa) are used to compress the powder. Soft powder particles can be pressed or keyed together readily and therefore do not require as high a pressure as the harder particles. The density and hardness of the finished product increases with the amount of pressure used to compact or briquette the powder. However, for every case there is an optimum pressure above which very little improvement in qualities and properties can be obtained.

Sintering

After compacting, the "green part" must be heated sufficiently in order to effect permanent cohesion of the metal particles into a solid. This operation of heating is known as *sintering*. The parts are passed through controlled protective-atmosphere furnaces which are maintained at a temperature approximately one-third below the melting point of the principal powder. The carefully controlled atmosphere and temperature during the sintering operation permits particle bonding and recrystallization across the particle interfaces. Most sintering is done in a hydrogen atmosphere, producing parts with no scaling or discoloration.

The temperature within the furnaces varies from 1600 to 2700°F (871.1 to 1482.2°C), depending on the type of metal powder being sintered. The time required to sinter a part varies, depending on it shape and size; normally, however, it is from 15 to 45 min.

The purpose of the sintering operation is to bond the powder particles together to form a strong homogeneous part having the desired physical characteristics.

FINISHING OPERATIONS

After sintering, most parts are ready for service. However, some parts requiring very close tolerances or other qualities may require some additional operations, such as sizing and repressing, impregnation, infiltration, plating, heat treating, and machining.

Sizing and Repressing

Parts requiring close tolerances or increased density must be *sized* or *repressed*. This involves putting the part into a die which is similar to the one used for compacting and repressing it. This operation improves the surface finish, the dimensional accuracy, increases the density, and gives added strength to the part.

Impregnation

This is the process of filling the pores of a sintered part with a lubricant or nonmetallic material. Oilite bearings are a good example of a part impregnated with oil to overcome the necessity for constant lubrication and maintenance. Parts may be impregnated by a vacuum process or by soaking the parts in oil for several hours.

Infiltration

This is the process of filling the pores of a part with a metal or alloy having a lower melting point than the sintered piece. The purpose of this operation is to increase the density, strength, hardness, impact resistance, and ductility of the manufactured part.

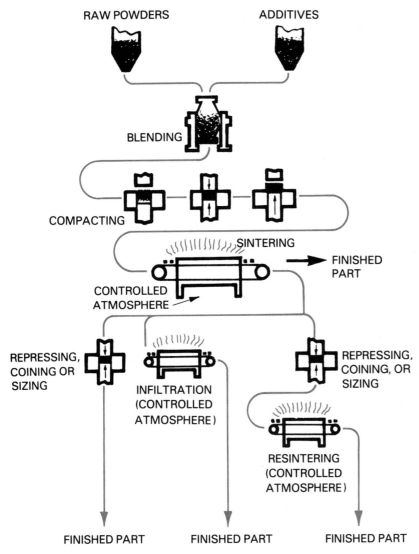

Fig. 92-7   The powder metallurgy process.

STEPS IN THE OPERATING
CYCLE OF FORMING A PART

1. THE DIE CAVITY IS FILLED WITH BLENDED POWDER.
2. THE TOP AND BOTTOM PUNCHES COMPRESS THE POWDER
   TO SHAPE.
3. THE COMPACTED PART IS EJECTED FROM THE DIE BY THE
   BOTTOM PUNCH.
4. THE COMPACTED PART IS MOVED CLEAR OF THE DIE AREA
   AND MORE BLENDED POWDER IS FED INTO THE DIE.

Fig. 92-8   Steps in compacting metal powders to shape.

Fig. 92-9 Examples of products available through powder metallurgy. (Courtesy The Powder Metallurgy Parts Mfg. Association.)

Slugs of the material to be infiltrated are placed on the compacted parts and passed through the sintering furnace. Because of its lower melting point, the infiltrated material melts and penetrates the pores of the compacted part by capillary action.

Plating, Heat Treating, Machining

Depending on the type of material, its application, and requirements, some powder metallurgy parts may be plated, heat treated, machined, brazed, or welded.

ADVANTAGES OF POWDER METALLURGY

The use of powder metallurgy is increasing rapidly, with many parts being made better and more economically than with previous manufacturing methods. Some of the advantages of the powder metallurgy process are:

1. Close dimensional tolerances of ±0.001 in. (0.02 mm) and smooth finishes can be obtained without costly secondary operations.
2. Complex or unusually shaped parts, which would be impractical to obtain by any other method, can be produced (Fig. 92-9).
3. It is capable of producing porous bearings and cemented-carbide tools.

4. The pores of a part can be infiltrated with other metals.
5. Surfaces of great wear resistance can be produced.
6. The porosity of the product can be accurately controlled.
7. Products made of extremely pure metals can be produced.
8. There is little waste in this process.
9. The operation can be automated and employ unskilled labor, keeping costs low.
10. The physical properties of the product can be minutely controlled.
11. Duplicate parts are accurate and unvarying.
12. Powders made from difficult-to-machine alloys can be formed into parts which would be difficult to produce by machining processes.

DISADVANTAGES OF POWDER METALLURGY

Although there are many advantages of the powder metallurgy process, there are also certain disadvantages or limitations.

1. Metal powders are quite expensive and must be carefully stored to avoid deterioration.
2. The cost of the presses and furnaces is high, and therefore it is not practical for short-run jobs.
3. The size of the part which can be produced is controlled

by the capacity of the press available and the compression ratio of the powders.

4. Parts requiring sharp corners or abrupt changes in thickness are very difficult to produce.

5. Parts requiring internal threads, grooves, or undercuts are impossible to produce.

6. Some of the low-melting-temperature powders may cause difficulties during the sintering operation.

Powder Metallurgy Part Design Rules

1. The shape of the part must permit it to be ejected from the die. The following cannot be molded and must be machined later: undercuts, grooves, internal threads, diamond knurls, holes at right angles, and reverse tapers.

2. Avoid part designs that would require powder to flow into thin walls, narrow splines, or sharp corners. Parts with these characteristics can be produced only with extreme difficulty and should be avoided whenever possible.

3. The die should be designed to provide the maximum stength to all the components.

4. The length of the part should not be longer than 2-½ times the diameter.

5. The part should be designed with as few steps or diameters as possible.

REVIEW QUESTIONS

HIGH-ENERGY FORMING

1. Explain why high-energy metal forming has become increasingly important in recent years.

2. Name three methods of explosive forming.

3. Explain briefly what occurs when a chemical explosive is detonated under water.

4. Name four advantages of explosive forming.

5. What are the disadvantages or limitations in using explosives for forming metals?

ELECTRO SPARK FORMING

6. How does electric spark discharge forming differ from explosive forming?

7. Describe briefly the process of electric spark discharge forming.

8. Explain what occurs when high energy is released between two submerged electrodes.

9. Why is it necessary to create a vacuum in the die cavity for explosive and electric spark discharge forming?

ELECTROMAGNETIC FORMING

10. On what principle does electromagnetic forming operate?

11. List four important factors which should be considered for electromagnetic forming.

12. Name five advantages of electromagnetic forming.

POWDER METALLURGY

13. List the four steps in the powder metallurgy process.

14. Name six products which are produced by powder metallurgy.

15. Compare conventional machining methods with powder metallurgy, and list four unique features of the process.

16. Name and briefly describe four methods of producing metal powders.

17. Describe the process of blending and state its importance to the finished product.

18. Describe the compacting operation with regard to the type of die used, pressures required, and the qualities obtained.

19. Fully describe and state the purposes of sintering.

20. Define and state the purpose of:
  (a) sizing and repressing
  (b) impregnation
  (c) infiltration

21. List six important advantages of powder metallurgy.

22. State four disadvantages or limitations of powder metallurgy.

THE LASER

UNIT 93

Since the first operational laser was built in 1960, laser technology has advanced at a rapid rate and is one of the most versatile developments at our disposal. The word laser is an abbreviation for light amplification by stimulated emission of radiation. Laser light can be used to cut holes in any material, including diamonds, or to perform microscopic surgery on the eye and many other parts of the human body. Laser may also be used to guide missiles and satellites, to trigger thermonuclear fusion, and to measure distances from milicrons (millionth of a meter) to hundreds of thousands of miles with extreme accuracy. For instance, scientists using lasers can determine the distance from the earth to the moon at any given time, to within an accuracy of about 1 ft.

The laser beam is a very narrow, intense, beam of monocolored light which can be controlled over a wide range of temperatures, ranging from one that would feel slightly warm to one which is several times hotter than the surface of the sun at the point of focus. Some lasers can instantaneously produce temperatures of 75,000°F (41, 650°C).

OBJECTIVES

After completing this unit you will be able to:
1. Identify and briefly describe the operation of four types of lasers
2. State the applications and advantages of industrial lasers
3. Describe the purpose and operation of the interferometer

TYPES OF LASERS

There are several types of lasers and each has a particular use, depending on the job required. Most lasers may be classified as follows: *solid, gas, liquid,* and *semiconductor.* Because the solid, gas, and liquid lasers employ basically the same principle, only these will be discussed. The main parts of a laser (Fig. 93-1) are basically the same for all types of lasers. These include a *power supply,* a *lasing medium* (suitably contained), and a pair of *precisely aligned mirrors.* One of the mirrors has a totally reflective surface, while the other is only partially reflective (about 96 percent).

THE SOLID LASER

The power source charge fires the large flash tube in a manner similar to a photographic electronic flash. The flash tube emits a burst of light sending or ''pumping'' energy into the lasing medium which could be a ruby rod. This excites the chromium atoms in the ruby rod to a high energy level. As these atoms fall back to their initial state, they emit heat and bundles of light energy called *photons.* The photons produced will strike other atoms causing other photons of identical wavelengths to be produced Fig. 93-2). These photons are reflected back and forth (oscillate) along the rod by the mirrored, square surface on each end of the rod. Consequently, the number of photons increase and produce more energy until some of them pass through the partially reflective mirror producing a laser beam which then passes through a lens onto the point of focus on the work (Fig. 93-3).

Solid lasers have proved very useful in the fields of surgery, atomic fusion, drilling diamond dies, measurement, and spot welding.

GAS LASERS

Probably the most common gas laser is the *helium-neon* (HeNe) *laser.* The gas, mixture of 10 parts helium to 1 of

neon, is contained in a sealed glass unit with angled ends to reduce reflection loss. As the power-pulsed circuit closes, an electron discharge takes place within the tube, exciting the helium atoms to higher energy levels. These atoms collide with the neon atoms bringing them up to the same level. When the neon atoms drop back to a lower level, they emit heat and red photons of laser light. The action continues to multiply and the atoms oscillate back and forth in the tube until some of the light escapes through the partial mirror in the form of a red laser beam (Fig. 93-4).

Fig. 93-1 The solid ruby laser is used in surgery, drilling, measurement, and spot welding.

WAVE GROWTH BY SIMULATED EMISSION

Fig. 93-2 Light waves traveling along the axis of the rod grow by stimulated emission of photons.

Fig. 93-3 Parallel light waves pass through the partially reflective mirror and the lense.

HeNe lasers are small, relatively inexpensive, and relatively safe. Consequently they are used in laboratories and schools for laser experiments.

Another form of gas laser is the *carbon dioxide (CO_2) laser* which uses CO_2 in the sealed tube. This device is capable of much higher output and is more efficient than the HeNe laser. If the CO_2 gas is pumped through the laser cavity or tube and cooled in a heat exchanger, it can operate continuously at very high power. Large units of this type are capable of vaporizing *any* material. CO_2 lasers are used for cutting organic materials such as rubber and leather, thus eliminating the need for punch presses. They are also used for hole production, perforating, welding, and oxygen-assisted metal cutting.

LIQUID LASERS

Liquid lasers employ an organic dye in a solvent as the lasing medium (Fig. 93-5). The liquid is circulated and the flash tube pumps the dye molecules to high energy levels and photons are produced. The process continues as the adjustable external mirrors create a feedback and the atoms oscillate in the tube until some of the light passes through the partial mirror as a laser beam. Liquid lasers are particularly suited to chemical analysis because of the tuning prism which can be rotated and which in turn produces a choice of several different colors and corresponding wavelengths.

INDUSTRIAL APPLICATIONS OF LASERS

Lasers may be used for:

☐ Cutting holes in any known substance. Ruby lasers are used to produce precise holes in diamonds for wire drawing.
☐ Producing holes in paper, plastic, rubber, etc. where a poor finish is left by conventional means. CO_2 lasers are used to punch holes in plastic pipes, baby bottle nipples, and aerosol can spray tips.
☐ Cutting shallow grooves and engraving measuring instruments and steel parts.
☐ Cutting metal. A laser beam leaves no tears or jagged edges like a saw does.
☐ Cutting hard quartz crystal. A laser beam will cut 100 times faster than a diamond wheel.
☐ Removing very narrow sections of unwanted metal and printed circuits.
☐ For welding which would be difficult by any other means such as welding the gears to the synchronizing mechanism in an automobile transmission.

Fig. 93-4 A sealed-tube gas laser.

Fig. 93-5 Liquid lasers are finding wide use in chemical laboratories.

☐ Machining microscopic parts.
☐ Heat treating, such as hardening the surfaces of gears or cylinders. The intense heat required for this is generated almost instantaneously. The beam may be aimed to harden specific areas and no further polishing is needed.
☐ Measuring with a very high degree of accuracy; within 0.000006 in. per foot of length (Fig. 93-6). (See Unit 17 for the principle of the laser interferometer.)

ADVANTAGES OF LASERS

1. Capable of producing a very narrow heat-affected zone which is suitable for welding next to a glass-metal seal.
2. Can be used in welding delicate electronic components that could not withstand resistance welding.
3. Can be used to weld in a vacuum.
4. Capable of producing holes in any known material.
5. Capable of producing extremely small and accurate holes.
6. There is no need to clamp a workpiece in position since there is no torque involved as in conventional drilling.
7. Ideal for tack-welding operations.
8. Very useful in welding parts together as no other material or equipment is needed.
9. Can weld dissimilar metals easily.

639

THE LASER

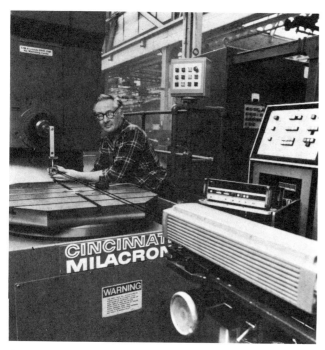

Fig. 93-6 A laser being used to check the alignment of machine parts.   (Courtesy Cincinnati Milacron Incorporated.)

10. Particularly suited for drilling paper, plastics, rubber, etc., where a poor finish is left by conventional means.

11. Can be used for localized heat treating on finished parts.

12. Useful for cutting shallow grooves and engraving letters and numbers in steel and measuring instruments.

REVIEW QUESTIONS

TYPES OF LASERS

1. Name four types of lasers.
2. Name the main parts of any laser.
3. Describe the operation of the solid laser.
4. Where are HeNe gas lasers used? Why?
5. State four uses for a CO_2 laser.
6. For what purpose are liquid lasers generally used?

INDUSTRIAL APPLICATIONS

7. State three major industrial laser applications.

ADVANTAGES OF LASERS

8. List six of the most important advantages of lasers.

TABLE 1: DECIMAL INCH, FRACTIONAL INCH, AND MILLIMETER EQUIVALENTS

Decimal inch	Fractional inch	Millimeter	Decimal inch	Fractional inch	Millimeter
.015625	1/64	0.397	.515625	33/64	13.097
.03125	1/32	0.794	.53125	17/32	13.494
.046875	3/64	1.191	.546875	35/64	13.891
.0625	1/16	1.588	.5625	9/16	14.288
.078125	5/64	1.984	.578125	37/64	14.684
.09375	3/32	2.381	.59375	19/32	15.081
.109375	7/64	2.778	.609375	39/64	15.478
.125	1/8	3.175	.625	5/8	15.875
.140625	9/64	3.572	.640625	41/64	16.272
.15625	5/32	3.969	.65625	21/32	16.669
.171875	11/64	4.366	.671875	43/64	17.066
.1875	3/16	4.762	.6875	11/16	17.462
.203125	13/64	5.159	.703125	45/64	17.859
.21875	7/32	5.556	.71875	23/32	18.256
.234375	15/64	5.953	.734375	47/64	18.653
.25	1/4	6.350	.75	3/4	19.05
.265625	17/64	6.747	.765625	49/64	19.447
.28125	9/32	7.144	.78125	25/32	19.844
.296875	19/64	7.541	.796875	51/64	20.241
.3125	5/16	7.938	.8125	13/16	20.638
.328125	21/64	8.334	.828125	53/64	21.034
.34375	11/32	8.731	.84375	27/32	21.431
.359375	23/64	9.128	.859375	55/64	21.828
.375	3/8	9.525	.875	7/8	22.225
.390625	25/64	9.922	.890625	57/64	22.622
.40625	13/32	10.319	.90625	29/32	23.019
.421875	27/64	10.716	.921875	59/64	23.416
.4375	7/16	11.112	.9375	15/16	23.812
.453125	29/64	11.509	.953125	61/64	24.209
.46875	15/32	11.906	.96875	31/32	24.606
.484375	31/64	12.303	.984375	63/64	25.003
.5	1/2	12.700	1.	1	25.400

TABLE 2:

CONVERSION OF INCHES TO MILLIMETERS						CONVERSION OF MILLIMETERS TO INCHES					
Inches	Millimeters	Inches	Millimeters	Inches	Millimeters	Millimeters	Inches	Millimeters	Inches	Millimeters	Inches
.001	0.025	.290	7.37	.660	16.76	0.01	.0004	0.35	.0138	0.68	.0268
.002	0.051	.300	7.62	.670	17.02	0.02	.0008	0.36	.0142	0.69	.0272
.003	0.076	.310	7.87	.680	17.27	0.03	.0012	0.37	.0146	0.70	.0276
.004	0.102	.320	8.13	.690	17.53	0.04	.0016	0.38	.0150	0.71	.0280
.005	0.127	.330	8.38	.700	17.78	0.05	.0020	0.39	.0154	0.72	.0283
.006	0.152	.340	8.64	.710	18.03	0.06	.0024	0.40	.0157	0.73	.0287
.007	0.178	.350	8.89	.720	18.29	0.07	.0028	0.41	.0161	0.74	.0291
.008	0.203	.360	9.14	.730	18.54	0.08	.0031	0.42	.0165	0.75	.0295
.009	0.229	.370	9.40	.740	18.80	0.09	.0035	0.43	.0169	0.76	.0299
.010	0.254	.380	9.65	.750	19.05	0.10	.0039	0.44	.0173	0.77	.0303
.020	0.508	.390	9.91	.760	19.30	0.11	.0043	0.45	.0177	0.78	.0307
.030	0.762	.400	10.16	.770	19.56	0.12	.0047	0.46	.0181	0.79	.0311
.040	1.016	.410	10.41	.780	19.81	0.13	.0051	0.47	.0185	0.80	.0315
.050	1.270	.420	10.67	.790	20.07	0.14	.0055	0.48	.0189	0.81	.0319
.060	1.524	.430	10.92	.800	20.32	0.15	.0059	0.49	.0193	0.82	.0323
.070	1.778	.440	11.18	.810	20.57	0.16	.0063	0.50	.0197	0.83	.0327
.080	2.032	.450	11.43	.820	20.83	0.17	.0067	0.51	.0201	0.84	.0331
.090	2.286	.460	11.68	.830	21.08	0.18	.0071	0.52	.0205	0.85	.0335
.100	2.540	.470	11.94	.840	21.34	0.19	.0075	0.53	.0209	0.86	.0339
.110	2.794	.480	12.19	.850	21.59	0.20	.0079	0.54	.0213	0.87	.0343
.120	3.048	.490	12.45	.860	21.84	0.21	.0083	0.55	.0217	0.88	.0346
.130	3.302	.500	12.70	.870	22.10	0.22	.0087	0.56	.0220	0.89	.0350
.140	3.56	.510	12.95	.880	22.35	0.23	.0091	0.57	.0224	0.90	.0354
.150	3.81	.520	13.21	.890	22.61	0.24	.0094	0.58	.0228	0.91	.0358
.160	4.06	.530	13.46	.900	22.86	0.25	.0098	0.59	.0232	0.92	.0362
.170	4.32	.540	13.72	.910	23.11	0.26	.0102	0.60	.0236	0.93	.0366
.180	4.57	.550	13.97	.920	23.37	0.27	.0106	0.61	.0240	0.94	.0370
.190	4.83	.560	14.22	.930	23.62	0.28	.0110	0.62	.0244	0.95	.0374
.200	5.08	.570	14.48	.940	23.88	0.29	.0114	0.63	.0248	0.96	.0378
.210	5.33	.580	14.73	.950	24.13	0.30	.0118	0.64	.0252	0.97	.0382
.220	5.59	.590	14.99	.960	24.38	0.31	.0122	0.65	.0256	0.98	.0386
.230	5.84	.600	15.24	.970	24.64	0.32	.0126	0.66	.0260	0.99	.0390
.240	6.10	.610	15.49	.980	24.89	0.33	.0130	0.67	.0264	1.00	.0394
.250	6.35	.620	15.75	.990	25.15	0.34	.0134				
.260	6.60	.630	16.00	1.000	25.40						
.270	6.86	.640	16.26								
.280	7.11	.650	16.51								

Courtesy Automatic Electric Company

TABLE 3: LETTER DRILL SIZES

Letter	mm	in.	Letter	mm	in.
A	5.9	.234	N	7.7	.302
B	6.0	.238	O	8.0	.316
C	6.1	.242	P	8.2	.323
D	6.2	.246	Q	8.4	.332
E	6.4	.250	R	8.6	.339
F	6.5	.257	S	8.8	.348
G	6.6	.261	T	9.1	.358
H	6.7	.266	U	9.3	.368
I	6.9	.272	V	9.5	.377
J	7.0	.277	W	9.8	.386
K	7.1	.281	X	10.1	.397
L	7.4	.290	Y	10.3	.404
M	7.5	.295	Z	10.5	.413

TABLE 4: DRILL GAUGE SIZES

No.	mm	inch	No.	mm	inch	No.	mm	inch
1	5.80	.2280	34	2.80	.1110	66	0.84	.0330
2	5.60	.2210	35	2.80	.1100	67	0.81	.0320
3	5.40	.2130	36	2.70	.1065	68	0.79	.0310
4	5.30	.2090	37	2.65	.1040	69	0.74	.0292
5	5.20	.2055	38	2.60	.1015	70	0.71	.0280
6	5.20	.2040	39	2.55	.0995	71	0.66	.0260
7	5.10	.2010	40	2.50	.0980	72	0.64	.0250
8	5.10	.1990	41	2.45	.0960	83	0.61	.0240
9	5.00	.1960	42	2.40	.0935	74	0.57	.0225
10	4.90	.1935	43	2.25	.0890	75	0.53	.0210
11	4.90	.1910	44	2.20	.0860	76	0.51	.0200
12	4.80	.1890	45	2.10	.0820	77	0.46	.0180
13	4.70	.1850	46	2.05	.0810	78	0.41	.0160
14	4.60	.1820	47	2.00	.0785	79	0.37	.0145
15	4.60	.1800	48	1.95	.0760	80	0.34	.0135
16	4.50	.1770	49	1.85	.0730	81	0.33	.0130
17	4.40	.1730	50	1.80	.0700	82	0.32	.0125
18	4.30	.1695	51	1.70	.0670	83	0.31	.0120
19	4.20	.1660	52	1.60	.0635	84	0.29	.0115
20	4.10	.1610	53	1.50	.0595	85	0.28	.0110
21	4.00	.1590	54	1.40	.0550	86	0.27	.0105
22	4.00	.1570	55	1.30	.0520	87	0.25	.0100
23	3.90	.1540	56	1.20	.0465	88	0.24	.0095
24	3.90	.1520	57	1.10	.0430	89	0.23	.0091
25	3.80	.1495	58	1.05	.0420	90	0.22	.0087
26	3.70	.1470	59	1.05	.0410	91	0.21	.0083
27	3.70	.1440	60	1.00	.0400	92	0.20	.0079
28	3.60	.1405	61	0.99	.0390	93	0.19	.0075
29	3.50	.1360	62	0.97	.0380	94	0.18	.0071
30	3.30	.1285	63	0.94	.0370	95	0.17	.0067
31	3.00	.1200	64	0.92	.0360	96	0.16	.0063
32	2.95	.1160	65	0.89	.0350	97	0.15	.0059
33	2.85	.1130						

TABLE 5: TAP DRILL SIZES

Nominal Diameter mm	Thread Pitch mm	Tap Drill Size mm	Nominal Diameter mm	Thread Pitch mm	Tap Drill Size mm
1.60	0.35	1.20	20.00	2.50	17.50
2.00	0.40	1.60	24.00	3.00	21.00
2.50	0.45	2.05	30.00	3.50	26.50
3.00	0.50	2.50	36.00	4.00	32.00
3.50	0.60	2.90	42.00	4.50	37.50
4.00	0.70	3.30	48.00	5.00	43.00
5.00	0.80	4.20	56.00	5.50	50.50
6.30	1.00	5.30	64.00	6.00	58.00
8.00	1.25	6.80	72.00	6.00	66.00
10.00	1.50	8.50	80.00	6.00	74.00
12.00	1.75	10.20	90.00	6.00	84.00
14.00	2.00	12.00	100.00	6.00	94.00
16.00	2.00	14.00			

TABLE 6: ISO METRIC PITCH & DIAMETER COMBINATIONS

Nominal Dia. (mm)	Thread Pitch (mm)	Nominal Dia. (mm)	Thread Pitch (mm)
1.6	0.35	20	2.5
2	0.40	24	3.0
2.5	0.45	30	3.5
3	0.50	36	4.0
3.5	0.60	42	4.5
4	0.70	48	5.0
5	0.80	56	5.5
6.3	1.00	64	6.0
8	1.25	72	6.0
10	1.50	80	6.0
12	1.75	90	6.0
14	2.00	100	6.0
16	2.00		

TABLE 7: TAP DRILL SIZES AMERICAN NATIONAL FORM THREAD

NC National Coarse			NF National Fine		
Tap Size	Threads per inch	Tap Drill Size	Tap Size	Threads per inch	Tap Drill Size
# 5	40	#38	# 5	44	#37
# 6	32	#36	# 6	40	#33
# 8	32	#29	# 8	36	#29
#10	24	#25	#10	32	#21
#12	24	#16	#12	28	#14
1/4	20	# 7	1/4	28	# 3
5/16	18	F	5/16	24	I
3/8	16	5/16	3/8	24	Q
7/16	14	U	7/16	20	25/64
1/2	13	27/64	1/2	20	29/64
9/16	12	31/64	9/16	18	33/64
5/8	11	17/32	5/8	18	37/64
3/4	10	21/32	3/4	16	11/16
7/8	9	49/64	7/8	14	13/16
1	8	7/8	1	14	15/16
1-1/8	7	63/64	1-1/8	12	1-3/64
1-1/4	7	1-7/64	1-1/4	12	1-11/64
1-3/8	6	1-7/32	1-3/8	12	1-19/64
1-1/2	6	1-11/32	1-1/2	12	1-27/64
1-3/4	5	1-9/16			
2	4-1/2	1-25/32			

NPT NATIONAL PIPE THREAD					
1/8	27	11/32	1	11-1/2	1-5/32
1/4	18	7/16	1-1/4	11-1/2	1-1/2
3/8	18	19/32	1-1/2	11-1/2	1-23/32
1/2	14	23/32	2	11-1/2	2-3/16
3/4	14	15/16	2-1/2	8	2-5/8

The major diameter of an NC or NF number size tap or screw = (N × .013) + .060

EXAMPLE: The major diameter of a #5 tap equals

(5 × .013) + .060 = .125 diameter

TABLE 8:

THREE WIRE THREAD MEASUREMENT
(60° Metric Thread)

$M = PD + C$ $PD = M - C$

M = Measurement over wires
PD = Pitch diameter
C = Constant

Pitch		Best Wire Size		Constant	
mm	Inches	mm	Inches	mm	Inches
0.2	.00787	0.1155	.00455	0.1732	.00682
0.225	.00886	0.1299	.00511	0.1949	.00767
0.25	.00934	0.1443	.00568	0.2165	.00852
0.3	.01181	0.1732	.00682	0.2598	.01023
0.35	.01378	0.2021	.00796	0.3031	.01193
0.4	.01575	0.2309	.00909	0.3464	.01364
0.45	.01772	0.2598	.01023	0.3897	.01534
0.5	.01969	0.2887	.01137	0.4330	.01705
0.6	.02362	0.3464	.01364	0.5196	.02046
0.7	.02756	0.4041	.01591	0.6062	.02387
0.75	.02953	0.4330	.01705	0.6495	.02557
0.8	.03150	0.4619	.01818	0.6928	.02728
0.9	.03543	0.5196	.02046	0.7794	.03069
1.0	.03937	0.5774	.02273	0.8660	.03410
1.25	.04921	0.7217	.02841	1.0825	.04262
1.5	.05906	0.8660	.03410	1.2990	.05114
1.75	.06890	1.0104	.03978	1.5155	.05967
2.0	.07874	1.1547	.04546	1.7321	.06819
2.5	.09843	1.4434	.05683	2.1651	.08524
3.0	.11811	1.7321	.06819	2.5981	.10229
3.5	.13780	2.0207	.07956	3.0311	.11933
4.0	.15748	2.3094	.09092	3.4641	.13638
4.5	.17717	2.5981	.10229	3.8971	.15343
5.0	.19685	2.8868	.11365	4.3301	.17048
5.5	.21654	3.1754	.12502	4.7631	.18753
6.0	.23622	3.4641	.13638	5.1962	.20457
7.0	.27559	4.0415	.15911	6.0622	.23867
8.0	.31496	4.6188	.18184	6.9282	.27276
9.0	.35433	5.1962	.20457	7.7942	.30686
10.0	.39370	5.7735	.22730	8.6603	.34095

TABLE 9: MORSE TAPERS

ANGLE OF KEY 8° 19' = TAPER 1-3/4 IN 12

Number of taper	Diameter of plug at small end	Diameter at end of socket	Whole length of shank	Shank depth	Depth of hole	Standard plug depth	Thickness of tongue	Length of tongue	Diameter of tongue	Width of keyway	Length of keyway	End of socket to keyway	Taper per foot
	D	A	B	S	H	P	t	T	d	w	L	K	
0	.252	.356	2-11/32	2-7/32	2-1/32	2	5/32	1/4	.235	.160	9/16	1-15/16	.6246
1	.369	.475	2-9/16	2-7/16	2-3/16	2-1/8	13/64	3/8	.343	.213	3/4	2-1/16	.5986
2	.572	.700	3-1/8	2-15/16	2-5/8	2-9/16	1/4	7/16	17/32	.260	7/8	2-1/2	.5994
3	.778	.938	3-7/8	3-11/16	3-1/4	3-3/16	5/16	9/16	23/32	.322	1-3/16	3-1/16	.6023
4	1.020	1.231	4-7/8	4-5/8	4-1/8	4-1/16	15/32	5/8	31/32	.478	1-1/4	3-7/8	.6232
5	1.475	1.748	6-1/8	5-7/8	5-1/4	5-3/16	5/8	3/4	1-13/32	.635	1-1/2	4-15/16	.6315
6	2.116	2.494	8-9/16	8-1/4	7-3/8	7-1/4	3/4	1-1/8	2	.760	1-3/4	7	.6256
7	2.750	3.270	11-1/4	11-5/8	10-1/8	10	1-1/8	1-3/8	2-5/8	1.135	2-5/8	9-1/2	.6240

Note: All measurements are in inches

TABLE 10: STANDARD MILLING MACHINE TAPER

Milling Machine Spindles	Milling Machine Arbors

Taper No.	A	B	C	D	L	N	Q	R	S	T	U	V	W
30	1.250	2.7493	.685 .692	21/32	2-7/8	1.250	1/2 – 13	.673 .675	13/16	1	2	2-3/4	1/16
40	1.750	3.4993	.997 1.005	21/32	3-7/8	1.750	5/8 – 11	.985 .987	1	1-1/8	2-5/16	3-3/4	1/16
50	2.750	5.0618	1.559 1.568	1-1/16	5-1/2	2.750	1 – 8	1.547 1.549	1	1-3/4	3-1/2	5-1/8	1/8
60	4.250	8.718	2.371 2.381	1-3/8	8-5/8	4.250	1-1/4 – 7	2.359 2.361	1-3/4	2-1/4	4-1/4	8-5/16	1/8

Note: All measurements are in inches

TABLE 11: TAPERS AND ANGLES						
Taper per Foot	Included Angle		With Center Line		Taper per Inch	Taper per Inch from Center Line
	Degree	Minute	Degree	Minute		
1/8	0	36	0	18	.010416	.005208
3/16	0	54	0	27	.015625	.007812
1/4	1	12	0	36	.020833	.010416
5/16	1	30	0	45	.026042	.013021
3/8	1	47	0	53	.031250	.015625
7/16	2	05	1	02	.036458	.018229
1/2	2	23	1	11	.041667	.020833
9/16	2	42	1	21	.046875	.023438
5/8	3	00	1	30	.052084	.026042
11/16	3	18	1	39	.057292	.028646
3/4	3	35	1	48	.062500	.031250
13/16	3	52	1	56	.067708	.033854
7/8	4	12	2	06	.072917	.036458
15/16	4	28	2	14	.078125	.039063
1	4	45	2	23	.083330	.041667
1-1/4	5	58	2	59	.104166	.052083
1-1/2	7	08	3	34	.125000	.062500
1-3/4	8	20	4	10	.145833	.072917
2	9	32	4	46	.166666	.083333
2-1/2	11	54	5	57	.208333	.104166
3	14	16	7	08	.250000	.125000
3-1/2	16	36	8	18	.291666	.145833
4	18	56	9	28	.333333	.166666
4-1/2	21	14	10	37	.375000	.187500
5	23	32	11	46	.416666	.208333
6	28	04	14	02	.500000	.250000

Courtesy Morse Twist Drill & Machine Co.

TABLE 12: ALLOWANCES FOR FITS	
RUNNING FITS	

Shaft Diameter	For Shafts with Speeds Under 600 r/min Ordinary Working Conditions	For Shafts with Speeds Over 600 r/min. Heavy Pressure – Severe Working Conditions
Up to 1/2	−.0005 to −.001	−.0005 to −.001
1/2 to 1	−.00075 to −.0015	−.001 to −.002
1 to 2	−.0015 to −.0025	−.002 to −.003
2 to 3-1/2	−.002 to −.003	−.003 to −.004
3-1/2 to 6	−.0025 to −.004	−.004 to −.005

SLIDING FITS	
Shaft Diameter	For Shafts with Gears, Clutches or Similar Parts which Must be Free to Slide
Up to 1/2	−.0005 to −.001
1/2 to 1	−.00075 to −.0015
1 to 2	−.0015 to −.0025
2 to 3-1/2	−.002 to −.003
3-1/2 to 6	−.0025 to −.004

PUSH FITS		
Shaft Diameter	For Light Service where Part is Keyed to Shaft and Clamped Endwise – No Fitting	With Play Eliminated – Part Should Assemble Readily – Some Fitting and Selecting May be Required
Up to 1/2	Standard to −.00025	Standard to +.00025
1/2 to 3-1/2	Standard to −.0005	Standard to +.0005
3-1/2 to 6	Standard to −.00075	Standard to +.00075

DRIVING FITS		
Shaft Diameter	For Permanent Assembly of Parts so Located that Driving Cannot be Done Readily	For Permanent Assembly and Severe Duty and Where There is Ample Room for Driving
Up to 1/2	Standard to +.00025	+.0005 to +.001
1/2 to 1	+.00025 to +.0005	+.0005 to +.001
1 to 2	+.0005 to +.00075	+.0005 to +.001
2 to 3-1/2	+.0005 to +.001	+.00075 to +.00125
3-1/2 to 6	+.0005 to +.001	+.001 to +.0015

FORCED FITS	
Shaft Diameter	For Permanent Assembly and Very Severe Service – Hydraulic Press Used for Larger Parts
Up to 1/2	+.00075 to +.001
1/2 to 1	+.001 to +.002
1 to 2	+.002 to +.003
2 to 3-1/2	+.003 to +.004
3-1/2 to 6	+.004 to +.005

Note: All measurements are in inches

TABLE 13: RULES FOR FINDING DIMENSIONS OF CIRCLES, SQUARES, ETC.
D is diameter of stock necessary to turn shape desired. E is distance "across flats," or diameter of inscribed circle. C is depth of cut into stock turned to correct diameter.

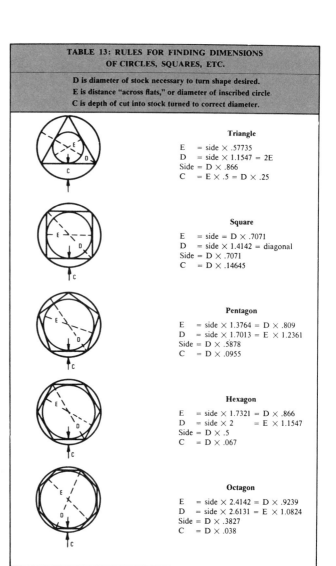

Triangle

$$E = \text{side} \times .57735$$
$$D = \text{side} \times 1.1547 = 2E$$
$$\text{Side} = D \times .866$$
$$C = E \times .5 = D \times .25$$

Square

$$E = \text{side} = D \times .7071$$
$$D = \text{side} \times 1.4142 = \text{diagonal}$$
$$\text{Side} = D \times .7071$$
$$C = D \times .14645$$

Pentagon

$$E = \text{side} \times 1.3764 = D \times .809$$
$$D = \text{side} \times 1.7013 = E \times 1.2361$$
$$\text{Side} = D \times .5878$$
$$C = D \times .0955$$

Hexagon

$$E = \text{side} \times 1.7321 = D \times .866$$
$$D = \text{side} \times 2 = E \times 1.1547$$
$$\text{Side} = D \times .5$$
$$C = D \times .067$$

Octagon

$$E = \text{side} \times 2.4142 = D \times .9239$$
$$D = \text{side} \times 2.6131 = E \times 1.0824$$
$$\text{Side} = D \times .3827$$
$$C = D \times .038$$

Courtesy Morse Twist Drill & Machine Co.

TABLE 14: HARDNESS CONVERSION CHART

10 mm Ball 3000 kg	120° Cone 150 kg	1/16 in. Ball 100 kg	Model C	Mpa	10 mm Ball 3000 kg	120° Cone 150 kg	1/16 in. Ball 100 kg	Model C	Mpa
Brinell	Rockwell C	Rockwell B	Shore Scleroscope	Tensile Strength	Brinell	Rockwell C	Rockwell B	Shore Scleroscope	Tensile Strength
800	72	.....	100	....	276	30	105	42	938
780	71	.....	99	....	269	29	104	41	910
760	70	.....	98	....	261	28	103	40	889
745	68	.....	97	2530	258	27	102	39	876
725	67		96	2460	255	26	102	39	862
712	66		95	2413	249	25	101	38	848
682	65		93	2324	245	24	100	37	820
668	64		91	2248	240	23	99	36	807
652	63		89	2193	237	23	99	35	793
626	62		87	2110	229	22	98	34	779
614	61		85	2062	224	21	97	33	758
601	60		83	2013	217	20	96	33	738
590	59		81	2000	211	19	95	32	717
576	57		79	1937	206	18	94	32	703
552	56		76	1862	203	17	94	31	689
545	55		75	1848	200	16	93	31	676
529	54		74	1786	196	15	92	30	662
514	53	120	72	1751	191	14	92	30	648
502	52	119	70	1703	187	13	91	29	634
495	51	119	69	1682	185	12	91	29	627
477	49	118	67	1606	183	11	90	28	621
461	48	117	66	1565	180	10	89	28	614
451	47	117	65	1538	175	9	88	27	593
444	46	116	64	1510	170	7	87	27	579
427	45	115	62	1441	167	6	87	27	565
415	44	115	60	1407	165	5	86	26	558
401	43	114	58	1351	163	4	85	26	552
388	42	114	57	1317	160	3	84	25	538
375	41	113	55	1269	156	2	83	25	524
370	40	112	54	1255	154	1	82	25	517
362	39	111	53	1234	152		82	24	510
351	38	111	51	1193	150		81	24	510
346	37	110	50	1172	147		80	24	496
341	37	110	49	1158	145		79	23	490
331	36	109	47	1124	143		79	23	483
323	35	109	46	1089	141		78	23	476
311	34	108	46	1055	140		77	22	476
301	33	107	45	1020	135		75	22	462
293	32	106	44	993	130		72	22	448
285	31	105	43	965					

TABLE 15: SOLUTIONS FOR RIGHT-ANGLED TRIANGLES

$$\text{Sine} \angle = \frac{\text{Side opposite}}{\text{Hypotenuse}} \qquad \text{Cosecant} \angle = \frac{\text{Hypotenuse}}{\text{Side opposite}}$$

$$\text{Cosine} \angle = \frac{\text{Side adjacent}}{\text{Hypotenuse}} \qquad \text{Secant} \angle = \frac{\text{Hypotenuse}}{\text{Side adjacent}}$$

$$\text{Tangent} \angle = \frac{\text{Side opposite}}{\text{Side adjacent}} \qquad \text{Cotangent} \angle = \frac{\text{Side adjacent}}{\text{Side opposite}}$$

Knowing	Formulas to find	
Sides a & b	$c = \sqrt{a^2 - b^2}$	$\sin B = \dfrac{b}{a}$
Side a & angle B	$b = a \times \sin B$	$c = a \times \cos B$
Sides a & c	$b = \sqrt{a^2 - c^2}$	$\sin C = \dfrac{c}{a}$
Side a & angle C	$b = a \times \cos C$	$c = a \times \sin C$
Sides b & c	$a = \sqrt{b^2 + c^2}$	$\tan B = \dfrac{b}{c}$
Side b & angle B	$a = \dfrac{b}{\sin B}$	$c = b \times \cot B$
Side b & angle C	$a = \dfrac{b}{\cos C}$	$c = b \times \tan C$
Side c & angle B	$a = \dfrac{c}{\cos B}$	$b = c \times \tan B$
Side c & angle C	$a = \dfrac{c}{\sin C}$	$b = c \times \cot C$

TABLE 16: SINE BAR CONSTANTS (5 in. BAR)
(Multiply Constants by Two for a 10 in. Sine Bar)

Min.	0°	1°	2°	3°	4°	5°	6°	7°	8°	9°	10°	11°	12°	13°	14°	15°	16°	17°	18°	19°	Min.
0	.00000	.08725	.17450	.26170	.34880	.43580	.52265	.60935	.69585	.78215	.86825	.95405	1.0395	1.1247	1.2096	1.2941	1.3782	1.4618	1.5451	1.6278	0
2	.00290	.09015	.17740	.26460	.35170	.43870	.52555	.61225	.69875	.78505	.87110	.95690	.0424	.1276	.2124	.2969	.3810	.4646	.5478	.6306	2
4	.00580	.09310	.18030	.26750	.35460	.44155	.52845	.61510	.70165	.78790	.87395	.95975	.0452	.1304	.2152	.2997	.3838	.4674	.5506	.6333	4
6	.00875	.09600	.18320	.27040	.35750	.44445	.53130	.61800	.70450	.79080	.87685	.96260	.0481	.1332	.2181	.3025	.3865	.4702	.5534	.6361	6
8	.01165	.09890	.18615	.27330	.36040	.44735	.53420	.62090	.70740	.79365	.87970	.96545	.0509	.1361	.2209	.3053	.3893	.4730	.5561	.6388	8
10	.01455	.10180	.18905	.27620	.36330	.45025	.53710	.62380	.71025	.79655	.88255	.96830	1.0538	1.1389	1.2237	1.3081	1.3921	1.4757	1.5589	1.6416	10
12	.01745	.10470	.19195	.27910	.36620	.45315	.54000	.62665	.71315	.79940	.88540	.97115	.0566	.1417	.2265	.3109	.3949	.4785	.5616	.6443	12
14	.02035	.10760	.19485	.28200	.36910	.45605	.54290	.62955	.71600	.80230	.88830	.97405	.0594	.1446	.2293	.3137	.3977	.4813	.5644	.6471	14
16	.02325	.11055	.19775	.28490	.37200	.45895	.54580	.63245	.71890	.80515	.89115	.97690	.0623	.1474	.2322	.3165	.4005	.4841	.5672	.6498	16
18	.02620	.11345	.20065	.28780	.37490	.46185	.54865	.63530	.72180	.80800	.89400	.97975	.0651	.1502	.2350	.3193	.4033	.4868	.5699	.6525	18
20	.02910	.11635	.20355	.29070	.37780	.46475	.55155	.63820	.72465	.81090	.89685	.98260	1.0680	1.1531	1.2378	1.3221	1.4061	1.4896	1.5727	1.6553	20
22	.03200	.11925	.20645	.29365	.38070	.46765	.55445	.64110	.72755	.81375	.89975	.98545	.0708	.1559	.2406	.3250	.4089	.4924	.5755	.6580	22
24	.03490	.12215	.20940	.29655	.38360	.47055	.55735	.64400	.73040	.81665	.90260	.98830	.0737	.1587	.2434	.3278	.4117	.4952	.5782	.6608	24
26	.03780	.12505	.21230	.29945	.38650	.47345	.56025	.64685	.73330	.81950	.90545	.99115	.0765	.1615	.2462	.3306	.4145	.4980	.5810	.6635	26
28	.04070	.12800	.21520	.30235	.38940	.47635	.56315	.64975	.73615	.82235	.90830	.99400	.0793	.1644	.2491	.3334	.4173	.5007	.5837	.6663	28
30	.04365	.13090	.21810	.30525	.39230	.47925	.56600	.65265	.73905	.82525	.91120	.99685	1.0822	1.1672	1.2519	1.3362	1.4201	1.5035	1.5865	1.6690	30
32	.04655	.13380	.22100	.30815	.39520	.48210	.56890	.65550	.74190	.82810	.91405	.99970	.0850	.1700	.2547	.3390	.4228	.5063	.5893	.6718	32
34	.04945	.13670	.22390	.31105	.39810	.48500	.57180	.65840	.74480	.83100	.91690	1.0016	.0879	.1729	.2575	.3418	.4256	.5091	.5920	.6745	34
36	.05235	.13960	.22680	.31395	.40100	.48790	.57470	.66130	.74770	.83385	.91975	.0054	.0907	.1757	.2603	.3446	.4284	.5118	.5948	.6772	36
38	.05525	.14250	.22970	.31685	.40390	.49080	.57760	.66415	.75055	.83670	.92260	.0082	.0935	.1785	.2631	.3474	.4312	.5146	.5975	.6800	38
40	.05820	.14540	.23265	.31975	.40680	.49370	.58045	.66705	.75345	.83960	.92545	1.0110	1.0964	1.1813	1.2660	1.3502	1.4340	1.5174	1.6003	1.6827	40
42	.06110	.14835	.23555	.32265	.40970	.49660	.58335	.66995	.75630	.84245	.92835	.0139	.0992	.1842	.2688	.3530	.4368	.5201	.6030	.6855	42
44	.06400	.15125	.23845	.32555	.41260	.49950	.58625	.67280	.75920	.84530	.93120	.0168	.1020	.1870	.2716	.3558	.4396	.5229	.6058	.6882	44
46	.06690	.15415	.24135	.32845	.41550	.50240	.58915	.67570	.76205	.84820	.93405	.0196	.1049	.1898	.2744	.3586	.4423	.5257	.6085	.6909	46
48	.06980	.15705	.24425	.33135	.41840	.50530	.59200	.67860	.76495	.85105	.93690	.0225	.1077	.1926	.2772	.3614	.4451	.5285	.6113	.6937	48
50	.07270	.15995	.24715	.33425	.42130	.50820	.59490	.68145	.76780	.85390	.93975	1.0253	1.1106	1.1955	1.2800	1.3642	1.4479	1.5312	1.6141	1.6964	50
52	.07565	.16285	.25005	.33715	.42420	.51105	.59780	.68435	.77070	.85680	.94260	.0281	.1134	.1983	.2828	.3670	.4507	.5340	.6168	.6991	52
54	.07855	.16580	.25295	.34010	.42710	.51395	.60070	.68720	.77355	.85965	.94550	.0310	.1162	.2011	.2856	.3698	.4535	.5368	.6196	.7019	54
56	.08145	.16870	.25585	.34300	.43000	.51685	.60355	.69010	.77645	.86250	.94835	.0338	.1191	.2039	.2884	.3726	.4563	.5395	.6223	.7046	56
58	.08435	.17160	.25875	.34590	.43290	.51975	.60645	.69300	.77930	.86540	.95120	.0367	.1219	.2068	.2913	.3754	.4591	.5423	.6251	.7073	58
60	.08725	.17450	.26170	.34880	.43580	.52265	.60935	.69585	.78215	.86825	.95405	1.0395	1.1247	1.2096	1.2941	1.3782	1.4618	1.5451	1.6278	1.7101	60

Courtesy Brown & Sharpe Mfg. Co.

Table 16 continued

TABLE 16: SINE BAR CONSTANTS (5 in. BAR)
(Multiply Constants by Two for a 10 in. Sine Bar)

Min.	20°	21°	22°	23°	24°	25°	26°	27°	28°	29°	30°	31°	32°	33°	34°	35°	36°	37°	38°	39°	Min.
0	1.7101	1.7918	1.8730	1.9536	2.0337	2.1131	2.1918	2.2699	2.3473	2.4240	2.5000	2.5752	2.6496	2.7232	2.7959	2.8679	2.9389	3.0091	3.0783	3.1466	0
2	.7128	.7945	.8757	.9563	.0363	.1157	.1944	.2725	.3499	.4266	.5025	.5777	.6520	.7256	.7984	.8702	.9413	.0114	.0806	.1488	2
4	.7155	.7972	.8784	.9590	.0390	.1183	.1971	.2751	.3525	.4291	.5050	.5802	.6545	.7280	.8008	.8726	.9436	.0137	.0829	.1511	4
6	.7183	.8000	.8811	.9617	.0416	.1210	.1997	.2777	.3550	.4317	.5075	.5826	.6570	.7305	.8032	.8750	.9460	.0160	.0852	.1534	6
8	.7210	.8027	.8838	.9643	.0443	.1236	.2023	.2803	.3576	.4342	.5100	.5851	.6594	.7329	.8056	.8774	.9483	.0183	.0874	.1556	8
10	1.7237	1.8054	1.8865	1.9670	2.0469	2.1262	2.2049	2.2829	2.3602	2.4367	2.5126	2.5876	2.6619	2.7354	2.8080	2.8798	2.9507	3.0207	3.0897	3.1579	10
12	.7265	.8081	.8892	.9697	.0496	.1289	.2075	.2855	.3627	.4393	.5151	.5901	.6644	.7378	.8104	.8821	.9530	.0230	.0920	.1601	12
14	.7292	.8108	.8919	.9724	.0522	.1315	.2101	.2881	.3653	.4418	.5176	.5926	.6668	.7402	.8128	.8845	.9554	.0253	.0943	.1624	14
16	.7319	.8135	.8946	.9750	.0549	.1341	.2127	.2906	.3679	.4444	.5201	.5951	.6693	.7427	.8152	.8869	.9577	.0276	.0966	.1646	16
18	.7347	.8162	.8973	.9777	.0575	.1368	.2153	.2932	.3704	.4469	.5226	.5976	.6717	.7451	.8176	.8893	.9600	.0299	.0989	.1669	18
20	1.7374	1.8189	1.8999	1.9804	2.0602	2.1394	2.2179	2.2958	2.3730	2.4494	2.5251	2.6001	2.6742	2.7475	2.8200	2.8916	2.9624	3.0322	3.1012	3.1691	20
22	.7401	.8217	.9026	.9830	.0628	.1420	.2205	.2984	.3755	.4520	.5276	.6025	.6767	.7499	.8224	.8940	.9647	.0345	.1034	.1714	22
24	.7428	.8244	.9053	.9857	.0655	.1447	.2232	.3010	.3781	.4545	.5301	.6050	.6791	.7524	.8248	.8964	.9671	.0369	.1057	.1736	24
26	.7456	.8271	.9080	.9884	.0681	.1473	.2258	.3036	.3807	.4570	.5327	.6075	.6816	.7548	.8272	.8988	.9694	.0392	.1080	.1759	26
28	.7483	.8298	.9107	.9911	.0708	.1499	.2284	.3061	.3832	.4596	.5352	.6100	.6840	.7572	.8296	.9011	.9718	.0415	.1103	.1781	28
30	1.7510	1.8325	1.9134	1.9937	2.0734	2.1525	2.2310	2.3087	2.3858	2.4621	2.5377	2.6125	2.6865	2.7597	2.8320	2.9035	2.9741	3.0438	3.1125	3.1804	30
32	.7537	.8352	.9161	.9964	.0761	.1552	.2336	.3113	.3883	.4646	.5402	.6149	.6889	.7621	.8344	.9059	.9764	.0461	.1148	.1826	32
34	.7565	.8379	.9188	.9991	.0787	.1578	.2362	.3139	.3909	.4672	.5427	.6174	.6914	.7645	.8368	.9082	.9788	.0484	.1171	.1849	34
36	.7592	.8406	.9215	2.0017	.0814	.1604	.2388	.3165	.3934	.4697	.5452	.6199	.6938	.7669	.8392	.9106	.9811	.0507	.1194	.1871	36
38	.7619	.8433	.9241	.0044	.0840	.1630	.2414	.3190	.3960	.4722	.5477	.6224	.6963	.7694	.8416	.9130	.9834	.0530	.1216	.1893	38
40	1.7646	1.8460	1.9268	2.0070	2.0867	2.1656	2.2440	2.3216	2.3985	2.4747	2.5502	2.6249	2.6987	2.7718	2.8440	2.9153	2.9858	3.0553	3.1239	3.1916	40
42	.7673	.8487	.9295	.0097	.0893	.1683	.2466	.3242	.4011	.4773	.5527	.6273	.7012	.7742	.8464	.9177	.9881	.0576	.1262	.1938	42
44	.7701	.8514	.9322	.0124	.0920	.1709	.2492	.3268	.4036	.4798	.5552	.6298	.7036	.7766	.8488	.9200	.9904	.0599	.1285	.1961	44
46	.7728	.8541	.9349	.0150	.0946	.1735	.2518	.3293	.4062	.4823	.5577	.6323	.7061	.7790	.8512	.9224	.9928	.0622	.1307	.1983	46
48	.7755	.8568	.9376	.0177	.0972	.1761	.2544	.3319	.4087	.4848	.5602	.6348	.7085	.7815	.8535	.9248	.9951	.0645	.1330	.2005	48
50	1.7782	1.8595	1.9402	2.0204	2.0999	2.1787	2.2570	2.3345	2.4113	2.4874	2.5627	2.6372	2.7110	2.7839	2.8559	2.9271	2.9974	3.0668	3.1353	3.2028	50
52	.7809	.8622	.9429	.0230	.1025	.1814	.2596	.3371	.4138	.4899	.5652	.6397	.7134	.7863	.8583	.9295	.9997	.0691	.1375	.2050	52
54	.7837	.8649	.9456	.0257	.1052	.1840	.2621	.3396	.4164	.4924	.5677	.6422	.7158	.7887	.8607	.9318	3.0021	.0714	.1398	.2072	54
56	.7864	.8676	.9483	.0283	.1078	.1866	.2647	.3422	.4189	.4949	.5702	.6446	.7183	.7911	.8631	.9342	.0044	.0737	.1421	.2095	56
58	.7891	.8703	.9510	.0310	.1104	.1892	.2673	.3448	.4215	.4975	.5727	.6471	.7207	.7935	.8655	.9365	.0067	.0760	.1443	.2117	58
60	1.7918	1.8730	1.9536	2.0337	2.1131	2.1918	2.2699	2.3473	2.4240	2.5000	2.5752	2.6496	2.7232	2.7959	2.8679	2.9389	3.0091	3.0783	3.1466	3.2139	60

Courtesy Brown & Sharpe Mfg. Co.

Table 16 continued

TABLE 16: SINE BAR CONSTANTS (5 in. BAR)
(Multiply Constants by Two for a 10 in. Sine Bar)

Min.	40°	41°	42°	43°	44°	45°	46°	47°	48°	49°	50°	51°	52°	53°	54°	55°	56°	57°	58°	59°	Min.
0	3.2139	3.2803	3.3456	3.4100	3.4733	3.5355	3.5967	3.6567	3.7157	3.7735	3.8302	3.8857	3.9400	3.9932	4.0451	4.0957	4.1452	4.1933	4.2402	4.2858	0
2	.2161	.2825	.3478	.4121	.4754	.5376	.5987	.6587	.7176	.7754	.8321	.8875	.9418	.9949	.0468	.0974	.1468	.1949	.2418	.2873	2
4	.2184	.2847	.3499	.4142	.4774	.5396	.6007	.6607	.7196	.7773	.8339	.8894	.9436	.9967	.0485	.0991	.1484	.1965	.2433	.2888	4
6	.2206	.2869	.3521	.4163	.4795	.5417	.6027	.6627	.7215	.7792	.8358	.8912	.9454	.9984	.0502	.1007	.1500	.1981	.2448	.2903	6
8	.2228	.2890	.3543	.4185	.4816	.5437	.6047	.6647	.7235	.7811	.8377	.8930	.9472	4.0001	.0519	.1024	.1517	.1997	.2464	.2918	8
10	3.2250	3.2912	3.3564	3.4206	3.4837	3.5458	3.6068	3.6666	3.7254	3.7830	3.8395	3.8948	3.9490	4.0019	4.0536	4.1041	4.1533	4.2012	4.2479	4.2933	10
12	.2273	.2934	.3586	.4227	.4858	.5478	.6088	.6686	.7274	.7850	.8414	.8967	.9508	.0036	.0553	.1057	.1549	.2028	.2494	.2948	12
14	.2295	.2956	.3607	.4248	.4879	.5499	.6108	.6706	.7293	.7869	.8433	.8985	.9525	.0054	.0570	.1074	.1565	.2044	.2510	.2963	14
16	.2317	.2978	.3629	.4269	.4900	.5519	.6128	.6726	.7312	.7887	.8451	.9003	.9543	.0071	.0587	.1090	.1581	.2060	.2525	.2978	16
18	.2339	.3000	.3650	.4291	.4921	.5540	.6148	.6745	.7332	.7906	.8470	.9021	.9561	.0089	.0604	.1107	.1597	.2075	.2540	.2992	18
20	3.2361	3.3022	3.3672	3.4312	3.4941	3.5560	3.6168	3.6765	3.7351	3.7925	3.8488	3.9039	3.9579	4.0106	4.0621	4.1124	4.1614	4.2091	4.2556	4.3007	20
22	.2384	.3044	.3693	.4333	.4962	.5581	.6188	.6785	.7370	.7944	.8507	.9058	.9596	.0123	.0638	.1140	.1630	.2107	.2571	.3022	22
24	.2406	.3065	.3715	.4354	.4983	.5601	.6208	.6805	.7390	.7963	.8525	.9076	.9614	.0141	.0655	.1157	.1646	.2122	.2586	.3037	24
26	.2428	.3087	.3736	.4375	.5004	.5621	.6228	.6824	.7409	.7982	.8544	.9094	.9632	.0158	.0672	.1173	.1662	.2138	.2601	.3052	26
28	.2450	.3109	.3758	.4396	.5024	.5642	.6248	.6844	.7428	.8001	.8562	.9112	.9650	.0175	.0689	.1190	.1678	.2154	.2617	.3066	28
30	3.2472	3.3131	3.3779	3.4417	3.5045	3.5662	3.6268	3.6864	3.7448	3.8020	3.8581	3.9130	3.9667	4.0193	4.0706	4.1206	4.1694	4.2169	4.2632	4.3081	30
32	.2494	.3153	.3801	.4439	.5066	.5683	.6288	.6883	.7467	.8039	.8599	.9148	.9685	.0210	.0722	.1223	.1710	.2185	.2647	.3096	32
34	.2516	.3174	.3822	.4460	.5087	.5703	.6308	.6903	.7486	.8058	.8618	.9166	.9703	.0227	.0739	.1239	.1726	.2201	.2662	.3111	34
36	.2538	.3196	.3844	.4481	.5107	.5723	.6328	.6923	.7505	.8077	.8636	.9184	.9720	.0244	.0756	.1255	.1742	.2216	.2677	.3125	36
38	.2561	.3218	.3865	.4502	.5128	.5744	.6348	.6942	.7525	.8096	.8655	.9202	.9738	.0262	.0773	.1272	.1758	.2232	.2692	.3140	38
40	3.2583	3.3240	3.3886	3.4523	3.5149	3.5764	3.6368	3.6962	3.7544	3.8114	3.8673	3.9221	3.9756	4.0279	4.0790	4.1288	4.1774	4.2247	4.2708	4.3155	40
42	.2605	.3261	.3908	.4544	.5169	.5784	.6388	.6981	.7563	.8133	.8692	.9239	.9773	.0296	.0807	.1305	.1790	.2263	.2723	.3170	42
44	.2627	.3283	.3929	.4565	.5190	.5805	.6408	.7001	.7582	.8152	.8710	.9257	.9791	.0313	.0823	.1321	.1806	.2278	.2738	.3184	44
46	.2649	.3305	.3950	.4586	.5211	.5825	.6428	.7020	.7601	.8171	.8729	.9275	.9809	.0331	.0840	.1337	.1822	.2294	.2753	.3199	46
48	.2671	.3326	.3972	.4607	.5231	.5845	.6448	.7040	.7620	.8190	.8747	.9293	.9826	.0348	.0857	.1354	.1838	.2309	.2768	.3213	48
50	3.2693	3.3348	3.3993	3.4628	3.5252	3.5866	3.6468	3.7060	3.7640	3.8208	3.8765	3.9311	3.9844	4.0365	4.0874	4.1370	4.1854	4.2325	4.2783	4.3228	50
52	.2715	.3370	.4014	.4649	.5273	.5886	.6488	.7079	.7659	.8227	.8784	.9329	.9861	.0382	.0891	.1386	.1870	.2340	.2798	.3243	52
54	.2737	.3391	.4036	.4670	.5293	.5906	.6508	.7099	.7678	.8246	.8802	.9347	.9879	.0399	.0907	.1403	.1886	.2356	.2813	.3257	54
56	.2759	.3413	.4057	.4691	.5314	.5926	.6528	.7118	.7697	.8265	.8820	.9364	.9896	.0416	.0924	.1419	.1902	.2371	.2828	.3272	56
58	.2781	.3435	.4078	.4712	.5335	.5947	.6548	.7138	.7716	.8283	.8839	.9382	.9914	.0433	.0941	.1435	.1917	.2387	.2843	.3286	58
60	3.2803	3.3456	3.4100	3.4733	3.5355	3.5967	3.6567	3.7157	3.7735	3.8302	3.8857	3.9400	3.9932	4.0451	4.0957	4.1452	4.1933	4.2402	4.2858	4.3301	60

Courtesy Brown & Sharpe Mfg. Co.

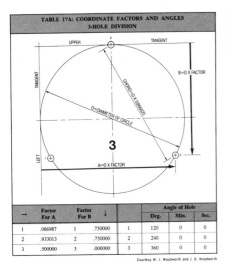

TABLE 17A: COORDINATE FACTORS AND ANGLES 3-HOLE DIVISION

→	Factor For A	↓	Factor For B		Deg.	Min.	Sec.
1	.066987	1	.750000	1	120	0	0
2	.933013	2	.750000	2	240	0	0
3	.500000	3	.000000	3	360	0	0

Courtesy W. J. Woodworth and J. D. Woodworth

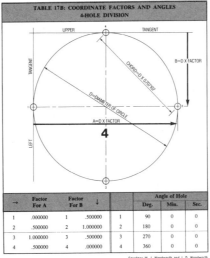

TABLE 17B: COORDINATE FACTORS AND ANGLES 4-HOLE DIVISION

→	Factor For A	↓	Factor For B		Deg.	Min.	Sec.
1	.000000	1	.500000	1	90	0	0
2	.500000	2	1.000000	2	180	0	0
3	1.000000	3	.500000	3	270	0	0
4	.500000	4	.000000	4	360	0	0

Courtesy W. J. Woodworth and J. D. Woodworth

TABLE 17C: COORDINATE FACTORS AND ANGLES 5-HOLE DIVISION

→	Factor For A	↓	Factor For B		Deg.	Min.	Sec.
1	.024472	1	.345492	1	72	0	0
2	.206107	2	.904508	2	144	0	0
3	.793893	3	.904508	3	216	0	0
4	.975528	4	.345492	4	288	0	0
5	.500000	5	.000000	5	360	0	0

Courtesy W. J. Woodworth and J. D. Woodworth

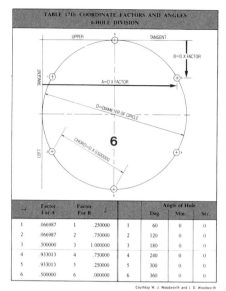

→	Factor For A		Factor For B	↓	Deg.	Min.	Sec.
1	.066987	1	.250000	1	60	0	0
2	.066987	2	.750000	2	120	0	0
3	.500000	3	1.000000	3	180	0	0
4	.933013	4	.750000	4	240	0	0
5	.933013	5	.250000	5	300	0	0
6	.500000	6	.000000	6	360	0	0

Courtesy W. J. Woodworth and J. D. Woodworth

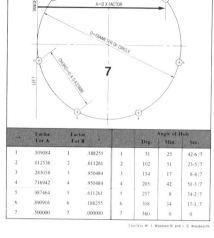

→	Factor For A		Factor For B	↓	Deg.	Min.	Sec.
1	.109084	1	.188255	1	51	25	42-6/7
2	.012536	2	.611261	2	102	51	23-5/7
3	.283058	3	.950484	3	154	17	8-4/7
4	.716942	4	.950484	4	205	42	51-3/7
5	.987464	5	.611261	5	257	8	34-2/7
6	.890916	6	.188255	6	308	34	17-1/7
7	.500000	7	.000000	7	360	0	0

Courtesy W. J. Woodworth and J. D. Woodworth

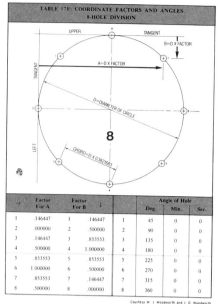

→	Factor For A		Factor For B	↓	Deg.	Min.	Sec.
1	.146447	1	.146447	1	45	0	0
2	.000000	2	.500000	2	90	0	0
3	.146447	3	.853553	3	135	0	0
4	.500000	4	1.000000	4	180	0	0
5	.853553	5	.853553	5	225	0	0
6	1.000000	6	.500000	6	270	0	0
7	.853553	7	.146447	7	315	0	0
8	.500000	8	.000000	8	360	0	0

Courtesy W. J. Woodworth and J. D. Woodworth

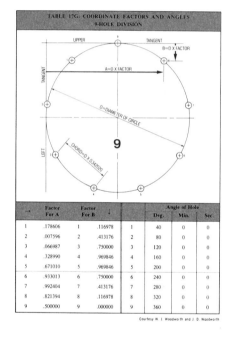

→	Factor For A		Factor For B	↓	Deg.	Min.	Sec.
1	.178606	1	.116978	1	40	0	0
2	.007596	2	.413176	2	80	0	0
3	.066987	3	.750000	3	120	0	0
4	.328990	4	.969846	4	160	0	0
5	.671010	5	.969846	5	200	0	0
6	.933013	6	.750000	6	240	0	0
7	.992404	7	.413176	7	280	0	0
8	.821394	8	.116978	8	320	0	0
9	.500000	9	.000000	9	360	0	0

Courtesy W. J. Woodworth and J. D. Woodworth

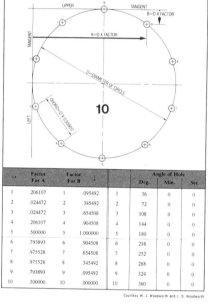

→	Factor For A		Factor For B	↓	Deg.	Min.	Sec.
1	.206107	1	.095492	1	36	0	0
2	.024472	2	.345492	2	72	0	0
3	.024472	3	.654508	3	108	0	0
4	.206107	4	.904508	4	144	0	0
5	.500000	5	1.000000	5	180	0	0
6	.793893	6	.904508	6	216	0	0
7	.975528	7	.654508	7	252	0	0
8	.975528	8	.345492	8	288	0	0
9	.793893	9	.095492	9	324	0	0
10	.500000	10	.000000	10	360	0	0

Courtesy W. J. Woodworth and J. D. Woodworth

→	Factor For A		Factor For B	↓	Deg.	Min.	Sec.
1	.229680	1	.079373	1	32	43	38-2/11
2	.045184	2	.292293	2	65	27	16-4/11
3	.005089	3	.571157	3	98	10	54-6/11
4	.122125	4	.827430	4	130	54	32-8/11
5	.359134	5	.979746	5	163	38	10-10/11
6	.640866	6	.979746	6	196	21	49-1/11
7	.877875	7	.827430	7	229	5	27-3/11
8	.994911	8	.571157	8	261	49	5-5/11
9	.954816	9	.292293	9	294	32	43-7/11
10	.770320	10	.079373	10	327	16	21-9/11
11	.500000	11	.000000	11	360	0	0

Courtesy W. J. Woodworth and J. D. Woodworth

NATURAL TRIGONOMETRIC FUNCTIONS

0° 89°

′	sin	tan	cot	cos	sec	cosec	′
0	.00000	.00000	Infinite	1.0000	1.0000	Infinite	60
1	.00029	.00029	3437.7	1.0000	1.0000	3437.7	59
2	.00058	.00058	1718.9	1.0000	1.0000	1718.9	58
3	.00087	.00087	1145.9	1.0000	1.0000	1145.9	57
4	.00116	.00116	859.44	1.0000	1.0000	859.44	56
5	.00145	.00145	687.55	1.0000	.0000	687.55	55
6	.00174	.00174	572.96	.99999	.0000	572.96	54
7	.00204	.00204	491.11	.99999	.0000	491.11	53
8	.00233	.00233	429.72	.99999	.0000	429.72	52
9	.00262	.00262	381.97	.99999	.0000	381.97	51
10	.00291	.00291	343.77	.99999	.0000	343.77	50
11	.00320	.00320	312.52	.99999	.0000	312.52	49
12	.00349	.00349	286.48	.99999	.0000	286.48	48
13	.00378	.00378	264.44	.99999	.0000	264.44	47
14	.00407	.00407	245.55	.99999	.0000	245.55	46
15	.00436	.00436	229.18	.99999	.0000	229.18	45
16	.00465	.00465	214.86	.99999	.0000	214.86	44
17	.00494	.00494	202.22	.99999	.0000	202.22	43
18	.00524	.00524	190.98	.99999	.0000	190.98	42
19	.00553	.00553	180.93	.99998	.0000	180.93	41
20	.00582	.00582	171.88	.99998	.0000	171.88	40
21	.00611	.00611	163.70	.99998	.0000	163.70	39
22	.00640	.00640	156.26	.99998	.0000	156.26	38
23	.00669	.00669	149.46	.99998	.0000	149.47	37
24	.00698	.00698	143.24	.99997	.0000	143.24	36
25	.00727	.00727	137.51	.99997	.0000	137.51	35
26	.00756	.00756	132.22	.99997	.0000	132.22	34
27	.00785	.00785	127.32	.99997	.0000	127.32	33
28	.00814	.00814	122.77	.99996	.0000	122.78	32
29	.00843	.00844	118.54	.99996	.0000	118.54	31
30	.00873	.00873	114.59	.99996	.0000	114.59	30
31	.00902	.00902	110.89	.99996	.0000	110.90	29
32	.00931	.00931	107.43	.99995	.0000	107.43	28
33	.00960	.00960	104.17	.99995	.0000	104.11	27
34	.00989	.00989	101.11	.99995	.0000	101.11	26
35	.01018	.01018	98.218	.99995	.0001	98.223	25
36	.01047	.01047	95.489	.99994	.0001	95.495	24
37	.01076	.01076	92.908	.99994	.0001	92.914	23
38	.01105	.01105	90.463	.99994	.0001	90.469	22
39	.01134	.01134	88.143	.99993	.0001	88.149	21
40	.01163	.01164	85.940	.99993	.0001	85.946	20
41	.01193	.01193	83.843	.99993	.0001	83.849	19
42	.01222	.01222	81.847	.99992	.0001	81.853	18
43	.01251	.01251	79.943	.99992	.0001	79.950	17
44	.01280	.01280	78.126	.99992	.0001	78.133	16
45	.01309	.01309	76.390	.99991	.0001	76.396	15
46	.01338	.01338	74.729	.99991	.0001	74.736	14
47	.01367	.01367	73.139	.99991	.0001	73.146	13
48	.01396	.01396	71.615	.99990	.0001	71.622	12
49	.01425	.01425	70.153	.99990	.0001	70.160	11
50	.01454	.01454	68.750	.99989	.0001	68.757	10
51	.01483	.01484	67.402	.99989	.0001	67.409	9
52	.01512	.01513	66.105	.99988	.0001	66.113	8
53	.01542	.01542	64.858	.99988	.0001	64.866	7
54	.01571	.01571	63.657	.99988	.0001	63.664	6
55	.01600	.01600	62.499	.99987	.0001	62.507	5
56	.01629	.01629	61.383	.99987	.0001	61.391	4
57	.01658	.01658	60.306	.99986	.0001	60.314	3
58	.01687	.01687	59.266	.99986	.0001	59.274	2
59	.01716	.01716	58.261	.99985	.0001	58.270	1
60	.01745	.01745	57.290	.99985	.0001	57.299	0

1° (bottom: 88°)

′	sin	tan	cot	cos	sec	cosec	′
0	.01745	.01745	57.290	.99985	.0002	57.299	60
1	.01774	.01775	56.350	.99984	.0002	56.359	59
2	.01803	.01804	55.441	.99984	.0002	55.450	58
3	.01832	.01833	54.561	.99983	.0002	54.570	57
4	.01861	.01862	53.708	.99983	.0002	53.718	56
5	.01891	.01891	52.882	.99982	.0002	52.891	55
6	.01920	.01920	52.081	.99981	.0002	52.090	54
7	.01949	.01949	51.303	.99981	.0002	51.313	53
8	.01978	.01978	50.548	.99980	.0002	50.558	52
9	.02007	.02007	49.816	.99980	.0002	49.826	51
10	.02036	.02036	49.104	.99979	.0002	49.114	50
11	.02065	.02066	48.412	.99979	.0002	48.422	49
12	.02094	.02095	47.739	.99978	.0002	47.750	48
13	.02123	.02124	47.085	.99977	.0002	47.096	47
14	.02152	.02153	46.449	.99977	.0002	46.460	46
15	.02181	.02182	45.829	.99976	.0002	45.840	45
16	.02210	.02211	45.226	.99975	.0002	45.237	44
17	.02240	.02240	44.638	.99974	.0002	44.650	43
18	.02269	.02269	44.066	.99974	.0002	44.077	42
19	.02298	.02298	43.508	.99973	.0003	43.520	41
20	.02326	.02327	42.964	.99973	.0003	42.976	40
21	.02356	.02357	42.433	.99972	.0003	42.445	39
22	.02385	.02386	41.916	.99971	.0003	41.928	38
23	.02414	.02415	41.410	.99971	.0003	41.423	37
24	.02443	.02444	40.917	.99970	.0003	40.930	36
25	.02472	.02473	40.436	.99969	.0003	40.448	35
26	.02501	.02502	39.965	.99969	.0003	39.978	34
27	.02530	.02531	39.506	.99968	.0003	39.518	33
28	.02560	.02560	39.057	.99967	.0003	39.069	32
29	.02589	.02589	38.618	.99966	.0003	38.631	31
30	.02618	.02618	38.188	.99966	.0003	38.201	30
31	.02647	.02648	37.769	.99965	.0003	37.782	29
32	.02676	.02677	37.358	.99964	.0004	37.371	28
33	.02705	.02706	36.956	.99963	.0004	36.969	27
34	.02734	.02735	36.563	.99963	.0004	36.576	26
35	.02763	.02764	36.177	.99962	.0004	36.191	25
36	.02792	.02793	35.800	.99961	.0004	35.814	24
37	.02821	.02822	35.431	.99960	.0004	35.445	23
38	.02850	.02851	35.069	.99959	.0004	35.084	22
39	.02879	.02880	34.715	.99958	.0004	34.729	21
40	.02908	.02910	34.368	.99958	.0004	34.382	20
41	.02937	.02939	34.027	.99957	.0004	34.042	19
42	.02967	.02968	33.693	.99956	.0004	33.708	18
43	.02996	.02997	33.366	.99955	.0004	33.381	17
44	.03025	.03026	33.045	.99954	.0004	33.060	16
45	.03054	.03055	32.730	.99953	.0005	32.745	15
46	.03083	.03084	32.421	.99952	.0005	32.437	14
47	.03112	.03113	32.118	.99952	.0005	32.134	13
48	.03141	.03143	31.820	.99951	.0005	31.836	12
49	.03170	.03172	31.528	.99950	.0005	31.544	11
50	.03199	.03201	31.241	.99949	.0005	31.257	10
51	.03228	.03230	30.960	.99948	.0005	30.976	9
52	.03257	.03259	30.683	.99947	.0005	30.699	8
53	.03286	.03288	30.411	.99946	.0005	30.428	7
54	.03315	.03317	30.145	.99945	.0005	30.161	6
55	.03344	.03346	29.882	.99944	.0006	29.899	5
56	.03374	.03375	29.624	.99942	.0006	29.641	4
57	.03403	.03405	29.371	.99941	.0006	29.388	3
58	.03432	.03434	29.122	.99940	.0006	29.139	2
59	.03461	.03463	28.877	.99939	.0006	28.894	1
60	.03490	.03492	28.636	.99939	.0006	28.654	0

2° (bottom: 87°)

′	sin	tan	cot	cos	sec	cosec	′
0	.03490	.03492	28.636	.99939	.0006	28.654	60
1	.03519	.03521	28.399	.99938	.0006	28.417	59
2	.03548	.03550	28.166	.99937	.0006	28.184	58
3	.03577	.03579	27.937	.99936	.0006	27.955	57
4	.03606	.03608	27.712	.99935	.0007	27.730	56
5	.03635	.03638	27.490	.99934	.0007	27.508	55
6	.03664	.03667	27.271	.99933	.0007	27.290	54
7	.03693	.03696	27.056	.99932	.0007	27.075	53
8	.03722	.03725	26.845	.99931	.0007	26.864	52
9	.03751	.03754	26.637	.99930	.0007	26.655	51
10	.03781	.03783	26.432	.99928	.0007	26.450	50
11	.03810	.03812	26.230	.99927	.0007	26.249	49
12	.03839	.03842	26.031	.99926	.0007	26.050	48
13	.03868	.03871	25.835	.99925	.0007	25.854	47
14	.03897	.03900	25.642	.99924	.0008	25.661	46
15	.03926	.03929	25.452	.99923	.0008	25.471	45
16	.03955	.03958	25.264	.99922	.0008	25.284	44
17	.03984	.03987	25.080	.99921	.0008	25.100	43
18	.04013	.04016	24.898	.99919	.0008	24.918	42
19	.04042	.04045	24.718	.99918	.0008	24.739	41
20	.04071	.04075	24.542	.99917	.0008	24.562	40
21	.04100	.04104	24.367	.99916	.0008	24.388	39
22	.04129	.04133	24.196	.99915	.0008	24.216	38
23	.04158	.04162	24.026	.99913	.0009	24.047	37
24	.04187	.04191	23.859	.99912	.0009	23.880	36
25	.04217	.04220	23.694	.99911	.0009	23.716	35
26	.04246	.04249	23.532	.99910	.0009	23.553	34
27	.04275	.04279	23.372	.99908	.0009	23.393	33
28	.04304	.04308	23.214	.99907	.0009	23.235	32
29	.04333	.04337	23.058	.99906	.0009	23.079	31
30	.04362	.04366	22.904	.99905	.0009	22.925	30
31	.04391	.04395	22.752	.99904	.0010	22.774	29
32	.04420	.04424	22.602	.99902	.0010	22.624	28
33	.04449	.04453	22.454	.99901	.0010	22.476	27
34	.04478	.04483	22.308	.99900	.0010	22.330	26
35	.04507	.04512	22.164	.99898	.0010	22.186	25
36	.04536	.04541	22.022	.99897	.0010	22.044	24
37	.04565	.04570	21.881	.99896	.0010	21.904	23
38	.04594	.04599	21.742	.99894	.0011	21.765	22
39	.04623	.04628	21.606	.99893	.0011	21.629	21
40	.04652	.04657	21.470	.99892	.0011	21.494	20
41	.04681	.04687	21.337	.99890	.0011	21.360	19
42	.04711	.04716	21.205	.99889	.0011	21.228	18
43	.04740	.04745	21.075	.99888	.0011	21.098	17
44	.04769	.04774	20.946	.99886	.0011	20.970	16
45	.04798	.04803	20.819	.99885	.0012	20.843	15
46	.04827	.04832	20.693	.99883	.0012	20.717	14
47	.04856	.04862	20.569	.99882	.0012	20.593	13
48	.04885	.04891	20.446	.99881	.0012	20.471	12
49	.04914	.04920	20.325	.99879	.0012	20.350	11
50	.04943	.04949	20.205	.99878	.0012	20.230	10
51	.04972	.04978	20.087	.99876	.0012	20.112	9
52	.05001	.05007	19.970	.99875	.0012	19.995	8
53	.05030	.05037	19.854	.99873	.0013	19.880	7
54	.05059	.05066	19.740	.99872	.0013	19.766	6
55	.05088	.05095	19.627	.99870	.0013	19.653	5
56	.05117	.05124	19.516	.99869	.0013	19.541	4
57	.05146	.05153	19.405	.99867	.0013	19.431	3
58	.05175	.05182	19.296	.99866	.0013	19.322	2
59	.05204	.05212	19.188	.99864	.0013	19.214	1
60	.05234	.05241	19.081	.99863	.0014	19.107	0

3° (bottom: 86°)

′	sin	tan	cot	cos	sec	cosec	′
0	.05234	.05241	19.081	.99863	.0014	19.107	60
1	.05263	.05270	18.975	.99861	.0014	19.002	59
2	.05292	.05299	18.871	.99860	.0014	18.897	58
3	.05321	.05328	18.768	.99858	.0014	18.794	57
4	.05350	.05357	18.665	.99857	.0014	18.692	56
5	.05379	.05387	18.564	.99855	.0014	18.591	55
6	.05408	.05416	18.464	.99854	.0015	18.491	54
7	.05437	.05445	18.365	.99852	.0015	18.393	53
8	.05466	.05474	18.268	.99850	.0015	18.295	52
9	.05495	.05503	18.171	.99849	.0015	18.198	51
10	.05524	.05532	18.075	.99847	.0015	18.103	50
11	.05553	.05562	17.980	.99846	.0016	18.008	49
12	.05582	.05591	17.886	.99844	.0016	17.914	48
13	.05611	.05620	17.793	.99842	.0016	17.821	47
14	.05640	.05649	17.701	.99841	.0016	17.730	46
15	.05669	.05678	17.610	.99839	.0016	17.639	45
16	.05698	.05707	17.520	.99838	.0016	17.549	44
17	.05727	.05737	17.431	.99836	.0017	17.460	43
18	.05756	.05766	17.343	.99834	.0017	17.372	42
19	.05785	.05795	17.256	.99832	.0017	17.285	41
20	.05814	.05824	17.169	.99831	.0017	17.198	40
21	.05843	.05853	17.084	.99829	.0017	17.113	39
22	.05872	.05883	16.999	.99827	.0017	17.028	38
23	.05902	.05912	16.915	.99826	.0018	16.944	37
24	.05931	.05941	16.832	.99824	.0018	16.861	36
25	.05960	.05970	16.750	.99822	.0018	16.779	35
26	.05989	.05999	16.668	.99820	.0018	16.698	34
27	.06018	.06029	16.587	.99819	.0018	16.617	33
28	.06047	.06058	16.507	.99817	.0018	16.538	32
29	.06076	.06087	16.428	.99815	.0019	16.459	31
30	.06105	.06116	16.350	.99813	.0019	16.380	30
31	.06134	.06145	16.272	.99812	.0019	16.303	29
32	.06163	.06175	16.195	.99810	.0019	16.226	28
33	.06192	.06204	16.119	.99808	.0019	16.150	27
34	.06221	.06233	16.043	.99806	.0019	16.075	26
35	.06250	.06262	15.969	.99804	.0020	16.000	25
36	.06279	.06291	15.894	.99803	.0020	15.926	24
37	.06308	.06321	15.821	.99801	.0020	15.853	23
38	.06337	.06350	15.748	.99799	.0020	15.780	22
39	.06366	.06379	15.676	.99797	.0020	15.708	21
40	.06395	.06408	15.605	.99795	.0020	15.637	20
41	.06424	.06437	15.534	.99793	.0021	15.566	19
42	.06453	.06467	15.464	.99792	.0021	15.496	18
43	.06482	.06496	15.394	.99790	.0021	15.427	17
44	.06511	.06525	15.325	.99788	.0021	15.358	16
45	.06540	.06554	15.257	.99786	.0021	15.290	15
46	.06569	.06583	15.189	.99784	.0022	15.222	14
47	.06598	.06613	15.122	.99782	.0022	15.155	13
48	.06627	.06642	15.056	.99780	.0022	15.089	12
49	.06656	.06671	14.990	.99778	.0022	15.023	11
50	.06685	.06700	14.924	.99776	.0022	14.958	10
51	.06714	.06730	14.860	.99774	.0023	14.893	9
52	.06743	.06759	14.795	.99772	.0023	14.829	8
53	.06773	.06788	14.732	.99770	.0023	14.765	7
54	.06802	.06817	14.668	.99768	.0023	14.702	6
55	.06830	.06846	14.606	.99766	.0023	14.640	5
56	.06859	.06876	14.543	.99764	.0024	14.578	4
57	.06888	.06905	14.482	.99762	.0024	14.517	3
58	.06918	.06934	14.421	.99760	.0024	14.456	2
59	.06947	.06963	14.361	.99758	.0024	14.395	1
60	.06976	.06993	14.301	.99756	.0024	14.335	0

Courtesy Bethlehem Steel Corporation

NATURAL TRIGONOMETRIC FUNCTIONS

7°

'	sin	cos	tan	cot	sec	cosec	'
	.12187	.99255	.12278	8.1443	1.0075	8.2055	60
	.12216	.99251	.12308	.1248	.0075	.1861	59
	.12245	.99247	.12337	.1053	.0076	.1668	58
	.12273	.99244	.12367	.0800	.0076	.1476	57
	.12302	.99240	.12396	.067	.0076	.1285	56
	.12331	.99237	.12426	8.0476	.0077	8.1094	55
	.12360	.99233	.12456	.0285	.0077	.0905	54
	.12389	.99229	.12485	.0095	.0078	.0717	53
	.12418	.99226	.12515	7.9906	.0078	.0529	52
	.12447	.99222	.12544	.9717	.0078	.0342	51
	.12476	.99219	.12574	7.9530	.0079	8.0156	50
	.12504	.99215	.12603	.9344	.0079	7.9971	49
	.12533	.99211	.12633	.9158	.0079	.9787	48
	.12562	.99208	.12662	.8973	.0080	.9604	47
	.12591	.99204	.12692	.8789	.0080	.9421	46
	.12620	.99200	.12722	7.8606	.0080	7.9240	45
	.12649	.99197	.12751	.8424	.0081	.9059	44
	.12678	.99193	.12781	.8243	.0081	.8879	43
	.12706	.99189	.12810	.8062	.0082	.8700	42
	.12735	.99186	.12840	.7882	.0082	.8522	41
	.12764	.99182	.12869	7.7703	.0082	7.8344	40
	.12793	.99178	.12899	.7525	.0083	.8168	39
	.12822	.99174	.12928	.7348	.0083	.7992	38
	.12851	.99171	.12958	.7171	.0084	.7817	37
	.12880	.99167	.12988	.6996	.0084	.7642	36
	.12908	.99163	.13017	7.6821	.0084	7.7469	35
	.12937	.99160	.13047	.6646	.0085	.7296	34
	.12966	.99156	.13076	.6473	.0085	.7124	33
	.12995	.99152	.13106	.6300	.0085	.6953	32
	.13024	.99148	.13136	.6129	.0086	.6783	31
	.13053	.99144	.13165	7.5957	.0086	7.6613	30
	.13081	.99141	.13195	.5787	.0087	.6444	29
	.13110	.99137	.13224	.5617	.0087	.6276	28
	.13139	.99133	.13254	.5449	.0087	.6108	27
	.13168	.99130	.13284	.5280	.0088	.5942	26
	.13197	.99125	.13313	7.5113	.0088	7.5776	25
	.13226	.99121	.13343	.4946	.0089	.5611	24
	.13254	.99118	.13372	.4780	.0089	.5446	23
	.13283	.99114	.13402	.4615	.0089	.5282	22
	.13312	.99110	.13432	.4451	.0090	.5119	21
	.13341	.99106	.13461	7.4287	.0090	7.4957	20
	.13370	.99102	.13491	.4124	.0090	.4795	19
	.13399	.99098	.13520	.3961	.0091	.4634	18
	.13427	.99094	.13550	.3800	.0091	.4474	17
	.13456	.99091	.13580	.3639	.0092	.4315	16
	.13485	.99086	.13609	7.3479	.0092	7.4156	15
	.13514	.99083	.13639	.3319	.0092	.3998	14
	.13543	.99079	.13669	.3160	.0093	.3840	13
	.13571	.99075	.13698	.3002	.0093	.3683	12
	.13600	.99071	.13728	.2844	.0094	.3527	11
	.13629	.99067	.13757	7.2687	.0094	7.3372	10
	.13658	.99063	.13787	.2531	.0094	.3217	9
	.13687	.99059	.13817	.2375	.0095	.3063	8
	.13716	.99055	.13846	.2220	.0095	.2909	7
	.13744	.99051	.13876	.2066	.0096	.2757	6
	.13773	.99047	.13906	7.1912	.0096	7.2604	5
	.13802	.99043	.13935	.1759	.0097	.2453	4
	.13831	.99039	.13965	.1607	.0097	.2302	3
	.13860	.99035	.13995	.1455	.0097	.2152	2
	.13888	.99031	.14024	.1304	.0098	.2002	1
	.13917	.99027	.14054	7.1154	.0098	7.1853	0
'	cos	sin	cot	tan	cosec	sec	'

82°

6°

'	sin	cos	tan	cot	sec	cosec	'
	.10453	.99452	.10510	9.5144	1.0055	9.5668	60
	.10482	.99449	.10540	.4878	.0055	.5404	59
	.10511	.99446	.10569	.4614	.0056	.5141	58
	.10540	.99443	.10599	.4351	.0056	.4880	57
	.10568	.99440	.10628	.4090	.0056	.4620	56
	.10597	.99437	.10657	9.3831	.0057	9.4362	55
	.10626	.99434	.10687	.3572	.0057	.4105	54
	.10655	.99431	.10716	.3315	.0057	.3850	53
	.10684	.99428	.10746	.3060	.0057	.3596	52
	.10713	.99424	.10775	.2806	.0058	.3343	51
	.10742	.99421	.10805	9.2553	.0058	9.3092	50
	.10771	.99418	.10834	.2302	.0059	.2842	49
	.10800	.99415	.10863	.2051	.0059	.2593	48
	.10829	.99412	.10893	.1803	.0059	.2346	47
	.10858	.99409	.10922	.1555	.0059	.2100	46
	.10887	.99406	.10952	9.1309	.0060	9.1855	45
	.10916	.99402	.10981	.1064	.0060	.1612	44
	.10944	.99399	.11011	.0821	.0060	.1370	43
	.10973	.99396	.11040	.0579	.0061	.1129	42
	.11002	.99393	.11069	.0338	.0061	.0890	41
	.11031	.99390	.11099	9.0098	.0061	9.0651	40
	.11060	.99386	.11128	8.9860	.0062	.0414	39
	.11089	.99383	.11158	.9623	.0062	.0179	38
	.11118	.99380	.11187	.9387	.0062	8.9944	37
	.11147	.99377	.11217	.9152	.0063	.9711	36
	.11176	.99373	.11246	8.8918	.0063	8.9479	35
	.11205	.99370	.11276	.8686	.0063	.9248	34
	.11234	.99367	.11305	.8455	.0064	.9018	33
	.11262	.99364	.11335	.8225	.0064	.8790	32
	.11291	.99360	.11364	.7996	.0064	.8563	31
	.11320	.99357	.11393	8.7769	.0065	8.8337	30
	.11349	.99354	.11423	.7542	.0065	.8112	29
	.11378	.99351	.11452	.7317	.0065	.7888	28
	.11407	.99347	.11482	.7093	.0066	.7665	27
	.11436	.99344	.11511	.6870	.0066	.7444	26
	.11465	.99341	.11541	8.6648	.0066	8.7223	25
	.11494	.99337	.11570	.6427	.0067	.7004	24
	.11523	.99334	.11600	.6208	.0067	.6786	23
	.11551	.99330	.11629	.5989	.0067	.6569	22
	.11580	.99327	.11659	.5772	.0068	.6353	21
	.11609	.99324	.11688	8.5555	.0068	8.6138	20
	.11638	.99320	.11718	.5340	.0069	.5924	19
	.11667	.99316	.11747	.5126	.0069	.5711	18
	.11696	.99313	.11777	.4913	.0069	.5499	17
	.11725	.99310	.11806	.4701	.0069	.5289	16
	.11754	.99307	.11836	8.4489	.0070	8.5079	15
	.11783	.99303	.11865	.4279	.0070	.4871	14
	.11811	.99300	.11895	.4070	.0070	.4663	13
	.11840	.99296	.11924	.3862	.0071	.4457	12
	.11869	.99293	.11954	.3655	.0071	.4251	11
	.11898	.99290	.11983	8.3449	.0071	8.4046	10
	.11927	.99286	.12013	.3244	.0072	.3843	9
	.11956	.99283	.12042	.3040	.0072	.3640	8
	.11985	.99279	.12072	.2837	.0073	.3439	7
	.12014	.99276	.12101	.2635	.0073	.3238	6
	.12042	.99272	.12131	8.2434	.0073	8.3039	5
	.12071	.99269	.12160	.2234	.0074	.2840	4
	.12100	.99265	.12190	.2035	.0074	.2642	3
	.12129	.99262	.12219	.1837	.0074	.2446	2
	.12158	.99258	.12249	.1640	.0075	.2250	1
	.12187	.99255	.12278	8.1443	.0075	8.2055	0
'	cos	sin	cot	tan	cosec	sec	'

83°

5°

'	sin	cos	tan	cot	sec	cosec	'
	.08715	.99619	.08749	11.430	1.0038	11.474	60
	.08744	.99617	.08778	.392	.0038	.436	59
	.08773	.99614	.08807	.354	.0039	.398	58
	.08802	.99612	.08837	.316	.0039	.360	57
	.08831	.99609	.08866	.279	.0039	.323	56
	.08860	.99607	.08895	11.242	.0039	11.286	55
	.08889	.99604	.08925	.205	.0040	.249	54
	.08918	.99601	.08954	.168	.0040	.213	53
	.08947	.99599	.08983	.132	.0040	.176	52
	.08976	.99596	.09013	.095	.0040	.140	51
	.09005	.99594	.09042	11.059	.0041	11.104	50
	.09034	.99591	.09071	.024	.0041	.069	49
	.09063	.99588	.09101	10.988	.0041	.033	48
	.09092	.99586	.09130	.953	.0041	10.998	47
	.09121	.99583	.09159	.918	.0042	.963	46
	.09150	.99580	.09189	10.883	.0042	10.929	45
	.09179	.99578	.09218	.848	.0042	.894	44
	.09208	.99575	.09247	.814	.0043	.860	43
	.09237	.99572	.09277	.780	.0043	.826	42
	.09266	.99569	.09306	.746	.0043	.792	41
	.09295	.99567	.09335	10.712	.0043	10.758	40
	.09324	.99564	.09365	.678	.0044	.725	39
	.09353	.99562	.09394	.645	.0044	.692	38
	.09382	.99559	.09423	.612	.0044	.659	37
	.09411	.99556	.09453	.579	.0044	.626	36
	.09440	.99553	.09482	10.546	.0045	10.593	35
	.09469	.99551	.09511	.514	.0045	.561	34
	.09498	.99548	.09541	.481	.0045	.529	33
	.09527	.99545	.09570	.449	.0046	.497	32
	.09556	.99542	.09599	.417	.0046	.465	31
	.09584	.99540	.09629	10.385	.0046	10.433	30
	.09613	.99537	.09658	.354	.0046	.402	29
	.09642	.99534	.09688	.322	.0047	.371	28
	.09671	.99531	.09717	.291	.0047	.340	27
	.09700	.99528	.09746	.260	.0047	.309	26
	.09729	.99525	.09776	10.229	.0048	10.278	25
	.09758	.99523	.09805	.199	.0048	.248	24
	.09787	.99520	.09834	.168	.0048	.217	23
	.09816	.99517	.09864	.138	.0048	.187	22
	.09845	.99514	.09893	.108	.0049	.157	21
	.09874	.99511	.09922	10.078	.0049	10.127	20
	.09903	.99508	.09952	.048	.0049	.098	19
	.09932	.99505	.09981	.019	.0050	.068	18
	.09961	.99503	.10011	9.9893	.0050	.039	17
	.09990	.99500	.10040	.9601	.0050	.010	16
	.10019	.99497	.10069	9.9310	.0050	9.9812	15
	.10048	.99494	.10099	.9021	.0051	.9525	14
	.10077	.99491	.10128	.8734	.0051	.9239	13
	.10106	.99488	.10158	.8448	.0051	.8955	12
	.10134	.99485	.10187	.8164	.0052	.8672	11
	.10163	.99482	.10216	9.7882	.0052	9.8391	10
	.10192	.99479	.10246	.7601	.0052	.8112	9
	.10221	.99476	.10275	.7322	.0053	.7834	8
	.10250	.99473	.10305	.7044	.0053	.7558	7
	.10279	.99470	.10334	.6768	.0053	.7283	6
	.10308	.99467	.10363	9.6493	.0053	9.7010	5
	.10337	.99464	.10393	.6220	.0054	.6739	4
	.10366	.99461	.10422	.5949	.0054	.6469	3
	.10395	.99458	.10452	.5679	.0054	.6200	2
	.10424	.99455	.10481	.5411	.0055	.5933	1
	.10453	.99452	.10510	9.5144	.0055	9.5668	0
'	cos	sin	cot	tan	cosec	sec	'

84°

4°

'	sin	cos	tan	cot	sec	cosec	'
0	.06976	.99756	.06993	14.301	1.0024	14.335	60
1	.07005	.99754	.07022	14.241	.0025	14.276	59
2	.07034	.99752	.07051	14.217	.0025	14.217	58
3	.07063	.99750	.07080	14.123	.0025	14.159	57
4	.07092	.99748	.07110	14.065	.0025	14.101	56
5	.07121	.99746	.07139	14.008	.0025	14.043	55
6	.07150	.99744	.07168	13.951	.0026	13.986	54
7	.07179	.99742	.07197	13.894	.0026	13.930	53
8	.07208	.99740	.07226	13.838	.0026	13.874	52
9	.07237	.99738	.07256	13.782	.0026	13.818	51
10	.07266	.99736	.07285	13.727	.0026	13.763	50
11	.07295	.99733	.07314	13.672	.0027	13.708	49
12	.07324	.99731	.07343	13.617	.0027	13.654	48
13	.07353	.99729	.07373	13.563	.0027	13.600	47
14	.07382	.99727	.07402	13.510	.0027	13.547	46
15	.07411	.99725	.07431	13.457	.0027	13.494	45
16	.07440	.99723	.07460	13.404	.0028	13.441	44
17	.07469	.99721	.07490	13.351	.0028	13.337	43
18	.07498	.99718	.07519	13.299	.0028	13.337	42
19	.07527	.99716	.07548	13.248	.0028	13.286	41
20	.07556	.99714	.07577	13.197	.0029	13.235	40
21	.07585	.99712	.07607	13.146	.0029	13.184	39
22	.07614	.99710	.07636	13.096	.0029	13.134	38
23	.07643	.99707	.07665	13.046	.0029	13.084	37
24	.07672	.99705	.07694	12.996	.0029	12.996	36
25	.07701	.99703	.07724	12.947	.0030	12.985	35
26	.07730	.99701	.07753	12.898	.0030	12.937	34
27	.07759	.99698	.07782	12.850	.0030	12.888	33
28	.07788	.99696	.07812	12.801	.0030	12.840	32
29	.07817	.99694	.07841	12.754	.0031	12.793	31
30	.07846	.99692	.07870	12.706	.0031	12.745	30
31	.07875	.99689	.07899	12.659	.0031	12.698	29
32	.07904	.99687	.07929	12.612	.0031	12.652	28
33	.07933	.99685	.07958	12.566	.0032	12.606	27
34	.07962	.99682	.07987	12.520	.0032	12.560	26
35	.07991	.99680	.08016	12.474	.0032	12.514	25
36	.08020	.99678	.08046	12.429	.0032	12.469	24
37	.08049	.99675	.08075	12.384	.0033	12.424	23
38	.08078	.99673	.08104	12.339	.0033	12.379	22
39	.08107	.99671	.08134	12.295	.0033	12.335	21
40	.08136	.99668	.08163	12.250	.0033	12.291	20
41	.08165	.99666	.08192	12.207	.0034	12.248	19
42	.08194	.99664	.08221	12.163	.0034	12.204	18
43	.08223	.99661	.08251	12.120	.0034	12.161	17
44	.08252	.99659	.08280	12.077	.0034	12.118	16
45	.08281	.99656	.08309	11.826	.0036	11.868	15
46	.08310	.99654	.08339	11.785	.0036	11.828	14
47	.08339	.99652	.08368	11.745	.0036	11.787	13
48	.08368	.99649	.08397	11.704	.0036	11.747	12
49	.08397	.99647	.08426	11.664	.0037	11.707	11
50	.08426	.99644	.08456	11.625	.0036	11.668	10
51	.08455	.99642	.08485	11.585	.0036	11.628	9
52	.08484	.99639	.08514	11.546	.0036	11.580	8
53	.08513	.99637	.08544	11.507	.0036	11.550	7
54	.08542	.99634	.08573	11.468	.0037	11.512	6
55	.08571	.99632	.08602	11.625	.0037	11.668	5
56	.08600	.99629	.08632	11.585	.0037	11.628	4
57	.08629	.99627	.08661	11.346	.0037	11.580	3
58	.08658	.99624	.08690	11.507	.0038	11.550	2
59	.08687	.99622	.08719	11.468	.0038	11.512	1
60	.08715	.99619	.08749	11.430	.0038	11.474	0
'	cos	sin	cot	tan	cosec	sec	'

85°

NATURAL TRIGONOMETRIC FUNCTIONS

11° (complement **78°**)

′	sin	cos	tan	cot	sec	cosec	′
0	.19081	.98163	.19438	5.1445	1.0187	5.2408	60
1	.19109	.98157	.19468	5.1366	.0188	.2330	59
2	.19138	.98152	.19498	5.1286	.0188	.2252	58
3	.19166	.98146	.19529	5.1207	.0189	.2174	57
4	.19195	.98140	.19559	5.1128	.0189	.2097	56
5	.19224	.98135	.19589	5.1049	.0190	5.2019	55
6	.19252	.98129	.19619	5.0970	.0191	.1942	54
7	.19281	.98124	.19649	5.0892	.0191	.1865	53
8	.19309	.98118	.19680	5.0814	.0192	.1788	52
9	.19338	.98112	.19710	5.0736	.0192	.1712	51
10	.19366	.98107	.19740	5.0658	.0193	5.1636	50
11	.19395	.98101	.19770	5.0581	.0193	.1560	49
12	.19423	.98095	.19800	5.0504	.0194	.1484	48
13	.19452	.98090	.19831	5.0427	.0195	.1409	47
14	.19480	.98084	.19861	5.0350	.0195	.1333	46
15	.19509	.98078	.19891	5.0273	.0196	5.1258	45
16	.19537	.98073	.19921	5.0197	.0196	.1183	44
17	.19566	.98067	.19952	5.0127	.0197	.1109	43
18	.19595	.98061	.19982	5.0045	.0198	.1034	42
19	.19623	.98056	.20012	4.9969	.0198	.0960	41
20	.19652	.98050	.20042	4.9894	.0199	5.0886	40
21	.19680	.98044	.20073	4.9819	.0199	.0812	39
22	.19709	.98039	.20103	4.9744	.0200	.0739	38
23	.19737	.98033	.20133	4.9669	.0201	.0666	37
24	.19766	.98027	.20163	4.9594	.0201	.0593	36
25	.19794	.98021	.20194	4.9520	.0202	5.0520	35
26	.19823	.98016	.20224	4.9446	.0202	.0447	34
27	.19851	.98010	.20254	4.9372	.0203	.0375	33
28	.19880	.98004	.20285	4.9298	.0204	.0302	32
29	.19908	.97998	.20315	4.9225	.0204	.0230	31
30	.19937	.97992	.20345	4.9151	.0205	5.0158	30
31	.19965	.97987	.20375	4.9078	.0205	.0087	29
32	.19994	.97981	.20406	4.9006	.0206	5.0015	28
33	.20022	.97975	.20436	4.8933	.0207	4.9944	27
34	.20051	.97969	.20466	4.8860	.0207	.9873	26
35	.20079	.97963	.20497	4.8788	.0208	4.9802	25
36	.20108	.97952	.20527	4.8716	.0208	.9732	24
37	.20136	.97952	.20557	4.8644	.0209	.9661	23
38	.20165	.97946	.20588	4.8573	.0210	.9591	22
39	.20193	.97940	.20618	4.8501	.0210	.9521	21
40	.20222	.97934	.20648	4.8430	.0211	4.9452	20
41	.20250	.97928	.20679	4.8359	.0211	.9382	19
42	.20279	.97922	.20709	4.8288	.0212	.9313	18
43	.20307	.97916	.20739	4.8217	.0213	.9243	17
44	.20336	.97910	.20770	4.8147	.0213	.9175	16
45	.20364	.97904	.20800	4.8077	.0214	4.9106	15
46	.20393	.97899	.20830	4.8007	.0214	.9037	14
47	.20421	.97893	.20861	4.7937	.0215	.8969	13
48	.20450	.97887	.20891	4.7867	.0216	.8901	12
49	.20478	.97881	.20921	4.7798	.0216	.8833	11
50	.20506	.97875	.20952	4.7728	.0218	4.8765	10
51	.20535	.97869	.20982	4.7659	.0218	.8697	9
52	.20563	.97863	.21012	4.7591	.0219	.8630	8
53	.20592	.97857	.21043	4.7522	.0219	.8563	7
54	.20620	.97851	.21073	4.7453	.0220	.8496	6
55	.20649	.97845	.21104	4.7385	.0220	4.8429	5
56	.20677	.97839	.21134	4.7317	.0221	.8362	4
57	.20706	.97833	.21164	4.7249	.0221	.8290	3
58	.20734	.97827	.21195	4.7181	.0222	.8229	2
59	.20763	.97821	.21225	4.7114	.0223	.8163	1
60	.20791	.97815	.21256	4.7046	.0223	4.8097	0
′	cos	sin	cot	tan	cosec	sec	′

(78°)

10° (complement **79°**)

′	sin	cos	tan	cot	sec	cosec	′
0	.17365	.98481	.17633	5.6713	1.0154	5.7588	60
60	.19081	.98163	.19438	5.1445	.0187	5.2408	0
′	cos	sin	cot	tan	cosec	sec	′

(79°)

9° (complement **80°**)

′	sin	cos	tan	cot	sec	cosec	′
0	.15643	.98769	.15838	6.3137	1.0125	6.3924	60
60	.17365	.98481	.17633	5.7588	.0154	5.7588	0
′	cos	sin	cot	tan	cosec	sec	′

(80°)

8° (complement **81°**)

′	sin	cos	tan	cot	sec	cosec	′
0	.13917	.99027	.14054	7.1154	1.0098	7.1853	60
60	.15643	.98769	.15838	6.3137	.0125	6.3924	0
′	cos	sin	cot	tan	cosec	sec	′

(81°)

NATURAL TRIGONOMETRIC FUNCTIONS

12°

′	sin	cos	tan	cot	sec	cosec
0	.20791	.97815	.21256	4.7046	1.0223	4.8097
1	.20820	.97803	.21286	.6979	.0224	.8032
2	.20848	.97797	.21316	.6912	.0225	.7966
3	.20876	.97791	.21347	.6845	.0225	.7901
4	.20905	.97784	.21377	.6778	.0226	.7835
5	.20933	.97778	.21408	4.6712	.0226	4.7770
6	.20962	.97772	.21438	.6646	.0227	.7706
7	.20990	.97766	.21469	.6580	.0228	.7641
8	.21019	.97760	.21499	.6514	.0228	.7576
9	.21047	.97754	.21529	.6448	.0229	.7512
10	.21076	.97748	.21560	4.6382	.0230	4.7448
11	.21104	.97742	.21590	.6317	.0230	.7384
12	.21132	.97735	.21621	.6252	.0231	.7320
13	.21161	.97729	.21651	.6187	.0232	.7257
14	.21189	.97723	.21682	.6122	.0232	.7193
15	.21218	.97717	.21712	4.6057	.0233	4.7130
16	.21246	.97711	.21742	.5993	.0234	.7067
17	.21275	.97704	.21773	.5928	.0234	.7004
18	.21303	.97698	.21803	.5864	.0235	.6942
19	.21331	.97692	.21834	.5800	.0235	.6879
20	.21360	.97686	.21864	4.5736	.0236	4.6817
21	.21388	.97680	.21895	.5673	.0237	.6754
22	.21417	.97673	.21925	.5609	.0237	.6692
23	.21445	.97667	.21956	.5546	.0238	.6631
24	.21473	.97661	.21986	.5483	.0239	.6569
25	.21502	.97655	.22017	4.5420	.0239	4.6507
26	.21530	.97648	.22047	.5357	.0240	.6446
27	.21559	.97642	.22078	.5294	.0241	.6385
28	.21587	.97636	.22108	.5232	.0241	.6324
29	.21615	.97630	.22139	.5169	.0242	.6263
30	.21644	.97623	.22169	4.5107	.0243	4.6201
31	.21672	.97617	.22200	.5045	.0243	.6142
32	.21701	.97611	.22230	.4983	.0244	.6081
33	.21729	.97604	.22261	.4921	.0245	.6021
34	.21757	.97598	.22291	.4860	.0245	.5961
35	.21786	.97592	.22322	4.4799	.0246	4.5901
36	.21814	.97585	.22353	.4737	.0247	.5841
37	.21843	.97579	.22383	.4676	.0248	.5782
38	.21871	.97573	.22414	.4615	.0248	.5722
39	.21899	.97566	.22444	.4555	.0249	.5663
40	.21928	.97560	.22475	4.4494	.0249	4.5604
41	.21956	.97553	.22505	.4434	.0250	.5546
42	.21985	.97547	.22536	.4373	.0251	.5486
43	.22013	.97541	.22566	.4313	.0251	.5428
44	.22041	.97534	.22597	.4253	.0252	.5369
45	.22070	.97528	.22628	4.4194	.0253	4.5311
46	.22098	.97521	.22658	.4134	.0253	.5254
47	.22126	.97515	.22689	.4074	.0254	.5195
48	.22155	.97508	.22719	.4015	.0255	.5137
49	.22183	.97502	.22750	.3956	.0255	.5079
50	.22211	.97496	.22781	4.3897	.0256	4.5021
51	.22240	.97489	.22811	.3838	.0257	.4964
52	.22268	.97483	.22842	.3779	.0258	.4907
53	.22297	.97476	.22872	.3721	.0259	.4850
54	.22325	.97470	.22903	.3662	.0259	.4793
55	.22353	.97463	.22934	4.3604	.0260	4.4736
56	.22382	.97457	.22964	.3546	.0261	.4679
57	.22410	.97450	.22995	.3488	.0262	.4623
58	.22438	.97444	.23025	.3430	.0262	.4566
59	.22467	.97437	.23056	.3372	.0263	.4510
60	.22495	.97437	.23087	4.3315	.0263	4.4454

(bottom: **77°** — cos, sin, cot, tan, cosec, sec)

13°

′	sin	cos	tan	cot	sec	cosec
0	.22495	.97437	.23087	4.3315	1.0263	4.4454
1	.22523	.97430	.23117	.3257	.0264	.4398
2	.22552	.97424	.23148	.3200	.0264	.4342
3	.22580	.97417	.23179	.3143	.0265	.4287
4	.22608	.97411	.23209	.3086	.0266	.4231
5	.22637	.97404	.23240	4.3029	.0266	4.4176
6	.22665	.97398	.23270	.2972	.0267	.4121
7	.22693	.97391	.23301	.2916	.0268	.4065
8	.22722	.97384	.23332	.2859	.0268	.4011
9	.22750	.97378	.23363	.2803	.0269	.3956
10	.22778	.97371	.23393	4.2747	.0270	4.3901
11	.22807	.97364	.23424	.2691	.0271	.3847
12	.22835	.97358	.23455	.2635	.0271	.3792
13	.22863	.97351	.23485	.2579	.0272	.3738
14	.22892	.97344	.23516	.2524	.0273	.3684
15	.22920	.97338	.23547	4.2468	.0273	4.3630
16	.22948	.97331	.23577	.2413	.0274	.3576
17	.22977	.97324	.23608	.2358	.0275	.3522
18	.23005	.97318	.23639	.2303	.0276	.3469
19	.23033	.97311	.23670	.2248	.0276	.3415
20	.23061	.97304	.23700	4.2193	.0277	4.3362
21	.23090	.97298	.23731	.2139	.0278	.3309
22	.23118	.97291	.23762	.2084	.0278	.3256
23	.23146	.97284	.23793	.2030	.0279	.3203
24	.23175	.97277	.23823	.1976	.0280	.3150
25	.23202	.97271	.23854	4.1921	.0280	4.3098
26	.23231	.97264	.23885	.1867	.0281	.3045
27	.23260	.97257	.23916	.1814	.0282	.2993
28	.23288	.97250	.23946	.1760	.0283	.2941
29	.23316	.97244	.23977	.1706	.0283	.2888
30	.23344	.97237	.24008	4.1653	.0284	4.2836
31	.23373	.97230	.24039	.1600	.0285	.2785
32	.23401	.97223	.24069	.1546	.0285	.2733
33	.23429	.97216	.24100	.1493	.0286	.2681
34	.23458	.97210	.24131	.1440	.0287	.2630
35	.23486	.97203	.24162	4.1388	.0288	4.2579
36	.23514	.97196	.24192	.1335	.0288	.2527
37	.23542	.97189	.24223	.1282	.0289	.2476
38	.23571	.97182	.24254	.1230	.0290	.2425
39	.23599	.97175	.24285	.1178	.0291	.2375
40	.23627	.97169	.24316	4.1126	.0291	4.2324
41	.23655	.97162	.24347	.1073	.0292	.2273
42	.23684	.97155	.24377	.1022	.0293	.2223
43	.23712	.97148	.24408	.0970	.0293	.2173
44	.23740	.97141	.24439	.0918	.0294	.2122
45	.23768	.97134	.24470	4.0867	.0295	4.2072
46	.23797	.97127	.24501	.0815	.0296	.2022
47	.23825	.97120	.24531	.0764	.0296	.1972
48	.23853	.97113	.24562	.0713	.0297	.1923
49	.23881	.97106	.24593	.0662	.0298	.1873
50	.23910	.97099	.24624	4.0611	.0299	4.1824
51	.23938	.97092	.24655	.0560	.0299	.1774
52	.23966	.97086	.24686	.0509	.0300	.1725
53	.23994	.97079	.24717	.0458	.0301	.1676
54	.24023	.97072	.24747	.0408	.0302	.1627
55	.24051	.97065	.24778	4.0358	.0302	4.1578
56	.24079	.97058	.24809	.0307	.0303	.1529
57	.24107	.97051	.24840	.0257	.0304	.1481
58	.24136	.97044	.24871	.0207	.0305	.1432
59	.24164	.97037	.24902	.0157	.0305	.1384
60	.24192	.97029	.24933	4.0108	.0306	4.1336

(bottom: **76°** — cos, sin, cot, tan, cosec, sec)

14°

′	sin	cos	tan	cot	sec	cosec
0	.24192	.97029	.24933	4.0108	1.0306	4.1336
1	.24220	.97022	.24964	.0058	.0307	.1287
2	.24249	.97015	.24995	.0009	.0308	.1239
3	.24277	.97008	.25025	3.9959	.0308	.1191
4	.24305	.97001	.25056	.9910	.0309	.1144
5	.24333	.96994	.25087	3.9861	.0310	4.1096
6	.24361	.96987	.25118	.9812	.0311	.1048
7	.24390	.96980	.25149	.9763	.0311	.1001
8	.24418	.96973	.25180	.9714	.0312	.0953
9	.24446	.96966	.25211	.9665	.0313	.0906
10	.24474	.96959	.25242	3.9616	.0314	4.0859
11	.24502	.96952	.25273	.9568	.0314	.0812
12	.24531	.96944	.25304	.9520	.0315	.0765
13	.24559	.96937	.25335	.9471	.0316	.0718
14	.24587	.96930	.25366	.9423	.0317	.0672
15	.24615	.96923	.25397	3.9375	.0317	4.0625
16	.24643	.96916	.25428	.9327	.0318	.0579
17	.24672	.96909	.25459	.9279	.0319	.0532
18	.24700	.96901	.25490	.9231	.0320	.0486
19	.24728	.96894	.25521	.9184	.0320	.0440
20	.24756	.96887	.25552	3.9136	.0321	4.0394
21	.24784	.96880	.25583	.9089	.0322	.0348
22	.24813	.96873	.25614	.9042	.0323	.0302
23	.24841	.96866	.25645	.8994	.0323	.0256
24	.24869	.96858	.25676	.8947	.0324	.0211
25	.24897	.96851	.25707	3.8900	.0325	4.0165
26	.24925	.96844	.25738	.8853	.0326	.0120
27	.24953	.96836	.25769	.8807	.0327	.0074
28	.24982	.96829	.25800	.8760	.0327	.0029
29	.25010	.96822	.25831	.8713	.0328	3.9984
30	.25038	.96815	.25862	3.8667	.0329	3.9939
31	.25066	.96807	.25893	.8621	.0330	.9894
32	.25094	.96800	.25924	.8574	.0331	.9850
33	.25122	.96793	.25955	.8528	.0331	.9805
34	.25151	.96785	.25986	.8482	.0332	.9760
35	.25179	.96778	.26017	3.8436	.0333	3.9716
36	.25207	.96771	.26048	.8390	.0334	.9672
37	.25235	.96763	.26079	.8345	.0334	.9627
38	.25263	.96756	.26110	.8299	.0335	.9583
39	.25291	.96749	.26141	.8254	.0336	.9539
40	.25319	.96741	.26172	3.8208	.0337	3.9495
41	.25348	.96734	.26203	.8163	.0338	.9451
42	.25376	.96727	.26234	.8118	.0338	.9408
43	.25404	.96719	.26266	.8073	.0339	.9364
44	.25432	.96712	.26297	.8027	.0340	.9320
45	.25460	.96704	.26328	3.7983	.0341	3.9277
46	.25488	.96697	.26359	.7938	.0341	.9234
47	.25516	.96690	.26390	.7893	.0342	.9190
48	.25544	.96682	.26421	.7848	.0343	.9147
49	.25573	.96675	.26452	.7804	.0344	.9104
50	.25601	.96667	.26483	3.7759	.0345	3.9061
51	.25629	.96660	.26514	.7715	.0345	.9018
52	.25657	.96652	.26546	.7671	.0346	.8976
53	.25685	.96645	.26577	.7627	.0347	.8933
54	.25713	.96638	.26608	.7583	.0348	.8890
55	.25741	.96630	.26639	3.7539	.0349	3.8848
56	.25769	.96623	.26670	.7495	.0349	.8805
57	.25798	.96615	.26701	.7451	.0350	.8763
58	.25826	.96608	.26732	.7407	.0351	.8721
59	.25854	.96600	.26764	.7364	.0352	.8679
60	.25882	.96592	.26795	3.7320	.0353	3.8637

(bottom: **75°** — cos, sin, cot, tan, cosec, sec)

15°

′	sin	cos	tan	cot	sec	cosec
0	.25882	.96592	.26795	3.7320	1.0353	3.8637
1	.25910	.96585	.26826	.7277	.0353	.8595
2	.25938	.96577	.26857	.7234	.0354	.8553
3	.25966	.96570	.26888	.7191	.0355	.8512
4	.25994	.96562	.26920	.7147	.0356	.8470
5	.26022	.96555	.26951	3.7104	.0357	3.8428
6	.26050	.96547	.26982	.7062	.0358	.8387
7	.26078	.96540	.27013	.7019	.0358	.8346
8	.26107	.96532	.27044	.6976	.0359	.8304
9	.26135	.96524	.27076	.6933	.0360	.8263
10	.26163	.96517	.27107	3.6891	.0361	3.8222
11	.26191	.96509	.27138	.6848	.0362	.8181
12	.26219	.96502	.27169	.6806	.0362	.8140
13	.26247	.96494	.27201	.6764	.0363	.8100
14	.26275	.96486	.27232	.6722	.0364	.8059
15	.26303	.96479	.27263	3.6679	.0365	3.8018
16	.26331	.96471	.27294	.6637	.0366	.7978
17	.26359	.96463	.27326	.6596	.0367	.7937
18	.26387	.96456	.27357	.6554	.0367	.7897
19	.26415	.96448	.27388	.6512	.0368	.7857
20	.26443	.96440	.27419	3.6470	.0369	3.7816
21	.26471	.96433	.27451	.6429	.0370	.7776
22	.26499	.96425	.27482	.6387	.0371	.7736
23	.26527	.96417	.27513	.6346	.0371	.7697
24	.26556	.96409	.27544	.6305	.0372	.7657
25	.26584	.96402	.27576	3.6263	.0373	3.7617
26	.26612	.96394	.27607	.6222	.0374	.7577
27	.26640	.96386	.27638	.6181	.0375	.7538
28	.26668	.96378	.27670	.6140	.0376	.7498
29	.26696	.96371	.27701	.6100	.0376	.7459
30	.26724	.96363	.27732	3.6059	.0377	3.7420
31	.26752	.96355	.27764	.6018	.0378	.7380
32	.26780	.96347	.27795	.5977	.0379	.7341
33	.26808	.96340	.27826	.5937	.0380	.7302
34	.26836	.96332	.27858	.5896	.0381	.7263
35	.26864	.96324	.27889	3.5856	.0382	3.7224
36	.26892	.96316	.27920	.5816	.0382	.7186
37	.26920	.96308	.27952	.5776	.0383	.7147
38	.26948	.96301	.27983	.5736	.0384	.7108
39	.26976	.96293	.28014	.5696	.0385	.7070
40	.27004	.96285	.28046	3.5656	.0386	3.7031
41	.27032	.96277	.28077	.5616	.0387	.6993
42	.27060	.96269	.28109	.5576	.0387	.6955
43	.27088	.96261	.28140	.5536	.0388	.6917
44	.27116	.96253	.28172	.5497	.0389	.6878
45	.27144	.96245	.28203	3.5457	.0390	3.6840
46	.27172	.96238	.28234	.5418	.0391	.6802
47	.27200	.96230	.28266	.5378	.0392	.6765
48	.27228	.96222	.28297	.5339	.0392	.6727
49	.27256	.96214	.28328	.5300	.0393	.6689
50	.27284	.96206	.28360	3.5261	.0394	3.6651
51	.27312	.96198	.28391	.5222	.0395	.6614
52	.27340	.96190	.28423	.5183	.0396	.6576
53	.27368	.96182	.28454	.5144	.0397	.6539
54	.27396	.96174	.28486	.5105	.0398	.6502
55	.27424	.96166	.28517	3.5066	.0399	3.6464
56	.27452	.96158	.28549	.5028	.0399	.6427
57	.27480	.96150	.28580	.4989	.0400	.6390
58	.27508	.96142	.28611	.4951	.0401	.6353
59	.27536	.96134	.28643	.4912	.0402	.6316
60	.27564	.96126	.28674	3.4874	.0403	3.6279

(bottom: **74°** — cos, sin, cot, tan, cosec, sec)

16°

′	sin	cos	tan	cot	sec	cosec	′
0	.27564	.96126	.28674	3.4874	1.0403	3.6279	60
1	.27592	.96118	.28706	.4836	.0404	.6243	59
2	.27620	.96110	.28737	.4798	.0405	.6206	58
3	.27648	.96102	.28769	.4760	.0406	.6169	57
4	.27675	.96094	.28800	.4722	.0406	.6133	56
5	.27703	.96086	.28832	3.4684	.0407	3.6096	55
6	.27731	.96078	.28863	.4646	.0408	.6060	54
7	.27759	.96070	.28895	.4608	.0409	.6024	53
8	.27787	.96062	.28926	.4570	.0410	.5987	52
9	.27815	.96054	.28958	.4533	.0411	.5951	51
10	.27843	.96045	.28990	3.4495	.0412	3.5915	50
11	.27871	.96037	.29021	.4458	.0413	.5879	49
12	.27899	.96029	.29053	.4420	.0413	.5843	48
13	.27927	.96021	.29084	.4383	.0414	.5807	47
14	.27955	.96013	.29116	.4346	.0415	.5772	46
15	.27983	.96005	.29147	3.4308	.0416	3.5736	45
16	.28011	.95997	.29179	.4271	.0417	.5700	44
17	.28039	.95989	.29210	.4234	.0418	.5665	43
18	.28067	.95981	.29242	.4197	.0419	.5629	42
19	.28094	.95972	.29274	.4160	.0420	.5594	41
20	.28122	.95964	.29305	3.4124	.0420	3.5559	40
21	.28150	.95956	.29337	.4087	.0421	.5523	39
22	.28178	.95948	.29368	.4050	.0422	.5488	38
23	.28206	.95940	.29400	.4014	.0423	.5453	37
24	.28234	.95931	.29432	.3977	.0424	.5418	36
25	.28262	.95923	.29463	3.3941	.0425	3.5383	35
26	.28290	.95915	.29495	.3904	.0426	.5348	34
27	.28318	.95907	.29526	.3868	.0427	.5313	33
28	.28346	.95898	.29558	.3832	.0428	.5279	32
29	.28374	.95890	.29590	.3795	.0428	.5244	31
30	.28401	.95882	.29621	3.3759	.0429	3.5209	30
31	.28429	.95874	.29653	.3723	.0430	.5175	29
32	.28457	.95865	.29685	.3687	.0431	.5140	28
33	.28485	.95857	.29716	.3651	.0432	.5106	27
34	.28513	.95849	.29748	.3616	.0433	.5072	26
35	.28541	.95840	.29780	3.3580	.0434	3.5037	25
36	.28569	.95832	.29811	.3544	.0435	.5003	24
37	.28597	.95824	.29843	.3509	.0436	.4969	23
38	.28624	.95816	.29875	.3473	.0437	.4935	22
39	.28652	.95807	.29906	.3438	.0438	.4901	21
40	.28680	.95799	.29938	3.3402	.0439	3.4867	20
41	.28708	.95791	.29970	.3367	.0440	.4833	19
42	.28736	.95782	.30001	.3332	.0441	.4799	18
43	.28764	.95774	.30033	.3296	.0442	.4766	17
44	.28792	.95765	.30065	.3261	.0443	.4732	16
45	.28820	.95757	.30096	3.3226	.0444	3.4698	15
46	.28847	.95749	.30128	.3191	.0445	.4665	14
47	.28875	.95740	.30160	.3156	.0446	.4632	13
48	.28903	.95732	.30192	.3121	.0447	.4598	12
49	.28931	.95723	.30223	.3087	.0448	.4565	11
50	.28959	.95715	.30255	3.3052	.0448	3.4532	10
51	.28987	.95707	.30287	.3017	.0449	.4498	9
52	.29014	.95698	.30319	.2983	.0450	.4465	8
53	.29042	.95690	.30350	.2948	.0451	.4432	7
54	.29070	.95681	.30382	.2914	.0452	.4399	6
55	.29098	.95673	.30414	3.2879	.0452	3.4366	5
56	.29126	.95664	.30446	.2845	.0453	.4334	4
57	.29154	.95656	.30478	.2811	.0454	.4301	3
58	.29181	.95647	.30509	.2777	.0455	.4268	2
59	.29209	.95639	.30541	.2742	.0456	.4235	1
60	.29237	.95630	.30573	3.2708	.0457	3.4203	0
′	cos	sin	cot	tan	cosec	sec	′

73°

17°

′	sin	cos	tan	cot	sec	cosec	′
0	.29237	.95630	.30573	3.2708	1.0457	3.4203	60
1	.29265	.95622	.30605	.2674	.0458	.4170	59
2	.29293	.95613	.30637	.2640	.0459	.4138	58
3	.29321	.95605	.30668	.2607	.0460	.4106	57
4	.29348	.95596	.30700	.2573	.0461	.4073	56
5	.29376	.95588	.30732	3.2539	.0461	3.4041	55
6	.29404	.95579	.30764	.2505	.0462	.4009	54
7	.29432	.95571	.30796	.2472	.0463	.3977	53
8	.29460	.95562	.30828	.2438	.0464	.3945	52
9	.29487	.95554	.30859	.2405	.0465	.3913	51
10	.29515	.95545	.30891	3.2371	.0466	3.3881	50
11	.29543	.95536	.30923	.2338	.0467	.3849	49
12	.29571	.95528	.30955	.2305	.0468	.3817	48
13	.29598	.95519	.30987	.2271	.0469	.3785	47
14	.29626	.95511	.31019	.2238	.0470	.3754	46
15	.29654	.95502	.31051	3.2205	.0471	3.3722	45
16	.29682	.95493	.31083	.2172	.0472	.3690	44
17	.29710	.95485	.31115	.2139	.0473	.3659	43
18	.29737	.95476	.31146	.2106	.0474	.3627	42
19	.29765	.95467	.31178	.2073	.0475	.3596	41
20	.29793	.95459	.31210	3.2041	.0476	3.3565	40
21	.29821	.95450	.31242	.2008	.0477	.3534	39
22	.29848	.95441	.31274	.1975	.0478	.3502	38
23	.29876	.95433	.31306	.1942	.0478	.3471	37
24	.29904	.95424	.31338	.1910	.0479	.3440	36
25	.29932	.95415	.31370	3.1877	.0480	3.3409	35
26	.29959	.95407	.31402	.1845	.0481	.3378	34
27	.29987	.95398	.31434	.1813	.0482	.3347	33
28	.30015	.95389	.31466	.1780	.0483	.3316	32
29	.30043	.95380	.31498	.1748	.0483	.3286	31
30	.30070	.95372	.31530	3.1716	.0485	3.3255	30
31	.30098	.95363	.31562	.1684	.0486	.3224	29
32	.30126	.95354	.31594	.1652	.0487	.3194	28
33	.30154	.95345	.31626	.1620	.0488	.3163	27
34	.30181	.95337	.31658	.1588	.0489	.3133	26
35	.30209	.95328	.31690	3.1556	.0490	3.3102	25
36	.30237	.95319	.31722	.1524	.0491	.3072	24
37	.30265	.95310	.31754	.1492	.0492	.3042	23
38	.30292	.95301	.31786	.1460	.0493	.3011	22
39	.30320	.95293	.31818	.1429	.0494	.2981	21
40	.30348	.95284	.31850	3.1397	.0495	3.2951	20
41	.30375	.95275	.31882	.1366	.0496	.2921	19
42	.30403	.95266	.31914	.1334	.0497	.2891	18
43	.30431	.95257	.31946	.1303	.0498	.2861	17
44	.30459	.95248	.31978	.1271	.0499	.2831	16
45	.30486	.95240	.32010	3.1240	.0500	3.2801	15
46	.30514	.95231	.32042	.1209	.0501	.2772	14
47	.30542	.95222	.32074	.1178	.0502	.2742	13
48	.30569	.95213	.32106	.1146	.0503	.2712	12
49	.30597	.95204	.32138	.1115	.0504	.2683	11
50	.30625	.95195	.32171	3.1084	.0505	3.2653	10
51	.30653	.95186	.32203	.1053	.0506	.2624	9
52	.30680	.95177	.32235	.1022	.0507	.2594	8
53	.30708	.95168	.32267	.0991	.0508	.2565	7
54	.30736	.95159	.32299	.0960	.0509	.2535	6
55	.30763	.95150	.32331	3.0930	.0510	3.2506	5
56	.30791	.95141	.32363	.0899	.0511	.2477	4
57	.30819	.95132	.32396	.0868	.0512	.2448	3
58	.30846	.95124	.32428	.0838	.0513	.2419	2
59	.30874	.95115	.32460	.0807	.0514	.2390	1
60	.30902	.95106	.32492	3.0777	.0515	3.2361	0
′	cos	sin	cot	tan	cosec	sec	′

72°

18°

′	sin	cos	tan	cot	sec	cosec	′
0	.30902	.95106	.32492	3.0777	1.0515	3.2361	60
1	.30929	.95097	.32524	.0746	.0516	.2332	59
2	.30957	.95088	.32556	.0716	.0517	.2303	58
3	.30985	.95079	.32588	.0686	.0518	.2274	57
4	.31012	.95070	.32621	.0655	.0519	.2245	56
5	.31040	.95061	.32653	3.0625	.0520	3.2216	55
6	.31068	.95051	.32685	.0595	.0521	.2188	54
7	.31095	.95042	.32717	.0565	.0522	.2159	53
8	.31123	.95033	.32749	.0535	.0523	.2131	52
9	.31150	.95024	.32782	.0505	.0524	.2102	51
10	.31178	.95015	.32814	3.0475	.0525	3.2074	50
11	.31206	.95006	.32846	.0445	.0526	.2045	49
12	.31233	.94997	.32878	.0415	.0527	.2017	48
13	.31261	.94988	.32910	.0385	.0528	.1989	47
14	.31289	.94979	.32943	.0356	.0529	.1960	46
15	.31316	.94970	.32975	3.0326	.0530	3.1932	45
16	.31344	.94961	.33007	.0296	.0531	.1904	44
17	.31372	.94952	.33039	.0267	.0532	.1876	43
18	.31399	.94942	.33072	.0237	.0533	.1848	42
19	.31427	.94933	.33104	.0208	.0534	.1820	41
20	.31454	.94924	.33136	3.0178	.0535	3.1792	40
21	.31482	.94915	.33169	.0149	.0536	.1764	39
22	.31510	.94906	.33201	.0120	.0537	.1736	38
23	.31537	.94897	.33233	.0090	.0538	.1708	37
24	.31565	.94888	.33265	.0061	.0539	.1681	36
25	.31592	.94878	.33298	3.0032	.0540	3.1653	35
26	.31620	.94869	.33330	.0003	.0541	.1625	34
27	.31648	.94860	.33362	2.9974	.0542	.1598	33
28	.31675	.94851	.33395	.9945	.0543	.1570	32
29	.31703	.94842	.33427	.9916	.0544	.1543	31
30	.31730	.94832	.33459	2.9887	.0545	3.1515	30
31	.31758	.94823	.33492	.9858	.0546	.1488	29
32	.31786	.94814	.33524	.9829	.0547	.1461	28
33	.31813	.94805	.33557	.9800	.0548	.1433	27
34	.31841	.94795	.33589	.9772	.0549	.1406	26
35	.31868	.94786	.33621	2.9743	.0550	3.1379	25
36	.31896	.94777	.33654	.9714	.0551	.1352	24
37	.31923	.94767	.33686	.9686	.0552	.1325	23
38	.31951	.94758	.33718	.9657	.0553	.1298	22
39	.31978	.94749	.33751	.9629	.0554	.1271	21
40	.32006	.94740	.33783	2.9600	.0555	3.1244	20
41	.32034	.94730	.33816	.9572	.0556	.1217	19
42	.32061	.94721	.33848	.9544	.0557	.1190	18
43	.32089	.94712	.33880	.9515	.0558	.1163	17
44	.32116	.94702	.33913	.9487	.0559	.1137	16
45	.32144	.94693	.33945	2.9459	.0560	3.1110	15
46	.32171	.94684	.33978	.9431	.0561	.1083	14
47	.32199	.94674	.34010	.9403	.0562	.1057	13
48	.32227	.94665	.34043	.9375	.0563	.1030	12
49	.32254	.94656	.34075	.9347	.0565	.1004	11
50	.32282	.94646	.34108	2.9319	.0566	3.0977	10
51	.32309	.94637	.34140	.9291	.0567	.0951	9
52	.32336	.94627	.34173	.9263	.0568	.0925	8
53	.32364	.94618	.34205	.9235	.0569	.0898	7
54	.32392	.94609	.34238	.9208	.0570	.0872	6
55	.32419	.94599	.34270	2.9180	.0571	3.0846	5
56	.32447	.94590	.34303	.9152	.0572	.0820	4
57	.32474	.94580	.34335	.9125	.0573	.0793	3
58	.32502	.94571	.34368	.9097	.0574	.0767	2
59	.32529	.94561	.34400	.9069	.0575	.0741	1
60	.32557	.94552	.34433	2.9042	.0576	3.0715	0
′	cos	sin	cot	tan	cosec	sec	′

71°

19°

′	sin	cos	tan	cot	sec	cosec	′
0	.32557	.94552	.34433	2.9042	1.0576	3.0715	60
1	.32584	.94542	.34465	.9015	.0577	.0690	59
2	.32612	.94533	.34498	.8987	.0578	.0664	58
3	.32639	.94523	.34530	.8960	.0579	.0638	57
4	.32667	.94514	.34563	.8933	.0580	.0612	56
5	.32694	.94504	.34595	2.8905	.0581	3.0586	55
6	.32722	.94495	.34628	.8878	.0582	.0561	54
7	.32749	.94485	.34661	.8851	.0584	.0535	53
8	.32777	.94476	.34693	.8824	.0585	.0509	52
9	.32804	.94466	.34726	.8797	.0586	.0484	51
10	.32832	.94457	.34758	2.8770	.0587	3.0458	50
11	.32859	.94447	.34791	.8743	.0588	.0433	49
12	.32887	.94438	.34824	.8716	.0589	.0407	48
13	.32914	.94428	.34856	.8689	.0590	.0382	47
14	.32942	.94418	.34889	.8662	.0591	.0357	46
15	.32969	.94409	.34921	2.8636	.0592	3.0331	45
16	.32996	.94399	.34954	.8609	.0593	.0306	44
17	.33024	.94390	.34987	.8582	.0594	.0281	43
18	.33051	.94380	.35019	.8555	.0595	.0256	42
19	.33079	.94370	.35052	.8529	.0596	.0231	41
20	.33106	.94361	.35085	2.8502	.0598	3.0206	40
21	.33134	.94351	.35117	.8476	.0599	.0181	39
22	.33161	.94342	.35150	.8449	.0600	.0156	38
23	.33180	.94332	.35183	.8423	.0601	.0131	37
24	.33216	.94322	.35215	.8396	.0602	.0106	36
25	.33243	.94313	.35248	2.8370	.0603	3.0081	35
26	.33271	.94303	.35281	.8344	.0604	.0056	34
27	.33298	.94293	.35314	.8318	.0605	.0031	33
28	.33326	.94283	.35346	.8291	.0606	.0007	32
29	.33353	.94274	.35379	.8265	.0607	2.9982	31
30	.33381	.94264	.35412	2.8239	.0608	2.9957	30
31	.33408	.94254	.35445	.8213	.0609	.9933	29
32	.33435	.94245	.35477	.8187	.0611	.9908	28
33	.33463	.94235	.35510	.8161	.0612	.9884	27
34	.33490	.94225	.35543	.8135	.0613	.9859	26
35	.33518	.94215	.35576	2.8109	.0614	2.9835	25
36	.33545	.94206	.35608	.8083	.0615	.9810	24
37	.33572	.94196	.35641	.8057	.0616	.9786	23
38	.33600	.94186	.35674	.8032	.0617	.9762	22
39	.33627	.94176	.35707	.8006	.0618	.9738	21
40	.33655	.94167	.35739	2.7980	.0619	2.9713	20
41	.33682	.94157	.35772	.7954	.0620	.9689	19
42	.33709	.94147	.35805	.7929	.0622	.9665	18
43	.33737	.94137	.35838	.7903	.0623	.9641	17
44	.33764	.94127	.35871	.7878	.0624	.9617	16
45	.33792	.94118	.35904	2.7852	.0625	2.9593	15
46	.33819	.94108	.35936	.7827	.0626	.9569	14
47	.33846	.94098	.35969	.7801	.0627	.9545	13
48	.33874	.94088	.36002	.7776	.0628	.9521	12
49	.33901	.94078	.36035	.7751	.0629	.9497	11
50	.33928	.94068	.36068	2.7725	.0630	2.9474	10
51	.33956	.94058	.36101	.7700	.0632	.9450	9
52	.33983	.94049	.36134	.7675	.0633	.9426	8
53	.34011	.94039	.36167	.7650	.0634	.9402	7
54	.34038	.94029	.36199	.7625	.0635	.9379	6
55	.34065	.94019	.36232	2.7600	.0636	2.9355	5
56	.34093	.94009	.36265	.7575	.0637	.9332	4
57	.34120	.93999	.36298	.7550	.0638	.9308	3
58	.34147	.93989	.36331	.7525	.0639	.9285	2
59	.34175	.93979	.36364	.7500	.0641	.9261	1
60	.34202	.93969	.36397	2.7475	.0642	2.9238	0
′	cos	sin	cot	tan	cosec	sec	′

70°

NATURAL TRIGONOMETRIC FUNCTIONS

20° (bottom: 69°)

'	sin	cos	tan	cot	sec	cosec	'
0	.34202	.93969	.36397	2.7475	1.0642	2.9238	60
1	.34229	.93959	.36430	.7450	.0643	.9215	59
2	.34257	.93949	.36463	.7425	.0644	.9191	58
3	.34284	.93939	.36496	.7400	.0645	.9168	57
4	.34311	.93929	.36529	.7376	.0646	.9145	56
5	.34339	.93919	.36562	2.7351	.0647	2.9122	55
6	.34366	.93909	.36595	.7326	.0648	.9098	54
7	.34393	.93899	.36628	.7302	.0650	.9075	53
8	.34421	.93889	.36661	.7277	.0651	.9052	52
9	.34448	.93879	.36694	.7252	.0652	.9029	51
10	.34475	.93869	.36727	2.7228	.0653	2.9006	50
11	.34502	.93859	.36760	.7203	.0654	.8983	49
12	.34530	.93849	.36793	.7179	.0655	.8960	48
13	.34557	.93839	.36826	.7155	.0656	.8937	47
14	.34584	.93829	.36859	.7130	.0658	.8915	46
15	.34612	.93819	.36892	2.7106	.0659	2.8892	45
16	.34639	.93809	.36925	.7082	.0660	.8869	44
17	.34666	.93799	.36958	.7058	.0661	.8846	43
18	.34693	.93789	.36991	.7033	.0662	.8824	42
19	.34721	.93779	.37024	.7009	.0663	.8801	41
20	.34748	.93769	.37057	2.6985	.0664	2.8778	40
21	.34775	.93759	.37090	.6961	.0666	.8756	39
22	.34803	.93748	.37123	.6937	.0667	.8733	38
23	.34830	.93738	.37156	.6913	.0668	.8711	37
24	.34857	.93728	.37190	.6889	.0669	.8688	36
25	.34884	.93718	.37223	2.6865	.0670	2.8666	35
26	.34912	.93708	.37256	.6841	.0671	.8643	34
27	.34939	.93698	.37289	.6817	.0673	.8621	33
28	.34966	.93688	.37322	.6794	.0674	.8599	32
29	.34993	.93677	.37355	.6770	.0675	.8577	31
30	.35021	.93667	.37388	2.6746	.0677	2.8554	30
31	.35048	.93657	.37422	.6722	.0678	.8532	29
32	.35075	.93647	.37455	.6699	.0679	.8510	28
33	.35102	.93637	.37488	.6675	.0680	.8488	27
34	.35130	.93626	.37521	.6652	.0681	.8466	26
35	.35157	.93616	.37554	2.6628	.0682	2.8444	25
36	.35184	.93606	.37587	.6604	.0683	.8422	24
37	.35211	.93596	.37621	.6581	.0684	.8400	23
38	.35239	.93585	.37654	.6558	.0685	.8378	22
39	.35266	.93575	.37687	.6534	.0686	.8356	21
40	.35293	.93565	.37720	2.6511	.0688	2.8334	20
41	.35320	.93555	.37754	.6487	.0689	.8312	19
42	.35347	.93544	.37787	.6464	.0690	.8290	18
43	.35375	.93534	.37820	.6441	.0691	.8269	17
44	.35402	.93524	.37853	.6418	.0692	.8247	16
45	.35429	.93513	.37887	2.6394	.0694	2.8225	15
46	.35456	.93503	.37920	.6371	.0695	.8204	14
47	.35483	.93493	.37953	.6348	.0696	.8182	13
48	.35511	.93482	.37986	.6325	.0697	.8160	12
49	.35538	.93472	.38020	.6302	.0698	.8139	11
50	.35565	.93462	.38053	2.6279	.0699	2.8117	10
51	.35592	.93451	.38086	.6256	.0701	.8096	9
52	.35619	.93441	.38120	.6233	.0702	.8074	8
53	.35647	.93431	.38153	.6210	.0703	.8053	7
54	.35674	.93420	.38186	.6187	.0704	.8032	6
55	.35701	.93410	.38220	2.6164	.0705	2.8010	5
56	.35728	.93400	.38253	.6142	.0707	.7989	4
57	.35755	.93389	.38286	.6119	.0708	.7947	3
58	.35782	.93379	.38320	.6096	.0709	.7948	2
59	.35810	.93368	.38353	.6073	.0710	.7925	1
60	.35837	.93358	.38386	2.6051	.0711	2.7904	0
'	cos	sin	cot	tan	cosec	sec	'

(69°)

21° (bottom: 68°)

'	sin	cos	tan	cot	sec	cosec	'
0	.35837	.93358	.38386	2.6051	1.0711	2.7904	60
1	.35864	.93348	.38420	.6028	.0713	.7883	59
2	.35891	.93337	.38453	.6006	.0714	.7862	58
3	.35918	.93327	.38486	.5983	.0715	.7841	57
4	.35945	.93316	.38520	.5960	.0716	.7820	56
5	.35972	.93306	.38553	2.5938	.0717	2.7799	55
6	.36000	.93295	.38587	.5916	.0718	.7778	54
7	.36027	.93285	.38620	.5893	.0720	.7757	53
8	.36054	.93274	.38654	.5871	.0721	.7736	52
9	.36081	.93264	.38687	.5848	.0722	.7715	51
10	.36108	.93253	.38720	2.5826	.0723	2.7694	50
11	.36135	.93243	.38754	.5804	.0725	.7674	49
12	.36162	.93232	.38787	.5782	.0726	.7653	48
13	.36189	.93222	.38821	.5759	.0727	.7632	47
14	.36217	.93211	.38854	.5737	.0728	.7611	46
15	.36244	.93201	.38888	2.5715	.0729	2.7591	45
16	.36271	.93190	.38921	.5693	.0731	.7570	44
17	.36298	.93180	.38955	.5671	.0732	.7550	43
18	.36325	.93169	.38988	.5649	.0733	.7529	42
19	.36352	.93158	.39022	.5627	.0734	.7509	41
20	.36379	.93148	.39055	2.5605	.0736	2.7488	40
21	.36406	.93137	.39089	.5583	.0737	.7468	39
22	.36433	.93127	.39122	.5561	.0738	.7447	38
23	.36460	.93116	.39156	.5539	.0739	.7427	37
24	.36488	.93106	.39189	.5517	.0740	.7406	36
25	.36515	.93095	.39223	2.5495	.0742	2.7386	35
26	.36542	.93084	.39257	.5473	.0743	.7366	34
27	.36569	.93074	.39290	.5451	.0745	.7346	33
28	.36596	.93063	.39324	.5430	.0746	.7325	32
29	.36623	.93052	.39357	.5408	.0747	.7305	31
30	.36650	.93042	.39391	2.5386	.0748	2.7285	30
31	.36677	.93031	.39425	.5365	.0749	.7265	29
32	.36704	.93020	.39458	.5343	.0750	.7245	28
33	.36731	.93010	.39492	.5322	.0751	.7225	27
34	.36758	.92999	.39525	.5300	.0753	.7205	26
35	.36785	.92988	.39559	2.5278	.0754	2.7185	25
36	.36812	.92978	.39593	.5257	.0755	.7165	24
37	.36839	.92967	.39626	.5236	.0756	.7145	23
38	.36867	.92956	.39660	.5214	.0758	.7125	22
39	.36893	.92945	.39694	.5193	.0759	.7105	21
40	.36921	.92935	.39727	2.5171	.0760	2.7085	20
41	.36948	.92924	.39761	.5150	.0761	.7065	19
42	.36975	.92913	.39795	.5129	.0763	.7045	18
43	.37002	.92902	.39829	.5108	.0764	.7026	17
44	.37029	.92892	.39862	.5086	.0765	.7006	16
45	.37056	.92881	.39896	2.5065	.0766	2.6986	15
46	.37083	.92870	.39930	.5044	.0768	.6967	14
47	.37110	.92859	.39963	.5023	.0769	.6947	13
48	.37137	.92848	.39997	.5002	.0770	.6927	12
49	.37164	.92838	.40031	.4981	.0771	.6908	11
50	.37191	.92827	.40065	2.4960	.0773	2.6888	10
51	.37218	.92816	.40098	.4939	.0774	.6869	9
52	.37245	.92805	.40132	.4918	.0775	.6849	8
53	.37272	.92794	.40166	.4897	.0776	.6830	7
54	.37299	.92784	.40200	.4876	.0778	.6810	6
55	.37326	.92773	.40233	2.4855	.0779	2.6791	5
56	.37353	.92762	.40267	.4834	.0780	.6772	4
57	.37380	.92751	.40301	.4813	.0781	.6752	3
58	.37407	.92740	.40335	.4792	.0783	.6733	2
59	.37434	.92729	.40369	.4772	.0784	.6714	1
60	.37461	.92718	.40403	2.4751	.0785	2.6695	0
'	cos	sin	cot	tan	cosec	sec	'

(68°)

22° (bottom: 67°)

'	sin	cos	tan	cot	sec	cosec	'
0	.37461	.92718	.40403	2.4751	1.0785	2.6695	60
1	.37488	.92707	.40436	.4730	.0787	.6675	59
2	.37514	.92696	.40470	.4709	.0788	.6656	58
3	.37541	.92686	.40504	.4689	.0789	.6637	57
4	.37568	.92675	.40538	.4668	.0790	.6618	56
5	.37595	.92664	.40572	2.4647	.0792	2.6599	55
6	.37622	.92653	.40606	.4627	.0793	.6580	54
7	.37649	.92642	.40640	.4606	.0794	.6561	53
8	.37676	.92631	.40674	.4586	.0795	.6542	52
9	.37703	.92620	.40707	.4565	.0797	.6523	51
10	.37730	.92609	.40741	2.4545	.0798	2.6504	50
11	.37757	.92598	.40775	.4525	.0799	.6485	49
12	.37784	.92587	.40809	.4504	.0801	.6466	48
13	.37811	.92576	.40843	.4484	.0802	.6447	47
14	.37838	.92565	.40877	.4463	.0803	.6428	46
15	.37865	.92554	.40911	2.4443	.0804	2.6410	45
16	.37892	.92543	.40945	.4423	.0806	.6391	44
17	.37919	.92532	.40979	.4403	.0807	.6372	43
18	.37946	.92521	.41013	.4382	.0808	.6353	42
19	.37972	.92510	.41047	.4362	.0810	.6335	41
20	.37999	.92499	.41081	2.4342	.0811	2.6316	40
21	.38026	.92488	.41115	.4322	.0812	.6297	39
22	.38053	.92477	.41149	.4302	.0813	.6279	38
23	.38080	.92466	.41183	.4282	.0815	.6260	37
24	.38107	.92455	.41217	.4262	.0816	.6242	36
25	.38134	.92443	.41251	2.4242	.0817	2.6223	35
26	.38161	.92432	.41285	.4222	.0819	.6205	34
27	.38188	.92421	.41319	.4202	.0820	.6186	33
28	.38214	.92410	.41353	.4182	.0821	.6168	32
29	.38241	.92399	.41387	.4162	.0823	.6150	31
30	.38268	.92388	.41421	2.4142	.0824	2.6131	30
31	.38295	.92377	.41455	.4122	.0825	.6113	29
32	.38322	.92366	.41489	.4102	.0826	.6095	28
33	.38349	.92355	.41524	.4083	.0828	.6076	27
34	.38376	.92343	.41558	.4063	.0829	.6058	26
35	.38403	.92332	.41592	2.4043	.0830	2.6040	25
36	.38429	.92321	.41626	.4023	.0832	.6022	24
37	.38456	.92310	.41660	.4004	.0833	.6003	23
38	.38483	.92299	.41694	.3984	.0834	.5985	22
39	.38510	.92287	.41728	.3964	.0836	.5967	21
40	.38537	.92276	.41762	2.3945	.0837	2.5949	20
41	.38564	.92265	.41797	.3925	.0838	.5931	19
42	.38591	.92254	.41831	.3906	.0840	.5913	18
43	.38617	.92243	.41865	.3886	.0841	.5895	17
44	.38644	.92231	.41899	.3867	.0842	.5877	16
45	.38671	.92220	.41933	2.3847	.0844	2.5859	15
46	.38698	.92209	.41968	.3828	.0845	.5841	14
47	.38725	.92197	.42002	.3808	.0846	.5823	13
48	.38751	.92186	.42036	.3789	.0847	.5805	12
49	.38778	.92175	.42070	.3770	.0849	.5787	11
50	.38805	.92164	.42105	2.3751	.0850	2.5770	10
51	.38832	.92152	.42139	.3731	.0851	.5752	9
52	.38859	.92141	.42173	.3712	.0853	.5734	8
53	.38886	.92130	.42207	.3693	.0854	.5716	7
54	.38912	.92119	.42242	.3673	.0855	.5699	6
55	.38939	.92107	.42276	2.3654	.0857	2.5681	5
56	.38966	.92096	.42310	.3635	.0858	.5663	4
57	.38993	.92085	.42344	.3616	.0859	.5646	3
58	.39020	.92073	.42379	.3597	.0861	.5628	2
59	.39046	.92062	.42413	.3577	.0862	.5610	1
60	.39073	.92050	.42447	2.3558	.0864	2.5593	0
'	cos	sin	cot	tan	cosec	sec	'

(67°)

23° (bottom: 66°)

'	sin	cos	tan	cot	sec	cosec	'
0	.39073	.92050	.42447	2.3558	1.0864	2.5593	60
1	.39100	.92039	.42482	.3539	.0865	.5575	59
2	.39125	.92028	.42516	.3520	.0866	.5558	58
3	.39153	.92016	.42550	.3501	.0868	.5540	57
4	.39180	.92005	.42585	.3482	.0869	.5523	56
5	.39207	.91993	.42619	2.3463	.0870	2.5506	55
6	.39234	.91982	.42654	.3445	.0872	.5488	54
7	.39260	.91971	.42688	.3426	.0873	.5471	53
8	.39287	.91959	.42722	.3407	.0874	.5453	52
9	.39314	.91948	.42757	.3388	.0876	.5436	51
10	.39341	.91936	.42791	2.3369	.0877	2.5419	50
11	.39367	.91925	.42826	.3350	.0878	.5402	49
12	.39394	.91913	.42860	.3332	.0880	.5384	48
13	.39421	.91902	.42894	.3313	.0881	.5367	47
14	.39448	.91891	.42929	.3294	.0882	.5350	46
15	.39474	.91879	.42963	2.3276	.0884	2.5333	45
16	.39501	.91868	.42998	.3257	.0885	.5316	44
17	.39528	.91856	.43032	.3238	.0886	.5299	43
18	.39554	.91845	.43067	.3220	.0888	.5281	42
19	.39581	.91833	.43101	.3201	.0889	.5264	41
20	.39608	.91822	.43136	2.3183	.0891	2.5247	40
21	.39635	.91810	.43170	.3164	.0892	.5230	39
22	.39661	.91798	.43205	.3145	.0893	.5213	38
23	.39688	.91787	.43239	.3127	.0895	.5196	37
24	.39715	.91775	.43274	.3109	.0896	.5179	36
25	.39741	.91764	.43308	2.3090	.0897	2.5163	35
26	.39768	.91752	.43343	.3072	.0899	.5146	34
27	.39795	.91741	.43377	.3053	.0900	.5129	33
28	.39822	.91729	.43412	.3035	.0902	.5112	32
29	.39848	.91718	.43447	.3017	.0903	.5095	31
30	.39875	.91706	.43481	2.2998	.0904	2.5078	30
31	.39902	.91694	.43516	.2980	.0906	.5062	29
32	.39928	.91683	.43550	.2962	.0907	.5045	28
33	.39955	.91671	.43585	.2944	.0908	.5028	27
34	.39981	.91659	.43620	.2925	.0910	.5011	26
35	.40008	.91648	.43654	2.2907	.0911	2.4995	25
36	.40035	.91636	.43689	.2889	.0913	.4978	24
37	.40061	.91625	.43723	.2871	.0914	.4961	23
38	.40088	.91613	.43758	.2853	.0915	.4945	22
39	.40115	.91601	.43793	.2835	.0917	.4928	21
40	.40141	.91590	.43827	2.2817	.0918	2.4912	20
41	.40168	.91578	.43862	.2799	.0920	.4895	19
42	.40195	.91566	.43897	.2781	.0921	.4879	18
43	.40221	.91555	.43932	.2763	.0922	.4862	17
44	.40248	.91543	.43966	.2745	.0924	.4846	16
45	.40275	.91531	.44001	2.2727	.0925	2.4829	15
46	.40302	.91519	.44036	.2709	.0927	.4813	14
47	.40328	.91508	.44070	.2691	.0928	.4797	13
48	.40355	.91496	.44105	.2673	.0929	.4780	12
49	.40381	.91484	.44140	.2655	.0931	.4764	11
50	.40408	.91472	.44175	2.2637	.0932	2.4748	10
51	.40434	.91461	.44209	.2619	.0934	.4731	9
52	.40461	.91449	.44244	.2602	.0935	.4715	8
53	.40488	.91437	.44279	.2584	.0936	.4699	7
54	.40514	.91425	.44314	.2566	.0938	.4683	6
55	.40541	.91414	.44349	2.2548	.0939	2.4666	5
56	.40567	.91402	.44383	.2531	.0941	.4650	4
57	.40594	.91390	.44418	.2513	.0942	.4634	3
58	.40620	.91378	.44453	.2495	.0943	.4618	2
59	.40647	.91366	.44488	.2478	.0945	.4602	1
60	.40674	.91354	.44523	2.2460	.0946	2.4586	0
'	cos	sin	cot	tan	cosec	sec	'

(66°)

NATURAL TRIGONOMETRIC FUNCTIONS

The page is a full-page numeric table of natural trigonometric functions. The table is divided into blocks for **24°**, **25°**, **26°**, and **27°** (read from the top), with complementary angles **65°**, **64°**, **63°**, and **62°** (read from the bottom).

Each block has the column headings (left to right at the top): `'  sin  cos  tan  cot  sec  cosec  '` and at the bottom (for the complementary angle): `'  cos  sin  cot  tan  cosec  sec  '`.

The leftmost and rightmost columns give the minutes `'` running 0–60 (top block) and 60–0 (with the complementary scale reversed).

NATURAL TRIGONOMETRIC FUNCTIONS

31°

′	cosec	sec	cot	tan	cos	sin
60	1.9416	1.6666	1.6643	.60086	.85717	.51504
59	.9407	1668	6632	60126	85702	51529
58	.9397	1670	6621	60165	85687	51554
57	.9388	1672	6610	60205	85672	51578
56	.9378	1674	6599	60244	85657	51603
55	1.9369	1676	1.6588	60284	85642	51628
54	.9360	1678	6577	60324	85627	51653
53	.9350	1681	6566	60363	85612	51678
52	.9341	1683	6555	60403	85597	51703
51	.9332	1685	6544	60443	85582	51728
50	1.9322	1687	1.6534	60483	85566	51753
49	.9313	1689	6523	60522	85551	51778
48	.9304	1691	6512	60562	85536	51803
47	.9295	1693	6501	60602	85521	51827
46	.9285	1695	6490	60642	85506	51852
45	1.9276	1697	1.6479	60681	85491	51877
44	.9267	1699	6469	60721	85476	51902
43	.9258	1701	6458	60761	85461	51927
42	.9248	1703	6447	60801	85446	51952
41	.9239	1705	6436	60841	85431	51977
40	1.9230	1707	1.6425	60881	85416	52002
39	.9221	1709	6415	60920	85400	52026
38	.9212	1712	6404	60960	85385	52051
37	.9203	1714	6393	61000	85370	52076
36	.9193	1716	6383	61040	85355	52101
35	1.9184	1718	1.6372	61080	85340	52126
34	.9175	1720	6361	61120	85325	52151
33	.9166	1722	6350	61160	85309	52175
32	.9157	1724	6340	61200	85294	52200
31	.9148	1726	6329	61240	85279	52225
30	1.9139	1728	1.6318	61280	85264	52250
29	.9130	1730	6308	61320	85249	52275
28	.9121	1732	6297	61360	85234	52299
27	.9112	1734	6286	61400	85218	52324
26	.9102	1737	6276	61440	85203	52349
25	1.9093	1739	1.6265	61480	85188	52374
24	.9084	1741	6255	61520	85173	52398
23	.9075	1743	6244	61560	85157	52423
22	.9066	1745	6233	61601	85142	52448
21	.9057	1747	6223	61641	85127	52473
20	1.9048	1749	1.6212	61681	85112	52498
19	.9039	1751	6202	61721	85096	52522
18	.9030	1753	6191	61761	85081	52547
17	.9021	1756	6181	61801	85066	52572
16	.9013	1758	6170	61842	85050	52597
15	1.9004	1760	1.6160	61882	85035	52621
14	.8995	1762	6149	61922	85020	52646
13	.8986	1764	6139	61962	85004	52671
12	.8977	1766	6128	62003	84989	52695
11	.8968	1768	6118	62043	84974	52720
10	1.8959	1770	1.6107	62083	84959	52745
9	.8950	1772	6097	62123	84943	52770
8	.8941	1775	6086	62164	84928	52794
7	.8932	1777	6076	62204	84912	52819
6	.8924	1779	6066	62244	84897	52844
5	1.8915	1781	1.6055	62285	84882	52868
4	.8906	1783	6045	62325	84866	52893
3	.8897	1785	6034	62366	84851	52918
2	.8888	1787	6024	62406	84836	52942
1	.8879	1790	6014	62446	84820	52967
0	1.8871	.1792	1.6003	62487	84805	52992
′	sec	cosec	tan	cot	sin	cos

58°

30°

′	cosec	sec	cot	tan	cos	sin
60	2.0000	1.1547	1.7320	.57735	.86603	.50000
59	1.9990	.1549	7309	57774	86588	50025
58	.9980	.1551	7297	57813	86573	50050
57	.9970	.1553	7286	57851	86559	50075
56	.9960	.1555	7274	57890	86544	50101
55	1.9950	1557	1.7262	57929	86530	50126
54	.9940	1559	7251	57968	86515	50151
53	.9930	1561	7239	58007	86500	50176
52	.9920	1562	7228	58046	86486	50201
51	.9910	1564	7216	58085	86471	50226
50	1.9900	1566	1.7205	58123	86457	50252
49	.9890	1568	7193	58162	86442	50277
48	.9880	1570	7182	58201	86427	50302
47	.9870	1572	7170	58240	86413	50327
46	.9860	1574	7159	58279	86398	50352
45	1.9850	1576	1.7147	58318	86383	50377
44	.9840	1578	7136	58357	86369	50402
43	.9830	1580	7124	58396	86354	50428
42	.9820	1582	7113	58435	86339	50453
41	.9811	1584	7101	58474	86325	50478
40	1.9801	1586	1.7090	58513	86310	50503
39	.9791	1588	7079	58552	86295	50528
38	.9781	1590	7067	58591	86281	50553
37	.9771	1592	7056	58630	86266	50578
36	.9761	1594	7044	58670	86251	50603
35	1.9752	1596	1.7033	58709	86237	50628
34	.9742	1598	7022	58748	86222	50653
33	.9732	1600	7010	58787	86207	50679
32	.9722	1602	6999	58826	86192	50704
31	.9713	1604	6988	58865	86178	50729
30	1.9703	1606	1.6977	58904	86163	50754
29	.9693	1608	6965	58944	86148	50779
28	.9683	1610	6954	58983	86133	50804
27	.9674	1612	6943	59022	86118	50829
26	.9664	1614	6931	59061	86104	50854
25	1.9654	1616	1.6920	59100	86089	50879
24	.9645	1618	6909	59140	86074	50904
23	.9635	1620	6898	59179	86059	50929
22	.9625	1622	6887	59218	86044	50954
21	.9616	1624	6875	59258	86030	50979
20	1.9606	1626	1.6864	59297	86015	51004
19	.9596	1628	6853	59336	86000	51029
18	.9587	1630	6842	59376	85985	51054
17	.9577	1632	6831	59415	85970	51079
16	.9568	1634	6820	59454	85955	51104
15	1.9558	1636	1.6808	59494	85941	51129
14	.9549	1638	6797	59533	85926	51154
13	.9539	1640	6786	59572	85911	51179
12	.9530	1642	6775	59612	85896	51204
11	.9520	1644	6764	59651	85881	51229
10	1.9501	1646	1.6753	59691	85866	51254
9	.9501	1648	6742	59730	85851	51279
8	.9491	1650	6731	59770	85836	51304
7	.9482	1652	6720	59809	85821	51329
6	.9473	1654	6709	59849	85806	51354
5	1.9463	1656	1.6698	59888	85791	51379
4	.9454	1658	6687	59928	85777	51404
3	.9444	1660	6676	59967	85762	51429
2	.9435	1662	6665	60007	85747	51454
1	.9425	1664	6654	60046	85732	51479
0	1.9416	.1666	1.6643	60086	85717	51504
′	sec	cosec	tan	cot	sin	cos

59°

29°

′	cosec	sec	cot	tan	cos	sin
60	2.0627	1.1433	1.8040	.55431	.87462	.48481
59	.0616	.1435	8028	55469	87448	48506
58	.0605	.1437	8016	55507	87434	48532
57	.0594	.1439	8003	55545	87420	48557
56	.0583	.1441	.7991	55583	87405	48583
55	2.0573	1443	1.7979	55621	87391	48608
54	.0562	1445	7966	55659	87377	48634
53	.0551	1446	7954	55697	87363	48659
52	.0540	1448	7942	55735	87349	48684
51	.0530	1450	7930	55774	87335	48710
50	2.0519	1452	1.7917	55812	87320	48735
49	.0508	1454	7905	55850	87306	48760
48	.0498	1456	7893	55888	87292	48786
47	.0487	1458	7881	55926	87278	48811
46	.0476	1459	7868	55964	87264	48837
45	2.0466	1461	1.7856	56003	87250	48862
44	.0455	1463	7843	56041	87235	48887
43	.0444	1465	7831	56079	87221	48913
42	.0434	1467	7820	56117	87207	48938
41	.0423	1469	7808	56156	87193	48964
40	2.0413	1471	1.7795	56194	87178	48989
39	.0402	1474	7783	56232	87164	49014
38	.0392	1476	7771	56270	87150	49040
37	.0381	1476	7759	56309	87136	49065
36	.0370	1478	7747	56347	87121	49090
35	2.0360	1480	1.7735	56385	87107	49116
34	.0349	1482	7723	56424	87093	49141
33	.0339	1484	7711	56462	87078	49166
32	.0329	1486	7699	56501	87064	49192
31	.0318	1488	7687	56539	87050	49217
30	2.0308	1489	1.7675	56577	87035	49242
29	.0297	1491	7663	56616	87021	49268
28	.0287	1493	7651	56654	87007	49293
27	.0276	1495	7639	56692	86992	49318
26	.0266	1497	7627	56731	86978	49343
25	2.0256	1499	1.7615	56769	86964	49369
24	.0245	1501	7603	56808	86949	49394
23	.0235	1503	7591	56846	86935	49419
22	.0224	1505	7579	56885	86921	49445
21	.0214	1507	7567	56923	86906	49470
20	2.0204	1508	1.7555	56962	86892	49495
19	.0194	1510	7544	57000	86878	49521
18	.0183	1512	7532	57039	86863	49546
17	.0173	1514	7520	57078	86849	49571
16	.0163	1516	7508	57116	86834	49596
15	2.0152	1518	1.7496	57155	86820	49622
14	.0142	1520	7484	57193	86805	49647
13	.0132	1522	7473	57232	86791	49672
12	.0122	1524	7461	57270	86776	49697
11	.0111	1526	7449	57309	86762	49723
10	2.0101	1528	1.7437	57348	86748	49748
9	.0091	1530	7426	57386	86733	49773
8	.0081	1531	7414	57425	86719	49798
7	.0071	1533	7402	57464	86704	49823
6	.0061	1535	7390	57502	86690	49849
5	2.0050	1537	1.7379	57541	86675	49874
4	.0040	1539	7367	57580	86661	49899
3	.0030	1541	7355	57619	86646	49924
2	.0020	1543	7344	57657	86632	49950
1	.0010	1545	7332	57696	86617	49975
0	2.0000	.1547	1.7320	57735	86603	50000
′	sec	cosec	tan	cot	sin	cos

60°

28°

′	cosec	sec	cot	tan	cos	sin
60	2.1300	1.1326	1.8807	.53171	.88295	.46947
59	.1289	.1327	8794	53208	88281	46973
58	.1277	.1329	8781	53246	88267	46998
57	.1266	.1331	8768	53283	88254	47024
56	.1254	.1333	8754	53320	88240	47050
55	2.1242	1334	1.8741	53358	88226	47075
54	.1231	1336	8728	53395	88213	47101
53	.1219	1338	8715	53432	88199	47127
52	.1208	1340	8702	53470	88185	47152
51	.1196	1341	8689	53507	88171	47178
50	2.1185	1343	1.8676	53545	88158	47204
49	.1173	1345	8663	53582	88144	47229
48	.1162	1347	8650	53619	88130	47255
47	.1150	1349	8637	53657	88117	47281
46	.1139	1350	8624	53694	88103	47306
45	2.1127	1352	1.8611	53732	88089	47332
44	.1116	1354	8598	53769	88075	47357
43	.1104	1356	8585	53807	88061	47383
42	.1093	1358	8572	53844	88048	47409
41	.1082	1359	8559	53882	88034	47434
40	2.1070	1361	1.8546	53919	88020	47460
39	.1059	1363	8533	53957	88006	47486
38	.1048	1365	8520	53995	87992	47511
37	.1036	1366	8507	54032	87979	47537
36	.1025	1368	8495	54070	87965	47562
35	2.1014	1370	1.8482	54107	87951	47588
34	.0991	1373	8469	54145	87937	47613
33	.0991	1373	8456	54183	87923	47639
32	.0980	1375	8443	54220	87909	47665
31	.0969	1377	8430	54258	87895	47690
30	2.0957	1379	1.8418	54295	87882	47716
29	.0946	1381	8405	54333	87868	47741
28	.0935	1382	8392	54371	87854	47767
27	.0924	1384	8379	54409	87840	47792
26	.0912	1386	8367	54446	87826	47818
25	2.0901	1388	1.8354	54484	87812	47844
24	.0890	1390	8341	54522	87798	47869
23	.0879	1391	8329	54559	87784	47895
22	.0868	1393	8316	54597	87770	47920
21	.0857	1395	8303	54635	87756	47946
20	2.0846	1397	1.8291	54673	87742	47971
19	.0835	1399	8278	54711	87728	47997
18	.0824	1401	8265	54748	87715	48022
17	.0813	1403	8253	54786	87701	48048
16	.0801	1404	8240	54824	87687	48073
15	2.0790	1406	1.8227	54862	87673	48099
14	.0779	1408	8215	54900	87659	48124
13	.0768	1410	8202	54937	87645	48150
12	.0757	1411	8190	54975	87631	48175
11	.0746	1413	8177	55013	87617	48201
10	2.0735	1415	1.8165	55051	87603	48226
9	.0725	1417	8152	55089	87588	48252
8	.0714	1419	8140	55127	87574	48277
7	.0703	1421	8127	55165	87560	48303
6	.0692	1422	8115	55203	87546	48328
5	2.0681	1424	1.8102	55241	87532	48354
4	.0670	1426	8090	55279	87518	48379
3	.0659	1428	8078	55317	87504	48405
2	.0648	1430	8065	55355	87490	48430
1	.0637	1432	8053	55393	87476	48455
0	2.0627	.1433	1.8040	55431	87462	48481
′	sec	cosec	tan	cot	sin	cos

61°

NATURAL TRIGONOMETRIC FUNCTIONS

32° / 57° (left block)

′	sin	cos	tan	cot	sec	cosec
0	.52992	.84805	.62487	1.6003	1.1792	1.8871
1	.53016	.84789	.62527	.5993	.1794	.8862
2	.53041	.84774	.62568	.5983	.1796	.8853
3	.53066	.84758	.62608	.5972	.1798	.8844
4	.53090	.84743	.62649	.5962	.1800	.8836
5	.53115	.84728	.62689	1.5952	1.1802	1.8827
6	.53140	.84712	.62730	.5941	.1805	.8818
7	.53164	.84697	.62770	.5931	.1807	.8809
8	.53189	.84681	.62811	.5921	.1809	.8801
9	.53214	.84666	.62851	.5910	.1811	.8792
10	.53238	.84650	.62892	1.5900	1.1813	1.8783
11	.53263	.84635	.62933	.5890	.1815	.8775
12	.53288	.84619	.62973	.5880	.1818	.8766
13	.53312	.84604	.63014	.5869	.1820	.8757
14	.53337	.84588	.63055	.5859	.1822	.8749
15	.53361	.84573	.63095	1.5849	1.1824	1.8740
16	.53386	.84557	.63136	.5839	.1826	.8731
17	.53411	.84542	.63177	.5829	.1828	.8723
18	.53435	.84526	.63217	.5818	.1831	.8714
19	.53460	.84511	.63258	.5808	.1833	.8706
20	.53484	.84495	.63299	1.5798	1.1835	1.8697
21	.53509	.84479	.63339	.5788	.1837	.8688
22	.53533	.84464	.63380	.5778	.1839	.8680
23	.53558	.84448	.63421	.5768	.1841	.8671
24	.53583	.84433	.63462	.5757	.1844	.8663
25	.53607	.84417	.63503	1.5747	1.1846	1.8654
26	.53632	.84402	.63543	.5737	.1848	.8646
27	.53656	.84386	.63584	.5727	.1850	.8637
28	.53681	.84370	.63625	.5717	.1852	.8629
29	.53705	.84355	.63666	.5707	.1855	.8620
30	.53730	.84339	.63707	1.5697	1.1857	1.8611
31	.53754	.84323	.63748	.5687	.1859	.8603
32	.53779	.84308	.63789	.5677	.1861	.8595
33	.53803	.84292	.63830	.5667	.1863	.8586
34	.53828	.84276	.63871	.5657	.1866	.8578
35	.53852	.84261	.63912	1.5646	1.1868	1.8569
36	.53877	.84245	.63953	.5636	.1870	.8561
37	.53901	.84229	.63994	.5626	.1872	.8552
38	.53926	.84214	.64035	.5616	.1874	.8544
39	.53950	.84198	.64076	.5606	.1877	.8535
40	.53975	.84182	.64117	1.5596	1.1879	1.8527
41	.53999	.84167	.64158	.5586	.1881	.8519
42	.54024	.84151	.64199	.5577	.1883	.8510
43	.54048	.84135	.64240	.5567	.1886	.8502
44	.54073	.84120	.64281	.5557	.1888	.8493
45	.54097	.84104	.64322	1.5547	1.1890	1.8485
46	.54122	.84088	.64363	.5537	.1892	.8477
47	.54146	.84072	.64404	.5527	.1894	.8468
48	.54171	.84057	.64446	.5517	.1897	.8460
49	.54195	.84041	.64487	.5507	.1899	.8452
50	.54220	.84025	.64528	1.5497	1.1901	1.8443
51	.54244	.84009	.64569	.5487	.1903	.8435
52	.54268	.83993	.64610	.5477	.1906	.8427
53	.54293	.83978	.64652	.5467	.1908	.8418
54	.54317	.83962	.64693	.5458	.1910	.8410
55	.54342	.83946	.64734	1.5448	1.1912	1.8402
56	.54366	.83930	.64775	.5438	.1915	.8394
57	.54391	.83914	.64817	.5428	.1917	.8385
58	.54415	.83899	.64858	.5418	.1919	.8377
59	.54439	.83883	.64899	.5408	.1921	.8369
60	.54464	.83867	.64941	1.5399	1.1922	1.8361
′	cos	sin	cot	tan	cosec	sec

(bottom heading: 57°)

33° / 56° (center-left block)

′	sin	cos	tan	cot	sec	cosec
0	.54464	.83867	.64941	1.5399	1.1924	1.8361
1	.54488	.83851	.64982	.5389	.1926	.8352
2	.54513	.83835	.65023	.5379	.1928	.8344
3	.54537	.83819	.65065	.5369	.1930	.8336
4	.54561	.83804	.65106	.5359	.1933	.8328
5	.54586	.83788	.65148	1.5350	1.1935	1.8320
6	.54610	.83772	.65189	.5340	.1937	.8311
7	.54634	.83756	.65231	.5330	.1939	.8303
8	.54659	.83740	.65272	.5320	.1942	.8295
9	.54683	.83724	.65314	.5311	.1944	.8287
10	.54708	.83708	.65355	1.5301	1.1946	1.8279
11	.54732	.83692	.65397	.5291	.1948	.8271
12	.54756	.83676	.65438	.5282	.1951	.8263
13	.54781	.83660	.65480	.5272	.1953	.8255
14	.54805	.83644	.65521	.5262	.1955	.8246
15	.54829	.83629	.65563	1.5252	1.1955	1.8238
16	.54854	.83613	.65604	.5234	.1960	.8230
17	.54878	.83597	.65646	.5229	.1962	.8222
18	.54902	.83581	.65688	.5223	.1965	.8214
19	.54926	.83565	.65729	.5214	.1967	.8206
20	.54951	.83549	.65771	1.5204	1.1969	1.8198
21	.54975	.83533	.65813	.5195	.1971	.8190
22	.54999	.83517	.65854	.5185	.1974	.8182
23	.55024	.83501	.65896	.5175	.1976	.8174
24	.55048	.83485	.65938	.5166	.1978	.8166
25	.55002	.83469	.65980	1.5156	1.1980	1.8158
26	.55090	.83453	.66021	.5147	.1983	.8150
27	.55121	.83437	.66063	.5137	.1985	.8142
28	.55145	.83421	.66105	.5127	.1987	.8134
29	.55169	.83405	.66147	.5117	.1990	.8126
30	.55194	.83388	.66188	1.5108	1.1992	1.8118
31	.55218	.83372	.66230	.5099	.1994	.8110
32	.55242	.83356	.66272	.5089	.1997	.8102
33	.55266	.83340	.66314	.5080	.1999	.8094
34	.55291	.83324	.66356	.5070	.2001	.8086
35	.55315	.83308	.66398	1.5061	1.2004	1.8078
36	.55339	.83292	.66440	.5051	.2006	.8070
37	.55363	.83276	.66482	.5042	.2008	.8062
38	.55388	.83260	.66524	.5032	.2010	.8054
39	.55412	.83244	.66566	.5023	.2013	.8047
40	.55436	.83228	.66608	1.5013	1.2015	1.8039
41	.55460	.83211	.66650	.5004	.2017	.8031
42	.55484	.83195	.66692	.4994	.2020	.8023
43	.55509	.83179	.66734	.4985	.2022	.8015
44	.55533	.83163	.66776	.4975	.2024	.8007
45	.55557	.83147	.66818	1.4966	1.2027	1.7999
46	.55581	.83131	.66860	.4957	.2029	.7992
47	.55605	.83115	.66902	.4947	.2031	.7984
48	.55629	.83098	.66944	.4938	.2034	.7976
49	.55654	.83082	.66986	.4928	.2036	.7968
50	.55678	.83066	.67028	1.4919	1.2039	1.7960
51	.55702	.83050	.67071	.4910	.2041	.7953
52	.55726	.83034	.67113	.4900	.2043	.7945
53	.55750	.83017	.67155	.4891	.2046	.7937
54	.55774	.83001	.67197	.4881	.2048	.7929
55	.55799	.82985	.67239	1.4872	1.2050	1.7921
56	.55823	.82969	.67282	.4863	.2053	.7914
57	.55847	.82952	.67324	.4853	.2055	.7906
58	.55871	.82936	.67366	.4844	.2057	.7898
59	.55895	.82920	.67408	.4835	.2060	.7891
60	.55919	.82904	.67451	1.4826	1.2062	1.7883
′	cos	sin	cot	tan	cosec	sec

56°

34° / 55°

′	sin	cos	tan	cot	sec	cosec
0	.55919	.82904	.67451	1.4826	1.2062	1.7883
1	.55943	.82887	.67493	.4816	.2064	.7875
2	.55967	.82871	.67535	.4807	.2067	.7867
3	.55992	.82855	.67578	.4798	.2069	.7860
4	.56016	.82839	.67620	.4788	.2072	.7852
5	.56040	.82822	.67663	1.4779	1.2074	1.7844
6	.56064	.82806	.67705	.4770	.2076	.7837
7	.56088	.82790	.67747	.4761	.2079	.7829
8	.56112	.82773	.67790	.4751	.2081	.7821
9	.56136	.82757	.67832	.4742	.2083	.7814
10	.56160	.82741	.67875	1.4733	1.2086	1.7806
11	.56184	.82724	.67917	.4724	.2088	.7798
12	.56208	.82708	.67960	.4714	.2091	.7791
13	.56232	.82692	.68002	.4705	.2093	.7783
14	.56256	.82675	.68045	.4696	.2095	.7776
15	.56280	.82659	.68087	1.4687	1.2098	1.7768
16	.56304	.82643	.68130	.4678	.2100	.7760
17	.56328	.82626	.68173	.4669	.2103	.7753
18	.56353	.82610	.68215	.4659	.2105	.7745
19	.56377	.82593	.68258	.4650	.2107	.7738
20	.56401	.82577	.68301	1.4641	1.2110	1.7730
21	.56425	.82561	.68343	.4632	.2112	.7723
22	.56449	.82544	.68386	.4623	.2115	.7715
23	.56473	.82528	.68429	.4614	.2117	.7708
24	.56497	.82511	.68471	.4605	.2119	.7700
25	.56521	.82495	.68514	1.4595	1.2122	1.7693
26	.56545	.82478	.68557	.4586	.2124	.7685
27	.56569	.82462	.68600	.4577	.2127	.7678
28	.56593	.82445	.68642	.4568	.2129	.7670
29	.56617	.82429	.68685	.4559	.2132	.7663
30	.56641	.82413	.68728	1.4550	1.2134	1.7655
31	.56664	.82396	.68771	.4541	.2136	.7648
32	.56688	.82380	.68814	.4532	.2139	.7640
33	.56712	.82363	.68857	.4523	.2141	.7633
34	.56736	.82347	.68899	.4514	.2144	.7625
35	.56760	.82330	.68942	1.4505	1.2146	1.7618
36	.56784	.82314	.68985	.4496	.2149	.7610
37	.56808	.82297	.69028	.4487	.2151	.7603
38	.56832	.82280	.69071	.4478	.2153	.7596
39	.56856	.82264	.69114	.4469	.2156	.7588
40	.56880	.82247	.69157	1.4460	1.2158	1.7581
41	.56904	.82231	.69200	.4451	.2161	.7573
42	.56928	.82214	.69243	.4442	.2163	.7566
43	.56952	.82198	.69286	.4433	.2166	.7559
44	.56976	.82181	.69329	.4424	.2168	.7551
45	.57000	.82165	.69372	1.4415	1.2171	1.7544
46	.57023	.82148	.69415	.4406	.2173	.7537
47	.57047	.82131	.69459	.4397	.2175	.7529
48	.57071	.82115	.69502	.4388	.2178	.7522
49	.57095	.82098	.69545	.4379	.2180	.7514
50	.57119	.82082	.69588	1.4370	1.2183	1.7507
51	.57143	.82065	.69631	.4361	.2185	.7500
52	.57167	.82048	.69674	.4352	.2188	.7493
53	.57191	.82032	.69718	.4343	.2190	.7485
54	.57214	.82015	.69761	.4335	.2193	.7478
55	.57238	.81998	.69804	1.4326	1.2195	1.7471
56	.57262	.81982	.69847	.4317	.2198	.7463
57	.57286	.81965	.69891	.4308	.2200	.7456
58	.57310	.81948	.69934	.4299	.2203	.7449
59	.57334	.81932	.69977	.4290	.2205	.7442
60	.57358	.81915	.70021	1.4281	1.2208	1.7434
′	cos	sin	cot	tan	cosec	sec

(bottom heading: 55°)

35° / 54° (right block)

′	sin	cos	tan	cot	sec	cosec
60	.57358	.81915	.70021	1.4281	1.2208	1.7434
59	.57381	.81898	.70064	.4273	.2210	.7427
58	.57405	.81882	.70107	.4264	.2213	.7420
57	.57429	.81865	.70151	.4255	.2215	.7413
56	.57453	.81848	.70194	.4246	.2218	.7405
55	.57477	.81832	.70238	1.4237	1.2220	1.7398
54	.57500	.81815	.70281	.4228	.2223	.7391
53	.57524	.81798	.70325	.4220	.2225	.7384
52	.57548	.81781	.70368	.4211	.2228	.7377
51	.57572	.81765	.70412	.4202	.2230	.7369
50	.57596	.81748	.70455	1.4193	1.2233	1.7362
49	.57619	.81731	.70499	.4185	.2235	.7355
48	.57643	.81714	.70542	.4176	.2238	.7348
47	.57667	.81698	.70586	.4167	.2240	.7341
46	.57691	.81681	.70629	.4158	.2243	.7334
45	.57714	.81664	.70673	1.4150	1.2245	1.7327
44	.57738	.81647	.70717	.4141	.2248	.7319
43	.57762	.81630	.70760	.4132	.2250	.7312
42	.57786	.81614	.70804	.4123	.2253	.7305
41	.57809	.81597	.70848	.4115	.2255	.7298
40	.57833	.81580	.70891	1.4106	1.2258	1.7291
39	.57857	.81563	.70935	.4097	.2260	.7284
38	.57881	.81546	.70979	.4089	.2263	.7277
37	.57904	.81530	.71022	.4080	.2265	.7270
36	.57928	.81513	.71066	.4071	.2268	.7263
35	.57952	.81496	.71110	1.4063	1.2270	1.7256
34	.57975	.81479	.71154	.4054	.2273	.7249
33	.57999	.81462	.71198	.4045	.2276	.7242
32	.58023	.81445	.71241	.4037	.2278	.7234
31	.58047	.81428	.71285	.4028	.2281	.7227
30	.58070	.81411	.71329	1.4019	1.2283	1.7220
29	.58094	.81395	.71373	.4011	.2286	.7213
28	.58118	.81378	.71417	.4002	.2288	.7206
27	.58141	.81361	.71461	.3994	.2291	.7199
26	.58165	.81344	.71505	.3985	.2293	.7192
25	.58189	.81327	.71549	1.3976	1.2296	1.7185
24	.58212	.81310	.71593	.3968	.2298	.7178
23	.58236	.81293	.71637	.3959	.2301	.7171
22	.58259	.81276	.71681	.3951	.2304	.7164
21	.58283	.81259	.71725	.3942	.2306	.7157
20	.58307	.81242	.71769	1.3933	1.2309	1.7151
19	.58330	.81225	.71813	.3925	.2311	.7144
18	.58354	.81208	.71857	.3916	.2314	.7137
17	.58378	.81191	.71901	.3908	.2316	.7130
16	.58401	.81174	.71945	.3899	.2319	.7123
15	.58425	.81157	.71990	1.3891	1.2322	1.7116
14	.58448	.81140	.72034	.3882	.2324	.7109
13	.58472	.81123	.72078	.3874	.2327	.7102
12	.58496	.81106	.72122	.3865	.2329	.7095
11	.58519	.81089	.72166	.3857	.2332	.7088
10	.58543	.81072	.72211	1.3848	1.2335	1.7081
9	.58566	.81055	.72255	.3840	.2337	.7075
8	.58590	.81038	.72299	.3831	.2340	.7068
7	.58614	.81021	.72344	.3823	.2342	.7061
6	.58637	.81004	.72388	.3814	.2345	.7054
5	.58661	.80987	.72432	1.3806	1.2348	1.7047
4	.58684	.80970	.72477	.3797	.2350	.7040
3	.58708	.80953	.72521	.3789	.2353	.7033
2	.58731	.80936	.72565	.3781	.2355	.7027
1	.58755	.80919	.72610	.3772	.2358	.7020
0	.58778	.80902	.72654	1.3764	1.2361	1.7013
′	cos	sin	cot	tan	cosec	sec

(bottom heading: 54°)

NATURAL TRIGONOMETRIC FUNCTIONS

36°

'	sin	cos	tan	cot	sec	cosec
0	.58778	.80902	.72654	1.3764	1.2361	1.7013
1	.58802	.80885	.72743	1.3747	1.2363	1.7006
2	.58825	.80867	.72788	1.3730	1.2366	1.6999
3	.58849	.80850	.72832	1.3730	1.2368	1.6993
4	.58873	.80833			1.2371	1.6986

(Table continues — 36° / 53°, 37° / 52°, 38° / 51°, 39° / 50°)

37° — columns: sin, cos, tan, cot, sec, cosec

38° — columns: sin, cos, tan, cot, sec, cosec

39° — columns: sin, cos, tan, cot, sec, cosec

Complementary angle labels at bottom: **53°**, **52°**, **51°**, **50°**

NATURAL TRIGONOMETRIC FUNCTIONS

43°

′	cosec	sec	cot	tan	cos	sin	cosec	sec	cot	tan	cos	sin	′
60	1.4663	.3673	1.0724	.93251	.73135	.68200	1.4945	.3456	1.1106	.90040	.74314	.66913	60
59	.4658	.3677	.0717	.93306	.73115	.68221	.4940	.3460	.1100	.90093	.74295	.66935	59
58	.4654	.3681	.0711	.93360	.73096	.68242	.4935	.3463	.1093	.90146	.74276	.66956	58
57	.4649	.3684	.0705	.93415	.73076	.68264	.4930	.3467	.1086	.90198	.74256	.66978	57
56	.4644	.3688	.0699	.93469	.73056	.68285	.4925	.3470	.1080	.90251	.74236	.66999	56
55	1.4640	.3692	1.0692	.93524	.73036	.68306	1.4921	.3474	1.1074	.90304	.74217	.67021	55
54	.4635	.3695	.0686	.93578	.73016	.68327	.4916	.3477	.1067	.90357	.74197	.67043	54
53	.4631	.3699	.0680	.93633	.72996	.68349	.4911	.3481	.1061	.90410	.74178	.67064	53
52	.4626	.3703	.0680	.93687	.72976	.68370	.4906	.3485	.1054	.90463	.74158	.67086	52
51	.4622	.3707	.0667	.93742	.72956	.68391	.4901	.3488	.1048	.90515	.74139	.67107	51
50	1.4617	.3710	1.0661	.93797	.72937	.68412	1.4897	.3492	1.1041	.90568	.74119	.67129	50
49	.4613	.3714	.0655	.93851	.72917	.68433	.4892	.3495	.1035	.90621	.74100	.67150	49
48	.4608	.3718	.0649	.93906	.72897	.68455	.4887	.3499	.1028	.90674	.74080	.67172	48
47	.4604	.3722	.0643	.93961	.72877	.68476	.4887	.3502	.1022	.90727	.74061	.67194	47
46	.4599	.3725	.0636	.94016	.72857	.68497	.4877	.3506	.1015	.90780	.74041	.67215	46
45	1.4595	.3729	1.0630	.94071	.72837	.68518	1.4873	.3509	1.1009	.90834	.74022	.67237	45
44	.4590	.3733	.0624	.94125	.72817	.68539	.4868	.3513	.1003	.90887	.74002	.67258	44
43	.4586	.3737	.0618	.94180	.72797	.68561	.4863	.3517	.0996	.90940	.73983	.67280	43
42	.4581	.3740	.0612	.94235	.72777	.68582	.4858	.3520	.0990	.90993	.73963	.67301	42
41	.4577	.3744	.0605	.94290	.72757	.68603	.4854	.3524	.0983	.91046	.73943	.67323	41
40	1.4572	.3748	1.0599	.94345	.72737	.68624	1.4849	.3527	1.0977	.91099	.73924	.67344	40
39	.4568	.3752	.0593	.94400	.72717	.68645	.4844	.3531	.0971	.91153	.73904	.67366	39
38	.4563	.3756	.0587	.94455	.72697	.68666	.4839	.3534	.0964	.91206	.73885	.67387	38
37	.4559	.3759	.0581	.94510	.72677	.68688	.4835	.3538	.0958	.91259	.73865	.67409	37
36	.4554	.3763	.0575	.94565	.72657	.68709	.4830	.3542	.0951	.91312	.73845	.67430	36
35	1.4550	.3767	1.0568	.94620	.72637	.68730	1.4825	.3545	1.0945	.91366	.73826	.67452	35
34	.4545	.3771	.0562	.94675	.72617	.68751	.4821	.3549	.0939	.91419	.73806	.67473	34
33	.4541	.3774	.0556	.94731	.72597	.68772	.4816	.3552	.0932	.91473	.73787	.67495	33
32	.4536	.3778	.0550	.94786	.72577	.68793	.4811	.3556	.0926	.91526	.73767	.67516	32
31	.4532	.3782	.0544	.94841	.72557	.68814	.4806	.3560	.0919	.91580	.73747	.67537	31
30	1.4527	.3786	1.0538	.94896	.72537	.68835	1.4802	.3563	1.0913	.91633	.73728	.67559	30
29	.4523	.3790	.0532	.94952	.72517	.68856	.4797	.3567	.0907	.91687	.73708	.67580	29
28	.4518	.3794	.0525	.95007	.72497	.68878	.4792	.3571	.0900	.91740	.73688	.67602	28
27	.4514	.3797	.0519	.95062	.72477	.68899	.4788	.3574	.0894	.91794	.73669	.67623	27
26	.4510	.3801	.0513	.95118	.72457	.68920	.4783	.3578	.0888	.91847	.73649	.67645	26
25	1.4505	.3805	1.0507	.95173	.72437	.68941	1.4778	.3581	1.0881	.91901	.73629	.67666	25
24	.4501	.3809	.0501	.95229	.72417	.68962	.4774	.3585	.0875	.91955	.73610	.67688	24
23	.4496	.3813	.0495	.95284	.72397	.68983	.4769	.3589	.0868	.92008	.73590	.67709	23
22	.4492	.3816	.0489	.95340	.72377	.69004	.4764	.3592	.0862	.92062	.73570	.67730	22
21	.4487	.3820	.0483	.95395	.72357	.69025	.4760	.3596	.0856	.92116	.73551	.67752	21
20	1.4483	.3824	1.0476	.95451	.72337	.69046	1.4755	.3600	1.0849	.92170	.73531	.67773	20
19	.4479	.3828	.0470	.95506	.72317	.69067	.4750	.3603	.0843	.92223	.73511	.67794	19
18	.4474	.3832	.0464	.95562	.72296	.69088	.4746	.3607	.0837	.92277	.73491	.67816	18
17	.4470	.3836	.0458	.95618	.72276	.69109	.4741	.3611	.0830	.92331	.73472	.67837	17
16	.4465	.3839	.0452	.95673	.72256	.69130	.4736	.3614	.0824	.92385	.73452	.67859	16
15	1.4461	.3843	1.0446	.95785	.72236	.69151	1.4732	.3618	1.0818	.92439	.73432	.67880	15
14	.4457	.3847	.0440	.95785	.72216	.69172	.4727	.3622	.0811	.92493	.73412	.67901	14
13	.4452	.3851	.0434	.95841	.72196	.69193	.4723	.3625	.0805	.92547	.73393	.67923	13
12	.4448	.3855	.0428	.95896	.72176	.69214	.4718	.3629	.0799	.92601	.73373	.67944	12
11	.4443	.3859	.0422	.95952	.72156	.69235	.4713	.3633	.0793	.92655	.73353	.67965	11
10	1.4439	.3863	1.0416	.96008	.72136	.69256	1.4709	.3636	1.0786	.92709	.73333	.67987	10
9	.4435	.3867	.0410	.96064	.72115	.69277	.4704	.3640	.0780	.92763	.73314	.68008	9
8	.4430	.3870	.0404	.96120	.72095	.69298	.4699	.3644	.0774	.92817	.73294	.68029	8
7	.4426	.3874	.0398	.96176	.72075	.69319	.4695	.3647	.0767	.92871	.73274	.68051	7
6	.4422	.3878	.0391	.96232	.72055	.69340	.4690	.3651	.0761	.92926	.73254	.68072	6
5	1.4417	.3882	1.0385	.96288	.72035	.69361	1.4686	.3655	1.0755	.92980	.73234	.68093	5
4	.4413	.3886	.0379	.96344	.72015	.69382	.4681	.3658	.0749	.93034	.73215	.68115	4
3	.4408	.3890	.0373	.96400	.71994	.69403	.4676	.3662	.0742	.93088	.73195	.68136	3
2	.4404	.3894	.0367	.96456	.71974	.69424	.4672	.3666	.0736	.93143	.73175	.68157	2
1	.4400	.3898	.0361	.96513	.71954	.69445	.4667	.3669	.0730	.93197	.73155	.68178	1
0	1.4395	.3902	1.0355	.96569	.71934	.69466	1.4663	.3673	1.0724	.93251	.73135	.68200	0
′	sec	cosec	tan	cot	sin	cos	sec	cosec	tan	cot	sin	cos	′

46° / 47° (bottom headers for above)

42°

(See combined columns above under headers sec / cosec / tan / cot / sin / cos for 42°)

41° and 40°

′	cosec	sec	cot	tan	cos	sin	cosec	sec	cot	tan	cos	sin	′
0	1.5557	.3054	1.1917	.83910	.76604	.64279	1.5242	.3250	1.1504	.86929	.75471	.65606	0
1	.5552	.3057	.1910	.83959	.76586	.64301	.5237	.3253	.1497	.86980	.75452	.65628	1
2	.5546	.3060	.1903	.84009	.76567	.64323	.5232	.3257	.1490	.87031	.75433	.65650	2
3	.5541	.3064	.1896	.84059	.76548	.64345	.5227	.3260	.1483	.87082	.75414	.65672	3
4	.5536	.3067	.1889	.84108	.76530	.64368	.5222	.3263	.1477	.87133	.75394	.65694	4
5	1.5530	.3070	1.1882	.84158	.76511	.64390	1.5217	.3267	1.1470	.87184	.75375	.65716	5
6	.5525	.3073	.1875	.84208	.76492	.64412	.5212	.3270	.1463	.87235	.75356	.65738	6
7	.5520	.3076	.1868	.84257	.76473	.64435	.5207	.3274	.1456	.87287	.75337	.65759	7
8	.5514	.3080	.1861	.84307	.76455	.64457	.5202	.3277	.1450	.87338	.75318	.65781	8
9	.5509	.3083	.1854	.84357	.76436	.64479	.5197	.3280	.1443	.87389	.75299	.65803	9
10	1.5503	.3086	1.1847	.84407	.76417	.64501	1.5192	.3284	1.1436	.87441	.75280	.65825	10
11	.5498	.3089	.1840	.84457	.76398	.64523	.5187	.3287	.1430	.87492	.75261	.65847	11
12	.5493	.3092	.1833	.84506	.76380	.64546	.5182	.3290	.1423	.87543	.75241	.65869	12
13	.5487	.3096	.1826	.84556	.76361	.64568	.5177	.3294	.1416	.87595	.75222	.65891	13
14	.5482	.3099	.1819	.84606	.76342	.64590	.5171	.3297	.1409	.87646	.75203	.65913	14
15	1.5477	.3102	1.1812	.84656	.76323	.64612	1.5166	.3301	1.1403	.87698	.75184	.65934	15
16	.5471	.3105	.1805	.84706	.76304	.64635	.5161	.3304	.1396	.87749	.75165	.65956	16
17	.5466	.3109	.1798	.84756	.76286	.64657	.5156	.3307	.1389	.87801	.75146	.65978	17
18	.5461	.3112	.1791	.84806	.76267	.64679	.5151	.3311	.1383	.87852	.75126	.66000	18
19	.5456	.3115	.1785	.84856	.76248	.64701	.5146	.3314	.1376	.87904	.75107	.66022	19
20	1.5450	.3118	1.1778	.84906	.76229	.64723	1.5141	.3318	1.1369	.87955	.75088	.66044	20
21	.5445	.3121	.1771	.84956	.76210	.64745	.5136	.3321	.1363	.88007	.75069	.66066	21
22	.5440	.3125	.1764	.85006	.76191	.64768	.5131	.3324	.1356	.88058	.75050	.66088	22
23	.5434	.3128	.1757	.85056	.76173	.64790	.5126	.3328	.1349	.88110	.75030	.66109	23
24	.5429	.3131	.1750	.85107	.76154	.64812	.5121	.3331	.1343	.88162	.75011	.66131	24
25	1.5424	.3134	1.1743	.85157	.76135	.64834	1.5116	.3335	1.1336	.88213	.74992	.66153	25
26	.5419	.3138	.1736	.85207	.76116	.64856	.5111	.3338	.1329	.88265	.74973	.66175	26
27	.5413	.3141	.1729	.85257	.76097	.64878	.5106	.3342	.1323	.88317	.74953	.66197	27
28	.5408	.3144	.1722	.85307	.76078	.64900	.5101	.3345	.1316	.88369	.74934	.66218	28
29	.5403	.3148	.1715	.85358	.76059	.64923	.5096	.3348	.1309	.88421	.74915	.66240	29
30	1.5398	.3151	1.1708	.85408	.76041	.64945	1.5092	.3352	1.1303	.88472	.74895	.66262	30
31	.5392	.3154	.1702	.85458	.76022	.64967	.5087	.3355	.1296	.88524	.74876	.66284	31
32	.5387	.3157	.1695	.85509	.76003	.64989	.5082	.3359	.1290	.88576	.74857	.66306	32
33	.5382	.3161	.1688	.85559	.75984	.65011	.5077	.3362	.1283	.88628	.74838	.66327	33
34	.5377	.3164	.1681	.85609	.75965	.65033	.5072	.3366	.1276	.88680	.74818	.66349	34
35	1.5371	.3167	1.1674	.85660	.75946	.65055	1.5067	.3369	1.1270	.88732	.74799	.66371	35
36	.5366	.3170	.1667	.85710	.75927	.65077	.5062	.3372	.1263	.88784	.74780	.66393	36
37	.5361	.3174	.1660	.85761	.75908	.65100	.5057	.3376	.1257	.88836	.74760	.66414	37
38	.5356	.3177	.1653	.85811	.75889	.65122	.5052	.3379	.1250	.88888	.74741	.66436	38
39	.5351	.3180	.1647	.85862	.75870	.65144	.5047	.3383	.1243	.88940	.74722	.66458	39
40	1.5345	.3184	1.1640	.85912	.75851	.65166	1.5042	.3386	1.1237	.88992	.74702	.66479	40
41	.5340	.3187	.1633	.85963	.75832	.65188	.5037	.3390	.1230	.89044	.74683	.66501	41
42	.5335	.3190	.1626	.86013	.75813	.65210	.5032	.3393	.1224	.89097	.74664	.66523	42
43	.5330	.3193	.1619	.86064	.75794	.65232	.5027	.3397	.1217	.89149	.74644	.66545	43
44	.5325	.3197	.1612	.86115	.75775	.65254	.5022	.3400	.1211	.89201	.74625	.66566	44
45	1.5319	.3200	1.1605	.86165	.75756	.65276	1.5018	.3404	1.1204	.89253	.74606	.66588	45
46	.5314	.3203	.1599	.86216	.75737	.65298	.5013	.3407	.1197	.89306	.74586	.66610	46
47	.5309	.3207	.1592	.86267	.75718	.65320	.5008	.3411	.1191	.89358	.74567	.66632	47
48	.5304	.3210	.1585	.86318	.75700	.65342	.5003	.3414	.1184	.89410	.74548	.66653	48
49	.5299	.3213	.1578	.86368	.75680	.65364	.4998	.3418	.1178	.89463	.74528	.66675	49
50	1.5294	.3217	1.1571	.86419	.75661	.65386	1.4993	.3421	1.1171	.89515	.74509	.66697	50
51	.5289	.3220	.1565	.86470	.75642	.65408	.4988	.3425	.1165	.89567	.74489	.66718	51
52	.5283	.3223	.1558	.86521	.75623	.65430	.4983	.3428	.1158	.89620	.74470	.66740	52
53	.5278	.3227	.1551	.86572	.75604	.65452	.4979	.3432	.1152	.89672	.74451	.66762	53
54	.5273	.3230	.1544	.86623	.75585	.65474	.4974	.3435	.1145	.89725	.74431	.66783	54
55	1.5268	.3233	1.1537	.86674	.75566	.65496	1.4969	.3439	1.1139	.89777	.74412	.66805	55
56	.5263	.3237	.1531	.86725	.75547	.65518	.4964	.3442	.1132	.89830	.74392	.66826	56
57	.5258	.3240	.1524	.86775	.75528	.65540	.4959	.3446	.1126	.89882	.74373	.66848	57
58	.5253	.3243	.1517	.86826	.75509	.65562	.4954	.3449	.1119	.89935	.74353	.66870	58
59	.5248	.3247	.1510	.86878	.75490	.65584	.4949	.3453	.1113	.89988	.74334	.66891	59
60	1.5242	.3250	1.1504	.86929	.75471	.65606	1.4945	.3456	1.1106	.90040	.74314	.66913	60
′	sec	cosec	tan	cot	sin	cos	sec	cosec	tan	cot	sin	cos	′

49° / 48°

NATURAL TRIGONOMETRIC FUNCTIONS

44°

′	sin	cos	tan	cot	sec	cosec	′
0	.69466	.71934	.96569	1.0355	1.3902	1.4395	60
1	.69487	.71914	.96625	.0349	.3905	.4391	59
2	.69508	.71893	.96681	.0343	.3909	.4387	58
3	.69528	.71873	.96738	.0337	.3913	.4382	57
4	.69549	.71853	.96794	.0331	.3917	.4378	56
5	.69570	.71833	.96850	1.0325	.3921	1.4374	55
6	.69591	.71813	.96907	.0319	.3925	.4370	54
7	.69612	.71792	.96963	.0313	.3929	.4365	53
8	.69633	.71772	.97020	.0307	.3933	.4361	52
9	.69654	.71752	.97076	.0301	.3937	.4357	51
10	.69675	.71732	.97133	1.0295	.3941	1.4352	50
11	.69696	.71711	.97189	.0289	.3945	.4348	49
12	.69716	.71691	.97246	.0283	.3949	.4344	48
13	.69737	.71671	.97302	.0277	.3953	.4339	47
14	.69758	.71650	.97359	.0271	.3957	.4335	46
15	.69779	.71630	.97416	1.0265	.3960	1.4331	45
16	.69800	.71610	.97472	.0259	.3964	.4327	44
17	.69821	.71589	.97529	.0253	.3968	.4322	43
18	.69841	.71569	.97586	.0247	.3972	.4318	42
19	.69862	.71549	.97643	.0241	.3976	.4314	41
20	.69883	.71529	.97700	1.0235	.3980	1.4310	40
21	.69904	.71508	.97756	.0229	.3984	.4305	39
22	.69925	.71488	.97813	.0223	.3988	.4301	38
23	.69945	.71468	.97870	.0218	.3992	.4297	37
24	.69966	.71447	.97927	.0212	.3996	.4292	36
25	.69987	.71427	.97984	1.0206	.4000	1.4288	35
26	.70008	.71406	.98041	.0200	.4004	.4284	34
27	.70029	.71386	.98098	.0194	.4008	.4280	33
28	.70049	.71366	.98155	.0188	.4012	.4276	32
29	.70070	.71345	.98212	.0182	.4016	.4271	31
30	.70091	.71325	.98270	1.0176	.4020	1.4267	30
31	.70112	.71305	.98327	.0170	.4024	.4263	29
32	.70132	.71284	.98384	.0164	.4028	.4259	28
33	.70153	.71264	.98441	.0158	.4032	.4254	27
34	.70174	.71243	.98499	.0152	.4036	.4250	26
35	.70194	.71223	.98556	1.0146	.4040	1.4246	25
36	.70215	.71203	.98613	.0141	.4044	.4242	24
37	.70236	.71182	.98671	.0135	.4048	.4238	23
38	.70257	.71162	.98728	.0129	.4052	.4233	22
39	.70277	.71141	.98786	.0123	.4056	.4229	21
40	.70298	.71121	.98843	1.0117	.4060	1.4225	20
41	.70319	.71100	.98901	.0111	.4065	.4221	19
42	.70339	.71080	.98958	.0105	.4069	.4217	18
43	.70360	.71059	.99016	.0099	.4073	.4212	17
44	.70381	.71039	.99073	.0093	.4077	.4208	16
45	.70401	.71018	.99131	1.0088	.4081	1.4204	15
46	.70422	.70998	.99189	.0082	.4085	.4200	14
47	.70443	.70977	.99246	.0076	.4089	.4196	13
48	.70463	.70957	.99304	.0070	.4093	.4192	12
49	.70484	.70936	.99362	.0064	.4097	.4188	11
50	.70505	.70916	.99420	1.0058	.4101	1.4183	10
51	.70525	.70895	.99478	.0052	.4105	.4179	9
52	.70546	.70875	.99536	.0047	.4109	.4175	8
53	.70566	.70854	.99593	.0041	.4113	.4171	7
54	.70587	.70834	.99651	.0035	.4117	.4167	6
55	.70608	.70813	.99709	1.0029	.4122	1.4163	5
56	.70628	.70793	.99767	.0023	.4126	.4159	4
57	.70649	.70772	.99826	.0017	.4130	.4154	3
58	.70669	.70752	.99884	.0012	.4134	.4150	2
59	.70690	.70731	.99942	.0006	.4138	.4146	1
60	.70711	.70711	1.00000	1.0000	.4142	1.4142	0

′	cos	sin	cot	tan	cosec	sec	′

45°

INDEX